LIPOIC ACID

Energy Production, Antioxidant Activity and Health Effects

OXIDATIVE STRESS AND DISEASE

Series Editors

LESTER PACKER, PH.D.
ENRIQUE CADENAS, M.D., PH.D.
University of Southern California School of Pharmacy
Los Angeles, California

1. Oxidative Stress in Cancer, AIDS, and Neurodegenerative Diseases, *edited by Luc Montagnier, René Olivier, and Catherine Pasquier*
2. Understanding the Process of Aging: The Roles of Mitochondria, Free Radicals, and Antioxidants, *edited by Enrique Cadenas and Lester Packer*
3. Redox Regulation of Cell Signaling and Its Clinical Application, *edited by Lester Packer and Junji Yodoi*
4. Antioxidants in Diabetes Management, *edited by Lester Packer, Peter Rösen, Hans J. Tritschler, George L. King, and Angelo Azzi*
5. Free Radicals in Brain Pathophysiology, *edited by Giuseppe Poli, Enrique Cadenas, and Lester Packer*
6. Nutraceuticals in Health and Disease Prevention, *edited by Klaus Krämer, Peter-Paul Hoppe, and Lester Packer*
7. Environmental Stressors in Health and Disease, *edited by Jürgen Fuchs and Lester Packer*
8. Handbook of Antioxidants: Second Edition, Revised and Expanded, *edited by Enrique Cadenas and Lester Packer*
9. Flavonoids in Health and Disease: Second Edition, Revised and Expanded, *edited by Catherine A. Rice-Evans and Lester Packer*
10. Redox–Genome Interactions in Health and Disease, *edited by Jürgen Fuchs, Maurizio Podda, and Lester Packer*
11. Thiamine: Catalytic Mechanisms in Normal and Disease States, *edited by Frank Jordan and Mulchand S. Patel*
12. Phytochemicals in Health and Disease, *edited by Yongping Bao and Roger Fenwick*
13. Carotenoids in Health and Disease, *edited by Norman I. Krinsky, Susan T. Mayne, and Helmut Sies*

14. Herbal and Traditional Medicine: Molecular Aspects of Health, *edited by Lester Packer, Choon Nam Ong, and Barry Halliwell*
15. Nutrients and Cell Signaling, *edited by Janos Zempleni and Krishnamurti Dakshinamurti*
16. Mitochondria in Health and Disease, *edited by Carolyn D. Berdanier*
17. Nutrigenomics, *edited by Gerald Rimbach, Jürgen Fuchs, and Lester Packer*
18. Oxidative Stress, Inflammation, and Health, *edited by Young-Joon Surh and Lester Packer*
19. Nitric Oxide, Cell Signaling, and Gene Expression, *edited by Santiago Lamas and Enrique Cadenas*
20. Resveratrol in Health and Disease, *edited by Bharat B. Aggarwal and Shishir Shishodia*
21. Oxidative Stress and Age-Related Neurodegeneration, *edited by Yuan Luo and Lester Packer*
22. Molecular Interventions in Lifestyle-Related Diseases, *edited by Midori Hiramatsu, Toshikazu Yoshikawa, and Lester Packer*
23. Oxidative Stress and Inflammatory Mechanisms in Obesity, Diabetes, and the Metabolic Syndrome, *edited by Lester Packer and Helmut Sies*
24. Lipoic Acid: Energy Production, Antioxidant Activity and Health Effects, *edited by Mulchand S. Patel and Lester Packer*

LIPOIC ACID

Energy Production, Antioxidant Activity and Health Effects

Edited by
Mulchand S. Patel *and* Lester Packer

CRC Press
Taylor & Francis Group
Boca Raton London New York

CRC Press is an imprint of the
Taylor & Francis Group, an **informa** business

CRC Press
Taylor & Francis Group
6000 Broken Sound Parkway NW, Suite 300
Boca Raton, FL 33487-2742

Printed in the United States of America on acid-free paper
10 9 8 7 6 5 4 3 2 1

International Standard Book Number-13: 978-1-4200-4537-6 (Hardcover)

Library of Congress Cataloging-in-Publication Data

Lipoic acid : energy production, antioxidant activity and health effects / edited by Mulchand S. Patel and Lester Packer.
p. ; cm. -- (Oxidative stress and disease ; 24)
Includes bibliographical references and index.
ISBN-13: 978-1-4200-4537-6 (hardcover : alk. paper)
ISBN-10: 1-4200-4537-7 (hardcover : alk. paper)
1. Thioctic acid. 2. Thioctic acid--Health aspects. 3. Thioctic acid--Therapeutic use. I. Patel, Mulchand S. II. Packer, Lester. III. Title. IV. Series.
[DNLM: 1. Thioctic Acid--metabolism. 2. Thioctic Acid--pharmacology. 3. Thioctic Acid--therapeutic use. W1 OX626 v.24 2008 / QU 135 A457 2008]

QP772.T54A47 2008
615'.35--dc22 2007035854

Visit the Taylor & Francis Web site at
http://www.taylorandfrancis.com

and the CRC Press Web site at
http://www.crcpress.com

Contents

Series Preface

OXYGEN BIOLOGY AND MEDICINE

Through evolution, oxygen—itself a free radical—was chosen as the terminal electron acceptor for respiration. The two unpaired electrons of oxygen spin in the same direction; thus, oxygen is a biradical. Other oxygen-derived free radicals such as superoxide anion or hydroxyl radicals formed during metabolism or by ionizing radiation are stronger *oxidants*, i.e., endowed with higher chemical reactivities. Oxygen-derived free radicals are generated during metabolism and energy production in the body and are involved in regulation of signal transduction and gene expression, activation of receptors and nuclear transcription factors, oxidative damage to cell components, antimicrobial and cytotoxic actions of immune system cells, as well as in aging and age-related degenerative diseases. Conversely, cells conserve antioxidant mechanisms to counteract the effects of oxidants; these *antioxidants* may remove oxidants either in a highly specific manner (for example, by superoxide dismutases) or in a less specific manner (for example, through small molecules such as vitamin E, vitamin C, and glutathione). *Oxidative stress* as classically defined is an *imbalance between oxidants and antioxidants*. Overwhelming evidence indicates that oxidative stress can lead to cell and tissue injury. However, the same free radicals that are generated during oxidative stress are produced during normal metabolism and, as a corollary, are involved in both human health and disease.

UNDERSTANDING OXIDATIVE STRESS

In recent years, the research disciplines interested in oxidative stress have grown and enormously increased our knowledge of the importance of the cell redox status and the recognition of oxidative stress as a process with implications for many pathophysiological states. From this multi- and inter-disciplinary interest in oxidative stress emerges a concept that attests to the vast consequences of the complex and dynamic interplay of oxidants and antioxidants in cellular and tissue settings. Consequently, our view of oxidative stress is growing in scope and new future directions. Likewise, the term *reactive oxygen species*—adopted at some stage in order to highlight non-radical oxidants such as H_2O_2 and $^1O_2$—now fails to reflect the rich variety of other reactive species in free radical biology and medicine encompassing nitrogen-, sulfur-, oxygen-, and carbon-centered radicals. With the discovery of nitric oxide, nitrogen-centered radicals gathered momentum and have matured into an area of enormous importance in biology and medicine. Nitric oxide or nitrogen monoxide (NO), a free radical generated in a variety of cell types by nitric oxide synthases (NOSs), is involved in a wide array of physiological and pathophysiological phenomena such as vasodilation, neuronal signaling, and inflammation. Of great importance is the radical–radical reaction of nitric oxide with superoxide anion. This is among the most rapid non-enzymatic reactions in biology (well over the diffusion-controlled limits) and yields the potent non-radical oxidant, peroxynitrite. The involvement of this species in tissue injury through oxidation and nitration reactions is well documented.

Virtually all diseases thus far examined involve free radicals. In most cases, free radicals are secondary to the disease process, but in some instances causality is established by free radicals. Thus, there is a delicate balance between oxidants and antioxidants in health and disease. Their proper balance is essential for ensuring healthy aging.

Both reactive oxygen and nitrogen species are involved in the redox regulation of cell functions. Oxidative stress is increasingly viewed as a major upstream component in the signaling cascade involved in inflammatory responses, stimulation of cell adhesion molecules, and chemoattractant production and as an early component in age-related neurodegenerative disorders such as Alzheimer's, Parkinson's, and Huntington's diseases, and amyotrophic lateral sclerosis. Hydrogen peroxide is probably the most important redox signaling molecule that, among others, can activate NFκB, Nrf2, and other universal transcription factors. Increasing steady-state levels of hydrogen peroxide have been linked to a cell's redox status with clear involvement in adaptation, proliferation, differentiation, apoptosis, and necrosis.

The identification of oxidants in regulation of redox cell signaling and gene expression was a significant breakthrough in the field of oxidative stress: the classical definition of oxidative stress as an *imbalance between the production of oxidants and the occurrence of cell antioxidant defenses* proposed by Sies in 1985 now seems to provide a limited concept of oxidative stress, but it emphasizes the significance of cell redox status. Because individual signaling and control events occur through discrete redox pathways rather than through global balances, a new definition of oxidative stress was advanced by Dean P. Jones (*Antioxidants & Redox Signaling* [2006]) as a disruption of redox signaling and control that recognizes the occurrence of compartmentalized cellular redox circuits. Recognition of discrete thiol redox circuits led Jones to provide this new definition of oxidative stress. Measurements of GSH/GSSG, cysteine/cystine, or thio-redoxin$_{\text{reduced}}$/thioredoxin$_{\text{oxidized}}$ provide a quantitative definition of oxidative stress. Redox status is thus dependent on the degree to which tissue-specific cell components are in the oxidized state.

In general, the reducing environments inside cells help to prevent oxidative damage. In this reducing environment, disulfide bonds (S–S) do not spontaneously form because sulfhydryl groups are maintained in the reduced state (SH), thus preventing protein misfolding or aggregation. The reducing environment is maintained by metabolism and by the enzymes involved in maintenance of thiol/disulfide balance and substances such as glutathione, thioredoxin, vitamins E and C, and enzymes such as superoxide dismutases, catalase, and the selenium-dependent glutathione reductase and glutathione and thioredoxin-dependent hydroperoxidases (periredoxins) that serve to remove reactive oxygen species (hydroperoxides). Also of importance is the existence of many tissue- and cell compartment-specific isoforms of antioxidant enzymes and proteins.

Compelling support for the involvement of free radicals in disease development originates from epidemiological studies showing that enhanced antioxidant status is associated with reduced risk of several diseases. Of great significance is the role that micronutrients play in modulation of redox cell signaling; this establishes a strong linking of diet and health and disease centered on the abilities of micronutrients to regulate redox cell signaling and modify gene expression.

These concepts are anticipated to serve as platforms for the development of tissue-specific therapeutics tailored to discrete, compartmentalized redox circuits. This, in essence, dictates principles of drug development-guided knowledge of mechanisms of oxidative stress. Hence, successful interventions will take advantage of new knowledge of compartmentalized redox control and free radical scavenging.

OXIDATIVE STRESS IN HEALTH AND DISEASE

Oxidative stress is an underlying factor in health and disease. In this series of books, the importance of oxidative stress and diseases associated with organ systems of the body is highlighted by exploring the scientific evidence and clinical applications of this knowledge. This series is intended for researchers in the basic biomedical sciences and clinicians. The potential of such knowledge for healthy aging and disease prevention warrants further knowledge about how oxidants and antioxidants modulate cell and tissue function.

Lester Packer
Enrique Cadenas

Preface

Oxidative stress is a major factor in health, aging, and disease and is often defined by the redox balance established by cellular oxidants and antioxidants. The ratio of reduced to oxidized levels of major low-molecular weight protein thiols, such as cysteine and glutathione or thioredoxin and glutaredoxin, is often used as a measure of oxidative stress. Among naturally occurring thiols, the discovery of α-lipoic acid has led to an unprecedented interest in basic research because of its role in energy metabolism and as an antioxidant for the treatment of certain diseases.

α-Lipoic acid has acquired importance because of its role as a coenzyme in the maintenance of energy metabolism, on one hand, and as an antioxidant and cell redox modulator, on the other hand. The former was supported by the discovery of the essential role of lipoic acid in mitochondrial dehydrogenase-driven reactions inherent in energy metabolism and the latter is supported by reduction of supplemental α-lipoic acid to dihydrolipoic acid in mammalian cells conferring further antioxidant protection. For example, lipoic acid is an essential cofactor of α-keto acid dehydrogenase complexes in eukaryotic cells as part of the lipoyl domain of the E2 enzymes (dihydrolipoamide acyltransferases). Also, exogenous *R*- (or *RS*)-lipoic acid is reduced by a flavoenzyme, dihydrolipoamide dehydrogenase, of the complex in mitochondria. Hence lipoic acid may be a mitochondrial-targeted nutrient. The cell redox status is a critical factor regulating cell proliferation, differentiation, apoptosis (programmed cell death), and necrosis, and the mitochondrion is a center of interest for its role in the generation of signaling molecules (e.g., hydrogen peroxide and nitric oxide) that, in turn, modulates the redox status of the cell.

This book is a state of the art compilation of information, which covers both historical aspects and new information on the chemistry, biological action, and significance of α-lipoic acid in energy production, antioxidant activity, and health. The chapters in the book are divided into three sections.

The chapters in Section I, "Discovery and Molecular Structure," elaborate upon the early studies leading to its discovery, molecular structure and biosynthesis, and characterization of novel chemical derivatives. Lipoic acid first was isolated and chemically identified in 1951 by Lester Reed and colleagues after numerous reports of the existence of a low-molecular weight factor required for microbial activity.

Specific chapters in Section II, "Metabolic Aspects," describe in detail the pyruvate dehydrogenase and other α-keto acid dehydrogenase complexes and the glycine cleavage system and their roles in pathways of carbohydrate, protein, and fat metabolism, the structure and function of lipoate-protein ligase A, human pharmacokinetics, metabolism and excretion of α-lipoic acid and metabolites, tissue-specific enzymes and thioredoxin in reduction of α-lipoic acid, and how lipoic acid acts as an inducer of phase II detoxification enzymes through activation of Nrf2-dependent gene expression.

Section III, "Clinical Aspects," includes chapters on enzyme deficiency disorders, primary biliary cirrhosis, effects of lipoic acid on insulin action in animal models of insulin resistance, action of lipoic acid on signaling pathways, topical antioxidant activity of selenium derivatives, effects of lipoic acid on AMP-activated kinase, and therapeutic implications for treatment of the metabolic syndrome, and how the age-related loss of mitochondrial function is improved by feeding lipoic acid to older animals.

The editors gratefully appreciate all of the authors who have shared their knowledge and who have written outstanding and scholarly accounts of the biological and biomedical actions of lipoic acid. The chapters in this book represent important contributions to the ongoing scientific investigation. Moreover, the book will serve as a valuable reference to researchers in academia, industry, and clinical medicine.

Mulchand S. Patel
Lester Packer

Editors

Mulchand S. Patel received his PhD in animal science from the University of Illinois, Urbana–Champaign. He is a UB Distinguished Professor, professor of biochemistry and associate dean for research and biomedical education, School of Medicine and Biomedical Sciences, University at Buffalo, State University of New York. Earlier he served as chairman of the Department of Biochemistry at the same university. Dr. Patel has been a member of the American Society for Biochemistry and Molecular Biology and the American Society for Nutrition for more than 35 years.

Professor Patel's research interests focus on the structure–function relationship of the mammalian pyruvate dehydrogenase complex and metabolic programming due to early-life nutritional interventions. He has more than 200 professional publications to his credit. He has received several scientific achievement awards including a Fulbright research scholar award to India.

Lester Packer received his PhD in microbiology and biochemistry from Yale University. For many years he was a professor and a senior researcher at the University of California at Berkeley. Currently he is an adjunct professor in the Department of Pharmacology and Pharmaceutical Sciences at the University of Southern California. Recently, he was appointed distinguished professor at the Institute of Nutritional Sciences of the Chinese Academy of Sciences, Shanghai. His research interests are related to the molecular, cellular, and physiological role of oxidants, free radicals, antioxidants, and redox regulation in health and disease.

Professor Packer is the recipient of numerous scientific achievement awards, including three honorary doctoral degrees. He has served as president of the International Society of Free Radical Research (SFRRI), president of the Oxygen Club of California (OCC), and vice-president of UNESCO's Molecular and Cell Biology Network (MCBN).

Contributors

Haiying Bao
Department of Biochemistry
Kansas State University
Manhattan, Kansas

José Antonio Bárcena
Department of Biochemistry and Molecular Biology
University of Córdoba
Córdoba, Spain

Moriya Ben Yakir
Departments of Pharmacology
School of Pharmacy
The Hebrew University
Jerusalem, Israel

Squire J. Booker
Department of Biochemistry and Molecular Biology and Department of Chemistry
The Pennsylvania State University
University Park, Pennsylvania

Christopher L. Bowlus
UCDavis Medical Center
University of California
Sacramento, California

and

Division of Internal Medicine and Liver Unit
School of Medicine
San Paolo Hospital
University of Milan
Milan, Italy

Erdem Buyukbingol
Department of Pharmaceutical Chemistry
Faculty of Pharmacy
Ankara University
Ankara, Turkey

Jessie M. Cameron
Genetics and Genome Biology
The Hospital for Sick Children
Toronto, Ontario, Canada

David A. Carlson
GeroNova Research, Inc.
Richmond, California

Eun Hee Cho
Department of Internal Medicine
University of Ulsan College of Medicine
Seoul, South Korea

Robert M. Cicchillo
Department of Biochemistry and Molecular Biology
The Pennsylvania State University
University Park, Pennsylvania

Britta Diesel
Department of Pharmacy
Saarland University
Saarbrücken, Germany

Sarah J. Fischer
GeroNova Research, Inc.
Richmond, California

Kazuko Fujiwara
Institute for Enzyme Research
The University of Tokushima
Tokushima, Japan

M. Eric Gershwin
Division of Rheumatology, Allergy
and Clinical Immunology
School of Medicine
University of California
Davis, California

Xioaming Gong
Department of Medicine
Rhode Island Hospital

and

Brown Medical School
Providence, Rhode Island

Tyler L. Grove
Department of Chemistry
The Pennsylvania State University
University Park, Pennsylvania

A. Selen Gurkan
Department of Pharmaceutical
Chemistry
Ankara University
Ankara, Turkey

Tory M. Hagen
Linus Pauling Institute
Oregon State University
Corvallis, Oregon

Ryan T. Hamilton
Department of Pharmacology and
Pharmaceutical Sciences
School of Pharmacy
University of Southern California
Los Angeles, California

Derick Han
University of Southern California
Research Center for Liver Diseases
Keck School of Medicine
University of Southern California
Los Angeles, California

Robert A. Harris
Department of Biochemistry and
Molecular Biology
Indiana University School of Medicine
Indianapolis, Indiana

Xiao-Song He
Division of Rheumatology, Allergy,
and Clinical Immunology
School of Medicine
University of California
Davis, California

Erik J. Henriksen
Department of Physiology
University of Arizona College
of Medicine
Tucson, Arizona

Yasuaki Hiromasa
Department of Biochemistry
Kansas State University
Manhattan, Kansas

Harumi Hosaka
Institute for Protein Research
Osaka University
Osaka, Japan

Liangyan Hu
Department of Biochemistry
Kansas State University
Manhattan, Kansas

Kenneth M. Humphries
Oklahoma Medical Research
Foundation
Oklahoma City, Oklahoma

Byounghoon Hwang
Department of Biochemistry
and Molecular Biology
Indiana University School
of Medicine
Indianapolis, Indiana

Stephan Jacob
Internist, Endocrinologist, and
Diabetologist
Villingen-Schwenningen, Germany

Nam Ho Jeoung
Department of Biochemistry and
Molecular Biology
Indiana University School
of Medicine
Indianapolis, Indiana

Mandar Joshi
Department of Biochemistry
and Molecular Biology
Indiana University School
of Medicine
Indianapolis, Indiana

Yehoshua Katzhendler
Department of Medicinal Chemistry
School of Pharmacy
The Hebrew University
Jerusalem, Israel

Alexandra K. Kiemer
Department of Pharmacy
Saarland University
Saarbrücken, Germany

Min-Seon Kim
Department of Internal Medicine
University of Ulsan College
of Medicine
Seoul, South Korea

Eun Hee Koh
Department of Internal Medicine
University of Ulsan College
of Medicine
Seoul, South Korea

Lioubov G. Korotchkina
Department of Biochemistry
School of Medicine and Biomedical
Sciences
University at Buffalo
State University of New York
Buffalo, New York

Philip Y. Lam
Department of Pharmacology
and Pharmaceutical Sciences
School of Pharmacy
University of Southern California
Los Angeles, California

Ki-Up Lee
Department of Internal Medicine
University of Ulsan College
of Medicine
Seoul, South Korea

Kyung-Hoon Lee
Department of Biochemistry
and Molecular Biology
Pennsylvania State University
University Park, Pennsylvania

Jiankang Liu
Institute for Brain Aging
and Dementia
University of California
Irvine, California

Mary C. Maj
Genetics and Genome Biology
The Hospital for Sick Children
Toronto, Ontario, Canada

Emilia Martínez-Galisteo
Department of Biochemistry and Molecular Biology
University of Córdoba
Córdoba, Spain

Yutaro Motokawa
Institute for Enzyme Research
The University of Tokushima
Tokushima, Japan

Atsushi Nakagawa
Institute for Protein Research
Osaka University
Osaka, Japan

Natasha M. Nesbitt
Department of Biochemistry and Molecular Biology
The Pennsylvania State University
University Park, Pennsylvania

Amy C. Nulton-Persson
Department of Physiology and Biophysics
Case Western Reserve University
Cleveland, Ohio

Lester Packer
Department of Pharmacology and Pharmaceutical Sciences
School of Pharmacy
University of Southern California
Los Angeles, California

Carmen Alicia Padilla
Department of Biochemistry and Molecular Biology
University of Córdoba
Córdoba, Spain

Joong-Yeol Park
Department of Internal Medicine
University of Ulsan College of Medicine
Seoul, South Korea

Mulchand S. Patel
Department of Biochemistry
School of Medicine and Biomedical Sciences
University at Buffalo
State University of New York
Buffalo, New York

José Rafael Pedrajas
Department of Experimental Biology
University of Jaén
Jaén, Spain

José Peinado
Department of Biochemistry and Molecular Biology
University of Córdoba
Córdoba, Spain

Tao Peng
Department of Chemistry and Biochemistry
Nieuwland Science Hall
University of Notre Dame
Notre Dame, Indiana

Pablo Porras
Department of Biochemistry and Molecular Biology
University of Córdoba
Córdoba, Spain

Rainer Preiss
Institute of Clinical Pharmacology
University of Leipzig
Leipzig, Germany

Lester J. Reed
Department of Chemistry and Biochemistry
The University of Texas at Austin
Austin, Texas

Raquel Requejo
Department of Biochemistry and Molecular Biology
University of Córdoba
Córdoba, Spain

Brian H. Robinson
Genetics and Genome Biology
The Hospital for Sick Children
Toronto, Ontario, Canada

Thomas E. Roche
Department of Biochemistry
Kansas State University
Manhattan, Kansas

Shlomo Sasson
Department of Pharmacology
School of Pharmacy
The Hebrew University
Jerusalem, Israel

William T. Self
Department of Molecular Biology and Microbiology
College of Medicine
University of Central Florida
Orlando, Florida

Carlo Selmi
Division of Rheumatology, Allergy, and Clinical Immunology
School of Medicine
University of California
Davis, California

Kate Petersen Shay
Linus Pauling Institute
Oregon State University
Corvallis, Oregon

Swapna Shenvi
Linus Pauling Institute
Oregon State University
Corvallis, Oregon

Luke I. Szweda
Oklahoma Medical Research Foundation
Oklahoma City, Oklahoma

Jens Teichert
Institute of Clinical Pharmacology
University of Leipzig
Leipzig, Germany

Heinz Ulrich
Private Medical Practice
Niedernberg, Germany

Karyn L. Young
GeroNova Research, Inc.
Richmond, California

Section I

Discovery and Molecular Structure

1 A Trail of Research on Lipoic Acid

Lester J. Reed

CONTENTS

My research on lipoic acid began with its isolation and characterization and then went on to identification of its functional form. The trail then led to elucidation of major features of the structure, function, and regulation of α-keto acid dehydrogenase multienzyme complexes. This trail of discovery started in the spring of 1949, about 6 months after I joined the faculty of the Department of Chemistry at the University of Texas at Austin. At that time I started working on the isolation of a factor that stimulated growth of *Lactobacillus casei* on an acetate-free synthetic medium. Research on the "acetate-replacing factor" was initiated by Esmond Snell and associates at the University of Wisconsin and then at the University of Texas at Austin. I inherited this project in the spring of 1949. We established that this factor is widely distributed in animal, plant, and microbial cells and that animal liver is a rich source. The factor is tightly bound to liver protein, and it is released by proteolysis or by acid hydrolysis. At that time, pharmaceutical companies were processing large amounts of pork and beef liver to obtain extracts suitable for the treatment of pernicious anemia. The active principle was shown later to be vitamin B_{12}. Fresh liver was extracted with warm water, and the residual liver proteins and fatty material were dried and sold as an animal feed supplement. Arrangements were made with Eli Lilly and Co. to obtain liver residue, and we developed procedures for extracting and purifying the acetate-replacing factor. We were eventually able to process about 6 lb of liver residue at a time. A 16,000- to 50,000-fold purification was achieved.

In the late 1940s and early 1950s, several other groups were trying to isolate factors that were similar to, if not identical with, the acetate-replacing factor. These factors included the "pyruvate oxidation factor" of D.J. O'Kane and I.C. Gunsalus that was essential for oxidation of pyruvate to acetate and carbon dioxide by *Streptococcus faecalis* cells grown in a synthetic medium; "protogen," an unidentified growth factor for a protozoan, *Tetrahymena geleii*, that was being purified by E.L.R. Stokstad, T.H. Jukes, and their collaborators at Lederle

Research Laboratories; and the "B.R. Factor" of L. Kline and H.A. Barker that was required for growth of *Butyribacterium rettgeri* on a medium containing lactate as the fermentable carbon source.

In the fall of 1950, a collaboration between Reed, Gunsalus, and the Lilly Research Laboratories was undertaken to isolate the acetate-replacing/pyruvate oxidation factor. The Lilly group adapted and scaled up isolation procedures developed by Reed and coworkers. Concentrates of the factor that were 0.1%–1% pure were sent to the Reed laboratory for further processing. I obtained the first pale-yellow crystals of the factor, about 3 mg, on or about March 15, 1951, a truly memorable occasion. It was partially characterized and given the trivial name α-lipoic acid [1]. The prefix α (alpha) was not used in a chemical sense, but to distinguish the isolated compound from structurally related substances present in extracts of biological materials. The isolation procedure involved an approximately 300,000-fold purification. A total of approximately 30 mg of crystalline lipoic acid was eventually isolated. We estimated that approximately 10 tons of liver residue were processed to obtain this small amount of the pure substance. NMR and mass spectrometers were not available in those days, but it was possible to establish that lipoic acid is a cyclic disulfide, either 6,8-, 5,8-, or 4,8-dithiooctanoic acid. That the correct structure is 6,8-dithiooctanoic acid (1,2-dithiolane-3-valeric acid) was established by synthesis of DL-lipoic acid, first achieved by E.L.R. Stokstad and associates at Lederle Laboratories. This research group proposed the name "6-thioctic acid" for the parent substance (protogen-A, α-lipoic acid). In 1955, the American Society of Biological Chemists recognized priority of the name "lipoic acid" and adopted it as the trivial designation of 1,2-dithiolane-3-valeric acid.

I was intrigued by this simple yet unique substance and wanted to know more about its biological function, i.e., with what and how does it function in living cells. We, therefore, set out to explore this trail, which turned out to be even more exciting than the isolation and characterization of lipoic acid. Elucidation of the mechanism of oxidative decarboxylation of α-keto acids is a fascinating chapter in modern biochemistry. I shall review briefly the major developments in this story. Before the isolation of lipoic acid and its characterization as a cyclic disulfide, contributions from several laboratories had established several cofactor requirements for the oxidative decarboxylation of α-keto acids represented by the following equation:

$$\mathrm{RCOCO_2^-} + \mathrm{CoASH} + \mathrm{NAD^+} \rightarrow \mathrm{RCOSCoA} + \mathrm{CO_2} + \mathrm{NADH}$$

In addition to CoA and NAD^+, thiamin diphosphate, a divalent metal ion, and protein-bound lipoic acid are required. A requirement for FAD was demonstrated later. The presence of a disulfide linkage in lipoic acid led Gunsalus to propose that lipoic acid underwent a cycle of reactions in α-keto acid oxidation comprising reductive acylation, acyl transfer, and electron transfer. Lipoic acid was visualized as functioning after thiamin diphosphate and before CoA and NAD^+. Gunsalus, Lowell Hager, and associates obtained evidence for this proposal using lipoic acid and derivatives thereof in substrate amounts. However, the physiological reactions presumably involve catalytic amounts of protein-bound lipoic acid. Elucidation of the nature of the functional form of lipoic acid was essential for verification of the postulated reactions and for further clarification of mechanism.

Reed, Hayao Nawa, and coworkers showed in the late 1950s that the lipoyl moiety in the *E. coli* pyruvate and α-ketoglutarate dehydrogenase complexes is attached in amide linkage to the ε-amino group of a lysine residue [2] (Figure 1.1). An enzyme that hydrolyzes the lipoyllysyl linkage, lipoamidase, as well as an ATP-dependent enzyme that reincorporates the lipoyl moiety, a lipoate-protein ligase, were detected in *S. faecalis* extracts and partially purified. These two enzymes, together with radioactive lipoic acid-$^{35}S_2$ proved to be invaluable in providing direct, unequivocal evidence of the involvement of protein-bound lipoic acid in the CoA- and NAD^+-linked oxidative decarboxylation of pyruvate and α-ketoglutarate and in providing clarification of the mechanism of model reactions catalyzed by the pyruvate and α-ketoglutarate dehydrogenase complexes and components thereof. This evidence comprised a demonstration of inactivation and reactivation of the enzyme or enzyme complex accompanying, respectively, release and reincorporation of the radioactive lipoyl moiety.

Before 1950, pyruvate and α-ketoglutarate oxidation had been studied mainly with particulate preparations that were unsuitable for detailed analysis. Solubilization of bacterial and animal α-keto acid oxidation systems in the early 1950s in the laboratories of Severo Ochoa and David Green was a significant advance. Seymour Korkes, Gunsalus, and Ochoa succeeded in separating the pyruvate oxidation system of *E. coli* into two components, designated fraction A and fraction B. Using lipoic acid and dihydrolipoic acid in substrate amounts, Hager and Gunsalus showed that fraction A contained a lipoyl transacetylase and that fraction B contained a lipoyl dehydrogenase. Richard Schweet and associates isolated a CoA- and NAD^+-linked pyruvate oxidation system from pigeon breast muscle in a highly purified state, with an apparent molecular weight of about 4 million. D.R. Sanadi and associates isolated a CoA- and NAD^+-linked α-ketoglutarate oxidation system from pig heart with an apparent molecular weight of 2 million.

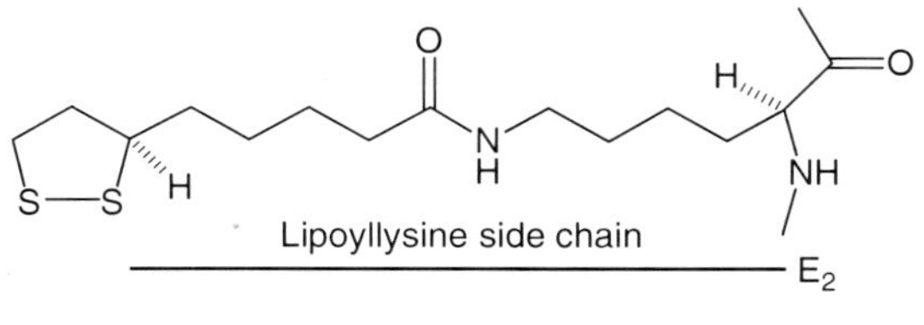

FIGURE 1.1 Functional form of lipoic acid. The carboxyl group of lipoic acid is bound in amide linkage to the ε-amino group of a lysine residue in the acyltransferase component (E_2) of the α-keto acid dehydrogenase complexes. This linkage may provide a "swinging arm" that facilitates communication between active sites. (From Reed, L.J., *J. Biol. Chem.*, 276, 38329, 2001. With permission.)

In my laboratory, we developed mild procedures for purification of the pyruvate and α-ketoglutarate oxidation systems from *E. coli* (Crookes strain). By the late 1950s, Reed, Masahiko Koike, and associates [3] succeeded in isolating these enzyme systems as highly purified functional multienzyme units with molecular weights of 4.8 million and 2.4 million, respectively. By careful and persistent work over a period of several years, we dissected the PDH and KGDH complexes into their component enzymes, characterized them, and reassembled the large functional units from the isolated enzymes [4]. We demonstrated that each of these functional units is composed of multiple copies of three enzymes: a pyruvate or α-ketoglutarate decarboxylase-dehydrogenase (E_1), a dihydrolipoamide acetyltransferase or a succinyltransferase (E_2), and a flavoprotein, dihydrolipoamide dehydrogenase (E_3). These three enzymes, acting in sequence, catalyze the reactions shown in Figure 1.2. E_1 catalyzes both the decarboxylation of the α-keto acid (Reaction 1) and the subsequent reductive acylation of the lipoyl moiety, which is covalently bound to E_2 (Reaction 2). E_2 catalyzes the acyl transfer to CoA (Reaction 3), and E_3 catalyzes the reoxidation of the dihydrolipoyl moiety with NAD^+ as the ultimate electron acceptor (Reactions 4 and 5). Binding experiments showed that the acetyltransferase serves a dual function, a catalytic function and a structural function, i.e., a scaffold for binding and localizing E_1 and E_3. In dilute acetic acid (0.83 M, pH 2.6) the acetyltransferase component dissociated into inactive subunits with a molecular weight of about 70,000. Dilution of the acidic subunit into suitable buffers resulted in restoration of enzymatic activity and the characteristic structure of the native acetyltransferase unit. The acetyltransferase appeared to be a self-assembling system. It was evident that the lipoyl moiety undergoes a cycle of

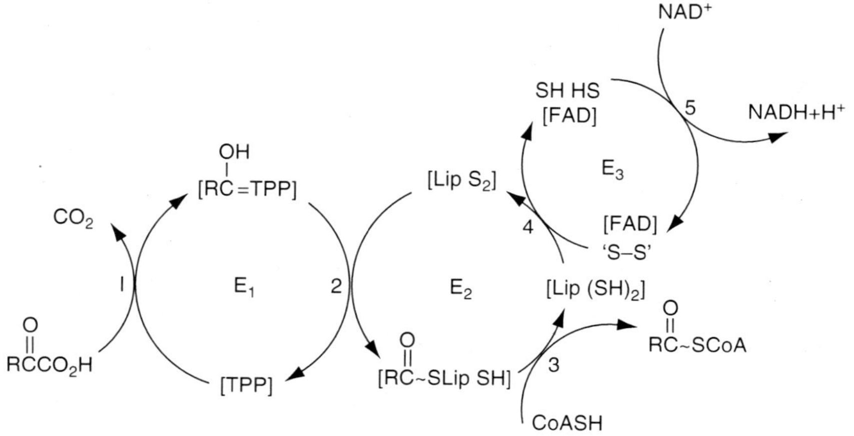

FIGURE 1.2 Reaction sequence in α-keto acid oxidation. TPP, thiamin diphosphate; $LipS_2$ and $Lip(SH)_2$, lipoyl moiety and its reduced form. (From Reed, L.J., *J. Biol. Chem.*, 276, 38329, 2001. With permission.)

transformations, i.e., reductive acylation, acyl transfer, and electron transfer, involving three separate enzymes within a complex in which movement of the individual enzymes is restricted and from which intermediates do not dissociate.

These were exciting times for us in the late 1950s and early 1960s. Our concept of the macromolecular organization of the PDH complex that emerged from these biochemical studies is that of an organized mosaic of enzymes in which each of the component enzymes is uniquely located to permit efficient coupling of the individual reactions catalyzed by these enzymes. This concept was confirmed and extended by electron microscopy studies conducted by my associate Robert Oliver. Electron micrographs of the *E. coli* PDH complex and its component enzymes negatively stained with phosphotungstate revealed that the complex had a polyhedral structure with a diameter of about 300 Å, that the acetyltransferase (E_2) occupied the center of the polyhedron, and that the molecules of E_1 and E_3 were distributed on its surface. The shape of the acetyltransferase indicated that it had a cube-like structure. The shape of the succinyltransferase component of the *E. coli* KGDH complex was very similar. These results, together with biochemical data, demonstrated that both E_2s consist of 24 identical polypeptide chains arranged as eight trimers (morphological subunits) at the vertices of a cube (Figure 1.3a). This proposed structure was confirmed later by x-ray diffraction analyses carried out by collaborators David DeRosier and Marvin Hackert demonstrating that both acyltransferases possess 432 molecular symmetry. Our interpretative model of the macromolecular organization of the *E. coli* PDH complex depicted 12 E_1 dimers and 6 E_3 dimers arranged, respectively, on the 12 edges and in the 6 faces of the cube-like E_2.

All dihydrolipoamide acyltransferases possess a unique multidomain structure. This architectural feature was revealed initially in my laboratory in the late 1970s by limited proteolysis studies of the *E. coli* dihydrolipoamide acetyltransferase containing [2-^{3}H]-lipoyl moieties. Reed, Dennis Bleile, and associates [5] found

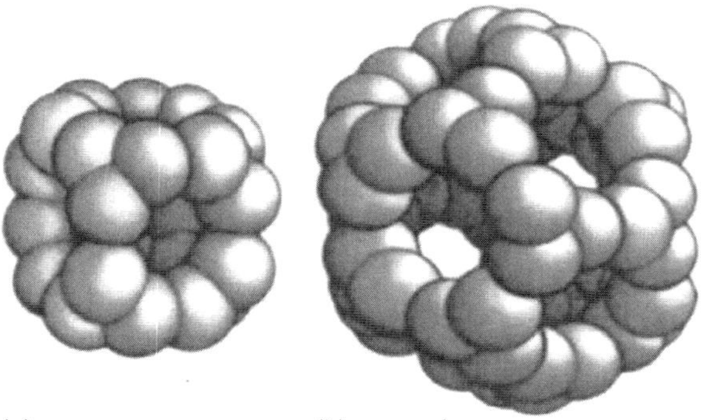

(a) (b)

FIGURE 1.3 Schematic representations of the 24- and 60-mer polyhedra of E_2 cores. (From Reed, L.J., *J. Biol. Chem.*, 276, 38329, 2001. With permission.)

that limited tryptic digestion at pH 7.0 and 4°C cleaved the E_2 subunits ($M_r \sim 64{,}500$) into two large fragments, an outer lipoyl-bearing domain and an inner catalytic and subunit binding domain. The latter fragment ($M_r \sim 29{,}600$) had a compact structure, and it possessed the inter-subunit binding sites of the acetyltransferase, the binding sites for E_1 and E_3, and the catalytic site for acetyl transfer. The assemblage of compact catalytic and subunit binding domains constitutes the inner core of the acetyltransferase, conferring the cube-like appearance of this E_2 seen with the electron microscope. The other tryptic fragment ($M_r \sim 31{,}600$), designated the lipoyl domain, contained the covalently bound lipoyl moieties and had an extended structure. We suggested that the two domains are connected by a trypsin-sensitive hinge region. These early findings on the domain structure of dihydrolipoamide acyltransferases were confirmed and extended by studies involving molecular genetics, limited proteolysis, and proton NMR spectroscopy in the laboratories of John Guest and Richard Perham. Briefly, the amino-terminal part of the acyltransferases contains one, two, or three highly similar lipoyl domains, each of about 80 amino acid residues. The lipoyl domain (or domains) is followed by another structurally distinct segment that is involved in binding E_3 and/or E_1. These domains are linked to each other and to the carboxyl-terminal part of the polypeptide chain (catalytic domain) by flexible segments (hinge regions) that are rich in alanine, proline, and charged amino acid residues. These segments are thought to provide flexibility to the lipoyl domains, facilitating active site coupling within the multienzyme complexes.

In the late 1960s, part of our research effort was directed toward isolation and characterization of the mammalian PDH and KGDH complexes, which are localized to mitochondria within the inner membrane-matrix compartment. Procedures were developed for preparation of mitochondria on a large scale from bovine kidney and heart, and relatively mild procedures were developed to isolate the PDH and KGDH complexes from the mitochondrial extracts. In the course of attempts to stabilize these complexes in crude extracts of bovine kidney mitochondria, Tracy Linn observed that the PDH complex, but not the KGDH complex, underwent a time-dependent inactivation in the presence of ATP. A systematic investigation by Linn, Flora Pettit, and coworkers revealed that the bovine kidney and heart PDH complexes are regulated by a phosphorylation–dephosphorylation cycle [6]. Phosphorylation and concomitant inactivation of the complex is catalyzed by an ATP-dependent kinase, which is tightly bound to the complex, and dephosphorylation and concomitant reactivation are catalyzed by an Mg^{2+}-dependent phosphatase, which is loosely attached to the complex. It seemed curious at the time (1968) that inactivation of the PDH complex by phosphorylation had not been detected earlier. The explanation may lie in a remark by Henry Lardy after receiving a preprint of our paper on the phosphorylation and inactivation of the mammalian PDH complex. This finding "explains why we have never been able to get pyruvate to be oxidized in submitochondrial particles, because we invariably add ATP to keep things in the 'optimum' state."

Over a period of several years, our group separated the bovine kidney and heart PDH complexes into their component enzymes (E_1 ($\alpha_2\beta_2$), E_2, E_3, PDH kinase, and PDH phosphatase) and characterized the individual enzymes [7]. We showed that the $E_1\alpha$ subunit undergoes phosphorylation on three seryl residues. We were surprised by the appearance in the electron microscope of negatively stained preparations of the mammalian dihydrolipoamide acetyltransferase (E_2). Its morphological subunits appeared to be located at the vertices of a pentagonal dodecahedron instead of at the vertices of a cube. Thus, our electron microscope studies revealed that there are two distinct polyhedral forms of E_2, the cube and the pentagonal dodecahedron (Figure 1.3). The former design, a cube, consists of 24 E_2 subunits (eight trimers) arranged with octahedral (432) symmetry. This design is exhibited by the E_2 components of the *E. coli* PDH and KGDH complexes and by the E_2 components of the mammalian KGDH and branched chain α-keto acid dehydrogenase complexes. The latter design, a pentagonal dodecahedron, consists of 60 E_2 subunits (20 trimers) arranged with icosahedral (532) symmetry. E_2 components of this morphology are found in the PDH complexes from mammalian and avian tissues, fungi, and the Grampositive bacterium *Bacillus stearothermophilus*. A morphological unit consisting of three E_2 subunits appeared to be important in the assembly of both types of polyhedral forms. These conclusions were confirmed and extended by results from x-ray diffraction analysis by collaborators David DeRosier and Marvin Hackert and by Wim Hol and associates. The bovine heart PDH complex has a molecular weight of about 9.5 million. Its subunit composition is 60 E_2 subunits, 30 E_1 tetramers, and 12 E_3 dimers, which are positioned on the E_2 core by 12 E_3-binding protein monomers. We proposed that the E_1 tetramers are located on the 30 edges and the E_3 dimers in the 12 faces of the E_2 pentagonal dodecahedron.

Subsequent research on the bovine PDH phosphatase and on structure–function relationships in the PDH complex from *Streptococcus cerevisiae* is described elsewhere [8].

REFERENCES

1. Reed, L.J., DeBusk, B.G., Gunsalus, I.C., and Hornberger, C.S., Jr., Crystalline α-lipoic acid: A catalytic agent associated with pyruvate dehydrogenase, *Science*, 114, 93–94, 1951.
2. Nawa, H., Brady, W.T., Koike, M., and Reed, L.J., Studies on the nature of protein-bound lipoic acid, *J. Am. Chem. Soc.*, 82, 896–903, 1960.
3. Koike, M., Reed, L.J., and Carroll, W.R., α-Keto acid dehydrogenation complexes. I. Purification and properties of pyruvate and α-ketoglutarate dehydrogenation complexes of *Escherichia coli*, *J. Biol. Chem.*, 235, 1924–1930, 1960.
4. Koike, M., Reed, L.J., and Carroll, W.R., α-Keto acid dehydrogenation complexes. IV. Resolution and reconstitution of the *Escherichia coli* pyruvate dehydrogenation complex, *J. Biol. Chem.*, 238, 30–39, 1963.
5. Bleile, D.M., Munk, P., Oliver, R.M., and Reed, L.J., Subunit structure of dihydrolipoyl transacetylase component of pyruvate dehydrogenase complex from *Escherichia coli*, *Proc. Natl. Acad. Sci. USA*, 76, 4385–4389, 1979.
6. Linn, T.C., Pettit, F.H., and Reed, L.J., α-Keto acid dehydrogenase complexes. X. Regulation of the activity of the pyruvate dehydrogenase complex from beef kidney

mitochondria by phosphorylation and dephosphorylation, *Proc. Natl. Acad. Sci. USA*, 62, 234–241, 1969.

7. Linn, T.C., Pelley, J.W., Pettit, F.H., Hucho, F., Randall, D.D., and Reed, L.J., α-Keto acid dehydrogenase complexes. XV. Purification and properties of the component enzymes of the pyruvate dehydrogenase complexes from bovine kidney and heart, *Arch. Biochem. Biophys.*, 148, 327–342, 1972.
8. Reed, L.J., A trail of research from lipoic acid to α-keto acid dehydrogenase complexes, *J. Biol. Chem.*, 276, 38329–38336, 2001.

2 Lipoic Acid Biosynthesis

Natasha M. Nesbitt, Robert M. Cicchillo, Kyung-Hoon Lee, Tyler L. Grove, and Squire J. Booker

CONTENTS

LIPOIC ACID AND LIPOYL CARRIER PROTEINS

α-Lipoic acid (1,2-dithiolane-3-pentanoic acid, or 6,8-thioctic acid) is a sulfur-containing cofactor found in most prokaryotic and eukaryotic microorganisms, as well as plant and animal tissue (Figure 2.1A) [1]. It is best known as a requisite component of several multienzyme complexes that function in the oxidative decarboxylation of various α-keto acids and glycine, as well as the oxidative cleavage of 3-hydroxy-2-butanone (acetoin) to acetaldehyde and acetyl-coenzyme A [2–7]. In its capacity as a cofactor, it is bound covalently in an amide linkage to the N^6-amino group of a lysine residue on one of the proteins of the complex, producing a long (14 Å) tether (Figure 2.1B), which facilitates the direct

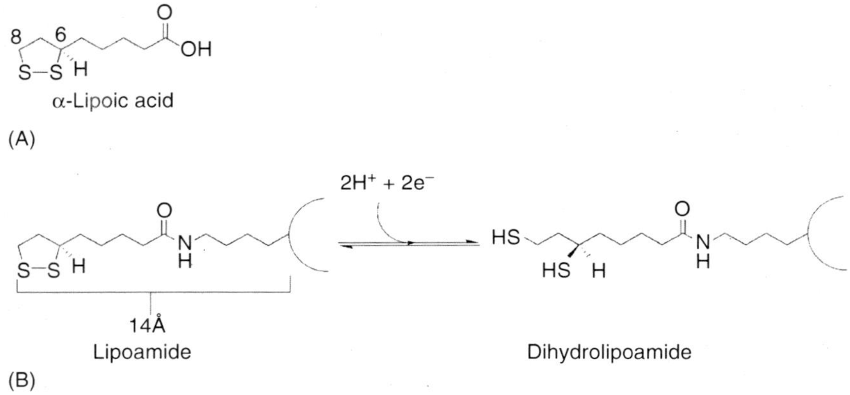

FIGURE 2.1 Structures of various forms of lipoic acid: (A) free acid; (B) lipoamide (left) and dihydrolipoamide (right) drawn attached to a lysine residue on a lipoyl carrier protein.

channeling of intermediates among the various active sites located on different subunits [8–10]. The key functional property of the lipoyl cofactor is its ability to undergo redox chemistry, interchanging between the cyclic disulfide (lipoamide) and reduced dithiol (dihydrolipoamide) forms (Figure 2.1B). The disulfide form of the cofactor acts as an electron acceptor in each of the multienzyme complexes that employ it, becoming reduced by two-electrons during turnover. Reoxidation of dihydrolipoamide must take place for additional rounds of turnover to occur, and is catalyzed by the flavoenzyme lipoamide dehydrogenase. The reducing equivalents generated in each overall reaction are ultimately transferred to NAD^+, affording NADH and a proton.

The role of the lipoyl cofactor in the pyruvate and α-ketoglutarate dehydrogenase complexes (PDC and KDC, respectively) is well established [11–14]. Other complexes in which it is known to function in a similar capacity are the branched-chain α-keto acid dehydrogenase complex (BCKDC), which catalyzes the oxidative decarboxylation of the branched-chain α-keto acids that are derived from the transamination of valine, leucine, and isoleucine; the glycine cleavage system (GCS), which degrades glycine to NH_4^+ and CO_2 with transfer of its α carbon to tetrahydrofolate, yielding N^5,N^{10}-methylene tetrahydrofolate; and the acetoin dehydrogenase complex (ADC), which catalyzes the oxidative cleavage of acetoin to acetaldehyde and acetyl-CoA, a reaction that allows certain bacteria to use acetoin as their sole carbon source [3–7]. The PDC, KDC, BCKDC, and ADC are all composed of multiple copies of three distinct proteins: E_1, E_2, and E_3. Each E_1 subunit requires thiamin diphosphate for its catalytic activity, which it uses to catalyze the decarboxylation of the appropriate α-keto acid and the subsequent reductive modification (e.g., acetylation in the PDC and succinylation in the KDC) of the lipoyl cofactor. The lipoyl cofactor is covalently attached to a conserved lysine residue contained within the E_2 subunit, which catalyzes a transthioesterification reaction in which the modification is transferred to coenzyme A (CoA). The resulting dihydrolipoamide group is reoxidized to its functional form by E_3, a flavin adenine dinucleotide (FAD)-containing dihydrolipoamide dehydrogenase. The E_1 and E_2 subunits in each of these complexes are products of different genes, and are specific for the dehydrogenase complex in which they participate. The E_3 subunits, however, are products of the same gene in *E. coli* (*lpd*) and many, but not all, other organisms [15–19]. One exception appears to be the bacterium *Pseudomonas putida*, which encodes three or perhaps four different E_3 subunits in its genome; one for the KDC and PDC, and one each for the remaining lipoyl-utilizing multienzyme complexes (BCKDC, GCS, and ADC) [17,20,21].

In the KDC, PDC, BCKDC, and ADC, the E_2 subunit serves as the core of the complex. It is composed, from N-terminus to C-terminus, of one to three lipoyl domains, each of which can bind one lipoyl cofactor, and a peripheral subunit-binding domain, which is involved in the binding of the E_1 or E_3 subunits to the E_2 core. A large core-forming acyltransferase (catalytic) domain resides at the C-terminus of the E_2 polypeptide [8,10,22]. These domains are linked by a flexible polypeptide chain approximately 25–30 residues in length and rich in alanine, proline, and charged amino acids [23–27].

Unlike the KDC, PDC, BCKDC, and ADC, the glycine cleavage system is not a multienzyme complex, but a system of independent proteins that loosely associate. The GCS is composed of four proteins—P, H, T, and L—that in conjunction catalyze both the decarboxylation and synthesis of glycine [7,28,29]. The homodimeric P-protein, which is functionally equivalent to the E_1 subunits of the KDC, PDC, BCKDC, and ADC, catalyzes the pyridoxal 5′-phosphate (PLP)-dependent decarboxylation of glycine, yielding CO_2 [30–32]. The resulting aminomethyl group is transferred to the lipoyl-bearing H-protein with concomitant reduction of the lipoyl group, forming an aminomethylthio intermediate [33–37]. The T-protein catalyzes the cleavage of the aminomethylthio intermediate, releasing ammonia and transferring the carbon atom to tetrahydrofolate with concomitant production of N^5,N^{10}-methylenetetrahydrofolate [34,38]. The lipoyl group is then reoxidized by the L-protein, which in most organisms is also the E_3 subunit of the lipoyl-containing dehydrogenase complexes [16,32,39].

Three-dimensional structures of lipoyl domains from the KDC of *E. coli* [40] and *Azotobacter vinelandii* [41]; the PDC from *E. coli* [42], *A. vinelandii* [43], and *Bacillus stearothermophilus* [44]; and the human BCKDC [45] have been determined by nuclear magnetic resonance (NMR) spectroscopy. The prototype structure is that of the lipoyl domain from *B. stearothermophilus*, which is composed of residues 1–79 of the dihydrolipoamide acetyltransferase subunit (Figure 2.2, left) [44]. The overall structure resembles a flattened eight-stranded β-barrel that wraps around a well-defined core of hydrophobic residues. The lipoyllysine is located at the tip of one β-turn, and the N- and C-termini are in close proximity to each other in adjacent β-strands that are on the opposite side of

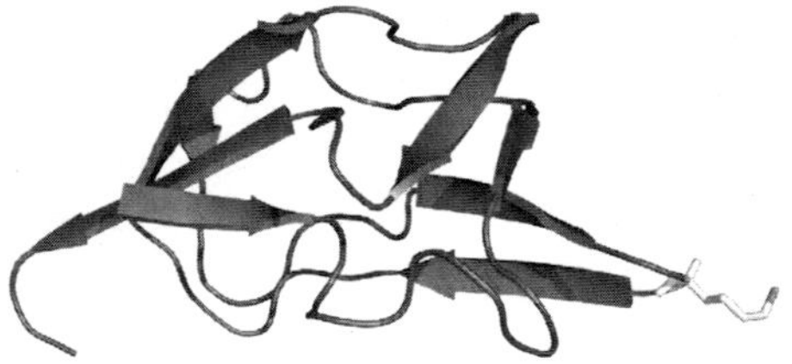

Lipoyl domain from PDC
of *Bacillus stearothermophilus*

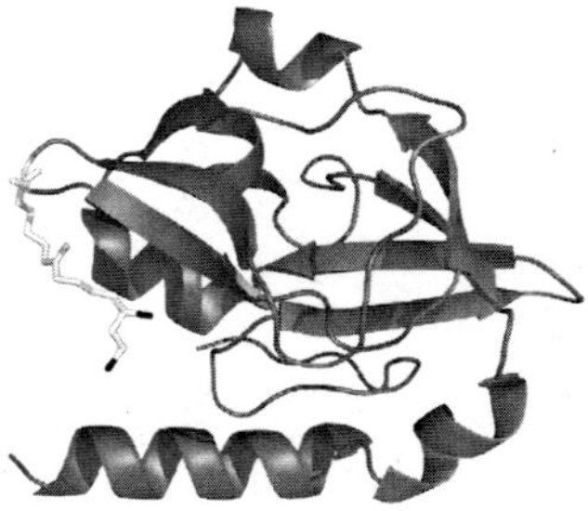

H-protein from *Pisum sativum*

FIGURE 2.2 Structures of lipoyl carrier proteins: (Left) Structure of the lipoyl domain of the E_2 subunit of the PDC from *Bacillus stearothermophilus* solved by NMR (PDB 1LAB). The lysine group that becomes lipoylated (Lys42) is shown in stick format, and is at the tip of a β-hairpin structure. (Right) Structure of the H-protein of the glycine cleavage system from *Pisum sativum* (PDB 1DXM). The lipoyllysine group (Lys63) is shown in stick format and is found at the tip of a β-hairpin structure. The lipoyl group is shown in its reduced form, and is believed to be freely swinging in this state. Both structures were prepared using the Pymol Molecular Graphics System (http://www.pymol.org).

the domain from the lipoyllysine group. The lipoyl-accepting lysine is flanked by an aspartate residue on its N-terminal side and an alanine residue on its C-terminal side. When the Asp_{41}→Lys, Lys_{42}→Ala (Lys42 is the site of lipoylation) double variant was constructed on the lipoyl domain of the *E. coli* PDC, no lipoylation was observed, suggesting that the absolute positioning of the target lysine is essential for proper recognition [46].

Structures of the H-protein from *Pisum sativum* and *Thermus thermophilus* have been solved by x-ray diffraction and resemble the structures of the lipoyl domains of the PDC, KDC, and BCKDC [5,47,48]. The structure of the *P. sativum* protein containing a reduced lipoyl cofactor is shown in Figure 2.2 (right). Similar to the lipoyl domains of the E_2 proteins, it is composed of 131 amino acids consisting of seven β-strands arranged in two antiparallel β-sheets that form a "sandwich." Again, the lipoyl group is covalently attached to a lysine residue (Lys63) that resides at the tip of a β-hairpin structure. In addition, the cofactor appears to be freely exposed to the solvent in its reduced form, whereas in the structure containing the aminomethylthio adduct, the lipoyl group is locked into a cleft on the protein [49]. One unique aspect of the H-protein structure is that the N-terminal exposed loop found in lipoyl domains of E_2 proteins is replaced by a helix. The H-protein also contains two additional β-strands at the N-terminus and a short helix at the C-terminus [5,48], rendering it significantly larger than the lipoyl domains of the E_2 subunits of the PDC and KDC. The structure of the H-protein from *T. thermophilus* resembles that of *P. sativum*, and was solved by molecular replacement using the latter structure [47].

LIPOIC ACID BIOSYNTHESIS: EARLY METABOLIC FEEDING STUDIES

A great deal of what is currently known about the details of lipoic acid biosynthesis at the level of bond making and bond breaking derives from in vivo feeding studies in *E. coli* that were conducted in the 1970s and early 1980s. An excellent and comprehensive review of these studies has been published, which notably highlights the syntheses of various potential in vivo precursors to lipoic acid and the elegant chemical methods used to interrogate various mechanistic scenarios [50]. One of the striking insights from this review are the strong parallels between the biosynthesis of lipoic acid and the biosynthesis of biotin, another sulfur-containing biological cofactor, which now extends even to the primary structures and reaction mechanisms of lipoyl synthase and biotin synthase, the enzymes that catalyze the sulfur insertion steps in the biosynthesis of these two cofactors [51–62].

Lipoic acid is one of nature's least complex cofactors with respect to chemical structure; it merely contains two sulfur atoms inserted into C–H bonds at carbons 6 and 8 of *n*-octanoic acid, which induced speculation that this short-chain fatty acid may be a direct biosynthetic precursor to lipoic acid. Evidence to support this hypothesis was provided in 1964 by Lester Reed in an abstract from the Third International Symposium on the Chemistry of Natural Products held in Kyoto, Japan, but related experimental details were never published in a mainstream journal. According to the abstract, [1-^{14}C]lipoic acid could be isolated from *E. coli* when its growth media was supplemented with [1-^{14}C]octanoic acid, but not with [1-^{14}C]hexanoic acid or [1,6-^{14}C]adipic acid [63]. Almost 10 years later, the subject of octanoic acid as a precursor to lipoic acid was revisited in two independent studies by the laboratories of Ronald Parry and Robert White. In one study, [1-^{14}C]octanoate was administered to shake cultures of *E. coli* with the goal of assessing whether [1-^{14}C]lipoic acid is produced. The cultures were harvested and protein-bound lipoic acid was liberated by acid hydrolysis. After addition of carrier lipoic acid and proper derivatization, the product was recrystallized to constant radioactivity, resulting in an incorporation of 0.17%. Further degradation studies, in which all of the incorporated radioactivity was shown to be in the carboxylate group of the recrystallized lipoic acid, provided firm evidence that octanoic acid served as a direct precursor to lipoic acid and that the radioactivity was not incorporated via processes that are related to metabolism of the administered fatty acid (Figure 2.3A) [64].

The synthesis and use of regiospecific and stereospecific isotopically labeled forms of octanoic acid yielded further insights into the biosynthesis of lipoic acid. The coadministration of [1-^{14}C]octanoate with [5(*R,S*)-^{3}H]octanoate (Figure 2.3B), [7(*R,S*)-^{3}H]octanoate (Figure 2.3C), or [8-^{3}H]octanoate (Figure 2.3D), and subsequent isolation and determination of the ^{3}H/^{14}C ratio in the isolates showed that the ratio in the lipoic acid product was essentially unchanged from that in the precursor compounds, suggesting that hydrogens at carbons 5 and 7 are not removed during the process of sulfur insertion [64]. This finding would also suggest that unsaturated species such as 4-, 5-, 6-, or 7-octenoic acid are not

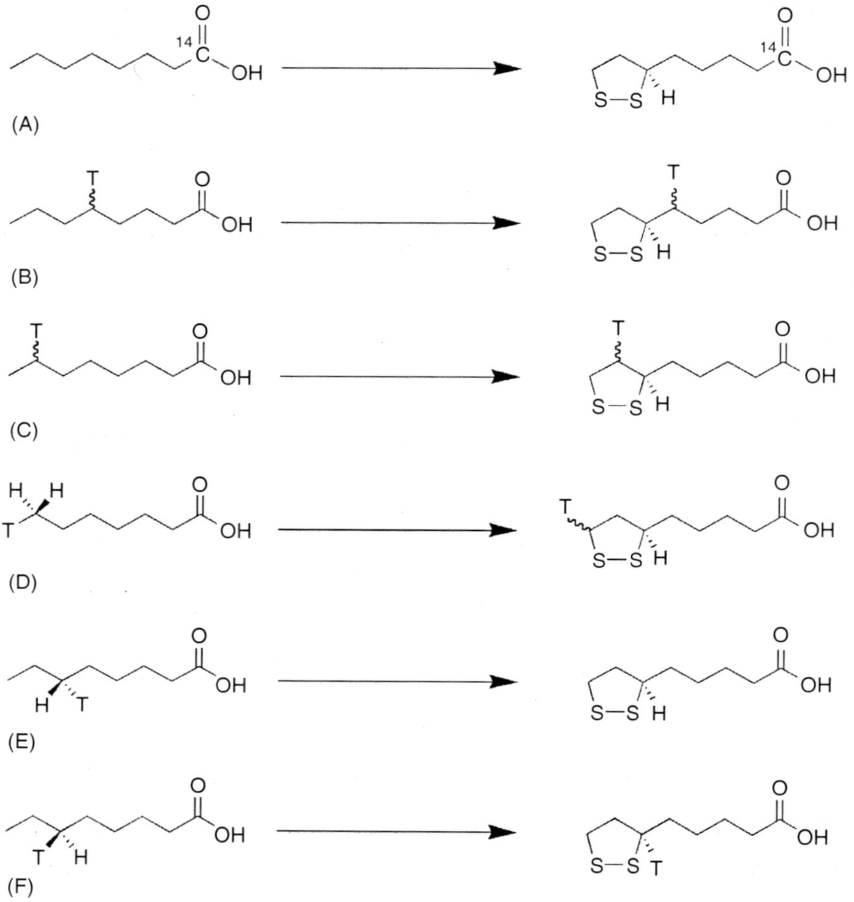

FIGURE 2.3 Structures of molecules used in metabolic labeling studies to probe the mechanism of lipoic acid biosynthesis: (A) [1-^{14}C]octanoic acid; (B) [5(*R*,*S*)-^{3}H]octanoic acid; (C) [7(*R*,*S*)-^{3}H]octanoic acid; (D) [8-^{3}H]octanoic acid; (E) [6(*R*)-^{3}H]octanoic acid; (F) [6(*S*)-^{3}H]octanoic acid.

Note: "T" is used to denote tritium (^{3}H).

intermediates in the biosynthesis of lipoic acid. The observation that the ^{3}H to ^{14}C ratio is unchanged with the [8-^{3}H]octanoate precursor compound was attributed to a significant k_H/k_T kinetic isotope effect at C-8, since the structure of lipoic acid necessitates that a hydrogen atom be removed from this position in octanoic acid to allow for sulfur insertion. As might be expected, when [1-^{14}C] octanoate was coadministered with [6(*R*,*S*)-^{3}H]octanoate, the percent of tritium retained in lipoic acid was 49.8, consistent with the stereoselective removal of only one of the two hydrogens at C-6. To resolve which of the two hydrogens (6-*proS* or 6-*proR*) is removed during turnover, [6(*S*)-^{3}H]- and [6(*R*)-^{3}H]octanoic acids were synthesized and used to supplement the growth media. Coadministration of [6(*R*)-^{3}H]octanoic acid with [1-^{14}C]octanoic acid resulted in 10.9% tritium retention in isolated lipoic acid (Figure 2.3E), while coadministration of [6(*S*)-^{3}H] octanoic acid with [1-^{14}C]octanoic acid resulted in 84% tritium retention (Figure 2.3F), indicating that removal of the 6-*proR* hydrogen is the predominant pathway for formation of lipoic acid, although removal of the 6-*proS* hydrogen might exist as a minor pathway [65]. Since the natural isomer of lipoic acid is of the *R* configuration at C-6, it follows that sulfur insertion takes place with inversion of configuration at this carbon. Note that there is a difference in the Cahn–Ingold–Prelog assignment at C-6 between octanoic acid and lipoic acid, which accounts for the inversion of configuration although the stereochemical assignment is the same.

Several of the above findings were reached independently in stable isotope metabolic feeding studies that were initially designed to probe the origins, stoichiometries, and stereochemistries of hydrogen atoms incorporated into fatty acids during their biosynthesis [66,67]. When *E. coli* B cells were cultured in a defined medium in the presence of [*methyl*-$^{2}H_{3}$]acetate, and their lipoic acid content analyzed by gas chromatography–mass spectrometry (GC–MS) after appropriate extraction and derivatization, the distribution of deuterium in the sample suggested that the cofactor was biosynthesized directly from a fatty acid. In fact, octanoic acid that was extracted from the same cultures produced a mass chromatogram that exhibited the same relative intensities for the various fragment ions as lipoic acid, except they were shifted upward by one mass unit. Moreover, GC–MS analysis of lipoic acid extracted from cells that were fed [U-$^{2}H_{15}$]octanoate indicated that 90.3% of the lipoic acid had been biosynthesized directly from the compound. The labeled compound also contained 13 deuterium atoms, corroborating the earlier studies that indicated that only two hydrogens are removed from octanoic acid during lipoic acid biosynthesis [66,67].

It was suggested that a mechanism that could account for the observed inversion of stereochemistry at C-6 during the biosynthesis of lipoic acid might involve intermediate hydroxylation at C-6 (or both C-6 and C-8) with retention of configuration followed by appropriate activation and displacement of the activated hydroxyl group by an appropriate sulfur nucleophile [50,67]. To address this possibility, [6(*R,S*)-$^{2}H_{1}$]-6-hydroxyoctanoic acid (Figure 2.4A), [8-$^{2}H_{2}$]-8-hydroxyoctanoic acid (Figure 2.4B), and [8-$^{2}H_{2}$]-(±)-6,8-dihydroxyoctanoic acid (Figure 2.4C) were synthesized and tested for their ability to be converted into lipoic acid in growing *E. coli*. None of the hydroxylated compounds served as precursors to lipoic acid. By contrast, both [6(*R,S*)-$^{2}H_{1}$]-mercaptooctanoic acid (Figure 2.4D) and [8-$^{2}H_{2}$]-8-mercaptooctanoic acid (Figure 2.4E) served as precursors, with the latter compound being converted almost 10 times more efficiently than the former [67]. Interestingly, metabolic feeding studies conducted with two different strains of *E. coli* (W1485-*lip2* and JRG33-*lip9*) that are auxotrophic for lipoic acid because of point mutations in the *lipA* gene showed that 8-mercaptooctanoic acid could support growth of both *lipA* auxotrophs, but 6-mercaptooctanoic acid could not [68].

A separate and independent study using lipoic acid auxotrophs that contained transposon insertions in the *lipA* gene produced somewhat similar results, except

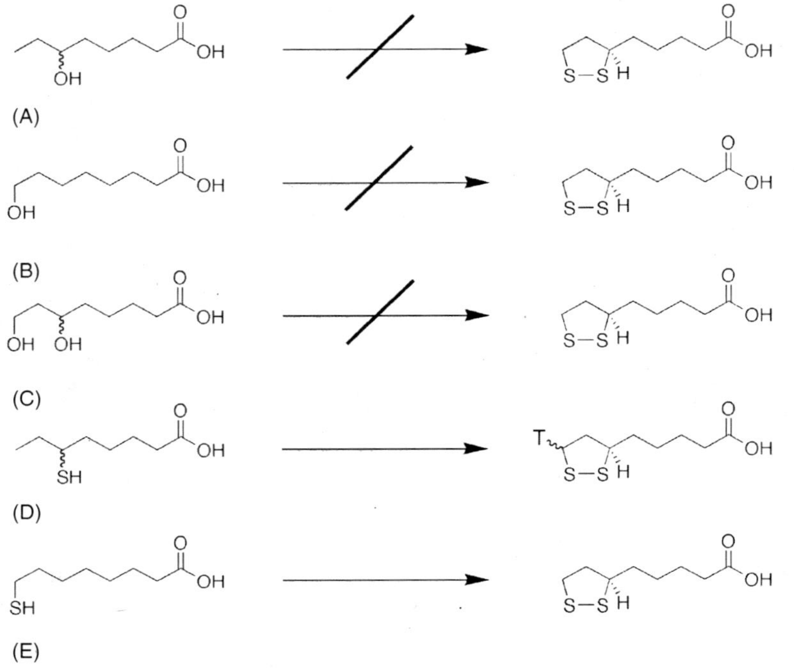

FIGURE 2.4 Structures of potential intermediates in the biosynthesis of lipoic acid: (A) 6(*R,S*)-hydroxyoctanoic acid; (B) 8-hydroxyoctanoic acid; (C) 6(*R,S*),8-dihydroxyoctanoic acid; (D) 6(*R,S*)-mercaptooctanoic acid; (E) 8-mercaptooctanoic acid.

Note: Each of the compounds also contains deuterium labels as indicated in the text; however, these labels were purposely omitted from the structures for brevity and clarity.

that 6-mercaptooctanoic acid could also support growth of the auxotroph to the same extent as 8-mercaptooctanoic acid when the former compound was included in the growth medium in a 17-fold excess of the amount used with the latter compound. Both these compounds, however, were significantly poorer than lipoic acid itself in supporting growth, which required supplementation at a concentration that was ~1000-fold less than that for 8-mercaptooctanoic acid to achieve a similar level of growth [58]. The observation that strains of *E. coli* containing null mutations in the *lipA* gene can grow on 8- and 6-mercaptooctanoic acid suggests the presence of another enzyme that can catalyze sulfur insertion, which may "moonlight" as a lipoic acid synthase if given an appropriate substrate.

PATHWAYS OF PROTEIN LIPOYLATION

In the cell, lipoic acid exists primarily as lipoamide, in an amide linkage to the ε-amino group of a specific lysine residue on a lipoyl carrier protein (LCP). In fact, very little free lipoic acid is found in cells that have not been treated with substantial amounts of the molecule from exogenous sources [2,69–71]. In *E. coli*, two mechanisms have been identified by which LCPs become lipoylated, an exogenous pathway and an endogenous pathway (Figure 2.5). In the exogenous pathway, lipoic acid obtained from nutrient sources is activated in an ATP-dependent process and then transferred to the appropriate LCP. In *E. coli*, both activation and transfer are catalyzed by the same protein, lipoate protein ligase A [72,73]. In the endogenous pathway, an octanoyl chain generated via type II fatty acid biosynthesis is transferred from octanoyl-ACP (acyl carrier protein) to the appropriate LCP in a reaction catalyzed by octanoyl-[acyl carrier protein]-protein transferase (octanoyltransferase), designated LipB in *E. coli* [58,74–76]. A subsequent enzyme, lipoyl synthase (LipA), catalyzes insertion of two sulfur atoms into C–H bonds at carbons 6 and 8, producing lipoic acid in its cofactor form (lipoamide) [51,58,61,76]. These two pathways readily account for the metabolic feeding studies discussed above, wherein intact *n*-octanoic acid was found to serve as a direct precursor to lipoic acid. The octanoate administered to the *E. coli* cells was most likely transferred to LCPs by the action of LplA,

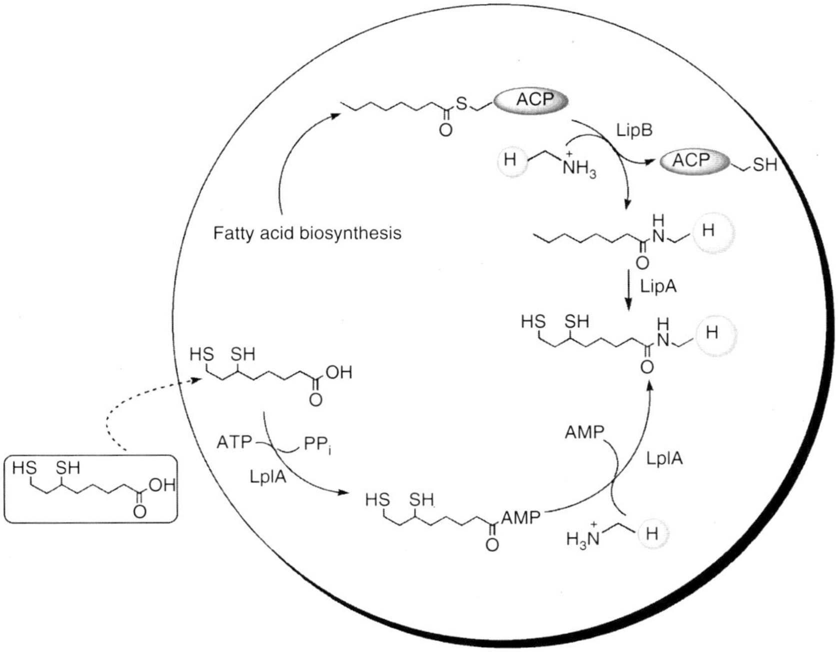

FIGURE 2.5 Pathways of protein lipoylation: (Top) endogenous pathway, which is an offshoot of fatty acid biosynthesis, occurring on the acyl carrier protein (ACP). LipB is octanoyltransferase and LipA is lipoyl synthase. (Bottom) exogenous pathway, wherein lipoic acid is obtained from nutrient sources. LplA is lipoate protein ligase, which in *Escherichia coli* catalyzes both the activation of lipoic acid to lipoyl-AMP and the transfer of the lipoyl group to lipoyl-accepting proteins.

whereupon LipA catalyzed sulfur insertion into the protein-bound octanoyl chain. As discussed in the following sections, LplA can catalyze the activation and transfer of several analogs of lipoic acid, including octanoic acid, albeit with varying catalytic efficiencies [42]. An excellent and comprehensive review of lipoic acid biosynthesis that details the genetic and biochemical approaches used to establish the presence of these two pathways has recently been published [17]. Key aspects of the strategies used to identify the genes that encode the participant enzymes in these two pathways are described below.

Cloning of *Escherichia coli lipA* and *lipB* Genes

The sequence of the *Escherichia coli lipA* gene was first reported in 1992 by Ashley and coworkers [54,77]. The gene was cloned by complementation, exploiting the ability of plasmid pLC15-5 from the Clark-Carbon colony library to confer prototrophy on the *E. coli* lipoic acid auxotrophs W1485-*lip2* and JRG33-*lip9*, which were generated by Herbert and Guest in the late 1960s in one of the seminal genetic studies on lipoic acid biosynthesis [54,77]. Further restriction analysis of the plasmid allowed localization of the gene to a 1.9 kb *Eco*R1/*Hin*dIII fragment, which was subsequently sequenced; however, the start site for the gene was not determined experimentally, and was tentatively assigned to a methionine residue (Met41) that would predict a protein of 281 amino acids and having a molecular mass of 31,350 Da. This molecular mass was smaller than that predicted in an earlier study conducted to identify the *E. coli lip* locus, wherein in vitro transcription and translation of the *lipA* gene suggested that it was a protein of a molecular mass of ~36 kDa [29].

In an independent study, Cronan and coworkers sequenced the *lip* locus, which they had previously identified [29], and experimentally determined the translational initiation site of the *lipA* gene. This was accomplished by amino acid sequencing of [^{35}S]-labeled LipA via Edman degradation and counting the number of cycles necessary to result in release of radioactivity; LipA labeled with [^{35}S] was produced in an in vitro transcription–translation system containing [^{35}S]methionine. This study showed that *E. coli* LipA is a protein of 321 amino acids and has a molecular mass of 36,072 Da [58], and that the true translational start site is 40 codons upstream of the start site that was previously suggested [54]. Efforts to characterize the LipA protein were met with difficulty; overexpression of the *lipA* gene from a T7 promoter resulted in production of the protein primarily as insoluble aggregates. Resolubilization of the inclusion bodies allowed partial purification of LipA to ~80% homogeneity; however, demonstration of turnover was not reported. Interestingly, both *lipA* and *lipB* null mutants could grow when the medium was supplemented with 6-mercaptooctanoic acid or 8-mercaptooctanoic acid in place of lipoic acid, which led the authors to suggest that LipA is required for insertion of the first sulfur atom into the octanoyl chain [58].

Transposon mutagenesis was also used to identify another gene within the *lip* locus, designated *lipB*, and which unlike the *lipA* gene could not rescue the previously described *lipA2* defect, suggesting that it was distinct from *lipA* [29].

Sequencing of the *lipB* gene revealed an open reading frame that encoded a protein of 191 amino acids and a molecular mass of 21,339 Da, which is not the full-length native protein (discussed below). In vitro transcription–translation analysis of a plasmid that carried the *lipB* gene under the control of a *tac* promoter revealed the production of a 25 kDa protein that was not produced in the vector control. Attempts to overproduce the *lipB* gene product were initially unsuccessful, however, suggesting that the *lipB* gene is poorly expressed or that its protein product is highly labile [58].

Similar results were obtained in a subsequent study, in which the *lipB* gene was found to be a negative regulator of *dam* gene expression in *E. coli* [78]. An N-terminal hexahistidine-tagged form of LipB, initiating with the previously suggested N-terminal methionine residue, was produced at very low concentrations and was unstable. By contrast, when transcription of the *lipB* gene was initiated from an upstream TTG start codon, a stable protein of the predicted size was produced. In addition, the transcript generated from the TTG start site allowed in vivo complementation of two different *lipB* mutant strains of *E. coli*. The hexahistidine-tagged protein was purified by immobilized metal affinity chromatography (IMAC) and shown to have a molecular mass that was consistent with the predicted size of 23,882 Da (213 amino acids). As expected from its predicted TTG start codon, the level of LipB protein in a culture of wild-type *E. coli* was found to be near the detection limit of quantitative Western analysis using an antibody that was generated against the purified protein, indicating that its in vivo concentrations are very low [78].

Cloning of the *Escherichia coli lplA* Gene

Transposon mutagenesis using strains of *E. coli* with *lipA* null mutations was used to identify the *lplA* gene, selecting for *E. coli* mutants that could grow only in the presence of acetate plus succinate, conditions that bypass the requirement for lipoate-dependent enzymes [72]. Strains of *E. coli* with null mutations in the *lplA* gene were severely defective in the ligation of exogenous lipoic acid, octanoic acid, and selenolipoic acid to apo-LCPs; however, the extent and rate of lipoic acid transported into the cell was identical in both the wild-type and the *lipA*/*lplA* double mutant, suggesting that the lack of ligation was not a result of a defect in transport into the cytoplasm [72]. DNA sequencing of the *lplA* gene showed an open reading frame of 1014 bp, which encoded a protein of 338 amino acids and a molecular mass of 37,926 Da. Analysis of the protein by electrospray ionization (EI) and matrix-assisted laser desorption ionization (MALDI) mass spectrometry afforded molecular masses of 37,795 and 37,791 Da, respectively for each method, which is in agreement with the molecular mass of the protein lacking the N-terminal formylmethionine group as predicted from its gene sequence.

Role of the Acyl Carrier Protein in Lipoic Acid Biosynthesis

A number of experimental observations were consistent with the participation of the ACP in lipoic acid biosynthesis. ACPs are versatile proteins that are involved in a number of biochemical processes, including acyl homoserine lactone formation [79], the biosynthesis of lipids and phospholipids [80,81], and the activation of protein toxins [82]. Their most central role in biochemistry, however, is their participation in the initiation, elongation, and transfer of acyl groups during fatty acid biosynthesis. The *E. coli* ACP has a molecular mass of 8640 Da and is one of the most abundant proteins in the cell [83–85]. It is an acidic protein, possessing an isoelectric point of 3.98, and contains no cysteine residues in its primary structure, which consists of 77 amino acids [85]. For *E. coli* ACP to fulfill its biochemical functions, the apo form of the protein must first be phosphopantetheinylated [86]. This is accomplished via the action of the enzyme holo-ACP synthase, which catalyzes the transfer of the 4′-phosphopantetheine moiety from CoA to the hydroxyl group of Ser36 (*E. coli* numbering) of ACP, resulting in the formation of a phosphodiester linkage (Figure 2.6) [87,88]. The resulting free sulfhydryl group exists in a thioester linkage with growing fatty acyl chains.

Structures of ACPs have been determined by both NMR and x-ray crystallography and include among others: the apo form from *E. coli* [89], *Bacillus subtilis* [90], *Mycobacterium tuberculosis* [91], and *Nitrosomonas europaea* [92]; the holo forms from *M. tuberculosis* [91] and *B. subtilis* [90]; and acylated forms from *E. coli* (butyryl-, hexanoyl-, heptanoyl-, and decanoyl-ACP) [93,94] and spinach (decanoyl- and stearoyl-ACP) [95]. All are very similar and are composed primarily of four α-helices connected by loops and arranged in a bundle. At the core of the bundle is a hydrophobic cleft, which accommodates the growing fatty acyl chain. A structure of a heptanoylated (C-7) form of *E. coli* ACP has been determined by x-ray crystallography and is shown in Figure 2.6. The 4′-phosphopantetheine prosthetic group (Figure 2.6, bottom) is attached to Ser36 in a phosphodiester linkage, and terminates with a β-mercaptoethylamine group to which the fatty acyl chain is attached.

Initial insight that ultimately led to the finding that ACP is involved in de novo biosynthesis of the lipoyl cofactor grew out of studies that were designed to characterize *lplA* null mutants in *E. coli*. It had been shown that *lipA* null mutants accumulated LCPs that were octanoylated if lipoic acid was not provided in the medium [96,97], and that this aberrant octanoylation could be reproduced under conditions in which LCPs were produced at high levels in the cell [98]. Subsequent studies showed that *lplA* null mutants grew fine on minimal medium unless they were transduced into strains that carried null mutations in the *lipA* or *lipB* genes [58,72,73,99]. Moreover, wild-type strains or strains with null mutations in *lplA* accumulated exclusively lipoylated LCPs, while strains with null mutations in *lipA* accumulated octanoylated LCPs unless they were supplemented with lipoic acid. Interestingly, strains with null mutations in *lipB* accumulated meager amounts of octanoylated and lipoylated LCPs in the absence of lipoate, but accumulated wild-type levels of lipoylated LCPs in the presence of excess lipoate [73]. Strains that carried null mutations in *lipB* were found to be leaky, in that slow growth occurred in the absence of added lipoic acid. However, growth was abolished in *lipB*/*lplA* null mutants, and normal growth was restored in the absence of added lipoate under conditions in which the *lplA* gene

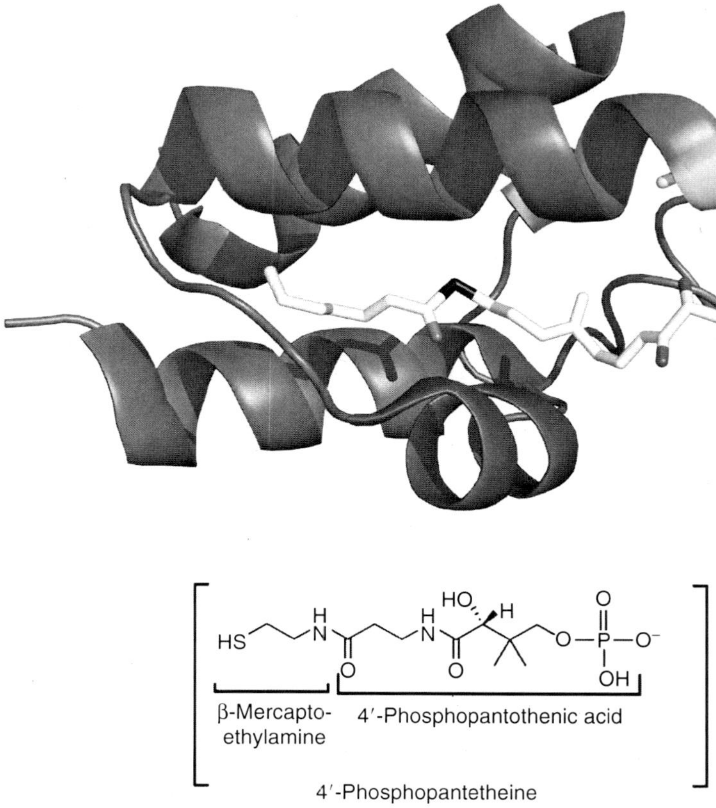

FIGURE 2.6 Structure of *Escherichia coli* acyl carrier protein containing a heptanoyl fatty acyl chain (PDB 2FAD): (Top) *E. coli* heptanoyl-ACP (acyl carrier protein). Shown in stick format is the heptanoyl fatty acyl chain in a thioester linkage to the 4′-phosphopantetheine prosthetic group, which in turn is in a phosphodiester linkage with Ser36. Oxygen atoms are shown in gray, while the sulfur atom of the β-mercaptoethylamine group is shown in black. The structure was prepared using the Pymol Molecular Graphics System (http://www.pymol.org). (Bottom) Structure of 4′-phosphopantetheine.

was overexpressed in a *lipB* mutant background [58,73]. These genetic studies provided strong evidence that the *lipA* and *lipB* genes were key players in the endogenous pathway of lipoic acid production.

Hypothesizing that the LipB-dependent pathway might involve an octanoyl chain generated from fatty acid biosynthesis, Jordan and Cronan showed that in extracts of *E. coli*, LipB does indeed catalyze the transfer of octanoyl or lipoyl chains from octanoyl- or lipoyl-ACP to the lipoyl domains of the PDC and KDC. Extracts of strains carrying null mutations in *lipB* were ineffective in catalyzing the reaction, whereas extracts of strains carrying null mutations in *lplA* showed enhanced transfer from lipoyl-ACP over the wild-type strain, presumably because of a lack of competition for the lipoyl domain from the exogenous pathway [75]. The same study found this LipB-dependent activity to be present in mitochondria of *Neurospora crassa* and *Pisum sativum* [75].

In mammals, plants, fungi, and certain mycobacteria, fatty acid synthesis is carried out by a large cytosolic multifunctional enzyme known as a type I fatty acid synthase (FAS), in which all of the active sites necessary to construct fatty acids are located on a single polypeptide [100]. In most bacteria, fatty acid biosynthesis is carried out by a series of small discrete proteins, which is termed type II fatty acid synthesis [83]. Because it is well accepted that most or all mitochondrial membrane fatty acids are synthesized outside of the mitochondrion and then imported, the finding of ACP and type II FAS enzymes in the mitochondria of *Neurospora crassa* [101,102], *Saccharomyces cerevisiae* [103], plants [104,105], and bovine heart muscle [106] was perplexing. In experiments that were designed to shed light on this issue, Ohlrogge and his colleagues incubated purified pea leaf mitochondria with [2-^{14}C]malonic acid and observed radioactivity to be incorporated into fatty acids, establishing that mitochondria contained all of the enzymes necessary to synthesize fatty acids. [1-^{14}C]Acetate was significantly less effective as a precursor for fatty acid synthesis, suggesting that mitochondria do not possess acetyl-CoA carboxylase [107]. When the labeled fatty acids were saponified and analyzed by thin-layer chromatography, most of the radiolabel was found incorporated into hydroxymyristic, lauric, and palmitic acids. Analysis of radiolabeled proteins by sodium dodecyl sulfate–polyacrylamide gel electrophoresis (SDS–PAGE) showed radioactivity to be associated with mitochondrial ACP as well as the H-protein of the GCS. Further analysis of the H-protein was consistent with the attachment of an eight-carbon (octanoyl) chain to it in an amide linkage. Since it was known that the H-protein is the lipoyl-bearing subunit of the GCS, this study suggested that acyl-ACPs are intermediates in the biosynthesis of the lipoyl cofactor [107].

Experiments conducted by the Bourguignon laboratory further corroborated the involvement of mitochondrial ACP in lipoic acid biosynthesis. The authors were able to show that soluble extracts from pea leaf mitochondria contained all enzymes required to synthesize fatty acids from malonate, including a malonyl-CoA synthetase and malonyl-CoA:ACP transacylase, which allow conversion of malonate into malonyl-ACP, the building blocks of fatty acids. Moreover, using MALDI mass spectrometry they found three major fatty acids to be synthesized as an appendage to *E. coli* ACP: octanoic acid (C8), hexadecanoic acid (C16), and octadecanoic acid (C18), and suggested that octanoyl-ACP was a precursor to lipoic acid. Indeed, upon the addition of apo-H-protein, *E. coli* ACP, and [2-^{14}C] malonic acid to a soluble mitochondrial extract, radioactivity was found to be transferred both to acyl-ACP and the H-protein. To assess whether this pathway in mitochondria is responsible for the biosynthesis of lipoic acid, a variant of the H-protein containing a $Glu_{14}{\rightarrow}Ala$ substitution was employed to allow distinction by MALDI-MS between endogenous H-protein and newly synthesized octanoylated or lipoylated H-protein. Upon incubation of the variant with mitochondrial matrix extract, *E. coli* ACP, *S*-adenosylmethionine, and potential sulfur donors (sodium sulfide and cysteine), peaks that corresponded to octanoylated and lipoylated forms of the Glu_{14}→Ala variant of the H-protein were observed, in addition to peaks for the lipoylated form of the wild-type H-protein and the apo form of the variant H-protein. Incubation of the lipoylated form of the variant H-protein with glycine, pyridoxal 5′-phosphate, and the P-protein of the GCS resulted in the decrease of the lipoylated form of the variant H-protein, as observed by MALDI-MS, with a concomitant increase in a form that had a methylamine group attached to the lipoyl cofactor, verifying that the appendage connected to the variant H-protein was indeed a lipoyl group [108].

Yet another study using *S. cerevisiae* showed that when the gene that encodes the mitochondrial type II acyl carrier protein (ACP1) is deleted, the resulting mutants had approximately 5%–10% of the typical wild-type content of lipoic acid and exhibited a respiratory deficient phenotype. Moreover, the lipoate deficiency co-segregated with the deletion [109]. Similar results were obtained when the yeast gene PPT2 was deleted, which encodes for a phosphopantetheine: protein transferase that catalyzes the phosphopantetheinylation of apo-ACP, converting it into its holo form [88]. Mitchondrial ACPs have also been found in other eukaryotic organisms, including mammals [110]. In fact, the ACP component of the membrane-bound NADH:ubiquinone oxidoreductase from bovine heart mitochondria is found mostly in the mitochondrial matrix and apparently can participate in type II fatty acid biosynthesis, since preparations of it have been found to contain appended octanoyl chains [110].

CHARACTERIZATION OF LIPOATE PROTEIN LIGASE A

Lipoate protein ligase A (LplA) is the enzyme in *E. coli* that catalyzes the activation of exogenous lipoic acid and its transfer to appropriate LCPs. Seminal studies by Reed and coworkers showed that the pyruvate dehydrogenase activating system in *Streptococcus faecalis* required lipoic acid, ATP, Mg^{2+}, and two separable components, only one of which was heat labile. Other nucleoside triphosphates could also support turnover; CTP, GTP, ITP, and UTP afforded 40%, 25%, 15%, and 50% of the activity, respectively, of that afforded by ATP [111]. In contrast to the *S. faecalis* lipoate-activating system, these also demonstrated that a fraction from *E. coli* that was capable of activating lipoic acid could not be further fractionated, suggesting that only one protein or a tightly bound complex was responsible for catalyzing both the activation and transfer of lipoic acid to LCPs in this organism [111]. Arsenite is known to form an adduct with the sulfhydryl groups of reduced lipoic acid, which is reversed upon treatment with 2,3-mercaptopropanol (BAL). Addition of arsenite to cellular extracts of *S. faecalis* either before or after incubation with lipoic acid led to inhibition that was reversible upon addition of BAL [111]. Since it was known that activation involves generation of a stable protein-bound form of lipoic acid, it followed that the activation event must have been a result of modifying the carboxyl end of the molecule rather than the sulfhydryl groups [14,111]. Consistent with this premise, activation in the presence of hydroxylamine resulted in formation of lipohydroxamic acid and one equivalent of pyrophosphate, suggesting that lipoyl adenylate (lipoyl-AMP) served as an intermediate in the transfer of exogenous lipoic acid to LCPs. In support of this premise, chemically synthesized lipoyl-AMP could substitute for lipoic acid and ATP in the activating systems. Further studies demonstrated that the heat labile component of the extract from *S. faecalis* catalyzed formation of lipoyl-AMP, while the heat stable component catalyzed the transfer of the lipoyl group from lipoyl-AMP to LCPs.

As previously described, the *lplA* gene from *E. coli* has been successfully overexpressed and its protein product purified [72,99]. In addition to DL-lipoic acid, LplA could also catalyze the activation and transfer of dihydro-DL-lipoic acid, D-lipoic acid, octanoic acid, and methyl-lipoic acid with activities that were 80%, 83%, 3%, and 73% of the activity with DL-lipoic acid. *E. coli* LplA could also use L-lipoic acid as a substrate, although only the physiologically relevant D form is transferred to the PDC in vivo [112]. In addition to the compounds above, 6- and 8-mercaptooctanoic acid, 6- and 8-selenooctanoic acid, and selenolipoic acid were also tested as substrates for LplA. Only 6-mercaptooctanoic acid and 6-selenolipoic acid served as reasonable substrates, affording specific activities for LplA that were 150% and 6%, respectively, of the specific activity in the presence of DL-lipoic acid [99].

A number of structures of lipoate protein ligases A have been determined by x-ray crystallography, including structures from *Thermoplasma acidophilum* and *E. coli*. The structure of the *E. coli* enzyme was determined in its apo form and with lipoic acid bound (Figure 2.7, Left) [113], while several structures of LplA from *T. acidophilum* were determined: the apo form [114,115], in complex with ATP [114], and in complex with lipoyl-AMP (Figure 2.7, Right) [114]. The overall folds of the two proteins are similar, and also bear some resemblance to the structure of the biotinyl protein ligase module of BirA, the *E. coli* biotin holoenzyme synthetase/*bio* repressor. There are two significant differences between the two LplA structures, however. The *E. coli* enzyme possesses a long C-terminal extension, which is not present in the enzyme from *T. acidophilum* (Figure 2.7, left, shown in light gray), and the binding sites for lipoic acid are slightly different in the two structures. In the structure of LplA from *T. acidophilum* (Figure 2.7, right), the lipoyl–AMP group is bound in a U-shape, and the phosphate group of the molecule is accessible from bulk solvent by a tunnel that is approximately 10 Å deep. This tunnel has dimensions that would accommodate the target lysine side chain of LCPs. The lipoyl–AMP structure was generated by soaking crystals of the apo form of LplA with ATP followed by lipoic acid, clearly showing that this enzyme can catalyze activation of lipoic acid to lipoyl–AMP. The structure of the enzyme in the presence of ATP shows that ATP binds to the same site to which the AMP moiety of lipoyl–AMP binds [114].

There are three regions of high sequence conservation in LplAs: motif I ($\mathbf{R}_{71}$**RXXGGGXV**[F/Y]$\mathbf{HD}_{82}$), motif II ($\mathbf{K}_{145}$hX**GXA**$_{150}$), and motif III ($\mathbf{H}_{161}$XX[L/M]**L**XXX[D/N]**L**XX**L**XXh$\mathbf{L}_{177}$). Note that each amino acid is denoted by its one-letter code and boldface amino acids indicate those that are

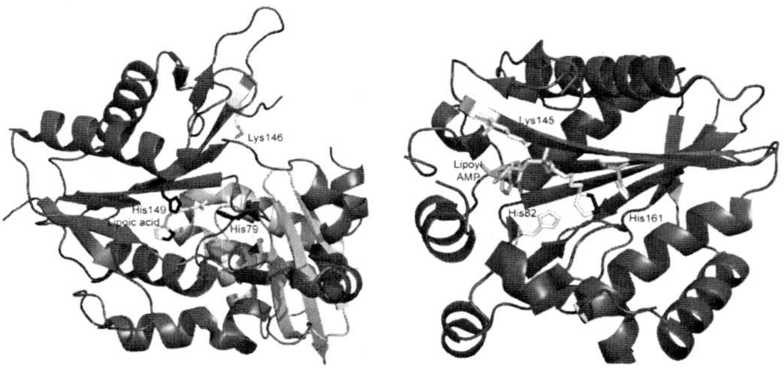

Escherichia coli LplA   *Thermoplasma acidolphilum* LplA

FIGURE 2.7 Structures of lipoate protein ligase A: (Left) LplA from *Escherichia coli* complexed with lipoic acid (PDB 1X2H). The C-terminal extension that is not found in the structure from *Thermoplasma acidophilum* is shown in light gray. (Right) LplA from *T. acidophilum* complexed with lipoyl-AMP (PDB 2ART). In this structure, lipoyl-AMP is bound in a "U" shape with Lys145 pointing at the phosphate group. In both structures, the lysine residue that is strictly conserved in all LplAs, LipB, and biotin ligase is marked, as are the histidine residues that are strictly conserved in LplAs and LipBs. Both structures were prepared using the Pymol Molecular Graphics System (http://www.pymol.org).

strictly conserved. In addition, X indicates any amino acid, [X/Y] indicates that either of the two amino acids may occur at the indicated position, and h denotes any hydrophobic amino acid. Lysine145 in motif II of the *T. acidophilum* LplA makes contact with the phosphate group in the structure with lipoyl–AMP and the carbonyl group of lipoic acid in the structure with lipoic acid. This amino acid is also conserved in LipB and biotin ligase, suggesting that it may play a role in the transfer of the lipoyl or biotinyl groups from the activated intermediate to the appropriate acceptor proteins. Two conserved histidine amino acids, His81 from motif I and His161 from motif III, are also found in LipB, but not in biotin ligase, and form part of the binding pocket for the dithiolane ring of the lipoyl group.

Models of the *T. acidophilum* LplA docked with the lipoyl domain of the PDC from *A. vinelandii* and the H-protein of the GCS from *Thermus thermophilus* have been constructed [114]. The target lysine of the LCPs could be accommodated in the tunnel in LplA, and excellent shape and charge complementarity were obtained between the two proteins. Interestingly, although LplA from *T. acidophilum* catalyzed formation of lipoyl–AMP in the crystal, there was no indication as to whether the protein could catalyze the subsequent transfer of the lipoyl group to LCPs. In fact, an independent determination of the x-ray structure of the *T. acidophilum* LplA by Perham and coworkers showed that the structures were essentially identical, but that the enzyme was inactive. Moreover, the authors of this study argued that the *T. acidophilum* genome encodes for another protein of ~10 kDa that is immediately adjacent to the *lplA* gene, and which is predicted to have a significant degree of structural homology with the C-terminal domain of LplA from *Streptococcus pneumoniae* as well as similar domains from a variety of other bacteria [115]. They suggest that this protein may be required to constitute a functional LplA in this organism.

LplA homologues have been identified in a number of organisms, including mammals, yeast, and bacterial pathogens. In fact, *Listeria monocytogenes*, a Gram-positive intracytosolic pathogen that causes severe disease in pregnant or immunocompromised individuals, contains two *lplA* or *lplA*-like genes, *lplA1* and *lplA2*. The *lplA1* gene apparently encodes for a protein that scavenges lipoic acid from the cytosol of its host, and uses it primarily to activate its own PDC. Mutant strains of *L. monocytogenes* with null mutations in *lplA1* were defective for growth in the host cytosol, and were 300-fold less virulent in intravenous mouse models of infection [116].

In mammals, as is seen in *S. faecalis*, activation of lipoic acid and its subsequent transfer are catalyzed by two separate enzymes, both of which are located in the mitochondrion [117–120]. A lipoate-activating enzyme (LAE) from bovine liver has been purified by conventional chromatography and characterized [120]. The protein was predicted to have a molecular mass of 61 kDa by SDS–PAGE, but migrated as a 49 kDa protein by molecular sieve chromatography. Steady-state kinetic analysis of the reaction revealed that the turnover number of the enzyme (k_{cat}) was similar with GTP, CTP, and UTP as the nucleoside monophosphate donor, but was reduced 1000-fold in the presence of ATP. The second-order rate constants ($k_{cat}K_M^{-1}$) for the reaction were 2275, 687, and 98 $M^{-1}s^{-1}$, respectively for GTP, CTP, and UTP, indicating that GTP is the best substrate. In addition, the protein was specific for *R*-lipoic acid; its activity with *S*-lipoic acid was less than 1% of that with the *R* form. The subsequent cloning of the full-length cDNA for LAE showed that it is a protein of 577 amino acids, and that it contains a 31 amino acid mitochondrial targeting sequence. Therefore, the mature protein is predicted to be 546 amino acids and to have a molecular mass of 61,146 Da. Interestingly, a BLAST search against the sequence of LAE showed it to be identical to that of xenobiotic-metabolizing/medium-chain fatty acid:CoA ligase-III from bovine liver [120,121]. Consistent with this finding, the protein catalyzed formation of the acyl–CoA and acyl–GMP derivatives of butyric acid, hexanoic acid, octanoic acid, decanoic acid, and dodecanoic acid; and, lipoic acid acted as a competitive inhibitor of the reactions.

Two isoforms of lipoyl-AMP:N^{ε}–lysine lipoyltransferase—the enzyme that catalyzes the second step in lipoic acid ligation to LCPs in mammals—have been purified from bovine liver mitochondria [117]. Neither of these were able to use lipoic acid plus MgATP as substrates in place of lipoyl-AMP, and similar to LAE, these were active with other fatty acyl groups, including hexanoyl-, octanoyl-, and decanoyl-AMP. Studies with other lipoyl-nucleoside triphosphates have not been reported, which is surprising since the enzyme should rarely "see" lipoyl-AMP, as suggested by the kinetic studies with LAE. The subsequent cloning of one of the isoforms, lipoyltransferase II, revealed it to be a protein of 373 amino acids, and to contain a 26 amino acid mitochondrial targeting sequence. Therefore, the mature protein is predicted to be 347 amino acids and to have a molecular mass of 39,137 Da. Although the cloning of lipoyltransferase I has not been reported, an antibody raised against recombinant lipoyltransferase II cross-reacted with lipoyltransferase I, suggesting that the two proteins are similar in sequence [119].

CHARACTERIZATION OF OCTANOYL-[ACYL CARRIER PROTEIN]-PROTEIN TRANSFERASE

Conducting a detailed mechanistic characterization of octanoyl-[acyl carrier protein]-protein transferase (LipB) is a tedious undertaking because the enzyme catalyzes the transfer of an appendage from one macromolecule to the other, which necessitates the synthesis of homogeneous protein-bound substrates and the development of quantitative methods to analyze the protein-bound products of the reaction, adding an additional level of difficulty associated with kinetic analysis of the enzyme. Nonetheless, important in vitro studies using purified protein have been conducted. Jordan and Cronan showed that the purified N-terminal hexahistidine-tagged enzyme could transfer a C-8 chain from either octanoyl-ACP or lipoyl-ACP, and that it was not active with lipoate plus ATP. The K_M value for lipoyl-ACP was estimated to be ~1 μM [74]. Interestingly, they also showed that LplA could catalyze the transfer of a C-8 chain from either octanoyl-ACP or lipoyl-ACP, albeit with much lower efficiencies as compared to LipB, which explains the "leakiness" observed in *lipB* mutants under lipoate deficient conditions that is corrected upon overproduction of LplA [29,58]. At high concentrations of LplA, sufficient transfer of the octanoyl chain from octanoyl-ACP to LCPs takes place to meet the cell's requirement for the lipoyl cofactor.

In an independent study, Nesbitt et al. [122] reported the purification of native *E. coli* LipB by conventional chromatographic methods and the development of alternative methods for quantifying its kinetic parameters. Native LipB catalyzed transfer of the octanoyl chain from octanoyl-LipB to the *E. coli* H-protein with a turnover number of $0.2\ s^{-1}$, and K_M values for octanoyl-ACP and apo-H-protein of 10.2 ± 4.4 μM and 13.2 ± 2.9 μM, respectively. The protein was shown by molecular sieve chromatography to exist in solution primarily as a monomer with a small amount of trimer, and not to require any common metal ions for catalysis. *E. coli* LipB contains three cysteine residues, only one of which is conserved (Cys169). Treatment of the protein with 5,5′-dithiobis-(2-nitrobenzoic acid) (DTNB) while kinetically analyzing the extent of cysteine modification showed that all three cysteines reacted in two distinct phases. One cysteine reacted with a rate constant that was 50-fold greater than that of the other two cysteines, suggesting that it might play a role in covalent catalysis [122].

Covalent catalysis in the LipB reaction was subsequently demonstrated by Zhao, Miller, and Cronan [76]. When [1-^{14}C]octanoyl-ACP was treated with LipB and either of the E_2 subunits of the PDC and KDC, radioactivity was found associated with all three proteins. Radioactivity was also shown to be transferred to LipB upon incubation of [1-^{14}C]octanoyl-ACP with LipB alone. After separating [1-^{14}C]octanoyl-ACP from [1-^{14}C]octanoyl-LipB by ion-exchange chromatography and incubating [1-^{14}C]octanoyl-LipB with the apo forms of E_2 from the PDC or KDC, radioactivity was found associated with the LCPs, showing that transfer of the octanoyl appendage from octanoyl-LipB is chemically competent. Moreover, when [1-^{14}C]octanoyl-LipB was incubated with holo-ACP, the octanoyl appendage was transferred to holo-ACP, indicating that the first step in the reaction, formation of octanoyl-LipB from octanoyl-ACP, is reversible [76].

Several lines of evidence were consistent with the presence of a thioester bond between the octanoyl chain and Cys169. Octanoyl-LipB was greatly sensitive to hydrolysis, especially under basic conditions, and was also sensitive to treatment with neutral hydroxylamine, affording the hydroxamate adduct. Substitution of Cys169 with Ser or Ala abolished turnover in vitro, and the corresponding mutants were unable to rescue *lipB* null mutants when grown on minimal medium. A small amount of radioactivity from octanoyl-ACP was shown to be transferred to the variant proteins, but this activity was not on the catalytic pathway [76].

Recently, a structure of LipB from *Mycobacterium tuberculosis* (MTB) was solved to 1.08 Å resolution by x-ray diffraction. The structure contained unexpected electron density, which was shown to be decanoic acid bound in the active site of the protein and covalently attached at C-3 in a thioether linkage to Cys176 (Cys169 in *E. coli* LipB) (Figure 2.8) [123]. The decanoic acid was

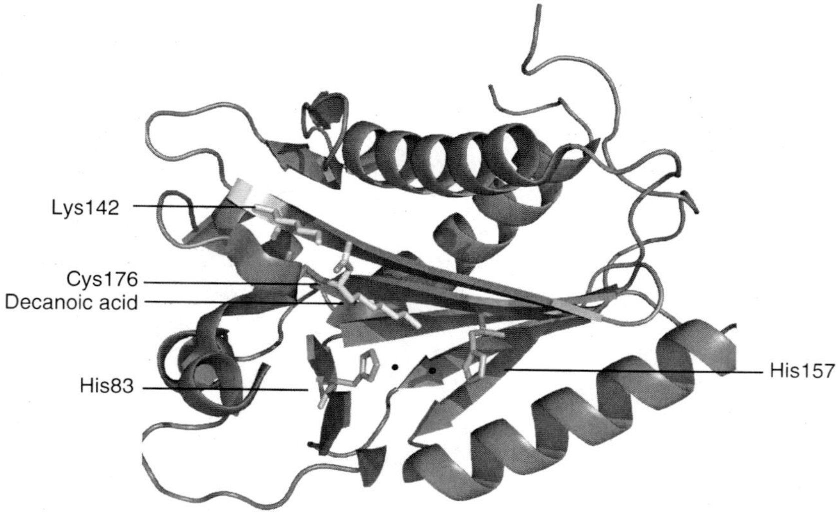

FIGURE 2.8 Structure of octanoyl-[acyl carrier protein]-protein transferase (LipB) from *Mycobacterium tuberculosis* (PDB 1W66). The lysine residue (Lys142) that is strictly conserved in all LipBs, as well as LplAs and biotin ligases, is depicted in stick format, as are the histidine residues (His83 and His157) that are conserved in LipBs and LplAs. Also shown is the decanoic acid that forms an adduct with Cys176, the amino acid that participates in covalent catalysis during the reaction. The two black spheres represent water molecules that are in hydrogen bonding distance with the indicated histidines. The structure was prepared using the Pymol Molecular Graphics System (http://www.pymol.org).

proposed to be an artifact of heterologous expression of the MTB *lipB* gene in *E. coli*, arising from a possible Michael addition of Cys176 onto the unsaturated bond of *trans*-2-decanoyl-ACP or *cis*-3-decanoyl-ACP, which are intermediates in fatty acid biosynthesis in *E. coli*. An additional unexpected finding was that the ε-amino group of Lys142, another strictly conserved residue among octanoyltransferases, was located 3.4 Å away from the sulfhydryl group of Cys176, suggesting that these two amino acid residues might form an ion pair in the absence of substrate (Figure 2.8). It was therefore proposed that LipB functions as a cysteine/lysine dyad acyltransferase [123]. In this mechanism, Cys176 of LipB, existing as a cysteine thiolate, initiates a *trans*-thioesterification reaction, resulting in the transfer of the octanoyl chain from octanoyl-ACP to Cys176 of LipB. After release of product from the first half-reaction (holo-ACP), and binding of an LCP for the second half-reaction, the relevant lysine residue from the second substrate attacks the thioester of the octanoyl-LipB intermediate, forming an amide linkage with concomitant regeneration of the LipB cysteine thiolate (Figure 2.9).

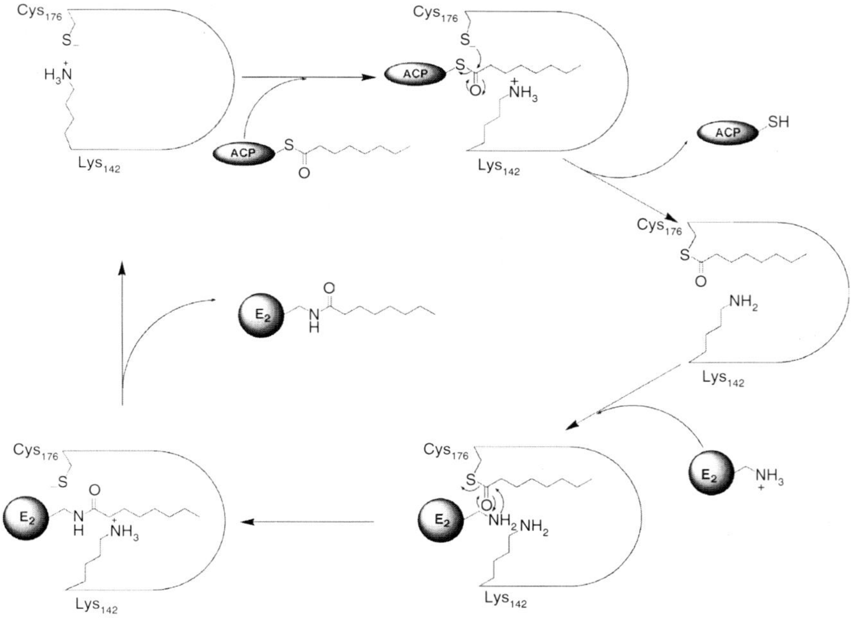

FIGURE 2.9 Postulated reaction mechanism of LipB. ACP denotes acyl carrier protein; E_2 denotes the lipoyl domains of lipoyl carrier proteins. Cys176 and Lys142 are proposed to exist in an ion pair in the resting state of the enzyme and to act as a Cys/Lys dyad. (Mechanism adapted from Ma, Q., Zhao, X., Eddine, A.N., Geerlof, A., Li, X., Cronan, J.E., Kaufmann, S.H.E., and Wilmanns, M., *Proc. Natl. Acad. Sci. USA*, 103, 8662, 2006.)

The structure of *M. tuberculosis* LipB is distantly related to the known structures of LplA. The Cα atoms of the structures can be superimposed on each other with a root mean square deviation of ~2.5 Å. In addition, the topology of most of the elements of secondary structure in the two classes of proteins is similar, and they seem to share similarities in the binding site for the fatty acyl chain to be transferred. In fact, the terminal methyl group of the decanoyl adduct in the LipB structure superimposes on the dithiolane ring of lipoic acid in the LplA structure in complex with lipoyl-AMP [123].

CHARACTERIZATION OF LIPOYL SYNTHASE

Radical SAM Superfamily of Enzymes

From a mechanistic and synthetic standpoint, the most intriguing step in lipoic acid biosynthesis is the reaction catalyzed by lipoyl synthase (LipA). This reaction entails insertion of two sulfur atoms into unactivated alkyl chains, and parallels the insertion of oxygen atoms into similar species by enzymes that use molecular oxygen as a co-substrate [51,52,61,62,124,125]. At about the time of the determination of the primary structure of lipoyl synthase, studies on a new superfamily of metalloenzymes were coming to light. These enzymes all contained [4Fe–4S] clusters and used *S*-adenosyl-L-methionine (SAM) as a source of a 5′-deoxy 5′-adenosyl radical (5′-dA$^{\bullet}$), previously postulated to be an intermediate in enzymatic reactions that require coenzyme-B_{12} [56,126–130]. In each system, the role of this radical was to initiate catalysis by removing a hydrogen atom (H$^{\bullet}$) from the appropriate protein or small molecule substrate, forming a protein- or substrate-based radical. In a special class of radical SAM enzymes, a hydrogen atom is removed from the α-carbon of a glycine residue located on a cognate protein, creating a protein glycyl radical cofactor. These glycyl radical cofactors have been shown to be stable in the absence of oxygen, and to support multiple turnovers [131–134]. Substrates for this superfamily of proteins include various small molecules, proteins, and nucleic acids, and very often the hydrogen atom to be removed is unactivated, as in the case with lipoyl synthase [127,129,130,135,136].

The hallmark of radical SAM enzymes is the presence of a [4Fe–4S] cluster that is obligate for turnover. This cluster is bound to its host protein via three cysteine ligands that reside in a CXXXCXXC motif, which is a signature sequence for radical SAM enzymes [135]. There are enzymes that have minor permutations of this signature motif and that have been included as members of the radical SAM superfamily, but none of these enzymes have been proven unambiguously to generate a 5′-deoxyadenosyl radical from SAM and use it in catalysis [135]. In the functional resting state of most, if not all, enzymes within the radical SAM superfamily, the [4Fe–4S] cluster exists in the +2 oxidation state. However, to achieve turnover, the $[4Fe–4S]^{+2}$ cluster must be transiently reduced by one electron, generating the $[4Fe–4S]^{+}$ state. In *E. coli*, the flavodoxin/flavodoxin:NADP oxidoreductase system [137–139] supplies the requisite electron; however, turnover can also be achieved in vitro with small molecule reductants such as sodium hydrosulfite (dithionite) or 5-deazariboflavin in the presence of light [126,140]. A series of elegant spectroscopic studies on the radical SAM enzymes, pyruvate–formate lyase (PFL) activase [141,142] and lysine 2,3-aminomutase (LAM) [143], followed by x-ray structures of coproporphyrinogen III oxidase (HemN) [144], biotin synthase (BioB) [145], MoaA [146], and LAM [147], indicate that SAM binds in contact with the [4Fe–4S] cluster via coordination of its α-amino and α-carboxylate groups to the unique iron site (i.e., the Fe atom that is not ligated by a cysteine residue) (Figure 2.10). In its reduced form, $[4Fe–4S]^+$, it is believed that the cluster transfers an electron into the sulfonium of SAM, creating a quasi-stable species, which leads to fragmentation of the molecule into L-methionine and the high-energy primary alkyl radical intermediate (Figure 2.11).

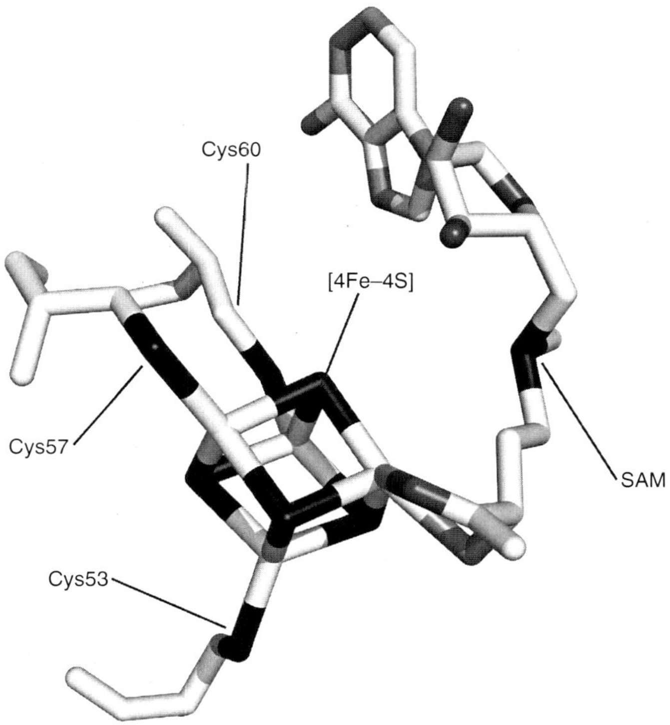

FIGURE 2.10 Representative structure of the binding mode of *S*-adenosylmethionine in the radical SAM superfamily of enzymes. The figure was prepared from the PDB file for biotin synthase (PDB 1R30). Shown in black are the sulfur atoms of SAM, the [4Fe–4S] cluster, and the cysteine ligands to the [4Fe–4S] cluster. Shown in gray are other heteroatoms (O and N), particularly the oxygen and nitrogen atoms of the α-carboxylate and α-amino groups of SAM, which coordinate to one of the iron atoms of the [4Fe–4S] cluster. The structure was prepared using the Pymol Molecular Graphics System (http://www.pymol.org).

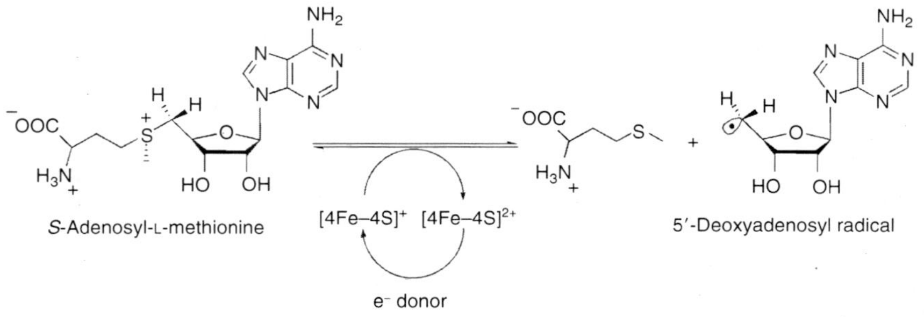

FIGURE 2.11 Reductive cleavage of *S*-adenosylmethionine catalyzed by the radical SAM superfamily of enzymes. The cleavage reaction generates L-methionine and the 5′-deoxyadenosyl radical intermediate. The electron (e^-) can be supplied by the flavodoxin:flavodoxin:$NADP^+$ oxidoreductase system, or by sodium dithionite or 5-deazaflavin in the presence of light.

Isolation and Characterization of Lipoyl Synthase from *Escherichia coli*

The isolation of *E. coli* LipA was first reported by Michael Marletta's laboratory and collaborators [148], and subsequently by Marc Fontecave's laboratory and collaborators [149]. In the former study, a clear description of the purification of the protein was not provided; however, it was reddish-brown in color and contained 1.8 ± 0.2 mol of iron and 2.2 ± 0.4 mol of acid-labile sulfide per mol of protein. The UV–visible and resonance Raman spectra of LipA were consistent with the presence of [4Fe–4S] clusters, which were not reduced upon addition of dithionite as noted from the absence of an EPR spectrum. Note that $[4Fe–4S]^{+2}$ clusters are diamagnetic, while $[4Fe–4S]^{+}$ clusters are paramagnetic, having an $S = 1/2$ spin ground state, and therefore are observable by EPR. The stoichiometry of iron and sulfide would indicate that a substantial fraction of the protein did not have iron–sulfur (Fe/S) clusters associated with them [148]. In the latter study, overexpression of the *lipA* gene in *E. coli* BL21(DE3) resulted in the production of the protein as insoluble aggregates. The inclusion bodies were resolubilized with 6 M guanidine hydrochloride and the protein was reconstituted with iron and sulfide to generate the expected [4Fe–4S] cluster. The reconstituted protein contained 1.8–2.3 mol of iron and sulfide per mole of polypeptide, and had a spectrum that was similar to that of proteins that contained [2Fe–2S] clusters. Upon reduction of the protein with 5-deazaflavin and light, the resulting EPR spectrum displayed temperature dependence and microwave power saturation properties that were consistent with a reduced $[4Fe–4S]^{+}$ cluster rather than a reduced $[2Fe–2S]^{+}$ cluster [149].

One study, carried out by Peter Roach and colleagues, attempted to define some of the determinants of protein solubility during in vivo expression of the *E. coli lipA* gene [150]. It was reasoned that inclusion bodies might form because of misfolding as a result of a deficiency in chaperone-like proteins to assist in the folding process, or because of inefficient incorporation of the Fe/S cluster during or after translation. Several systems have been shown to be important in protein folding in vivo. Co-expression with the gene for thioredoxin, for example, is thought to increase solubility of certain proteins by altering the redox potential of the cytosol [151]. The Chaperonin GroESL has been shown to assist in general in vivo folding as well as cluster insertion into Fe/S proteins [151,152]. When the *E. coli lipA* gene was co-expressed with the genes that encode GroESL or thioredoxin, no increase in expression was observed as compared to expression of *lipA* alone. By contrast, when *lipA* was co-expressed with an operon (*isc*) that encodes proteins known to be involved in Fe/S cluster biosynthesis, soluble protein production increased about threefold [150]. Proteins encoded by the *isc* operon include: IscS, a cysteine desulfurase, which liberates sulfur from cysteine to be used in Fe/S cluster assembly; IscU and IscA, which are thought to serve as scaffolds upon which precursor clusters are built; HscB and HscA, which are molecular chaperones believed to facilitate Fe/S cluster insertion; and Fdx, a ferredoxin believed to be involved in maintaining redox balance during the process [153–155]. In none of the above studies on the characterization of LipA was the protein reported to catalyze turnover. Although octanoic acid was shown to be a precursor to lipoic acid in vivo, it did not induce turnover in vitro.

The first demonstration of in vitro turnover was reported by Cronan and Marletta and their coworkers and collaborators, using a spectrophotometric assay to monitor LipA-dependent production of lipoyl-PDC [61]. The assay exploited the ability of lipoyl-PDC to catalyze reduction of 3-acetylpyridine adenine dinucleotide (APAD)—an analog of NAD^+ with a higher molar absorptivity in its reduced form—in the presence of pyruvate, CoA, thiamin pyrophosphate, and cysteine, the rate of which was directly proportional to the concentration of lipoyl-PDC generated in the LipA reaction. At the time, the authors thought that octanoyl-ACP, and not an octanoyl-LCP, was the substrate for LipA, and that the true role of LipB was that of a lipoyltransferase rather than an octanoyltransferase. Reactions therefore included octanoyl-ACP in addition to LipA, SAM, and sodium dithionite as the source of the requisite electron. After a given incubation period, the lipoyl group from the suspected lipoyl-ACP product was then transferred to apo-PDC by addition of LipB to the reaction, and the extent of lipoylation was subsequently quantified as described above using a standard curve of known concentrations of lipoyl-PDC. Using this assay, LipA was reported to catalyze formation of 0.032 mol of product per mol of LipA polypeptide [61].

Interestingly, the authors could never detect a lipoyl-ACP species using a variety of methods, which included polyacrylamide gel electrophoresis in the presence of urea, hydrophobic chromatography, immobilized metal affinity chromatography, size exclusion chromatography, and MALDI mass spectrometry. By contrast, the lipoylated E_2 domain of the PDC could, in fact, be observed by MALDI mass spectrometry. It is clear from the aforementioned description of the pathways of lipoylation that lipoyl-ACP is never generated, because octanoyl-ACP is not a substrate for LipA. The ability of the authors to detect formation of lipoyl-PDC most likely resulted from the subsequent addition of LipB and apo-PDC to the LipA reaction, which generated the octanoyl-PDC substrate in the presence of reduced LipA and *S*-adenosylmethionine. A subsequent study by Cronan, Marletta, and their coworkers provided firm biochemical evidence that octanoyl-PDC and octanoyl-KDC, but not octanoyl-ACP, are the true substrates for LipA. It was found that *lipB* mutant strains of *E. coli* were able to grow in minimal media supplemented with octanoate in place of lipoate, and that this growth required a functional copy of the *lplA* gene and higher concentrations of octanoate as compared to lipoate [62]. This experiment suggested, therefore, that LplA could catalyze the attachment of octanoate onto LCPs in vivo, whereupon LipA would insert sulfur atoms into the appended octanoyl chain. To test this hypothesis, a *lipA lipB fadE* null mutant (λ) strain of *E. coli* was created, which allowed accumulation of a d_{15}-labeled octanoyl-E_2 domain when cultured in the presence of d_{15}-octanoic acid; the mutation in *lipA* blocked lipoic acid synthesis, the mutation in *lipB* blocked transfer of octanoate from fatty acid biosynthesis, and the mutation in *fadE* blocked β-oxidative degradation of the added octanoate. Upon in vivo generation of the labeled E_2 domains, lipoyl synthase activity was induced by transduction of the cells with phage λ particles containing a *lipA* cosmid. Subsequent isolation of the E_2 domains and their analysis by ES–MS showed that the d_{15}-labeled octanoyl–E_2 domains had been converted into d_{13}-labeled lipoyl–E_2 domains, demonstrating that lipoyl synthase could indeed use octanoylated LCPs as substrates [62].

In vitro studies, assessing the LipA-dependent cleavage of *S*-adenosylmethionine by monitoring the production of 5′-deoxyadenosine, were consistent with the in vivo studies. Significant cleavage of SAM was observed in the presence of octanoyl-E_2 of the PDC but not in the presence of octanoyl-ACP [62]. Similar results were obtained subsequently in studies by the Booker laboratory, wherein it was shown that octanoyl-H protein served as a substrate for LipA [51].

E. coli LipA Contains Two [4Fe–4S] Clusters per Polypeptide in Its Active Form

Further spectroscopic characterization of LipA was carried out in a study by the Booker and Krebs laboratories [52]. In this study, the *E. coli lipA* gene was co-expressed with plasmid pDB1282, which encoded genes from the *isc* operon from *A. vinelandii* under the control of an arabinose-inducible promoter. The *lipA* gene was cloned into a pET-28a-derived vector, and was under the control of an IPTG (isopropyl-β-D-thiogalactopyranoside)-inducible promoter. Similarly to that previously reported in Ref. [150], co-expression of the *lipA* gene with an *isc* operon dramatically increased the ratio of soluble to insoluble LipA in the crude extract, which allowed the isolation of hundreds of milligram quantities of the protein. Moreover, the expression system worked well in minimal medium, which allowed supplementation with ^{57}Fe in the absence of significant concentrations of contaminating natural abundance iron so that the isolated and ^{57}Fe-labeled protein could be analyzed by Mössbauer spectroscopy [52]. The details of Mössbauer spectroscopy are beyond the scope of this review. However, it is important to note that this technique can yield a tremendous amount of information about the oxidation state and configuration of iron in most molecules that contain ^{57}Fe, especially when used in conjunction with electron paramagnetic resonance spectroscopy. The caveat, however, is that only the ^{57}Fe isotope is Mössbauer active, which must be introduced into proteins by reconstitution methods, or by producing the protein in media in which natural abundance iron has been replaced with the ^{57}Fe isotope. Several excellent reviews about the use of Mössbauer spectroscopy to elucidate the structures of biologically relevant iron-containing molecules have been published [156–159], and the reader is encouraged to refer to them for a deeper understanding of the technique.

LipA was purified by immobilized metal affinity chromatography, exploiting an N-terminal hexahistidine tag, and the isolated protein was characterized by UV–visible, electron paramagnetic resonance (EPR), and Mössbauer spectroscopies, which were used in conjunction with chemical analyses for iron and sulfide to determine the number and configuration of associated Fe/S clusters. The as-isolated protein contained 6.9 ± 0.5 equiv of iron and 6.4 ± 0.9 equiv of sulfide per polypeptide, considerably more than what was reported from previous isolations [61,148,150] or even after reconstitution [149,160]. The UV–visible, EPR, and Mössbauer spectra of the as-isolated protein were all consistent with the presence of $[4Fe\text{–}4S]^{+2}$ clusters, which were determined by Mössbauer spectroscopy to represent >97% of all of the iron associated with the protein [52]. The stoichiometry of iron and sulfide and the failure to detect the presence of significant amounts of $[2Fe\text{–}2S]^{+2}$ clusters by Mössbauer spectroscopy suggested that the protein contained more than one $[4Fe\text{–}4S]^{+2}$ cluster, and that a small percentage of polypeptides had no $[4Fe\text{–}4S]^{+2}$ clusters associated with them. When the as-isolated protein was reconstituted with additional iron and sulfide, the stoichiometry of iron and sulfide associated with the protein subsequent to gel-filtration increased dramatically to 13.8 ± 0.6 and 13.1 ± 0.2 per polypeptide. Complementary Mössbauer studies showed, however, that only $67\% \pm 6\%$ of that iron was in the configuration $[4Fe\text{–}4S]^{+2}$, while the remaining 33% was adventitiously bound ferrous iron, which is referred to as "junk" iron. Therefore, quantification of iron species by Mössbauer showed that the reconstituted protein contained 9.2 ± 0.4 irons and 8.8 ± 0.1 sulfides per polypeptide in the configuration $[4Fe\text{–}4S]^{+2}$, indicating the presence of two $[4Fe\text{–}4S]^{+2}$ clusters per polypeptide [52].

E. coli LipA contains eight cysteine residues, six of which are strictly conserved. Three of the conserved cysteines reside in the $Cys_{94}XXXCys_{98}XXCys_{101}$ radical SAM signature motif, while the remaining three reside in a $Cys_{68}XXXXCys_{73}XXXXXCys_{79}$ motif, which is found only in lipoyl synthases. When Cys→Ala substitutions were made for cysteines lying in the two indicated motifs, the isolated proteins were inactive and contained reduced amounts of iron and sulfide per polypeptide [52]. Triple variants containing Cys→Ala substitutions at each of the cysteines in the two conserved motifs were constructed and analyzed in detail by UV–visible, EPR, and Mössbauer spectroscopies, as well as by chemical analyses for iron and sulfide. The $Cys_{68}Ala–Cys_{73}Ala–Cys_{79}Ala$ triple variant contained 3.0 ± 0.1 irons and 3.6 ± 0.4 sulfides per polypeptide, while the $Cys_{94}Ala–Cys_{98}Ala–Cys_{101}Ala$ triple variant contained 4.2 ± 0.1 irons and 4.7 ± 0.8 sulfides per polypeptide, which in conjunction with Mössbauer spectroscopy indicated the presence of only one $[4Fe–4S]^{+2}$ cluster in each of the triple variants. Moreover, the stoichiometry of $[4Fe–4S]^{+2}$ clusters per polypeptide did not increase upon reconstitution, which is consistent with a model in which the cysteines within the two conserved motifs serve as ligands to two $[4Fe–4S]^{+2}$ clusters [52]. The [4Fe–4S] cluster in the radical SAM motif, $Cys_{94}XXXCys_{98}XXCys_{101}$ (Cluster A), was proposed to interact with, and donate an electron to, SAM for generation of the 5′-dA$^{\bullet}$ intermediate required for catalysis, while the [4Fe–4S] cluster in the lipoyl synthase motif, $Cys_{68}XXXXCys_{73}XXXXXCys_{79}$ (Cluster B), was proposed to act as the sulfur donor [52].

Mechanistic Characterization of the LipA Reaction

The early in vivo feeding studies in *E. coli*, which indicated that lipoic acid is synthesized from octanoic acid with removal of only those hydrogen atoms that are replaced by sulfur atoms, and recognition of LipA as a member of the radical SAM superfamily of enzymes, allows formulation of a working hypothesis for how sulfur insertion might take place (Figure 2.12). An electron is transferred to Cluster A from reduced flavodoxin (or dithionite, or 5-deazaflavin and light in vitro), and then subsequently to SAM, inducing its fragmentation into a 5′-dA$^{\bullet}$ and L-methionine. The 5′-dA$^{\bullet}$ abstracts a hydrogen atom from C-6 of the octanoyl-LCP substrate, affording an alkyl radical, which attacks one of the bridging sulfido ligands of Cluster B (Figure 2.12A). The second step of the reaction is a repeat of the first step, except that the 5′-dA$^{\bullet}$ abstracts a hydrogen atom from C-8, affording an alkyl radical that attacks another of the bridging sulfido ligands of Cluster B. The addition of two protons allows dissociation of the lipoyl product in the form of dihydrolipoamide (Figure 2.12B). The mechanism shown in Figure 2.12 predicts that two moles of SAM must be expended per mol of lipoamide generated. Analysis of the time-dependent production of 5′-dA, which reflects the irreversible cleavage of SAM, and comparison with the time-dependent production of lipoamide shows that each reaction takes place with similar pseudo first-order rate constants (0.144 ± 0.005 min^{-1} and 0.175 ± 0.010 min^{-1}, respectively). Moreover, at each time point, the ratio of 5′-dA to lipoamide varies from 2.38 to 2.71. The excess 5′-dA was attributed to abortive cleavage of SAM, which could arise if the 5′-dA$^{\bullet}$ removes a hydrogen atom from some species other than substrate, or if sulfur insertion takes place at one carbon but not the other, leaving a monothiolated species [51].

The use of a substrate containing a perdeuterated octanoyl chain, [octanoyl-d_{15}]-H-protein, allowed verification of the abortive cleavage of SAM, and also yielded other interesting results. 5′-Deoxyadenosine was isolated as a function of

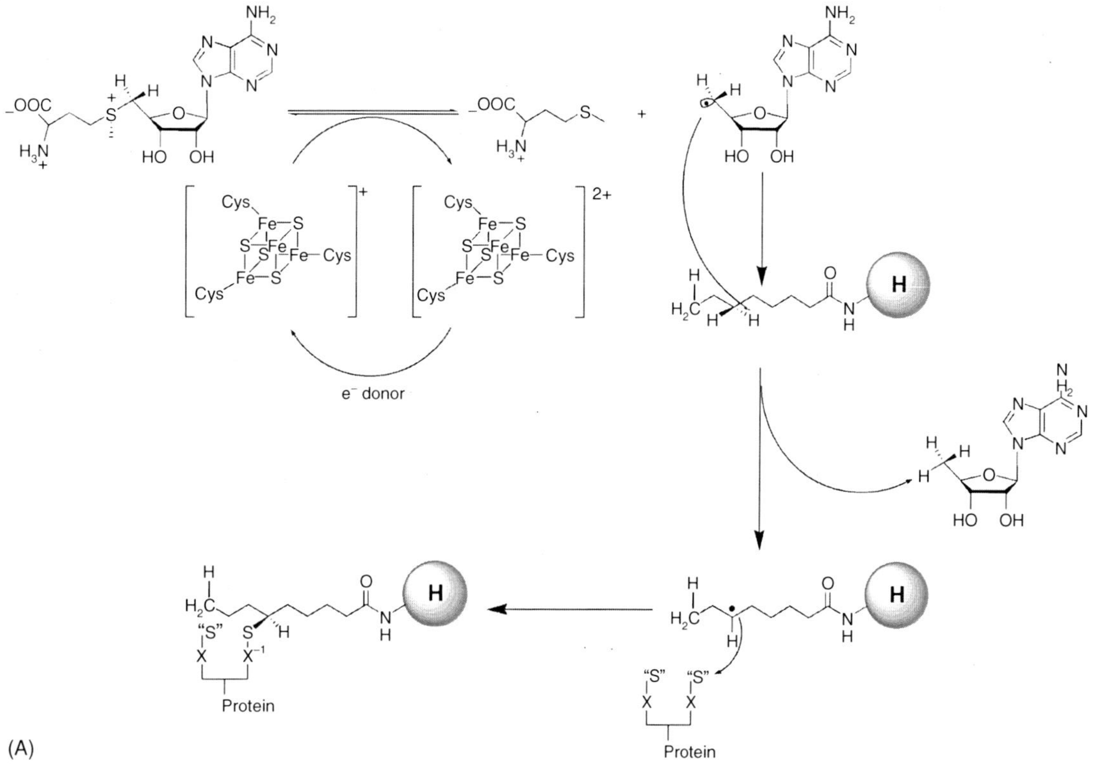

(A)

FIGURE 2.12 Working hypothesis for the reaction mechanism of LipA. (A) First half of the reaction mechanism, which involves hydrogen atom abstraction from C-6 and subsequent radical addition to a sulfur source, indicated as "S," but believed to be the [4Fe–4S] cluster residing in the CXXXXCXXXXXC motif.

(continued)

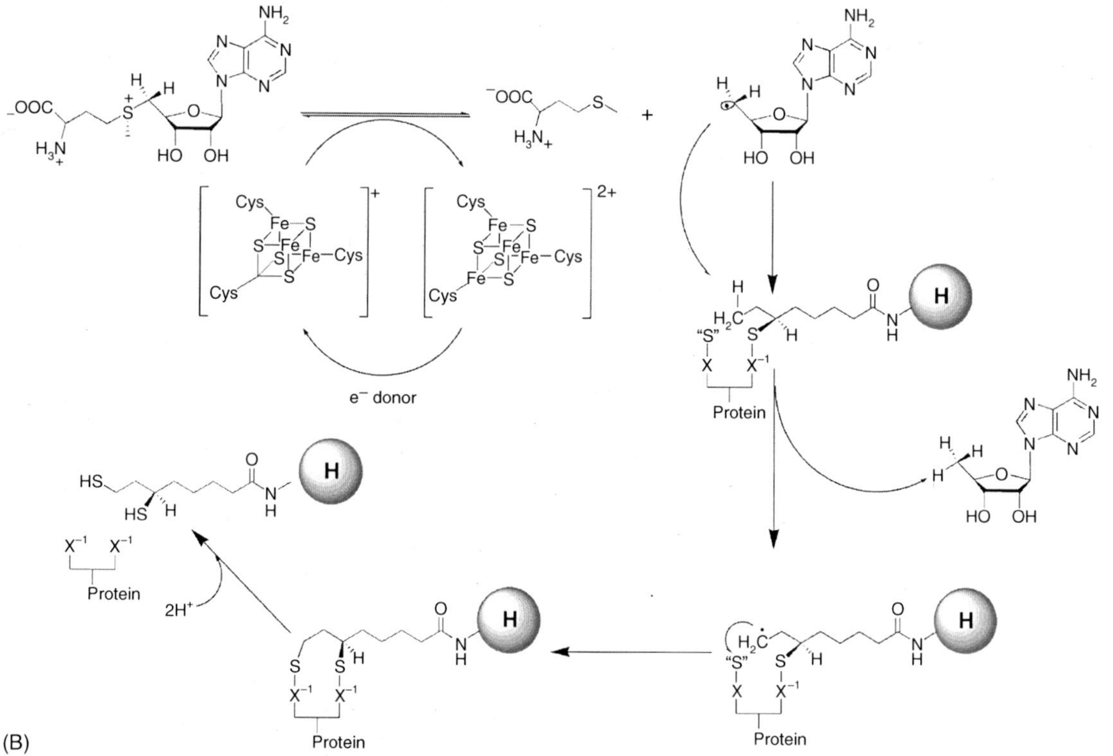

(B)

FIGURE 2.12 (continued) (B) Second half of the reaction mechanism, which involves hydrogen atom abstraction from C-8 and subsequent radical addition to a sulfur source. The subsequent addition of two protons results in the production of lipoamide in its reduced form. The sphere containing "H" represents the H-protein of the glycine cleavage system, an LCP.

time and analyzed by GC–MS to determine its deuterium content. Both 5′-dA containing one deuterium atom (5′-dA-d_1) and 5′-dA containing no deuterium atoms (5′-dA-d_0) were observed; the latter ranged from 13% to 22% of the total 5′-dA (abortive cleavage). Interestingly, the maximum amount of 5′-dA-d_1 generated was ~50% of that generated when unlabeled substrate was used, and assays for lipoamide indicated that none was formed above the detection limit of the assay, suggesting that abstraction of a hydrogen atom at either C-6 or C-8 proceeded with a large isotope effect. This inhibition of lipoamide formation in the presence of a perdeuterated substrate was also observed by electrospray ionization mass spectrometry [51].

The large isotope effect observed upon deuterium substitution in the substrate is interesting in light of the aforementioned metabolic feeding studies by Parry, wherein [8-^{3}H]octanoate was converted into lipoic acid with essentially no loss in specific radioactivity, suggestive of a significant k_H/k_T kinetic isotope effect at C-8 [64]. Studies by Peter Roach and colleagues, using a lipoyl synthase from *Sulfolobus solfataricus* and an octanoylated tripeptide substrate (glutamyl-octanoyllysyl-isoleucine), verified that deuterium substitution at C-8 and not C-6 dramatically impedes in vitro formation of lipoamide [124]. When the tripeptide containing an [8-$^2H_3$]octanoyl chain was used as substrate, a monothiolated species containing a sulfur atom at C-6 was observed by liquid chromatography–mass spectrometry (LC–MS), while very little lipoamide was formed relative to unlabeled substrate. To exclude the possibility that exclusive sulfur insertion at C-6 was not attributable to discrimination as a result of isotopic labeling at C-8, and to obtain insight into the order of sulfur insertion, reactions were conducted with unlabeled substrate and quenched after 20 min of incubation in an effort to show the presence of an intermediate species. The reaction products were purified by HPLC after reduction and treatment with iodoacetamide, and then analyzed by NMR to identify protons attached to C-6, C-7, and C-8 of the monothiolated species. The only monothiolated species observed contained a sulfur atom at C-6, suggesting that sulfur insertion at C-6 occurs before insertion at C-8 [124]. It must be mentioned that the conclusion reached in this study is in direct contrast to the aforementioned metabolic feeding studies, which indicated that 8-mercaptooctanoic acid is a much better substrate than 6-mercaptooctanoic acid for the enzyme that catalyzes sulfur insertion [67,68].

Unequivocal evidence that LipA is both a catalyst and a substrate was provided by Cicchillo and Booker [161]. LipA was overproduced in minimal media in which added $Na_2^{34}S$ (sodium sulfide) prepared from elemental ^{34}S was the only source of sulfur available. The protein was isolated and used in a reaction in which the only other compounds that contained sulfur were the octanoyl-H-protein substrate and SAM. Moreover, these compounds contained natural abundance sulfur, which is ~95% ^{32}S. Upon completion of the reaction, lipoic acid was cleaved from the lipoyl-H-protein product and analyzed by GC–MS. The sulfur atoms in the lipoic acid product were found to be of the ^{34}S isotope, containing the same percentage of ^{34}S that was used in the growth media for production of the labeled LipA, establishing that LipA itself contributes the sulfur atoms for the synthesis of lipoamide [161]. To assess whether both sulfur atoms are derived from the same LipA subunit, or whether each sulfur atom in the final product is derived from a different LipA polypeptide, the above experiment was repeated with equimolar concentrations of ^{32}S-labeled (i.e., natural abundance) LipA and ^{34}S-labeled LipA. If both sulfur atoms are derived from the same polypeptide, it would be expected that the isolated lipoic acid should contain either two atoms of ^{32}S or two atoms of ^{34}S. By contrast, if each LipA polypeptide contributes only one sulfur atom in the lipoyl product, the product distribution would be expected to be a 1:2:1 ratio of products containing two ^{32}S atoms, one ^{32}S and one ^{34}S atom, and two ^{34}S atoms. GC–MS analysis of lipoic acid produced by an equimolar mixture of these two differentially labeled proteins showed two prominent peaks at m/z values that were consistent with two ^{32}S atoms and two ^{34}S atoms, establishing that both sulfur atoms are contributed by the same LipA polypeptide [161].

SIMILARITIES BETWEEN LIPOYL SYNTHASE AND BIOTIN SYNTHASE

There are clear similarities between the chemistry of sulfur insertion in lipoic acid biosynthesis and the chemistry of sulfur insertion in the biosynthesis of biotin, another important biological cofactor. Biotin is formed by insertion of one sulfur atom between C-6 and C-9 of dethiobiotin, its direct precursor (Figure 2.13). As previously stated, two sulfur atoms are inserted into a protein-bound octanoyl derivative to generate lipoamide [50,51,56,61,124]. In vivo metabolic labeling studies in *E. coli* and *Aspergillus niger* that were similar to those carried out in the study of lipoic acid biosynthesis showed that biotin formation requires the removal of only two hydrogens from dethiobiotin, one from C-6 and one from C-9, and that neither [1-^{3}H]-6-hydroxydethiobiotin nor [9(*R,S*)-^{3}H]-9-hydroxydethiobiotin is a precursor to biotin [50,162]. In analogy to lipoic acid biosynthesis, analogs of dethiobiotin containing sulfur (^{34}S or ^{35}S) at C-9 were capable of supporting growth of a strain of *E. coli* that is auxotrophic for biotin, and were directly converted to biotin by *Bacillus sphaericus* [163]. In addition,

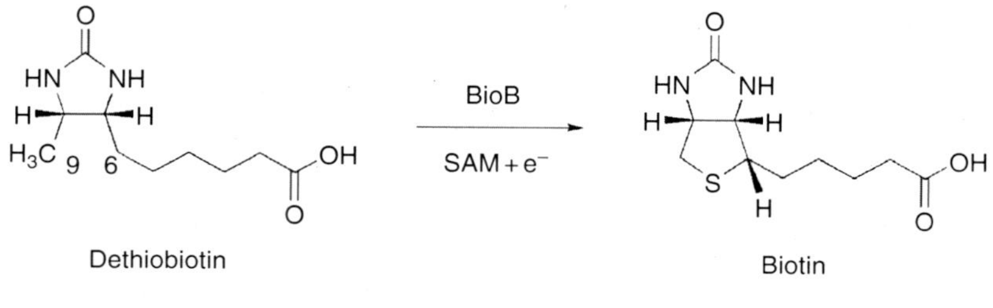

FIGURE 2.13 Biotin synthase reaction.

biotin synthase (BioB) was shown to be able to convert 9-mercaptodethiobiotin into biotin in vitro at a rate that was similar to that in the presence of dethiobiotin itself [164].

BioB also contains six conserved cysteine residues. Three of the cysteines lie in the conserved $C_{53}XXXC_{57}XXC_{60}$ radical SAM motif, while the remaining

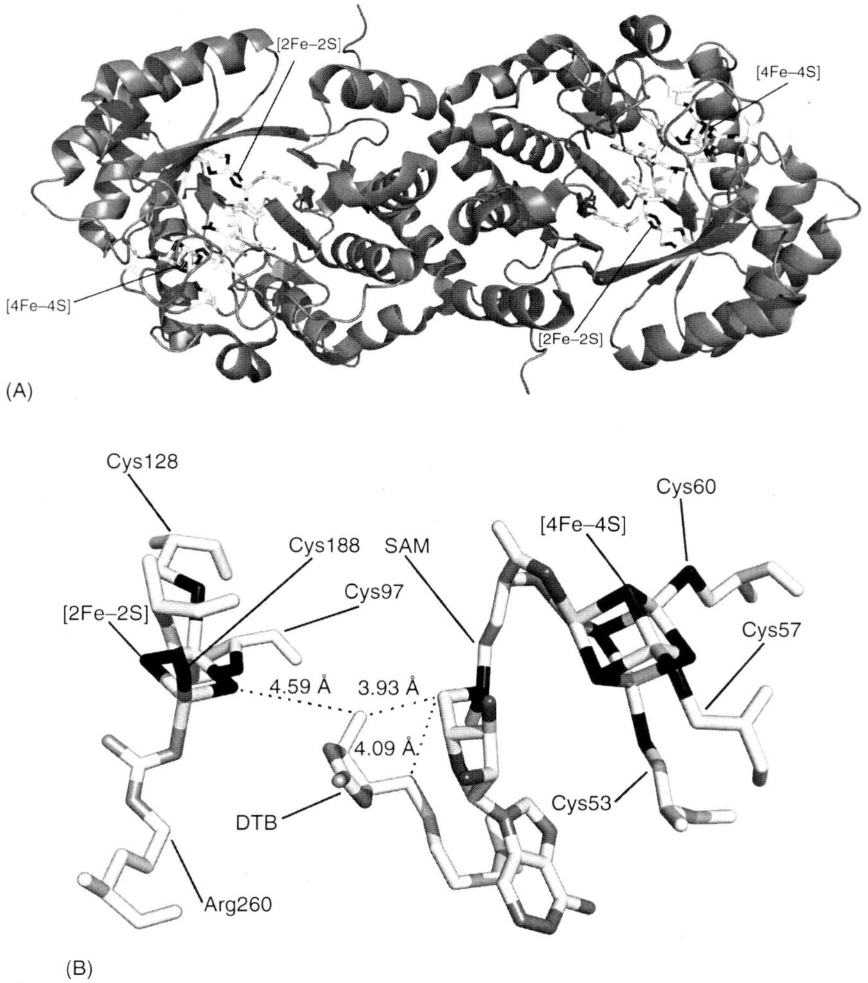

FIGURE 2.14 Structure of biotin synthase (PDB 1R30). (A) Overall fold of the homodimer. The positions of the [4Fe–4S] and [2Fe–2S] clusters are indicated. (B) View of the active site. SAM and dethiobiotin are sandwiched between the [2Fe–2S] and [4Fe–4S] clusters. Sulfur atoms are shown in black. Other heteroatoms are shown in gray. The [2Fe–2S] cluster has three Cys ligands and one Arg ligand. DTB is dethiobiotin. Structures were prepared using the Pymol Molecular Graphics System (http://www.pymol.org).

three conserved cysteines, Cys97, Cys128, and Cys188, reside predominantly in the C-terminal half of the protein, and are not found in a compact sequence motif as is seen in LipA. Similarly to LipA, this second set of cysteines coordinates a second Fe/S cluster that is required for activity, and the substitution of any of the cysteine ligands with Ala or Ser results in loss of activity. In contrast to LipA, this second Fe/S cluster is a [2Fe–2S] cluster rather than a [4Fe–4S] cluster [59,165–167].

An x-ray structure of BioB was solved to 3.4 Å resolution and includes both SAM and dethiobiotin bound to the enzyme [145]. The structure shows that the protein is homodimeric, composed of subunits that are in a triosephosphate isomerase (TIM) type $(\alpha/\beta)_8$ barrel. There are also two additional helices at the N-terminus and a disordered region at the C-terminus. The [4Fe–4S] clusters are at the opposite end of the dimer interface, and the [2Fe–2S] clusters are buried deep inside the barrel (Figure 2.14A). A close-up view of the active site of BioB shows that the two substrates are sandwiched between the [2Fe–2S] and [4Fe–4S] clusters (Figure 2.14B), predicted to be present on the enzyme by various spectroscopic methods [59,168,169]. In the structure, the α-amino and α-carboxylate groups of SAM are coordinated to the unique iron of the [4Fe–4S] cluster, and C-5′ of SAM is ~4 Å away from C-6 and C-9 of dethiobiotin. Moreover, C-9 of dethiobiotin is 4.59 Å away from one of the bridging sulfurs of the [2Fe–2S] cluster, which appears to contain a conserved Arg amino acid as a ligand in addition to the three conserved cysteine amino acids that act as ligands. The structure is consistent with the proposed model that the [2Fe–2S] cluster participates somehow in sulfur mobilization [60]. Both the [2Fe–2S] and [4Fe–4S] clusters contribute features to the UV–visible spectrum of biotin synthase. During turnover, loss of features in the spectrum that are ascribed to the [2Fe–2S] cluster have been reported to correlate with formation of biotin [60]. A similar study, in which Mössbauer spectroscopy was used to monitor turnover-dependent changes in the Fe/S clusters of BioB, indicated that loss of the [2Fe–2S] cluster occurred on a timescale that was significantly faster than biotin production [55].

CONCLUSION

The current picture of the pathways of lipoamide biosynthesis is a result of well-designed genetic and biochemical studies conducted primarily in, or with, *Escherichia coli*, or using *E. coli* proteins [17,50]. It is clear, however, that similar, if not exact, pathways are operative in many other organisms, and that some of the participant enzymes might serve as targets for the design of antibacterial or antiparasitic agents, or maybe even antineoplastic agents. The importance of lipoate protein ligase in *Listeria monocytogenes* has already been described, wherein this pathogen uses this particular ligase to obtain lipoic acid from the cytosol of its host. The finding that *L. monocytogenes* LplA1 is not required when the bacterium is grown in broth culture, and the understanding that very little free lipoic acid exists in the mammalian cytosol [2,69–71], suggests that this protein can somehow remove lipoate groups from host LCPs and use them to modify its own LCPs, particularly the PDC [116]. Details of how this is accomplished are yet to be revealed. Other parasites, such as *Plasmodium falciparum* and *Toxoplasma gondii* possess lipoylation pathways that are organelle specific. For example in *T. gondii*, de novo biosynthesis of lipoamide occurs in the organism's apicoplast, an endosymbiotic organelle that performs a number of important biosynthetic functions for the organism. For its mitochondrial functions, however, the organism cannot use apicoplast-derived lipoic acid, but must scavenge it from its host via the action of a lipoate protein ligase [170]. Similar pathways are operative in *Plasmodium falciparum*, one of the species of *Plasmodium* that causes human malaria. In this organism, de novo synthesis of lipoamide takes place in the apicoplast, whereas its lipoate protein ligase is located in its mitochondrion [171,172]. Efforts to generate null mutations in the LipB gene of *Mycobacterium tuberculosis* have been unsuccessful [123], which is consistent with the finding that LipB is essential for growth of the organism [173]. Moreover, the expression of the gene for LipB is strongly upregulated in the lungs of patients with pulmonary multiple-drug-resistant tuberculosis [123,174]. Mice containing heterozygous null mutations in the Lias (lipoyl synthase) gene develop normally but have markedly reduced levels of erythrocyte glutathione, indicating that they have reduced antioxidant capacity. Mice containing homozygous null mutations in the Lias gene do not develop past the blastocyst stage, even when the diets of their mothers are supplemented with lipoic acid [175]. Therefore, even though mammalian cells contain both pathways for lipoamide formation, it appears that the ability to endogenously synthesize the cofactor is critical at least for development. At present, there is very little understanding of the pathways of lipoylation in humans, and which pathways predominate in particular cell or tissue types

The proteins that constitute the pathways for lipoamide biosynthesis are, themselves, targets for further mechanistic characterization. LipA is the most intriguing of the proteins from a mechanistic standpoint. There is circumstantial evidence that the sulfur atoms inserted during catalysis originate from the [4Fe–4S] cluster that resides in the $C_{68}XXXXC_{73}XXXXXC_{79}$ motif, which is common to all lipoyl synthases [52,161]; however, much of this evidence derives from work done on the related protein, biotin synthase [55,60,176–178]. There is need for structural studies on this enzyme, and the design of experiments that will test specifically the hypothesis that the Fe/S cluster is the direct donor of the sulfur atoms. The other major question that relates to the mechanism of action of lipoyl synthase is whether the active form of the protein is regenerated after each turnover in vivo. There is evidence that *E. coli* biotin synthase catalyzes multiple turnovers in vivo, and that catalysis (i.e., loss of the [2Fe–2S]) renders the protein susceptible to proteolysis [179]. There is still very little insight as to how the [2Fe–2S] cluster is regenerated after each turnover; however, evidence suggests that IscS or an IscS-related protein may be involved [177,180–182]. One major take-home message from this review on lipoic acid biosynthesis is that there is still much to learn about these incredibly exciting pathways!

ACKNOWLEDGMENTS

Our work on lipoic acid biosynthesis was supported by the National Institutes of Health (GM-63847), and an NIH minority predoctoral fellowship to N.M.N. (GM-64033).

REFERENCES

1. Herbert, A.A. and J.R. Guest, Lipoic acid content of *Escherichia coli* and other microorganisms. *Arch. Biochem. Biophys.*, 1975. **106**: 259–266.
2. Biewenga, G.P., G.R.M.M. Haenen, and A. Bast, An overview of lipoate chemistry, in *Lipoic Acid in Health and Disease*, J. Fuchs, L. Packer, and G. Zimmer, Editors. 1997, Marcel Dekker, Inc.: New York. pp. 1–32.
3. Douce, R., et al., The glycine decarboxylase system in higher plant mitochondria: Structure, function and biogenesis. *Biochem. Soc. Trans.*, 1994. **22**: 184–188.
4. Kruger, N., et al., Biochemical and molecular characterization of the Clostridium magnum acetoin dehydrogenase enzyme system. *J. Bacteriol.*, 1994. **176**: 3614–3630.
5. Pares, S., et al., X-ray structure determination at 2.6-Å resolution of a lipoate-containing protein: The H-protein of the glycine decarboxylase complex from pea leaves. *Proc. Natl. Acad. Sci. USA*, 1994. **91**: 4850–4853.
6. Priefer, H., et al., Identification and molecular characterization of the *Alcaligenes eutrophus* H16 aco operon genes involved in acetoin catabolism. *J. Bacteriol.*, 1991. **173**: 4056–4071.
7. Reed, L., Multienzyme complexes. *Acc. Chem. Res.*, 1974. **7**: 40–46.
8. Perham, R.N., Domains, motifs, and linkers in 2-oxo acid dehydrogenase multienzyme complexes: A paradigm in the design of a multifunctional protein. *Biochemistry*, 1991. **30**: 8501–8512.
9. Perham, R.N., Swinging arms and swinging domains in mulftifunctional enzymes: Catalytic machines for multistep reactions. *Annu. Rev. Biochem.*, 2000. **69**: 961–1004.
10. Reed, L.J. and M.L. Hackert, Structure–function relationships in dihydrolipoamide acyltransferases. *J. Biol. Chem.*, 1990. **265**: 8971–8974.
11. Ambrose-Griffen, M.C., et al., Kinetic analysis of the role of lipoic acid residues in the pyruvate dehydrogenase multienzyme complex of *Escherichia coli. Biochem. J.*, 1980. **187**: 393–401.
12. Angelides, K.J. and G.G. Hammes, Mechanism of action of the pyruvate dehydrogenase multienzyme complex from *Escherichia coli. Proc. Natl. Acad. Sci.* USA, 1978. **75**: 4877–4880.
13. Koike, M. and L.J. Reed, α-Keto acid dehydrogenation complexes. II. The role of protein-bound lipoic acid and flavin adenine dinucleotide. *J. Biol. Chem.*, 1960. **235**: 1931–1938.
14. Reed, L.J., et al., Studies on the nature and reactions of protein-bound lipoic acid. *J. Biol. Chem.*, 1958. **232**: 143–158.
15. Bourguignon, J., et al., Isolation, characterization, and sequence analysis of a cDNA clone encoding L-protein, the dihydrolipoamide dehydrogenase component of the glycine cleavage system from pea-leaf mitochondria. *Eur. J. Biochem.*, 1992. **204**: 865–873.

16. Bourguignon, J., et al., Glycine decarboxylase and pyruvate dehydrogenase complexes share the same dihydrolipoamide dehydrogenase in pea leaf mitochondria: Evidence from mass spectrometry and primary-structure analysis. *Biochem. J.*, 1996. **313**: 229–234.
17. Cronan, J.E., X. Zhao, and Y. Jiang, Function, attachment and synthesis of lipoic acid in *Escherichia coli. Adv. Microb. Physiol.*, 2005. **50**: 103–146.
18. Otulakowski, G. and B.H. Robinson, Isolation and sequence determination of cDNA clones for porcine and human lipoamide dehydrogenase. Homology to other disulfide oxidoreductases. *J. Biol. Chem.*, 1987. **262**: 17313–17318.
19. Turner, S.R., R. Ireland, and S. Rawsthorne, Purification and primary amino acid sequence of the L subunit of glycine decarboxylase. Evidence for a single lipoamide dehydrogenase in plant mitochondria. *J. Biol. Chem.*, 1992. **267**: 7745–7750.
20. Oppermann, F.B. and A. Steinbuchel, Identification and molecular characterization of the aco genes encoding the *Pelobacter carbinolicus* acetoin dehydrogenase enzyme system. *J. Bacteriol.*, 1994. **176**: 469–485.
21. Palmer, J.A., et al., Cloning, sequence and transcriptional analysis of the structural gene for LPD-3, the third lipoamide dehydrogenase of *Pseudomonas putida. Eur. J. Biochem.*, 1991. **202**: 231–240.
22. Berg, A. and A. de Kok, 2-Oxo acid dehydrogenase multienzyme complexes. The central role of the lipoyl domain. *Biol. Chem.*, 1997. **378**: 617–634.
23. Green, J.D.F., et al., Conformational studies of the interdomain linker peptides in the dihydrolipoyl acetyltransferase component of the pyruvate dehydrogenase multienzyme complex of *E. coli. J. Biol. Chem.*, 1992. **267**: 23484–23488.
24. Stephens, P.E., et al., The pyruvate dehydrogenase complex of *Escherichia coli* K12. Nucleotide sequence encoding the pyruvate dehydrogenase component. *Eur. J. Biochem.*, 1983. **133**: 155–162.
25. Stephens, P.E., et al., The pyruvate dehydrogenase complex of *Escherichia coli* K12. Nucleotide sequence encoding the dihydrolipoamide acetyltransferase component. *Eur. J. Biochem.*, 1983. **133**: 481–489.
26. Stephens, P.E., et al., Nucleotide sequence of the lipoamide dehydrogenase gene of *Escherichia coli* K12. *Eur. J. Biochem.*, 1983. **135**: 519–527.
27. Turner, S.L., et al., Restructuring an interdomain linker in the dihydrolipoamide acetyltransferase component of the pyruvate dehydrogenase complex of *Escherichia coli. Protein Eng.*, 1993. **6**: 101–108.
28. Sagers, R.D. and I.C. Gunsalus, Intermediary metabolism of *Diplococcus glycinophilus. I.* Glycine cleavage and one-carbon interconversions. *J. Bacteriol.*, 1961. **81**: 541–549.
29. Vanden Boom, T.J., K.E. Reed, and J.J.E. Cronan, Lipoic acid metabolism in *Escherichia coli:* Isolation of null mutants defective in lipoic acid biosynthesis, molecular cloning and characterization of the *E. coli* lip locus, and identification of the lipoylated protein of the glycine cleavage system. *J. Bacteriol.*, 1991. **173**(20): 6411–6420.
30. Fujiwara, K., K. Okamura-Ikeda, and Y. Motokawa, Mechanism of the glycine cleavage reaction. Further characterization of the intermediate attached to H-protein and of the reaction catalyzed by T-protein. *J. Biol. Chem.*, 1984. **259**: 10664–10668.
31. Hiraga, K. and G. Kikuchi, The mitochondrial glycine cleavage system. Purification and properties of glycine decarboxylase from chicken liver mitochondria. *J. Biol. Chem.*, 1980. **255**: 11664–11670.

32. Klein, S.M. and R.D. Sagers, Glycine metabolism. III. A flavin-linked dehydrogenase associated with the glycine cleavage system in *Peptococcus glycinophilus*. *J. Biol. Chem.*, 1967. **242**: 297–300.
33. Fujiwara, K., K. Okamura, and Y. Motokawa, Hydrogen carrier protein from chicken liver: Purification, characterization, and role of its prosthetic group, lipoic acid, in the glycine cleavage reaction. *Arch. Biochem. Biophys.*, 1979. **197**: 454–462.
34. Kochi, H. and G. Kikuchi, Mechanism of the reversible glycine cleavage reaction in *Arthrobacter globiformis*. I. Purification and function of protein components required for the reaction. *J. Biochem.*, 1974. **75**: 1113–1127.
35. Kochi, H. and G. Kikuchi, Mechanism of reversible glycine cleavage reaction in *Arthrobacter globiformis*. Function of lipoic acid in the cleavage and synthesis of glycine. *Arch. Biochem. Biophys.*, 1976. **173**: 71–81.
36. Motokawa, Y. and G. Kikuchi, Glycine metabolism by rat liver mitochondria. IV. Isolation and characterization of hydrogen carrier protein, and essential factor for glycine metabolism. *Arch. Biochem. Biophys.*, 1969. **135**: 402–409.
37. Robinson, J.R., S.M. Klein, and R.D. Sagers, Glycine metabolism. Lipoic acid as the prosthetic group in the electron transfer protein P2 from *Peptococcus glycinophilus*. *J. Biol. Chem.*, 1973. **248**: 5319–5323.
38. Motokawa, Y. and G. Kikuchi, Glycine metabolism by rat liver mitochondria. II. Methylene tetrahydrofolate as the direct one carbon donor in the reaction of glycine synthesis. *J. Biochem.*, 1969. **65**: 71–75.
39. Baginsky, M.L. and F.M. Huennekens, Further studies on the electron transport proteins involved in the oxdative decarboxylation of glycine. *Arch. Biochem. Biophys.*, 1967. **120**: 703–711.
40. Ricaud, P.M., et al., Three-dimensional structure of the lipoyl domain from the dihydrolipoyl succinyltransferase component of the 2-oxoglutarate dehydrogenase multienzyme complex of *Escherichia coli*. *J. Mol. Biol.*, 1996. **264**: 179–190.
41. Berg, A., J. Vervoort, and A. de Kok, Solution structure of the lipoyl domain of the 2-oxoglutarate dehydrogenase complex from *Azotobacter vinelandii*. *J. Mol. Biol.*, 1996. **261**: 432–442.
42. Green, J.D.F., et al., Three-dimensional structure of a lipoyl domain from the dihydrolipoyl acetyltransferase component of the pyruvate dehydrogenase multienzyme complex of *Escherichia coli*. *J. Mol. Biol.*, 1995. **248**: 328–343.
43. Berg, A., J. Vervoort, and A. de Kok, Three-dimensional structure in solution of the N-terminal lipoyl domain of the pyruvate dehydrogenase complex from *Azotobacter vinelandii*. *Eur. J. Biochem.*, 1997. **244**: 352–360.
44. Dardel, F., et al., Three-dimensional structure of the lipoyl domain from *Bacillus stearothermophilus* pyruvate dehydrogenase multienzyme complex. *J. Mol. Biol.*, 1993. **229**: 1037–1048.
45. Chang, C.-F., et al., Solution structure and dynamics of the lipoic acid-bearing domain of human mitochondrial branched chain alpha-keto acid dehydrogenase complex. *J. Biol. Chem.*, 2002. **277**: 15865–15873.
46. Wallis, N.G. and R.N. Perham, Structural dependence of post-translational modification and reductive acetylation of the lipoyl domain of the pyruvate dehydrogenase multienzyme complex. *J. Mol. Biol.*, 1994. **236**: 209–216.
47. Nakai, T., et al., Structure of *Thermus thermophilus* HB8 H-protein of the glycine-cleavage system, resolved by a six-dimensional molecular-replacement method. *Acta Crystallogr.*, Sect. D, 2003. **59**: 1610–1618.

48. Pares, S., et al., Refined structures at 2 and 2.2-angstrom resolution of 2 forms of the H-protein, a lipoamide-containing protein of the glycine decarboxylase complex. *Acta Crystallogr. D Biol. Crystallogr.*, 1995. **51**: 1041–1051.
49. Cohen-Addad, C., et al., The lipoamide arm in the glycine decarboxylase complex is not freely swinging. *Nat. Struct. Biol.*, 1995. **2**: 63–68.
50. Parry, R.J., Biosynthesis of some sulfur-containing natural products. Investigations of the mechanism of carbon–sulfur bond formation. *Tetrahedron*, 1983. **39**: 1215–1238.
51. Cicchillo, R.M., et al., Lipoyl synthase requires two equivalents of *S*-adenosyl-L-methionine to synthesize one equivalent of lipoic acid. *Biochemistry*, 2004. **43**: 6378–6386.
52. Cicchillo, R.M., et al., *Escherichia coli* lipoyl synthase binds two distinct [4Fe–4S] clusters per polypeptide. *Biochemistry*, 2004. **43**: 11770–11781.
53. Fontecave, M., S. Ollagnier-de Choudens, and E. Mulliez, Biological radical sulfur insertion reactions. *Chem. Rev.*, 2003. **103**: 2149–2166.
54. Hayden, M.A., et al., The biosynthesis of lipoic acid: Cloning of lip, a lipoate biosynthetic locus of *Escherichia coli. J. Biol. Chem.*, 1992. **267**: 9512–9515.
55. Jameson, G.N.L., et al., Role of the [2Fe–2S] cluster in recombinant *Escherichia coli* biotin synthase. *Biochemistry*, 2004. **43**: 2022–2031.
56. Marquet, A., Enzymology of carbon–sulfur bond formation. *Curr. Opin. Chem. Biol.*, 2001. **5**: 541–549.
57. Otsuka, A.J., et al., The *Escherichia coli* biotin biosynthetic enzyme sequences predicted from the nucleotide sequence of the bio operon. *J. Biol. Chem.*, 1988. **263**: 19577–19585.
58. Reed, K.E. and J.J.E. Cronan, Lipoic acid metabolism in *Escherichia coli*: Sequencing and functional characterization of the lipA and lipB genes. *J. Bacteriol.*, 1993. **175**: 1325–1336.
59. Ugulava, N.B., B.R. Gibney, and J.T. Jarrett, Biotin synthase contains two distinct iron–sulfur binding sites: Chemical and spectroelectrochemical analysis of iron–sulfur cluster interconversions. *Biochemistry*, 2001. **40**: 8343–8351.
60. Ugulava, N.B., C.J. Sacanell, and J.T. Jarrett, Spectroscopic changes during a single turnover of biotin synthase: Destruction of a [2Fe–2S] cluster accompanies sulfur insertion. *Biochemistry*, 2001. **40**: 8352–8358.
61. Miller, J.R., et al., *Escherichia coli* LipA is a lipoyl synthase: In vitro biosynthesis of lipoylated pyruvate dehydrogenase complex from octanoyl-acyl carrier protein. *Biochemistry*, 2000. **39**: 15166–15178.
62. Zhao, S., et al., Assembly of the covalent linkage between lipoic acid and its cognate enzymes. *Chem. Biol.*, 2003. **10**: 1293–1302.
63. Reed, L.J., T. Okaichi, and I. Nakanishi, Studies on the biosynthesis of lipoic acid. *Abstr. Int. Symp. Chem. Nat. Prod. (Kyoto)*, 1964: 218–220.
64. Parry, R.J., Biosynthesis of lipoic acid. 1. Incorporation of specifically tritiated octanoic acid into lipoic acid. *J. Am. Chem. Soc.*, 1977. **99**: 6464–6466.
65. Parry, R.J. and D.A. Trainor, Biosynthesis of lipoic acid. 2. Stereochemistry of sulfur introduction at C-6 of octanoic acid. *J. Am. Chem. Soc.*, 1978. **100**: 5243–5244.
66. White, R.H., Stable isotope studies on the biosynthesis of lipoic acid in *Escherichia coli. Biochemistry*, 1980. **19**: 15–19.
67. White, R.H., Biosynthesis of lipoic acid: Extent of incorporation of deuterated hydroxy- and thiooctanoic acids into lipoic acid. *J. Am. Chem. Soc.*, 1980. **102**: 6605–6607.

68. Hayden, M.A., et al., Biosynthesis of lipoic acid: Characterization of the lipoic acid auxotrophs *Escherichia coli W1485-lip2 and JRG33-lip9. Biochemistry*, 1993. **32**: 3778–3782.
69. Kamata, K. and K. Akiyama, High-performance liquid chromatography with electrochemical detection for the determination of thioctic acid and thioctic acid amide. *J. Pharm. Biomed. Anal.*, 1990. **8**: 453–456.
70. Packer, L., E.H. Witt, and H.J. Tritschler, Alpha-Lipoic acid as a biological oxidant. *Free Radic. Biol. Med.*, 1995. **19**: 227–250.
71. Podda, M., et al., Alpha-lipoic acid supplementation prevents symptoms of vitamin E deficiency. *Biochem. Biophys. Res. Commun.*, 1994. **204**: 98–104.
72. Morris, T.W., K.E. Reed, and J.E. Cronan Jr., Identification of the gene encoding lipoate-protein ligase of *Escherichia coli*. Molecular cloning and characterization of the *lplA* gene and gene product. *J. Biol. Chem.*, 1994. **269**: 16091–16100.
73. Morris, T.W., K.E. Reed, and J.J.E. Cronan, Lipoic acid metabolism in *Escherichia coli:* The *lplA* and *lipB* genes define redundant pathways for ligation of lipoyl groups to apoprotein. *J. Bacteriol.*, 1995. **177**: 1–10.
74. Jordan, S.W. and J.E. Cronan Jr., The *Escherichia coli lipB* gene encodes lipoyl (octanoyl)-acyl carrier protein:protein transferase. *J. Bacteriol.*, 2003. **185**: 1582–1589.
75. Jordan, S.W. and J.J.E. Cronan, A new metabolic link. The acyl carrier protein of lipid synthesis donates lipoic acid to the pyruvate dehydrogenase complex in *Escherichia coli* and mitochondria. *J. Biol. Chem.*, 1997. **272**: 17903–17906.
76. Zhao, X., J.R. Miller, and J.E. Cronan, The reation of LipB, the octanoyl-[acyl carrier protein]:protein *N*-octanoyltransferase of lipoic acid synthesis, proceeds through an acyl-enzyme intermediate. *Biochemistry*, 2005. **44**: 16737–16746.
77. Hebert, A.A. and J.R. Guest, Biochemical and genetic studies with lysine + methionine mutants of *Escherichia coli:* Lipoic acid and a-ketoglutarate dehydrogenase-less mutants. *J. Gen. Microbiol.*, 1968. **53**: 363–381.
78. Vaisvila, R., et al., The LipB protein is a negative regulator of dam gene expression in *Escherichia coli. Biochim. Biophys. Acta*, 2000. **1494**: 43–53.
79. Schaefer, A.L., et al., Generation of cell-to-cell signals in quorum sensing: Acyl homoserine lactone synthase activity of a purified *Vibrio fischeri* LuxI protein. *Proc. Natl. Acad. Sci.* USA, 1996. **93**: 9505–9509.
80. Anderson, M.S. and C.R. Raetz, Biosynthesis of lipid A precursors in *Escherichia coli*. A cytoplasmic acyltransferase that converts UDP-*N*-acetylglucosamine to UDP-3-O-(*R*-3-hydroxymyristoyl)-*N*-acetylglucosamine. *J. Biol. Chem.*, 1987. **262**: 5159–5169.
81. Rock, C.O. and S. Jordan, Regulation of phospholipid synthesis in *Escherichia coli. J. Biol. Chem.*, 1982. **257**: 10759–10765.
82. Issartel, J.P., V. Koronakis, and C. Hughes, Activation of *Escherichia coli* prohaemolysin to the mature toxin by acyl carrier protein-dependent fatty acylation. *Nature*, 1991. **351**: 759–761.
83. Magnuson, K., et al., Regulation of fatty acid biosynthesis in *Escherichia coli. Microbiol. Rev.*, 1993. **57**: 522–542.
84. Rock, C.O. and J.E. Cronan Jr., Solubilization, purification, and salt activation of acyl–acyl carrier protein synthetase from *Escherichia coli. J. Biol. Chem.*, 1979. **254**: 7116–7122.
85. Vanaman, T.C., S.J. Wakil, and R.L. Hill, The complete amino acid sequence of the acyl carrier protein of *Escherichia coli. J. Biol. Chem.*, 1968. **243**: 6420–6431.

86. Majerus, P.W., A.W. Alberts, and P.R. Vagelos, The acyl carrier protein of fatty acid synthesis: purification, physical properties, and substrate binding site. *Proc. Natl. Acad. Sci.* USA, 1964. **51**: 1231–1238.
87. Elovson, J. and P.R. Vagelos, Acyl carrier protein. X. Acyl carrier protein synthetase. *J. Biol. Chem.*, 1968. **243**: 3603–3611.
88. Stuible, H.P., et al., A novel phosphopantetheine: Protein transferase activating yeast mitochondrial acyl carrier protein. *J. Biol. Chem.*, 1998. **273**: 22334–22339.
89. Kim, R., et al., Overexpression of archaeal proteins in *Escherichia coli. Biotechnol. Lett.*, 1998. **20**: 207–210.
90. Xu, G.Y., et al., Solution structure of *B. subtilis* acyl carrier protein. *Structure*, 2001. **9**: 277–287.
91. Wong, H.C., et al., The solution structure of acyl carrier protein from *Mycobacterium tuberculosis. J. Biol. Chem.*, 2002. **277**: 15874–15880.
92. Srisailam, S., et al., Solution structure of acyl carrier protein from *Nitrosomonas europaea. Proteins*, 2006. **64**: 800–803.
93. Roujeinikova, A., et al., X-ray crystallographic studies on butyryl-ACP reveal flexibility of the structure around a putative acyl chain binding site. *Structure*, 2002. **10**: 825–835.
94. Roujeinikova, A., et al., Structural studies of fatty acyl-(acyl carrier protein) thioesters reveal a hydrophobic binding cavity that can expand to fit longer substrates. *J. Mol. Biol.*, 2007. **365**: 135–145.
95. Zornetzer, G.A., B.G. Fox, and J.L. Markely, Solution structures of spinach acyl carrier protein with decanoate and stearate. *Biochemistry*, 2006. **45**: 5217–5227.
96. Ali, S.T., et al., Octanoylation of the lipoyl domains of the pyruvate dehydrogenase complex in a lipoyl-deficient strain of *Escherichia coli. Mol. Microbiol.*, 1990. **4**: 943–950.
97. Dardel, F., L.C. Packman, and R.N. Perham, Expression in *Escherichia coli* of a subgene encoding the lipoyl domain of the pyruvate dehydrogenase complex of *Bacillus stearothermophilus. FEBS Lett.*, 1990. **264**: 206–210.
98. Fujiwara, K., K. Okamura-Ikeda, and Y. Motokawat, Expression of mature bovine H-protein of the glycine cleavage system in *Escherichia coli* and in vitro lipoylation of the apoform. *J. Biol. Chem.*, 1992. **267**: 20011–20016.
99. Green, D.E., et al., Purification and properties of the lipoate protein ligase of *Escherichia coli. Biochem. J.*, 1995. **309**: 853–862.
100. Wakil, S.J., Fatty acid synthase, a proficient multifunctional enzyme. *Biochemistry*, 1989. **28**: 4523–4530.
101. Brody, S. and S. Mikolajczyk, *Neurospora* mitochondria contain an acyl carrier protein. *Eur. J. Biochem.*, 1988. **173**: 353–359.
102. Sackmann, U., et al., The acyl-carrier protein in *Neurospora crassa* mitochondria is a subunit of NADH:ubiquinone reductase (complex I). *Eur. J. Biochem.*, 1991. **200**: 463–469.
103. Schneider, R., et al., Different respiratory-defective phenotypes of *Neurospora crassa* and *Saccharomyces cerevisiae* after inactivation of the gene encoding the mitochondrial acyl carrier protein. *Curr. Genet.*, 1995. **29**: 10–17.
104. Chuman, L. and S. Brody, Acyl carrier protein is present in the mitochondria of plants and eucaryotic micro-organisms. *Eur. J. Biochem.*, 1989. **184**: 643–649.
105. Shintani, D.K. and J.B. Ohlrogge, The characterization of a mitochondrial acyl carrier protein isoform isolated from *Arabidopsis thaliana. Plant Physiol.*, 1994. **104**: 1221–1229.

106. Runswick, M.J., et al., Presence of an acyl carrier protein in NADH:ubiquinone oxidoreductase from bovine heart mitochondria. *FEBS Lett.*, 1991. **286**: 121–124.
107. Wada, H., D. Shintani, and J. Ohlrogge, Why do mitochondria synthesize fatty acids? Evidence for involvement in lipoic acid production. *Proc. Natl. Acad. Sci.* USA, 1997. **94**: 1591–1596.
108. Gueguen, V., et al., Fatty acid and lipoic acid biosynthesis in higher plant mitochondria. *J. Biol. Chem.*, 2000. **275**: 5016–5025.
109. Brody, S., C. Oh, and U.H.E. Schweizer, Mitochondrial acyl carrier protein is involved in lipoic acid synthesis in *Saccharomyces cerevisiae. FEBS Lett.*, 1997. **408**: 217–220.
110. Cronan, J.E., I.M. Fearnley, and J.E. Walker, Mammalian mitochondria contain a soluble acyl carrier protein. *FEBS Lett.*, 2005. **579**: 4892–4896.
111. Reed, L.J., F.R. Leach, and M. Koike, Studies on a lipoic acid-activating system. *J. Biol. Chem.*, 1958. **232**: 123–142.
112. Oehring, R. and H. Bisswanger, Incorporation of the enantiomers of lipoic acid into the pyruvate dehydrogenase complex from *Escherichia coli. Biol. Chem. Hoppe-Seyler*, 1992. **373**: 333–335.
113. Fujiwara, K., et al., Crystal structure of lipoate-protein ligase A from *Escherichia coli. J. Biol. Chem.*, 2005. **280**: 33645–33651.
114. Kim, D.J., et al., Crystal structure of lipoate-protein ligase A bound with the activated intermediate. *J. Biol. Chem.*, 2005. **280**: 38081–38089.
115. McManus, E., B.F. Luisi, and R.N. Perham, Structure of a putative lipoate protein ligase from *Thermoplasma acidophilum* and the mechanism of target selection for post-translational modification. *J. Mol. Biol.*, 2006. **356**: 625–637.
116. O'Riordan, M., M.A. Moors, and D.A. Portnoy, Listeria intracellular growth and virulence require host-derived lipoic acid. *Science*, 2003. **302**: 462–464.
117. Fujiwara, K., K. Okamura-Ikeda, and Y. Motokawa, Purification and characterization of lipoyl-AMP:N^{ϵ}-lysine lipoyltransferase from bovine liver mitochondria. *J. Biol. Chem.*, 1994. **269**: 16605–16609.
118. Fujiwara, K., K. Okamura-Ikeda, and Y. Motokawa, Lipoylation of acyltransferase components of a-ketoacid dehydrogenase complexes. *J. Biol. Chem.*, 1996. **271**: 12932–12936.
119. Fujiwara, K., et al., Molecular cloning, structural characterization and chromosomal localization of human lipoyltransferase. *Eur. J. Biochem.*, 1999. **260**: 761–767.
120. Fujiwara, K., et al., Purification, characterization, and cDNA cloning of lipoate-activating enzyme from bovine liver. *J. Biol. Chem.*, 2001. **276**: 28819–28823.
121. Vessey, D.A., E. Lau, and M. Kelley, Isolation and sequencing of cDNAs for the XL-1 and XL-III forms of bovine liver xenobiotic-metabolizing medium-chain fatty acid:CoA ligase. *J. Biochem. Mol. Toxicol.*, 2000. **14**: 11–19.
122. Nesbitt, N.M., et al., Expression, purification, and physical characterization of *Escherichia coli* lipoyl(octanoyl)transferase. *Protein Expr. Purif.*, 2005. **39**: 269–282.
123. Ma, Q., et al., The *Mycobacterium tuberculosis* LipB enzyme functions as a cysteine/lysine dyad acyltransferase. *Proc. Natl. Acad. Sci. USA*, 2006. **103**: 8662–8667.
124. Douglas, P., et al., Lipoyl synthase inserts sulfur atoms into an octanoyl substrate in a stepwise manner. *Angew. Chem.*, 2006. **118**: 5321–5323.
125. Frey, P.A., Importance of organic radicals in enzymic cleavage of unactivated carbon-hydrogen bonds. *Chem. Rev.*, 1990. **90**: 1343–1357.
126. Cheek, J. and J.B. Broderick, Adenosylmethionine-dependent iron-sulfur enzymes: Versatile clusters in a radical new role. *J. Biol. Inorg. Chem.*, 2001. **6**: 209–226.

127. Fontecave, M., E. Mulliez, and S. Ollagnier-de Choudens, Adenosylmethionine as a source of 5′-deoxyadenosyl radicals. *Curr. Opin. Chem. Biol.*, 2001. **5**: 506–511.
128. Frey, P.A. and S. Booker, Radical intermediates in the reaction of lysine 2,3-aminomutase, in *Advances in Free Radical Chemistry*, S.Z. Zard, Editor. 1999, JAI Press Inc.: Stamford, CT. pp. 1–43.
129. Frey, P.A. and S.J. Booker, Radical mechanisms of *S*-adenosylmethionine-dependent enzymes. *Adv. Protein Chem.*, 2001. **58**: 1–45.
130. Frey, P.A. and O.T. Magnusson, *S*-Adenosylmethionine: A wolf in sheep's clothing, or a rich man's adenosylcobalamin? *Chem. Rev.*, 2003. **103**: 2129–2148.
131. Knappe, J. and A.F. Wagner, Stable glycyl radical from pyruvate formate-lyase and ribonucleotide reductase (III). *Adv. Protein Chem.*, 2001. **58**: 277–315.
132. Mulliez, E., et al., An iron-sulfur center and a free radical in the active anaerobic ribonucleotide reductase of *Escherichia coli*. *J. Biol. Chem.*, 1993. **268**: 2296–2299.
133. Sun, X., et al., The free radical of the anaerobic ribonucleotide reductase from *Escherichia coli* is at glycine 681. *J. Biol. Chem.*, 1996. **271**: 6827–6831.
134. Wagner, A.F., et al., The free radical in pyruvate formate-lyase is located on glycine-734. *Proc. Natl. Acad. Sci.* USA, 1992. **89**: 996–1000.
135. Sofia, H.J., et al., Radical SAM, a novel protein superfamily linking unresolved steps in familiar biosynthetic pathways with radical mechanisms: functional characterization using new analysis and information visualization methods. *Nucleic Acids Res.*, 2001. **29**: 1097–1106.
136. Wang, S.C. and P.A. Frey, *S*-adenosylmethionine as an oxidant: The radical SAM superfamily. *Trends Biochem. Sci.*, 2007. **32**: 101–110.
137. Bianchi, V., et al., *Escherichia coli* ferredoxin $NADP^+$ reductase: Activation of *E. coli* anaerobic ribonucleotide reduction, cloning of the gene (fpr), and over-expression of the protein. *J. Bacteriol.*, 1993. **175**: 1590–1595.
138. Osborne, C., et al., Isolation, cloning, mapping, and nucleotide sequencing of the gene encoding flavodoxin in *Escherichia coli*. *J. Bacteriol.*, 1991. **173**: 1729–1737.
139. Bianchi, V., et al., Flavodoxin is required for the activation of the anaerobic ribonucleotide reductase. *Biochem. Biophys. Res. Commun.*, 1993. **197**: 792–797.
140. Walsby, C.J., et al., Spectroscopic approaches to elucidating novel iron-sulfur chemistry in the "radical-SAM" protein superfamily. *Inorg. Chem.*, 2005. **44**: 727–741.
141. Walsby, C.J., et al., Electron-nuclear double resonance spectroscopic evidence that *S*-adenosylmethionine binds in contact with the catalytically active $[4Fe–4S]^+$ cluster of pyruvate formate-ayase activating enzyme. *J. Am. Chem. Soc.*, 2002. **124**: 3143–3151.
142. Walsby, C.J., et al., An anchoring role for FeS clusters: Chelation of the amino acid moiety of *S*-adenosylmethionine to the unique iron site of the [4Fe–4S] cluster of pyruvate formate–lyase activating enzyme. *J. Am. Chem. Soc.*, 2002. **124**: 11270–11271.
143. Chen, D., et al., Coordination and mechanism of reversible cleavage of *S*-adenosylmethionine by the [4Fe–4S] center in lysine 2,3-aminomutase. *J. Am. Chem. Soc.*, 2003. **125**: 11788–11789.
144. Layer, G., et al., Crystal structure of coproporphyrinogen III oxidase reveals cofactor geometry of radical SAM enzymes. *EMBO J.*, 2003. **22**: 6214–6224.
145. Berkovitch, F., et al., Crystal structure of biotin synthase, an *S*-adenosylmethionine-dependent radical enzyme. *Science*, 2004. **303**: 76–79.

146. Hanzelmann, P. and H. Schindelin, Crystal structure of the *S*-adenosylmethionine-dependent enzyme MoaA and its implications for molybdenum cofactor deficiency in humans. *Proc. Natl. Acad. Sci. USA*, 2004. **101**: 12870–12875.
147. Lepore, B.W., et al., The x-ray crystal structure of lysine-2,3-aminomutase from *Clostridium subterminale. Proc. Natl. Acad. Sci. USA*, 2005. **102**: 13819–13824.
148. Busby, R.W., et al., Lipoic acid biosynthesis: LipA is an iron sulfur protein. *J. Am. Chem. Soc.*, 1999. **121**: 4706–4707.
149. Ollagnier-de Choudens, S. and M. Fontecave, The lipoate synthase from *Escherichia coli* is an iron-sulfur protein. *FEBS Lett.*, 1999. **453**: 25–28.
150. Kriek, M., et al., Effect of iron–sulfur cluster assembly proteins on the expression of *Escherichia coli* lipoic acid synthase. *Protein Expr. Purif.*, 2003. **28**: 241–245.
151. Yasukawa, T., et al., Increase of solubility of foreign proteins in *Escherichia coli* by coproduction of the bacterial thioredoxin. *J. Biol. Chem.*, 1995. **270**: 25328–25331.
152. Iametti, S., et al., GroEL-assisted refolding of adrenodoxin during chemical cluster insertion. *Eur. J. Biochem.*, 2001. **268**: 2421–2429.
153. Johnson, D.C., et al., Structure, function, and formation of biological iron–sulfur clusters. *Annu. Rev. Biochem.*, 2005. **74**: 247–281.
154. Takahashi, Y. and M. Nakamura, Functional assignment of the ORF2-iscS-iscU-iscA-hscB-hscA-fdx-ORF3 gene cluster involved in the assembly of Fe–S clusters in *Escherichia coli. J. Biochem. (Tokyo)*, 1999. **126**: 917–926.
155. Zheng, L.M., et al., Assembly of iron–sulfur clusters. Identification of an iscSUA-hscBA-fdx gene cluster from *Azotobacter vinelandii. J. Biol. Chem.*, 1998. **273**: 13264–13272.
156. Debrunner, P.G., Mössbauer spectroscopy of iron proteins, in *Biological Magnetic Resonance*, L.J. Berliner and J. Reuber, Editors. 1993, Plenum: New York. pp. 59–101.
157. Krebs, C., et al., Rapid Freeze-quench ^{57}Fe Mössbauer spectroscopy: Monitoring changes of an iron-containing active site during a biochemical reaction. *Inorg. Chem.*, 2005. **44**: 742–757.
158. Münck, E., Mössbauer spectroscopy of proteins: Electron carriers. *Methods Enzymol.*, 1978. **54**: 346–379.
159. Münck, E., Aspects of ^{57}Fe Mössbauer spectroscopy, in *Physical Methods in Bioinorganic Chemistry: Spectroscopy and Magnetism*, J. Lawrence Que, Editor. 2000, University Science Books: Sausalito, CA. pp. 287–319.
160. Ollagnier-de Choudens, S., et al., Iron-sulfur center of biotin synthase and lipoate synthase. *Biochemistry*, 2000. **39**: 4165–4173.
161. Cicchillo, R.M. and S.J. Booker, Mechanistic investigations of lipoic acid biosynthesis in *Escherichia coli*: Both sulfur atoms in lipoic acid are contributed by the same lipoyl synthase polypeptide. *J. Am. Chem. Soc.*, 2005. **127**: 2860–2861.
162. Frappier, F., et al., On the mechanism of conversion of dethiobiotin to biotin in *Escherichia coli.* Discussion of the occurrence of an intermediate hydroxylation. *Biochem. Biophys. Res. Commun.*, 1979. **91**: 521–527.
163. Marquet, A., et al., Biotin biosynthesis: Synthesis and biological evaluation of the putative intermediate thiols. *J. Am. Chem. Soc.*, 1993. **115**: 2139–2145.
164. Tse Sum Bui, B., et al., Further investigation on the turnover of *Escherichia coli* biotin synthase with dethiobiotin and 9-mercaptodethiobiotin as substrates. *Biochemistry*, 2004. **43**: 16432–16441.
165. Cosper, M.M., et al., Characterization of the cofactor composition of *Escherichia coli* biotin synthase. *Biochemistry*, 2004. **43**: 2007–2021.

166. Hewitson, K.S., et al., Mutagenesis of the proposed iron-sulfur cluster binding ligands in *Escherichia coli* biotin synthase. *FEBS Lett.*, 2000. **466**: 372–376.
167. Hewitson, K.S., et al., The iron-sulfur center of biotin synthase: Site-directed mutants. *J. Biol. Inorg. Chem.*, 2002. **7**: 83–93.
168. Duin, E.C., et al., [2Fe–2S] to [4Fe–4S] cluster conversion in *Escherichia coli* biotin synthase. *Biochemistry*, 1997. **36**: 11811–11820.
169. Ugulava, N.B., et al., Evidence from Mössbauer spectroscopy for distinct $[2Fe–2S]^{2+}$ and $[4Fe–4S]^{2+}$ cluster binding sites in biotin synthase from *Escherichia coli*. *J. Am. Chem. Soc.*, 2002. **124**: 9050–9051.
170. Crawford, M.J., et al., *Toxoplasma gondii* scavenges host-derived lipoic acid despite its de novo synthesis in the apicoplast. *EMBO J.*, 2006. **25**: 3214–3222.
171. Allary, M., et al., Scavenging of the cofactor lipoate is essential for the survival of the malaria parasite *Plasmodium falciparum*. *Mol. Microbiol.*, 2007. **63**: 1331–1344.
172. Wrenger, C. and S. Müller, The human malaria parasite *Plasmodium falciparum* has distinct organelle-specific lipoylation pathways. *Mol. Microbiol.*, 2004. **53**: 103–113.
173. Sassetti, C.M., D.H. Boyd, and E.J. Rubin, Genes required for mycobacterial growth defined by high density mutagenesis. *Mol. Microbiol.*, 2003. **48**: 77–84.
174. Rachman, H., et al., Unique transcriptome signature of *Mycobacterium tuberculosis* in pulmonary tuberculosis. *Infect. Immun.*, 2006. **74**: 1233–1242.
175. Yi, X. and N. Maeda, Endogenous production of lipoic acid is essential for mouse development. *Mol. Cell. Biol.*, 2005. **25**: 8387–8392.
176. Tse Sum Bui, B., et al., Fate of the $[2Fe–2S]^{2+}$ cluster of *Escherichia coli* biotin synthase during reaction: A Mössbauer characterization. *Biochemistry*, 2003. **42**: 8791–8798.
177. Tse Sum Bui, B., et al., Enzyme-mediated sulfide production for the reconstitution of [2Fe–2S] clusters into apo-biotin synthase of *Escherichia coli*: Sulfide transfer from cysteine to biotin. *Eur. J. Biochem.*, 2000. **267**: 2688–2694.
178. Tse Sum Bui, B., et al., *Escherichia coli* biotin synthase produces selenobiotin. Further evidence of the involvement of the $[2Fe–2S]^{2+}$ cluster in the sulfur insertion step. *Biochemistry*, 2006. **45**: 3824–3834.
179. Choi-Rhee, E. and J.E. Cronan, Biotin synthase is catalytic in vivo, but catalysis engenders destruction of the protein. *Chem. Biol.*, 2005. **12**: 461–468.
180. Kiyasu, T., et al., Contribution of cysteine desulfurase (NifS protein) to the biotin synthase reaction of *Escherichia coli*. *J. Bacteriol.*, 2000. **182**: 2879–2885.
181. Önder, Ö., et al., Modifications of the lipoamide-containing mitochondrial subproteome in a yeast mutant defective in cysteine desulfurase. *Mol. Cell. Proteomics*, 2006. **5**: 1426–1436.
182. Picciocchi, A., R. Douce, and C. Alban, The plant biotin synthase reaction: Identification and characterization of essential mitochondrial accessory protein components. *J. Biol. Chem.*, 2003. **278**: 24966–24975.

3 The Search for Potent Alpha-Lipoic Acid Derivatives: Chemical and Pharmacological Aspects

Moriya Ben Yakir, Yehoshua Katzhendler, and Shlomo Sasson

CONTENTS

INTRODUCTION

Two cellular systems combine to protect cells against damaging effects of free radicals that are generated under normal metabolic conditions or in excess in pathological processes: Antioxidant enzymes (i.e., superoxide dismutase, catalase) eliminate free radicals by enzymatic reactions whereas low-molecular weight antioxidants neutralize radicals by direct chemical interactions (i.e., vitamin C, vitamin E, glutathione). α-Lipoic acid [LA, CAS 62-46-4, thioctic acid, 5-(1,2-dithiolan-3-yl), pentanoic acid] which belongs to the second group, is an endogenous cofactor for several 2-oxoacid dehydrogenase multienzyme complexes. For instance, it is covalently linked to a lysine residue of dihydrolipoamide acetyltransferase in the pyruvate dehydrogenase (PDH) complex to accept an acyl intermediate from the dehydrogenase components and transfer it to coenzyme-A. The resulting reduced form of LA, dihydrolipoic acid (DHLA), is then reoxidized by lipoamide dehydrogenase (Reed et al. 1958; Loffelhardt et al. 1995; Biewenga et al. 1996). Free LA is considered a therapeutically potent thiol antioxidant due to its reductive power (Packer et al. 1995). The low oxidation potential (−0.29 V) of the reduced form of DHLA results from the two vicinal thiol groups within the molecule that enable efficient scavenging of free radicals, such as superoxide, hydroxyl, and peroxyl radicals (Packer et al. 1995; Biewenga et al. 1997; Bustamante et al. 1998). These interactions also contribute to the regeneration of other low-molecular weight antioxidants, such as vitamin C, glutathione, and vitamin E (Packer et al. 1995; Biewenga et al. 1997). In addition, the metal-chelating activity of LA also contributes to its antioxidative activity through the inhibition of metal ion–dependent formation of free radicals in cells (Muller and Menzel 1990; Packer et al. 1995). Another mechanism suggests that dihydrolipoamide dehydrogenase that reduces LA to DHLA consumes cellular NAD(P)H, which is an essential cofactor to several free radical–producing enzymes (e.g., nitric oxide synthase) (Guo et al. 2001). Many reports and trials have suggested that by virtue of these antioxidative interactions LA may ameliorate various symptoms associated with oxidative stress, such as diabetes-induced neuropathy (Packer et al. 1995; Nickander et al. 1996; Bustamante et al. 1998; Packer et al. 2001; Ziegler 2004; Ziegler et al. 2004; Bilska and Wlodek 2005).

α-Lipoic acid exists as *R*(+)- and *S*(−)-enantiomers due to the presence of an asymmetric carbon C3 (Figure 3.1). Analyses of bacterial and mammalian PDH complexes have revealed that the natural cofactor of the complex is the *R*(+) enantiomer. Moreover, *S*(−)-LA acts either as a poor substrate or as an inhibitor of *R*(+)-LA in its interaction with 2-oxoacid dehydrogenase multienzyme complexes. However, both free LA enantiomers are reduced intracellularly to their respective reduced forms, albeit by different enzymatic interactions: *R*(+)-LA

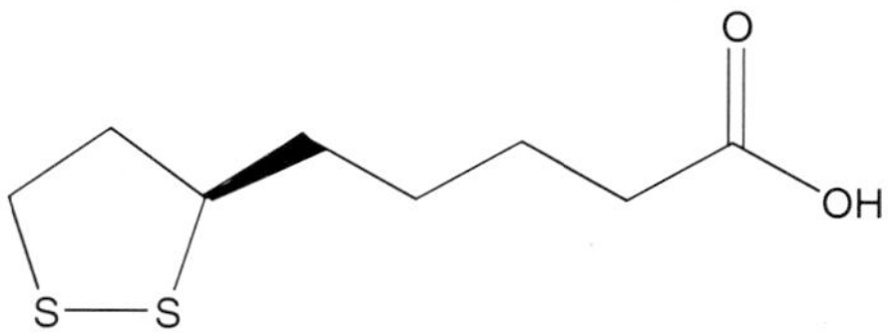

R(+)-LA

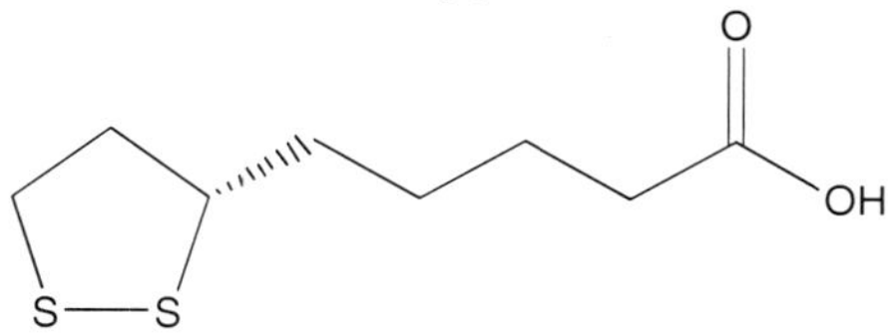

S(–)-LA

FIGURE 3.1 Structure of *R*(+)- and *S*(−)-lipoic acid (LA).

by dihydrolipoamide dehydrogenase (the E3 enzyme in the PDH complex) and *S*(−)-LA by glutathione reductase (Pick et al. 1995; Haramaki et al. 1997). Consequently, *R*(+)-LA/*R*(+)-DHLA and *S*(−)-LA/*S*(−)-DHLA act as redox couples and free-radical scavengers in cells (Suzuki et al. 1991; Hermann et al. 1996). Although some reports assign only the *R*(+) enantiomer cellular functions (Maitra et al. 1996; Frolich et al. 2004), other studies show similar cellular activities for both the enantiomers (Moini et al. 2002). For example, in vivo studies in insulin resistant rats showed that *R*(+)-LA (i.p. administration) was significantly more effective than *S*(−)-LA in protecting cells against oxidative stress; also, *R*(+)-LA, but not *S*(−)-LA, enhanced insulin-stimulated glucose transport and metabolism in skeletal muscles (Streeper et al. 1997), protected hepatocytes against *tert*-butylhydroperoxide-induced stress (Hagen et al. 2000), and quenched NO radicals in cultured macrophages (Guo et al. 2001). Similarly, *R*(+)-LA was more protective than *S*(−)-LA against buthionine sulfoximine-induced cataract in lenses of newborn rats. Yet, administration of racemic LA in this animal model protected the lenses better than treatments with each pure enantiomer (Maitra et al. 1996). Similarly, both enantiomers equipotently reduced in vitro lipid peroxidation of both brain and nerve preparations (Nickander et al. 1996).

These antioxidant properties led to a widespread use of racemic LA or *R*(+)-LA as dietary supplements (Berkson 1998; Sosin and Ley-Jacobs 1998; Packer and Colman 2000). Yet, results of numerous studies and clinical trials on the preferred mode of administration, effective dosage, potency, efficacy, and the pharmacokinetics of LA often remain inconsistent (Muller and Menzel 1990; Jacob et al. 1996; Streeper et al. 1997; Hagen et al. 1999; Jacob et al. 1999; Konrad et al. 1999; Ruhnau et al. 1999; Evans and Goldfine 2000; Suh et al. 2001; Moini et al. 2002; Frolich et al. 2004; Hahm et al. 2004; Ziegler 2004; Ziegler et al. 2004). Generally, the pharmacokinetic analysis of LA suggests that intravenous administration of LA reaches and sustains effective concentrations of the compounds in the plasma of treated individuals better than oral administration of similar or even larger doses (Evans and Goldfine 2000). Therefore, much effort has been invested in designing and synthesizing novel LA derivatives devoid of such unfavorable properties and more suitable for oral administration. This chapter describes main chemical pharmaceutical approaches that have served in preparing such novel LA derivatives, and summarizes their pharmacological properties.

PHARMACOKINETICS OF α-LIPOIC ACID

α-Lipoic acid is transported into cells via the short-chain fatty acid carrier and is reduced intracellularly to DHLA. Efflux of the latter allows it to exert antioxidant effects also extracellularly (Peinado et al. 1989; Bustamante et al. 1998). In contrast to the distinct distribution of other water- or lipid-soluble antioxidants to their respective subcellular compartments and organelles, LA can interact in both milieus (Roy and Packer 1998). This amphiphilic character of LA is attributed to the copresence of the alkyl-substituted 1,2-dithiolane ring and the carboxylic group (Roy and Packer 1998). There are inconsistent reports on the therapeutic effectiveness of LA when administered orally or in a parenteral manner in man. For instance, oral consumption of 600 mg of LA up to three times a day for a period of four weeks improved moderately insulin-stimulated glucose disposal in type-2 diabetic patients, but failed to reduce fasting plasma glucose levels (Jacob et al. 1999). Similarly, daily intravenous infusions of 500 mg of LA for 10 days to such patients augmented insulin-stimulated glucose disposal by merely 28% (Jacob et al. 1996). Yet, an acute parenteral administration of 1 g of LA to other diabetic individuals increased this disposal parameter to 55% (Jacob et al. 1995). A four week oral treatment of LA in lean and obese diabetic patients reduced fasting plasma glucose levels in the lean group, but not in obese patients (Konrad et al. 1999). Evans and Goldfine (2000) reviewed the studies on the metabolic clearance rate of glucose following oral and intravenous administration of LA in man. In the latter case LA increased the rate of glucose disposal up to 55% above the control level in a dose-dependent manner, whereas a moderate effect (18%) was observed following the oral administration of LA, even at doses as high as 1.8 g. Likewise, beneficial effects of intravenously administered LA on glucose disposal in type-2 diabetic patients were evident within a 10 day treatment period, whereas oral treatments, up to 30 days, failed to produce significant effects (Evans and Goldfine 2000). Unlike these disparate effects of an orally- or intravenously administered LA in man, both routes of administration of daily doses of 600 mg of LA improved symptoms of diabetic polyneuropathy in diabetic patients (Bustamante et al. 1998; Hahm et al. 2004; Ziegler 2004; Ziegler et al. 2006).

The limited potency and efficacy of LA when given orally may result from the pharmacokinetics of the compound. The bioavailability of LA is rather low

(0.2–0.3) (Biewenga et al. 1997; Teichert et al. 1998; Evans and Goldfine 2000). Animal studies showed that over 90% of an oral dose of LA was readily absorbed from the digestive tract of rat (Peter and Borbe 1995). Similarly, gastrointestinal absorption of LA in man is considered fast (<1 h) and efficient (Packer et al. 1995; Packer et al. 2001). The low bioavailability of LA results from its extensive hepatic first-pass metabolism through β-oxidation of the carboxylic acid side chain and S-methylation of the dithiolane substructure (Teichert et al. 2003). Likewise, the short half-life (<1 h) of LA reflects its high rate of elimination due to its biotransformation in the liver and peripheral tissues (Biewenga et al. 1997; Evans and Goldfine 2000; Teichert et al. 2003). The renal clearance of untransformed LA, on the other hand, is limited and does not contribute significantly to its fast clearance. The range of effective concentrations of LA, determined in different in vitro assays, is surprisingly wide (0.01–2.5 mM) (Tirosh et al. 1999; Gruzman et al. 2004). The peak plasma concentrations of LA following a single or repeated oral administration of 600 mg ranges from 10 to 50 μM in man (Breithaupt-Grogler et al. 1999; Teichert et al. 2003). The pharmacokinetic analyses described above were based on the use of racemic LA. Detailed pharmacokinetic analysis of LA enantiomers in man revealed that the bioavailability of the *R*(+)-enantiomer was relatively higher than that of the *S*(−)-enantiomer (0.38 ± 0.15 and 0.28 ± 0.14, for solutions given p.o.). As expected, the respective values of total body clearance of the two enantiomers showed a reciprocal relationship (12.2 ± 2.6 and 15.6 ± 4.2 mL/min/kg, respectively) (Hermann et al. 1996). Similar observations were made in other independent studies (Breithaupt-Grogler et al. 1999). These differences were attributed to a preferential hepatic extraction and metabolism of *S*(−)-LA in comparison with *R*(+)-LA (Hermann et al. 1996; Bustamante et al. 1998; Hermann et al. 1998). It should be noted, however, that despite these differences, the half-life of both enantiomers remained short (<1 h) and similar to that of racemic LA (Hermann et al. 1996; Evans and Goldfine 2000; Packer et al. 2001).

The limited bioavailability and short half-life of racemic LA and its enantiomers following their oral administration seem to produce peak plasma concentrations below the minimal effective threshold. A lower volume of distribution (V_d) of LA in obese individuals, due to fat depot-specific accumulation of the compound, may explain the reported lower efficacy of the compound in improving glucose disposal rates in obese diabetic patients than in the lean group (Konrad et al. 1999).

Such drawbacks in xenobiotic, naturaceutical, or drug actions that are inherent to their chemical entities are usually resolved by chemical modifications in the structure of the molecules to improve their pharmacokinetic parameters while maintaining or improving their potency and efficacy. Such modifications have been made in LA and in its enantiomers in order to extend their half-life, and reduce the rate of their elimination.

α-LIPOIC ACID-BASED DERIVATIVES AND PRO-DRUGS

Pharmaceutical chemistry provides numerous methods to alter and improve physicochemical, pharmacological, and delivery properties of bioactive molecules. For instance, chemical modifications made in the structure of small molecules to enhance their lipid solubility, without altering their biological action, may improve their bioavailability by augmenting their membrane permeability and absorption through the gastrointestinal system to the hepatic circulation (Pagliara et al. 1999). Such modifications may result in a slower and even different hepatic metabolism, improved bioavailability and longer half-life, than that of the parent molecule, while maintaining similar biological functions. Another line of chemical synthesis consists of pro-drugs containing inactive lipophilic dissociable or readily hydrolyzable groups covalently bound to the active molecule. Such pro-drugs may exhibit preferential lipid/water partition. Spontaneous or enzyme-mediated release of the active moiety may then release the active drug from this complex during or after first-pass metabolism in the liver (Liederer and Borchardt 2006).

Such chemical modifications were made in LA. Structure–activity relationship analysis of LA points to the critical function of the 1,2-dithiolane ring in mediating antioxidant interactions of the molecule. Hence, covalent and non-dissociable modifications in this ring may produce inactive derivatives. The terminal carboxyl group, however, is a preferred target to chemical modifications. α-Lipoic acid has also been coupled covalently to various drugs and bioactive molecules in attempts to improve the pharmacokinetics of the partner drug, while introducing potential antioxidant group to such conjugates.

Seleno-α-Lipoic Acid Derivatives

The advantage of the amphiphilic character of LA is demonstrated when compared with its seleno-analogue [1,2]-diselenolane-3-pentanoic acid (Figure 3.2, compound 1) (Matsugo et al. 1997). α-Lipoic acid protected BSA and LDL ApoB as well as salicylate from hydroxyl radical attack. Compound 1 protected salicylate from hydroxylation, but in contrast to LA, was unable to protect both BSA and LDL ApoB. The different lipid/water solubility of these two compounds (5:1 for LA and 18:1 for compound 1) allows the former to interact in both

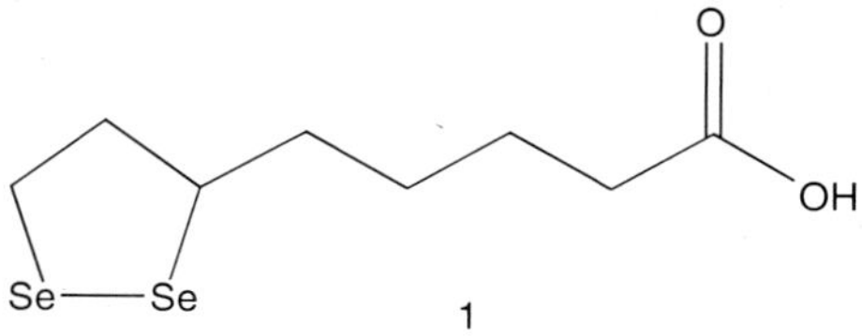

1

FIGURE 3.2 Structure of compound 1 (1,2-diselenolane-3-pentanoic acid).

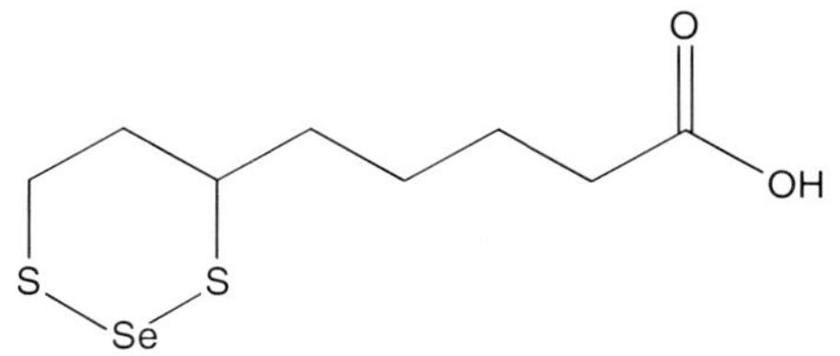

2

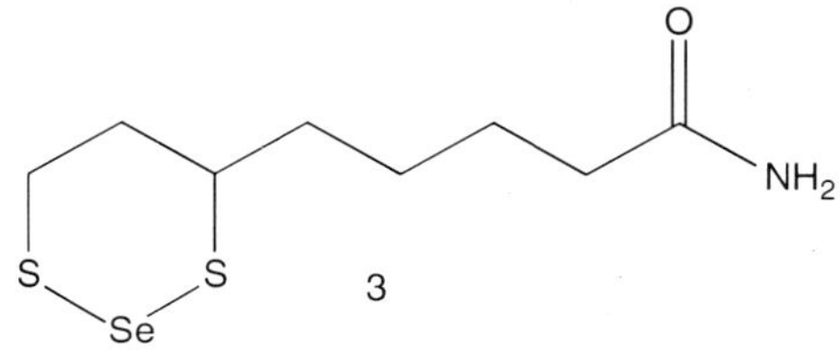

3

FIGURE 3.3 Structures of compounds 2 and 3 (selenotrisulfides).

aqueous and lipid compartments, whereas the latter is restricted to the lipid phase (Matsugo et al. 1997). Disulfide dithioselenan derivatives of LA (Figure 3.3, compounds 2 and 3) are synthetic analogues of LA that were found to be ligands to thioredoxin reductase. Since these compounds are more stable than LA due to the formation of a six-membered ring, it is suggested that reduction and oxidation of the S–Se bonds, as in S–S bonds, may effectively affect the overall redox state of treated cells (Self et al. 2000).

Amide and Ester Derivatives of Bis-α-Lipoic Acid

An example for rational design and synthesis of LA-based pro-drugs with an increased membrane permeability and potential for enzymatic hydrolysis to release free LA has been provided by Gruzman et al. (2004). They synthesized several LA derivatives, in which two LA molecules were linked with various linkers through their carboxyl groups. Of these novel compounds, only two, an amide- and an ester-based derivative (Figure 3.4, compounds 4 and 5, respectively) were more effective and potent than LA in enhancing the rate of hexose uptake in cultured L6 myotubes. The maximal effect of free LA (2.5 mM) was rather moderate: $39\% \pm 9\%$ and $51\% \pm 6\%$ increase in myotubes maintained at 5 and 23 mM glucose, respectively. Compounds 4 and 5 (200 μM) increased the rate of hexose transport by $92\% \pm 11\%$ and $82\% \pm 27\%$ in myotubes exposed to 5 mM glucose, and by $120\% \pm 09\%$ and $93\% \pm 11\%$, respectively, at 23 mM. Interestingly, although compounds 4 and 5 are more lipophilic and cell-permeable than free LA, they required 22 and 36 h, respectively, to produce maximal stimulation, whereas free LA-induced maximal effect was obtained within 2 h. It is suggested that this extended induction period reflects a slow enzymatic release of active LA molecules from both derivatives by nonspecific amidases and esterases, respectively. Of interest is the finding that the amide derivative

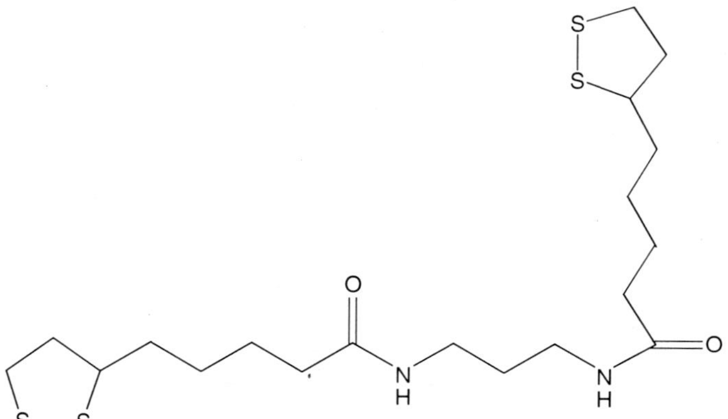

4

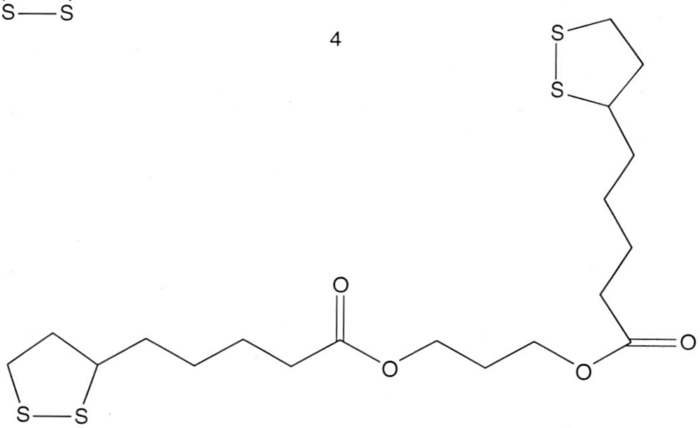

5

FIGURE 3.4 Structures of compound 4 (5-[1,2]-dithiolan-3-yl-pentanoic acid 3-(5-[1,2]-dithiolan-3-yl-pentanoylamino)-propyl-amide), and compound 5 (5-[1,2]-dithiolan-3-yl-pentanoic acid 3-(5-[1,2]-dithiolan-3-yl-pentanoyloxy)-propyl ester).

(compound 4), but not the ester derivative (compound 5), lowered blood glucose levels in diabetic mice, being nearly 10-fold more potent than racemic LA. These results indicate that the amide bond in compound 4 was less susceptible to enzymatic hydrolysis than the ester bond in compound 5, and therefore could accumulate in peripheral tissues and produce LA-like effects. Conversely, compound 5 was most likely subjected to rapid hydrolysis by esterases in plasma and cells and therefore could not accumulate in target tissue to reach biologically active concentrations (Liederer and Borchardt 2006).

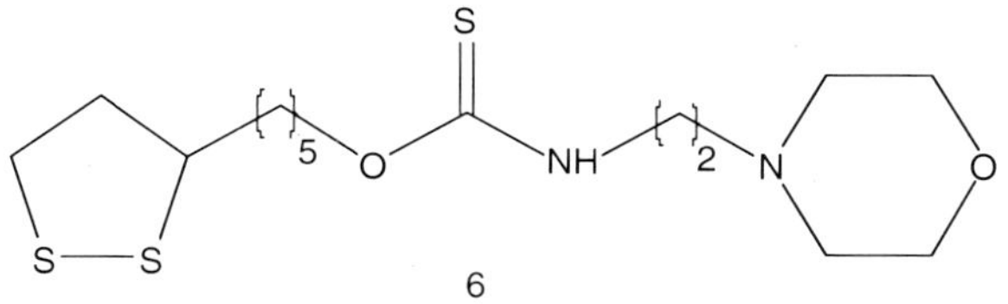

6

FIGURE 3.5 Structure of compound 6 ((2-morpholin-4-yl-ethyl)-thiocarbamic acid *O*-(5-[1,2]-dithiolan-3-yl-pentyl) ester hydrochloride).

Morpholine-α-Lipoic Acid Derivative

Guillonneau et al. (2003) prepared novel LA derivatives by connecting lipolol (the alcohol form of LA) to various alkyl substituted morpholine ring via amide, thioamide, carbamate, or thiocarbamate bonds. Compound 6 (Figure 3.5), in which the linker group was thiocarbamate coupled to the morpholine ring via an ethylene group, was two- to threefold more active than racemic LA in inhibiting in vitro lipid peroxidation and in protecting mice against *tert*-butylhydroperoxide-induced lethality or against alloxan-induced hyperglycemia. Interestingly, the *R*(+)- and *S*(−)-derivatives of LA were nearly equipotent and only slightly less active than the racemic derivatives. Dithiocarbamate derivatives, such as pyrrolidine dithiocarbamate, are potent antioxidants due to their metal ion–chelating capacity (Schreck et al. 1992). It has been hypothesized that the thiocarbamate moiety in compound 6 also possesses similar antioxidant potential. It should be noted, however, that the chelating and antioxidant capacity of thiocarbamate might be lower or even insignificant in comparison with dithiocarbamates due to the absence of a free thiol group in the former. The metabolism and the pharmacokinetic parameters of compound 6 have not yet been determined. Whether such lipolol conjugates act as novel antioxidants or merely as LA-pro-drug capable of releasing the lipolol moiety following enzymatic degradation remains to be investigated. Hitherto, the antioxidant potential of lipolol has not been established and it is not clear whether it is also reduced enzymatically to form lipolol/dihydrolipolol redox couple, in a similar manner to the enzymatic reduction of LA to DHLA.

Fluorinated Amphiphilic-α-Lipoic Acid Derivative

Ortial et al. (2006), who conjugated the antioxidant α-phenyl-*N*-*tert*-butylnitrone (PBN) to a fluorinated amphiphilic carrier (Durand et al. 2003), applied the same principle to synthesize analogous derivatives of LA. The procedure combined an introduction of a perfluoroalkylamino chain to a protected lysine, condensation of a lactobionic acid to the α-amino group of this lysine derivative following the fixation of an LA moiety to the lysine side chain. The resulting amphiphilic compound 7 (Figure 3.6) provided somewhat better protection against hydrogen peroxide, peroxynitrite, and doxorubicin-induced cell death. Unlike these modest effects of compound 7 and similar derivatives, the corresponding PBN derivatives

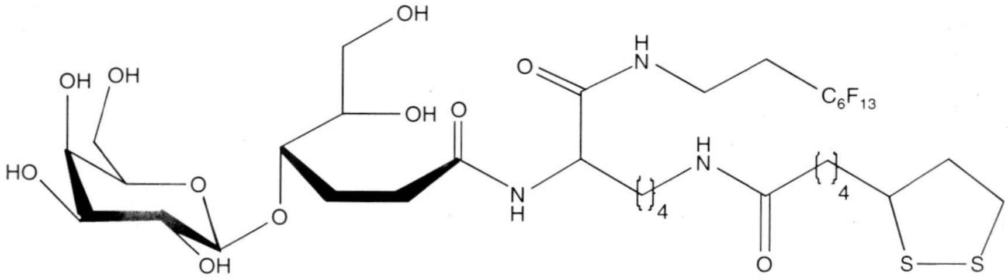

7

FIGURE 3.6 Structure of compound 7 (*N*-lactobionyl-N^{ε}-(5-[1,2]-dithiolan-3-yl-penoyl)-L-lysinyl-1*H*,1*H*,2*H*,2*H*-perfluorooctylamide).

were significantly more potent than PBN itself in the same assays. It appears that the amide bond linking LA to the bulky amphiphilic carrier was relatively resistant to amidase-induced hydrolysis in cells due to steric hindrance. Therefore, mitochondrial or cytoplasmic enzymatic reduction of LA to DHLA by dihydrolipoamide dehydrogenase or glutathione reductase was slowed (Pick et al. 1995; Haramaki et al. 1997). Consequently, a limited generation of functional LA/DHLA redox couple in cells may explain the marginal antioxidative effects of this derivative.

Indole-α-Lipoic Acid Derivatives

The rationale that directed Gurkan et al. (2005) to design and synthesize an indole-lipoic acid derivative was the documented reactive oxygen species scavenging activity of other indole-containing molecules, such as melatonin and serotonin (Stolc 1999). Three different conjugates of LA were synthesized through the introduction of an amine group at position 1 or 3 of the indole ring followed by conjugation with LA, or by amidation of a substituted 5-amino indole compound with LA. Compound 8 (Figure 3.7), which belongs to the last group, was more potent than LA and the other derivatives in inhibiting lipid peroxidation

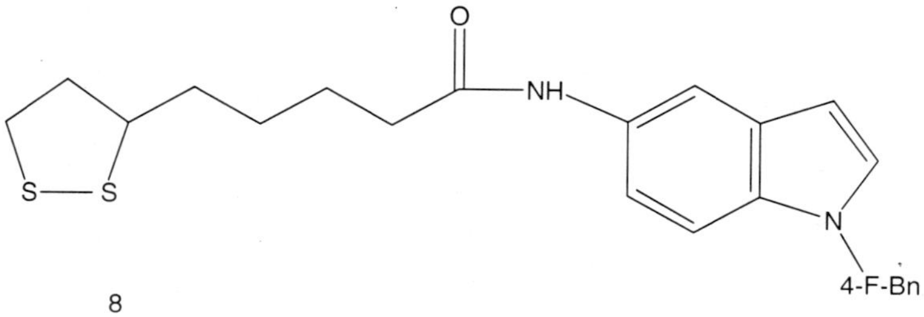

8

FIGURE 3.7 Structure of compound 8 (5-[1,2]dithiolan-3-yl-pentanoic acid [1-(4-fluorobenzyl)-1*H*-indole-5-yl]-amide).

in vitro. The in vivo antioxidant potential of these derivatives has not been evaluated yet. Further studies are required to examine the hypothesis that a slow amidase-dependent hydrolysis of this amide bond in compound 8 might improve the pharmacokinetics and efficacy of this indolic-LA derivative.

LA-Plus: *N,N*-Dimethyl, *N′*-2-Amidoethyl-Lipoate

Sen et al. (1998) planned a chemical strategy to reduce the efflux of DHLA from cells by introducing in the molecule a tertiary amine group that is predominantly protonated under physiological conditions. The carboxyl group in LA was covalently linked to *N,N*-dimethylethylenediamine and the product, *N,N*-dimethyl, *N′*-2-amidoethyl-lipoate, (Figure 3.8, compound 9) was designated LA-plus to indicate the protonated form of the compound. The hypothesis that LA-plus was better reduced to its DHLA-plus form than LA by mitochondrial dihydrolipoamide dehydrogenase was confirmed in isolated mitochondria from rat liver (Tirosh et al. 2003). The intracellular accumulation of LA and LA-plus and their corresponding reduced forms were determined and compared in cultured human lymphoma Wurzburg cells. Although the level of DHLA was merely 10% of that of LA, due to the efflux of the former, DHLA-plus was retained in cells and accumulated to a similar level as LA-plus. Moreover, the total accumulated content of both oxidized and reduced forms of LA-plus in the cells was nearly 3.5-fold higher than that of the corresponding forms of LA in the cells that were treated similarly with LA. Consequently, it has been proposed that the LA-plus/DHLA-plus redox couple is more effective in neutralizing free radicals than the LA/DHLA couple. Indeed, LA-plus (100 μM), but not LA, inhibited hydrogen peroxide or TNFα-induced activation of NF-κB in Wurzburg cells. Similarly, LA-plus (5 μM) protected rat thymocytes against spontaneous and etoposide-induced DNA fragmentation.

Guo et al. (2001) synthesized the *R*(+) enantiomer of RA-plus and found less striking ratios of intracellular accumulation of its oxidized and reduced forms in RAW 264.7 macrophages, in comparison with the accumulation of the corresponding forms of LA (10:1 and 25:1, respectively). In Wurzburg cells, however, the corresponding ratios of oxidized and reduced RA-plus was nearly 1:1 (Sen et al. 1998). When treated with 100 μM of LA or *R*(+)-LA-plus for

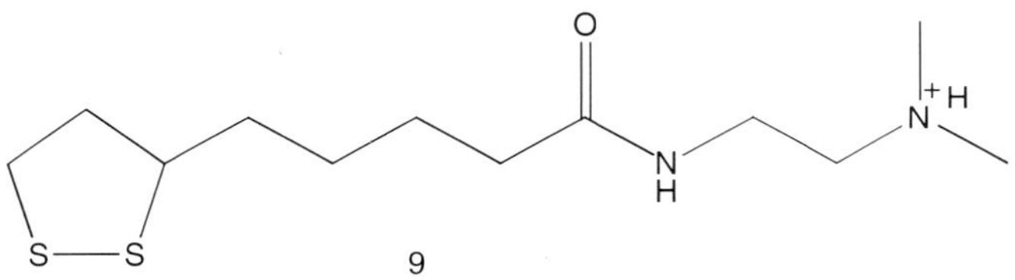

9

FIGURE 3.8 Structure of compound 9 (a protonated form of *N,N*-dimethyl,*N′*-2-amidoethyl-lipoate).

90 min these cells accumulated higher amounts of *R*(+)-LA and *R*(+)-DHLA (245 ± 16 and 4.5 ± 0.3 pmol/10^6 cells, respectively) than the corresponding forms of racemic LA-plus (158 ± 8 and 15 ± 3 pmol/10^6 cells). Yet, even under this less favorable accumulation of *R*(+)-LA-plus, it still inhibited basal- and rotenone-induced NO production in cultured macrophages better than LA. Since these compounds had no direct NO-scavenging activity in a cell-free system it has been suggested that they lack direct free-radical scavenging activities. Rather, *R*(+)-LA and *R*(+)-LA-plus depleted cellular NAD(P)H during their reduction, thus indirectly reduced NO production, because an appropriate NADPH supply was required to support NO production in macrophages. This proposed mechanism may also explain the negligible effects of both *R*(+)-LA and *R*(+)-LA-plus on glucose-induced production of NO in these cells, due to sufficient replenishment of pyridine nucleotides during the metabolism of glucose.

Lipoamide

Persson et al. (2001) aimed at reducing cellular injury induced by oxidative stress reaction and lysosomal rupture by targeting LA into lysosomes (Ollinger and Brunk 1995). α-Lipoic acid may prevent generation of free radicals and maintain lysosomal integrity by chelating iron ions and consequently prevent Fenton type reactions intralysosomally. α-Lipoic acid is effectively reduced in lysosomes to DHLA due to the abundance of reducing equivalents and cysteine and to the presence of lysosomal thiol-reductase (Collins et al. 1991; Arunachalam et al. 2000). Yet, because cytoplasmic LA is predominantly ionized ($pK_a = 5.4$), its diffusional permeation into lysosomes is limited. Moreover, the acidic milieu in the lysosomes readily protonates the ionized LA to its neutral form, permitting its efflux back to the cytoplasm. This obstacle can be chemically overcome by converting the carboxylic moiety into a nonionizable amide. The resulting compound 10 is lipoamide (Figure 3.9), which remains uncharged at physiological and acidic pH. This property of lipoamide resulted in a twofold higher efficiency in protecting macrophage-like L774 cells against hydrogen peroxide-induced apoptosis.

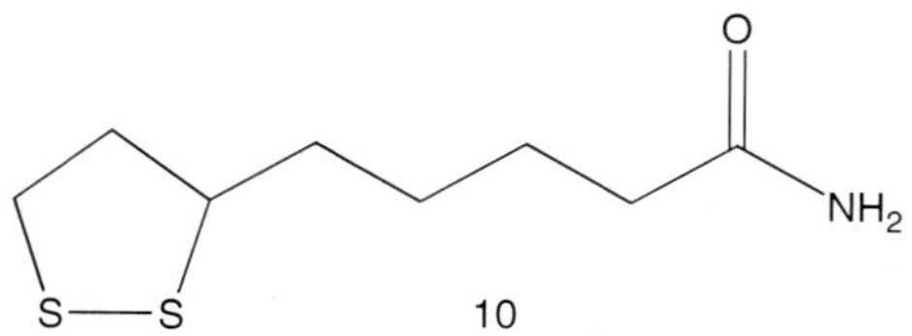

10

FIGURE 3.9 Structure of compound 10 (α-lipoamide; 5-[1,2]Dithiolan-3-yl-pentanoic acid amide).

α-LIPOIC ACID IN CO-DRUGS

Covalent binding of two different pharmacophores to produce chimeric or hybrid compounds is a classic approach in pharmaceutical chemistry to (1) produce novel molecules with unique functions, (2) combine effects of each of the paired molecules into one pharmaceutical, or (3) improve the delivery of one compound by covalent binding with a cell-permeable partner.

Trolox/α-Lipoic Acid Co-Drug

Koufaki et al. (2001) designed and synthesized a series of amide-based hybrid molecules of lipoic acid and trolox (6-hydroxy-2,5,7,8-tetramethylchroman-2-carboxylic acid; a cell-permeable and water-soluble derivative of vitamin E). The leading principle was that the co-drug might exhibit combined antioxidant effects of LA and vitamin E (Thurich et al. 1998). The resulting compound 11 (Figure 3.10) was the most potent inhibitor of lipid peroxidation in vitro (IC_{50} 0.06 ± 0.01 versus >1000 and 24.8 ± 1.0 μM for LA and trolox, respectively) and completely suppressed arrhythmias following reoxygenation of isolated rat hearts. Also, the level of the malondialdehyde (an end product of lipid peroxidation by reactive oxygen species) in these reperfused hearts was halved following treatments with trolox, lipoic acid, or compound 11. Other amide and amine derivatives of trolox that were synthesized in this study or by others were 5–1000-fold less potent than compound 11 in inhibiting lipid peroxidation (Jacobsen et al. 1992; Vajragupta et al. 2000). This enhanced potency of the lipoic acid-trolox hybrid has been attributed specifically to the oxidized form of LA, because a similar DHLA-trolox pair was found to be 10-fold less active than the former. It remains to be investigated whether free trolox and LA released

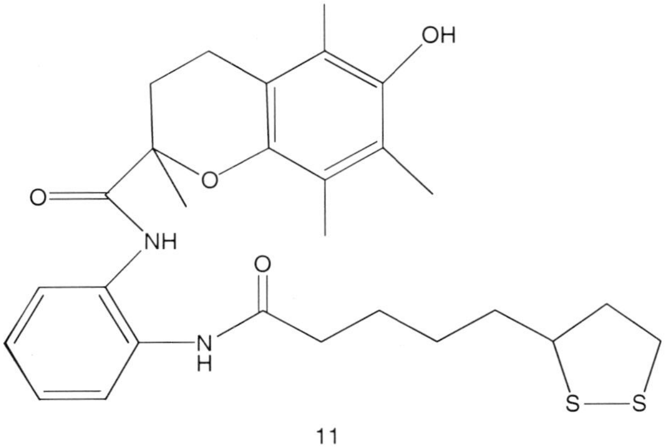

11

FIGURE 3.10 Structure of compound 11 (*N*-(3,4-dihydro-6-hydroxy-2,5,7,8-tetramethyl-2*H*-1-benzopyran-2-carbonyl)-*N*′-(1,2-dithiolan-3-pentanoyl)-1,2-phenylenediamine).

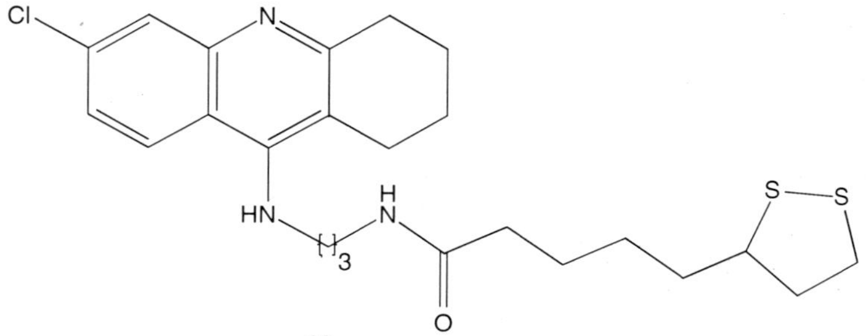

12

FIGURE 3.11 Structure of compound 12 (1,2,3,4-tetrahydro-6-chloro-9-(*N*-*R*,*S*-α-lipoylamidopropyl)-amino-acridine).

following an enzymatic hydrolysis of the amide bonds in compound 11, or the intact hybrid molecule mediate these beneficial effects. The enhanced lipophilicity of compound 11 (ClogP = 5.99) may also explain its increased potency in comparison with free lipoic acid and trolox (ClogP = 2.39 and 3.10, respectively).

Tacrine/α-Lipoic Acid Co-Drug (Lipocrine)

The idea to use LA for treatment of neurodegenerative diseases emerged from studies that assigned LA a protective role against cytotoxic effects induced by amyloid-β (Aβ) and LA-dependent inhibition of formation and maturation of Aβ, which is common in Alzheimer's disease (Zhang et al. 2001; Ono et al. 2006). In addition, LA has been shown to ameliorate several neurotoxic effects induced by various agents in animal models for neurodegenerative diseases (Collins et al. 1991; Andreassen et al. 2001; Virmani et al. 2005). Compound 12 (lipocrine, Figure 3.11) is an amide-based hybrid of tacrine and LA. Tacrine is an acetylcholine esterase (AchE) inhibitor approved for the treatment of Alzheimer's disease (Davis and Powchik 1995). Compound 12 was a more potent inhibitor of human AchE than tacrine ($IC_{50} = 0.25$ and 424 μM, respectively), yet LA exhibited no such inhibitory effects. Neither tacrine nor LA inhibited AchE-induced Aβ aggregation induced by AchE, whereas compound 12 caused such inhibition ($IC_{50} = 45$ μM). Moreover, the viability of neuronal-like cells (SH-SY5Y) was not compromised following exposure to compound 12 or to LA (0.1–50 μM). Finally, the antioxidative effects of compound 12 were compared to that of LA in these cells following exposure to *tert*-butylhydroperoxide that induces an oxidative stress: LA (50 μM) decreased free-radical formation by 32%, whereas half maximal inhibitory effect of compound 12 was already obtained at 20 μM. It is conceivable that compound 12 might be better absorbed than tacrine, whose oral bioavailability is limited, and possesses a longer half-life than LA, which undergoes an intensive first-pass metabolism. Further studies on the ability of compound 12 to cross the blood–brain barrier and accumulate in the

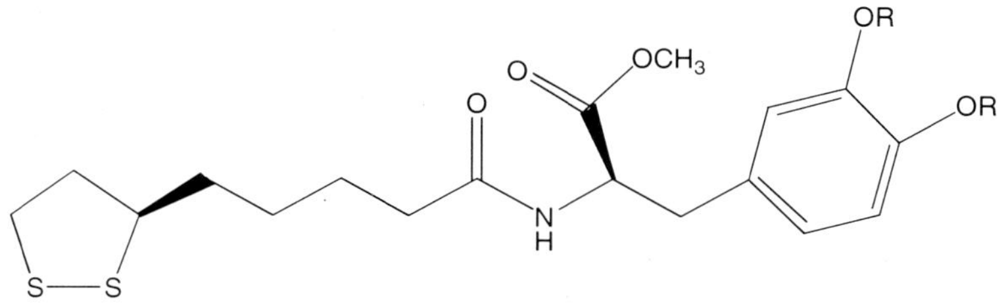

13: R=COCH$_3$
14: R=H

FIGURE 3.12 Structures of compound 13 (methyl-*O*-acetyl-3-(acetyloxy)-*N*-{5-[(3*R*)-1,2-dithiolan-3-yl]-pentanoyl}-L-tyrosinate) and compound 14 (methyl *N*-{5-[(3*R*)-1,2-dithiolan-3-yl]-pentanoyl}-3-hydroxy-L-tyrosinate).

central nervous system, its pharmacokinetic parameters and rate of hydrolysis and metabolism are required to ascertain the feasibility of this compound to be a potential pharmacological agent. Following the synthesis of this tacrine-LA co-drug, LA has become a leading molecule in designing, synthesizing, and testing other novel multitargeted co-drugs for the treatment of neurodegenerative diseases (Bolognesi et al. 2006; Holzgrabe et al. 2007).

L-DOPA/α-Lipoic Acid Co-Drug

Oxidative stress is a major detrimental factor in the etiology of Parkinson's disease and degradation of dopaminergic neurons in the substantia nigra. Lowering the oxidative burden in these neurons might be beneficial when combined with other traditional pharmacological treatments of the disease, such as L-DOPA. The beneficial effects of L-DOPA need to be weighted against its cytotoxic effects that are associated with free radicals formed as a result of autooxidation of the catechol moiety (Basma et al. 1995). Moreover, L-DOPA-induced neurotoxicity is also caused by generation of highly reactive dopamine- and DOPA-quinines, which are dopaminergic neuron-specific cytotoxic molecules. Thus, a co-drug linking L-DOPA and LA might be a useful compound devoid of these pro-oxidant effects. Compounds 13 and 14 (Figure 3.12) are linked *R*(+)-LA amide to 3,4-diacetyloxy-L-DOPA or to L-DOPA, respectively. Both resulting co-drugs were readily hydrolyzed in rat plasma, yet, their half-life values in human plasma were nearly 50 and 100 min, respectively. These findings along with some indications for an in vivo dopaminergic activity of these hybrid molecules point to the potential advantage of using L-DOPA/LA conjugates or other L-DOPA-based co-drugs for the treatment of Parkinson's disease (Asanuma et al. 2003). It should also be noted that neuromelanin is formed by nonenzymatic autooxidation of cytoplasmic dopamine to dopamine-quinone that subsequently cyclizes and oxidatively polymerizes (Graham 1978; Tsuji-Naito et al. 2006).

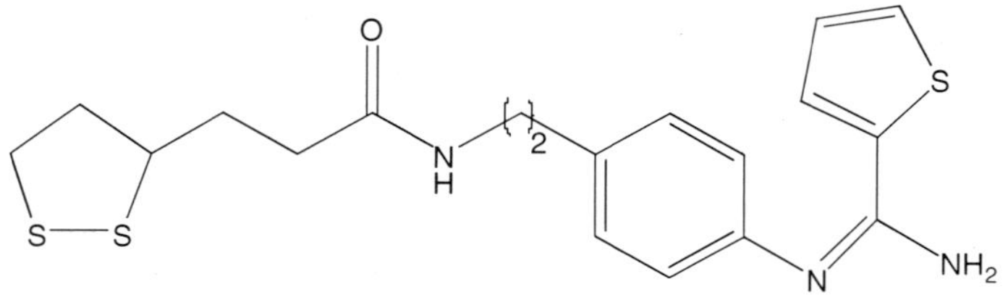

FIGURE 3.13 Structure of compound 15 (3-[1,2]Dithiolan-3-yl-*N*-(2-{4-[(thiophene-2-carboximidoyl)-amino]-phenyl]-phenyl}-ethyl)-propionamide).

L-cysteine and glutathione scavenge these quinones and form soluble 5-*S*-cysteinyl-catecholamine and 5-*S*-glutathionyl-catecholamine adducts reducing the formation of neuromelanin polymer. It has been shown that DHLA, but not LA, competes with L-cysteine and glutathione to form 5-*S*-lipoyl-DOPA adducts, in a system generating superoxide radicals, proposing this as a model of oxidative stress for the nervous system (Tsuji-Naito et al. 2006).

NOS Inhibitor/α-Lipoic Acid Co-Drug

Another approach to protect neural cells against oxidative stress is to coadminister LA and a nitric oxide synthase (NOS) inhibitor. This strategy that was proven effective against transient focal ischemia in rat (Spinnewyn et al. 1999) led to the synthesis of a series of LA-arylthiophene amidine co-drugs, linked by a carboxyamide bond. Compound 15 (Figure 3.13), a representative of this series, was (1) fourfold more potent than the free NOS inhibitor in inhibiting nNOS (respective IC_{50} values were 1.1 and 4.0 μM), (2) 10- and 6-fold more effective than the NOS inhibitor and LA, respectively, in protecting hippocampal neural HT22 cells from glutamate-induced oxidative stress, and (3) 10- and 4-fold more potent than the respective parent compounds in preventing glutathione depletion following glutamate-induced inhibition of cysteine transport. Following this report Chabrier and colleagues have further proved the feasibility of dual nNOS/antioxidant co-drugs by synthesizing other novel nNOS inhibitors/antioxidant pairs (Auvin et al. 2003).

Thiazolidinedione/α-Lipoic Acid Co-Drug

Members of the thiazolidinedione (TZD) group are synthetic ligands of peroxisome proliferator-activated receptor-γ (PPARγ) and principally act as peripheral insulin sensitizers for the treatment of type-2 diabetes. In addition, some anti-inflammatory effects of TZDs have been documented (Delerive et al. 2001). In particular, troglitazone improved psoriasis and ameliorated proliferative skin disease (Pershadsingh et al. 1998; Pershadsingh 1999; Ellis et al. 2000). α-Lipoic acid also significantly inhibits the consumption of vitamin E and ubiquinone in

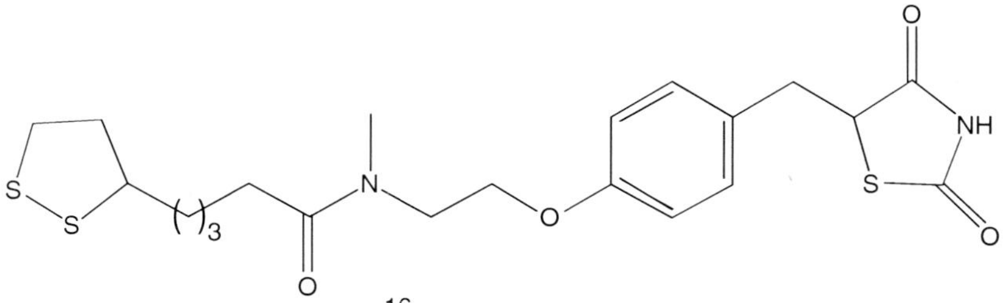

16

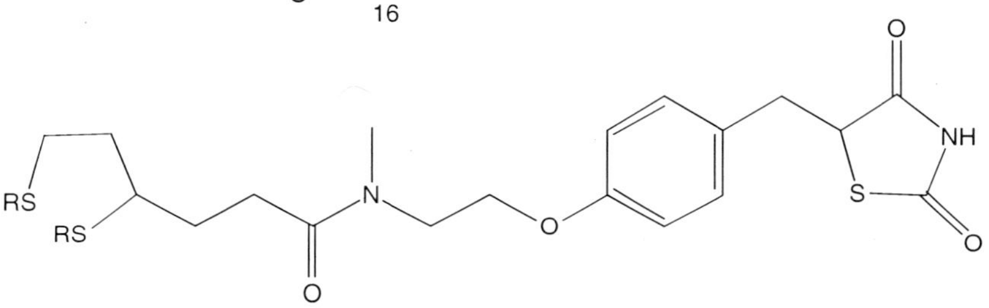

17: R ═ $COCH_2NH_2$. HC1

FIGURE 3.14 Structures of compound 16 (*N*-(2-{4-[2,4-dioxo(1,3-thiazolidin-5-yl) methyl] phenoxy}ethyl)-5-(1,2-dithiolan-3-yl)-*N*-methylpentanamide) and compound 17 (2,2′-[{8-[(2-{4-[(2,4-dioxo-1,3-thiazolodin-5-yl)methyl]phenoxy}-ethyl)(methyl)amino]-8-oxooctane-1,3-diyl}bis(thio)bis(2-oxo-ethanaminium) dichloride).

skin after the UV radiation (Podda et al. 2001). Thus, it has been assumed that LA provides protection against oxidative stress in skin (Podda and Grundmann-Kollmann 2001). α-Lipoic acid-based TZD co-drug derivatives were synthesized by Venkatraman et al. (2004) and tested for their ability to slow keratinocyte proliferation while maintaining their other functions, such as the ability to induce differentiation of preadipocytes to mature, fat-laden adipocytes. Compound 16 (Figure 3.14), in which the pyridine head of rosiglitazone was removed and α-LA was conjugated to the resulting 2-aminoethoxy-benzyl-2-4-thiazolidinedione moiety via an amide bond, was the most active derivative. Compound 16 was more potent than rosiglitazone in inhibiting keratinocyte proliferation in vitro (IC_{50} values of 0.1 and ~10 μM, respectively). Interestingly, despite the modification in the chemical backbone of rosiglitazone, the hybrid molecule was 18-fold more potent than free rosiglitazone in activating PPARγ in a cell-based heterologous transactivation assay (using a GAL4-PPARγ chimeric receptor), while producing a similar adipogenic response in murine 3T3L1 preadipocytes. This enhanced activity has been attributed to the optimal tetramethylene moiety spacer between the carboxamide moiety and the 1,2-dithiolane ring, whereas shorter spacers produced inactive or less active derivatives. In a mouse model, however, orally or topically administered compound 16 did not show significant anti-inflammatory effects, whereas the dithioglycinate derivative (compound 17) did. However, in the in vitro assays described above, compound 16 was 10–40-fold more potent than compound 17. This seems related to the vast differences in the lipid/water solubility of the two hybrid molecules. Compound 17 is water soluble (ClogP = 1.23) whereas compound 16 is lipophilic (ClogP = 4.00). The lack of effect of an orally administered compound 16 might indicate limited absorption from the gastrointestinal tract and/or intensive first-pass metabolism. Nevertheless, it was also found that when compound 16 was administered by gavage to Zucker rats, it effectively reduced plasma insulin and triglyceride levels (Chittiboyina et al. 2006). Therefore, the basis of these inconsistent results requires a thorough determination of species-specific pharmacokinetic parameters and metabolic pathways of these hybrid molecules.

Unlike the LA moiety in compound 16, compound 17 is resistant to thioreduction due to the linking of the glycinate groups to the thiol groups. Therefore, it is reasonable to assume that upon hydrolysis of the amide bond in compound 17, the released and modified LA and rosiglitazone are inactive due to chemical modifications in their primary structures. Crystallographic analysis of the ligand-binding domain of PPARγ lends support to the view that the entire LA-rosiglitazone hybrid molecule interacts with this receptor. The co-crystal structure of rosiglitazone with the ligand-binding domain of PPARγ reveals the presence of an unoccupied hydrophobic domain in the ligand-binding pocket (Nolte et al. 1998). It is suggested that the LA moieties in compounds 16 and 17 interact with this hydrophobic portion within the ligand-binding pocket and improves the hybrid's binding affinity in comparison with that of rosiglitazone (Chittiboyina et al. 2006). To explain the compound 16-mediated plasma triglyceride-, but not glucose, lowering effect in Zucker rats, it has been suggested that compound 16 may also bind and activate PPARα (Chittiboyina et al. 2006). Hitherto, it is unclear whether this LA-based compound could reduce the oxidative stress, which is common in this animal model of type-2 diabetes (Laight et al. 1999).

γ-Linoleic Acid/α-Lipoic Acid Co-Drug

α-Lipoic acid was combined with γ-linoleic acid to produce a co-drug aimed at improving whole-body and skeletal muscle insulin action on glucose disposal in insulin-resistant Zucker rats (Peth et al. 2000). The rationale for synthesizing such co-drug is the observation that like LA, γ-linoleic acid also seems to relieve symptoms of diabetic polyneuropathy in diabetic patients (Keen et al. 1993). Chronic treatments with various doses of LA, γ-linoleic acid, and the hybrid molecule showed that the latter improved the rate of glucose disposal and augmented insulin-mediated glucose transport in skeletal muscles better than free LA or γ-linoleic acid. However, the routes of administration of these compounds were not uniform: LA, dissolved in Tris buffer, was injected intraperitoneally; γ-linoleic acid and the conjugated molecule, dissolved in corn oil, were administered by gavage. Thus, it is difficult to assess the impact of absorption and first-pass metabolism of the last two compounds in comparison with the injected LA.

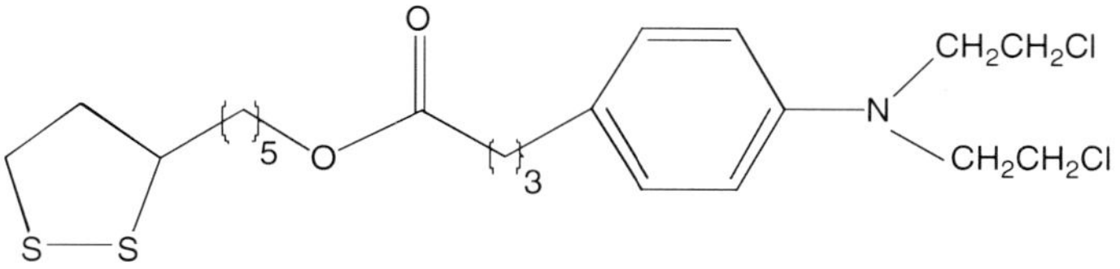

18

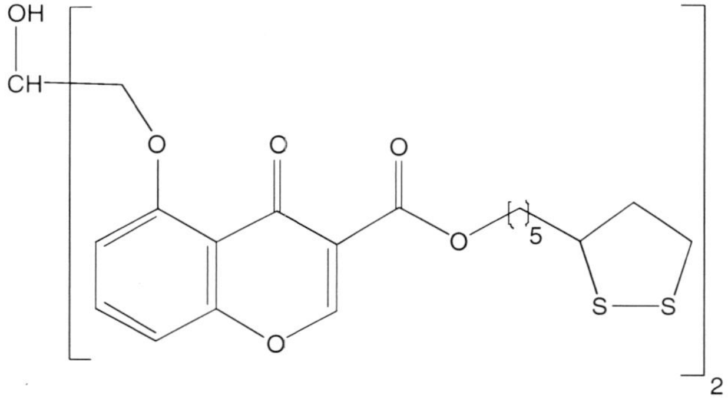

19

FIGURE 3.15 Structures of compound 18 (1,2-dithiolan-3-ylpent-5-yl 4-{p-[bis(2-chloroethyl)amino]phenyl}-butyrate) and compound 19 (1,3-bis[[2-[[[5-(1,2-dithiolan-3-yl)pentyl]oxy]carbonyl]chromon-5-yl]oxy]-2-hydroxypropane).

Chlorambucil- and Cromolyn/α-Lipoic Acid Co-Drugs

Saah et al. (1996) aimed at solving major problems associated with the clinical use of the alkylating agent chlorambucil and the antiasthmatic, mast cell-stabilizer cromolyn, by synthesizing respective co-drugs with LA. Cell membrane permeability of chlorambucil is limited due to the presence of the carboxyl group ($pK_a = 2.3$) (Cullis et al. 1995) that is predominantly ionized at physiological pH. Similarly, the bioavailability of cromolyn, which is polar and water soluble, is poor. Compound 18 and compound 19 (Figure 3.15) are the respective chlorambucil- and cromolyn-LA conjugates in which the ionizable carboxyl groups were masked by forming ester bonds with lipoyl moieties. The lipid solubility of the conjugated molecule is significantly higher than that of parent compounds (ClogP values: conjugated versus free chlorambucil, 2.7 and −0.5; conjugated versus free cromolyn, 2.9 and −1.7). Both conjugates were found highly susceptible to degradation in rat and rabbit plasma but remained relatively stable in human plasma. The delivery of these hybrid compounds to various tissues and most notably to the lungs of rabbits and rats following intravenous administration was enhanced in comparison with the parent drugs. The view that the preferential distribution of these co-drugs to lungs resulted from the presence of the lipoyl moieties in the hybrid needs to be investigated. Of note are reported effects of lipoic acid in the pulmonary system: inhibition of airway inflammation and hyperresponsiveness (Cho et al. 2004), on one hand, and pro-oxidative induction of apoptosis in lung epithelial cells, on the other (Moungjaroen et al. 2006). Thus, it remains to be determined whether the clinical use of such LA-based co-drugs would be beneficial or harmful due to their preferential accumulation in lungs.

Other α-Lipoic Acid-Based Co-Drugs

A series of co-drugs linking LA to the α1-adrenoreceptor antagonist prazosin were synthesized (Antonello et al. 2005). In addition to a potent inhibition of the receptor these co-drugs also inhibited intracellular oxidative stress and some exerted antiproliferative effects. α-Lipoic acid was also covalently bound to the ε-amino group of Lys^{B29} in insulin to produce lipoyl insulin (Huang and Huang 2006). This derivative was biologically active and induced hypoglycemia in mice to the same extent as insulin, but showed longer duration of action. One of the trypsin cleavage sites in insulin is located between Lys^{B29} and Ala^{B30}. It has been suggested that the extended duration of action of lipoyl insulin results from a slower proteolytic degradation due to a partial protection of the latter cleavage site. Coumarin compounds have been found to inhibit various stages in the HIV replication cycle (Taddeo et al. 1994). α-Lipoic acid was therefore covalently linked to biscoumarin analogues. Some of these new hybrid molecules inhibited HIV-1 integrase in an in vitro assay (Su et al. 2006). In addition, LA was also linked to cationic amphiphiles to produce an efficient redox-controlled DNA delivery system (Balakirev et al. 2000).

SUMMARY

The antioxidant properties of the LA/DHLA redox couple have made LA a popular dietary supplement for over four decades. Several clinical trials indicate LA-dependent melioration of symptoms of polyneuropathy, especially in diabetic patients. Other studies indicate that LA may also have some antihyperglycemic properties by augmenting basal and insulin-mediated glucose transport in peripheral tissues, notably in skeletal muscles. Unlike the cellular LA, which is bound to proteins, exogenous LA remains mostly free. The low bioavailability and the short half-life of orally administered LA result predominantly from an extensive rate of metabolism of the compound. Some chemical modifications in LA that have been performed to solve these problems are reviewed in this chapter. It seems that derivatives, in which an amide bond was formed to link two LA molecules or to add hydrophobic moieties, were less susceptible to rapid enzymatic hydrolysis than ester-based derivatives. Covalent linking of these chemical groups to the carboxylic carbon of LA modify the original amphiphilic character of LA and the resulting derivatives are usually more lipid soluble. The dithiolane moiety in LA is usually left intact to allow it for the reduction and formation of an active redox couple. In another case, the carboxylic group in LA was blocked by linking it to an amine moiety (LA-plus). Unlike the substantial efflux of DHLA from cells, the reduced amine-derivatives, which are predominantly protonated at physiological pH, are retained intracellularly and become more effective due to an extended duration of action. In the case of lipoamide, the carboxylic group of LA was exchanged with a nonionizable amide moiety. It has been argued that substantial antioxidative effects of free LA occur in lysosomes due to metal ions chelation and inhibition of Fenton type reactions. However, the acidic pH in lysosomes interferes with their capacity to retain LA, which is predominantly neutral at this pH. In contrast, derivatives containing nonionizable amine groups, such as LA-plus, may accumulate in lysosomes due to an efficient protonation of the amine moiety, which results in a reduced efflux to the cytoplasm. There are more reported chemical modifications that were made in LA. This chapter reviews several procedures to illustrate fundamental chemical approaches and the complexity in designing novel LA derivatives.

This chapter also discusses strategies to synthesize various LA-based co-drugs. Two main reasons underlie the use of LA in the synthesis of co-drugs: First, to link LA to another molecule via its carboxyl group and improve the lipid solubility of the partner drug in order to increase its absorption and uptake into cells, while slowing its first-pass metabolism. Second, to introduce an antioxidant moiety coupled to the other active drug. This chapter reviews several types of such LA-based co-drugs for the treatment of neurodegenerative diseases, hyperglycemia, and skin and pulmonary diseases. In view of these and other formulations of LA-co-drugs that escaped the scope of this review, LA seems to takes a center role in co-drug design and synthesis.

ACKNOWLEDGMENTS

M. Ben Yakir received a fellowship from the Diabetes Research Center of the Hebrew University. S. Sasson and J. Katzhendler are members of the David R. Bloom Center for Pharmacy at the Hebrew University of Jerusalem. We are grateful to the support of the Alex Grass Center for Drug Design and Synthesis of Novel Therapeutics.

REFERENCES

Andreassen, O.A., R.J. Ferrante, A. Dedeoglu, and M.F. Beal. 2001. Lipoic acid improves survival in transgenic mouse models of Huntington's disease. *Neuroreport* 12(15):3371–3373.

Antonello, A., P. Hrelia, A. Leonardi, G. Marucci, M. Rosini, A. Tarozzi et al. 2005. Design, synthesis, and biological evaluation of prazosin-related derivatives as multipotent compounds. *J Med Chem* 48(1):28–31.

Arunachalam, B., U.T. Phan, H.J. Geuze, and P. Cresswell. 2000. Enzymatic reduction of disulfide bonds in lysosomes: Characterization of a gamma-interferon-inducible lysosomal thiol reductase (GILT). *Proc Natl Acad Sci U S A* 97(2):745–750.

Asanuma, M., I. Miyazaki, and N. Ogawa. 2003. Dopamine-or L-DOPA-induced neurotoxicity: the role of quinone formation and tyrosinase in a model of Parkinson's disease. *Neurotox Res* (5):165–176.

Auvin, S., M. Auguet, E. Navet, J.J. Harnett, I. Viossat, J. Schulz et al. 2003. Novel inhibitors of neuronal nitric oxide synthase with potent antioxidant properties. *Bioorg Med Chem Lett* 13(2):209–212.

Balakirev, M., G. Schoehn, and J. Chroboczek. 2000. Lipoic acid-derived amphiphiles for redox-controlled DNA delivery. *Chem Biol* 7(10):813–819.

Basma, A.N., E.J. Morris, W.J. Nicklas, and H.M. Geller. 1995. L-DOPA cytotoxicity to PC12 cells in culture is via its autoxidation. *J Neurochem* 64(2):825–832.

Berkson, B. 1998. *The Alpha-Lipoic Acid Breakthrough*: Three River Press.

Biewenga, G.P., M.A. Dorstijn, J.V. Verhagen, G.R. Haenen, and A. Bast. 1996. Reduction of lipoic acid by lipoamide dehydrogenase. *Biochem Pharmacol* 51(3):233–238.

Biewenga, G.P., G.R. Haenen, and A. Bast. 1997. The pharmacology of the antioxidant lipoic acid. *Gen Pharmacol* 29(3):315–331.

Bilska, A. and L. Wlodek. 2005. Lipoic acid—the drug of the future? *Pharmacol Rep* 57(5):570–577.

Bolognesi, M.L., A. Minarini, V. Tumiatti, and C. Melchiorre. 2006. Lipoic acid, a lead structure for multi-target-directed drugs for neurodegradation. *Mini-Reviews Med Chem* 6:1269–1274.

Breithaupt-Grogler, K., G. Niebch, E. Schneider, K. Erb, R. Hermann, H.H. Blume et al. 1999. Dose-proportionality of oral thioctic acid- coincidence of assessments via pooled plasma and individual data. *Eur J Pharm Sci* 8:57–65.

Bustamante, J., J.K. Lodge, L. Marcocci, H.J. Tritschler, L. Packer, and B.H. Rihn. 1998. Alpha-lipoic acid in liver metabolism and disease. *Free Radic Biol Med* 24(6):1023–1039.

Chittiboyina, A.G., M.S. Venkatraman, C.S. Mizuno, P.V. Desai, A. Patny, S.C. Benson et al. 2006. Design and synthesis of the first generation of dithiolane thiazolidinedione- and phenylacetic acid-based PPARgamma agonists. *J Med Chem* 49(14): 4072–4084.

Cho, Y.S., J. Lee, T.H. Lee, E.Y. Lee, K.U. Lee, J.Y. Park et al. 2004. Alpha-Lipoic acid inhibits airway inflammation and hyperresponsiveness in a mouse model of asthma. *J Allergy Clin Immunol* 114(2):429–435.

Collins, D.S., E.R. Unanue, and C.V. Harding. 1991. Reduction of disulfide bonds within lysosomes is a key step in antigen processing. *J Immunol* 147(12):4054–4059.

Cullis, P.M., R.E. Green, and M.E. Malone. 1995. Mechanism and reactivity of chlorambucil and chlorambucil-spermidine conjugate. *J Chem Soc Perkin Trans* 2:1503–1511.

Davis, K.L. and P. Powchik. 1995. Tacrine. *Lancet* 345(8950):625–630.

Delerive, P., J.C. Fruchart, and B. Staels. 2001. Peroxisome proliferator-activated receptors in inflammation control. *J Endocrinol* 169(3):453–459.

Durand, G., A. Polidori, O. Ouari, P. Tordo, V. Geromel, P. Rustin et al. 2003. Synthesis and preliminary biological evaluations of ionic and nonionic amphiphilic alpha-phenyl-N-tert-butylnitrone derivatives. *J Med Chem* 46(24):5230–5237.

Ellis, C.N., J. Varani, G.J. Fisher, M.E. Zeigler, H.A. Pershadsingh, S.C. Benson et al. 2000. Troglitazone improves psoriasis and normalizes models of proliferative skin disease: Ligands for peroxisome proliferator-activated receptor-gamma inhibit keratinocyte proliferation. *Arch Dermatol* 136(5):609–616.

Evans, J.L. and I.D. Goldfine. 2000. α-Lipoic acid: A multifunctional antioxidant that improves insulin sensitivity in patients with type 2 diabetes. *Diabetes Technol Ther* 2(3):401–413.

Frolich, L., M.E. Gotz, M. Weinmuller, M.B.H. Youdim, N. Barth, A. Dirr et al. 2004. (r)-, but not (s)-alpha lipoic acid stimulates deficient brain pyruvate dehydrogenase complex in vascular dementia, but not in Alzheimer dementia. *J Neur Trans* 111:295–310.

Graham, D.G. 1978. Oxidative pathways for catecholamines in the genesis of neuromelanin and cytotoxic quinones. *Mol Pharmacol* 14(4):633–643.

Gruzman, A., A. Hidmi, J. Katzhendler, A. Haj-Yehie, and S. Sasson. 2004. Synthesis and characterization of new and potent alpha-lipoic acid derivatives. *Bioorg Med Chem* 12(5):1183–1190.

Guillonneau, C., Y. Charton, Y.M. Ginot, M.V. Fouquier-d'Herouel, M. Bertrand, B. Lockhart et al. 2003. Synthesis and pharmacological evaluation of new 1,2-dithiolane based antioxidants. *Eur J Med Chem* 38(1):1–11.

Guo, Q., O. Tirosh, and L. Packer. 2001. Inhibitory effect of alpha-lipoic acid and its positively charged amide analogue on nitric oxide production in RAW 264.7 macrophages. *Biochem Pharmacol* 61(5):547–554.

Gurkan, A.S., A. Karabay, Z. Buyukbingol, A. Adejare, and E. Buyukbingol. 2005. Syntheses of novel indole lipoic acid derivatives and their antioxidant effects on lipid peroxidation. *Arch Pharm (Weinheim)* 338(2–3):67–73.

Hagen, T.M., R.T. Ingersoll, J. Lykkesfeldt, J. Liu, C.M. Wehr, V. Vinarsky et al. 1999. (R)-alpha-lipoic acid-supplemented old rats have improved mitochondrial function, decreased oxidative damage, and increased metabolic rate. *Faseb J* 13(2):411–418.

Hagen, T.M., V. Vinarsky, C.M. Wehr, and B.N. Ames. 2000. (R)-alpha-lipoic acid reverses the age-associated increase in susceptibility of hepatocytes to tert-butylhydroperoxide both in vitro and in vivo. *Antioxid Redox Signal* 2(3):473–483.

Hahm, J.R., B.J. Kim, and K.W. Kim. 2004. Clinical experience with thioctacid (thioctic acid) in the treatment of distal symmetric polyneuropathy in Korean diabetic patients. *J Diabetes Complications* 18(2):79–85.

Haramaki, N., D. Han, G.J. Handelman, H.J. Tritschler, and L. Packer. 1997. Cytosolic and mitochondrial systems for NADH- and NADPH-dependent reduction of alpha-lipoic acid. *Free Radic Biol Med* 22(3):535–542.

Hermann, R., G. Niebch, H.O. Borbe, H. Fieger-Buschges, P. Ruus, H. Nowak, H. Riethmuller-Winzen, M. Peukert, and H.H. Blume. 1996. Enantioselsctive pharmacokinetics and bioavailibility of different racemic α-lipoic acid formulations in healthy volunteers. *Eur J Pharm Sci* 1996.

Hermann, R., H.J. Wildgrube, P. Ruus, G. Niebch, H. Nowak, and C.H. Gleiter. 1998. Gastric emptying in patients with insulin dependent diabetes mellitus and bioavailability of thioctic acid-enantiomers. *Eur J Pharm Sci* 6(1):27–37.

Holzgrabe, U., P. Kapkova, V. Alptuzun, J. Scheiber, and E. Kugelmann. 2007. Targeting acetylcholinesterase to treat neurodegeneration. *Expert Opin Ther Targets* 11(2):161–179.

Huang, T. and K. Huang. 2006. Synthesis, characterization and biological activity of chemically modified insulin derivative with alpha lipoic acid. *Protein Pept Lett* 13(2):135–142.

Jacob, S., E.J. Henriksen, A.L. Schiemann, I. Simon, D.E. Clancy, H.J. Tritschler et al. 1995. Enhancement of glucose disposal in patients with type 2 diabetes by alpha-lipoic acid. *Arzneimittelforschung* 45(8):872–874.

Jacob, S., E.J. Henriksen, H.J. Tritschler, H.J. Augustin, and G.J. Dietze. 1996. Improvement of insulin-stimulated glucose-disposal in type 2 diabetes after repeated parenteral administration of thioctic acid. *Exp Clin Endocrinol Diabetes* 104(3):284–288.

Jacob, S., P. Ruus, R. Hermann, H.J. Tritschler, E. Maerker, W. Renn et al. 1999. Oral administration of RAC-α-lipoic acid modulates insulin sensitivity in patients with type-2 diabetes mellitus: A placebo-controlled pilot trial. *Free Radic Biol Med* 27(3–4):309–314.

Jacobsen, E.J., F.J. VanDoornik, D.E. Ayer, K.L. Belonga, J.M. Braughler, E.D. Hall et al. 1992. 2-(Aminomethyl)chromans that inhibit iron-dependent lipid peroxidation and protect against central nervous system trauma and ischemia. *J Med Chem* 35(23):4464–4472.

Keen, H., J. Payan, J. Allawi, J. Walker, G.A. Jamal, A.I. Weir et al. 1993. Treatment of diabetic neuropathy with gamma-linolenic acid. The gamma-Linolenic Acid Multicenter Trial Group. *Diabetes Care* 16(1):8–15.

Konrad, T., P. Vicini, K. Kusterer, A. Hoflich, A. Assadkhani, H.J. Bohles et al. 1999. alpha-Lipoic acid treatment decreases serum lactate and pyruvate concentrations and improves glucose effectiveness in lean and obese patients with type 2 diabetes. *Diabetes Care* 22(2):280–287.

Koufaki, M., T. Calogeropoulou, A. Detsi, A. Roditis, A.P. Kourounakis, P. Papazafiri et al. 2001. Novel potent inhibitors of lipid peroxidation with protective effects against reperfusion arrhythmias. *J Med Chem* 44(24):4300–4303.

Laight, D.W., K.M. Desai, N.K. Gopaul, E.E. Anggard, and M.J. Carrier. 1999. F2-isoprostane evidence of oxidant stress in the insulin resistant, obese Zucker rat: Effects of vitamin E. *Eur J Pharmacol* 377(1):89–92.

Liederer, B.M. and R.T. Borchardt. 2006. Enzymes involved in the bioconversion of ester-based prodrugs. *J Pharm Sci* 95(6):1177–1195.

Loffelhardt, S., C. Bonaventura, M. Locher, H.O. Borbe, and H. Bisswanger. 1995. Interaction of alpha-lipoic acid enantiomers and homologues with the enzyme components of the mammalian pyruvate dehydrogenase complex. *Biochem Pharmacol* 50(5):637–646.

Maitra, I., E. Serbinova, H.J. Tritschler, and L. Packer. 1996. Stereospecific effects of R-lipoic acid on buthionine sulfoximine-induced cataract formation in newborn rats. *Biochem Biophys Res Commun* 221(2):422–429.

Matsugo, S., L.J. Yan, T. Konishi, H.D. Youn, J.K. Lodge, H. Ulrich et al. 1997. The lipoic acid analogue 1,2-diselenolane-3-pentanoic acid protects human low density lipoprotein against oxidative modification mediated by copper ion. *Biochem Biophys Res Commun* 240(3):819–824.

Moini, H., O. Tirosh, Y.C. Park, K.J. Cho, and L. Packer. 2002. R-alpha-lipoic acid action on cell redox status, the insulin receptor, and glucose uptake in 3T3-L1 adipocytes. *Arch Biochem Biophys* 397(2):384–391.

Moungjaroen, J., U. Nimmannit, P.S. Callery, L. Wang, N. Azad, V. Lipipun et al. 2006. Reactive oxygen species mediate caspase activation and apoptosis induced by lipoic acid in human lung epithelial cancer cells through Bcl-2 down-regulation. *J Pharmacol Exp Ther* 319(3):1062–1069.

Muller, L. and H. Menzel. 1990. Studies on the efficacy of lipoate and dihydrolipoate in the alteration of cadmium2+ toxicity in isolated hepatocytes. *Biochim Biophys Acta* 1052(3):386–391.

Nickander, K.K., B.R. McPhee, P.A. Low, and H.J. Tritschler. 1996. Alpha-lipoic acid: Antioxidant potency against lipid peroxidation of neural tissues in vitro and implications for diabetic neuropathy. *Free Radic Biol Med* 21:631–639.

Nolte, R.T., G.B. Wisely, S. Westin, J.E. Cobb, M.H. Lambert, R. Kurokawa et al. 1998. Ligand binding and co-activator assembly of the peroxisome proliferator-activated receptor-gamma. *Nature* 395(6698):137–143.

Ollinger, K. and U.T. Brunk. 1995. Cellular injury induced by oxidative stress is mediated through lysosomal damage. *Free Radic Biol Med* 19(5):565–574.

Ono, K., M. Hirohata, and M. Yamada. 2006. Alpha-lipoic acid exhibits anti-amyloidogenicity for beta-amyloid fibrils in vitro. *Biochem Biophys Res Commun* 341(4):1046–1052.

Ortial, S., G. Durand, B. Poeggeler, A. Polidori, M.A. Pappolla, J. Boker et al. 2006. Fluorinated amphiphilic amino acid derivatives as antioxidant carriers: A new class of protective agents. *J Med Chem* 49(9):2812–2820.

Packer, L. and C. Colman. 2000. *The Antioxidant Miracle: Put Lipoic Acid, Pycnogenol, and Vitamin E and C to Work for You*: Wiley US.

Packer, L., K. Kraemer, and G. Rimbach. 2001. Molecular aspects of lipoic acid in the prevention of diabetes complications. *Nutrition* 17(10):888–895.

Packer, L., E.H. Witt, and H.J. Tritschler. 1995. Alpha-Lipoic acid as a biological antioxidant. *Free Radic Biol Med* 19(2):227–250.

Pagliara, A., M. Reist, S. Geinoz, P.A. Carrupt, and B. Testa. 1999. Evaluation and prediction of drug permeation. *J Pharm Pharmacol* 51(12):1339–1357.

Peinado, J., H. Sies, and T.P. Akerboom. 1989. Hepatic lipoate uptake. *Arch Biochem Biophys* 273(2):389–395.

Pershadsingh, H.A. 1999. Pharmacological peroxisome proliferator-activated receptor-gamma ligands: Emerging clinical indications beyond diabetes. *Expert Opin Investig Drugs* 8(11):1859–1872.

Pershadsingh, H.A., J.A. Sproul, E. Benjamin, J. Finnegan, and N.M. Amin. 1998. Treatment of psoriasis with troglitazone therapy. *Arch Dermatol* 134(10): 1304–1305.

Persson, H.L., A.I. Svensson, and U.T. Brunk. 2001. Alpha-lipoic acid and alpha-lipoamide prevent oxidant-induced lysosomal rupture and apoptosis. *Redox Rep* 6(5):327–334.

Peter, G. and H.O. Borbe. 1995. Absorption of [7,8-14C]rac-a-lipoic acid from in situ ligated segments of the gastrointestinal tract of the rat. *Arzneimittelforschung* 45(3):293–299.

Peth, J.A., T.R. Kinnick, E.B. Youngblood, H.J. Tritschler, and E.J. Henriksen. 2000. Effects of a unique conjugate of alpha-lipoic acid and gamma-linolenic acid on insulin action in obese Zucker rats. *Am J Physiol Regul Integr Comp Physiol* 278(2):R453–R459.

Pick, U., N. Haramaki, A. Constantinescu, G.J. Handelman, H.J. Tritschler, and L. Packer. 1995. Glutathione reductase and lipoamide dehydrogenase have opposite stereospecificities for alpha-lipoic acid enantiomers. *Biochem Biophys Res Commun* 206(2):724–730.

Podda, M. and M. Grundmann-Kollmann. 2001. Low molecular weight antioxidants and their role in skin ageing. *Clin Exp Dermatol* 26(7):578–582.

Podda, M., T.M. Zollner, M. Grundmann-Kollmann, J.J. Thiele, L. Packer, and R. Kaufmann. 2001. Activity of alpha-lipoic acid in the protection against oxidative stress in skin. *Curr Probl Dermatol* 29:43–51.

Reed, L.J., M. Koike, M.E. Levitch, and F.R. Leach. 1958. Studies on the nature and reactions of protein-bound lipoic acid. *J Biol Chem* 232:143–158.

Roy, S. and L. Packer. 1998. Redox regulation of cell functions by alpha-lipoate: Biochemical and molecular aspects. *Biofactors* 8(1–2):17–21.

Ruhnau, K.J., H.P. Meissner, J.R. Finn, M. Reljanovic, M. Lobisch, K. Schutte et al. 1999. Effects of 3-week oral treatment with the antioxidant thioctic acid (alpha-lipoic acid) in symptomatic diabetic polyneuropathy. *Diabet Med* 16(12):1040–1043.

Saah, M., W.M. Wu, K. Eberst, E. Marvanyos, and N. Bodor. 1996. Design, synthesis, and pharmacokinetic evaluation of a chemical delivery system for drug targeting to lung tissue. *J Pharm Sci* 85(5):496–504.

Schreck, R., R. Grassmann, B. Fleckenstein, and P.A. Baeuerle. 1992. Antioxidants selectively suppress activation of NF-kappa B by human T-cell leukemia virus type I Tax protein. *J Virol* 66(11):6288–6293.

Self, W.T., L. Tsai, and T.C. Stadtman. 2000. Synthesis and characterization of selenotrisulfide-derivatives of lipoic acid and lipoamide. *Proc Natl Acad Sci U S A* 97(23):12481–12486.

Sen, C.K., O. Tirosh, S. Roy, M.S. Kobayashi, and L. Packer. 1998. A positively charged alpha-lipoic acid analogue with increased cellular uptake and more potent immunomodulatory activity. *Biochem Biophys Res Commun* 247(2):223–228.

Sosin, A.E. and B.M. Ley-Jacobs. 1998. *Alpha-Lipoic Acid: Nature's Ultimate Antioxidant.* Kensington Publishing.

Spinnewyn, B., S. Cornet, M. Auguet, and P.E. Chabrier. 1999. Synergistic protective effects of antioxidant and nitric oxide synthase inhibitor in transient focal ischemia. *J Cereb Blood Flow Metab* 19(2):139–143.

Stolc, S. 1999. Indole derivatives as neuroprotectants. *Life Sci* 65(18–19):1943–1950.

Streeper, R.S., E.J. Henriksen, S. Jacob, J.Y. Hokama, D.L. Fogt, and H.J. Tritschler. 1997. Differential effects of lipoic acid stereoisomers on glucose metabolism in insulin-resistant skeletal muscle. *Am J Physiol* 273(1 Pt 1):E185–E191.

Su, C.X., J.F. Mouscadet, C.C. Chiang, H.J. Tsai, and L.Y. Hsu. 2006. HIV-1 integrase inhibition of biscoumarin analogues. *Chem Pharm Bull (Tokyo)* 54(5):682–686.

Suh, J.H., E.T. Shigeno, J.D. Morrow, B. Cox, A.E. Rocha, B. Frei et al. 2001. Oxidative stress in the aging rat heart is reversed by dietary supplementation with (R)-(alpha)-lipoic acid. *Faseb J* 15(3):700–706.

Suzuki, Y.J., M. Tsuchiya, and L. Packer. 1991. Thioctic acid and dihydrolipoic acid are novel antioxidants which interact with reactive oxygen species. *Free Radic Res Commun* 15(5):255–263.

Taddeo, B., W.A. Haseltine, and C.M. Farnet. 1994. Integrase mutants of human immunodeficiency virus type 1 with a specific defect in integration. *J Virol* 68(12):8401–8405.

Teichert, J., R. Hermann, P. Ruus, and R. Preiss. 2003. Plasma kinetics, metabolism, and urinary excretion of alpha-lipoic acid following oral administration in healthy volunteers. *J Clin Pharmacol* 43(11):1257–1267.

Teichert, J., J. Kern, H.J. Tritschler, H. Ulrich, and R. Preiss. 1998. Investigations on the pharmacokinetics of alpha-lipoic acid in healthy volunteers. *Int J Clin Pharmacol* 36:625.

Thurich, T., J. Bereiter-Hahn, M. Schneider, and G. Zimmer. 1998. Cardioprotective effects of dihydrolipoic acid and tocopherol in right heart hypertrophy during oxidative stress. *Arzneimittelforschung* 48(1):13–21.

Tirosh, O., C.K. Sen, S. Roy, M.S. Kobayashi, and L. Packer. 1999. Neuroprotective effects of alpha-lipoic acid and its positively charged amide analogue. *Free Radic Biol Med* 26(11–12):1418–1426.

Tirosh, O., S. Shilo, A. Aronis, and C.K. Sen. 2003. Redox regulation of mitochondrial permeability transition: Effects of uncoupler, lipoic acid and its positively charged analog LA-plus and selenium. *Biofactors* 17(1–4):297–306.

Tsuji-Naito, K., T. Hatani, T. Okada, and T. Tehara. 2006. Evidence for covalent lipoyl adduction with dopaquinone following tyrosinase-catalyzed oxidation. *Biochem Biophys Res Commun* 343(1):15–20.

Vajragupta, O., S. Toasaksiri, C. Boonyarat, Y. Wongkrajang, P. Peungvicha, H. Watanabe et al. 2000. Chroman amide and nicotinyl amide derivatives: Inhibition of lipid peroxidation and protection against head trauma. *Free Radic Res* 32(2):145–155.

Venkatraman, M.S., A. Chittiboyina, J. Meingassner, C.I. Ho, J. Varani, C.N. Ellis et al. 2004. Alpha-Lipoic acid-based PPARgamma agonists for treating inflammatory skin diseases. *Arch Dermatol Res* 296(3):97–104.

Virmani, A., F. Gaetani, and Z. Binienda. 2005. Effects of metabolic modifiers such as carnitines, coenzyme Q10, and PUFAs against different forms of neurotoxic insults: metabolic inhibitors, MPTP, and methamphetamine. *Ann N Y Acad Sci* 1053:183–191.

Zhang, L., G.Q. Xing, J.L. Barker, Y. Chang, D. Maric, W. Ma et al. 2001. Alpha-lipoic acid protects rat cortical neurons against cell death induced by amyloid and hydrogen peroxide through the Akt signalling pathway. *Neurosci Lett* 312(3):125–128.

Ziegler, D. 2004. Thioctic acid for patients with symptomatic diabetic polyneuropathy: A critical review. *Treat Endocrinol* 3(3):173–189.

Ziegler, D., A. Ametov, A. Barinov, P.J. Dyck, I. Gurieva, P.A. Low et al. 2006. Oral treatment with alpha-lipoic acid improves symptomatic diabetic polyneuropathy: The SYDNEY 2 trial. *Diabetes Care* 29(11):2365–2370.

Ziegler, D., H. Nowak, P. Kempler, P. Vargha, and P.A. Low. 2004. Treatment of symptomatic diabetic polyneuropathy with the antioxidant alpha-lipoic acid: A meta-analysis. *Diabet Med* 21(2):114–121.

4 Novel Indole Lipoic Acid Derivatives: Synthesis and Their Antioxidant Effects

A. Selen Gurkan and Erdem Buyukbingol

CONTENTS

INTRODUCTION

Impaired antioxidant defense mechanisms are important features in the pathogenesis of various diseases including diabetes, cancer, and neurological disorders. Antioxidants are required to prevent cellular damage observed in many of these diseases. Impairment of these defense mechanisms may be due to increased oxidative stress. α-Lipoic acid (α-LA, thioctic acid, 1,2-dithiolane-3-pentanoic acid) is a natural multifunctional antioxidant [1]. The two enantiomers of this substance are the *S* form and the *R* form (Figure 4.1); *R*-α-LA acid is a naturally occuring form. α-LA acts as a coenzyme in biological group transfer reactions [2–4]. Its main function is to increase production of glutathione, which eliminates toxic substances in the liver.

α-Lipoic acid is found endogenously as lipoamide in animals and plants [5]. Its carboxylic group is covalently bound by amide linkage to the ε-amino group of lysine residues. It functions as a cofactor of mitochondrial enzymes in catalyzing oxidative decarboxylation of α-keto acids such as pyruvate, α-ketoglutarate, and branched-chain α-keto acids [6]. α-Lipoic acid is readily absorbed and rapidly converted to reduced form, dihydrolipoic acid (DHLA) in many tissues

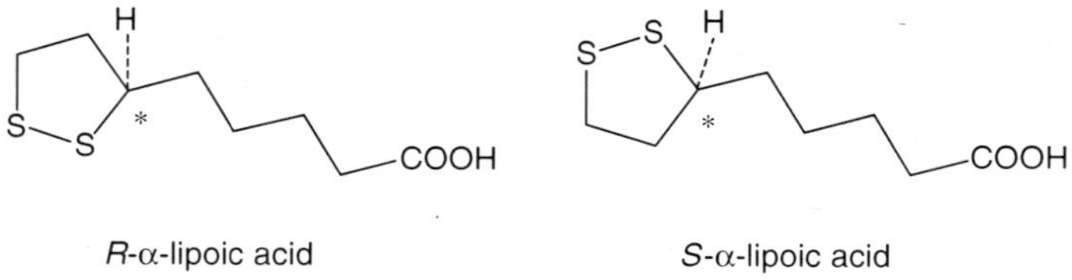

FIGURE 4.1 Chemical structures of *R*- and *S*-α-lipoic acid.

(Figure 4.2) [7]. Usually, antioxidant substances exhibit antioxidant properties in their reduced form. α-Lipoic acid is unique, because it retains protective functions in both its reduced and oxidized forms, although DHLA is the more effective than α-LA [8]. The antioxidant properties of α-LA can be related to four categories: its ability to scavenge free radicals, metal ion–chelating activity, capacity to regenerate endogenous antioxidants for example glutathione and tocopherol, and ability to repair oxidative damage in macromolecules [9,10].

On the other hand, the indole nucleus is an important moiety in many pharmacologically active endogenous and exogenous compounds. Indole-3-carbinol (I3C) (Figure 4.3), a compound naturally occurring in cruciferous vegetables, has been demonstrated to be a modulator of carcinogenesis in various models and suggested that it plays a role in the prevention of cancer [11–14]. It was reported I3C induces a strong inhibition of CDK2 specific kinase activity as part of a G1 cell cycle arrest of human breast cancer cells [15].

Tryptophan is an indole amino acid found in the constitution of numerous proteins and it is the precursor of several substances, such as serotonin and melatonin (Figure 4.4). It is evaluated for its possible antioxidant activity [16–18]. It was also studied for its antioxidant effects. However, due to the obtained results from such study, tryptophan was shown to having very slightly antioxidant properties because the oral administration of tryptophan did not show antioxidant activity on impaired endogenous antioxidant defense system [19]. As will be explained below, the proposed mechanism of the indole ring with its capability of capturing the electrons leading to scavenging of certain free radicals, will be supposed to have the peripheric groups may play important roles in enhancing the electron-capturing capacity of indole ring, e.g., electron-

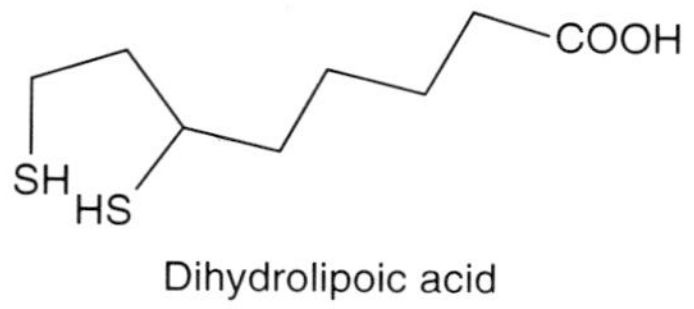

FIGURE 4.2 Chemical structure of dihydrolipoic acid.

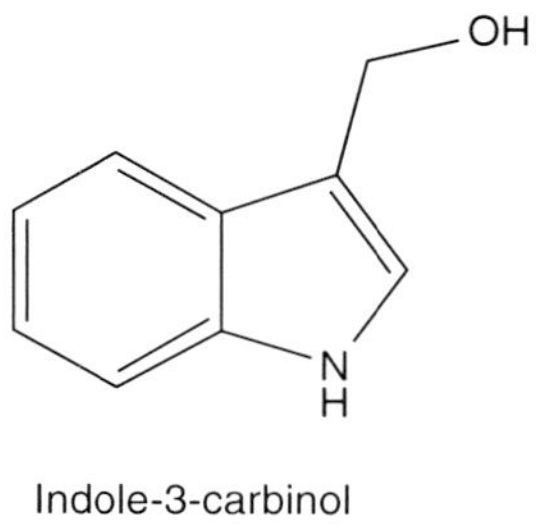

FIGURE 4.3 Molecular structure of indole-3-carbinol.

donating groups such as $-OCH_3$ at fifth position and side chain acetylation in melatonin.

Another well-known indole derivative is the neurotransmitter serotonin (Figure 4.4), which displays a wide range of physiological actions including protection of biological tissues against radiation injury [20]. The mechanism for this protection is thought to include inhibition of excessive free-radical formation.

Melatonin (*N*-acetyl-5-methoxytryptamine) is also a well-known indole compound produced endogenously by the pineal gland, having the properties of potent antioxidative efficacy [21] and scavenging free radicals [20], that protects against DNA strand breaks and lipid peroxidation [21]. *In vivo*, it is an effective pharmacological agent in reducing oxidative damage under conditions in which excessive free-radical generation is believed to be involved. *In vitro*, the data related to the direct scavenging of highly toxic reactive oxygen and reactive nitrogen species by melatonin have been obtained. Melatonin was discovered to directly scavenge the highly toxic hydroxyl radicals ($^{\bullet}OH$). Additionally, melatonin in cell-free systems has been shown to directly scavenge H_2O_2, singlet oxygen ($^1O_2$), and nitric oxide ($NO^{\bullet}$), but little or no ability to scavenge the superoxide anion radical ($O_2^{\bullet-}$). *In vitro*, melatonin also

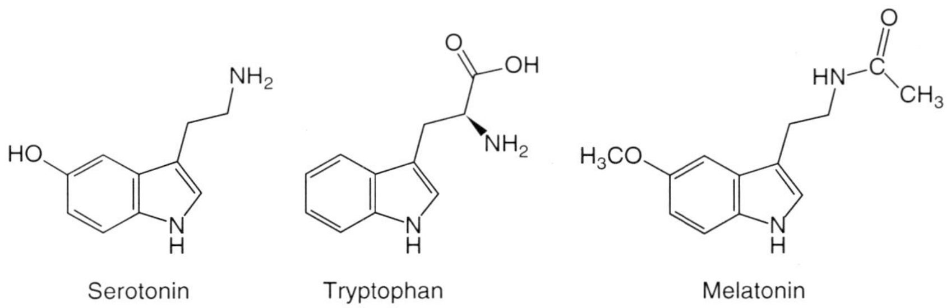

FIGURE 4.4 Molecular structures of serotonin, tryptophan, and melatonin.

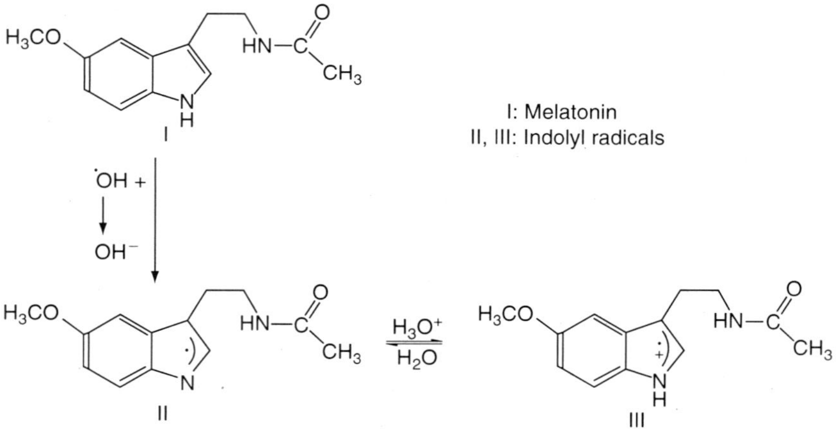

SCHEME 4.1 Interactions of melatonin with the hydroxyl radical. (From Reiter R.J., *Prog. Neurobiol.*, 56, 359, 1998. With permission.)

directly detoxifies the peroxynitrite anion ($ONOO^-$) or peroxynitrous acid (ONOOH) or both. The ability of melatonin to scavenge the lipid peroxyl radical ($LOO^•$) is debated. The evidence is that melatonin is probably not a classic chain-breaking antioxidant, because its ability to scavenge the $LOO^•$ seems weak. Its ability to reduce lipid peroxidation may stem from its function as a preventive antioxidant (scavenging initiating radicals), or yet unidentified actions [22].

Melatonin is believed to detoxify these toxic radicals by electron donation. In doing so, it becomes a radical, indolyl cation (or melatonyl) radical, of which there may be several isoforms (Scheme 4.1) and each melatonin molecule apparently has the capability to scavenge two radicals [21,22].

In the process, melatonin initially donates an electron to a $^•OH$ and it becomes the melatonyl radical, which has low reactivity. The melatonyl radical then removes a second $^•OH$ to produce cyclic 3-hydroxymelatonin which is transported to the kidney where it is excreted into the urine. It was confirmed that the loss of H_2O_2 in the presence of melatonin in a purely chemical system and identified one of the resulting products as N^1-acetyl-N^2-formyl-5-methoxykynuramine (AFMK), which is a potent scavenger. Thus, when melatonin interacts with H_2O_2 it generates a molecule, AFMK, which like the parent compound, melatonin, neutralizes the $^•OH$. Additionally, AFMK may be removed from cells by catalase; the resulting product, N^1-acetyl-5-methoxy-kynuramine (AMK), can also be found in the urine (Scheme 4.2). This effect presumably contributes to the high efficacy of melatonin in protecting against free-radical damage [23].

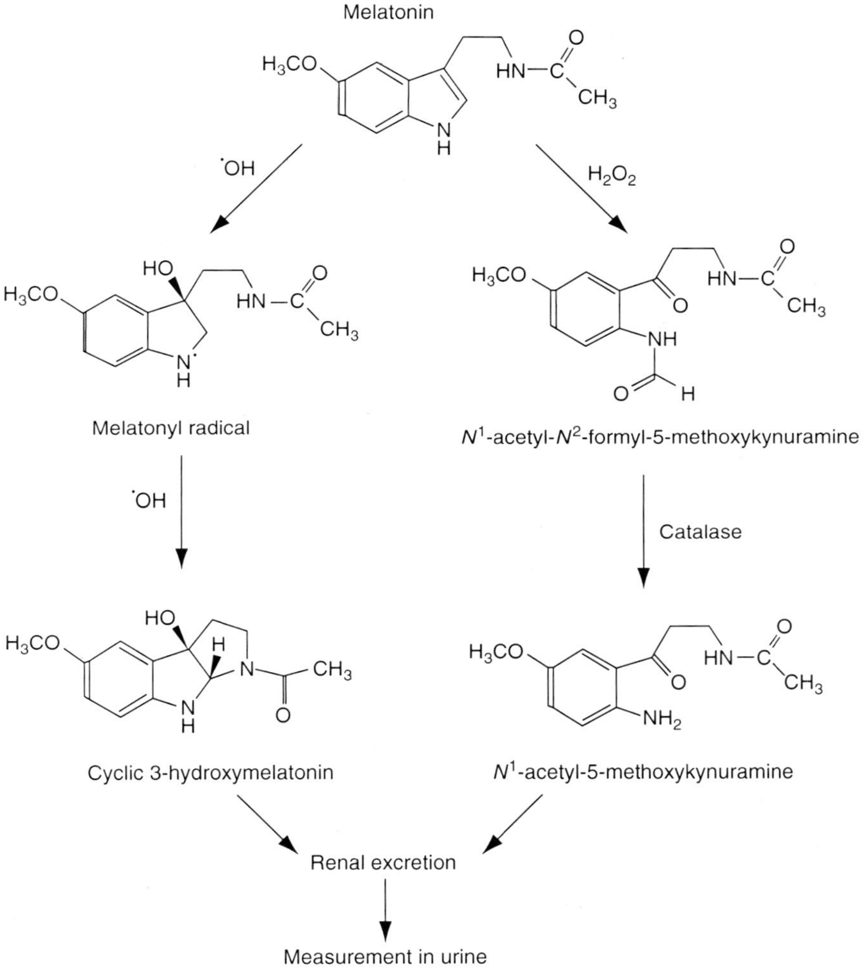

SCHEME 4.2 The interactions of melatonin with the $^{\bullet}OH$ (left) and with H_2O_2 (right). (From Reiter, J., Burkhardt, S., Cabrera, J., and Garcia, J.J., *Curr. Med. Chem.: Cent. Nerv. Syst. Agents*, 2, 45, 2002. With permission.)

Peroxynitrite ($ONOOH/ONOO^-$), which is produced by the reaction of superoxide anion with nitric oxide [24], is capable of hydroxylating and nitrating aromatic species. However, nitromelatonin is not found as a final product when melatonin was allowed to react with peroxynitrite. The melatonyl radical cation is also formed in the reaction of melatonin with peroxynitrite, providing direct evidence that peroxynitrite is capable of one-electron oxidation [25].

Besides the direct detoxification of reactive oxygen and nitrogen species by melatonin, it was also shown to have indirect antioxidative actions, through the stimulation of several antioxidative enzymes and the stabilization of membrane fluidity. In both *in vitro* and *in vivo* studies, melatonin reduced free-radical damage to lipids, proteins, and DNA. Because of the role of oxidative damage in disease processes, antioxidants may help to resist the development of various pathophysiologies [22].

Antioxidant compounds have been used to protect against radiation toxicity as irradiation produces toxic free radicals. The recent studies support that melatonin may be used as an anti-irradiation drug due to its potent free-radical scavenging and antioxidant activity. The antioxidative properties of melatonin resulting in its prophylactic property against acute radiation-induced biochemical and cellular alterations in the cerebellum were indicated [26]. It was suggested that supplementing cancer patients with adjuvant therapy of melatonin may be beneficial in alleviating the genitourinary complications because melatonin ameliorates radiation-induced injury to these organs [27]. The findings indicate the antioxidative properties of melatonin against the gamma radiation. Pretreatment with the lower concentration of melatonin affords potential protective effect against radiation-induced oxidative stress and mortality [28]. However, in a study it was demonstrated the application of melatonin conferred no protection against immune suppression or sunburn when applied topically to human skin immediately after irradiation. Melatonin may prove more effective following oral administration [29].

On the other hand, supplementation with thiols causes radioprotection, and depletion of thiols causes UV sensitization. Results of the studies on radioprotection by lipoic acid indicate beneficial biological effects. The protective effect of lipoic acid against radiation injury to mice hematopoietic tissues was showed [30]. It was also showed that lipoic acid protects the eye from the damaging effect of ultraviolet radiation [31], prevents radiation damage, and normalizes organ function after radiation exposure [32].

The basis of the approaches indicated above led to the consideration of the integration study of α-LA with melatonin-type compounds which are supposed to possess relatively potent antioxidant properties. Regarding through this consideration, we synthesized novel indole-lipoic acid derivatives which were found to have certain antioxidant activities when compared to α-LA which has been shown to have substantial multifunctional free-radical scavenging properties. The antioxidant properties of target compounds were investigated using rat liver microsomal, NADPH-dependent lipid peroxidation inhibition [33–35]. Some of the target compounds especially those containing amide linker at position 5 of indole ring proved to be highly effective in inhibiting lipid peroxidation as compared to α-LA [36].

Subsequently, some of the melatonin–lipoic acid derivatives which were synthesized in our laboratory were also applied to mice followed by exposing total body radiation. The results indicate that the compounds significantly protect tissues from radiation injury as well as extending the survival time of the mice [37]. One of these compounds (III-5b) recently synthesized by Fujita and Yokoyoma [38] and also in our laboratory [36] was also reported by Venkatachalam for its *in vitro* radioprotection activity [39].

CHEMISTRY AND RESULTS

Several novel α-LA/indole derivatives were synthesized [36]. Commercially available α-LA (racemic mixture) and appropriate indole compounds served as starting materials. Fifteen α-LA derivatives were designed and synthesized, composing of three series bearing substituents at positions 1, 3, and 5 of the indole ring. Conjugation of the 2 moieties was through amide bond which is between the carboxyl group of α-LA and amine group on position 1, 3, or 5 of indole ring (Table 4.1). Fourteen out of fifteen conjugates are novel α-LA compounds. Designed compounds as 3 groups are shown in Schemes 4.3 through 4.5 [36]. All synthesized compounds were purified with column chromatography and their melting points were determined and uncorrected. The structures of all synthesized compounds were assigned on the basis of IR, ^{1}H-NMR, and mass spectra (ES$^+$) analyses.

The target compounds were evaluated for antioxidant properties using *in vitro* nonenzymatic lipid peroxidation of rat hepatic microsomal membranes by measurement of the formation of 2-thiobarbituric acid (TBA) reactive substances. Lipid peroxidation consists of a radical-initiated reaction that serves for the evaluation of the antioxidant properties of a compound. The compounds exhibited significant lipid peroxidation inhibitory effects at the concentration of 1.0 mM

TABLE 4.1
Physicochemical Properties of Synthesized Compounds as Group First, Second, Third, and Their Inhibition of Lipid Peroxidation (LPO)

Compound (10^{-4}M)	R	R_f^a	Yield, %	mp,°C	LPO[b] % inh.
I-4a	H	0.62	62	96–98	82.1
I-4b	CH_3	0.68	36	116–117	60.4
I-4c	C_2H_5	0.76	9	93–95	77.4
I-4d	*i*-C_3H_7	0.82	52	130	88.7
I-4e	*n*-C_4H_9	0.89	69	65–66	78.1
I-4f	Bn	0.85	61	105	14.7
I-4g	4-F-Bn	0.86	51	111	90.7
I-4h	2,4-diCl-Bn	0.94	29	69–71	64.6
II-3a	H	0.88	32	105–107	89.6
II-3b	OCH_3	0.84	25	88–90	78.2
II-3c	Br	0.82	27	118	79.3
II-3d	NO_2	0.54	21	124–126	16.1
II-3e	CH_3	0.83	50	43–44	63.4
III-5a	H	0.16	54	58–60	83.3
III-5b	OCH_3	0.16	38	oil	75.7

Source: From Gurkan, A.S., Karabay, A., Buyukbingol, Z., Adejare, A., and Buyukbingol, E., *Arch. Pharm. Chem. Life Sci.*, 338, 67, 2005. With permission.

[a] Ethylacetate/*n*-Hexane (1:1).

[b] α-LA; LPO inh.: 45%.

with exception of compounds I-4f and II-3d, which showed 14.7% and 16.1% inhibition, respectively.

The conjugation of α-LA to indole moiety resulted in substitution patterns which were classified in three groups. For the first series of compounds (Scheme 4.3), substitution at the amine of 5-nitroindole followed by reduction generated amine intermediate. Conjugation of this with α-LA through amide bond formation was then conducted. The second series (Scheme 4.4) involved introduction of amine at position 1 followed by conjugation. The third series (Scheme 4.5) involved introduction of ethylamine at position 3 via formylation and methyl nitrite followed by conjugation to the nitrogen.

Synthesis of Compounds I-4 (a-h) Comprising the First Group

Designed compounds bearing substituent at position 1 were synthesized in three steps (Scheme 4.3) and involved amidation of 5-amino indoles with α-LA. The substitution pattern for this group lies on the indole nitrogen where alkyl homologues and various benzyl substituents were employed. Synthesis of 1-substituted-5-nitro-1*H*-indole derivatives I-2(b-h) were obtained according to the procedure described by Mor et al. [40], starting from I-1 and using an alkylating agent. Catalytic hydrogenation of the aromatic nitro group in ethanol was conducted at room temperature and under a pressure of 35 psi with 10% Pd/C as catalyst. The compounds (I-3 (c-h)) obtained as brown oils were used *in situ* and without purification in the next step. The amide compounds I-4 (b-h) (Table 4.1) were prepared from α-LA and appropriate amine as illustrated above after activation of *N,N'*-carbonyldiimidazole (*N,N'*-CDI) according to the method described by Fujita and Yokoyoma [38]. The same method was used to synthesize the compound I-4a starting from α-LA and 5-amino-indole.

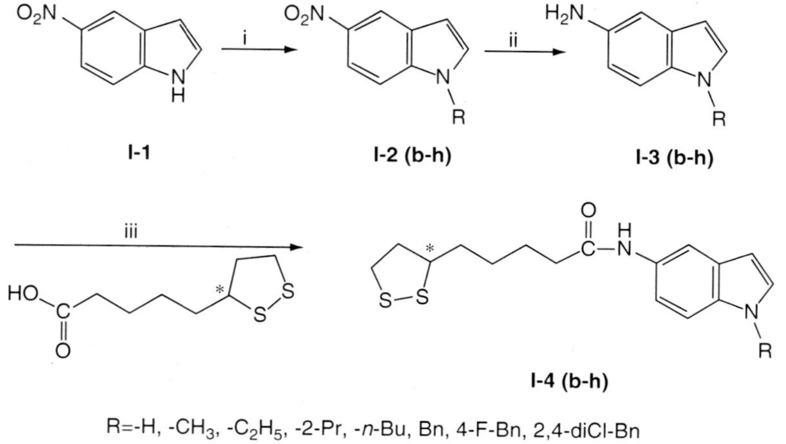

SCHEME 4.3 Synthesis of compounds I-4 (b-h) comprising the first group (i) alkyl/benzyl/substituted benzyl halogenid, NaH, DMF, r.t.; (ii) 10% Pd/C, H_2;(iii) *N,N'*-CDI, DMF, r.t. (From Gurkan, A.S., Karabay, A., Buyukbingol, Z., Adejare, A., and Buyukbingol, E., *Arch. Pharm. Chem. Life Sci.*, 338, 67, 2005. With permission.)

The most active compound within the first group was a 4-fluorobenzyl derivative (compound I-4g), which exerted lipid peroxidation (LPO) inhibitory activity as 90.7% inhibition. However, inconsistency was found with the compound I-4f which exhibited 14.7% inhibition. This compound has a bare benzyl substitution and the activity observed was significantly less than the fluorobenzyl compound. This difference in terms of inhibitory levels for two similar substitution patterns generated some interest and assays were repeated several times. However, the difference seems to be due to the significance of fluoro group being located at the para position of the benzyl ring. The 2,4-dichlorobenzyl derivative in this series also gave moderate inhibitory activity with confirmation of the important role of halogen substitution on the benzyl ring. A feasible approach to continuing studies in this area might be rationalized based on the difference between the halogens, including size and electronegativity, and the location. Removal of alkyl substitution (except I-4d) from the indole nitrogen resulted in a slight drop in the inhibitory activity. The i-propyl derivative (I-4d) might contain the optimal branching feature for the alkyl substitution. In order to perceive the activity of α-LA conjugation to indole moiety at first (1) position, we synthesized α-LA derivatives of indole in which the indole nitrogen was maintained to establish the α-LA integration while at position 5 of indole moiety was further modified by substituting electron donating (OCH_3, Br, CH_3) and electron withdrawing (NO_2) groups in order to optimize the electronic and lipophilic characteristics. The amino linker at the position 1 of indole ring was required for the conjugation of α-LA, due to the failure of direct bonding of α-LA to indole nitrogen.

Synthesis of Compounds II-3 (a-e) Comprising the Second Group

The reaction depicted in Scheme 4.4 was used for preparation of the second series of α-LA/indole conjugates bearing substituent at position 5. Synthesis of 5-substituted-1*H*-indole-1-amine derivatives II-2 (a-e) were achieved following the procedure described by Somei and Natsume [41], starting from 5-substituted-1*H*-indoles II-1 (a-e) and using hydroxylamine-*O*-sulfonic acid (HOSA) as an amination agent. The amide compounds II-3 (a-e) (Table 4.1) were prepared from α-LA and appropriate amines after activation of *N,N′*-CDI in accordance with the procedure described in the literature [38].

In this group, all compounds showed good inhibitory activity except compound II-3d (16.1% LPO inhibition) which bears a nitro substituent at position 5. The nonsubstituted derivative, II-3a, showed good activity (89.6%), higher than those with electron-donating substituents. These data illustrate that the presence of substituents might not be necessary to obtain good potency while electron-withdrawing substitution might be involved in significant alteration of the electron density of the indole ring to cause major loss in LPO inhibition activity. However, further studies are needed to elucidate the impact of such modifications or the influence of such substitution patterns for α-LA/indole compounds on LPO inhibitory activity.

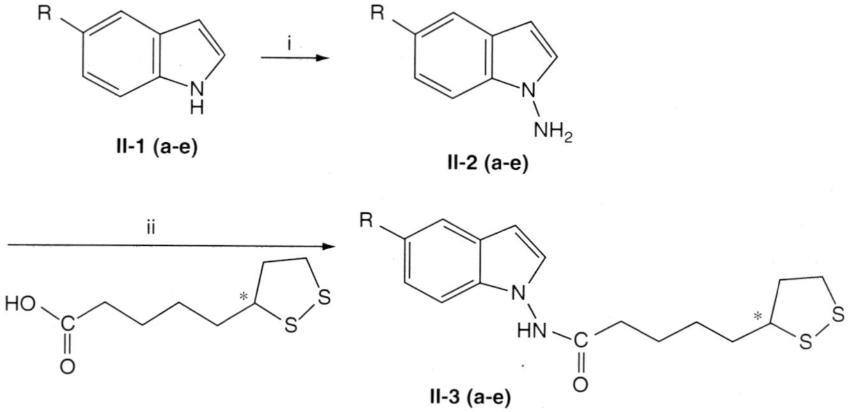

R= -H, -OCH3, -Br, -NO2, -CH3

SCHEME 4.4 Synthesis of compounds II-3 (a-e) comprising the second group (i) H-*O*-SA, KOH, DMF, 0°C; (ii) *N,N'*-CDI, DMF, r.t. (From Gurkan, A.S., Karabay, A., Buyukbingol, Z., Adejare, A., and Buyukbingol, E., *Arch. Pharm. Chem. Life Sci.*, 338, 67, 2005. With permission.)

Synthesis of Compounds III-5 (a-b) Comprising the Third Group

Designed compounds bearing substituent at position 5 were synthesized in four steps (Scheme 4.5). By treating III-1 (a-b) with DMF and $POCl_3$ [40], the corresponding aldehydes III-2 (a-b) were obtained. These aldehydes were condensed

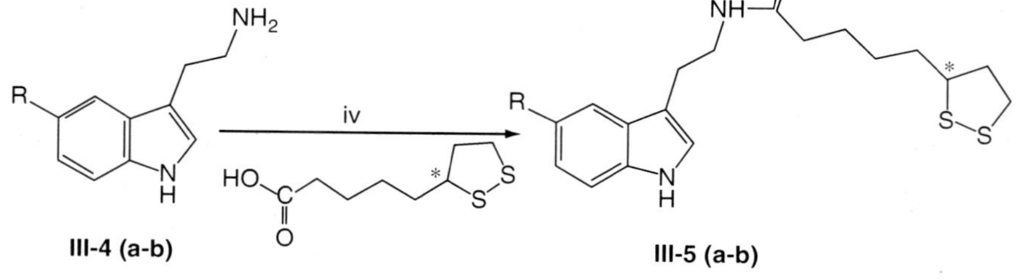

R = -H, -OCH3

SCHEME 4.5 Synthesis of compounds III-5 (a-b) comprising the third group (i) $POCl_3$, DMF, 0°C; (ii) CH_3NO_2, CH_3COONH_4, 70°C, reflux; (iii) $LiAlH_4$, THF, reflux; (iv) *N,N'*-CDI, DMF, r.t. (From Gurkan, A.S., Karabay, A., Buyukbingol, Z., Adejare, A., and Buyukbingol, E., *Arch. Pharm. Chem. Life Sci.*, 338, 67, 2005. With permission.)

with nitromethane to give III-3 (a-b) as described by Mor et al. [40]. 3–2′-Vinylindoles were reduced to the subsequent aminoethylindoles III-4 (a-b) by $LiAlH_4$ [42]. The amide compounds III-5 (a-b) (Table 4.1) were prepared from α-LA and aforementioned amine as described above after activation of *N,N′*-CDI [38].

The 5-substituted tryptamine conjugates synthesized as shown in Scheme 4.5 also exhibited greater than 75% inhibition.

The compound III-5b, a conjugate of melatonin and α-LA named melatoninolipoamide, was prepared using DCC mediated coupling. The conjugate has the ability to react with both oxidizing and reducing free radicals. On the basis of the pulse radiolysis studies of the conjugate carried out in aqueous solutions with these toxic radicals, the conjugate has two parts: the melatonin moiety and the lipoic acid moiety. The melatonin moiety of the conjugate reacts preferably with oxidizing radicals and the lipoic acid moiety exhibits preferential reaction with reducing radicals. While melatonin is less reactive toward hydrated electron, the hydrated electron preferentially attacks the lipoic acid part of the conjugate and reacts with the disulfide bond of α-LA-forming anion radical [39].

CONCLUSION

The most unique and important character of α-LA is its ability to reduce free radicals in both aqueous and lipid environments situations. This signifies the antioxidant activity of α-LA is effective in the extracellular fluid and also within the cell. It also has metal-chelating ability, helping the body rid itself of accumulated ingested toxins. Indeed, α-LA also protects against cholesterol oxidation and the consequent atherosclerosis in diabetics and also those in others at risk of cardiovascular diseases [1] as well as protecting chemically induced cataract formation related to diabetic complications which are resulted mostly aldose reductase enzyme induction [43,44].

Since both melatonin and α-LA are known to exhibit good radioprotecting ability and are excellent free-radical scavengers, it is expected that the conjugates derived from several combinations can be supposed to prove good radioprotecting ability, in which the rationally designed conjugates are anticipated to react with all types of free radicals, oxidizing and reducing radicals, formed during the radiation injury. The *in vitro* radioprotection ability of the compound was examined by γ-radiation-induced LPO in liposomes and hemolysis of erythrocytes, and compared the results with the parent compounds, melatonin and α-LA. It was found to be superior in protecting erythrocytes from radiation-induced hemolysis and is a moderate inhibitor of radiation-induced LPO. The studies suggest that the conjugate can be regarded as a promising radioprotector [39].

In conclusion, this study designed and synthesized α-LA/indole conjugates. These compounds possess good inhibitory activities on lipid peroxidation. These findings become significant in that α-LA is less active than some of the synthesized compounds. This could indicate a new strategy in designing α-LA derivatives with improved antioxidant capacity which could have an indirect beneficial effect on the human antioxidant defense system.

REFERENCES

1. L. Packer, E.H. Witt, and H.J. Tritschler. 1995. Alpha-lipoic acid as a biological antioxidant. *Free Radic. Biol. Med.*, 19, 227–250.
2. J.L. Evans and I.D. Goldfine. 2000. α-Lipoic acid: A multifunctional antioxidant that improves insulin sensitivity in patients with type 2 diabetes. *Diabetes Technol. Ther.*, 2, 401–413.
3. D. Voet and J.G. Voet. 1995. *Biochemistry*, Wiley, New York, 2nd edn., pp. 541ff and 736ff.
4. A. Berg and A. de Kok. 1997. 2-Oxo acid dehydrogenase multienzyme complexes. The central role of the lipoyl domain. *J. Biol. Chem.*, 378, 617–634.
5. A.A. Herbert and J.R. Guest. 1975. Lipoic acid content of *Escherichia coli* and other microorganisms. *Arch. Microbiol.*, 106, 259–266.
6. L.J. Reed. 1974. Multienzyme complexes. *Acc. Chem. Res.*, 7, 40–46.
7. L. Packer and H.J. Tritschler. 1996. Alpha-lipoic acid: The metabolic antioxidant. *Free Radic. Biol. Med.*, 20, 625–626.
8. F. Navari-Izzo, M.F. Quartacci, and C. Sgherri. 2002. Lipoic acid: A unique antioxidant in the detoxification of activated oxygen species. *Plant Physiol. Biochem.*, 40, 463–470.
9. A. Gruzman, A. Hidmi, J. Katzhendler, A. Haj-Yehie, and S. Sason. 2004. Synthesis and characterization of new and potent α-lipoic acid derivatives. *Bioorg. Med. Chem.*, 12, 1183–1190.
10. G. Biewenga, G.R.M.M. Haenen, and A. Bast. 1997. The pharmacology of the antioxidant lipoic acid. *Gen. Pharm.*, 29, 315–331.
11. A. Oganesian, J.D. Hendricks, and D.E. Williams. 1997. Long term dietary indole-3-carbinol inhibits diethylnitrosamine initiated hepatocarcinogenesis in the infant mouse model. *Cancer Lett.*, 118, 87–94.
12. C.Y. Zhu and S. Loft. 2003. Effect of chemopreventive compounds from Brassica vegetables on NAD(P)H:Quinone reductase and induction of DNA strand breaks in murine hepa1c1c7 cells. *Food Chem. Toxicol.*, 41, 455–462.
13. C.W. Nho and E. Jeffery. 2004. Crambene, a bioactive nitrile derived from glucosinolate hydrolysis, acts via the antioxidant response element to upregulate quinone reductase alone or synergistically with indole-3-carbinol. *Toxicol. Appl. Pharmacol.*, 198, 40–48.
14. K. Niva, K. Tagami, Z. Lian, J. Gao, Y. Wu, K. Onogi, H. Mori, and T. Tamaya. 2004. Preventive effects of indole-3-carbinol on endometrial carcinogenesis in mice. *Jpn. J. Reprod. Endocrinol.*, 9, 61–65.
15. H.H. Garcia, G.A. Brar, D.H.H. Nguyen, L.F. Bjeldanes, and G.L. Firestone. 2005. Indole-3-carbinol (I3C) inhibits cyclin-dependent kinase-2 function in human breast cancer cells by regulating the size distribution, associated cyclin E forms, and subcellular localization of the CDK2 protein complex. *J. Biol. Chem.*, 280, 8756–8764.
16. T. Brzozowski, P.C. Konturek, P.S.J. Konturek et al. 1997. The role of melatonin and L-tryptophan in prevention of acute gastric lesions induced by stress, ethanol, ischemia, and aspirin. *J. Pineal Res.*, 23, 79–89.
17. B. Moosmann and C. Behl. 2000. Cytoprotective antioxidant function of tyrosine and tryptophan residues in transmembrane proteins. *Eur. J. Biochem.*, 267, 5687–5692.
18. Y. Noda, A. Mori, R. Liburdy, and L. Packer. 1999. Melatonin and its precursors scavenge nitric oxide. *J. Pineal Res.*, 27, 159–163.
19. A.Z. Karabay. 2005. Personal communication. Ankara University, Department of Biochemistry, MS science thesis. (Results will be published soon.)

20. S. Stolc. 1999. Indole derivatives as neuroprotectants. *Life Sciences*, 65, 1943–1950.
21. R.J. Reiter. 1998. Oxidative damage in the central nervous system: Protection by melatonin. *Prog. Neurobiol.*, 56, 359–384.
22. R.J. Reiter, D. Tan, L.C. Manchester, and W. Qi. 2001. Biochemical reactivity of melatonin with reactive oxygen and nitrogen species. *Cell Biochem. Biophys.*, 34, 237–256.
23. J. Reiter, S. Burkhardt, J. Cabrera, and J.J. Garcia. 2002. Beneficial neurobiological effects of melatonin under conditions of increased oxidative stress. *Curr. Med. Chem.: Cent. Nerv. Syst. Agents*, 2, 45–58.
24. M. Saran, C. Michel, and W. Bors. 1990. Reaction of NO with $O_2^{-\bullet}$: Implications for the action of endothelium-releasing factor. *Free Radic. Res. Commun.*, 10, 221–226.
25. H. Zhang, G.L. Squadrito, and W.A. Pryor. 1998. The reaction of melatonin with peroxynitrite: Formation of melatonin radical cation and absence of stable nitrated products. *Biochem. Biophys. Res. Commun.*, 251, 83–87.
26. R. Sisodia, S. Kumari, R.K. Verma, and A.L. Bhatia. 2006. Prophylactic role of melatonin against radiation induced damage in mouse cerebellum with special reference to Purkinje cells. *J. Radiol. Prot.*, 26, 227–234.
27. G. Sener, B.M. Atasoy, Y. Ersoy, S. Arbak, M. Sengoz, and B.C. Yegen. 2004. Melatonin protects against ionizing radiation-induced oxidative damage in corpus cavernosum and urinary bladder in rats. *J. Pineal Res.*, 37, 241–246.
28. A.L. Bhatia and K. Manda. 2004. Study on pre-treatment of melatonin against radiation-induced oxidative stress in mice. *Environ. Toxicol. Pharmacol.*, 18, 13–20.
29. R.A. Howes, G.M. Halliday, and D.L. Damian. 2006. Effect of topical melatonin on ultraviolet radiation-induced suppression of Mantoux reactions in humans. *Photodermatol Photoimmunol Photomed*, 22, 267–269.
30. N. Ramakrishnan, W.W. Wolfe, and G.N. Catravas. 1992. Radioprotection of hematopoietic tissues in mice by lipoic acid. *Radiat. Res.*, 130, 360–365.
31. U. Demir, T. Demir, and N. Ilhan. 2005. The protective effect of alpha-lipoic acid against oxidative damage in rabbit conjunctiva and cornea exposed to ultraviolet radiation. *Ophthalmologica*, 219, 49–53.
32. L.G. Korkina, I.B. Afanas'ef, and A.T. Diplock. 1993. Antioxidant therapy in children affected by irradiation from the Chernobyl nuclear accident. *Biochem. Soc. Trans.*, 21, 314S.
33. M. Tuncbilek, G. Ayhan-Kilcigil, R. Ertan, B. Can-Eke, and M. Iscan. 2000. Synthesis and antioxidant properties of some new flavone derivatives in rat liver. *Pharmazie*, 55, 359–361.
34. M. Iscan, E. Arinc, N. Vural, and M.Y. Iscan. 1984. *In vivo* effects of 3MC, PB, pyrethrum and 2,4,5-T isooctylester on liver, lung and kidney microsomal mixed-function oxidase system of guinea-pig: A comparative study. *Comp. Biochem. Physiol.*, 77C, 177–190.
35. O.H. Lowry, N.J. Rosebrough, A.L. Farr, and R.J. Randall. 1951. Protein measurement with the Folin phenol reagent. *J. Biol. Chem.*, 193, 265–275.
36. A.S. Gurkan, A. Karabay, Z. Buyukbingol, A. Adejare, and E. Buyukbingol. 2005. Syntheses of novel indole lipoic acid derivatives and their antioxidant effects on lipid peroxidation. *Arch. Pharm. Chem. Life Sci.*, 338, 67–73.
37. Z. Alagoz, E. Buyukbingol, and A.S. Gurkan. 2006. Unpublished results.
38. T. Fujita and T. Yokoyoma. 1998. Dithiolane derivatives, their preparation and their therapeutic effect. EP0869126.

39. S.R. Venkatachalam, A. Salaskar, A. Chattopadhyay, A. Barik, B. Mishra, R. Gangabhagirathic, and K.I. Priyadarsinib. 2006. Synthesis, pulse radiolysis, and *in vitro* radioprotection studies of melatoninolipoamide, a novel conjugate of melatonin and α-lipoic acid. *Bioorg. Med. Chem.*, 14, 6414–6419.
40. M. Mor, S. Rivara, C. Silva, F. Bordi, P.V. Plazzi, G. Spadoni, G. Diamantini, C. Balsamini, and G. Tarzia. 1998. Melatonin receptor ligans: Synthesis of new melatonin derivatives and comprehensive comparatavite moleculer field analysis (CoMFA) study. *J. Med. Chem.*, 41, 3831–3844.
41. M. Somei and M. Natsume. 1974. 1-Aminoindoles. *Tetrahedron Lett.*, 15(5), 461–462.
42. R. Faust, P.J. Garratt, R. Jones, and L.K. Yeh. 2000. Mapping the melatonin receptor 6 Mèlatonin agonists and antagonists derived from 6H-Isoindolo[2,1-a]indoles, 5,6-Dihydroindolo [2,1-a] isoquinolines, and 6, 7-dihydro-5H-benzo[c]azepino[2,1-a] indoles. *J. Med. Chem.*, 43, 1050–1061.
43. F. Kilic, G.J. Handelman, E. Serbinova, L. Packer, and J.R. Trevithick. 1995. Modelling cortical cataractogenesis 17: *In vitro* effect of α-lipoic acid on glucose-induced lens membrane damage, a model of diabetic cataractogenesis. *Biochem. Mol. Bio. Int.*, 37, 361–370.
44. S. Suzen and E. Buyukbingol. 2003. Recent studies of aldose reductase enzyme inhibition for diabetic complications. *Curr. Med. Chem.*, 10, 1329–1352.

Section II

Metabolic Aspects

5 Alpha-Keto Acid Dehydrogenase Complexes and Glycine Cleavage System: Their Regulation and Involvement in Pathways of Carbohydrate, Protein, and Fat Metabolism

Robert A. Harris, Nam Ho Jeoung, Mandar Joshi, and Byounghoon Hwang

CONTENTS

REACTIONS CATALYZED BY MITOCHONDRIAL ENZYMES THAT REQUIRE LIPOIC ACID

Four enzyme complexes requiring lipoic acid are located in the mitochondrial matrix space of higher eurkaryotes. Three of them are α-keto acid dehydrogenase complexes that catalyze the same overall general reaction—the oxidative decarboxylation of an α-keto acid with the formation of CO_2, NADH, and an acyl-CoA:

$$RC(=O)-CO_2^- + NAD^+ + CoASH \longrightarrow RC(=O)-SCoA + NADH + CO_2$$

The fourth is a related enzyme system for the cleavage of glycine with the formation of CO_2, NH_4^+, NADH, and N^5, N^{10}-CH_2-tetrahydrofolate (THF)

$$NH_3^+CH_2CO_2^- + NAD^+ + THF \longrightarrow CO_2 + NH_4^+ + NADH + H^+ + N^5,N^{10}-CH_2-THF$$

The α-keto acid dehydrogenase complexes are composed of multiple copies of three enzymes that catalyze a total of five reactions (Figure 5.1). The enzymes are

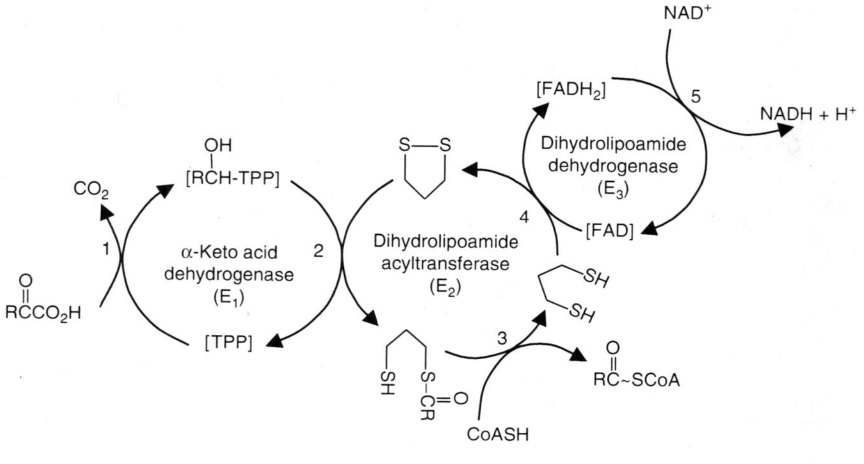

FIGURE 5.1 Reactions catalyzed by components of the α-keto acid dehydrogenase complexes. TPP, thiamine pyrophosphate; FAD, flavin adenine dinucleotide.

designated E_1, E_2, and E_3 in the order of the reactions they catalyze. The E_1 component, α-keto acid dehydrogenase (decarboxylase), uses thiamine pyrophosphate as its coenzyme for oxidative decarboxylation of an α-keto acid followed by reduction and acylation of lipoic acid covalently bound to the E_2 component, the dihydrolipoamide acyltransferase. E_2 transfers the acyl group from its lipoic acid moiety to CoA to form the acyl-CoA product. The reduced lipoyl group of E_2 is then used as a hydride ion source for the reduction of FAD bound to the E_3 component, the dihydrolipoamide dehydrogenase, which in turn binds and reduces NAD^+ to NADH. Specificity of the complexes for their respective α-keto acid substrates is conferred by the kinetic properties of the E_1 and E_2 components. All of the complexes bind the same E_3 component, which catalyzes the same reaction in each of them. Inner regions of the E_2 components associate noncovalently with each other to form 24 or 60 subunit structures that make up the central cores of the complexes. Outer regions of each E_2 subunit bearing one or two lipoyl groups extend outward from the surface of the central core to create swinging "super arms" that participate in catalysis of three of the five reactions carried out by the complex. Multiple copies of E_1 and E_3 associate noncovalently with specific binding domains on the super arms of E_2. The complexes function as large enzymatic machines (metabolons) in which the lipoyl groups of the swinging arms of the E_2 components interact with the catalytic sites of the E_1 and E_3 components of the complexes to carry out the overall reactions catalyzed by the complexes (Figure 5.1). The total molecular weights of the complexes vary because of differences in the number of E_1, E_2, and E_3 components and accessory proteins that associate with each complex.

α-Ketoglutarate Dehydrogenase Complex

The α-ketoglutarate dehydrogenase complex catalyzes the oxidative decarboxylation of α-ketoglutarate with the formation succinyl-CoA, NADH, and CO_2:

$$^{-}O-\overset{O}{\overset{\|}{C}}(CH_2)_2\overset{O}{\overset{\|}{C}}-CO_2^- + NAD^+ + CoASH \longrightarrow {}^{-}O-\overset{O}{\overset{\|}{C}}(CH_2)_2\overset{O}{\overset{\|}{C}}-SCoA + NADH + CO_2$$

The complex consists of a core of 24 dihydrolipoamide succinyltransferase (E_2) components to which 6 homodimeric E_1 components and 6 homodimeric E_3 components are attached. The E1 component of the α-ketoglutarate dehydrogenase complex is a homodimer (α_2 structure) with a subunit molecular weight of 113 kDa.

Pyruvate Dehydrogenase Complex

The pyruvate deydrogenase complex catalyzes the oxidative decarboxylation of pyruvate with the formation of acetyl-CoA, NADH, and CO_2:

$$CH_3\overset{O}{\overset{\|}{C}}-CO_2^- + NAD^+ + CoASH \longrightarrow CH_3\overset{O}{\overset{\|}{C}}-SCoA + NADH + CO_2$$

The mammalian pyruvate dehydrogenase complex is built around a 60 subunit inner core consisting of 48 tightly but noncovalently associated dihydrolipoamide acetyltransferase subunits (E_2) and 12 E_3 binding protein (E_3BP) components [1–4]. Each E_2 subunit has an inner region that forms the core and an outer region that extends as a swinging super arm with two lipoyl-bearing domains [1–4]. Each outer region consists of three self-folding domains connected in series by flexible linkers. The E_3BP is structurally related to E_2 but is not catalytically active and has only one lipoyl domain [5,6]. Sixty outer regions extend from a corresponding number of E_2 and E_3BP inner regions in the core. Pyruvate dehydrogenase (E_1), responsible for decarboxylation of pyruvate, is a heterotetrameric $\alpha_2\beta_2$ structure that binds noncovalently to an E_1-binding domain on E_2. A maximum of 30 E_1 tetramers can associate with each complex, although the number varies in different physiological conditions. Twelve E_3BPs anchor 12 E_3 homodimers. Dihydrolipoamide dehydrogenase (E_3), responsible for oxidation of the reduced lipoyl groups of E_2 and the subsequent production of NADH, binds to the complex by way of an E_3-binding protein (E_3BP). If the regulatory kinases and phosphatases are not included, each pyruvate dehydrogenase complex is composed of 216 subunits that sum up to about 9.5 million daltons, multiple copies of which exist in the matrix space of each mitochondrion. A small but uncertain number of pyruvate dehydrogenase kinase (PDK) molecules bind to each complex via the E_2 component. The kinase is believed to move hand over hand from lipoyl domain of one E_2 to lipoyl domain of an adjacent E_2 of the same complex to access sites of phosphorylation in the E_1 component [7]. Pyruvate dehydrogenase phosphatase (PDP) also binds to the complex, most likely via an anchor protein that binds it to the E_2 component.

Branched-Chain α-Keto Acid Dehydrogenase Complex

The branched-chain α-keto acid dehydrogenase complex catalyzes the oxidative decarboxylation of the branched-chain α-keto acids (α-ketoisocaproate, α-keto-β-methylvalerate, and α-ketoisovalerate) produced by transamination of the respective branched-chain amino acids (BCAAs) (leucine, isoleucine, and valine). The branched-chain α-keto acids are converted to isovaleryl-CoA, α-methylbutyryl-CoA, and isobutyryl-CoA, respectively with concurrent production of NADH and CO_2 by the following reactions:

$$CH_3CH_2CH(CH_3){-}\overset{O}{\overset{\|}{C}}{-}CO_2^- + NAD^+ + CoASH \longrightarrow CH_3CH_2CH(CH_3){-}\overset{O}{\overset{\|}{C}}{-}SCoA + NADH + CO_2$$

$$(CH_3)_2CHCH_2{-}\overset{O}{\overset{\|}{C}}{-}CO_2^- + NAD^+ + CoASH \longrightarrow (CH_3)_2CHCH_2{-}\overset{O}{\overset{\|}{C}}SCoA + NADH + CO_2$$

$$(CH_3)_2CH{-}\overset{O}{\overset{\|}{C}}{-}CO_2^- + NAD^+ + CoASH \longrightarrow (CH_3)_2CH{-}\overset{O}{\overset{\|}{C}}{-}SCoA + NADH + CO_2$$

The branched-chain α-keto acid dehydrogenase complex also decarboxylates α-ketobutyrate, the transamination product of threonine, and also γ-methylthio-α-ketobutyrate, the transamination product of methionine [8,9]:

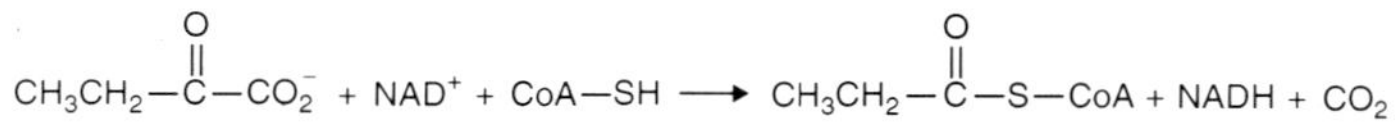

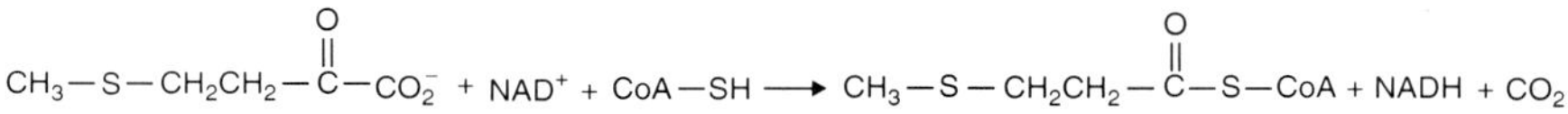

The decarboxylation of α-ketobutyrate by this reaction plays an important role in the catabolism of threonine [8,9]. Whether decarboxylation of γ-methylthio-α-ketobutyrate by this reaction is of physiological significance for the catabolism of methionine is uncertain [10].

The branched-chain α-keto acid dehydrogenase complex consists of a core of 24 dihydrolipoamide acyltransferase (E_2) to which 12 branched-chain α-keto acid dehydrogenase (E_1) components and 6 dihydrolipoamide dehydrogenase (E_3) components are attached to give a total molecular weight of 3.5 million daltons. The E_1 component of the branched-chain α-keto acid dehydrogenase complex is a heterotetramer ($\alpha_2\beta_2$ structure). An uncertain number of branched-chain α-keto acid dehydrogenase kinase molecules bind to each complex via the E_2 component. A phosphatase must also bind, but less is known about this enzyme, as discussed below.

A Hypothetical α-Ketoadipate Dehydrogenase Complex

Assuming it exists, the α-ketoadipate dehyrogenase complex catalyzes the oxidative decarboxylation of α-ketoadipate with the formation of glutaryl-CoA, NADH, and CO_2:

Although this reaction can be readily measured with isolated tissues, it remains uncertain whether a unique complex exists for this purpose. Indeed, the α-ketoglutarate dehydrogenase complex purified from mammalian tissues can use α-ketoadipate as a substrate [11], which is not surprising since α-ketoadipate differs from α-ketoglutarate by only one additional methylene carbon. The V_{max} of the complex for α-ketoglutarate is five times greater than for α-ketoadipate [11], consistent with the important role of the α-ketoglutarate dehydrogenase complex in the citric acid cycle. Nevertheless, a rare condition characterized by markedly elevated levels of α-ketoadipic acid, α-aminoadipic acid, and α-hydroxyadipic acid but surprisingly not α-ketoglutaric acid occurs in humans [12]. Fibroblasts grown from individuals with α-ketoadipic acidemia are almost completely void of α-ketoadipate dehydrogenase activity as measured by $^{14}CO_2$

production from [1-^{14}C]α-ketoadipate [13]. Reports that there are no significant metabolic or neurological consequences add to the puzzle [14]. Although it would seem that α-ketoadipic acidemia must be due to a genetic defect in a component of the α-ketoglutarate dehydrogenase complex, it is difficult to understand how the phenotype of individuals with a defect in an enzyme critically important for the citric acid cycle can be benign. The possibility that the α-ketoglutarate dehydrogenase complex harbors two E_1 components, one specific for the oxidation of α-ketoglutarate and the other specific for the oxidation of α-ketoadipate, has not been ruled out. Indeed, a sequence (NM_20025720) encoding a hypothetical protein with 40% identity to the E_1 component of the α-ketoglutarate dehydrogenase complex, including a well-conserved thiamine pyrophosphate binding domain, is present in the mammalian genome DNA databases (D.T. Johnson and R.A. Harris, unpublished observation). Preliminary experiments with recombinant protein have failed, however, to demonstrate enzyme activity with α-ketoadipate as substrate.

Glycine Cleavage System

Cleavage of glycine by the glycine cleavage system requires four enzyme components (P, H, T, and L) to catalyze the overall reaction (Figure 5.2). Although these enzymes likely associate to form a multimeric complex analogous to the α-keto acid dehydrogenase complexes, the interactions between the components are not retained during isolation and therefore must be weak. The only published stoichiometry is for the pea-leaf complex: 1 L-protein dimer, 2 P-protein dimers, 27 H-protein monomers, and 9 T-protein monomers [15], which seems an unlikely combination based on findings with the α-keto acid dehydrogenase

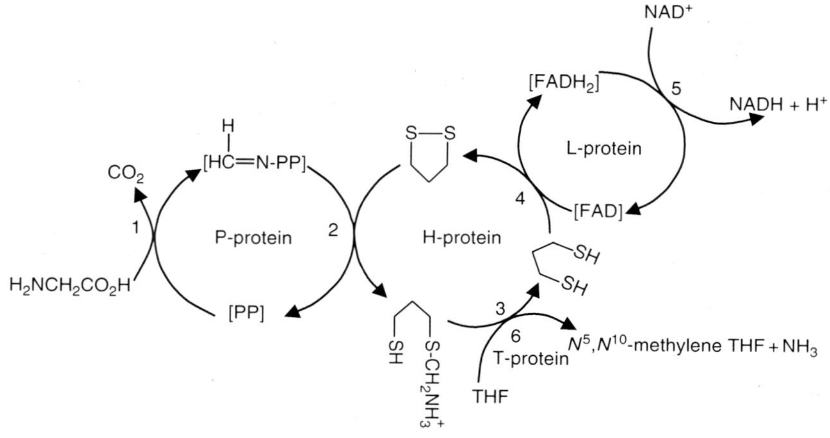

FIGURE 5.2 Reactions catalyzed by components of the glycine cleavage system. PP, pyridoxal phosphate; THF, tetrahydrofolate.

complexes. The glycine cleavage system uses pyridoxal phosphate, lipoic acid, FAD, NAD^+, and THF to carry out a redox reaction with formation of NADH and a one carbon THF derivative. The catalytic mechanism is similar in principle to that catalyzed by the α-keto acid dehydrogenase complexes (Figure 5.2), but involves six rather than five reactions. The P component of the system plays the same role as the E_1 component of the α-keto acid dehydrogenase complexes but uses pyridoxal phosphate rather than thiamine pyrophosphate for oxidative decarboxylation of glycine and the reductive aminomethylation of lipoic acid covalently attached to the H component. The H component plays the same role as the E_2 component of the α-keto acid dehydrogenase complexes but participates in the transfer of methylene groups from its reduced lipoate prosthetic group to THF to form N^5,N^{10}-CH_2-THF with concurrent release of ammonia. The latter reaction is catalyzed by the T component, an aminomethyltransferase enzyme unique to the glycine cleavage system. The L component, which corresponds to the E_3 component in the α-keto acid dehydrogenase complexes, uses the reduced lipoyl group of the H component as a hydride ion source for the reduction of NAD^+ to NADH. Although similar in many ways to the α-keto acid dehydrogenase complexes, the glycine cleavage system differs in its requirement for pyridoxal phosphate rather than thiamine pyrophosphate for the oxidative decarboxylation step, the involvement of an additional enzyme (T component) for the acyl transferase/deamination step, and the formation of methylene-THF rather than an acyl-CoA ester as its end product.

REGULATION OF MITOCHONDRIAL ENZYMES THAT REQUIRE LIPOIC ACID

α-Ketoglutarate Dehydrogenase Complex

Regulation by allosteric effectors. The α-ketoglutarate dehydrogenase complex is regulated by positive and negative allosteric effectors and feedback inhibitors (Figure 5.3). The effects of this diverse collection of compounds are all consistent

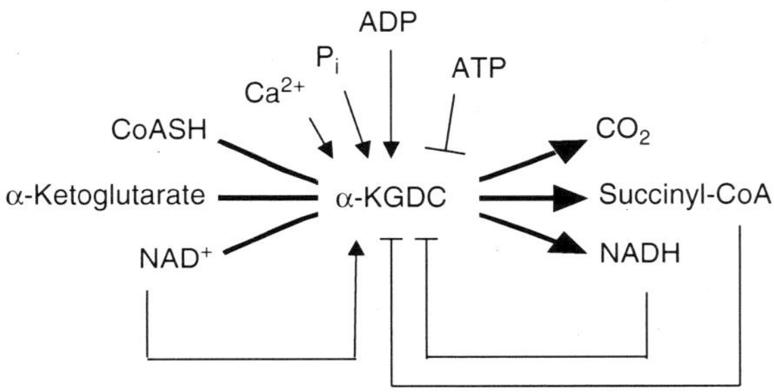

FIGURE 5.3 Regulation of α-ketoglutarate dehydrogenase complex (α-KGDC) by allosteric effectors.

with the role of the α-ketoglutarate dehydrogenase complex in the citric acid cycle. The positive allosteric effectors ADP, P_i, and Ca^{2+} decrease the K_m of the E_1 component for α-ketoglutarate; the negative allosteric effectors NADH and ATP have the opposite effect [16]. Feedback inhibitions exerted by NADH on E_3 and succinyl-CoA on E_2 are also important [16]. Long-chain acyl-CoA esters [17], and especially phytanoyl-CoA [18] produced from the phytol, the side chain of chlorophyll, are effective inhibitors of the complex, but it is uncertain whether the free concentrations of these compounds are ever high enough to be physiologically important.

Lack of regulation by phosphorylation. The primary sequence of the E_1 component of α-ketoglutarate dehydrogenase complex lacks serines at the sites subject to phosphorylation in the pyruvate and the branched-chain α-keto acid dehydrogenase complexes. The α-ketoglutarate dehydrogenase complex has therefore evolved in a manner that excludes regulation by phosphorylation, at least not by the same mechanism as the closely related pyruvate and branched-chain α-keto acid dehydrogenase complexes. This is surprising, considering what might be considered the hierarchy of importance of these complexes. However, as discussed below, complete inactivation of the pyruvate dehydrogenase complex by covalent modification is vital during starvation because it conserves three carbon compounds (lactate, pyruvate, and alanine) for glucose synthesis. Likewise, complete inactivation of the branched-chain α-keto acid dehydrogenase complex is necessary for the conservation of BCAAs for protein synthesis during periods of dietary protein deficiency. Lack of any condition under which it would be desirable to inhibit the citric acid cycle completely may explain why the α-ketoglutarate dehydrogenase complex is not regulated by phosphorylation. Indeed, it helps explain why the citric acid cycle is one of the few pathways of higher organisms that does not involve regulation of an enzyme by phosphorylation.

Pyruvate Dehydrogenase Complex

Regulation by allosteric effectors. The pyruvate dehydrogenase complex is regulated by feedback inhibition by acetyl-CoA and NADH (Figure 5.4). Acetyl-CoA inhibits the E_2 component by competing with CoA [19]; NADH inhibits E_3 by competing with NAD^+ [19]. The intramitochondrial NAD^+ redox state and acetyl-CoA/CoA ratio therefore directly impact pyruvate dehydrogenase complex activity. Although NADH and acetyl-CoA are the end products of reactions catalyzed by the pyruvate dehydrogenase complex, they are also produced by fatty acid oxidation. In starvation and diabetes, fatty acid oxidation is primarily responsible for producing the acetyl-CoA and NADH that inhibits flux through the pyruvate dehydrogenase complex.

Regulation by phosphorylation. The pyruvate dehydrogenase complex is also controlled by phosphorylation by kinases, which inactivate the complex, and by dephosphorylation by phosphatases, which activate the complex (Figure 5.4). Four pyruvate dehydrogenase kinase isoenzymes (PDK1, 2, 3, and 4) and two

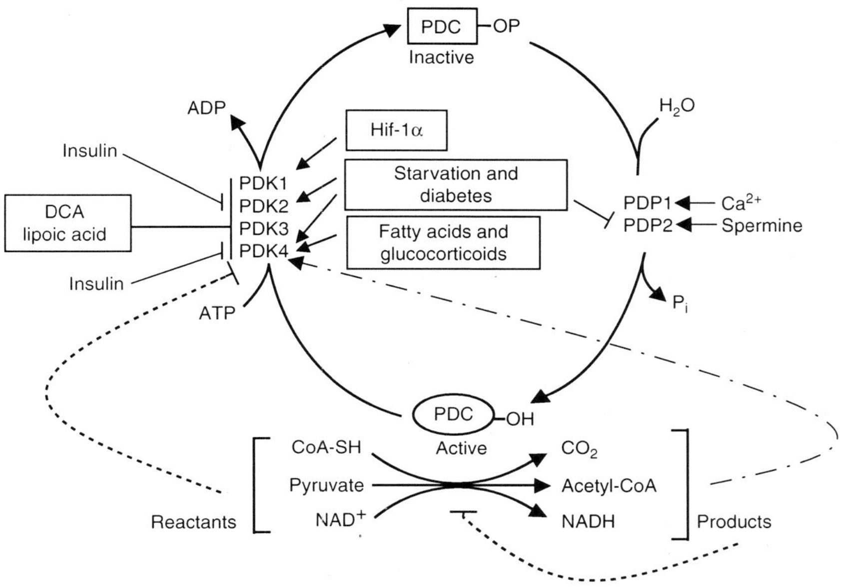

FIGURE 5.4 Regulation of pyruvate dehydrogenase complex (PDC) and its kinases (PDKs) and phosphatases (PDPs). DCA, dichloroacetate.

pyruvate dehydrogenase phosphatase isoenzymes (PDP1 and PDP2) are expressed in a tissue-specific manner. The primary sequences of the PDKs contain regions homologous to the bacterial histidine protein kinases [20] whereas the primary sequences of the PDPs place them in the PP2C family of phosphoprotein phosphatases [21], also found in bacteria. The PDKs and PDPs associate with the pyruvate dehydrogenase complex through the inner lipoyl-bearing domain of the E_2 super arm. Phosphorylation occurs on three serines in the $E_1\alpha$ subunit. Phosphorylation of site 1, corresponding to serine 264 in the $E_1\alpha$ subunit, almost completely inhibits the complex [22].

The pyruvate dehydrogenase complex does not exist in any tissue in its completely active, dephosphorylated state, nor in its completely inactive, phosphorylated state. The relative activities of the PDKs and the opposing PDPs determine the extent of phosphorylation and therefore the activity of the pyruvate dehydrogenase complex. Induced changes in the sum of the kinase activities relative to the sum of the phosphatase activities during transitions to different nutritional and hormonal states cause substantial change in the phosphorylation state and therefore activity of the pyruvate dehydrogenase complex.

Regulation of PDKs by allosteric effectors. The substrates and products of the reaction catalyzed by the pyruvate dehydrogenase complex are physiologically important effectors of the PDKs (Figure 5.4). The effects they exert upon the activity of the complex via their effects upon the PDKs occur almost instantaneously and make perfect physiological sense. Pyruvate, CoA, and NAD^+ are inhibitors of the PDKs; acetyl-CoA and NADH are activators [23,24]. The product of the kinase-catalyzed reaction, ADP, also inhibits the kinase by competing with the binding of ATP. Thus, the intramitochondrial NAD^+ redox state, acetyl-CoA/CoA ratio, and ATP/ADP ratio directly impact PDK activity and therefore indirectly pyruvate dehydrogenase complex activity. The sensitivities of the kinases to the effects of these compounds vary considerably, with the sensitivity of PDK2>PDK1>PDK4>PDK3 for activation by acetyl-CoA and NADH [25] and the sensitivity of PDK2>PDK4>PDK1>PDK3 for inhibition by the pyruvate analogue dichloroacetate [25,26]. PDK2 is therefore the most sensitive to regulation by these effectors; PDK3 the least sensitive. Fatty acid oxidation is the most important source of the acetyl-CoA and NADH for activation of the PDKs. The effects of acetyl-CoA and NADH upon the kinases are exerted through chemical modification of the lipoyl moieties where the kinases bind to the E_2 subunits [23,24]. A high NADH to NAD^+ ratio affects reduction of the lipoyl moieties of E_2 through a reversal of the reaction catalyzed by E_3. A concurrent high acetyl-CoA to CoA ratio promotes acetylation of the reduced lipoyl moiety. Binding of PDK to the reduced and acetylated lipoyl domain stimulates kinase activity [23,24], resulting in greater E_1 phosphorylation. NAD^+ inhibits the activation of the kinase by competing with the binding of NADH to E_3, thereby blocking reduction of the lipoyl groups of E_2. CoA likewise inhibits activation of the kinase by competing with acetyl-CoA for binding and acetylation of the lipoyl groups of E_2. Pyruvate mediates its effects by binding directly to the kinases. The pyruvate concentration fluctuates markedly in the effective inhibitory range in different nutritional states, making it a particularly important regulator of PDK activity. High concentrations of pyruvate produced when glucose is abundantly available promote its own oxidation by inhibiting PDK activity. Since PDK4 is less sensitive to pyruvate inhibition than PDK2, the shift to greater PDK4 expression during fasting and fat feeding decreases the effectiveness of pyruvate as an activator of the pyruvate dehydrogenase complex in these nutritional states [27].

Regulation of the PDKs by level of expression (Figure 5.4). This is an important mechanism for regulation of the activity of the pyruvate dehydrogenase complex in fasting, starvation, hibernation, heart disease, diabetes, and cancer. Induction of a stable increase in PDK activity in tissues of the rat by starvation and diabetes provided the first hint of this mechanism [28,29]. The discovery that four isoforms of PDK are expressed in tissues [20,30] led to the finding that starvation and diabetes increase the expression of PDK2 in liver and kidney and PDK4 in heart, skeletal muscle, kidney, lactating mammary gland, and liver [27,31–35]. It is now well established that increased expression of PDK2 and PDK4 contribute to the inactivation of the pyruvate dehydrogenase complex in starvation and diabetes. Reduced levels of insulin are largely responsible for the increases in both kinases, as evidenced by the effectiveness of insulin in decreasing their levels in diabetic animals [31]. In Pima Indians in whom type 2 diabetes and obesity are common, the level of muscle PDK4 mRNA is positively correlated with fasting insulin concentration and percentage of body fat, and negatively correlated with insulin-mediated glucose uptake rates [36]. Dexamethasone causes a rapid increase in PDK4 expression in several tissues of the rat [37], suggesting glucocorticoids are also important for induction of PDK4 expression during starvation and diabetes. Incubation of hepatoma cells [38] with dexamethasone causes a rapid increase in PDK4 expression as measured by changes in PDK4 message and protein. No effect on message stability is observed. The effect occurs at physiological concentrations of glucocorticoids and is inhibited by the glucocorticoid receptor antagonist RU486. The hyperglycemia characteristic of Cushing syndrome can be explained in part by the increase in PDK4 expression caused by the elevated blood levels of glucocorticoids. Insulin effectively blocks and reverses induction of PDK4 expression by dexamethasone in hepatoma cells [38]. The ability of insulin to dominate over glucocorticoids in this regard is similar to the negative effect that insulin has on glucocorticoid induction of the gluconeogenic enzyme phosphoenolpyruvate carboxykinase [39]. Euglycemic–hyperinsulinemic clamp studies suggest that insulin is the most important regulator of skeletal muscle PDK4 expression [40]. PDK4 is elevated in tissues of the diabetic-prone Otsuka Long-Evans Tokushima fatty rats [41] and insulin-resistant human subjects [42]. Improvement in insulin sensitivity has been shown to decrease expression of PDK4 mRNA in skeletal muscle of morbidly obese patients [42]. Hypoxia downregulates PDK4 expression in the heart [43], presumably to promote pyruvate oxidation and, therefore, maximize the efficiency of oxygen utilization by the heart in this condition.

PDK2 is constitutively expressed in hepatoma cells. Insulin very effectively downregulates PDK2 expression in hepatoma cells [38]. Whether this occurs by inhibition of transcription, reduced stability of its message, or both has not been established.

WY-14,643, a synthetic mimic of fatty acids and specific PPARα agonist, causes a rapid increase in PDK4 expression in several tissues [32]. WY-14,643 loses its effectiveness in PPARα-null mice [44], indicating that its effects are exerted via this receptor. GW0742, a specific PPARδ agonist, also effectively increases PDK4 expression in rodent and human myocytes [45,46]. Expression of PDK4 in response to starvation is partially attenuated in PPARα-null mice [44,47,48], suggesting that activation of the PPARα receptor is involved in upregulation of PDK4 in starvation. The increase in PDK4 that occurs in starved PPARα-null mice may be due in part to activation of the PPARδ receptor by endogenous agonists (free fatty acids). On the basis of the known activation of PPARα by free fatty acids and the increase in their blood levels that occur in response to starvation, diabetes, carnitine deficiency, hibernation, and high fat feeding, activation of both PPARα and PPARδ by free fatty acids is likely involved in upregulation of PDK4 expression [47,49,50]. Consistent with this idea, both WY-14,643 and free fatty acids are effective inducers of PDK4 expression in hepatoma cells [38]. Dexamethasone and WY-14,643 synergistically induce PDK4 expression, perhaps because glucocorticoids increase PPARα expression. Thyroid hormone [51] and bile acids [52] also increase PDK4 expression in liver.

PDK1 is constitutively expressed in some tumors [53] and upregulated by hypoxia [54]. Inactivation of the pyruvate dehydrogenase complex by PDK1 is important for the Warburg effect (aerobic glycolysis in the presence of oxygen), which is frequently observed in cancer cells [55], as discussed below.

A decreased level of PDK3 expression with concurrent greater activity of the pyruvate dehydrogenase complex activity occurs in the liver of knockout mice for the carbohydrate response element binding protein (ChREBP) (N.H. Jeoung, K. Uyeda, and R.A. Harris, unpublished observations). Why this occurs in mice lacking this glucose-sensitive transcription factor is not known.

Regulation of PDK4 gene expression in the human and rodents (Figure 5.5). The PDK4 gene is located on chromosome 7 in humans [56]. Transcription factors

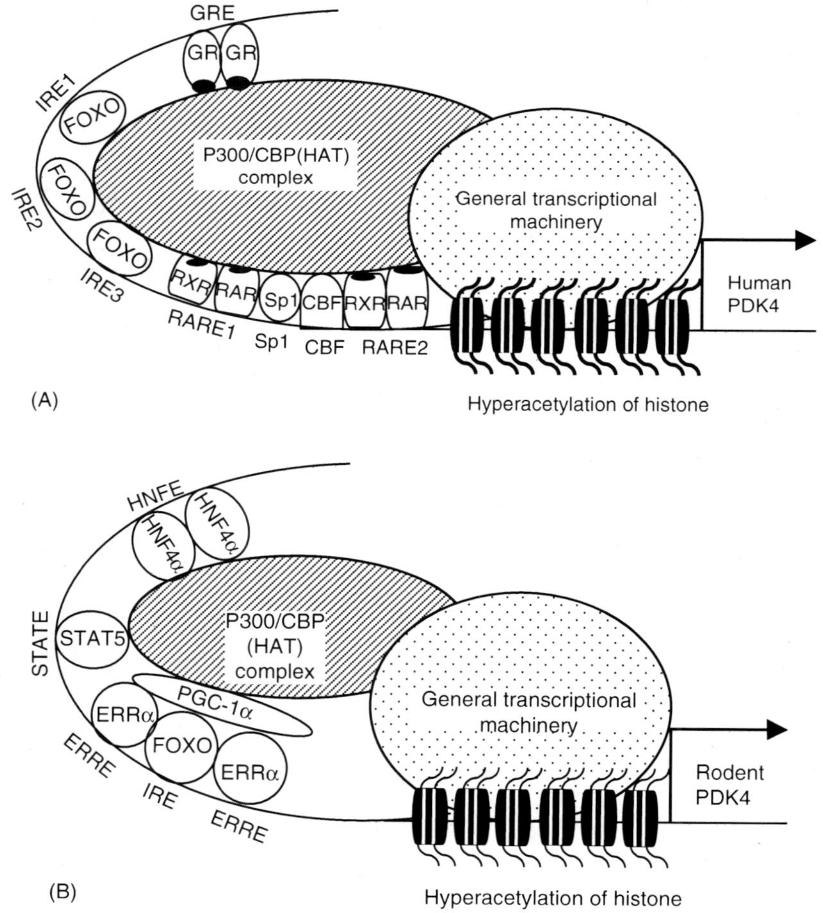

FIGURE 5.5 Regulation of PDK4 expression at the level of transcription. (A) Human PDK4 promoter, (B) rodent PDK4 promoter. HNFE, HNF4α response element; STATE, STAT5 response element; ERRE, ERRα response element.

implicated in its regulation include the glucocorticoid receptor (GR), FOXO, HNF4α, RARα, RXRα, and ERRα. Coactivators implicated include PGC-1α and p300/CBP. A glucocorticoid response element (GRE) is present in the −824/−809 promoter region of the human gene [57]. Dexamethasone activates reporter constructs in HepG2 cells that include this region of the promoter; glucocorticoid antagonist RU486 and mutations of the hypothetical GRE inhibit dexamethasone stimulation of PDK4 expression. Two retinoic acid response elements (RAREs) that bind retinoid X receptor α (RXRα) and retinoic acid receptor α (RARα) are also present in the human PDK4 promoter [58]. Three insulin response sequences (IRSs) are present in a region near the GRE in the human PDK4 gene [57]. Expression of FOXO1 and FOXO3a, transcription factors that interact with IRSs of other genes [59], increases PDK4 induction by dexamethasone in HepG2 cells. Mutation of the IRSs attenuates stimulation of PDK4 expression by dexamethasone. Constitutively active protein kinase B (PKB/Akt), a component of the insulin signaling pathway, represses stimulation of PDK4 expression by dexamethasone. Constitutively, active FOXO transcription factors that cannot be phosphorylated by PKB prevent PKB inhibition of the stimulatory effect of dexamethasone on PDK4 expression. On the basis of these findings, a model for human PDK4 gene regulation has been proposed in which activation of the GR by glucocorticoids recruits the p300/CBP coactivator complex that catalyzes histone acetylation [60]. RXRα/RARα and FOXO factors cooperate with GR and contribute to the formation of a stable complex that promotes transcription initiation (Figure 5.5A). Insulin inhibits transcription by activation of PKB, which phosphorylates the FOXO transcription factors, resulting in their export from the nucleus and destabilization of the transcription initiation complex on the PDK4 promoter.

The PDK4 genes for mouse and rat, located on chromosomes 6 and 4, respectively, are well conserved relative to each other [60]. Surprisingly, the rodent genes are only 40% identical in the first 4 kb to the sequence of the human gene. The GRE and the two RAREs identified in the human PDK4 gene are not conserved, and only one of the three IRSs of the human gene is present in the rodent genes (Figure 5.5B). A potential GRE is located 6–7 kb upstream in the rodent genes. Glucocorticoid receptor binds to this hypothetical GRE but it is not clear whether it is functional. The very rapid increase in PDK4 expression in rat hepatoma cells caused by dexamethasone [38] suggests a direct effect upon rodent PDK4 gene transcription but does not rule out indirect mechanisms. Increased expression of FOXO proteins by glucocorticoids [61] is involved in the long term. Indeed, induction of FOXO proteins in response to starvation or glucocorticoids correlates with induction of PDK4 gene with a 12 h delay in mouse skeletal muscle [61]. The IRS that contributes the most to PDK4 expression in the human promoter [57,60] is conserved in the rodent promoters [60]. Overexpression of constitutively active FOXO1 in C2C12 mouse myotubes increases expression of reporter constructs that include the intact IRS but not the mutated IRS, suggesting FOXO proteins promote transcription of the rodent PDK4 gene via this element. It is likely, therefore, that insulin inhibits expression of the rodent PDK4 gene by activating PKB-mediated phosphorylation of FOXO [61]. PGC-1α (peroxisome proliferators-activated receptor γ coactivator-1α) serves as coactivator for PDK4 transcription [62]. Upon binding to a transcription factor associated with a promoter, the conformation of PGC-1α changes to a form that recruits steroid receptor coactivator-1 (SRC-1) and CREB binding protein (CBP)/p300 to chromatin that acetylate histones, leading to transcriptional activation [63]. Overexpression of PGC-1α increases PDK4 message in primary rat hepatocytes and ventricular myocytes. The −578/−328 region of the rat PDK4 promoter is responsive to stimulation by PGC-1α [62]. PGC-1α is activated during starvation and interacts with an orphan member of the steroid/thyroid hormone receptor family, hepatic nuclear factor 4α (HNF4α), to induce the expression of phosphoenolpyruvate carboxykinase and glucose 6-phosphatase [64]. HNF4α also promotes PDK4 expression, but its effects may not be mediated by PGC-1α [62]. HNF4α binds to three regions of the rat PDK4 promoter, but its positive effects are lost only when the −1115/−1093 region is mutated. Estrogen-related receptor α (ERRα), orphan nuclear hormone receptor that promotes expression of fatty acid oxidation enzymes, induces PDK4 expression in skeletal muscle [65,66] and liver [67] via PGC-1α. A PGC-1α response region in mouse PDK4 promoter has been dissected and identified at −371/−308 [65]. A putative nuclear hormone receptor binding site (NR) exists within the PGC-1α response region. ERRα binds to this region (−327/−321) and promotes PDK4 gene expression in cooperation with PGC-1α. An additional ERRα binding region exists in the mouse PDK4 promoter at −359/−353 [66]. Cooperation between ERRα and PGC-1α has been confirmed with the rat PDK4 promoter. Insulin reduces the positive effects of ERRα and ERRα/PGC-1α on PDK4 expression [67].

The molecular mechanism responsible for the stimulation of PDK4 expression by WY-14,643 and fatty acids in cultured cells [38] and in vivo [32,44] has not been established. PDK4 promoter/reporter constructs are unresponsive to WY-14,643 and fatty acids. Rapid stimulation of transcription of the PDK4 gene by WY-14,643 in cultured cells favors the existence of one or more PPREs in the gene but none have been identified, suggesting that PPARα agonists may mediate their effects indirectly via other transcription factors.

Bile acids, long known to be important for cholesterol homeostasis, have recently been shown to also affect carbohydrate metabolism. The farnesoid X receptor (FXR), a nuclear receptor activated by bile acids, stimulates PDK4 as well as phosphoenolpyruvate carboxykinase expression [52,68]. This finding is complicated by the evidence that bile acids negatively regulate phosphoenolpyruvate carboxykinase expression by repressing the activities of HNF4α and FOXO1 [69].

Prolactin and growth hormone increase PDK4 message in 3T3-L1 adipocytes via STAT5 in an insulin-sensitive manner [70]. The binding region of STAT5 is located in the −389/−378 region of the mouse PDK4 promoter.

On the basis of these findings, a model for regulation of rodent gene expression can be proposed in which PGC-1α recruited by ERRα brings p300/CBP

(HAT) complex to the PDK4 promoter for histone acetylation (Figure 5.5B). HNF4α, FOXO, STAT5 also contribute to the formation of a stable transcription initiation complex.

Regulation of PDK2 gene expression. Starvation increases expression of PDK2 in liver and kidney but only modestly or not at all in other tissues. Nothing has been published on the PDK2 promoter. Downregulation of the PDK2 message by insulin in 7800C1 hepatoma cells may be due to decreased stability of the transcript rather than inhibition of transcription [38].

Regulation of PDK1 gene expression. Upregulation of PDK1 expression with concurrent inactivation of the pyruvate dehydrogenase complex is characteristic of some cancers [53]. This may be an important feature of the mechanism responsible for dominance of aerobic glycolysis (Warburg effect) in some tumors [71]. The mechanisms responsible for upregulation of PDK1 have been barely studied, but hypoxia inducible transcription factor-1α (HIF-1α), induced either constitutively or by hypoxia, is involved. HIF-1α responding elements are present in the promoter of the PDK1 gene [54,72].

Regulation of PDK3 gene expression. Although initial studies suggested expression of PDK3 is restricted to the testes [25], PDK3 is also expressed in skeletal muscle and heart [73], and very recent findings suggest a role for PDK3 in regulation of the pyruvate dehydrogenase complex in the liver (N.H. Jeoung, K. Uyeda, and R.A. Harris, unpublished observations).

Regulation of the phosphorylation state of the pyruvate dehydrogenase complex by the pyruvate dehydrogenase phosphatases. Two pyruvate dehydrogenase phosphatases (PDP1 and PDP2) are expressed in a tissue-specific manner in mammalian cells [21]. They are Mg^{2+}-dependent phosphatases of the PP2C phosphatase family, recently renamed the PPM1 phosphatase family. PDP1 is sensitive to stimulation by micromolar concentrations of Ca^{2+}. Activation of the pyruvate dehydrogenase complex during skeletal muscle contraction and by hormones that mobilize Ca^{2+} is believed due to activation of PDP1 by Ca^{2+}. PDP2 is closely related by sequence to PDP1 but its K_m for Mg^{2+} is much higher and Ca^{2+} has no effect upon its activity. PDP2 is stimulated by the polyamine spermine, which has no effect on PDP1. High expression in liver and adipose tissue [21] makes PDP2 a candidate target for the "mediator" of insulin action that activates the pyruvate dehydrogenase complex in these tissues [74]. Starvation and chemically induced diabetes decrease PDP2 in rat heart and kidney [75]. Refeeding and insulin treatment reverse these effects. Starvation also reduces PDP activity and the amount of PDP2 protein in skeletal muscle [75A].

Branched-Chain α-Keto Acid Dehydrogenase Complex

Regulation by allosteric effectors. The branched-chain α-keto acid dehydrogenase complex catalyzes the rate-limiting step in the catabolic pathways of the BCAAs [76]. Short-term control of the activity of the branched-chain α-keto acid dehydrogenase complex involves direct inhibition of the activity of the complex by

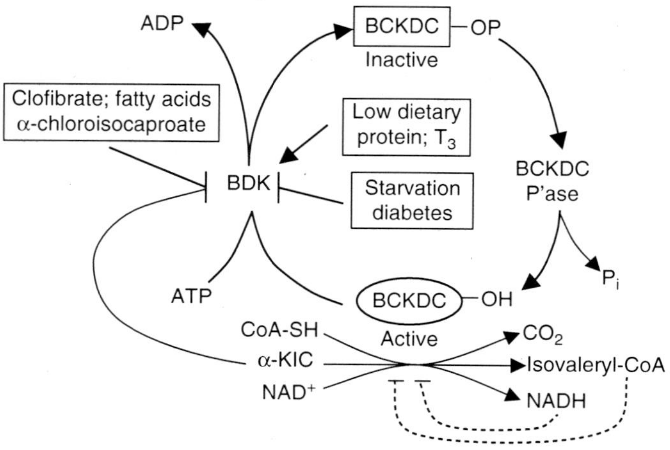

FIGURE 5.6 Regulation of branched-chain α-keto acid dehydrogenase complex (BCKDC) by phosphorylation and allosteric effectors. α-KIC, α-ketoisocaproate; BDK, branched-chain α-keto acid dehydrogenase kinase; P'ase, phosphatase; T_3, thyroid hormone.

NADH (competitive with NAD^+ at the level of E_3) and CoA esters derived from oxidation of BCAAs and fatty acids (competitive with acyl-CoA esters at the level of E_2) (Figure 5.6). Thus, like the other α-ketoacid complexes of this family of mitochondrial enzymes, the NAD^+-redox state and the acyl-CoA/CoA ratio influence the enzyme activity of the complex.

Regulation by phosphorylation. The branched-chain α-keto acid dehydrogenase complex is controlled by a kinase (BDK) that inactivates the complex and a phosphatase (BDP) that activates the complex. BDK phosphorylates the E1α subunit on serine 293 (site 1) and serine 303 (site 2) [77] (Figure 5.6). Phosphorylation at site 1 is physiologically more significant because it occurs more robustly and has a greater effect on the activity of the complex [77].

Regulation of BDK by allosteric effectors. α-Ketoisocaproate, the transamination product of leucine, is the most effective, naturally occurring inhibitor of BDK (Figure 5.6). Inhibition of BDK by this keto acid is believed to be the most important short-term control mechanism for the activity of the branched-chain α-keto acid dehydrogenase complex [78]. Structural analogues of α-ketoisocaproate, including octanoate [79], α-chloroisocaproate [80], and clofibric acid [79], also promote activation of the branched-chain α-keto acid dehydrogenase complex by inhibition of BDK. α-Keto-β-methylvalerate, transamination product of isoleucine, is less effective than α-ketoisocaproate, but may be physiologically important under some conditions. α-Ketoisovalerate, the transamination product of valine, is much less effective and not likely important. In contrast to the PDKs, convincing evidence has not been presented for regulation of BDK by NADH and CoA esters.

Regulation of BDK expression. Expression of BDK is sensitive to dietary protein content, starvation, female sex hormones, glucocorticoids, insulin, and PPARα agonists (Figure 5.6). Feeding rats a low-protein diet induces the expression of BDK message and protein in the liver [81]. The amount of BDK bound to the branched-chain α-keto acid dehydrogenase complex increases concurrently with the increase in the total amount of BDK [82], which in turn is inversely correlated with the activity of the branched-chain α-keto acid dehydrogenase complex. On the other hand, the expression of BDK as well as amount of BDK associated with branched-chain α-keto acid dehydrogenase complex is reduced in the liver of rats fed on a high-protein diet [81,82]. In female rats, the branched-chain α-keto acid dehydrogenase complex is mostly active in the beginning of the light cycle and mostly inactive at the end of the light cycle [83]. Changes in the amount of BDK associated with the branched-chain α-keto acid dehydrogenase complex at different times during the light cycle are responsible for these changes in activity, which are specific to females. Ovariectomy prevents these diurnal variations, suggesting that female sex hormones regulate the amount of BDK bound to the branched-chain α-keto acid dehydrogenase complex [84]. Starvation and diabetes increase the branched-chain α-keto acid dehydrogenase complex activity in liver by increasing the expression of its subunits [85] and reducing the expression of BDK [86] (Figure 5.6). Glucocorticoids activate the branched-chain α-keto acid dehydrogenase complex by inhibiting the expression of BDK [87]. Insulin inhibits the expression of branched-chain α-keto acid dehydrogenase complex subunits but increases the expression of BDK [88]. PPARα ligands such as clofibrate increase branched-chain α-keto acid dehydrogenase complex activity by increasing the expression of its subunits, decreasing the expression of BDK, and directly inhibiting the activity of BDK [89]. Free fatty acids, the physiological activators of PPARα, increase in starvation and diabetes. Free fatty acids along with glucocorticoids and low insulin may increase activity of the branched-chain α-keto acid dehydrogenase complex under such conditions. Thyroid hormone inhibits the branched-chain α-keto acid dehydrogenase complex activity in liver by upregulating expression of BDK [90]. Although this stands in contrast to the activating effect that thyroid hormone has on the basal metabolic rate, it may conserve BCAAs for protein synthesis in the hyperthyroid state.

Regulation of the phosphorylation state of the branched-chain α-keto acid dehydrogenase complex by the branched-chain α-keto acid dehydrogenase phosphatase. A protein phosphatase with specificity for the branched-chain α-keto acid dehydrogenase complex has been reported [91,92]. However, this enzyme has neither been cloned nor been definitively shown to be responsible for regulation of the branched-chain α-keto acid dehydrogenase complex. A third mitochondrial phosphatase, which belongs by sequence to the same phosphatase family (PPM1) as the mitochondrial pyruvate dehydrogenase phosphatases, has been cloned and partially characterized [93]. This enzyme dephosphorylates the branched-chain α-keto acid dehydrogenase complex [93], but its activity is low relative to the putative branched-chain α-keto acid dehydrogenase phosphatase previously isolated [92], making its role also uncertain. The difficulty in identifying the phosphatase for the branched-chain α-keto acid dehydrogenase complex has fueled speculation that such an enzyme does not exist. However, incubation of hepatocytes with the BDK inhibitor α-chloroisocaproate causes rapid and complete dephosphorylation of the branched-chain α-keto acid dehydrogenase complex [85]. Therefore, a phosphatase clearly exists for this complex—its identity just remains a mystery.

Regulation of the level of the subunits of the branched-chain α-keto acid dehydrogenase complex components. The subunits of the branched-chain α-keto acid dehydrogenase complex are subject to downregulation by low-protein feeding and insulin. Expression of E_2 and $E_1\beta$ generally show harmony between the mRNA and the protein levels under most but not all conditions. Expression of E_2 and $E_1\beta$ mRNA and protein is reduced in the liver of rats fed on protein-free diet. This effect can be reversed by switching rats to a chow diet with normal amounts of protein. $E_1\alpha$ protein levels show similar changes under these conditions but without changes in message levels. Likewise, E_2 and $E_1\beta$ mRNA and protein are increased whereas $E_1\alpha$ protein is increased without a change in message level in the liver of rats fed on a high-protein (60%) diet. Starvation, on the other hand, induces concurrent increases in the protein as well as mRNA levels of all the three subunits. Glucocorticoids induce the expression of E_2 mRNA in H4IIEC3 cells, suggesting that the increase in glucocorticoid levels during starvation may be responsible for the increase in E_2 levels. Expression of the E_2 subunit is inhibited by small inhibitory RNAs (miR), miR29a and miR29b [94]. Changes in the expression of miR29b may be responsible for alterations in the expression of E_2 mRNA and protein with variation in the protein content of the diet.

Destabilization of E_1 component by phosphorylation. Normally, the branched-chain α-keto acid dehydrogenase complex is only partially active in tissues because of extensive phosphorylation. However, in mice lacking the BDK protein ($BDK^{-/-}$ mice), the branched-chain α-keto acid dehydrogenase complex is completely dephosphorylated and therefore totally active. Surprisingly, the total activity of branched-chain α-keto acid dehydrogenase complex in $BDK^{-/-}$ mice is higher in most tissues than that of wild-type mice [95]. Since there is no physiological advantage, upregulation of the branched-chain α-keto acid dehydrogenase complex in the face of loss of control by phosphorylation is a paradox. A similar phenomenon occurs with rats treated with the BDK inhibitor clofibrate [89]. Clofibrate increases branched-chain α-keto acid dehydrogenase complex activity by inhibition of BDK and induction of E_1. Conversely, feeding rats on a low-protein diet not only causes branched-chain α-keto acid dehydrogenase complex to be highly phosphorylated but also reduces the amount of E_1 in the liver whereas feeding rats on a high-protein diet, causes the branched-chain α-keto acid dehydrogenase complex to be highly dephosphorylated and increases the amount of E_1 [96]. A difference in stability of phosphorylated E_1 compared to nonphosphorylated E_1 toward proteolytic degradation could explain these findings. The affinity of E_1 for the E_2 component of the branched-chain α-keto acid dehydrogenase complex is reduced by phosphorylation [97], perhaps making it more vulnerable to degradation by proteases. Thus, conditions that promote dephosphorylation may increase the stability and, therefore, increase the amount of E_1. In $BDK^{-/-}$ mice, E_1 is always dephosphorylated and may therefore be more stable, resulting in an increase in its amount.

A Hypothetical α-Ketoadipate Dehydrogenase Complex

Because the existence of this complex remains hypothetical, little can be said about its possible regulation. If this reaction is catalyzed by the α-ketoglutarate dehydrogenase complex, the factors that regulate the α-ketoglutarate dehydrogenase complex should also regulate α-ketoadipate oxidation. Glutaryl-CoA, the product of α-ketoadipate oxidation, should exert feedback inhibition of the α-ketoglutarate dehydrogenase complex, but this has not been documented.

Glycine Cleavage System

Regulation by allosteric effectors. The glycine cleavage system is sensitive to inhibition by NADH and NADPH [98] (Figure 5.7). End product inhibition by NADH is readily explained by competition with NAD^+ at the level of E_3. The basis and physiological reason for inhibition by NADPH have not been established.

Feeding rats a high-protein diet greatly increases flux through the glycine cleavage system as measured in intact liver mitochondria [99]. This finding fits with the observation that low dietary protein paradoxically increases blood glycine levels [100] while high dietary proteins causes a reduction [101]. Glucagon, which is elevated by high dietary protein, stimulates flux through the glycine cleavage system in isolated hepatocytes via a cAMP- and Ca^{2+}-mediated mechanism [102]. Although these effects have been known since the 1990s, how these messengers stimulate the glycine cleavage system remains an enigma. It cannot be explained by a change in the matrix space volume or an increase in glycine transport or a decrease in the redox states of the pyridine nucleotides. Although clearly mediated upon the glycine cleavage system, the stimulatory effects of

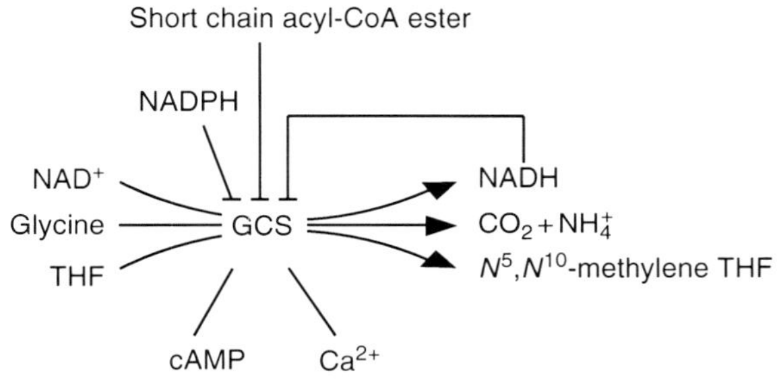

FIGURE 5.7 Regulation of glycine cleavage system (GCS) by allosteric effectors. THF, tetrahydrofolate.

these second messengers of glucagon action are lost upon disruption of the mitochondrial membrane system.

Short chain acyl-CoA esters derived from BCAA oxidation are inhibitors of the glycine cleavage system [103]. Enzymatic defects in the pathway for BCAA catabolism induce ketotic hyperglycinemia by the accumulation of short chain acyl-CoA esters or other intermediates [104].

Lack of regulation by phosphorylation. Although this would be an attractive mechanism for regulation of the glycine cleavage system, no evidence exists for phosphorylation of any component of the complex.

Regulation by level of expression of the subunits of the component enzymes. While it would be an attractive mechanism to explain long-term effects of different nutritional states and hormones on the capacity for glycine disposal, no evidence exists for regulation of the glycine cleavage system at the level of expression of its component enzymes.

PHYSIOLOGICAL ROLES OF MITOCHONDRIAL ENZYMES THAT REQUIRE LIPOIC ACID

α-Ketoglutarate Dehydrogenase Complex

One turn of the citric acid cycle results in the complete oxidation of the acetyl moiety of acetyl-CoA to two molecules of CO_2, three molecules of NADH, one molecule of $FADH_2$, and one molecule of ATP. Subsequent oxidation of the NADH and $FADH_2$ by the electron transport chain yields a total of nine molecules of ATP by oxidative phosphorylation. The α-ketoglutarate dehydrogenase complex is special among the enzymes required for the citric acid cycle because of what the reaction accomplishes and the importance of its regulation to citric acid cycle flux (Figure 5.8). The reaction not only produces NADH for the generation of ATP by oxidative phosphorylation but also the high-energy compound succinyl-CoA for the generation of GTP or ATP by substrate-level phosphorylation catalyzed by succinyl-CoA synthetase:

$$\text{Succinyl-CoA} + \text{GDP (ADP)} + P_i \rightarrow \text{Succinate} + \text{CoA} + \text{GTP (ATP)}$$

The sensitivity of the complex to the ratios of ADP to ATP and NAD^+ to NADH stimulates flux through the cycle when energy levels are low and reduces flux when energy levels are high. Moreover, the activation of the complex evoked by Ca^{2+} provides greater ATP production to meet the energy needs of muscle contraction and processes that are stimulated by Ca^{2+}-mobilizing hormones, e.g., hepatic gluconeogenesis [105].

Although most cells of the body employ a complete citric acid cycle to meet their energy needs, some cancer cells as well as some normal cells find it advantageous to use a truncated citric acid cycle. Such cells prefer to oxidize glutamine to alanine or aspartate by glutaminolysis [106], an ATP-yielding

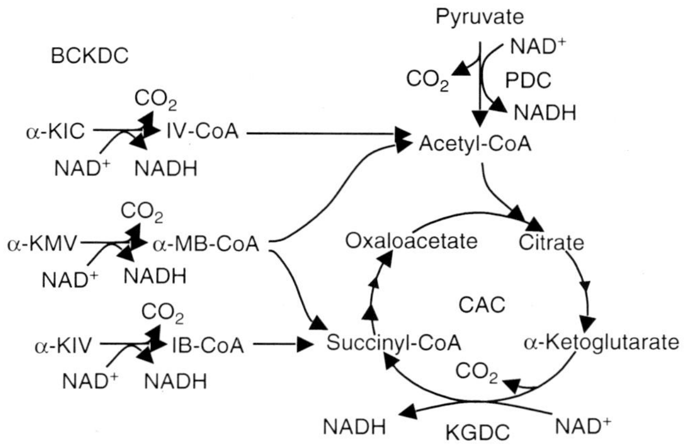

FIGURE 5.8 Physiological roles of α-keto acid dehydrogenase complexes. PDC, pyruvate dehydrogenase complex; KGDC, α-ketoglutarate dehydrogenase complex; BCKDC, branched-chain α-keto acid dehydrogenase complex; CAC, citric acid cycle; α-KIC, α-ketoisocaproate; α-KMV, α-keto-β-methylvalerate; α-ketoisovalerate; IV-CoA, isovaleryl-CoA; α-MB-CoA, α-methylbutyryl-CoA; α-ketoisovalerate; IB-CoA, isobutyryl-CoA.

pathway that relies upon the reaction catalyzed by the α-ketoglutarate dehydrogenase complex.

Loss of α-ketoglutarate dehydrogenase complex activity is particularly harmful to cells because of the importance of the reaction it catalyzes to ATP generation. The brain is most vulnerable because of its dependence upon glucose oxidation to meet its energy needs. Defects in the α-ketoglutarate dehydrogenase complex that accumulate with age may contribute to a number of neurodegenerative diseases [107,108]. Infantile lactic acidosis, Friedreich's, Alzheimer's, and Parkinson's may be caused or worsened by inactivation of the α-ketoglutarate dehydrogenase complex by ROS [107,108] produced by complexes I and III of the electron transport chain [109,110] and perhaps E_3 of the α-ketoacid dehydrogenase complexes [111–113]. Glutathionylation of sulfhydryl groups may be responsible for inactivation of the α-ketoglutarate dehydrogenase complex medicated by ROS [114].

Pyruvate Dehydrogenase Complex

The reaction catalyzed by the pyruvate dehydrogenase complex irreversibly binds the glycolytic pathway to the citric acid cycle (Figure 5.8). Since all cells can convert pyruvate to lactate and release lactate and since lactate can be converted back to glucose by gluconeogenesis in the liver, the pyruvate dehydrogenase complex occupies a key position in the pathway for glucose disposal. In most cells, pyruvate is transported into mitochondria for conversion to acetyl-CoA, the primary substrate of the citric acid cycle. The citric acid cycle generates electrons, which the electron transport chain uses to produce ATP by oxidative phosphorylation. Tight control of the activity of the pyruvate dehydrogenase is critical for numerous reasons. Unfettered pyruvate dehydrogenase complex activity would limit anaplerosis by lowering the pyruvate concentration below the K_m of pyruvate carboxylase, decrease the cytosolic NAD^+/NADH redox ratio by increasing the lactate/pyruvate ratio, and inhibit the oxidation of alternative fuels (fatty acids, ketone bodies, and amino acids). Excessive restraint of the pyruvate dehydrogenase complex would limit citric acid cycle activity for want of acetyl-CoA, reduce ATP production by oxidative phosphorylation for want of NADH, stimulate glycolysis and induce lactic acidosis for want of ATP, and promote utilization of fuels other than glucose. Regulation of the activity of the pyruvate dehydrogenase complex is therefore critical for balancing the rate of glycolysis with the rate of the citric acid cycle, fuel selection by tissues, disposal of excess glucose, and conservation of compounds that can be used to synthesize glucose.

Activation of the pyruvate dehydrogenase complex in the well-fed state helps dispose of excess glucose. Consumption of meals rich in carbohydrate induces the active, dephosphorylated state of the pyruvate dehydrogenase complex. Because high blood levels of glucose are toxic due to oxidative stress [115], all possible mechanisms that can restore euglycemia in the body are called into action. The synthesis of glycogen, which does not involve the pyruvate dehydrogenase complex, provides a temporary but limited solution. Glucose oxidation, which is dependent upon the pyruvate dehydrogenase complex, provides the only solution. As long as glucose is plentiful, the pyruvate dehydrogenase complex remains highly active and glucose oxidation dominates over the oxidation of alternative fuels. Inhibition of fatty acid oxidation by glucose oxidation directs dietary fatty acids into the formation of triacylglycerols for storage in adipose tissue. Inhibition of amino acid oxidation by glucose oxidation directs amino acids into the synthesis of proteins. The acetyl-CoA produced by the pyruvate dehydrogenase complex in excess of that needed by the citric acid cycle is used for the synthesis of fatty acids in the liver, a major pathway in rodents but a minor one in humans [116].

Although the pyruvate dehydrogenase complex is more dephosphorylated and therefore more active in the fed state than in the starved state, the complex is seldom, if ever, in its completely active, dephosphorylated state. A reserve of pyruvate dehydrogenase complex activity that can be called into play by dephosphorylation is always available for balancing the rates of glycolysis and the citric acid cycle. By operating at a submaximal activity, readily adjustable up or down in response to allosteric effectors and covalent modification, the pyruvate dehydrogenase complex responds to the relative rates of glycolysis and the citric acid cycle. The tendency for some tissues to overproduce lactate and pyruvate by aerobic glycolysis in the well-fed state is countered by pyruvate inhibition of PDK activity and transient activation of the pyruvate dehydrogenase complex via its phosphatase.

Inactivation of the pyruvate dehydrogenase complex during starvation conserves three carbon compounds for glucose synthesis. During fasting (going without food overnight) and starvation (fasting long enough to deplete liver glycogen), the pyruvate dehydrogenase complex is progressively shut down by phosphorylation in most tissues of the body (Figure 5.9A). The brain is a major exception. The pyruvate dehydrogenase complex remains active in neurons for utilization of

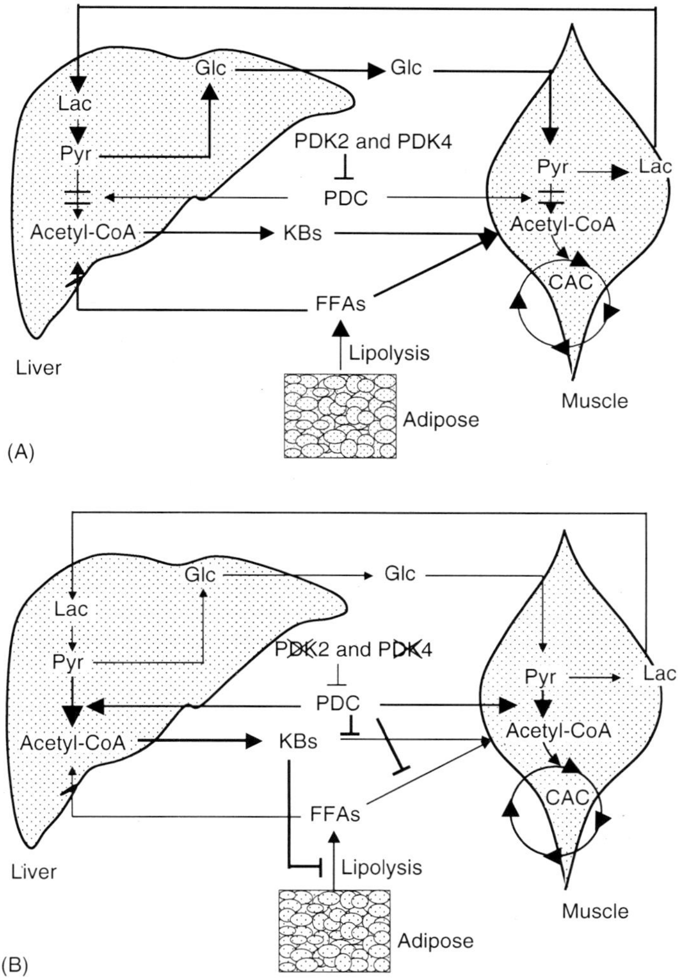

FIGURE 5.9 (A) Role of the pyruvate dehydrogenase complex in the starved state. (B) Effect of knocking out PDK2 and PDK4 on metabolic processes. Glc, glucose; Lac, lactate; Pyr, pyruvate; CAC, citric acid cycle; PDC, pyruvate dehydrogenase complex; KB, ketone bodies; FFAs, free fatty acids.

lactate, their preferred substrate, generated from glucose in astrocytes. High serum levels of free fatty acids generated by lipolysis in the adipose tissue promote fatty acid oxidation in almost every tissue of the body with the exception of the brain. The increase in mitochondrial concentrations of NADH and acetyl-CoA produced by fatty acid oxidation promotes inactivation of the pyruvate dehydrogenase complex by stimulating PDK activity. Reduced insulin levels and elevated levels of serum free fatty acids and glucocorticoids induce expression of PDK4 in muscle, kidney, and heart, and PDK2 in liver and kidney. The resulting increase in PDK activity brings about phosphorylation and inactivation of the pyruvate dehydrogenase complex. Inhibition of the conversion of the three-carbon compound pyruvate to the two-carbon compound acetyl-CoA in peripheral tissues, especially skeletal muscle, heart, and liver, conserves three carbon compounds (pyruvate, lactate, and alanine) that can be used by the liver to make glucose. Inactivation of the pyruvate dehydrogenase complex in the liver blocks ketone body synthesis from lactate, pyruvate, and alanine but allows their synthesis from fatty acids and ketogenic amino acids. Inhibition of ketogenesis from three-carbon compounds is critical because no pathway exists for the conversion of ketone bodies into glucose. Conservation of the compounds that can be used to synthesize glucose (lactate, pyruvate, and alanine) at the expense of compounds that cannot be converted to glucose (fatty acids, acetyl-CoA, and ketone bodies) helps maintain the blood glucose levels required by the brain. Inactivation of the pyruvate dehydrogenase complex in starvation indirectly conserves body protein because it minimizes the need for gluconeogenesis from glucogenic amino acids and prevents complete oxidation of the carbon skeletons of glucogenic amino acids that can be converted to glucose. Indeed, survival during long-term starvation depends upon inactivation of the pyruvate dehydrogenase complex. If the complex remained active in the starved state, the three carbon compounds needed for gluconeogenesis would be converted to CO_2 in peripheral tissues and to ketone bodies in the liver. Since survival requires maintenance of glucose levels, animals would have to consume their protein stores at a faster rate as a carbon source for the synthesis of glucose. Since pyruvate is an intermediate in the catabolism of several amino acids, much of the carbon coming from protein would be wasted. Thus, control of the pyruvate dehydrogenase complex plays an important role in dictating the fuel used by tissues of the body in different nutritional and hormonal states.

Inactivation of the pyruvate dehydrogenase complex contributes to the hyperglycemia in type 1 diabetes. The same mechanisms that suppress the pyruvate dehydrogenase complex in starvation are operative in uncontrolled type 1 diabetes mellitus. Increased expression of PDK2 and PDK4 and activation of these and the other PDKs by NADH and acetyl-CoA conserve three-carbon compounds for gluconeogenesis. Hyperglycemia occurs in the fed and fasted states because no insulin is produced to suppress gluconeogenesis. In a normal person, hyperglycemia occurs only transiently after meals and euglycemia persists during fasting because insulin produced by a functional pancreas keeps gluconeogenesis under control.

Activation of the pyruvate dehydrogenase complex by dichloroacetate lowers blood glucose levels in starvation and diabetes. Dichloroacetate, an inhibitor of pyruvate dehydrogenase kinase, is a useful tool for studies on regulation of the pyruvate dehydrogenase complex. Dichloroacetate indirectly activates the pyruvate dehydrogenase complex by inhibiting the pyruvate dehydrogenase kinases by the same mechanism as pyruvate. Although there are many compounds that inhibit metabolic processes, dichloroacetate is one of the few compounds that stimulates a metabolic process (pyruvate oxidation) by inhibiting an enzyme. Indeed, treatment of animals with dichloroacetate induces activation of the pyruvate dehydrogenase comlex in tissues and organs throughout the body [117]. The result in the fasted state is lowered blood levels of lactate, pyruvate, and alanine [118]. Decreased availability of these three-carbon compounds limits glucose synthesis by the liver [118] and, thereby, reduces blood glucose in the fasted state [119,120]. Dichloroacetate is only effective when blood glucose is being maintained by gluconeogenesis, i.e., in fasting and starvation where the complex is normally inactive and therefore limiting for oxidative disposal of pyruvate. In other words, dichloroacetate is without effect in situations in which the activity of the pyruvate dehydrogenase complex is not limiting for the disposal of pyruvate.

Knocking out PDK4 results in higher pyruvate dehydrogenase complex activities and lowers blood glucose levels in starvation. Strong upregulation of PDK4 during starvation suggests a dominant role for this PDK in regulation of fuel homeostasis in this condition. Studies with PDK4-knockout ($PDK4^{-/-}$) mice support this conclusion [121]. Lack of PDK4 during fasting and starvation of these mice results in higher than normal pyruvate dehydrogenase activities. As found in dichloroacetate-treated rats [118], gluconeogenic substrates are lower in the blood of starved $PDK4^{-/-}$ mice, consistent with reduced formation in peripheral tissues. Moreover, liver concentrations of intermediates of the gluconeogenic pathway are lower in starved $PDK4^{-/-}$ mice, consistent with lower rates of gluconeogenesis due to a limitation of substrate supply. Diaphragms from starved $PDK4^{-/-}$ mice accumulate less lactate and pyruvate because of a faster rate of pyruvate oxidation and a reduced rate of glycolysis [121], analogous to the effects of dichloroacetate [122]. Branched-chain amino acids are higher in the blood in starved $PDK4^{-/-}$ mice, consistent with lower blood alanine levels and the importance of BCAAs as a source of amino groups for alanine formation via these transamination reactions:

$$\text{BCAAs} + \alpha\text{-ketoglutarate} \rightarrow \text{branched-chain } \alpha\text{-keto acids} + \text{glutamate}$$

$$\text{Glutamate} + \text{pyruvate} \rightarrow \alpha\text{-ketoglutarate} + \text{alanine}$$

Free fatty acids and ketone bodies are also elevated more in the blood of starved $PDK4^{-/-}$ mice, consistent with slower rates of fatty acid oxidation due to increased rates of glucose and pyruvate oxidation due to greater pyruvate dehydrogenase complex activity. Upregulation of PDK4 in tissues is clearly important during starvation for regulation of pyruvate dehydrogenase complex activity and glucose homeostasis.

Knocking out PDK2 has no effects on the pyruvate dehydrogenase complex unless coupled with knocking out PDK4. Constitutive expression in most tissues and strong upregulation of PDK2 in liver and kidney during starvation have suggested an important role of this PDK in regulation of the pyruvate dehydrogenase complex. This was not confirmed, however, with PDK2-knockout ($PDK2^{-/-}$) mice (N.H. Jeoung and R.A. Harris, unpublished observations). $PDK2^{-/-}$ mice maintain normal pyruvate dehydrogenase activities in major tissues and normal blood glucose levels during fasting, most likely because upregulation of PDK4 compensates for lack of PDK2 in the tissues of these mice. This is supported by findings with PDK2/PDK4 double knockout mice. Overnight fasting of PDK2/PDK4 double knockout mice results in severe hypoglycemia (N.H. Jeoung and R.A. Harris, unpublished findings). Ketone bodies are markedly elevated by overnight fasting of these mice, most likely because of hepatic synthesis from lactate, pyruvate, and alanine (Figure 5.9B).

Are the PDKs a viable target for the treatment of type 2 diabetes? Upregulation of PDK4 in diabetes begs the question of whether PDK4 and the other PDKs should be considered therapeutic targets for the treatment of diabetes. $PDK4^{-/-}$ mice, generated in an attempt to answer this question, have lower than normal fasting blood glucose levels and slightly but significantly better glucose tolerance [121]. This is observed in both chow-fed $PDK4^{-/-}$ mice that have normal insulin sensitivity and diet-induced obese $PDK4^{-/-}$ mice that are insulin resistant (N.H. Jeoung and R.A. Harris, unpublished findings). PDK2/PDK4 double knockout mice develop hypoglycemia after overnight and are much more glucose tolerant than $PDK4^{-/-}$ mice (N.H. Jeoung and R.A. Harris, unpublished findings). Dichloroacetate decreases fasting blood glucose levels but has relatively low potency and long-term treatment induces peripheral neuropathy [120,123]. 2-Chloroproprionate also inhibits the PDKs, lowers fasting blood glucose levels, and induces peripheral neuropathy [124]. α-Lipoic acid also inhibits the PDKs [125], which explains at least in part the finding that lipoic acid stimulates pyruvate oxidation and inhibits fatty acid oxidation and gluconeogenesis by cultured hepatocytes [126]. These in turn may explain in part earlier findings that lipoic acid increases insulin-stimulated glucose disposal in patients with type 2 diabetes [127] and decreases blood levels of lactate and pyruvate in response to a glucose load [128]. A synthetic inhibitor of the PDKs, SDZ048-619, increases pyruvate dehydrogenase complex activity in tissues of the hyperglycemic Zucker diabetic rat and reduces blood lactate but, surprisingly, does not lower blood glucose [129,130]. AZD7545, a specific PDK2 inhibitor, markedly lowers blood glucose in hyperglycemic Zucker diabetic fatty rats [129,131].

Although the above findings are somewhat encouraging, there are potential negative consequences of inhibition of the PDKs that may preclude using this mechanism for lowering blood glucose levels. Ketosis, caused by conversion of

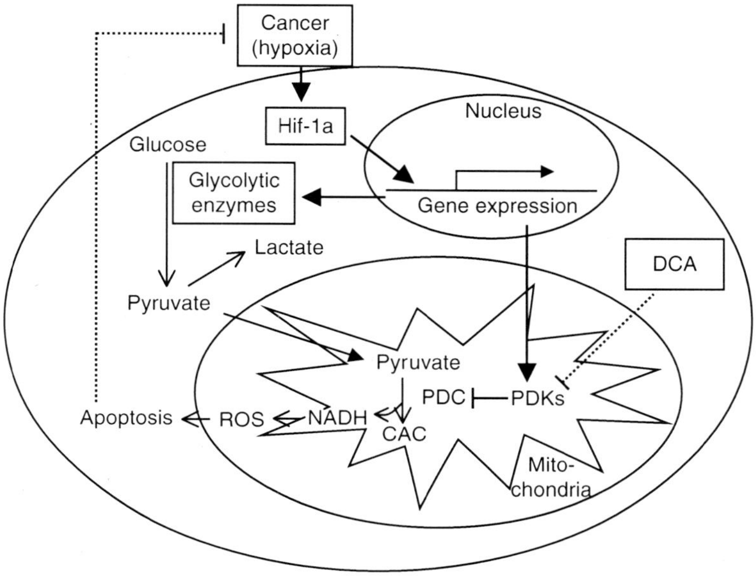

FIGURE 5.10 Role of the pyruvate dehydrogenase complex and its kinases in hypoxia and cancer. PDC, pyruvate dehydrogenase complex; DCA, dichloroacetate; CAC, citric acid cycle; ROS, reactive oxygen species.

pyruvate to ketone bodies by the liver and also inhibition of the oxidation of ketone bodies by peripheral tissues [121], could have a positive therapeutic potential if it is mild [132] but life threatening if it evolves into ketoacidosis [133]. Likewise, the inhibition of fatty acid oxidation caused by greater pyruvate oxidation results in an increase in blood free fatty acids [121] and perhaps tissue levels of fatty acids and fatty acid derivatives, harbingers for insulin resistance [134]. Interestingly, however, diet-induced obese PDK4$^{-/-}$ mice have lower fasting blood levels of insulin and are more rather than less insulin sensitive than control mice (N.H. Jeoung and R.A. Harris, unpublished findings).

Inactivation of the pyruvate dehydrogenase complex promotes cell survival in hypoxia (Figure 5.10). Out of the blue, at least for investigators studying the role of the PDKs in metabolism, the PDKs have emerged to be critically important for the survival of cells during hypoxia [54,71,72]. Conservation of three carbon compounds for glucose synthesis is not the reason. Oxidation of pyruvate during hypoxia promotes the generation of reactive oxygen species (ROS) by the mitochondrial electron transport chain to levels that can kill cells by apopotois. Inactivation of the pyruvate dehydrogenase complex by phosphorylation protects hypoxic cells against apoptosis by minimizing the availability of electrons for the electron transport chain and, therefore, the production of ROS. Hypoxia means oxygen is low but not absent, in contrast to anaerobiosis where oxygen is completely absent and production of ROS is impossible. Anaerobiosis induces robust rates of "anaerobic glycolysis," defined as the conversion of glucose to two molecules of lactic acid in the absence of oxygen. This is an emergency mechanism for the generation of ATP that cells must resort to for their survival in the absence of oxygen. Although life saving in the short term, anaerobic glycolysis produces only two ATP molecules per glucose and the decrease in pH caused by the accumulation of lactic acid can produce irreversible damage. For want of NAD^+, the pyruvate dehydrogenase complex is dead in the absence of oxygen and therefore plays no role in anaerobic glycolysis. Without oxygen, NADH generated by mitochondrial dehydrogenases accumulates in the matrix space and cannot be oxidized. The FAD bound to E_3 and the lipoyl groups of E_2 become fully reduced and cannot turn over in the pyruvate dehydrogenase complex. Introduction of oxygen inhibits anaerobic glycolysis by engaging the pyruvate dehydrogenase complex and the citric acid cycle in the production of ATP by oxidative phosphorylation. This is the "Pasteur effect," formally defined as the suppression of glucose utilization and lactate production when oxygen is reintroduced into a suspension of anaerobic cells. In effect, an abundant supply of oxygen induces a balance between the rates of glucose conversion to pyruvate and the complete oxidation of pyruvate to CO_2, maximizing the yield of ATP from glucose (~32/glucose) and minimizing the impact of the process upon cellular pH. Relative to aerobic conditions, where oxygen is abundantly available, and anaerobiosis, where oxygen is completely absent, hypoxia is a more complicated situation. In hypoxia, oxygen is present but the amount is far less than that required for maximum operation of the electron transport chain. Because oxygen is limiting, the carriers of electron transport chain are largely reduced. Paradoxically, in spite of its low concentration, oxygen can still tap into the electron transport chain at inappropriate sites and accept single electrons from carriers in Complex I and Complex III, resulting in the formation of superoxide radicals convertible by superoxide dismutase to H_2O_2, a diffusible second messenger [109,110]. E_3 of the α-ketoacid dehydrogenase complexes may also be a significant source of ROS [111–113]. Generation of ROS in moderation is good because H_2O_2 signals activation of hypoxia-inducible factor 1 (HIF-1), a transcription factor critical for the induction of survival mechanisms in hypoxia. However, unless rapidly squelched by defense mechanisms, ROS can accumulate to concentrations that induce cell death by apoptosis [71]. Macrophages and granulocytes are examples of "normal" cells that must function under hypoxic conditions in connective tissue created by inflammation and infections. Overproduction of ROS and apoptosis may be averted in such cells by HIF-1α mediated activation of genes that minimize the production and accumulation of ROS. These include upregulation of the enzymes required for glycolysis, increased expression of an isoform of lactate dehydrogenase with kinetic properties favoring conversion of pyruvate to lactate, and increased expression of PDK1, which inactivates the pyruvate dehydrogenase complex [54,71,72]. All of these favor glycolysis and minimize mitochondrial consumption of oxygen. Most important, inhibition of the pyruvate dehydrogenase complex and, consequently, the citric acid cycle minimize the production of toxic ROS by the mitochondrial electron transport chain and saves cells from apoptosis.

Inactivation of the pyruvate dehydrogenase complex contributes to the development of chronic hypoxic pulmonary hypertension (Figure 5.10). Chronic hypoxia can lead to the development of pulmonary hypertension as a result of damage to the arterial endothelium [135,136]. Smooth muscle cells of arteries proliferate excessively in this condition, in part because decreased expression or inhibition of plasma membrane voltage-gated potassium channels (K_v channels) and decreased activity of proapoptotic caspases. Under normal aerobic conditions, growth and proliferation of smooth muscle cells is kept under control by mitochondrial production of the second messenger H_2O_2, which activates and induces expression of Kv channels and proapoptotic caspases. Chronic hypoxia induces greater enzymatic capacity for aerobic glycolysis by the mechanisms discussed above, including upregulation of PDK1, resulting in reduced steady-state levels of H_2O_2 [136]. As a result, Kv channels are downregulated, causing depolarization of the plasma membrane and activation of voltage-gated L-type calcium channels. Increased uptake of Ca^{2+} promotes cell proliferation and growth [137]. The elevated K^+ concentration suppresses apoptosis by inhibition of the proapoptotic caspases. At least with rats, it has been found that oral treatment with dichloroacetate reverses chronic hypoxic pulmonary hypertension by inhibiting aerobic glycolysis and promoting mitochondrial generation of H_2O_2 [135,136,138].

Inactivation of the pyruvate dehydrogenase complex promotes growth, proliferation, and metastasis of some cancers (Figure 5.10). The blood vessels that provide nutrients and oxygen to solid tumors are often in a tangled mess. Tumor cells are adept at survival with or without oxygen. Like most cells, tumor cells are capable of robust rates of glycolysis under anaerobic conditions, and as a general rule, glycolytic enzymes are highly expressed in tumor cells. However, unlike most normal cells that suppress glycolysis under aerobic conditions, i.e., exhibit a strong Pasteur effect, tumor cells often exhibit a poor Pasteur effect. This is counterintuitive, since so much more ATP can be generated by complete oxidation of a molecule of glucose relative to the splitting of glucose to lactate by glycolysis. There is a rationale, however, to this madness of a tumor cell. Rather than shutting down glycolysis in response to oxygen, many tumor cells exhibit exceptional capacity for glycolysis with the production of lactate in the presence of oxygen. Championed as important in cancer cells by Otto Warburg, this phenomenon is termed "aerobic glycolysis" [55], or the Warburg effect, defined as rapid glycolysis with lactate as the end product in the presence of oxygen. The mechanism responsible for the Warburg effect is the same as that described above for normal cells, including high expression of the glycolytic enzymes and remarkable induction of PDK1 as a consequence of upregulation of HIF-1α by either hypoxia or oncogenes. High capacity for aerobic glycolysis during hypoxia allows cancer cells to maintain their ATP level in the face of decreased mitochondrial consumption of oxygen. Inhibition of the utilization of pyruvate carbon by the pyruvate dehydrogenase complex and subsequently by the citric acid cycle minimizes the production of toxic ROS by the electron transport chain. Expression of high levels of PDK1 serves as a protective mechanism against ROS-induced apoptosis of cancer cells during hypoxia. Recognition that cancer cells rely upon this mechanism for survival has opened up exciting new possibilities for cancer chemotherapy. Indeed, dichloroacetate increases ROS production by tumors, promotes apoptosis, inhibits proliferation, and suppresses tumor growth in rodents [139]. These findings have raised the hope that potent inhibitors of PDKs with less toxic side effects than dichloroacetate will be found effective against cancers.

Branched-Chain α-Keto Acid Dehydrogenase Complex

The branched-chain α-keto acid dehydrogenase complex catalyzes the most important regulatory step in the catabolic pathways for the BCAAs (Figure 5.8). The first step of the pathway, transamination of the BCAAs with α-ketoglutarate to produce the corresponding branched-chain α-keto acids, is a near equilibrium, reversible reaction. The branched-chain α-keto acid dehydrogenase complex catalyzes the second step, an irreversible reaction that commits the individual BCAAs to degradation. Unfettered branched-chain α-keto acid dehydrogenase activity lowers the concentration of the BCAAs and inhibits protein synthesis. Lack of sufficient branched-chain α-keto acid dehydrogenase activity causes accumulation of the BCAAs and keto acids that are toxic, as evidenced by the clinical experience with maple syrup urine disease [140]. Regulation of the activity of the branched-chain α-keto acid dehydrogenase complex is therefore of critical importance for conservation of the BCAAs for protein synthesis as well as disposal of the potentially toxic BCAAs when these are present in excess.

Activation of the branched-chain α-ketoacid dehydrogenase complex in the high-protein fed state helps dispose of excess BCAAs. Activation by dephosphorylation of the branched-chain α-keto acid dehydrogenase complex occurs when BCAAs are present in excess. The amount of protein consumed by humans varies greatly, not only among different individuals because of dietary preferences and financial resources but also within a particular individual from meal to meal and day to day. For the most part, the marked difference in BCAAs provided by the amount and quality of the protein of the diet can be readily dealt with by adjusting the activity of the branched-chain α-keto acid dehydrogenase complex. This is achieved in the short term by the extent of inhibition of BDK activity by α-ketoisocaproate, the transamination product of leucine, and in the long term by downregulation of the amount of BDK expressed in tissues and upregulation of components of branched-chain α-keto acid dehydrogenase complex.

Inactivation of the branched-chain α-keto acid dehydrogenase complex during protein starvation conserves BCAAs for protein synthesis. Phosphorylation of the $E_1\alpha$ subunit of the branched-chain α-keto acid dehydrogenase complex occurs when there is a need to conserve BCAAs for protein synthesis. Complete activation of branched-chain α-keto acid dehydrogenase in mice lacking BDK ($BDK^{-/-}$ mice) results in over-oxidation of BCAA and causes growth impairment, glucose intolerance, and neurological abnormalities [95]. Thus, over-oxidation of BCAA has detrimental effects, which in turn emphasizes the role played by BDK in the maintenance of optimal levels of BCAA. Animals cannot synthesize BCAAs. These amino acids are therefore essential and must be continuously supplied by the diet for growth and good health. A diet lacking just one essential amino acid will cause negative nitrogen balance and inhibit growth. Diets deficient in protein relative to energy, likewise, limit the supply of essential amino acids, restrict protein synthesis, and inhibit growth. Overall, the BCAAs are not different in this regard from other essential amino acids. However, the ability of leucine to stimulate protein translation to a greater extent than other amino acids [141–143] makes it special among the essential amino acids and therefore of greater interest.

Since BCAAs account for about 20% of our dietary protein and are required for protein synthesis and neurotransmitter synthesis, the catabolism of BCAAs must be tightly regulated. The activity of the branched-chain α-keto acid dehydrogenase complex is reduced in rats fed a low-protein diet [85] or treated with thyroid hormone [90] but increased in starvation [85], diabetes [144], sepsis [145,146], cancer [147,148], uremia [149,150], and infections and inflammatory disease caused by endotoxin and cytokines [151,152]. The BCAAs are also precursors of branched long-chain fatty acids found in rat skin surface lipids [153] and the covering of human eye retina. BCAA deficiency can induce anorexia [154].

Leucine is special because it stimulates protein synthesis [155]. Leucine stimulates protein translation by the mammalian target of rapamycin (mTOR) pathway. mTOR activation signals activation of eIF4E that binds to eIF4G to form eIF4F, the initiation complex required for protein translation [142,156,157]. mTOR also activates the S6 kinase that stimulates initiation of translation as well as elongation [158,159]. The 40S ribosomal protein S6 is phosphorylated by S6 kinase [160]. The enzymatic components responsible for translation are part of the biochemical machinery required to direct cell growth.

Consistent with this model, the administration of leucine to rats induces hyperphosphorylation of 4E-BP1, promotes formation of an eIF4F complex, causes hyperphosphorylation of S6 kinase, and stimulates protein synthesis [157]. Likewise, feeding rats a 20% protein diet stimulates protein synthesis in skeletal muscle and liver [161]. Stimulation of protein synthesis by the 20% protein diet is associated with reduced binding of eIF4E to 4EBP1, increased formation of eIF4F complex, and increased phosphorylation of 4EBP1. None of these effects are produced by a diet lacking in protein [160].

A dietary deficiency of leucine inhibits protein translation via a GCN2-mediated pathway [142,162]. A fall in the concentration of leucine below the K_m of its activating enzyme results in an increase in uncharged tRNA, an activator of GCN2. GCN2 phosphorylates eIF2α, which in turn is an inhibitor of the guanine nucleotide exchange activity of factor eIF2B. Inhibition of eIF2 causes inhibition of translation initiation and therefore global protein synthesis. However, not all proteins are downregulated in response to this mechanism. Translation of specific messages, e.g., that of the transcription factor ATF4, is increased. Greater expression of ATF4 promotes transcription of CHOP, another transcription factor that promotes transcription of growth-inhibitory genes.

$BDK^{-/-}$ mice provide a model for studying the impact of reduced leucine on basal levels of protein synthesis [95]. Hyperphosphorylation of eIF2α occurs in the brain but not in other tissues of $BDK^{-/-}$ mice, suggesting that regulation of translation via eIF2α-mediated pathway is particularly sensitive to the concentration of BCAAs in nervous tissue. No alterations in the phosphorylation state of downstream targets of the mTOR signaling pathway have been found in tissues of $BDK^{-/-}$ mice. In spite of this, knockdown of BDK expression with a small interfering RNA has been shown to suppress phosphorylation of S6 kinase in C2C12 cells [163].

Branched-chain amino acid metabolism is important for neurotransmitter synthesis in the brain. Leucine plays an important role in the metabolic interplay between neurons and astrocytes [164,165]. Leucine enters the brain more rapidly than any other amino acid. Transport into astrocytes is followed by transamination with α-ketoglutarate to form α-ketoisocaproate and glutamate. Leucine provides more of the amino groups present in glutamate than any other amino acid. Glutamate, an excitatory neurotransmitter, regulates 90% of the synapses in the brain. Its concentration must be kept low in the extracellular space to minimize the risk of excessive stimulation of susceptible neurons. Astrocytes release α-ketoisocaproate into the extracellular space from where it is picked up by neurons. Within neurons, α-ketoisocaproate transaminates with glutamate back to leucine and α-ketoglutarate. Release of leucine from the neurons and uptake by the astrocytes constitutes a "leucine/α-ketoisocaproate cycle" analogous to the "glutamate-glutamine cycle" [164,165]. The leucine/α-ketoisocaproate cycle provides a mechanism for buffering the glutamate concentration (by promoting net glutamate formation by transamination of α-ketoglutarate produced from glucose in the astrocytes and net glutamate utilization by transamination with α-ketoisocaproate in neurons) [164,165]. A sudden "surge" of extracellular glutamate is probably what initiates a seizure. The leucine/α-ketoisocaproate cycle is pictured to help maintain an appropriate concentration of glutamate in neurons, as part of a mechanism designed to avoid high concentrations of glutamate in the synaptic cleft where it could induce excitotoxic injury of other neurons. When the internal glutamate level increases beyond a defined limit (~40 nmol/mg protein in nerve endings), it tends to "spill" into the extracellular fluid, thereby increasing the risk of excitatory injury. By facilitating the rapid consumption of glutamate to produce leucine, the reamination of α-ketoisocaproate by glutamate may "buffer" the intraneuronal glutamate concentration.

The brain does not take up glutamate and therefore has to synthesize large amounts of this neurotransmitter from glucose. The brain can completely oxidize glutamate and also release glutamine in exchange for neutral amino acids, especially leucine. Indeed, transport of glutamate out of the brain as glutamine is another way that leucine may help "cleanse" the brain of excess glutamate [166].

The mitochondrial isoform of the branched-chain amino transferase is responsible for transmination of leucine to α-ketoisocaproate in astrocytes whereas the cytoplasmic isoform is responsible for conversion of α-ketoisocaproate back to leucine in neurons [167,168]. The branched-chain α-keto acid dehydrogenase complex lacks a direct role in the leucine/α-ketoisocaproate cycle, consistent with evidence that the branched-chain α-keto acid dehydrogenase complex activity is relatively low in the brain [169]. Significant capacity for oxidation of BCAAs exists in the brain [170], but the rate is kept low by keeping the branched-chain α-keto acid dehydrogenase complex predominately in its phosphorylated state [169]. Activation of the complex by dephosphorylation might disrupt the glutamate buffering effect by α-ketoisocaproate in neurons by lowering its concentration.

Genetic defects in the branched-chain α-keto acid dehydrogenase induce brain damage. Patients with maple syrup urine disease have markedly elevated levels of BCAAs and α-keto acids [171]. Severe mental retardation and physical dysfunction occur in untreated patients. Glutamate levels are decreased in the brain in maple syrup urine disease [172]. Yudkoff has suggested that an increase in α-ketoisocaproate causes a deficiency of glutamate by transamination [164]. As a consequence, the malate/aspartate shuttle is interrupted, resulting in an increase in lactate and diminished production of glutamate from glucose carbon. The addition of α-ketoisocaproate to brain slices [173] or synaptosomes [174] causes consumption of glutamate. In maple syrup urine disease, brain concentration of α-ketoisocaproate can increase 10–20 fold, which most likely accounts for glutamate depletion. These findings are consistent with an inverse relationship between brain leucine and glutamate levels. In ischemia, the level of leucine in the rat brain rises, and this occurs simultaneously with a fall in glutamate [175].

Metabolic and neurological defects are induced by knocking out BDK in mice. The effects of disruption of the BDK gene on growth and development of mice have helped elucidate the importance of regulation of the branched-chain α-keto acid dehydrogenase complex by phosphorylation [95]. Branched-chain α-keto acid dehydrogenase complex activity is much greater in most tissues of $BDK^{-/-}$ mice, in part because the E_1 component of the complex could not be phosphorylated due to the absence of BDK and in part because of the presence of larger than normal amounts of the E_1 component. Blood and tissue levels of the BCAAs and keto acids are markedly reduced in $BDK^{-/-}$ mice. The mice are smaller than wild-type and their fur lacked normal luster. Weights of the brain, muscle, and adipose tissue are reduced while weights of the liver and kidney are increased. Young $BDK^{-/-}$ mice exhibit neurological abnormalities, including hind limb clasping and shivering, whereas old mice develop epileptic seizures. Inhibition of protein synthesis in the brain due to hyperphosphorylation of eIF2α, increased glycine levels in the brain, or disturbance in the leucine-glutamate cycle may contribute to the development of the neurological abnormalities of $BDK^{-/-}$ mice. Growth, appearance, and neurological characteristics are improved in $BDK^{-/-}$ mice when a high-protein diet is provided, suggesting that higher amounts of dietary BCAA can partially compensate for increased oxidation in $BDK^{-/-}$ mice. Although both young and old $BDK^{-/-}$ mice are hypersensitive to insulin, elevated fasting blood glucose levels and glucose intolerance develop as the mice age. Release of insulin in response to glucose administered in vivo worsened as the mice age. These studies show that control of the activity of the BCKDH complex by phosphorylation is critically important for the regulation of oxidative disposal of BCAAs. The phenotype of the $BDK^{-/-}$ mice demonstrates the importance of tight regulation of oxidative disposal of BCAAs for normal growth, neurological function, and glucose homeostasis.

Branched-chain amino acid catabolism is accelerated in chronic kidney failure. Oxidation of BCAAs is elevated in patients with chronic renal failure and in rats with experimentally induced chronic renal failure or diabetes [176], resulting in reduced BCAA concentration [177]. Metabolic acidosis and increased glucocorticoids activate the branched-chain α-keto acid dehydrogenase complex [178]. Increased glucocorticoids and acidosis are the hallmark of chronic renal failure whereas reduced insulin and elevated glucocorticoids occur in diabetes. Reduction in insulin and elevation in glucocorticoids result in inhibition of the PI3 kinase pathway activity [179]. Glucocorticoids and acidosis induce E_2 expression in LLC-PK_1 cells regardless of the presence or absence of GRs, suggesting that glucocorticoids and acidosis independently induce the expression of E_2 [180]. Glucocortoids mediated increase in E_2 is attributed to the presence of GREs in E_2 promoter and displacement of NF-κB, proposed to be a corepressor for E_2 expression [181]. Induction of glucocorticoids and acidosis mediated expression of BCKDH E_2 may be responsible for increased BCAA oxidation in chronic renal failure.

Glycine Cleavage System

Glycine is a nonessential amino acid that can be synthesized from readily available intermediates. The numerous pathways in which glycine is involved are summarized in Figure 5.11. Although only a two-carbon compound, glycine is glucogenic via its conversion to serine by serine hydroxymethyltransferase. Glycine is also used in the synthesis of purines, heme, glutathione, and the conjugation of bile acids. Glycine is a major source of one carbon units via transfer of its α-carbon to THF to give methylene-THF. The conversion of glycine to sarcosine provides an important pathway for disposal of excess methyl groups originating from methionine:

$$\text{Glycine} + S\text{-adenosylmethionine (SAM)} \rightarrow \text{sarcosine} + S\text{-adenosylhomocysteine (SAH)}$$

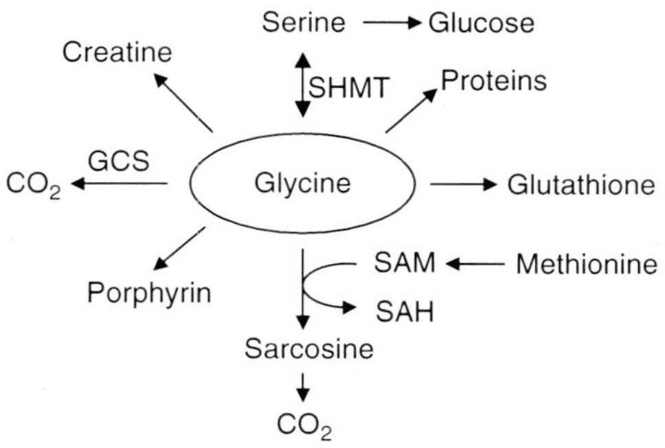

FIGURE 5.11 Pathways of glycine metabolism. GCS, glycine cleavage system; SAM, *S*-adenosylmethionine; SAH, *S*-adenosylhomocysteine; SHMT, serine hydroxymethyltransferase.

$$\text{Sarcosine} + \text{electron transfer factor (ETF)} \rightarrow \text{glycine} + \text{reduced ETF} + \text{formaldehyde}$$

$$\text{Sum: SAM} + \text{ETF} \rightarrow \text{SAH} + \text{reduced ETF} + \text{formaldehyde}$$

The glycine cleavage system catalyzes the first step in the major pathway for oxidative disposal of glycine [182]. Nonketotic hyperglycinemia is caused by inherited deficiencies in the glycine cleavage system. Ketotic hyperglycinemia results from defects in the oxidation of BCAAs cause. Both cause accumulation of glycine in the cerebral spinal fluid, which induces neurological dysfunction due to the function of glycine as inhibitory neurotransmitter and coactivator of the NMDA receptor. Housed within the mitochondrial matrix space, the glycine cleavage system is subject to activation by cAMP and Ca^{2+}-mediated mechanisms that have long been recognized but not explained at the molecular level.

ACKNOWLEDGMENTS

This work was supported by grants to R.A.H. from the American Diabetes Association and the U.S. Public Health Service (DK47844).

REFERENCES

1. S.J. Yeaman, The 2-oxo acid dehydrogenase complexes: Recent advances, *Biochem. J.* 257, 1989, 625–632.
2. L.J. Reed, A trail of research from lipoic acid to alpha-keto acid dehydrogenase complexes, *J. Biol. Chem.* 276, 2001, 38329–38336.
3. M.S. Patel and T.E. Roche, Molecular biology and biochemistry of pyruvate dehydrogenase complexes, *FASEB J.* 4, 1990, 3224–3233.
4. R.N. Perham, Swinging arms and swinging domains in multifunctional enzymes: Catalytic machines for multistep reactions, *Annu. Rev. Biochem.* 69, 2000, 961–1004.

5. J. Neagle, O. De Marcucci, B. Dunbar, and J.G. Lindsay, Component X of mammalian pyruvate dehydrogenase complex: Structural and functional relationship to the lipoate acetyltransferase (E2) component, *FEBS. Lett.* 253, 1989, 11–15.
6. R.A. Harris, M.M. Bowker-Kinley, P. Wu, J. Jeng, and K.M. Popov, Dihydrolipoamide dehydrogenase-binding protein of the human pyruvate dehydrogenase complex. DNA-derived amino acid sequence, expression, and reconstitution of the pyruvate dehydrogenase complex, *J. Biol. Chem.* 272, 1997, 19746–19751.
7. Y. Hiromasa and T.E. Roche, Facilitated interaction between the pyruvate dehydrogenase kinase isoform 2 and the dihydrolipoyl acetyltransferase, *J. Biol. Chem.* 278, 2003, 33681–33693.
8. R. Paxton, P.W. Scislowski, E.J. Davis, and R.A. Harris, Role of branched-chain 2-oxo acid dehydrogenase and pyruvate dehydrogenase in 2-oxobutyrate metabolism, *Biochem. J.* 234, 1986, 295–303.
9. S.M. Jones and S.J. Yeaman, Oxidative decarboxylation of 4-methylthio-2-oxobutyrate by branched-chain 2-oxo acid dehydrogenase complex, *Biochem. J.* 237, 1986, 621–623.
10. T.K. Makar, M. Nedergaard, A. Preuss, L. Hertz, and A.J. Cooper, Glutamine transaminase K and omega-amidase activities in primary cultures of astrocytes and neurons and in embryonic chick forebrain: Marked induction of brain glutamine transaminase K at time of hatching, *J. Neurochem.* 62, 1994, 1983–1988.
11. K. Koike, M. Hamada, N. Tanaka, K.I. Otsuka, K. Ogasahara, and M. Koike, Properties and subunit composition of the pig heart 2-oxoglutarate dehydrogenase, *J. Biol. Chem.* 249, 1974, 3836–3842.
12. F.E.S.I. Goodman and F.E. Freman, Organic acidemias due to defects in lysine oxidation: 2-ketoadipic acidemia and glutaric acidemia, in *The Metabolic and Molecular Bases of Inherited Diseases*, C.R. Scriver, A.L. Beaudet, W.S. Sly, and D. Valle (Eds.), McGraw-Hill, New York, 2001, pp. 2195–2204.
13. M. Duran, F.A. Beemer, S.K. Wadman, U. Wendel, and B. Janssen, A patient with alpha-ketoadipic and alpha-aminoadipic aciduria, *J. Inherit. Metab. Dis.* 7, 1984, 61.
14. B. Wilcken, A. Smith, and D.A. Brown, Urine screening for aminoacidopathies: Is it beneficial? Results of a long-term follow-up of cases detected bny screening one millon babies, *J. Pediatr.* 97, 1980, 492–497.
15. D.J. Oliver, M. Neuburger, J. Bourguignon, and R. Douce, Interaction between the component enzymes of the glycine decarboxylase multienzyme complex, *Plant Physiol.* 94, 1990, 833–839.
16. S. Strumilo, Short-term regulation of the alpha-ketoglutarate dehydrogenase complex by energy-linked and some other effectors, *Biochemistry (Mosc)* 70, 2005, 726–729.
17. J.D. Erfle and F. Sauer, The inhibitory effects of acyl-coenzyme A esters on the pyruvate and alpha-oxoglutarate dehydrogenase complexes, *Biochim. Biophys. Acta* 178, 1969, 441–452.
18. V.I. Bunik, G. Raddatz, R.J. Wanders, and G. Reiser, Brain pyruvate and 2-oxoglutarate dehydrogenase complexes are mitochondrial targets of the CoA ester of the Refsum disease marker phytanic acid, *FEBS Lett.* 580, 2006, 3551–3557.
19. R.H. Behal, D.B. Buxton, J.G. Robertson, and M.S. Olson, Regulation of the pyruvate dehydrogenase multienzyme complex, *Annu. Rev. Nutr.* 13, 1993, 497–520.
20. R.A. Harris, K.M. Popov, Y. Zhao, N.Y. Kedishvili, Y. Shimomura, and D.W. Crabb, A new family of protein kinases—the mitochondrial protein kinases, *Adv. Enzyme Regul.* 35, 1995, 147–162.

21. B. Huang, R. Gudi, P. Wu, R.A. Harris, J. Hamilton, and K.M. Popov, Isoenzymes of pyruvate dehydrogenase phosphatase. DNA-derived amino acid sequences, expression, and regulation, *J. Biol. Chem.* 273, 1998, 17680–17688.
22. L.G. Korotchkina and M.S. Patel, Probing the mechanism of inactivation of human pyruvate dehydrogenase by phosphorylation of three sites, *J. Biol. Chem.* 276, 2001, 5731–5738.
23. S. Ravindran, G.A. Radke, J.R. Guest, and T.E. Roche, Lipoyl domain-based mechanism for the integrated feedback control of the pyruvate dehydrogenase complex by enhancement of pyruvate dehydrogenase kinase activity, *J. Biol. Chem.* 271, 1996, 653–662.
24. D. Yang, X. Gong, A. Yakhnin, and T.E. Roche, Requirements for the adaptor protein role of dihydrolipoyl acetyltransferase in the up-regulated function of the pyruvate dehydrogenase kinase and pyruvate dehydrogenase phosphatase, *J. Biol. Chem.* 273, 1998, 14130–14137.
25. M.M. Bowker-Kinley, W.I. Davis, P. Wu, R.A. Harris, and K.M. Popov, Evidence for existence of tissue-specific regulation of the mammalian pyruvate dehydrogenase complex, *Biochem. J.* 329 (Pt 1), 1998, 191–196.
26. M.S. Patel and L.G. Korotchkina, Regulation of the pyruvate dehydrogenase complex, *Biochem. Soc. Trans.* 34, 2006, 217–222.
27. M.J. Holness, A. Kraus, R.A. Harris, and M.C. Sugden, Targeted upregulation of pyruvate dehydrogenase kinase (PDK)-4 in slow-twitch skeletal muscle underlies the stable modification of the regulatory characteristics of PDK induced by high-fat feeding, *Diabetes* 49, 2000, 775–781.
28. N.J. Hutson and P.J. Randle, Enhanced activity of pyruvate dehydrogenase kinase in rat heart mitochondria in alloxan-diabetes or starvation, *FEBS Lett.* 92, 1978, 73–76.
29. S.J. Fuller and P.J. Randle, Reversible phosphorylation of pyruvate dehydrogenase in rat skeletal-muscle mitochondria. Effects of starvation and diabetes, *Biochem. J.* 219, 1984, 635–646.
30. K.M. Popov, N.Y. Kedishvili, Y. Zhao, Y. Shimomura, D.W. Crabb, and R.A. Harris, Primary structure of pyruvate dehydrogenase kinase establishes a new family of eukaryotic protein kinases, *J. Biol. Chem.* 268, 1993, 26602–26606.
31. P. Wu, J. Sato, Y. Zhao, J. Jaskiewicz, K.M. Popov, and R.A. Harris, Starvation and diabetes increase the amount of pyruvate dehydrogenase kinase isoenzyme 4 in rat heart, *Biochem. J.* 329 (Pt 1), 1998, 197–201.
32. P. Wu, K. Inskeep, M.M. Bowker-Kinley, K.M. Popov, and R.A. Harris, Mechanism responsible for inactivation of skeletal muscle pyruvate dehydrogenase complex in starvation and diabetes, *Diabetes* 48, 1999, 1593–1599.
33. P. Wu, P.V. Blair, J. Sato, J. Jaskiewicz, K.M. Popov, and R.A. Harris, Starvation increases the amount of pyruvate dehydrogenase kinase in several mammalian tissues, *Arch. Biochem. Biophys.* 381, 2000, 1–7.
34. M.C. Sugden, A. Kraus, R.A. Harris, and M.J. Holness, Fibre-type specific modification of the activity and regulation of skeletal muscle pyruvate dehydrogenase kinase (PDK) by prolonged starvation and refeeding is associated with targeted regulation of PDK isoenzyme 4 expression, *Biochem. J.* 346 (Pt 3), 2000, 651–657.
35. S.J. Peters, R.A. Harris, G.J. Heigenhauser, and L.L. Spriet, Muscle fiber type comparison of PDH kinase activity and isoform expression in fed and fasted rats, *Am. J. Physiol. Regul. Integr. Comp. Physiol.* 280, 2001, R661–R668.
36. M. Majer, K.M. Popov, R.A. Harris, C. Bogardus, and M. Prochazka, Insulin downregulates pyruvate dehydrogenase kinase (PDK) mRNA: Potential mechanism

contributing to increased lipid oxidation in insulin-resistant subjects, *Mol. Genet. Metab.* 65, 1998, 181–186.
37. P. Wu, M. Bowker-Kinley, J. Jaskiewicz, and R.A. Harris, Dexamethasone induces PDK4 expression in rat peripheral tissues by not in liver, *Diabetes* 52 (Supplement 1), 2003, 1347P.
38. B. Huang, P. Wu, M.M. Bowker-Kinley, and R.A. Harris, Regulation of pyruvate dehydrogenase kinase expression by peroxisome proliferator-activated receptor-alpha ligands, glucocorticoids, and insulin, *Diabetes* 51, 2002, 276–283.
39. K. Sasaki, T.P. Cripe, S.R. Koch, T.L. Andreone, D.D. Petersen, E.G. Beale, and D.K. Granner, Multihormonal regulation of phosphoenolpyruvate carboxykinase gene transcription. The dominant role of insulin, *J. Biol. Chem.* 259, 1984, 15242–15251.
40. F. Lee, D. Zheng, L. Zhang, and J. Youn, Insulin suppresses PDK4 mRNA expression in skeletal muscle independent of plasma FFA or insulin resistance, *Diabetes* 52 (Supplement 1), 2003, 1292P.
41. G. Bajotto, T. Murakami, M. Nagasaki, T. Tamura, N. Tamura, R.A. Harris, Y. Shimomura, and Y. Sato, Downregulation of the skeletal muscle pyruvate dehydrogenase complex in the Otsuka Long-Evans Tokushima Fatty rat both before and after the onset of diabetes mellitus, *Life Sci.* 75, 2004, 2117–2130.
42. G. Rosa, P. Di Rocco, M. Manco, A.V. Greco, M. Castagneto, H. Vidal, and G. Mingrone, Reduced PDK4 expression associates with increased insulin sensitivity in postobese patients, *Obes. Res.* 11, 2003, 176–182.
43. P. Razeghi, M.E. Young, S. Abbasi, and H. Taegtmeyer, Hypoxia in vivo decreases peroxisome proliferator-activated receptor alpha-regulated gene expression in rat heart, *Biochem. Biophys. Res. Commun.* 287, 2001, 5–10.
44. P. Wu, J.M. Peters, and R.A. Harris, Adaptive increase in pyruvate dehydrogenase kinase 4 during starvation is mediated by peroxisome proliferator-activated receptor alpha, *Biochem. Biophys. Res. Commun.* 287, 2001, 391–396.
45. D.M. Muoio, P.S. MacLean, D.B. Lang, S. Li, J.A. Houmard, J.M. Way, D.A. Winegar, J.C. Corton, G.L. Dohm, and W.E. Kraus, Fatty acid homeostasis and induction of lipid regulatory genes in skeletal muscles of peroxisome proliferator-activated receptor (PPAR) alpha knock-out mice. Evidence for compensatory regulation by PPAR delta, *J. Biol. Chem.* 277, 2002, 26089–26097.
46. E.L. Abbot, J.G. McCormack, C. Reynet, D.G. Hassall, K.W. Buchan, and S.J. Yeaman, Diverging regulation of pyruvate dehydrogenase kinase isoform gene expression in cultured human muscle cells, *FEBS J.* 272, 2005, 3004–3014.
47. M.C. Sugden, K. Bulmer, G.F. Gibbons, and M.J. Holness, Role of peroxisome proliferator-activated receptor-alpha in the mechanism underlying changes in renal pyruvate dehydrogenase kinase isoform 4 protein expression in starvation and after refeeding, *Arch. Biochem. Biophys.* 395, 2001, 246–252.
48. M.C. Sugden, K. Bulmer, G.F. Gibbons, B.L. Knight, and M.J. Holness, Peroxisome-proliferator-activated receptor-alpha (PPARalpha) deficiency leads to dysregulation of hepatic lipid and carbohydrate metabolism by fatty acids and insulin, *Biochem. J.* 364, 2002, 361–368.
49. M. Horiuchi, K. Kobayashi, M. Masuda, H. Terazono, and T. Saheki, Pyruvate dehydrogenase kinase 4 mRNA is increased in the hypertrophied ventricles of carnitine-deficient juvenile visceral steatosis (JVS) mice, *Biofactors* 10, 1999, 301–309.

50. M.T. Andrews, T.L. Squire, C.M. Bowen, and M.B. Rollins, Low-temperature carbon utilization is regulated by novel gene activity in the heart of a hibernating mammal, *Proc. Natl. Acad. Sci. USA* 95, 1998, 8392–8397.
51. M.C. Sugden, H.S. Lall, R.A. Harris, and M.J. Holness, Selective modification of the pyruvate dehydrogenase kinase isoform profile in skeletal muscle in hyperthyroidism: Implications for the regulatory impact of glucose on fatty acid oxidation, *J. Endocrinol.* 167, 2000, 339–345.
52. R.S. Savkur, K.S. Bramlett, L.F. Michael, and T.P. Burris, Regulation of pyruvate dehydrogenase kinase expression by the farnesoid X receptor, *Biochem. Biophys. Res. Commun.* 329, 2005, 391–396.
53. M.I. Koukourakis, A. Giatromanolaki, E. Sivridis, K.C. Gatter, and A.L. Harris, Pyruvate dehydrogenase and pyruvate dehydrogenase kinase expression in non small cell lung cancer and tumor-associated stroma, *Neoplasia* 7, 2005, 1–6.
54. J.W. Kim, I. Tchernyshyov, G.L. Semenza, and C.V. Dang, HIF-1-mediated expression of pyruvate dehydrogenase kinase: A metabolic switch required for cellular adaptation to hypoxia, *Cell Metab.* 3, 2006, 177–185.
55. O. Warburg, On the origin of cancer cells, *Science* 123, 1956, 309–314.
56. J. Rowles, S.W. Scherer, T. Xi, M. Majer, D.C. Nickle, J.M. Rommens, K.M. Popov, R.A. Harris, N.L. Riebow, J. Xia, L.C. Tsui, C. Bogardus, and M. Prochazka, Cloning and characterization of PDK4 on 7q21.3 encoding a fourth pyruvate dehydrogenase kinase isoenzyme in human, *J. Biol. Chem.* 271, 1996, 22376–22382.
57. H.S. Kwon, B. Huang, T.G. Unterman, and R.A. Harris, Protein kinase B-alpha inhibits human pyruvate dehydrogenase kinase-4 gene induction by dexamethasone through inactivation of FOXO transcription factors, *Diabetes* 53, 2004, 899–910.
58. H.S. Kwon, B. Huang, N. Ho Jeoung, P. Wu, C.N. Steussy, and R.A. Harris, Retinoic acids and trichostatin A (TSA), a histone deacetylase inhibitor, induce human pyruvate dehydrogenase kinase 4 (PDK4) gene expression, *Biochim. Biophys. Acta* 1759, 2006, 141–151.
59. R.K. Hall, T. Yamasaki, T. Kucera, M. Waltner-Law, R. O'Brien, and D.K. Granner, Regulation of phosphoenolpyruvate carboxykinase and insulin-like growth factor-binding protein-1 gene expression by insulin. The role of winged helix/forkhead proteins, *J. Biol. Chem.* 275, 2000, 30169–30175.
60. H.S. Kwon and R.A. Harris, Mechanisms responsible for regulation of pyruvate dehydrogenase kinase 4 gene expression, *Adv. Enzyme Regul.* 44, 2004, 109–121.
61. T. Furuyama, K. Kitayama, H. Yamashita, and N. Mori, Forkhead transcription factor FOXO1 (FKHR)-dependent induction of PDK4 gene expression in skeletal muscle during energy deprivation, *Biochem. J.* 375, 2003, 365–371.
62. K. Ma, Y. Zhang, M.B. Elam, G.A. Cook, and E.A. Park, Cloning of the rat pyruvate dehydrogenase kinase 4 gene promoter: Activation of pyruvate dehydrogenase kinase 4 by the peroxisome proliferator-activated receptor gamma coactivator, *J. Biol. Chem.* 280, 2005, 29525–29532.
63. P. Puigserver, G. Adelmant, Z. Wu, M. Fan, J. Xu, B. O'Malley, and B.M. Spiegelman, Activation of PPARgamma coactivator-1 through transcription factor docking, *Science* 286, 1999, 1368–1371.
64. J. Rhee, Y. Inoue, J.C. Yoon, P. Puigserver, M. Fan, F.J. Gonzalez, and B.M. Spiegelman, Regulation of hepatic fasting response by PPARgamma coactivator-1alpha (PGC-1): Requirement for hepatocyte nuclear factor 4alpha in gluconeogenesis, *Proc. Natl. Acad. Sci. USA* 100, 2003, 4012–4017.

65. A.R. Wende, J.M. Huss, P.J. Schaeffer, V. Giguere, and D.P. Kelly, PGC-1alpha coactivates PDK4 gene expression via the orphan nuclear receptor ERRalpha: A mechanism for transcriptional control of muscle glucose metabolism, *Mol. Cell. Biol.* 25, 2005, 10684–10694.
66. M. Araki and K. Motojima, Identification of ERRalpha as a specific partner of PGC-1alpha for the activation of PDK4 gene expression in muscle, *FEBS J.* 273, 2006, 1669–1680.
67. Y. Zhang, K. Ma, P. Sadana, F. Chowdhury, S. Gaillard, F. Wang, D.P. McDonnell, T.G. Unterman, M.B. Elam, and E.A. Park, Estrogen-related receptors stimulate pyruvate dehydrogenase kinase isoform 4 gene expression, *J. Biol. Chem.* 281, 2006, 39897–39906.
68. K.R. Stayrook, K.S. Bramlett, R.S. Savkur, J. Ficorilli, T. Cook, M.E. Christe, L.F. Michael, and T.P. Burris, Regulation of carbohydrate metabolism by the farnesoid X receptor, *Endocrinology* 146, 2005, 984–991.
69. K. Yamagata, H. Daitoku, Y. Shimamoto, H. Matsuzaki, K. Hirota, J. Ishida, and A. Fukamizu, Bile acids regulate gluconeogenic gene expression via small heterodimer partner-mediated repression of hepatocyte nuclear factor 4 and Foxo1, *J. Biol. Chem.* 279, 2004, 23158–23165.
70. U.A. White, A.A. Coulter, T.K. Miles, and J.M. Stephens, The STAT5-mediated induction of pyruvate dehydrogenase kinase 4 expression by prolactin or growth hormone in adipocytes, *Diabetes* 56, 2007, 1623–1629.
71. G.L. Semenza, Oxygen-dependent regulation of mitochondrial respiration by hypoxia-inducible factor 1, *Biochem. J.* 405, 2007, 1–9.
72. I. Papandreou, R.A. Cairns, L. Fontana, A.L. Lim, and N.C. Denko, HIF-1 mediates adaptation to hypoxia by actively downregulating mitochondrial oxygen consumption, *Cell Metab.* 3, 2006, 187–197.
73. R. Gudi, M.M. Bowker-Kinley, N.Y. Kedishvili, Y. Zhao, and K.M. Popov, Diversity of the pyruvate dehydrogenase kinase gene family in humans, *J. Biol. Chem.* 270, 1995, 28989–28994.
74. J. Larner, L.C. Huang, C.F. Schwartz, A.S. Oswald, T.Y. Shen, M. Kinter, G.Z. Tang, and K. Zeller, Rat liver insulin mediator which stimulates pyruvate dehydrogenase phosphate contains galactosamine and D-chiroinositol, *Biochem. Biophys. Res. Commun.* 151, 1988, 1416–1426.
75. B. Huang, P. Wu, K.M. Popov, and R.A. Harris, Starvation and diabetes reduce the amount of pyruvate dehydrogenase phosphatase in rat heart and kidney, *Diabetes* 52, 2003, 1371–1376.
75A. P.J. Leblanc, R.A. Harris, and S.J. Peters, Skeletal muscle fiber type comparison of pyruvate dehydrogenase phosphatase activity and isoform expression in fed and food-deprived rats, *Am. J. Physiol. Endocrinol. Metab.* 292, 2007, E571–E576.
76. R.A. Harris, M. Joshi, and N.H. Jeoung, Mechanisms responsible for regulation of branched-chain amino acid catabolism, *Biochem. Biophys. Res. Commun.* 313, 2004, 391–396.
77. K.G. Cook, R. Lawson, and S.J. Yeaman, Multi-site phosphorylation of bovine kidney branched-chain 2-oxoacid dehydrogenase complex, *FEBS Lett.* 157, 1983, 59–62.
78. A.C. Han, G.W. Goodwin, R. Paxton, and R.A. Harris, Activation of branched-chain alpha-ketoacid dehydrogenase in isolated hepatocytes by branched-chain alpha-ketoacids, *Arch. Biochem. Biophys.* 258, 1987, 85–94.

79. R. Paxton and R.A. Harris, Clofibric acid, phenylpyruvate, and dichloroacetate inhibition of branched-chain alpha-ketoacid dehydrogenase kinase in vitro and in perfused rat heart, *Arch. Biochem. Biophys.* 231, 1984, 58–66.
80. R.A. Harris, R. Paxton, and A.A. DePaoli-Roach, Inhibition of branched chain α-ketoacid dehydrogenase kinase by α-chloroisocaproate, *J. Biol. Chem.* 257, 1982, 13913–13918.
81. K.M. Popov, Y. Zhao, Y. Shimomura, J. Jaskiewicz, N.Y. Kedishvili, J. Irwin, G.W. Goodwin, and R.A. Harris, Dietary control and tissue specific expression of branched-chain alpha-ketoacid dehydrogenase kinase, *Arch. Biochem. Biophys.* 316, 1995, 148–154.
82. M. Obayashi, Y. Sato, R.A. Harris, and Y. Shimomura, Regulation of the activity of branched-chain 2-oxo acid dehydrogenase (BCODH) complex by binding BCODH kinase, *FEBS Lett.* 491, 2001, 50–54.
83. R. Kobayashi, Y. Shimomura, T. Murakami, N. Nakai, N. Fujitsuka, M. Otsuka, N. Arakawa, K.M. Popov, and R.A. Harris, Gender difference in regulation of branched-chain amino acid catabolism, *Biochem. J.* 327 (Pt 2), 1997, 449–453.
84. M. Obayashi, Y. Shimomura, N. Nakai, N.H. Jeoung, M. Nagasaki, T. Murakami, Y. Sato, and R.A. Harris, Estrogen controls branched-chain amino acid catabolism in female rats, *J. Nutr.* 134, 2004, 2628–2633.
85. R.A. Harris, S.M. Powell, R. Paxton, S.E. Gillim, and H. Nagae, Physiological covalent regulation of rat liver branched-chain alpha-ketoacid dehydrogenase, *Arch. Biochem. Biophys.* 243, 1985, 542–555.
86. R. Kobayashi, Y. Shimomura, T. Murakami, N. Nakai, M. Otsuka, N. Arakawa, K. Shimizu, and R.A. Harris, Hepatic branched-chain alpha-keto acid dehydrogenase complex in female rats: Activation by exercise and starvation, *J. Nutr. Sci. Vitaminol.* (*Tokyo*) 45, 1999, 303–309.
87. Y.S. Huang and D.T. Chuang, Down-regulation of rat mitochondrial branched-chain 2-oxoacid dehydrogenase kinase gene expression by glucocorticoids, *Biochem. J.* 339 (Pt 3), 1999, 503–510.
88. M.M. Nellis, C.B. Doering, A. Kasinski, and D.J. Danner, Insulin increases branched-chain alpha-ketoacid dehydrogenase kinase expression in Clone 9 rat cells, *Am. J. Physiol. Endocrinol. Metab.* 283, 2002, E853–E860.
89. R. Kobayashi, T. Murakami, M. Obayashi, N. Nakai, J. Jaskiewicz, Y. Fujiwara, Y. Shimomura, and R.A. Harris, Clofibric acid stimulates branched-chain amino acid catabolism by three mechanisms, *Arch. Biochem. Biophys.* 407, 2002, 231–240.
90. R. Kobayashi, Y. Shimomura, M. Otsuka, K.M. Popov, and R.A. Harris, Experimental hyperthyroidism causes inactivation of the branched-chain alpha-ketoacid dehydrogenase complex in rat liver, *Arch. Biochem. Biophys.* 375, 2000, 55–61.
91. Z. Damuni, M.L. Merryfield, J.S. Humphreys, and L.J. Reed, Purification and properties of branched-chain alpha-keto acid dehydrogenase phosphatase from bovine kidney, *Proc. Natl. Acad. Sci. USA* 81, 1984, 4335–4338.
92. Z. Damuni and L.J. Reed, Purification and properties of the catalytic subunit of the branched-chain alpha-keto acid dehydrogenase phosphatase from bovine kidney mitochondria, *J. Biol. Chem.* 262, 1987, 5129–5132.
93. M. Joshi, N.H. Jeoung, K.M. Popov, and R.A. Harris, Identification of a novel PP2C-type mitochondrial phosphatase, *Biochem. Biophys. Res. Commun.* 356, 2007, 38–44.
94. B.D. Mersey, P. Jin, and D.J. Danner, Human microRNA (miR29b) expression controls the amount of branched chain alpha-ketoacid dehydrogenase complex in a cell, *Hum. Mol. Genet.* 14, 2005, 3371–3377.

95. M.A. Joshi, N.H. Jeoung, M. Obayashi, E.M. Hattab, E.G. Brocken, E.A. Liechty, M.J. Kubek, K.M. Vattem, R.C. Wek, and R.A. Harris, Impaired growth and neurological abnormalities in branched-chain alpha-keto acid dehydrogenase kinase-deficient mice, *Biochem. J.* 400, 2006, 153–162.
96. Y. Zhao, K.M. Popov, Y. Shimomura, N.Y. Kedishvili, J. Jaskiewicz, M.J. Kuntz, J. Kain, B. Zhang, and R.A. Harris, Effect of dietary protein on the liver content and subunit composition of the branched-chain alpha-ketoacid dehydrogenase complex, *Arch. Biochem. Biophys.* 308, 1994, 446–453.
97. K.G. Cook, A.P. Bradford, and S.J. Yeaman, Resolution and reconstitution of bovine kidney branched-chain 2-oxo acid dehydrogenase complex, *Biochem. J.* 225, 1985, 731–735.
98. R.K. Hampson, L.L. Barron, and M.S. Olson, Regulation of the glycine cleavage system in isolated rat liver mitochondria, *J. Biol. Chem.* 258, 1983, 2993–2999.
99. H.S. Ewart, M. Jois, and J.T. Brosnan, Rapid stimulation of the hepatic glycine-cleavage system in rats fed on a single high-protein meal, *Biochem. J.* 283 (Pt 2), 1992, 441–447.
100. S.E. Snyderman, L.E. Holt, Jr., P.M. Nortn, E. Roitman, and S.V. Phansalkar, The plasma aminogram. I. Influence of the level of protein intake and a comparison of whole protein and amino acid diets, *Pediatr. Res.* 2, 1968, 131–144.
101. E. Ishikawa, The regulation of uptake and output of amino acids by rat tissues, *Adv. Enzyme Regul.* 14, 1976, 117–136.
102. G.M. Mabrouk, M. Jois, and J.T. Brosnan, Cell signalling and the hormonal stimulation of the hepatic glycine cleavage enzyme system by glucagon, *Biochem. J.* 330 (Pt 2), 1998, 759–763.
103. R.E. Hillman and E.F. Otto, Inhibition of glycine-serine interconversion in cultured human fibroblasts by products of isoleucine catabolism, *Pediatr. Res.* 8, 1974, 941–945.
104. L. Sweetman and J.C. Williams, Branched chain organic aciduirias, in *The Metabolic and Molecular Bases of Inherited Disease*, C.R. Scriver, A.L. Beaudet, W.S. Sly, and D. Valle (Eds.), McGraw-Hill, New York, 2001, pp. 2125–2163.
105. B.J. Nichols and R.M. Denton, Towards the molecular basis for the regulation of mitochondrial dehydrogenases by calcium ions, *Mol. Cell. Biochem.* 149–150, 1995, 203–212.
106. E.A. Newsholme, B. Crabtree, and M.S. Ardawi, The role of high rates of glycolysis and glutamine utilization in rapidly dividing cells, *Biosci. Rep.* 5, 1985, 393–400.
107. K.F. Sheu and J.P. Blass, The alpha-ketoglutarate dehydrogenase complex, *Ann. NY. Acad. Sci.* 893, 1999, 61–78.
108. G.E. Gibson, J.P. Blass, M.F. Beal, and V. Bunik, The alpha-ketoglutarate-dehydrogenase complex: A mediator between mitochondria and oxidative stress in neurodegeneration, *Mol. Neurobiol.* 31, 2005, 43–63.
109. A.J. Kowaltowski and A.E. Vercesi, Mitochondrial damage induced by conditions of oxidative stress, *Free Radic. Biol. Med.* 26, 1999, 463–471.
110. R.S. Balaban, S. Nemoto, and T. Finkel, Mitochondria, oxidants, and aging, *Cell* 120, 2005, 483–495.
111. I.G. Gazaryan, B.F. Krasnikov, G.A. Ashby, R.N. Thorneley, B.S. Kristal, and A.M. Brown, Zinc is a potent inhibitor of thiol oxidoreductase activity and stimulates reactive oxygen species production by lipoamide dehydrogenase, *J. Biol. Chem.* 277, 2002, 10064–10072.

112. E.B. Tahara, M.H. Barros, G.A. Oliveira, L.E. Netto, and A.J. Kowaltowski, Dihydrolipoyl dehydrogenase as a source of reactive oxygen species inhibited by caloric restriction and involved in Saccharomyces cerevisiae aging, *FASEB J.* 21, 2007, 274–283.
113. A.A. Starkov, G. Fiskum, C. Chinopoulos, B.J. Lorenzo, S.E. Browne, M.S. Patel, and M.F. Beal, Mitochondrial alpha-ketoglutarate dehydrogenase complex generates reactive oxygen species, *J. Neurosci.* 24, 2004, 7779–7788.
114. A.C. Nulton-Persson, D.W. Starke, J.J. Mieyal, and L.I. Szweda, Reversible inactivation of alpha-ketoglutarate dehydrogenase in response to alterations in the mitochondrial glutathione status, *Biochemistry* 42, 2003, 4235–4242.
115. M. Brownlee, Biochemistry and molecular cell biology of diabetic complications, *Nature* 414, 2001, 813–820.
116. F.B. Hillgartner, L.M. Salati, and A.G. Goodridge, Physiological and molecular mechanisms involved in nutritional regulation of fatty acid synthesis, *Physiol. Rev.* 75, 1995, 47–76.
117. S. Whitehouse, R.H. Cooper, and P.J. Randle, Mechanism of activation of pyruvate dehydrogenase by dichloroacetate and other halogenated carboxylic acids, *Biochem. J.* 141, 1974, 761–774.
118. P.J. Blackshear, P.A. Holloway, and K.G. Alberti, The metabolic effects of sodium dichloroacetate in the starved rat, *Biochem. J.* 142, 1974, 279–286.
119. O.B. Evans and P.W. Stacpoole, Prolonged hypolactatemia and increased total pyruvate dehydrogenase activity by dichloroacetate, *Biochem. Pharmacol.* 31, 1982, 1295–1300.
120. P.W. Stacpoole, The pharmacology of dichloroacetate, *Metabolism* 38, 1989, 1124–1144.
121. N.H. Jeoung, P. Wu, M.A. Joshi, J. Jaskiewicz, C.B. Bock, A.A. Depaoli-Roach, and R.A. Harris, Role of pyruvate dehydrogenase kinase isoenzyme 4 (PDHK4) in glucose homoeostasis during starvation, *Biochem. J.* 397, 2006, 417–425.
122. A.S. Clark, W.E. Mitch, M.N. Goodman, J.M. Fagan, M.A. Goheer, and R.T. Curnow, Dichloroacetate inhibits glycolysis and augments insulin-stimulated glycogen synthesis in rat muscle, *J. Clin. Invest.* 79, 1987, 588–594.
123. D.W. Crabb, E.A. Yount, and R.A. Harris, The metabolic effects of dichloroacetate, *Metabolism* 30, 1981, 1024–1039.
124. E.A. Yount, S.Y. Felten, B.L. O'Connor, R.G. Peterson, R.S. Powell, M.N. Yum, and R.A. Harris, Comparison of the metabolic and toxic effects of 2-chloropropionate and dichloroacetate, *J. Pharmacol. Exp. Ther.* 222, 1982, 501–508.
125. L.G. Korotchkina, S. Sidhu, and M.S. Patel, R-lipoic acid inhibits mammalian pyruvate dehydrogenase kinase, *Free Radic. Res.* 38, 2004, 1083–1092.
126. J.L. Walgren, Z. Amani, J.M. McMillan, M. Locher, and M.G. Buse, Effect of R(+) alpha-lipoic acid on pyruvate metabolism and fatty acid oxidation in rat hepatocytes, *Metabolism* 53, 2004, 165–173.
127. S. Jacob, P. Ruus, R. Hermann, H.J. Tritschler, E. Maerker, W. Renn, H.J. Augustin, G.J. Dietze, and K. Rett, Oral administration of RAC-alpha-lipoic acid modulates insulin sensitivity in patients with type-2 diabetes mellitus: A placebo-controlled pilot trial, *Free Radic. Biol. Med.* 27, 1999, 309–314.
128. T. Konrad, P. Vicini, K. Kusterer, A. Hoflich, A. Assadkhani, H.J. Bohles, A. Sewell, H.J. Tritschler, C. Cobelli, and K.H. Usadel, alpha-Lipoic acid treatment decreases serum lactate and pyruvate concentrations and improves glucose

effectiveness in lean and obese patients with type 2 diabetes, *Diabetes Care* 22, 1999, 280–287.
129. G.R. Bebernitz, T.D. Aicher, J.L. Stanton, J. Gao, S.S. Shetty, D.C. Knorr, R.J. Strohschein, J. Tan, L.J. Brand, C. Liu, W.H. Wang, C.C. Vinluan, E.L. Kaplan, C.J. Dragland, D. DelGrande, A. Islam, R.J. Lozito, X. Liu, W.M. Maniara, and W.R. Mann, Anilides of (*R*)-trifluoro-2-hydroxy-2-methylpropionic acid as inhibitors of pyruvate dehydrogenase kinase, *J. Med. Chem.* 43, 2000, 2248–2257.
130. T.D. Aicher, R.C. Anderson, J. Gao, S.S. Shetty, G.M. Coppola, J.L. Stanton, D.C. Knorr, D.M. Sperbeck, L.J. Brand, C.C. Vinluan, E.L. Kaplan, C.J. Dragland, H.C. Tomaselli, A. Islam, R.J. Lozito, X. Liu, W.M. Maniara, W.S. Fillers, D. DelGrande, R.E. Walter, and W.R. Mann, Secondary amides of (*R*)-3,3,3-trifluoro-2-hydroxy-2-methylpropionic acid as inhibitors of pyruvate dehydrogenase kinase, *J. Med. Chem.* 43, 2000, 236–249.
131. R.M. Mayers, R.J. Butlin, E. Kilgour, B. Leighton, D. Martin, J. Myatt, J.P. Orme, and B.R. Holloway, AZD7545, a novel inhibitor of pyruvate dehydrogenase kinase 2 (PDHK2), activates pyruvate dehydrogenase in vivo and improves blood glucose control in obese (fa/fa) Zucker rats, *Biochem. Soc. Trans.* 31, 2003 1165–1167.
132. R.L. Veech, The therapeutic implications of ketone bodies: The effects of ketone bodies in pathological conditions: Ketosis, ketogenic diet, redox states, insulin resistance, and mitochondrial metabolism, *Prostaglandins Leukot Essent. Fatty Acids* 70, 2004, 309–319.
133. T.B. VanItallie and T.H. Nufert, Ketones: Metabolism's ugly duckling, *Nutr. Rev.* 61, 2003, 327–341.
134. G.I. Shulman, Cellular mechanisms of insulin resistance, *J. Clin. Invest.* 106, 2000, 171–176.
135. E.D. Michelakis, M.S. McMurtry, X.C. Wu, J.R. Dyck, R. Moudgil, T.A. Hopkins, G.D. Lopaschuk, L. Puttagunta, R. Waite, and S.L. Archer, Dichloroacetate, a metabolic modulator, prevents and reverses chronic hypoxic pulmonary hypertension in rats: Role of increased expression and activity of voltage-gated potassium channels, *Circulation* 105, 2002, 244–250.
136. S. Bonnet, E.D. Michelakis, C.J. Porter, M.A. Andrade-Navarro, B. Thebaud, S. Bonnet, A. Haromy, G. Harry, R. Moudgil, M.S. McMurtry, E.K. Weir, and S.L. Archer, An abnormal mitochondrial-hypoxia inducible factor-1alpha-Kv channel pathway disrupts oxygen sensing and triggers pulmonary arterial hypertension in fawn hooded rats: Similarities to human pulmonary arterial hypertension, *Circulation* 113, 2006, 2630–2641.
137. S. Sakao, L. Taraseviciene-Stewart, J.D. Lee, K. Wood, C.D. Cool, and N.F. Voelkel, Initial apoptosis is followed by increased proliferation of apoptosis-resistant endothelial cells, *FASEB J.* 19, 2005, 1178–1180.
138. M.S. McMurtry, S. Bonnet, X. Wu, J.R. Dyck, A. Haromy, K. Hashimoto, and E.D. Michelakis, Dichloroacetate prevents and reverses pulmonary hypertension by inducing pulmonary artery smooth muscle cell apoptosis, *Circ. Res.* 95, 2004, 830–840.
139. S. Bonnet, S.L. Archer, J. Allalunis-Turner, A. Haromy, C. Beaulieu, R. Thompson, C.T. Lee, G.D. Lopaschuk, L. Puttagunta, S. Bonnet, G. Harry, K. Hashimoto, C.J. Porter, M.A. Andrade, B. Thebaud, and E.D. Michelakis, A mitochondria-K+ channel axis is suppressed in cancer and its normalization promotes apoptosis and inhibits cancer growth, *Cancer Cell* 11, 2007, 37–51.

140. D.T. Chuang and V.E. Shih, Maple syrup urine disease, in *The Metabolic and Molecular Bases of Inherited Diseases*, C.R. Scriver, A.L. Beaudet, W.S. Sly, and D. Valle (Eds.), McGraw-Hill, New York, 1995, pp. 1239–1277.
141. J.C. Anthony, T.G. Anthony, S.R. Kimball, and L.S. Jefferson, Signaling pathways involved in translational control of protein synthesis in skeletal muscle by leucine, *J. Nutr.* 131, 2001, 856S–860S.
142. C.G. Proud, Regulation of mammalian translation factors by nutrients, *Eur. J. Biochem.* 269, 2002, 5338–5349.
143. S.R. Kimball and L.S. Jefferson, Regulation of protein synthesis by branched-chain amino acids, *Curr. Opin. Clin. Nutr. Metab. Care* 4, 2001, 39–43.
144. Y.B. Lombardo, M. Thamotharan, S.Z. Bawani, H.S. Paul, and S.A. Adibi, Post-transcriptional alterations in protein masses of hepatic branched-chain keto acid dehydrogenase and its associated kinase in diabetes, *Proc. Assoc. Am. Physicians.* 110, 1998, 40–49.
145. C. Garcia-Martinez, M. Llovera, F.J. Lopez-Soriano, and J.M. Argiles, The effects of endotoxin administration on blood amino acid concentrations: Similarities with sepsis, *Cell Mol. Biol. (Noisy-le-grand)* 39, 1993, 537–542.
146. N. Hayashi, D. Yoshihara, N. Kashiwabara, Y. Takeshita, H. Handa, and M. Yamakawa, Effect of carnitine on decrease of branched chain amino acids and glutamine in serum of septic rats, *Biol. Pharm. Bull.* 19, 1996, 157–159.
147. J.M. Argiles and F.J. Lopez-Soriano, The oxidation of leucine in tumour-bearing rats, *Biochem. J.* 268, 1990, 241–244.
148. P.A. Lazo, Tumour induction of host leucine starvation, *FEBS Lett.* 135, 1981, 229–231.
149. W.E. Mitch, Robert H. Herman Memorial Award in Clinical Nutrition Lecture, 1997. Mechanisms causing loss of lean body mass in kidney disease, *Am. J. Clin. Nutr.* 67, 1998, 359–366.
150. S.R. Price, D. Reaich, A.C. Marinovic, B.K. England, J.L. Bailey, R. Caban, W.E. Mitch, and B.J. Maroni, Mechanisms contributing to muscle-wasting in acute uremia: Activation of amino acid catabolism, *J. Am. Soc. Nephrol.* 9, 1998, 439–443.
151. M.D. Nawabi, K.P. Block, M.C. Chakrabarti, and M.G. Buse, Administration of endotoxin, tumor necrosis factor, or interleukin 1 to rats activates skeletal muscle branched-chain alpha-keto acid dehydrogenase, *J. Clin. Invest.* 85, 1990, 256–263.
152. O.E. Rooyackers, J.M. Senden, P.B. Soeters, W.H. Saris, and A.J. Wagenmakers, Prolonged activation of the branched-chain alpha-keto acid dehydrogenase complex in muscle of zymosan treated rats, *Eur. J. Clin. Invest.* 25, 1995, 548–552.
153. L.N. Jones and D.E. Rivett, The role of 18-methyleicosanoic acid in the structure and formation of mammalian hair fibres, *Micron* 28, 1997, 469–485.
154. D.W. Gietzen and L.J. Magrum, Molecular mechanisms in the brain involved in the anorexia of branched-chain amino acid deficiency, *J. Nutr.* 131, 2001, 851S–855S.
155. M.G. Buse and S.S. Reid, Leucine. A possible regulator of protein turnover in muscle, *J. Clin. Invest.* 56, 1975, 1250–1261.
156. A.K. Reiter, T.G. Anthony, J.C. Anthony, L.S. Jefferson, and S.R. Kimball, The mTOR signaling pathway mediates control of ribosomal protein mRNA translation in rat liver, *Int. J. Biochem. Cell Biol.* 36, 2004, 2169–2179.
157. J.C. Anthony, F. Yoshizawa, T.G. Anthony, T.C. Vary, L.S. Jefferson, and S.R. Kimball, Leucine stimulates translation initiation in skeletal muscle of postabsorptive rats via a rapamycin-sensitive pathway, *J. Nutr.* 130, 2000, 2413–2419.

158. D. Prevot, J.L. Darlix, and T. Ohlmann, Conducting the initiation of protein synthesis: The role of eIF4G, *Biol. Cell* 95, 2003, 141–156.
159. A.C. Gingras, B. Raught, and N. Sonenberg, eIF4 initiation factors: Effectors of mRNA recruitment to ribosomes and regulators of translation, *Annu. Rev. Biochem.* 68, 1999, 913–963.
160. S.R. Kimball and L.S. Jefferson, Signaling pathways and molecular mechanisms through which branched-chain amino acids mediate translational control of protein synthesis, *J. Nutr.* 136, 2006, 227S–231S.
161. F. Yoshizawa, S.R. Kimball, T.C. Vary, and L.S. Jefferson, Effect of dietary protein on translation initiation in rat skeletal muscle and liver, *Am. J. Physiol.* 275, 1998, E814–E820.
162. H. Kobayashi, E. Borsheim, T.G. Anthony, D.L. Traber, J. Badalamenti, S.R. Kimball, L.S. Jefferson, and R.R. Wolfe, Reduced amino acid availability inhibits muscle protein synthesis and decreases activity of initiation factor eIF2B, *Am. J. Physiol. Endocrinol. Metab.* 284, 2003, E488–E498.
163. N. Nakai, Y. Shimomura, T. Tamura, N. Tamura, K. Hamada, F. Kawano, and Y. Ohira, Leucine-induced activation of translational initiation is partly regulated by the branched-chain alpha-keto acid dehydrogenase complex in C2C12 cells, *Biochem. Biophys. Res. Commun.* 343, 2006, 1244–1250.
164. M. Yudkoff, Brain metabolism of branched-chain amino acids, *Glia* 21, 1997, 92–98.
165. S.M. Hutson, E. Lieth, and K.F. LaNoue, Function of leucine in excitatory neurotransmitter metabolism in the central nervous system, *J. Nutr.* 131, 2001, 846S–850S.
166. M. Yudkoff, Y. Daikhin, I. Nissim, A. Lazarow, and I. Nissim, Brain amino acid metabolism and ketosis, *J. Neurosci. Res.* 66, 2001, 272–281.
167. M. Bixel, Y. Shimomura, S. Hutson, and B. Hamprecht, Distribution of key enzymes of branched-chain amino acid metabolism in glial and neuronal cells in culture, *J. Histochem. Cytochem.* 49, 2001, 407–418.
168. E. Lieth, K.F. LaNoue, D.A. Berkich, B. Xu, M. Ratz, C. Taylor, and S.M. Hutson, Nitrogen shuttling between neurons and glial cells during glutamate synthesis, *J. Neurochem.* 76, 2001, 1712–1723.
169. M.E. Brosnan, A. Lowry, and J.T. Brosnan, Diabetes increases active fraction of branched-chain 2-oxoacid dehydrogenase in rat brain, *Diabetes* 35, 1986, 1041–1043.
170. E.R. Chaplin, A.L. Goldberg, and I. Diamond, Leucine oxidation in brain slices and nerve endings, *J. Neurochem.* 26, 1976, 701–707.
171. D.T. Chuang, J.L. Chuang, and R.M. Wynn, Lessons from genetic disorders of branched-chain amino acid metabolism, *J. Nutrition* 136, 2006, 243S–249S.
172. P.R. Dodd, S.H. Williams, A.L. Gundlach, P.A. Harper, P.J. Healy, J.A. Dennis, and G.A. Johnston, Glutamate and gamma-aminobutyric acid neurotransmitter systems in the acute phase of maple syrup urine disease and citrullinemia encephalopathies in newborn calves, *J. Neurochem.* 59, 1992, 582–590.
173. G.E. Shambaugh 3rd and R.A. Koehler, Fetal fuels VI. Metabolism of alpha-ketoisocaproic acid in fetal rat brain, *Metabolism* 32, 1983, 421–427.
174. M. Yudkoff, Y. Daikhin, I. Nissim, D. Pleasure, J. Stern, and I. Nissim, Inhibition of astrocyte glutamine production by alpha-ketoisocaproic acid, *J. Neurochem.* 63, 1994, 1508–1515.

175. M. Erecinska and D. Nelson, Activation of glutamate dehydrogenase by leucine and its nonmetabolizable analogue in rat brain synaptosomes, *J. Neurochem.* 54, 1990, 1335–1343.
176. Y. Hara, R.C. May, R.A. Kelly, and W.E. Mitch, Acidosis, not azotemia, stimulates branched-chain, amino acid catabolism in uremic rats, *Kidney Int.* 32, 1987, 808–814.
177. J. Bergstrom, A. Alvestrand, and P. Furst, Plasma and muscle free amino acids in maintenance hemodialysis patients without protein malnutrition, *Kidney Int.* 38, 1990, 108–114.
178. X. Wang, J.M. Chinsky, P.A. Costeas, and S.R. Price, Acidification and glucocorticoids independently regulate branched-chain alpha-ketoacid dehydrogenase subunit genes, *Am. J. Physiol. Cell Physiol.* 280, 2001, C1176–C1183.
179. X. Wang, J. Hu, and S.R. Price, Inhibition of PI3-kinase signaling by glucocorticoids results in increased branched-chain amino acid degradation in renal epithelial cells, *Am. J. Physiol. Cell Physiol.* 292, 2007, C1874–C1879.
180. X. Wang and S.R. Price, Differential regulation of branched-chain alpha-ketoacid dehydrogenase kinase expression by glucocorticoids and acidification in LLC-PK1-GR101 cells, *Am. J. Physiol. Renal. Physiol.* 286, 2004, F504–F508.
181. P.A. Costeas and J.M. Chinsky, Glucocorticoid regulation of branched-chain alpha-ketoacid dehydrogenase E2 subunit gene expression, *Biochem. J.* 347, 2000, 449–457.
182. T. Yoshida and G. Kikuchi, Comparative study on major pathways of glycine and serine catabolism in vertebrate livers, *J. Biochem.* (*Tokyo*) 72, 1972, 1503–1516.

6 Pyruvate Dehydrogenase Complex Regulation and Lipoic Acid

Lioubov G. Korotchkina and Mulchand S. Patel

CONTENTS

INTRODUCTION

Dietary carbohydrates provide at least one-third of daily caloric intake in Westernized societies. This calorie contribution of carbohydrates is even higher (>50%) in some populations in developing countries due to relatively low costs and availability. Hence the metabolism of glucose via the glycolytic pathway and the tricarboxylic acid cycle represents a major metabolic scheme for generation of energy. These two pathways are directly linked by the pyruvate dehydrogenase complex (PDC) localized in the mitochondria (Figure 6.1). Pyruvate derived from glucose via the glycolytic pathway is oxidatively decarboxylated to acetyl-CoA by PDC. Acetyl-CoA is further metabolized for

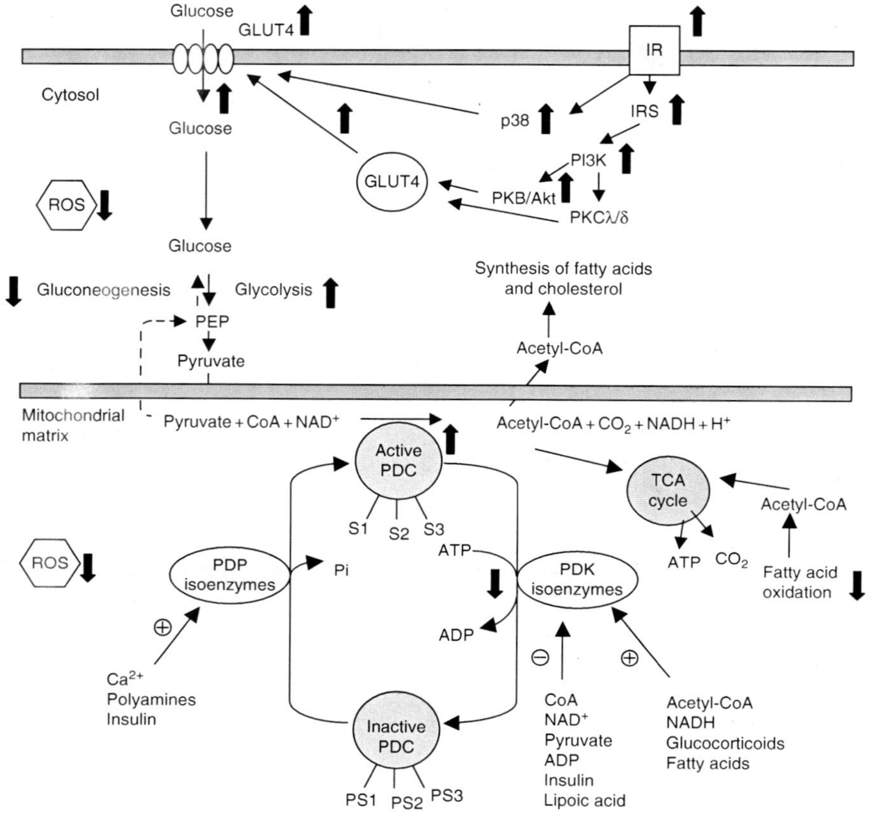

FIGURE 6.1 The suggested effects of lipoic acid on glucose metabolism. The positive (upward thick arrows) and negative (downward thick arrows) effects of lipoic acid in its oxidized or reduced form are shown as thick arrows. ROS, reactive oxygen species; IR, insulin receptor; IRS, insulin receptor substrate-1; GLUT4, glucose transporter 4; PI3K, phosphatidylinositol 3-kinase; PKB/Akt, protein kinase B/Akt1; p38, p38 mitogen-activated protein kinase; and PKCλ/δ, protein kinase C isoforms λ and δ.

energy production via the tricarboxylic acid cycle or utilized for biosynthesis of lipids. Pyruvate dehydrogenase complex catalyzes the only known reaction in higher eukaryotes to generate acetyl-CoA from pyruvate and since the PDC reaction is irreversible, the flux through PDC is highly regulated to maintain glucose homeostasis and to meet the metabolic needs of different tissues during the fed and fasted states. Regulation of PDC activity in mammals is accomplished by sophisticated mechanisms involving both covalent modification and transcriptional mechanisms. Insulin is a central hormone in regulating glucose homeostasis in mammals, and impairment in its levels and/or action results in the development of diabetes.

Lipoic acid plays an important coenzyme function in the multienzyme mitochondrial complexes referred to as the α-keto acid dehydrogenase complexes. Recent structural information about the component proteins of these complexes has provided insight about the roles of lipoyl moieties as well as the lipoyl domains of the components for enhancement of the catalytic function and protein–protein interactions in the complex. Additionally, lipoic acid in its free form has emerged as an antioxidant/prooxidant with beneficial effects for diabetic patients [1–3]. Lipoic acid is shown to affect glucose metabolism by increasing glucose uptake, enhancing the activity of PDC through inhibition of its phosphorylation, decreasing gluconeogenesis, and protecting proteins involved in glucose metabolism from oxidative stress [4–8]. Additionally, lipoic acid provides beneficiary effects in the prevention or reduction in some complications of type-2 diabetes mellitus [9]. These aspects are discussed in several chapters in this volume. In this chapter, we will discuss the aspects of lipoic acid as a coenzyme as well as an antioxidant using the pyruvate dehydrogenase complex as the center of this discussion.

STRUCTURE AND CATALYTIC MECHANISM OF PYRUVATE DEHYDROGENASE COMPLEX

Pyruvate dehydrogenase complex is one of the largest multienzyme complexes known in a living cell with the molecular mass of about 10 million daltons. The PDC reaction is catalyzed by the sequential action of the three catalytic components (Figure 6.2): pyruvate dehydrogenase (E1), dihydrolipoamide acetyltransferase (E2), and dihydrolipoamide dehydrogenase (E3). E1 carries out the rate-limiting reaction of the decarboxylation of pyruvate (Figure 6.2, reaction 1) and reductive acetylation of the lipoyl moieties covalently attached to E2 (Figure 6.2, reaction 2). E2 transfers the acetyl group to CoA with the formation of acetyl-CoA (Figure 6.2, reaction 3). E3 with FAD reoxidizes the reduced lipoyl moieties of E2 (Figure 6.2, reaction 4) with the consequent reduction of NAD^+ to NADH (Figure 6.2, reaction 5) [10].

The structural organization of PDC provides the effective function of its catalytic components. Pyruvate dehydrogenase complex from different organisms has a different morphology of the central PDC core formed by E2: 24-mer cube with octahedral symmetry for Gram-negative bacteria and 60-mer pentagonal dodecahedron with icosahedral symmetry for eukaryotes and some Gram-positive bacteria. In higher eukaryotes the central core of PDC is formed by E2 and the E3-binding protein (BP), which is a structural and not a catalytic component of PDC. E2 and BP have similar domain structures (Figure 6.3). These two proteins are composed of the lipoyl domains (two for mammalian E2 and one for mammalian BP), subunit-binding domains, and inner domains. Lipoyl domains have lipoyl groups covalently attached to a specific lysine residue(s) of E2 and

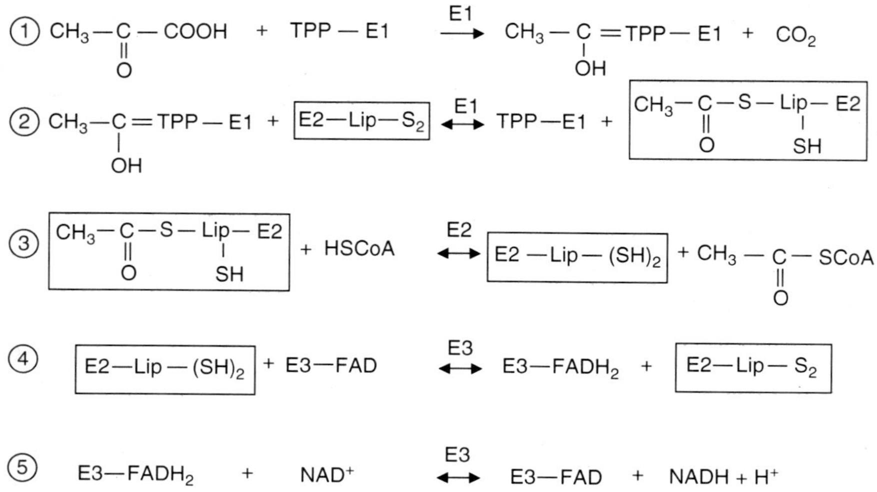

① $CH_3-C-COOH + TPP-E1 \xrightarrow{E1} CH_3-C=TPP-E1 + CO_2$
       $\|$ $|$
       O OH

② $CH_3-C=TPP-E1 + \boxed{E2-Lip-S_2} \xrightleftharpoons{E1} TPP-E1 + \boxed{\begin{array}{c} CH_3-C-S-Lip-E2 \\ \| \quad\quad\quad | \\ O \quad\quad\quad SH \end{array}}$
       $|$
       OH

③ $\boxed{\begin{array}{c} CH_3-C-S-Lip-E2 \\ \| \quad\quad\quad | \\ O \quad\quad\quad SH \end{array}} + HSCoA \xrightleftharpoons{E2} \boxed{E2-Lip-(SH)_2} + CH_3-C-SCoA$
                                                                                                                                                            $\|$
                                                                                                                                                            O

④ $\boxed{E2-Lip-(SH)_2} + E3-FAD \xrightleftharpoons{E3} E3-FADH_2 + \boxed{E2-Lip-S_2}$

⑤ $E3-FADH_2 + NAD^+ \xrightleftharpoons{E3} E3-FAD + NADH + H^+$

FIGURE 6.2 Sequence of partial reactions catalyzed by PDC components: E1 (reactions 1 and 2), E2 (reaction 3), and E3 (reactions 4 and 5). Lipoyl moieties attached to E2 participate in reactions 2, 3, and 4 and are shown in boxes.

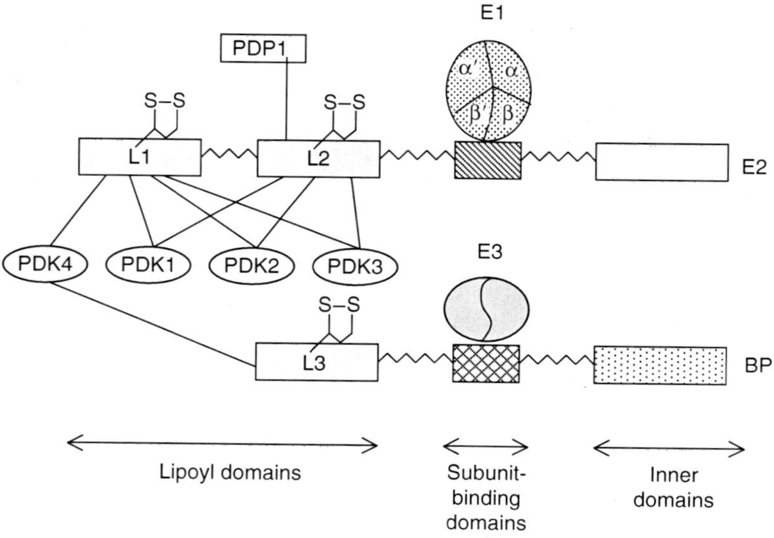

FIGURE 6.3 Structural domains of mammalian PDC E2 and BP and binding of E1, E3, PDKs, and PDP1 to E2 and BP. Lipoyl groups attached to lysine residues of the lipoyl domains in E2 and BP are shown. Binding of PDKs and PDP1 to the lipoyl domains of E2 and BP is indicated by lines.

BP. Lipoic acid is a prosthetic group of three mitochondrial α-keto acid dehydrogenase complexes, PDC, α-ketoglutarate dehydrogenase complex (KGDC), and branched-chain α-keto acid dehydrogenase complex (BCKDC) and is also a prosthetic group of the glycine cleavage system [11]. Lipoyl domains of E2 in PDC provide the coupling of the individual PDC reactions by transferring acetyl and reducing equivalents between active sites of E1, E2, and E3 (Figure 6.2). The subunit-binding domains of E2 and BP bind E1 and E3, respectively (Figure 6.3). The inner domains of E2 and BP form the central core, 60-mer pentagonal dodecahedron with icosahedral symmetry for eukaryotes composed of 48 subunits of E2 and 12 subunits of BP [12]. The inner domains of E2 catalyze the E2 reaction (Figure 6.2, reaction 3). The inner domains of BP cannot carry out the transacetylation reaction because they do not contain essential histidine in the active site [13]. All other components of PDC are connected with the central core: 20–30 tetramers ($\alpha_2\beta_2$) of E1 are bound to the E1-binding domains of E2 and 6–12 homodimers of E3 are bound to the E3-binding domains of BP (Figure 6.3). Two regulatory enzymes, pyruvate dehydrogenase kinase (PDK) and pyruvate dehydrogenase phosphatase (PDP), are bound to the lipoyl domains of E2 and BP [14].

The structural information about PDC is provided by cryoelectron microscopy and crystallography. The structure of the entire PDC complex is not determined but can be predicted based on the three-dimensional structures of individual PDC components and the central core of PDC. The three-dimensional structures are determined for the following individual components of mammalian PDCs or their fragments: human E1, lipoyl domain of human E2, human E3, subcomplex of human E3 with the E3-binding domain of BP, rat and human PDK2, and human PDK3 with the second lipoyl domain (L2) of E2 [15–21]. The three-dimensional structures of the bacterial core of PDC composed of truncated E2 (the inner domains only) revealed a hollow structure of the inner core of PDC with inner domains of E2 organized in trimers [22,23]. The suggested three-dimensional reconstruction model of bovine PDC based on cryoelectron microscopy shows that the E1s are not randomly distributed on the surface of PDC but organized in trimers, at a distance of about 100 Å above the core [24].

E1 catalyzes the first and rate-limiting step of PDC catalysis with participation of thiamin pyrophosphate (TPP) as a coenzyme. E1 has two active sites composed of amino acid residues of both α and β subunits. The two active sites of human E1 are proposed to interact with each other during catalysis by a flip-flop mechanism [15]. Half-of-the-site reactivity of E1 during its interaction with TPP and pyruvate was detected recently by spectral and NMR studies [25–27]. The interaction between the two active sites of human and mammalian E1 and the half-of-the-site reactivity was previously revealed during phosphorylation of E1 by PDK. Phosphorylation of a single phosphorylation site in one of the two active sites was sufficient for complete inactivation of E1 [28,29].

REGULATION OF PDC BY PHOSPHORYLATION/ DEPHOSPHORYLATION

Mechanism of Phosphorylation/Dephosphorylation of PDC

The major mechanism of regulation of PDC activity in higher eukaryotes is reversible phosphorylation/dephosphorylation of E1 by dedicated PDKs and PDPs (Figure 6.1). The targets of the phosphorylation are specific serine residues in the α subunit of E1. Different species have from one to three phosphorylation sites in each α subunit [30]. Mammalian E1 has three sites of phosphorylation: Ser264 (of mature human E1α sequence) as site 1, Ser271 as site 2, and Ser203 as site 3 [31,32]. Each of the three phosphorylation sites of mammalian E1 can be phosphorylated, and phosphorylation of any one site results in inactivation of E1 [29]. The mechanism of inactivation of human E1 is phosphorylation site-specific [33]. Site 1 is localized at the entrance to the active site [15]. On the basis of the analysis of mutants of site 1, phosphorylation of site 1 was suggested to interfere with the interaction of E1 with the lipoyl domains of E2 affecting the second reductive acetylation step of E1 reaction (Figure 6.2, reaction 2). The decarboxylation reaction was not affected significantly. The mechanism of inactivation by phosphorylation of sites 2 and 3 is different from that of site 1. Site 3 was protected from phosphorylation by TPP, and its modification could affect TPP binding [33]. Mechanism of inactivation by phosphorylation of site 2 is not known yet.

Phosphorylation of E1α is catalyzed by a family of PDKs (Table 6.1) [30]. A higher number of PDK isoenzymes provide higher level of tissue-specific and metabolic-state-specific regulation of PDC activity in mammals. Four isoenzymes of mammalian PDKs are distributed differently in tissues (Table 6.1) [34,35]. PDK1 is present in heart and to a lesser extent in skeletal muscle, liver, and pancreas. PDK2 is present in all tissues, with low levels in spleen and lung. PDK3 is highly expressed in testis and to a lesser extent in lung, kidney, and brain. PDK4 is detected mostly in heart and skeletal muscle and to a lesser extent in lungs, liver, and kidney.

Pyruvate dehydrogenase kinases have different activities and specificities for the three phosphorylation sites of mammalian PDC (Table 6.1). All four PDKs can phosphorylate sites 1 and 2, whereas only PDK1 is shown to phosphorylate site 3 [36,37]. PDK2 has highest activity for site 1 and PDK3 for site 2 in phosphate buffer. When PDKs phosphorylate free E1, the specificity remains the same; however, activities are dramatically less for sites 2 and 3. Only PDK4 has almost similar activities towards sites 1 and 2 with both PDC and free E1, and the highest activity towards site 2 of free E1 among four PDKs.

Pyruvate dehydrogenase phosphatase is a heterodimer composed of catalytic and regulatory subunits. Two isoenzymes of the catalytic subunit (PDP1 and PDP2) were identified in mammalian tissues with different tissue-specific distribution [14,38] (Table 6.1). PDP1 is expressed highly in heart and brain and less in other tissues, while PDP2 levels are high in liver, kidney, heart, and brain [39]. Activity of PDP1 is higher than PDP2, and both PDP1 and PDP2 can

TABLE 6.1
PDK Isoenzyme-Specific and PDP Isoenzyme-Specific Regulation of PDC Activity

Properties	PDK1	PDK2	PDK3	PDK4	PDP1	PDP2	References
Phosphorylation sites specificity	1 > 2 > 3	1 > 2	2 > 1	1 > 2	2 > 3 > 1	2 > 3 > 1	[36,37,40]
Binding to lipoyl domains	L1, L2	L2 > L1	L2 > L1	L3 > L1	L2		[14,36,49,50,52]
Tissue distribution							
Brain		+++	++		++++	++++	[34,35,38,39]
Heart	+++++	++++	+	++++	+++++	+++++	
Skeletal muscle	+++	++++		+++++	+		
Pancreas	+	+++					
Liver	++	+++		++		+++++	
Kidney		+++	++	+		+++++	
Testis		+++	++++		+++		
Spleen		+	+			+	
Lung		+	+++	+++	+		
Posttranslational regulation	Inhibited by ADP and pyruvate; activated by NADH and acetyl-CoA	Inhibited strongly by ADP and pyruvate; activated maximally by NADH and acetyl-CoA	Inhibited weakly by pyruvate	Inhibited by ADP and pyruvate; activated by NADH and acetyl-CoA	Activated by increases in Mg^{2+} and Ca^{2+}	Activated by spermine	[14,36,54–56]
Transcriptional regulation		Upregulated during starvation (L, K, MG, FTM); diabetes (SM); and hyperthyroidism (FTM)		Upregulated during starvation (L, K, muscle, MG, H); diabetes (H, SM); and hyperthyroidism (H, SM)	Downregulated during diabetes (SM)	Downregulated during starvation (H, K) and diabetes (H, K)	[39,41–45]
3D-structures		Rat and human	Human with L2				[19,20,21]

Abbreviations: L, liver; K, kidney; H, heart; MG, mammary gland; SM, skeletal muscle; FTM, fast-twitch muscle.

dephosphorylate all three sites of E1α with the following preference: site 2 > site 3 > site 1 [40].

Regulation of Phosphorylation/Dephosphorylation of PDC

Regulation of PDC activity in mammals by reversible phosphorylation involves a sophisticated mechanism. Not only does it involve three phosphorylation sites of E1α, four PDK isoenzymes, and two PDP isoenzymes with different activities and specificities as participants, but this mechanism is highly regulated at both transcriptional and posttranslational levels [14,41] (Table 6.1). At the level of transcription the expression of at least two PDKs (PDK4 and to a lesser extent PDK2) and both PDP1 and PDP2 changes in specific physiological conditions when it becomes necessary to conserve glucose and decrease PDC activity [39,42–45]. PDK4 is upregulated in several tissues (heart, skeletal muscle, and others) during starvation, diabetes, and hyperthyroidism. Pyruvate dehydrogenase phosphatases are shown to be downregulated in starvation and diabetes. Insulin downregulates PDK4 expression and increases the expression of PDPs reversing the inhibition and reactivating PDC [46–48].

Activities of PDKs are regulated by different physiological ligands which are substrates or products of either PDK and PDP (ADP, P_i) or the PDC reaction (pyruvate, NADH, acetyl-CoA) (Table 6.1, Figure 6.1). Pyruvate and ADP inhibit PDK thus resulting in increased PDC activity, whereas NADH and acetyl-CoA, the products of the PDC reaction, activate PDKs that leads to the reduction in PDC activity. In PDC, PDKs are bound to the outer (L1) or inner (L2) lipoyl domain of E2 or lipoyl domain (L3) of BP: PDK1 to L1 and L2; PDK2 and PDK3 to L2 > L1; and PDK4 to L3 of BP or L1 of E2 (Figure 6.3) [49,50]. Activities of PDKs are regulated through their binding to the lipoyl domains of E2: (1) PDKs are activated when bound to the lipoyl domain of E1 (PDK3 is activated more than other isoenzymes); (2) PDKs are transferred in PDC from one lipoyl domain to the other by the so-called hand-over-hand mechanism (dimeric PDK is held by two lipoyl domains, then it randomly lets go off one domain and binds to the new lipoyl domain) which allows PDKs translocation without complete dissociation from PDC and (3) PDKs activities are stimulated by the reduction and acetylation of the lipoyl domains of E2 (in the presence of high concentrations of NADH and acetyl-CoA) and this effect is due to increased dissociation of ADP [14,51–54]. Crystal structures of PDK2 and PDK3 with L2 [20,21] demonstrated that both PDKs form a "crossover" conformation of C-terminal tails which participate in binding to lipoyl domains. Binding to lipoyl domains provides PDK3 activation by releasing ADP, thus removing product inhibition [21]. The presence of pyruvate and ADP, inhibitors of PDK, changes the conformation of PDK2 weakening its association with lipoyl domain and reducing its activity [55].

Activities of PDPs are regulated by the concentrations of divalent ions [14]. Both PDP isoenzymes require Mg^{2+}; however, at different concentrations.

The regulatory subunit increases the K_m for Mg^{2+} for PDP1. Spermine and other polyamines decrease K_m of PDP1 for Mg^{2+} by binding to the regulatory subunit and of PDP2 by binding directly to the catalytic subunit. Binding of PDP2 with the regulatory subunit was not demonstrated [40]. Ca^{2+} participates in the binding of PDP1 to the L2 domain [56]. PDP2 is not shown to bind lipoyl domains.

EFFECT OF LIPOIC ACID ON GLUCOSE METABOLISM AND PDC

Effect of Lipoic Acid on Glucose Uptake

R-Lipoic acid was found to be beneficiary in the treatment of diabetes. Lipoic acid treatment was shown to decrease the serum lactate and pyruvate concentrations and improve insulin sensitivity in patients with type-2 diabetes [1,2]. In the cell lipoic acid is converted into dihydrolipoic acid. The presence of two forms of lipoic acid, oxidized and reduced, in the cell explains its dual action as a prooxidant (oxidized form) and as an antioxidant (reduced form) [3].

As a prooxidant lipoic acid enhances glucose uptake by stimulating the insulin-signaling pathway. Lipoic acid affects several components of the insulin-signaling pathway leading to the translocation and activation of glucose transporter-4: insulin receptor (IR), insulin receptor substrate-1, phosphatidylinositol 3-kinase, protein kinase B/Akt1, and p38 mitogen-activated protein kinase (Figure 6.1) [4,5]. The insulin-signaling pathway is regulated by redox signals. Lipoic acid through its oxidant properties can oxidize thiol groups in redox-sensitive components of insulin-signaling pathway. Lipoic acid was demonstrated to modify the thiol groups of insulin receptor and protein tyrosine phosphatase 1B, thus stimulating insulin receptor kinase activity and inhibiting protein tyrosine phosphatase 1B [57]. Both actions result in tyrosine phosphorylation of insulin receptor and its activation. Recently another mechanism was suggested for the lipoic acid activation of IR. That mechanism involves direct binding of lipoic acid to the tyrosine kinase domain of IR which stabilizes the active form of the kinase [58]. The mechanisms of the lipoic acid actions on other components of the insulin-signaling pathway are not yet known.

As an antioxidant the reduced form of lipoic acid (dihydrolipoic acid) protects the insulin-signaling pathway from oxidative stress (Figure 6.1). Dihydrolipoic acid is able to (1) directly scavenge reactive oxygen and nitrogen species, (2) chelate metals (iron, copper, zinc, and cadmium), and (3) regenerate oxidized antioxidants (ascorbate, vitamin E, glutathione, etc.). Several complications in diabetes are caused by oxidative stress, mostly due to hyperglycemia. Dihydrolipoic acid is shown to reduce the symptoms of diabetes complications such as cataract formation, vascular damage, and polyneuropathy [9,59–61]. Additionally, lipoic acid was shown to inhibit gluconeogenesis and free fatty acid oxidation possibly by decreasing the levels of free CoA in the cell [8].

Effect of Lipoic Acid on PDC Components

Reactive oxygen species (ROS) are thought to play roles in dysfunctions of the central nervous system in different diseases such as brain ischemia, and Alzheimer's and Wilson's diseases [62]. Reactive oxygen species lead to the reductions in several enzyme activities including mitochondrial dehydrogenases (such as PDC and KGDC) leading to neuronal death [62,63]. Lipoic acid due to its antioxidant properties can protect enzymes such as PDC from ROS. The reduction of lipoic acid to dihydrolipoic acid, that serves as an antioxidant is catalyzed by dihydrolipoamide dehydrogenase (E3), the common component of four mitochondrial complexes: PDC, KGDC, BCKDC, and the glycine cleavage system. E3 catalyzes the reoxidation of lipoyl moieties bound to other components of these complexes; however, it can use free lipoic acid as a substrate with low catalytic efficiency [64]. Purified E3 is inhibited by lipoic acid because of the partial reversal of the forward reaction of E3 (unpublished results). However, lipoic acid does not inhibit the PDC activity of cultured HepG2 cells [64].

Oxidative stress present under different pathophysiological conditions causes lipid peroxidation. The major product of lipid peroxidation, 4-hydroxy-2-nonenal (HNE), was shown to inhibit NADH-linked mitochondrial respiration through inactivation of the α-keto acid dehydrogenase complexes [7,65]. 4-Hydroxy-2-nonenal affects PDC, KGDC, and BCKDC by modifying the reduced lipoyl moieties of the E2s of these three complexes. Purified PDC is protected from HNE inactivation by thiol compounds, including lipoic acid and dihydrolipoic acid (Figure 6.4A) [7]. Treatment of cultured HepG2 cells with HNE results in a significant reduction of activities of the three mitochondrial α-keto acid dehydrogenase complexes in the following order: KGDC > PDC > BCKDC (Figure 6.4B). Addition of lipoic acid to the incubation medium significantly protects the activities of PDC, KGDC, and BCKDC (Figure 6.4B). Low-molecular weight thiols, cysteine and glutathione, afford even more protection than lipoic acid [7]. However, lipoic acid has an advantage of being a natural vitamin and can be used as a dietary supplement.

Administration of lipoic acid resulted in increased activity of PDC in liver mitochondria of rats [66]. Treatment of rat hepatocytes with *R*-LA was shown to increase PDC activity [8]. It was suggested that the activation of PDC could result from the decrease in acetyl-CoA levels which is an activator of PDK [8]. Recently the effect of lipoic acid on the phosphorylation/dephosphorylation of human PDC was investigated with the purified recombinant proteins [6]. It was observed that lipoic acid did not have any significant effect on the activities of PDPs. PDP1 was not stimulated by lipoic acid and PDP2 activity was instead inhibited to a small extent (Figure 6.5) [6]. However, lipoic acids (*R*- and *S*-lipoic acids) as well as *R*-dihydrolipoic acid caused significant inhibition of PDK isoenzymes (Figure 6.5), which may lead to lesser phosphorylation and hence activation of PDC (Figure 6.1). Inhibition was isoenzyme-specific. For *R*-lipoic acid and *R*-dihydrolipoic acid, the relative degree of inhibition was

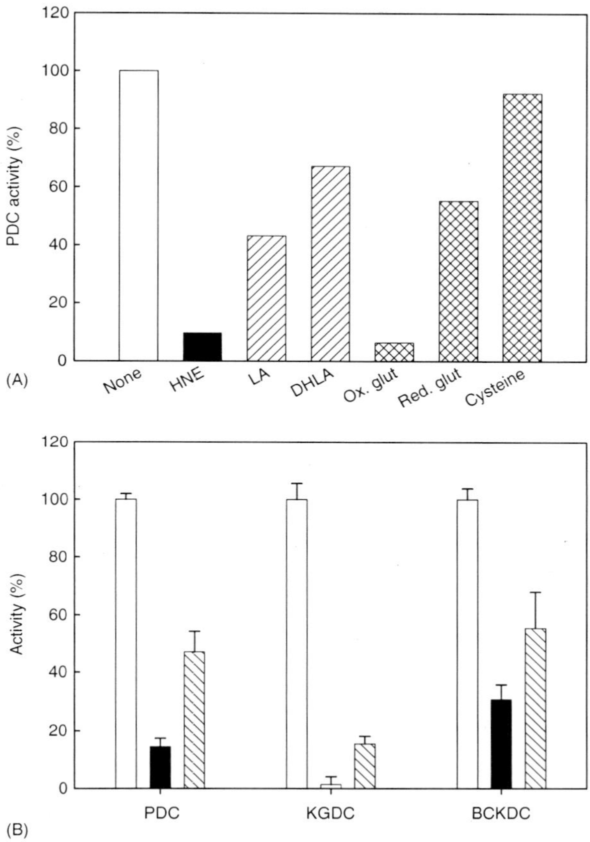

FIGURE 6.4 Protection of α-keto acid dehydrogenase complexes from 4-hydroxy-2-nonenal (HNE). (A) Human PDC reconstituted from purified recombinant E1, E2-BP, and E3 was preincubated in 50 mM potassium phosphate buffer, pH 7.5; 5 mM NADH in the presence or absence of thiol compounds for 10 min (white bar). 0.5 mM HNE was added and the time-course of inactivation was determined as described in [7]. On the basis of these results the averaged activity was calculated for 30 min incubation only for comparison. Black bar, in the absence of thiol compounds; hatched bars, in the presence of 3 mM racemic lipoic acid (LA) and dihydrolipoamide (DHLA); and cross-hatched bars, in the presence of oxidized glutathione (ox. glut.), reduced glutathione (red. glut.), and cysteine at 3 mM concentration as indicated. The correlation coefficients were at least 95%. (B) HepG2 cells were preincubated for 10 min in the presence or absence of 3 mM DL-lipoic acid, then treated with 1 mM HNE for 30 min at 37°C. The cells were then collected and activities of PDC, KGDC, and BCKDC were measured by $^{14}CO_2$ production from [1-^{14}C]pyruvate, [1-^{14}C]α-ketoglutarate, and [1-^{14}C]α-ketoisovalerate, respectively. White bars, no treatment (absence of DL-lipoic acid and HNE); black bars, cells were preincubated without addition of DL-lipoic acid for 10 min and then treated with HNE; and hatched bars, cells were preincubated in the presence of DL-lipoic acid for 10 min and then treated with HNE. Results are means ± SD ($n = 3$). Hundred percent of activity corresponds to 6.42 mU/mg of cellular protein for PDC, 1.48 mU/mg of cellular protein for KGDC, and 0.81 mU/mg of cellular protein for BCKDC [7]. (From Korotchkina, L.G. et al., *Free Radic. Biol. Med.*, 30, 992, 2001.)

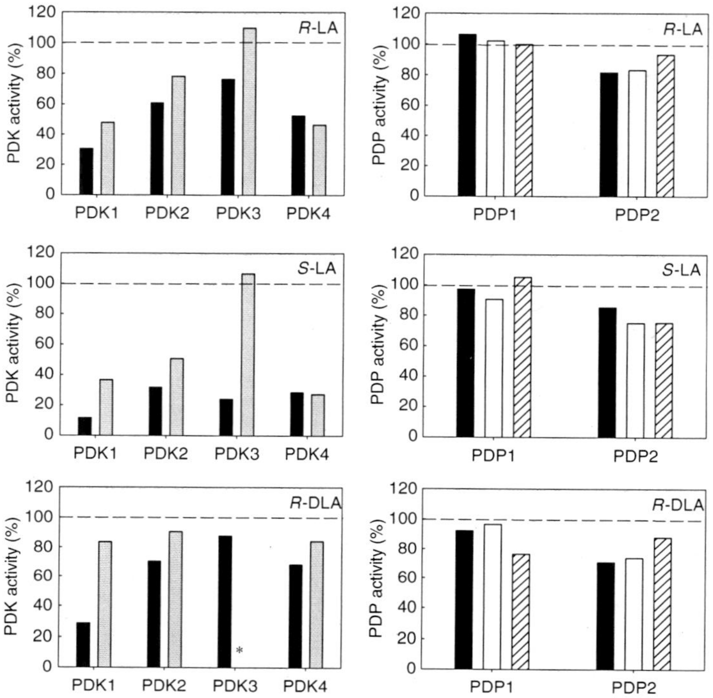

FIGURE 6.5 Effects of lipoic acids on the activities of PDKs and PDPs. Left panels: Inhibition of PDK isoenzymes by *R*-lipoic acid (*R*-LA), *S*-lipoic acid (*S*-LA), and *R*-dihydrolipoic acid (*R*-DLA). Activities of PDK isoenzymes were determined towards PDC (black bars) and E1 alone (gray bars) in the presence of 0–5 mM lipoic acids as described in [6]. On the basis of these results the averaged PDKs' activities were calculated for 3 mM concentration of lipoic acid compounds only for comparison. Results are expressed as percent of PDK activity in the absence of lipoic acid. Hundred percent of PDK activities towards reconstituted PDC corresponded to 56 mU/mg for PDK1, 87 mU/mg for PDK2, 64 mU/mg for PDK3, and 38 mU/mg for PDK4. Hundred percent of PDK activities towards E1 alone corresponded to 31 mU/mg for PDK1, 27 mU/mg for PDK2, 5.9 mU/mg for PDK3, and 87 mU/mg for PDK4. *, not determined [6]. Right panels: Effect of lipoic acids on activities of PDP isoenzymes. Activities of PDP isoenzymes were determined by dephosphorylation of E1 reconstituted in PDC and phosphorylated at site 1 (black bars), site 2 (white bars), or site 3 (hatched bars) in the presence of 0–5 mM of *R*-LA, *S*-LA, or *R*-DLA as described in [6]. On the basis of these results the averaged PDPs' activities were calculated for 3 mM concentration of lipoic acid compounds only for comparison. Results are expressed as percent of PDP activity in the absence of lipoic acid. Hundred percent of PDP1 and PDP2 activities corresponded to 9.4 and 5.5 mU/mg, respectively, for site 1; 18 and 23 mU/mg, respectively, for site 2; and 17 and 13 mU/mg, respectively, for site 3. E1 was added at the unsaturating concentration to reveal inhibition for both PDK and PDP reactions. Dashed lines represent 100% of activities of PDKs and PDPs. The correlation coefficients in all experiments used were at least 95% [6]. (From Korotchkina, L.G., Sidhu, S., and Patel, M.S., *Free Radic. Res.*, 38, 1083, 2004.)

PDK1 > PDK4 ~ PDK2 > PDK3. *S*-lipoic acid inhibited all four isoenzymes similarly. The effect of reduction and acetylation of the lipoyl domains of E2 binding PDKs on the inhibition of PDK2 by lipoic acid was further investigated. PDK2 was chosen because this isoenzyme exhibits the highest response to the reduction/acetylation of the lipoyl domains of E2 among the four PDK isoenzymes. The level of inhibition by *R*-lipoic acid and *R*-dihydrolipoic acid but not by *S*-lipoic acid increased with reduction and even more with reduction and acetylation of the lipoyl domain of E2 by incubation of PDC in the presence of the ratio (equivalent to 3) of NADH over NAD^+ and acetyl-CoA as described in Ref. [6].

The target of inhibition of the phosphorylation reaction by lipoic acid was suggested to be PDK itself based on the following observations: (1) the *R*-lipoic acid inhibition of phosphorylation of sites 2 or 3 by PDK1 (the only PDK isoenzyme that can phosphorylate each of the three sites of E1) was similar to the inhibition of site 1 phosphorylation. This observation argued against possible protection of E1 phosphorylation by binding of lipoic acid in the active site. Lipoic acid can bind in the active site of E1 in the vicinity of site 1 and not sites 2 and 3; (2) lipoic acid compounds were able to inhibit phosphorylation of free E1 by PDKs in the absence of E2 excluding competition between lipoic acid and the lipoyl domain of E2 for the binding of PDK as the possible mechanism of inhibition; and (3) the autophosphorylation reaction of PDK2 was inhibited by *R*-lipoic acid. PDK2 is the only protein participant in the autophosphorylation reaction. These results indicated that lipoic acid compounds can exert their inhibitory action mostly on PDK itself.

The mechanism by which PDK activity is inhibited by lipoic acid is not known. Oxidized and reduced forms of lipoic acid displayed similar degrees of inhibition. Additionally, oxidized glutathione did not have any effect on PDK indicating that the possible oxidation of essential SH groups of PDKs was not involved in the inhibitory mechanism. Inhibition of PDK1 by *R*-lipoic acid was not changed by ADP but decreased in the presence of pyruvate. Pyruvate is a noncompetitive inhibitor of PDK in the presence of ADP acting synergistically with ADP [67]. Binding of pyruvate and ADP were shown to induce a conformational change in PDK2 leading to reduction in its binding with the L2 of E2 [55]. This may explain how pyruvate can affect lipoic acid inhibition. Lipoic acid compounds most probably bind PDK at the binding site of the lipoyl moiety of lipoyl domain. However, the mechanism of PDK inactivation remains elusive at present.

CONCLUDING REMARKS

Lipoic acid can increase the activity of PDC in the cell by (1) scavenging ROS and products of lipid peroxidation which can cause PDC inhibition and (2) inhibition of PDK isoenzymes resulting in higher activity of PDC in the cell (Figure 6.1). In conditions with impaired glucose metabolism (such as type 2 diabetes mellitus), lipoic acid provides improvement by (1) increasing glucose uptake by stimulating components of the insulin-signaling pathway including glucose transporter-4, (2) protection of the components of insulin-signaling pathway involved in glucose metabolism and other proteins of glucose metabolism from oxidative stress, (3) activating PDC by reducing its phosphorylation, and (4) decreasing complications from diabetes through its antioxidant properties (Figure 6.1).

REFERENCES

1. Konrad, T. et al. Alpha-lipoic acid treatment decreases serum lactate and pyruvate concentrations and improves glucose effectiveness in lean and obese patients with type 2 diabetes, *Diabetes Care* 22, 280, 1999.
2. Jacob, S. et al. Oral administration of RAC-alpha-lipoic acid modulates insulin sensitivity in patients with type-2 diabetes mellitus: A placebo-controlled pilot trial, *Free Radic. Biol. Med.* 27, 309, 1999.
3. Moini, H., Packer, L., and Saris, N.E. Antioxidant and prooxidant activities of alpha-lipoic acid and dihydrolipoic acid, *Toxicol. Appl. Pharmacol.* 182, 84, 2002.
4. Konrad, D. et al. The antihyperglycemic drug alpha-lipoic acid stimulates glucose uptake via both GLUT4 translocation and GLUT4 activation: Potential role of p38 mitogen-activated protein kinase in GLUT4 activation, *Diabetes* 50, 1464, 2001.
5. Moini, H. et al. R-alpha-lipoic acid action on cell redox status, the insulin receptor, and glucose uptake in 3T3-L1 adipocytes, *Arch. Biochem. Biophys.* 397, 384, 2002.
6. Korotchkina, L.G., Sidhu, S., and Patel, M.S. R-lipoic acid inhibits mammalian pyruvate dehydrogenase kinase, *Free Radic Res.* 38, 1083, 2004.
7. Korotchkina, L.G. et al. Protection by thiols of the mitochondrial complexes from 4-hydroxy-2-nonenal, *Free Radic. Biol. Med.* 30, 992, 2001.
8. Walgren, J.L. et al. Effect of R(+)alpha-lipoic acid on pyruvate metabolism and fatty acid oxidation in rat hepatocytes, *Metabolism: Clin. & Exper.* 53, 16, 2004.
9. Packer, L., Kraemer, K., and Rimbach, G. Molecular aspects of lipoic acid in the prevention of diabetes complications, *Nutrition* 17, 888, 2001.
10. Patel, M.S. and Korotchkina, L.G. The biochemistry of the pyruvate dehydrogenase complex, *Biochem. Mol. Biol. Educ.* 31, 5, 2003.
11. Bustamante, J. et al. Alpha-lipoic acid in liver metabolism and disease, *Free Radic. Biol. Med.* 24, 1023, 1998.
12. Hiromasa, Y. et al. Organization of the cores of the mammalian pyruvate dehydrogenase complex formed by E2 and E2 plus the E3-binding protein and their capacities to bind the E1 and E3 components, *J. Biol. Chem.* 279, 6921, 2004.
13. Harris, R.A. et al. Dihydrolipoamide dehydrogenase-binding protein of the human pyruvate dehydrogenase complex. DNA-derived amino acid sequence, expression, and reconstitution of the pyruvate dehydrogenase complex, *J. Biol. Chem.* 272, 19746, 1997.
14. Roche, T.E. et al. Distinct regulatory properties of pyruvate dehydrogenase kinase and phosphatase isoforms, *Progr. Nucl. Acid Res. Molec. Biol.* 70, 33, 2001.
15. Ciszak, E.M. et al. Structural basis for flip-flop action of thiamin pyrophosphate-dependent enzymes revealed by human pyruvate dehydrogenase, *J. Biol. Chem.* 278, 21240, 2003.
16. Howard, M.J. et al. Three-dimensional structure of the major autoantigen in primary biliary cirrhosis, *Gastroenterology* 115, 139, 1998.

17. Brautigam, C.A. et al. Crystal structure of human dihydrolipoamide dehydrogenase: NAD^+/NADH binding and the structural basis of disease-causing mutations, *J. Mol. Biol.* 350, 543, 2005.
18. Ciszak, E.M. et al. How dihydrolipoamide dehydrogenase-binding protein binds dihydrolipoamide dehydrogenase in human pyruvate dehydrogenase complex, *J. Biol. Chem.* 281, 648, 2006.
19. Steussy, C.N. et al. Structure of pyruvate dehydrogenase kinase. Novel folding pattern for a serine protein kinase, *J. Biol. Chem.* 276, 37443, 2001.
20. Knoechel, T.R. et al. Regulatory roles of the N-terminal domain based on crystal structures of human pyruvate dehydrogenase kinase 2 containing physiological and synthetic ligands, *Biochemistry* 45, 402, 2006.
21. Kato, M. et al. Crystal structure of pyruvate dehydrogenase kinase 3 bound to lipoyl domain 2 of human pyruvate dehydrogenase complex, *EMBO J.* 24, 1763, 2005.
22. Mattevi, A. et al. Atomic structure of the cubic core of the pyruvate dehydrogenase multienzyme complex, *Science* 255, 1544, 1992.
23. Izard, T. et al. Principles of quasi-equivalence and Euclidean geometry govern the assembly of cubic and dodecahedral cores of pyruvate dehydrogenase complexes, *Proc. Natl. Acad. Sci. U.S. A.* 96, 1240, 1999.
24. Zhou, Z.H. et al. The remarkable structural and functional organization of the eukaryotic pyruvate dehydrogenase complexes, *Proc. Natl. Acad. Sci. U.S.A.* 98, 14802, 2001.
25. Nemeria, N.S. et al. Acetylphosphinate is the most potent mechanism-based substrate inhibitor of both the human and E. coli pyruvate dehydrogenase components of the pyruvate dehydrogenase complex, *Bioorg. Chem.* 34, 362, 2006.
26. Seifert, F. et al. Direct kinetic evidence for half-of-the-sites reactivity in the E1 component of the human pyruvate dehydrogenase multienzyme complex through alternating sites cofactor activation, *Biochemistry* 45, 12775, 2006.
27. Nemeria, N. et al. The 1′,4′-iminopyrimidine tautomer of thiamin diphosphate is poised for catalysis in asymmetric active centers on enzymes, *Proc. Natl. Acad. Sci. U.S.A.* 104, 78, 2007.
28. Sugden, P.H. and Randle, P.J. Regulation of pig heart pyruvate dehydrogenase by phosphorylation. Studies on the subunit and phosphorylation stoichiometries, *Biochem. J.* 173, 659, 1978.
29. Korotchkina, L.G. and Patel, M.S. Mutagenesis studies of the phosphorylation sites of recombinant human pyruvate dehydrogenase. Site-specific regulation, *J. Biol. Chem.* 270, 14297, 1995.
30. Patel, M.S. and Korotchkina, L.G. Regulation of the pyruvate dehydrogenase complex, *Biochem. Soc. Trans.* 34, 217, 2006.
31. Yeaman, S.J. et al. Sites of phosphorylation on pyruvate dehydrogenase from bovine kidney and heart, *Biochemistry* 17, 2364, 1978.
32. Dahl, H.H. et al. The human pyruvate dehydrogenase complex. Isolation of cDNA clones for the E1 alpha subunit, sequence analysis, and characterization of the mRNA, *J. Biol. Chem.*, 262, 7398, 1987.
33. Korotchkina, L.G. and Patel, M.S. Probing the mechanism of inactivation of human pyruvate dehydrogenase by phosphorylation of three sites, *J. Biol. Chem.* 276, 5731, 2001.
34. Bowker-Kinley, M.M. et al. Evidence for existence of tissue-specific regulation of the mammalian pyruvate dehydrogenase complex, *Biochem. J.* 329, 191, 1998.

35. Rowles, J. et al. Cloning and characterization of PDK4 on 7q21.3 encoding a fourth pyruvate dehydrogenase kinase isoenzyme in human, *J. Biol. Chem.* 271, 22376, 1996.
36. Korotchkina, L.G. and Patel, M.S. Site specificity of four pyruvate dehydrogenase kinase isoenzymes toward the three phosphorylation sites of human pyruvate dehydrogenase, *J. Biol. Chem.* 276, 37223, 2001.
37. Kolobova, E. et al. Regulation of pyruvate dehydrogenase activity through phosphorylation at multiple sites, *Biochem. J.* 358, 69, 2001.
38. Huang, B. et al. Isoenzymes of pyruvate dehydrogenase phosphatase. DNA-derived amino acid sequences, expression and regulation, *J. Biol. Chem.* 273, 17680, 1998.
39. Huang, B. et al. Starvation and diabetes reduce the amount of pyruvate dehydrogenase phosphatase in rat heart and kidney, *Diabetes* 52, 1371, 2003.
40. Karpova, T. et al. Characterization of the isozymes of pyruvate dehydrogenase phosphatase: Implications for the regulation of pyruvate dehydrogenase activity, *Biochim. Biophys. Acta* 1652, 126, 2003.
41. Harris, R.A., Huang, B., and Wu, P. Control of pyruvate dehydrogenase kinase gene expression, *Adv. Enzyme Regul.* 41, 269, 2001.
42. Wu, P. et al. Starvation and diabetes increase the amount of pyruvate dehydrogenase kinase isoenzyme 4 in rat heart, *Biochem. J.* 329, 197, 1998.
43. Wu, P. et al. Starvation increases the amount of pyruvate dehydrogenase kinase in several mammalian tissues, *Arch. Biochem. Biophys.* 381, 1, 2000.
44. Sugden, M.C. and Holness, M.J. Therapeutic potential of the mammalian pyruvate dehydrogenase kinases in the prevention of hyperglycaemia, *Curr. Drug Targets—Imm. Endocr. Met. Dis.* 2, 151, 2002.
45. Bajotto, G. et al. Downregulation of the skeletal muscle pyruvate dehydrogenase complex in the Otsuka Long-Evans Tokushima fatty rat both before and after the onset of diabetes mellitus, *Life Sci.* 75, 2117, 2004.
46. Kwon, H.S. and Harris, R.A. Mechanisms responsible for regulation of pyruvate dehydrogenase kinase 4 gene expression, *Adv. Enzyme Regul.* 44, 109, 2004.
47. Caruso, M. et al. Activation and mitochondrial translocation of protein kinase Cδ are necessary for insulin stimulation of pyruvate dehydrogenase complex activity in muscle and liver cells, *J. Biol. Chem.* 276, 45088, 2001.
48. Johnson, S.A. and Denton, R.M. Insulin stimulation of pyruvate dehydrogenase in adipocytes involves two distinct signalling pathways, *Biochem. J.* 369, 351, 2003.
49. Tuganova, A., Boulatnikov, I., and Popov, K.M. Interaction between the individual isoenzymes of pyruvate dehydrogenase kinase and the inner lipoyl-bearing domain of transacetylase component of pyruvate dehydrogenase complex, *Biochem. J.* 366, 129, 2002.
50. Roche, T.E. et al. Essential roles of lipoyl domains in the activated function and control of pyruvate dehydrogenase kinases and phosphatase isoform 1, *Eur. J. Biochem.* 270, 1050, 2003.
51. Tuganova, A. and Popov, K.M. Role of protein-protein interactions in the regulation of pyruvate dehydrogenase kinase activity, *Biochem. J.* 387, 147, 2005.
52. Baker, J.C. et al. Marked differences between two isoforms of human pyruvate dehydrogenase kinase, *J. Biol. Chem.* 275, 15773, 2000.
53. Hiromasa, Y. and Roche, T.E. Facilitated interaction between the pyruvate dehydrogenase kinase isoform 2 and the dihydrolipoyl acetyltransferase, *J. Biol. Chem.* 278, 33681, 2003.
54. Bao, H. et al. Pyruvate dehydrogenase kinase isoform 2 activity stimulated by speeding up the rate of dissociation of ADP, *Biochemistry* 43, 13442, 2004.

55. Hiromasa, Y., Hu, L., and Roche, T.E. Ligand-induced effects on pyruvate dehydrogenase kinase isoform 2, *J. Biol. Chem.* 281, 12568, 2006.
56. Turkan, A., Hiromasa, Y., and Roche, T.E. Formation of a complex of the catalytic subunit of pyruvate dehydrogenase phosphatase isoform 1 (PDP1c) and the L2 domain forms a Ca2+ binding site and captures PDP1c as a monomer, *Biochemistry* 43, 15073, 2004.
57. Cho, K.J. et al. Alpha-lipoic acid decreases thiol reactivity of the insulin receptor and protein tyrosine phosphatase 1B in 3T3-L1 adipocytes, *Biochem. Pharmacol.* 66, 849, 2003.
58. Diesel, B. et al. α-Lipoic acid as a directly binding activator of the insulin receptor: Protection from hepatocyte apoptosis, *Biochemistry* 46, 2146, 2007.
59. Maitra, I. et al. Alpha-lipoic acid prevents buthionine sulfoximine-induced cataract formation in newborn rats, *Free Radic. Biol. Med.* 18, 823, 1995.
60. Kunt, T. et al. Alpha-lipoic acid reduces expression of vascular cell adhesion molecule-1 and endothelial adhesion of human monocytes after stimulation with advanced glycation end products, *Clin. Sci. (Lond.)* 96, 75, 1999.
61. Haak, E. et al. Effects of alpha-lipoic acid on microcirculation in patients with peripheral diabetic neuropathy, *Exp. Clin. Endocrinol. Diabetes* 108, 168, 2000.
62. Sheline, C.T. and Wei, L. Free radical-mediated neurotoxicity may be caused by inhibition of mitochondrial dehydrogenases in vitro and in vivo, *Neuroscience* 140, 235, 2006.
63. Tabatabaie, T., Potts, J.D., and Floyd, R.A. Reactive oxygen species-mediated inactivation of pyruvate dehydrogenase, *Arch. Biochem. Biophys.* 336, 290, 1996.
64. Hong, Y.S. et al. The inhibitory effects of lipoic compounds on mammalian pyruvate dehydrogenase complex and its catalytic components, *Free Radic. Biol. Med.* 26, 685, 1999.
65. Humphries, K.M., Yoo, Y., and Szweda, L.I. Inhibition of NADH-linked mitochondrial respiration by 4-hydroxy-2-nonenal, *Biochemistry* 37, 552, 1998.
66. Gandhi, V.M. et al. Lipoic acid and diabetes II: Mode of action of lipoic acid, *J. Biosci.* 9, 117, 1985.
67. Bao, H. et al. Pyruvate dehydrogenase kinase isoform 2 activity limited and further inhibited by slowing down the rate of dissociation of ADP, *Biochemistry* 43, 13432, 2004.

7 Role of Lipoyl Domains in the Function and Regulation of Mammalian Pyruvate Dehydrogenase Complex

Thomas E. Roche, Tao Peng, Liangyan Hu, Yasuaki Hiromasa, Haiying Bao, and Xioaming Gong

CONTENTS

INTRODUCTION

The pyruvate dehydrogenase complex (PDC) catalyzes the irreversible conversion of pyruvate to acetyl-CoA and CO_2 coupled to the reduction of NAD^+ to NADH. This overall reaction (Figure 7.1) results from sequential catalysis by the pyruvate dehydrogenase (E1), the dihydrolipoyl acetyltransferase (E2), and the dihydrolipoyl dehydrogenase (E3) components. The five-step overall reaction is integrated by the three central steps that use the lipoyl prosthetic group as a substrate. The substrate lipoyl group is attached to specific lysine residues on each among a large set of mobile lipoyl domains that serve as intermediate carriers moving between the active sites of E1, E2, and E3 (Patel and Roche, 1990; Reed and Hackert, 1990; Perham, 2000; Roche et al., 2001). Depending on the source, E2 subunits have one to three lipoyl domains. The capacity of the lipoyl domains for traversing efficiently between the E1, E2, and E3 active sites is advanced by the mobility of the extended linker regions that connect these and other globular domains in the E2 structures (Figures 7.1 and 7.2). The C-terminal inner (I) domains of the multidomain E2 subunits from a wide range of organisms assemble into either 24-subunit cubic structures or 60-subunit structures with the appearance of a pentagonal dodecahedron (Oliver and Reed, 1982; Reed and Hackert, 1990; Mattevi et al., 1992; Perham, 2000). Figure 7.2 shows a slab cut through the inner core of human E2. The lipoyl and component binding domains of E2 surround this inner core structure. Here, we will consider the roles of E2 in the organization and function of PDC with emphasis on the distinct features (and components) in mammalian PDC. We will consider specificity-determining lipoyl domain structure in PDC catalysis with emphasis on the E1 reaction.

Mammalian PDC activity is primarily regulated by phosphorylation and dephosphorylation of the E1 component (Reed, 1974; Randle, 1986; Patel and Roche, 1990; Randle and Priestman, 1996; Roche et al., 2001; Patel and Korotchkina, 2006; Roche and Hiromasa, 2007). The activity of PDC is set by the proportion of nonphosphorylated (active) E1. Adaptable control of PDC

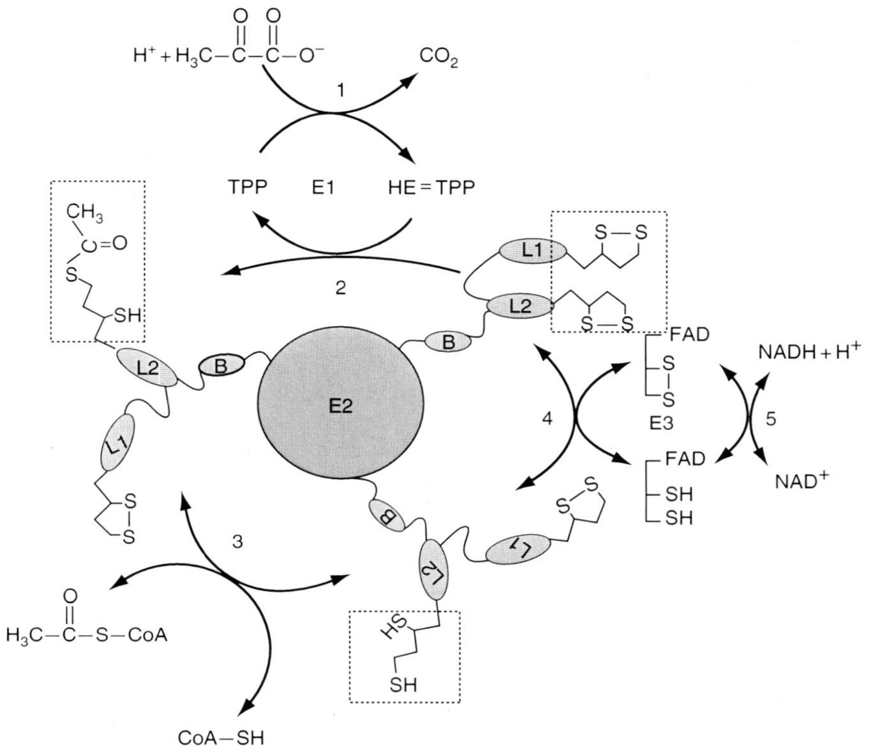

FIGURE 7.1 Overall PDC reaction. This scheme shows the steps in the overall reaction. HE = TPP is hydroxyethylidene thiamine pyrophosphate; this interconverts to C2 carbanion, which makes a nucleophilic attack on the dithiolane ring of lipoate at S8. Reaction steps 2 to 4 proceed counterclockwise for the forward direction with the lipoyl prosthetic group on the L2 domain of E2, dotted boxes, undergoing these reactions. This reaction series also models the domain structure for the outer domains of E2; see Figure 7.2 for further understanding of these domains and the structure of the inner core (labeled E2 here), which catalyzes step 3, the acetyltransferase reaction.

activity is needed to achieve the required variation in fuel consumption and storage in different tissues under different conditions (nutritional state, exercise, etc.) (Randle, 1986; Patel and Roche, 1990; Randle and Priestman, 1996; Roche et al., 2001; Sugden et al., 2001). When carbohydrate stores are reduced, PDC activity is downregulated to limit the oxidative utilization of glucose in most nonneural tissues. Extended starvation results in PDC activity being suppressed to a very low level in most tissues. Following the consumption of excess dietary carbohydrate, the activity of PDC in fat synthesizing tissues is increased to accelerate fatty acid biosynthesis from glucose and there is also increased oxidative use of glucose by heart, kidneys, lungs, and other active tissues. Four pyruvate dehydrogenase kinase (PDK) and two pyruvate dehydrogenase phosphatase (PDP) isoforms function in

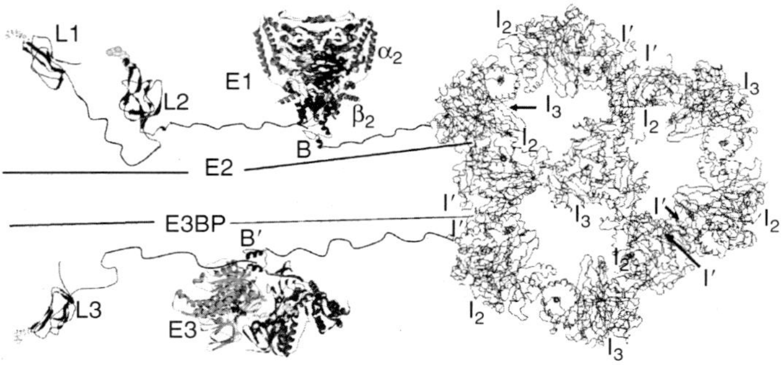

FIGURE 7.2 Domains of E2 and E3BP, inner core organization, and binding of E1 and E3. Ribbon structures are shown for the domains. The lipoyl domains are based on structures determined for the L2 domain. Specifically, L1 and L3 are based on NMR structure of L2 (Howard et al., 1998) but have an oxidized lipoyl groups added as stick structures; L2 is from the x-ray crystal structure of the PDK3-L2 complex (Kato et al., 2005) and has a space-filled lipoyl group in the reduced form but lacks H atoms on S atoms of the lipoyl group. The interaction of human E1 (Ciszak et al., 2003) is with the B domain of E2 component of the *Bacillus stearothermophilus* PDC (Mande et al., 1996; Perham, 2000). The interaction is based on x-ray crystal structures interaction of the E1 component of branched-chain α-ketoacid dehydrogenase complexes (Aevarsson et al., 1999; Aevarsson et al., 2000). The interaction of the B domain of E3BP and E3 is based on the structure crystal structures with human E3 and human B domain of E3BP (Brautigam et al., 2006; Ciszak et al., 2006). The inner core is based on the structure derived for the human E2 based on 8 Δ resolution structure obtained by cryoelectron microscopy coupled with reconstruction based on bacterial homolog (Yu et al., 2007). In this model, I domains of E2 are assigned as I′ domains based on the model proposed for the inner core of E2 E3BP (Hiromasa et al., 2004). The positioning of the linker regions connecting the domains is arbitrary; the length of the linker regions varies as found in human E2 and E3BP (see Supplement Hiromasa et al., 2004 for E2 and E3BP domain sequences).

adjusting the activation state of the PDC by determining the fraction of active (nonphosphorylated) E1 (Gudi et al., 1995; Rowles et al., 1996; Huang et al., 1998; Roche et al., 2001; Patel and Korotchkina, 2006; Roche and Hiromasa, 2007). The required versatility in the metabolic control of PDC in different tissues and cell types is met by the selective expression of the different PDK isoforms (PDK1, PDK2, PDK3, or PDK4) and PDP isoforms (PDP1 and PDP2) and by the pronounced variation in the inherent functional and regulatory properties of these enzymes (Randle and Priestman, 1996; Harris et al., 2001; Roche et al., 2001; Sugden et al., 2001; Roche and Hiromasa, 2007).

The lipoyl domains also play central roles in the function and the regulation of dedicated kinase and phosphatase components that carry out this regulatory interconversion (Roche et al., 1996; Roche et al., 2001; Roche et al., 2003a; Roche and Hiromasa, 2007). Here we will describe how the lipoyl domains function in the binding and delivery of the regulatory enzymes to their substrate and how they participate in effector regulation of these enzymes. These lipoyl domain-mediated processes result in the predominant changes in the activities of the regulatory enzymes. We will also delineate the mosaic of structure within the lipoyl domain and prosthetic group that impart the specificity for these activity-altering lipoyl domain interactions with the regulatory enzymes.

ORGANIZATION OF MAMMALIAN PYRUVATE DEHYDROGENASE COMPLEX

E2 and E3BP Domains and E1 and E3 Binding

In the case of mammalian PDC (and many but not all eukaryotic PDC), more than one component provides lipoyl domains and contributes to the central organization of the complex (Patel and Roche, 1990; Mattevi et al., 1992; Roche et al., 2001; Roche et al., 2003a; Roche et al., 2003b; Hiromasa et al., 2004). Mammalian PDC-E2 has four globular domains (Figure 7.2) (Thekkumkara et al., 1988). This includes two lipoyl domains (L1 and L2) and, when expressed by itself, E2 subunits assemble with an inner core shaped as a pentagonal dodecahedron. This 60mer structure is formed via association of 20 trimers of E2's C-terminal I domains at the 20 vertices of a dodecahedron (10 trimers shown in Figure 7.2). The trimer units of the inner core catalyze the transfer of the acetyl group between the dihydrolipoyl group and CoA. Each E2 subunit also includes an E1-binding (B) domain held by linker region connections between the lipoyl domains and the I domain. The E1 $\alpha_2\beta_2$ tetramer binds the B domain of E2 via its β subunits (Figure 7.2). The three intervening linker regions between the four domains of E2 are high in Ala and Pro residues, which contribute to maintaining separation while remaining highly mobile (Perham, 2000). Mammalian E3BP was first characterized as a new component (initially designated protein X) with a reactive lipoyl group on a single lipoyl domain (Jilka et al., 1986; Rahmatullah and Roche, 1987; Neagle et al., 1989; Rahmatullah et al., 1989a; Rahmatullah et al., 1989b; Rahmatullah et al., 1990) that, alone, could support the overall reaction (Rahmatullah et al., 1990). E3BP was shown to be tightly bound to E2 (Jilka et al., 1986) by its C-terminal I′ domain (Rahmatullah et al., 1989a). The E3BP component was then shown to contribute to the organization of the complex by binding the E3 component (Gopalakrishnan et al., 1989; Powers-Greenwood et al., 1989; Rahmatullah et al., 1989b; Sanderson et al., 1996a; Sanderson et al., 1996b). The N-terminal lipoyl domain, E3-binding (B′) domain, and C-terminal I′ domain (Figure 7.2) are each related to varying degrees to the corresponding domains in E2; the largest differences between homologous domains are with the sequences of the B and B′ domains (only 30% aligned identical residues) (Harris et al., 1997; Hiromasa et al., 2004). In Figure 7.2, the strands of the B and B′ domains can be observed as connecting to incoming and outgoing linker regions. Again the E3BP domains are connected by two Ala-Pro rich linker regions; these have indefinite sequence alignments and disparate lengths as compared to the linker regions in E2, with the linker region connecting the B′ and I′ domains being particularly long (48 amino acids as compared to 20 amino acids for the linker connecting the B to I domains in E2).

Inner Framework and Stoichiometry of E3BP to E3 Dimers

The mammalian PDC is, therefore, organized on a fixed inner framework and a mobile outer network formed by the assembled E2•E3BP subcomplex. On the basis of sedimentation (velocity and equilibrium) and small angle x-ray scattering studies on E2 and E2•E3BP and their fully loaded complexes with E1 and E3, E2•E3BP was shown to be smaller than E2 60mer, bind fewer E1 tetramers, and to bind ~12 E3 per complex (Hiromasa et al., 2004). An $E2_{48}$•$E3BP_{12}$ structure was proposed in which the related I′ domain of BP substitutes for E2's I domain at symmetric positions within the dodecahedron structure. In this model, the I′ domains are proposed to form 6 dimer edges (three I′–I′ dimers assigned as shown in Figure 7.2) and associate with two I domains. In the model, the I domains of each $I'I_2$ trimer only associate along twofold axis with I_3 trimers (Figure 7.2) (Hiromasa et al., 2004). This organization optimizes the maintenance of a critical capacity for expansion and contraction of the inner core (Zhou et al., 2001; Kong et al., 2003).

On the basis of variation in structure observed by cryo-electron microscopy, the dimer connections between I domains were deduced to extend and contract via a cantilever mechanism (Zhou et al., 2001; Kong et al., 2003). However, the I′ domains of E3BP lack critical residues needed to support this cantilever movement but retain varied sequence that is anticipated to form I′–I′ dimers (Hiromasa et al., 2004). The very similar I′ domains of E3BP are assumed to form dimers that maintain the same off-axis positioning as the I domains of E2 in the trimer connections along the twofold axis. This positioning places one of B′ domain facing into each of the 12 open faces (around the 12 fivefold symmetry axes) of the dodecahedron. This, in turn, would favor the two active sites of a bound E3 dimer catalytically servicing all the set of lipoyl groups located in one face—statistically and by favored positioning attached to nine lipoyl domains, eight from four E2 subunits and one from the E3BP subunit holding the E3 dimer. With E3's much higher catalytic rate than the E1 or E2, the E3 reaction under most conditions should approach equilibrium, thereby maintaining the ratio of oxidized to reduced lipoyl groups in near proportion to the intramitochondrial NAD^+ to NADH ratio.

X-ray crystal structures with the one B′ domain associated with an E3 dimer support the assumption of one-to-one binding (Brautigam et al., 2006; Ciszak et al., 2006). As with E1 $\alpha_2\beta_2$ tetramer binding to the B domain by its two β subunits (Aevarsson et al., 1999), these crystal structures establish that the single B′ domain binds in an asymmetric fashion in a space located between the subunits of the E3 homodimer. The B′ binding region is located in the center of the extended dimer interaction with the active sites at each end of the E3 dimer. Smolle et al. (2006) have presented evidence for a two binding domain-E3 dimer complex and interpreted this as the E3 dimer associating via equivalent interactions with two binding domains. Since there does not appear to be enough room for fitting two B′ domains within the region of the E3 dimer that holds the binding domain (Brautigam et al., 2006; Ciszak et al., 2006), a more likely explanation is that E3-binding domains can form dimers that remain self-associated while only one of those B′ domains binds the E3 dimer. Hiromasa et al. (2004) found that 12 and not 6 E3 dimers, as suggested by Smolle et al. (2006), can bind to E2•E3BP by analytical ultracentrifugation and small angle x-ray diffraction studies.

SPECIFICITY IN THE USE OF LIPOYL PROSTHETIC GROUP AND LIPOYL DOMAINS IN COMPONENT REACTIONS

E1 Specificity

Following E1-catalyzed decarboxylation of pyruvate to produce the tightly bound hydroxyethylidene-thiamine pyrophosphate intermediate, E1 catalyzes the reductive acetylation reaction that produces an acetylated dihydrolipoyl group (Figure 7.1). The PDC-E1 reaction or those by the E1 components of other α-keto acid dehydrogenase complexes require that the reacting lipoate be attached to a lipoyl domain of the cognate E2 component of that complex (Bleile et al., 1981; Graham et al., 1989; Berg et al., 1998a; Perham, 2000; Liu et al., 2001). This specificity is determined by interactions between the lipoyl domain and its cognate E1 that, in part, depend on the distinct steric structure of the lipoyl domain. The congruent β-barrel structures of different lipoyl domains present the lipoyl-lysine prosthetic group on a tight β turn at a narrow end of the kidney-shaped domain (Dardel et al., 1993; Berg et al., 1996; Berg et al., 1998b; Howard et al., 1998; Perham, 2000). Key E1-specificity residues are found near the prosthetic group although some are also located some distance away (Berg et al., 1995; Wallis et al., 1996; Berg et al., 1998a; Gong et al., 2000; Fries et al., 2007). The specificity residues for mammalian E1 reaction, which were determined for the inner lipoyl domain (L2) of the E2 component of mammalian PDC, are delineated in Figure 7.3 (Gong et al., 2000). Substitution for these residues substantially reduced use of the L2 domain as a substrate. Leu140, Ser141, and Thr143 are located in a loop designated the E1-specificity loop. Ala substitution for Asp197 (located on the opposite side of lipoyl group and not shown) actually enhanced the affinity of L2 for E1. Phosphorylation prevents use of L2 as a substrate (Liu et al., 2001). The acidic residues Asp172, near the lipoyl group, and Glu179 and Glu162, well removed from the lipoylated end of L2, contribute to specific binding. Phosphorylation of E1 may result in electrostatic repulsion with one or more of these anionic residues. E1-catalyzed reductive acetylation appears to be the rate-limiting step in the mammalian PDC reaction (Cate et al., 1980). Comparing E1 use of L2 to a lipoylated peptide containing 15 of the residues of the L2 domain and blocked termini, there was >1500-fold decrease in k_{cat}/K_m of E1 with the peptide substrate (Liu et al., 2001).

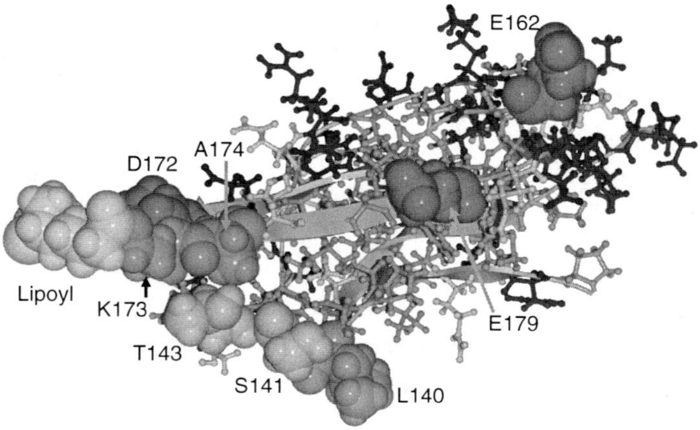

FIGURE 7.3 Specificity residues in the L2 domain for E1 catalyzed reductive acetylation. The lipoyl group and specificity residues in the L2 domain for the E1 reaction are shown with space-filled atoms and are labeled.

E2 and E3 Specificity

E2 catalyzes the transfer of the acetyl group to CoA and E3 catalyzes the transfer of the electrons on the resulting lipoyl group to NAD^+ (Figure 7.1). The specificities for these reactions are primarily for the reactive form of the lipoyl prosthetic group; indeed, the same E3 component is shared by different α-keto acid dehydrogenase complexes in many organisms including three complexes in mammalian mitochondria (PDC, α-ketoglutarate dehydrogenase, and branched-chain α-ketoacid dehydrogenase complex). There was greater than threefold increase in k_{cat}/K_m comparing the use of the free L2 domain to lipoamide as substrates in the human E3 reaction (Liu et al., 2001). The lipoamide used was a mixture of the *R*- and *S*-isomers; that may be contributing to the above difference since the E3 component has a higher activity with the natural *R* isomer (Pick et al., 1995). Only the *R* isomer is attached to the lipoyl domain substrate. E3-lipoyl domain interaction beyond the lipoyl-lysine was shown to be limited to the Asp and Ala that precede and follow the prosthetic group Lys based on NMR spectroscopy studies with the *Bacillus stearothermophilus* PDC system E3 and E2 lipoyl domain (Fries et al., 2007).

Relative Use of Different Lipoyl Domains in Overall PDC and E3 Reactions

We have prepared recombinant E2•E3BP structures containing just L1, L2, or lipoyl domain of E3BP with Lys that can be lipoylated; the other two Lys residues are converted by mutations to Ala residues. With excess E1 and E3, these E2•E3BP structures support per core ~72% (L1 lipoylated), ~84% (L2 lipoylated), and ~19% (L3 lipoylated) of the rate in the overall PDC reaction obtained with native E2•E3BP (Bao, 2004). As would be expected with the Lys of all three lipoyl domains converted to alanines, the resulting nonlipoylated E2•E3BP failed to support any activity in the overall PDC reaction. Therefore, all three lipoyl domains are effectively used by E1, E2, and E3, but L2 is used somewhat more effectively than L1. The E3BP lipoyl domain supports activity of $25 \pm 2.5\%$ when compared to just L1 or just L2 lipoylated. This is in accord with near equivalent use of 12 versus 48 lipoyl domains with just one of the three lipoyl domains carrying a lipoyl group in an $E2_{48}$•$E3BP_{12}$ structure. The structure with the lipoylated L2 was also used somewhat more effectively in an E3 cycling reaction (Liu et al., 1995b) performed with excess E3. The lipoylated E3BP lipoyl domain supported 25% or 21% as much activity as the E2•E3BP lipoylated only at L1 or L2, respectively (Bao, 2004). Under conditions that maximize incorporation, all the structures were acetylated to near the expected levels; the E2•E3BP subcomplex with L1 lipoylated had slightly more acetyl groups incorporated than the subcomplex with lipoylated L2 (Bao, 2004). Thus, the lesser support of the overall PDC reaction and the E3 cycling activity by the E2•E3BP lipoylated only on L1 than on L2 was not due to lack of lipoylation. The native E2•E3BP with all three sites carrying lipoyl groups supported only 16% higher PDC than the E2•E3BP with only L2 domain providing a lipoyl group. Therefore, there are sufficient lipoyl domains available in native human complex that PDC catalysis is not limited by access to these tethered substrates under most conditions. This conclusion is in accord with the finding that half of the bi-lipoyl domain regions of E2 subunits could be removed by selective cleavage on the N-terminal side of the B domain of bovine kidney E2•E3BP without any reduction in reconstituted PDC activity (Rahmatullah et al., 1990).

REGULATORY ENZYMES

Kinase Isoforms and Isoform Focus

The PDK are a unique class of Ser-kinases unrelated to the large class of standard Ser/Thr/Tyr kinases. The PDK have catalytic domains distantly related to ATP/ADP binding domains of bacterial histidine kinases, Hsp90, DNA gyrase, and other proteins, known as the GHKL superfamily (Bowker-Kinley and Popov, 1999; Wynn et al., 2000; Machius et al., 2001; Steussy et al., 2001). Among mammalian sources, there is invariant conservation of each PDK isoform (e.g., >94% sequence identity comparing human and rat sources), whereas the different isoforms differ in 35% of their aligned sequences in the same source. To exemplify the variation in PDK properties, we primarily compare PDK2 with PDK3 below. A more thorough description of kinase properties has been provided in a recent review (Roche and Hiromasa, 2007); by no means are PDK2 and PDK3 the most disparate among the PDKs. However, only in the case of PDK2 and PDK3 isoforms are the crystal structures available (Steussy et al., 2001; Kato et al., 2005; Knoechel et al., 2006) and these isoforms are the most extensively characterized among the PDKs. PDK2 is the most universally distributed in mammalian tissues of the PDKs (Bowker-Kinley et al., 1998; Wu et al., 2000). Human PDK2 responds to the full set of known regulatory responses (Baker et al., 2000; Bao et al., 2004a; Bao et al., 2004b). PDK3 is also widely distributed (kidneys, lungs, brain, testes, etc.) but is found only at very low levels in muscle tissues, liver, and adipose and some other tissues (Bowker-Kinley et al., 1998; Wu et al., 2000).

PDK2 Structure

PDK2 is a stable dimer (Steussy et al., 2001; Hiromasa and Roche, 2003); subunit association is via a β-sheet region in the C-terminal catalytic (Cat) domain as exhibited in PDK2 ribbon structure (Figure 7.4) (Steussy et al., 2001; Knoechel et al., 2006). Additional but variable dimer interactions are formed between subunits (below). The catalytic cavities are at each end of the dimer. A regulatory (R) domain sits above each Cat domain forming an extended interface between these domains at the center of the large open active site cavity (Figure 7.4). The large size of the cavity between the R and Cat domain appears to be needed for placing E1 in a position to undergo phosphorylation. The sites of phosphorylation are located on the α subunit of E1 at the active site entry, a central cleft between

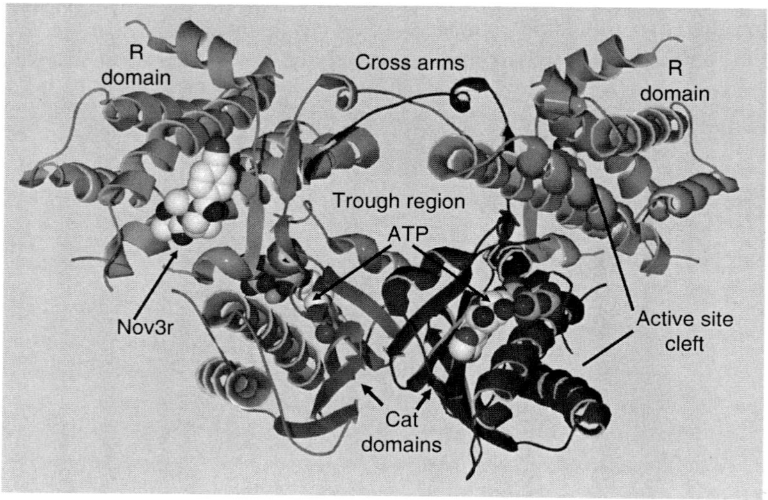

FIGURE 7.4 PDK2 dimer with ATP and Nov3r bound. The Cat domains associate by interactions of their β sheet regions, bind ATP, and form the base of the trough region. The inter-subunit cross arms also arise from the C-terminal end of the Cat domains (Knoechel et al., 2006). The R domain is shown binding Nov3r; most of the Nov3r is not visible behind the R domain on the right. On the ends, the R domains forms the upper side of the active site clefts (one on left is toward the back and not labeled) and, on the opposite sides, forms the sides of the trough regions that is spanned by the inter-subunit cross arms.

the α and β subunits of E1. On the central dimer side, rising from the base formed by the associating Cat domains, the R domains form the sides of skewed central trough region (Figure 7.4). Trough-spanning cross-arm structures create additional dimer interactions (Figure 7.4) (Kato et al., 2005; Knoechel et al., 2006) and play a critical role in E2 binding as described below.

PDP Isoforms, PDP1 Subunits, and Focus

The PDP are members of the divalent metal requiring 2C class of phosphatases (Teague et al., 1982; Huang et al., 1998) and specifically use Mg^{2+} in their di-metal active sites. A major feature in control of PDP activity is their effector-modified response to Mg^{2+} levels (Damuni et al., 1984; Thomas et al., 1986; Yan et al., 1996; Turkan et al., 2004). PDP1 has both a 50 kDa PP2C class catalytic subunit (Teague et al., 1982) and a large (96 kDa) regulatory subunit with a bound FAD (Teague et al., 1982; Lawson et al., 1997). Several structures of related proteins in the PP2C class have been described (e.g., Das et al., 1996; Pullen et al., 2004). The regulatory subunit is related to a series of proteins that carry out routine steps in catabolism. Crystal structures exist for some of these that are homologs of the N-terminal FAD-binding domain (e.g., monomeric sarcosine oxidase) (Wagner et al., 2000) and others that are homologs of the tetrahydrofolate-binding C-terminal half of PDP1r (e.g., glycine-T-cleavage enzyme) (Lee et al., 2004). There are also large proteins that contain both these large domains of PDP1r, including one, dimethylglycine oxidase, for which an x-ray crystal structure is available (Leys et al., 2003). This review will focus on PDP1 and its catalytic subunit, PDP1c, which bind to the L2 domain of E2 via an interaction that requires the lipoyl group of L2 (Chen et al., 1996; Turkan et al., 2002; Karpova et al., 2003; Turkan et al., 2004). This critical, activity-enhancing interaction is effector regulated.

ROLES OF LIPOYL DOMAINS IN THE REGULATION OF MAMMALIAN PDC

Enhanced PDK and PDP1 Function

The function and regulation of the PDK and PDP1 are dependent on regulated interactions with the lipoyl domains of the E2 component, primarily with the L2 domain (Roche et al., 1996; Roche et al., 2001; Roche et al., 2003a; Roche and Hiromasa, 2007). Indeed, the predominant changes in the PDK and PDP1 activities result from the dynamic, effector-modified interactions with L2 domain of E2. In comparison to use of the same level of free E1 substrate (nonphosphorylated- or phosphorylated-E1), the activities of PDK2, PDK3, or PDP1 are increased ~10-fold by E2 when assays are conducted with typical intramitochondrial concentrations of PDC (1.0–2 mg/mL PDC or 0.5–1 mg/mL E1) as a substrate (Chen et al., 1996; Baker et al., 2000; Bao et al., 2004b). The magnitude of these activity enhancements increases with dilution of these substrate sources (i.e., yet higher rates with E2•E1 than with free E1) (Hiromasa and Roche, unpublished). Enhanced activity is due to the greatly increased access that the regulatory enzymes, bound to and moving between the L2 domains of E2, have for the many E1 bound to the E2 oligomer.

E2 Constructs

As indicated above, this review will emphasize the structural basis within the lipoyl domains (domain and prosthetic group) and how these interactions participate in effector control of PDK and PDP activity. These properties have been uncovered using a variety of constructs of E2 or E2 domains, including free lipoyl domains, the dimer glutathione-*S*-transferase (GST) with a pair of lipoyl domains fused at the C-terminus, one or both lipoyl domains of E2 with the connecting E1 binding domain, E2 oligomers lacking one of specific lipoyl domains or the E1 binding domain, and E2 with the sites of lipoylation mutated as described above (Liu et al., 1995b; Yang et al., 1997; Yang et al., 1998; Baker et al., 2000; Bao et al., 2004b; Tuganova and Popov, 2005). The reaction state of the lipoyl prosthetic group (oxidized versus reduced versus acetylated) has been varied and prosthetic group removed and analogs were enzymatically substituted on lipoyl domains (Cate and Roche, 1978; Ravindran et al., 1996; Peng, 2001). Additionally, the effects of high levels of the free cofactor in different reaction states have been tested (Peng, 2001, see below). Before addressing the specific roles of lipoyl domains in PDK and PDP1 function, the general regulatory properties of these enzymes are described.

Regulatory Properties of PDK

Pyruvate dehydrogenase kinase activities are modulated by key metabolites that vary with fundamental changes in metabolic state (Randle, 1986; Randle and Priestman, 1996; Roche et al., 2001; Sugden et al., 2001; Roche and Hiromasa, 2007). Kinase activity is reduced by ADP, phosphate (P_i), and pyruvate. Elevation of ADP and P_i indicates an energy deficit. Pyruvate levels are elevated by hormone-induced or workload-fostered increases in glucose or glycogen breakdown. The most universal of the PDK isoforms, PDK2, is inhibited synergistically by the combination of these inhibitors (Baker et al., 2000; Bao et al., 2004a). In contrast PDK3 is insensitive to pyruvate inhibition but very sensitive to inhibition by ADP and P_i (Baker et al., 2000). Elevation of the NADH:NAD^+ and acetyl-CoA: CoA ratios stimulates PDK activity (Pettit et al., 1975; Cate and Roche, 1978; Ravindran et al., 1996; Baker et al., 2000; Bao et al., 2004b). NADH and acetyl-CoA are not only the products of the PDC reaction but also the catabolism of fatty acids, ketone bodies, and, even, amino acids. PDC activity is reduced by stimulation of PDK activity particularly when fatty acids are used as an energy source. This regulation of PDC activity provides critical intervention for conservation of body carbohydrate reserves. Again comparing PDK2 and PDK3 as examples, PDK2 activity is stimulated up to fourfold by acetyl-CoA and NADH whereas PDK3 activity is only modestly stimulated when overcoming inhibition by ADP and P_i (57). NADH and acetyl-CoA do not directly interact with PDK.

Increases in the intramitochondrial NADH/NAD$^+$ and acetyl-CoA/CoA ratios increase the proportion of oxidized, reduced, and acetylated lipoyl groups via the reversible E3 and E2 reactions (Cate and Roche, 1978; Rahmatullah and Roche, 1985; Ravindran et al., 1996; Popov, 1997; Baker et al., 2000; Bao et al., 2004b). Reduced and acetylated lipoyl domains bind to and enhance the PDK activity. Reductive acetylation of the L2 domain of PDK2 stimulates PDK2 activity by ~3.5-fold (Baker et al., 2000). Using E2 structures with lipoyl groups only on L1 or L2, reductive acetylation of just the L1 domain was much less effective in stimulating PDK2 activity (Baker et al., 2000). This specificity and other studies suggested that PDK2 dimer binds to the inner lipoyl domain (L2) via interacting with extended parts of the L2 domain structure and the lipoyl-lysine prosthetic group (Radke et al., 1993; Yang et al., 1998; Baker et al., 2000; Tuganova et al., 2002; Hiromasa and Roche, 2003; Tuganova and Popov, 2005). Tight binding analogs of acetylated dihydrolipoyl group have been developed (Aicher et al., 1999; Aicher et al., 2000; Bebernitz et al., 2000; Morrell et al., 2003). Crystal structures of PDK2 dimer with two of these analogs bound demonstrated binding at the lipoyl binding site on the trough side of the R domain of PDK2 (Knoechel et al., 2006).

Regulated PDK2 Interaction with L2

PDK2 has a weak affinity for the free monomeric L2 domain (Hiromasa and Roche, 2003). PDK2 binding is greatly enhanced with oligomeric L2 structures (E2 60mer but also the dimeric GST-L2) (Hiromasa and Roche, 2003). PDK2 binding affinity is enhanced by reduction and further strengthened by acetylation of the lipoyl domain (Hiromasa and Roche, 2003; Bao et al., 2004b). PDK2 binds GST-L2 with reduced lipoyl groups (i.e., GST-L2$_{red}$) with a $K_d = \sim 0.4\ \mu M$ (Hiromasa and Roche, 2003). The effects of ligands on this interaction have been evaluated by sedimentation velocity studies in potassium phosphate buffer (Hiromasa et al., 2006). ATP, ADP, or pyruvate reduces the binding of PDK2 to GST-L2$_{red}$ and the combination of ADP and pyruvate markedly reduces binding while causing the PDK2 dimer to associate and form a tetramer. When all the lipoyl groups of E2 are converted to the oxidized form, there is a marked reduction in the capacity of E2 to enhance PDK2 activity (Bao et al., 2004b). This state is only slowly reversed upon reduction of the lipoyl groups. We suggest this requires a change in the conformation of the L2 domain of E2 (Bao et al., 2004b). With bacterial lipoyl domains, the oxidized lipoyl group was shown to bind to the surface of the domain rather than persisting as an extended cofactor (Perham et al., 2002).

Effector Modification of PDK2 Activity and Structure

When PDK2 activity is aided by E2-enhanced access to E1, kinetic and other studies support ADP dissociation becoming the rate-limiting step in PDK2 catalysis (Bao et al., 2004a). Pyruvate further inhibits PDK2 catalysis by preferentially binding to PDK2•ADP (Bao et al., 2004a; Hiromasa et al., 2006). In the opposite direction, reductive acetylation of the L2 group of E2 leads to enhanced dissociation of ADP (Bao et al., 2004b). PDK2 structures demonstrate that the pyruvate and lipoyl binding sites are located in the R domain well separated from the active site (Knoechel et al., 2006). Therefore, allosteric coupling between these sites and the active site is required for these ligands influencing ADP dissociation. Similarly, allosteric coupling is needed for ATP/ADP binding at the active site greatly enhancing pyruvate binding and the combination greatly hindering binding of PDK2 to GST-$L2_{red}$.

Tryptophan fluorescence detects these conformational changes (Hiromasa et al., 2006). ATP/ADP or pyruvate quenches most of the fluorescence of Trp383. The concentration of pyruvate required for half-maximal quenching is lowered 150-fold in the presence of ADP (Hiromasa et al., 2006). Trp383 is located at the end of an inter-subunit cross arm, which arises from the Cat domain of one subunit and extends to the trough side interface where Trp383 inserts between the Cat and R domains of the other subunit. The capacity of ADP and pyruvate in potassium phosphate buffer to cause PDK2 to convert to a tetramer may result from release of the inter-subunit cross arm followed by direct involvement of the released cross arms in tetramer formation (Hiromasa et al., 2006). K and Pi are critical for coupled ligand binding, quenching of Trp fluorescence, and tetramer formation (Hiromasa and Roche, in press)

SPECIFIC NATURE OF PDK3 INTERACTION WITH L2

Tight PDK3-L2 Complex

Among the PDKs, PDK3 binds to the L2 domain with the highest affinity. Analytical ultracentrifugation studies (Hiromasa and Roche, in preparation) support an equilibrium dissociation constant of PDK3 for L2 in the low nanomolar range. A crystal structure of the PDK3 dimer in a complex with two L2 domains has been described (Figure 7.5A) (Kato et al., 2005). PDK3 is directly activated by the L2 domain (Baker et al., 2000) and that activity effect was used to evaluate the effects mutations in the L2 domain on this interaction (Peng, 2001; Roche et al., 2001; Roche et al., 2003a; Tso et al., 2006). PDK3 binding of the L2 domain engages the trough regions spanning the inter-subunit cross arm and beyond this a segment C-terminal tail that wraps around the lipoyl domain and moves up alongside the binding site of the lipoyl group in the R domain of the other subunit (Figure 7.5A) (Kato et al., 2005). As predicted from this 3-D structure, deletion of C-terminal residues in the C-terminus of PDK3 subunits interferes with binding to the L2 domain (Klyuyeva et al., 2005; Tso et al., 2006).

Tight binding of PDK to E2 led to the proposal that hand-over-hand movement of the PDK dimers between lipoyl domains of the E2 oligomer served as a mechanism for PDK accessing their E1 substrate (Ono et al., 1993; Liu et al., 1995a; Baker et al., 2000; Roche et al., 2001; Roche et al., 2003a). The very tight

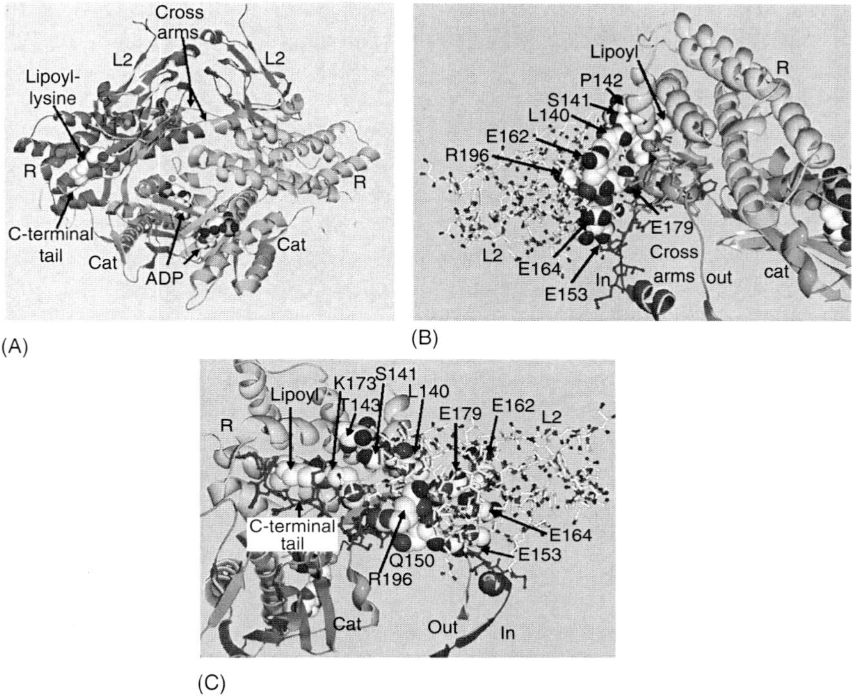

FIGURE 7.5 PDK3•L2 complex shown as dimer (A) with two L2 bound and with specificity residues of L2 (B and C). (A) Ribbon structures are shown for the PDK3 dimer and two bound L2 domains (Kato et al., 2005). Bound ADP in the Cat domains and lipoyl-lysine group inserted into R domain on the left are shown with space-filled atoms. (B and C) The L2 domains are shown as stick structures with varied shading of atoms except for the specificity residues that are shown with space-filled atoms and labeled. The PDK3 R and Cat domains are shown as ribbon structures except for parts of the incoming and outgoing cross arms/C-terminal tails that interact with L2, which are shown as stick structures with constant shading. Panel B shows the view where the cross arms enter and leave and panel C shows the other side of L2 and the end of the R domain where the lipoyl group inserts.

binding of PDK3 to E2 led to further studies that support this mechanism (Y. Hiromasa, S.A. Kasten, and T.E. Roche, unpublished). This includes evidence for PDK3 acting to efficiently phosphorylate E2-bound E1 in dilute complexes and evidence that the propensity of PDK3 to phosphorylate a second site is due to continued residency on E2 oligomer.

Critical L2 Domain Structure for Binding and Activation of PDK3

PDK3 undergoes an approximately eightfold enhancement in activity due to binding to the free L2 domain (Baker et al., 2000). Specific residues within the

L2 domain contribute to effective binding and activation of PDK3 (Peng, 2001; Roche et al., 2001; Roche et al., 2003a; Tso et al., 2006). The binding of L2 to the cross arm-containing C-terminal side of PDK3 appears to favor movement of the R domain that narrows the spacing in the central trough region while increasing the spacing in the exterior active site clefts (Kato et al., 2005). However, a similar trough closure with cross-arm formation was observed with PDK2 in the absence of L2 (Knoechel et al., 2006). L2 enhancement of PDK3 activity may in part result from the opening of the active site since this may aid binding of the large E1 substrate. Without L2, PDK3 dimers tend to self-associate, but binding of the L2 domain stabilizes PDK3 as a dimer (Peng, 2001; Roche et al., 2001; Roche et al., 2003a; Kato et al., 2005).

The effects of amino acid substitutions in the L2 domain on activation of PDK3 were characterized in our laboratory (Peng, 2001; Roche et al., 2001; Roche et al., 2003a) and by Tso et al. (2006). Among 12 amino acid substitutions near the lipoylated lysine, large effects in reducing L2 activation of PDK3 were observed with L140A, D172N, A174S as well as Ala, Met, or Gln substitutions for the lipoylated Lys173 (Figure 7.6). Smaller but significant effects resulted from mutation of S141A, T143A, Q150, R196Q, and mutations at the lipoylated

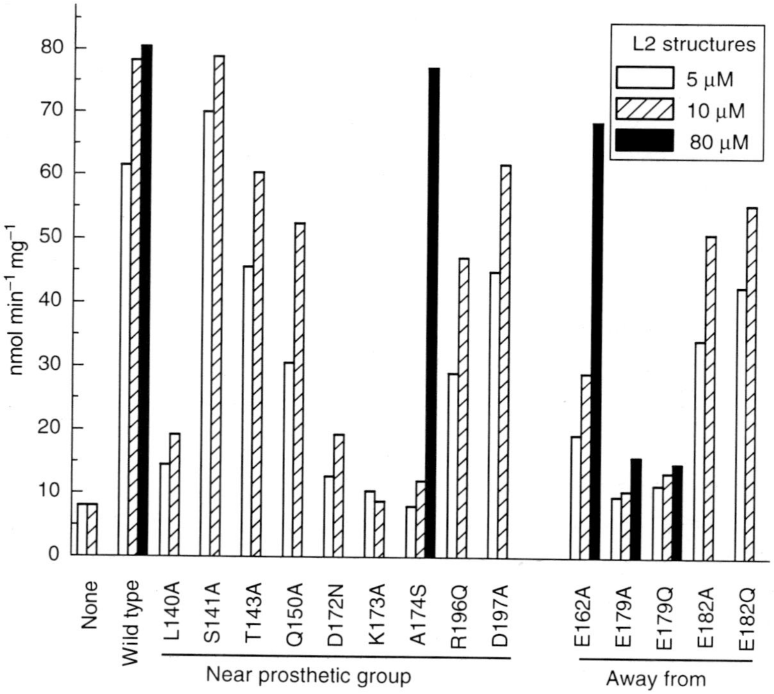

FIGURE 7.6 Effects of mutant L2 structures on human PDK3 activity. PDK3 activity was measured with the indicated level of L2 construct (Peng, 2001).

end of L2 (Figure 7.6). In addition to these mutant effects, Tso et al. (2006) found that D170A mutation at this end of L2 significantly affected PDK3 binding. The S141A mutant was unusual in that it had increased binding (and activation) PDK3 activity at 5 μM. At the lipoylated end of L2, all the L2 mutants that affected E1 use and binding to PDP1 (below) were located adjacent to the lipoylated Lys173 or in the E1-specificity loop. However, Arg196 and Thr143 substitutions, which affect PDK3 activation, are located on the opposite side of this end of L2. With the exception of mutants at Lys173, all the mutants at the lipoylated end of L2 that reduced activation at 5 μM had an increasing capacity to activate PDK3 as their concentrations were increased. Only in the case of L140A mutant did an 80 μM level not enhance PDK3 activity nearly as much as a 10 μM level wild type L2. Therefore, the substitution of residues near the lipoyl-lysine prosthetic group (Figure 7.5B and C) reduced the binding affinity but did not prevent L2 binding to PDK3 and when bound PDK3 was highly activated.

More than a dozen surface residues, not near the lipoyl prosthetic group, were also substituted and these mutant L2 domains were also evaluated for changes in their capacities to enhance PDK3 activity (Peng, 2001; Roche et al., 2001; Roche et al., 2003a; Tso et al., 2006). PDK3 activation was substantially reduced by substitutions for Glu162, Glu179 (Figure 7.6) (Peng, 2001; Roche et al., 2003a), as well as Glu153 and Glu164 (Tso et al., 2006) (Figure 7.5B and C). The conversion of Glu179 to an Ala or Gln markedly reduced PDK3 activation, suggesting a critical role for this residue. Increasing the concentration of E179A or E179Q from 5 to 80 μM did not significantly increase activation (Figure 7.6) and these mutant domains did not compete effectively in reducing activation by wild-type L2 (80 μM E179Q reduced the activation of PDK3 by 10 μM wild-type L2 by 11%). There was minimal binding of E179Q to PDK3 in sedimentation velocity studies. In contrast, at high levels there was significant binding by all other mutants except those substituted at the Lys173. Using an E179A mutant, these potent effects of mutating Glu179 were confirmed by Tso et al. (2006).

PDK3 has three structural regions that interact with the lipoyl domain: the trough-side of the R domain near the site of insertion of the lipoyl group, the outgoing beginning of the C-terminal cross arm, and the incoming C-terminal cross arm and its extended C-terminal tail (Figure 7.5B and C) (Kato et al., 2005; Tso et al., 2006). At the front of the lipoyl domain residues, Asp172, Arg196, and Glu170 interact with the C-terminal tail following the incoming cross arm and Thr143 and Leu140 interact with the R domain (Figure 7.5B and C). The S141A conversion may allow the side chain to gain an interaction with Phe22 of the R domain. Away from the lipoylated end of L2, Glu179 and Glu162 interact with the outgoing cross arm and Gln150, Glu153, and Glu164 interacting with the incoming cross arm (Figure 7.5B). N- and C-terminal regions (residues 128–130 and 214–220), which were in an unorganized state in the L2 structure determined by NMR (Howard et al., 1998), become organized in the PDK3•L2 complex (Kato et al., 2005), associating with each other at the distal end of the L2 domain from the lipoylated Lys173. However, none of these residues directly interact with PDK3.

L2 Prosthetic Group Structure for Binding and Activation of PDK3

We also evaluated the effects of enzymatic substitution of analogs for the lipoyl group of L2 on PDK3 activity (Figure 7.7). Octanoyl-L2 gave some enhancement of PDK3 activity whereas heptanoyl-L2 inhibited PDK3 activity. 8-Thiol-octanoyl-L2 was very effective in activating PDK3. Even though PDK3 is not particularly sensitive to stimulation by reductive-acetylation, E2 catalyzed acetylation of 8-thiol-octanoyl-L2 increased PDK3 activity as much as acetylation of L2's lipoyl group. This indicates that addition of the acetyl group to the 8-position is the primary change leading to direct stimulation of kinase activities. Therefore, neither the 6-SH of the reduced lipoyl group nor the formation of 6-*S*-acetyl group is needed for binding or activation. Reduction of the dithiolane ring and then acetylation sequentially extend the potential reach of the lipoyl prosthetic group, which introduces the capacities for additional interactions within the extended lipoyl binding site of PDK. The additional reach seems critical for interactions that generate conformational changes that stimulate PDK catalysis (Ravindran

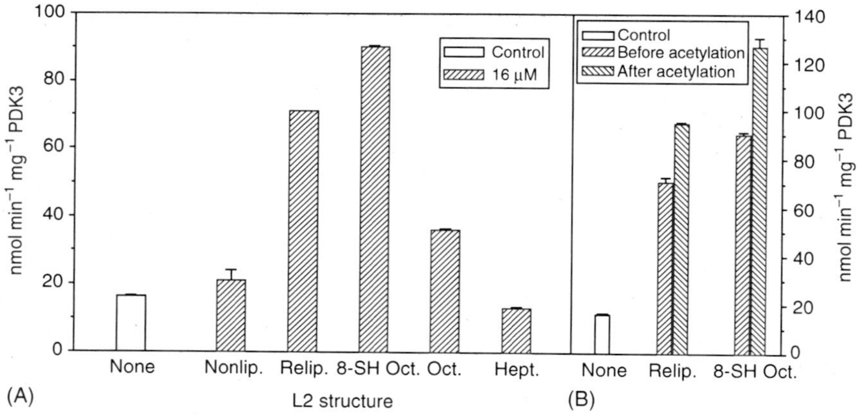

FIGURE 7.7 Effects of cofactor analogs and acetylation of 8-thiol-octanoate on human PDK3 activity. PDK3 activity was measured with 16 μM of the indicated L2 structures (Peng, 2001). Nonlipoylated L2 was prepared by expressing L2 domains without lipoyl supplementation, followed by separation of lipoylated and nonlipoylated L2 by hydrophobic interaction chromatography (Gong et al., 2000). Lipoate (relip) and lipoyl analogs were incorporated using *E. coli* lipoyl protein ligase (Gong et al., 2000). For acetylation, reaction mixtures included 150 μM NADH and 50 μM NAD^+, 1 μg E3 (for 60 s) and 0.5 μg tE2 (inner core of E2) (Hiromasa et al., 2004); tE2 construction (Peng, 2001) catalyzed acetylation with 30 μM acetyl-CoA for 25 s before initiating the kinase reaction (Ravindran et al., 1996).

et al., 1996; Roche et al., 2001). That is consistent with additional space within the lipoyl binding site (Knoechel et al., 2006) and effects of mutations of PDK2 to be described elsewhere (Hu, Hiromasa, and Roche, in preparation).

Effects of Free Lipoyl Group Structures on PDK3 Activity and L2 Binding

Free lipoamide and dihydrolipoamide at 0.5 mM gave greater than twofold enhancements of PDK3 activity but had no effect on PDK2 activity (Figure 7.8). However, activation of PDK3 was not increased in combination with L2 mutated at the site of lipoylation (e.g., 48 μM K173A-L2 or K173M-L2) (Figure 7.8). The activation of PDK3 activity by 10 μM wild-type L2 is reduced only slightly by 0.5 mM levels of lipoamide or dihydrolipoamide or by these those ligands in combination with 48 μM levels of the above L2 domains mutated at the site of lipoylation (Figure 7.8). Therefore, binding of a lipoyl group covalently attached to L2 is critical for activation. Heptanoyl-L2 did compete somewhat with wild-type L2 with a threefold higher level causing about a 50% reduction in the

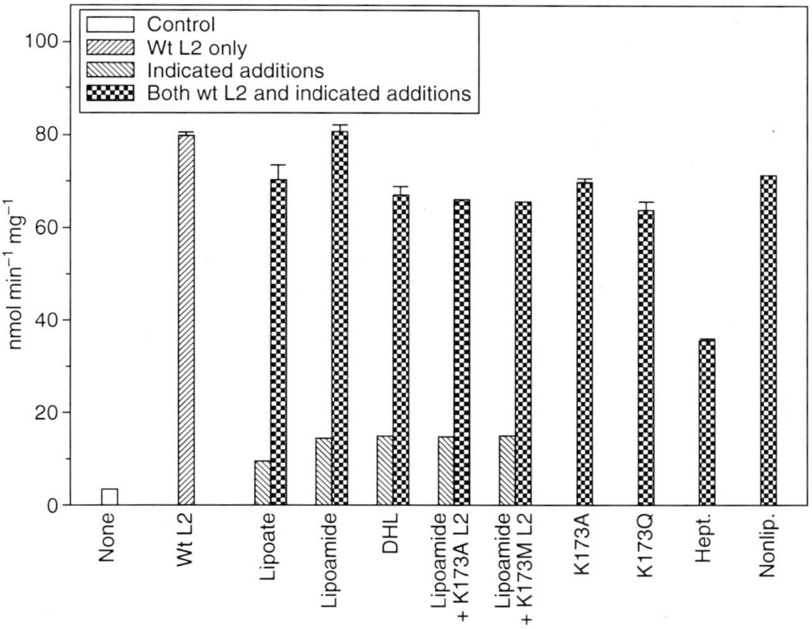

FIGURE 7.8 Interactions of PDK3 with elevated levels of free cofactor structures and of L2 structures mutated or modified at lipoylation site. Components were as shown in the figure inset and on *x*-axis (Peng, 2001). When included, native L2 was always at 10 μM, the other L2 constructs were at 50 μM, and free cofactor structures (lipoate, lipoamide, and dihydrolipoamide [DHL]) were added at 0.5 mM.

activation of PDK3. Thus, the hydrophobic character of the first seven carbons of the lipoyl group does make a significant contribution to lipoyl group binding.

The potent inhibitor, Nov3r, binds where the lipoyl group binds in the R domain of PDK2 and structurally appears to be related to acetyl-dihydrolipoamide (Knoechel et al., 2006; Roche and Hiromasa, 2007). The residues in the R domains of PDK2 and PDK3 that enclose the Nov3r/lipoyl group binding site are completely conserved (Knoechel et al., 2006) and yet there is no effect of the free lipoyl groups and very small effects of free lipoyl domains on PDK2 activity, even with the dimeric GST-L2 that binds well to PDK2 (Baker et al., 2000; Hiromasa and Roche, 2003). Reductive acetylation of GST-L2 does give about a twofold increase in PDK2 activity. Despite having complete conservation of the residues that form direct interactions in the lipoyl group binding site, the fact that just the free lipoyl group enhances PDK3 activity while not affecting PDK2 activity indicates that there is some difference in either the lipoyl group interaction or the capacity of lipoyl group binding to increase catalysis. It seems likely that a combination of differences in lipoyl group binding and in the interaction with L2 domain lead to the 1000-fold tighter binding of L2 to PDK3 than PDK2. PDK1 is also activated by free lipoyl domains but less so than PDK3 and similarly has an affinity in between PDK2 and PDK3 for the L2 domain. In marked contrast to the other isoforms, PDK4 activity is only modestly increased by binding to E2 except at low E1 levels but reductive acetylation of E2 does enhance PDK4 activity (Dong, 2001). Also uniquely, PDK4 activity is stimulated by Nov3r (Morrell et al., 2003). In the lipoyl-binding site, PDK4 differs from the other three PDK isoforms in having Leu32 replace the aligned Phe28 of PDK2.

REQUIREMENTS FOR L2 BINDING TO PDP1

Metal Requirements for PDP1 and PDP1c Binding to L2 Domain

The activities of PDP1 and PDP1c are Mg^{2+}-dependent with PDP1c having a K_m of 0.5 mM for Mg^{2+} (Hucho et al., 1972; Lawson et al., 1993). The PDP1r subunit increases the K_m for Mg^{2+} of PDP1 by approximately sixfold and polyamines like spermine reverse this increase (Damuni et al., 1984; Thomas et al., 1986). Like PDK, E2 increases PDP1 activity by ~10-fold via PDP1 binding to the L2 domain (Chen et al., 1996) by providing access to E2-bound phosphorylated E1 (Pettit et al., 1972). This interaction of PDP1 or just the PDP1c subunit with L2 domain of E2 requires and is regulated by Ca^{2+} (Pettit et al., 1972; Chen et al., 1996; Turkan et al., 2002). Half-maximal activation of PDP1 is attained with ~1.0 μM Ca^{2+} at near-saturating Mg^{2+} and with ~3.0 μM Ca^{2+} at 0.8 mM Mg^{2+} whereas activation of PDP1c required ~3.0 μM Ca^{2+} at high or low Mg^{2+} levels (Turkan et al., 2004). Cytoplasmic Ca^{2+} is increased by signaling, exercise, and other energy-demanding changes and then is actively taken up by mitochondria where Ca^{2+} enhances NADH production and consequently ATP production (Denton and McCormack, 1990; Hansford, 1994; Denton et al.,

1996). PDP1 and PDP1c bind in a Ca^{2+}-dependent manner to free L2 structures. This selective interaction has been used for affinity purification of PDP1 and PDP1c with gel-anchored GST-L2 (Chen et al., 1996; Turkan et al., 2004); tight binding of PDP1 (but not PDP1c) additionally requires elevated Mg^{2+} (Chen et al., 1996; Turkan et al., 2004). Thus, PDP1c•Ca^{2+}•L2 complex formation is supported by micromolar Ca^{2+} and binding by PDP1c becomes sensitive to the Mg^{2+} level when PDP1c is bound to PDP1r. Sedimentation velocity and equilibrium studies revealed that PDP1c exists as a reversible monomer–dimer mixture with an equilibrium dissociation constant of 8 ± 3.5 μM (Turkan et al., 2004). L2 binds tightly and preferentially to the PDP1c monomer (Turkan et al., 2004). About 45 PDP1c monomers bind to E2 60mer with K_d of ~0.3 μM (Turkan et al., 2004).

Isothermal titration calorimetry and $^{45}Ca^{2+}$-binding studies failed to detect Ca^{2+} (<100 μM) binding to L2 or PDP1c, alone, but readily observed binding to L2 plus PDP1c (Turkan et al., 2004). Therefore, both proteins are required to form a complex with tightly held Ca^{2+} and complex formation hinders the tendency of PDP1c to form a dimer. Neither PDP1c nor L2 exhibit sequences characteristic of normal Ca^{2+} binding domains (EF hands or C2 domains). Apparently, PDP1c•L2 complex formation creates a site for tightly binding Ca^{2+}.

L2 Domain Structure Needed for Binding PDP1c

Two regions in the L2 structure are required for binding of PDP1c (Turkan et al., 2002). The first again involves the lipoyl group and the lipoylated end of the L2. PDP1 and PDP1c binding to L2 was prevented by mutations of Lys173 or lack of lipoylation of this lysyl residue and is greatly hindered by substitutions, D172N and A174S, at the adjacent residues as well as by the L140A substitution in the nearby E1-specificity loop. T143A substitution in the E1-specificity loop enhanced L2 binding to PDP1 and PDP1c (Turkan et al., 2002). Octanoyl-L2 effectively binds PDP1c indicating the hydrophobic character of the lipoyl group is most important whereas the dithiolane ring and 8-thiol are needed for effective binding to E1 and PDK3, respectively (Turkan et al., 2002).

The second region of L2 involves a cluster of acidic residues (glutamates 162, 179, and 182 along with Gln181). Particularly large reductions in binding of PDP1 and PDP1c to the L2 ensued from mutation of Glu182 (Turkan et al., 2002). Since mutation of all these residues hindered L2 binding to PDP1, an appealing prospect is that these form a strong electrostatic interaction either with Ca^{2+} in combination with residues provided by PDP1c or that they interact with a positively charged region of PDP1c. Therefore, the Ca^{2+}-dependent binding of PDP1 to L2 requires domain-aided hydrophobic binding by the exterior lipoyl group at one end of L2 and electrostatic binding by the opposite end of the L2 domain. An appealing prospect is that the interaction of the lipoylated end of L2 promotes a conformational change in PDP1c which in turn fosters and stabilizes PDP1c participation in a Ca^{2+} binding site formed with the collaboration of the acidic end of L2. Understanding of this interaction will be aided by determination of the structure of PDP1c.

CONCLUSIONS

This chapter has focused on the use of lipoyl domains in the E1 reaction and in binding and supporting the catalytic function and regulation of PDK2, PDK3, and PDP1. In each lipoyl domain, role-specific surface structure in the protein domain works with the lipoyl prosthetic group in achieving the consequential functional roles. The domain structure gives specificity for the reductive acetylation reaction catalyzed by E1. In the case of the PDK and PDP1, the interactions make important contributions to function and regulation. Binding of PDP1 and PDK to the lipoyl domain of E2 enormously increases their activities by providing access to their E1 substrates. Feedback effector control of PDC is achieved by NADH and acetyl-CoA being used to reductively acetylate the L2 domain of E2; acetylated-L2 speeds up PDK-catalyzed inactivation of PDC. PDK2 activity is increased due to acetylated-L2 causing a conformational change that decreases ADP binding at the active site. Activation of PDC is achieved by enhanced PDP1 activity being enhanced by Ca^{2+}-regulated binding of PDP1 to the L2 domain of E2. Thus, these novel mechanisms utilize the mobile outer domains of E2 with specific involvement of the lipoyl group in the translation of key metabolic signals to upregulate or downregulate PDC activity.

ACKNOWLEDGMENT

This work was supported by NIH grant DK18320.

REFERENCES

Aevarsson, A., Seger, K., Turley, S., Sokatch, J.R., and Hol, W.G. 1999. Crystal structure of 2-oxoisovalerate and dehydrogenase and the architecture of 2-oxo acid dehydrogenase multienzyme complexes. *Nat. Struct. Biol.* 6:785–792.

Aevarsson, A., Chuang, J.L., Wynn, R.M., Turley, S., Chuang, D.T., and Hol, W.G. 2000. Crystal structure of human branched-chain alpha-ketoacid dehydrogenase and the molecular basis of multienzyme complex deficiency in maple syrup urine disease. *Structure* 8:277–291.

Aicher, T.D., Anderson, R.C., Beberntiz, G.R., Coppola, G.M., Jewell, C.F., Knorr, D.C., Liu, C., Sperbeck, D.M., Brand, L.J., Strohschein, R.J., Gao, J., Vinluan, C.C., Shetty, S.S., Dragland, C., Kaplan, E.L., DelGrande, D., Islam, A., Liu, X., Lozito, R.J., Maniara, W.M., Walter, R.E., and Mann, W.R. 1999. *R*-3,3,3-trifluoro-2-hydroxy-2-methylpropionamides are orally active inhibitors of pyruvate dehydrogenase kinase. *J. Med. Chem.* 42:2741–2746.

Aicher, T.D., Anderson, R.C., Gao, J., Shetty, S.S., Coppola, G.M., Stanton, J.L., Knorr, D.C., Sperbeck, D.M., Brand, L.J., Winluan, C.C., Kaplan, E.L., Dragland, C.J., Tomaselli, H.C., Islam, A., Lozito, R.J., Liu, X., Maniara, W.M., Fillers, W.S., DelGrande, D., Walter, R.E., and Mann, W.R. 2000. Secondary amides of *R*-3,3,3-trifluoro-2-hydroxy-2-methylpropionic acid as inhibitors of pyruvate dehydrogenase kinase. *J. Med. Chem.* 43:236–249.

Baker, J.C., Yan, X., Peng, T., Kasten, S.A., and Roche, T.E. 2000. Marked differences between two isoforms of human pyruvate dehydrogenase kinase. *J. Biol. Chem.* 275:15773–15781.

Bao, H. 2004. Regulation of pyruvate dehydrogenase kinase 2 and related functional properties of the human pyruvate dehydrogenase complex. MS Thesis, Kansas State University.

Bao, H., Kasten, S.A., Yan, X., and Roche, T.E. 2004a. Pyruvate dehydrogenase kinase isoform 2 activity limited and further inhibited by slowing down the rate of dissociation of ADP. *Biochemistry* 43:13432–13441.

Bao, H., Kasten, S.A., Yan, X., Hiromasa, Y., and Roche, T.E. 2004b. Pyruvate dehydrogenase kinase isoform 2 activity stimulated by speeding up the rate of dissociation of ADP. *Biochemistry* 43:13442–13451.

Bebernitz, G.R., Aicher, T.D., Stanton, J.L., Gao, J., Shetty, S.S., Knorr, D.C., Strohschein, R.J., Tan, J., Brand, L.J., Liu, C., Wang, W.H., Vinluan, C.C., Kaplan, E.L., Drangland, C.J., DelGrande, D., Islam, A., Lozito, R.J., Liu, X., Maniara, W.M., and Mann, W.R. 2000. Anilides of *R*-trifluoro-2-hydroxy-2-methylpropionic acid as inhibitors of pyruvate dehydrogenase kinase. *J. Med. Chem.* 43:2248–2257.

Berg, A., Vervoort, J., and de Kok, A. 1995. Sequential 1H and 15N nuclear magnetic resonance assignments and secondary structure of the lipoyl domain of the 2-oxoglutarate dehydrogenase complex from *Azotobacter vinelandii*. Evidence for high structural similarity with the lipoyl domain of the pyruvate dehydrogenase complex. *Eur. J. Biochem.* 234:148–159.

Berg, A., Vervoort, J., and deKok, A. 1996. Solution structure of the lipoyl domain of the 2-oxoglutarate dehydrogenase complex from *Azotobacter vinelandii*. *J. Mol. Biol.* 261:432–442.

Berg, A., Westphal, A.H., Bosma, H.J., and De Kok, A. 1998a. Kinetics and specificity of reductive acylation of wild-type and mutated lipoyl domains of 2-oxo-acid dehydrogenase complexes from *Azotobacter vinelandii*. *Eur. J. Biochem.* 252: 45–50.

Berg, A., Westphal, A.H., Bosma, H.J., and De Kok, A. 1998b. Three-dimensional structure in solution of the N-terminal lipoyl domain of the pyruvate dehydrogenase complex from *Azotobacter vinelandii*. *Eur. J. Biochem.* 252:45–50.

Bleile, D.M., Hackert, M.L., Pettit, F.H., and Reed, L.J. 1981. Subunit structure of dihydrolipoyl transacetylase component of pyruvate dehydrogenase complex from bovine heart. *J. Biol. Chem.* 256:514–519.

Bowker-Kinley, M. and Popov, K.M. 1999. Evidence that pyruvate dehydrogenase kinase belongs to the ATPase/kinase superfamily. *Biochem. J.* 344:47–53.

Bowker-Kinley, M.M., Davis, W.I., Wu, P., Harris, R.A., and Popov, K.M. 1998. Evidence for existence of tissue-specific regulation of the mammalian pyruvate dehydrogenase complex. *Biochem. J.* 329:191–196.

Brautigam, C.A., Wynn, R.M., Chuang, J.L., Machius, M., Tomchick, D.R., Chuang, D.T. 2006. Structural insight into interactions between dihydrolipoamide dehydrogenase E3 and E3 binding protein of human pyruvate dehydrogenase complex. *Structure* 14:611–621.

Cate, R.L. and Roche, T.E. 1978. A unifying mechanism for stimulation of mammalian pyruvate dehydrogenase kinase activity by NADH, dihydrolipoamide, acetyl coenzyme A, or pyruvate. *J. Biol. Chem.* 253:496–503.

Cate, R.L., Roche, T.E., and Davis, L.C. 1980. Rapid intersite transfer of acetyl groups and movement of pyruvate dehydrogenase component in the kidney pyruvate dehydrogenase complex. *J. Biol. Chem.* 255:7556–7562.

Chen, G., Wang, L., Liu, S., Chang, C., and Roche, T.E. 1996. Activated function of the pyruvate dehydrogenase phosphatase through Ca^{2+}-facilitated binding to the inner lipoyl domain of the dihydrolipoyl acetyltransferase. *J. Biol. Chem.* 271:28064–28070.

Ciszak, E.M., Korotchkina, L.G., Dominiak, M., Sidhu, S., and Patel, M.S. 2003. Structural basis for flip-flop action of thiamin pyrophosphate-dependent enzymes revealed by human pyruvate dehydrogenase. *J. Biol. Chem.* 278:21240–21246.

Ciszak, E.M., Makal, A., Hong, Y.S., Vettaikkorumakankauv, A.K., Korotchkina, L.G., and Patel, M.S. 2006. How dihydrolipoamide dehydrogenase-binding protein binds dihydrolipoamide dehydrogenase in the human pyruvate dehydrogenase complex. *J. Biol. Chem.* 281:648–655.

Damuni, Z., Humphreys, J.S., and Reed, L.J. 1984. Stimulation of pyruvate dehydrogenase phosphatase activity by polyamines. *Biochem. Biophys. Res. Commun.* 124:95–99.

Dardel, F., Davis, A.L., Laue, E.D., and Perham, R.N. 1993. Three-dimensional structure of the lipoyl domain from *Bacillus stearothermophilus* pyruvate dehydrogenase multienzyme complex. *J. Mol. Biol.* 229:1037–1048.

Das, A.K., Helps, N.R., Cohen, P.T.W., and Barford, D. 1996. Crystal structure of the protein serine/threonine phosphatase 2C at 2.0 Å resolution. *EMBO Journal* 15:6798–6809.

Denton, R.M. and McCormack, J.G. 1990. Ca^{2+} as a second messenger within mitochondria of the heart and other tissues. *Ann. Rev. Physiol.* 52:451–466.

Denton, R.M., McCormack, J.G., Rutter, G.A., Burnett, P., Edgell, N.J., Moule, S.K., and Diggle, T.A. 1996. The hormonal regulation of pyruvate dehydrogenase complex. *Advan. Enzyme Regul.* 36:183–198.

Dong, J. 2001. Expression, purification, and characterization of human pyruvate dehydrogenase kinase isoform 4. PhD Thesis, Kansas State University.

Fries, M., Stott, K.M., Reynolds, S., and Perham, R.N. 2007. Distinct modes of recognition of the lipoyl domain as substrate by the E1 and E3 components of the pyruvate dehydrogenase multienzyme complex. *J. Mol. Biol.* 366:132–139.

Gong, X., Peng, T., Yakhnin, A., Zolkiewski, M., Quinn, J., Yeaman, S.J., and Roche, T.E. 2000. Specificity determinants for the pyruvate dehydrogenase component reaction mapped with mutated and prosthetic group modified lipoyl domains. *J. Biol. Chem.* 275:13645–13653.

Gopalakrishnan, S., Rahmatullah, M., Radke, G.A., Powers-Greenwood, S.L., and Roche, T.E. 1989. Role of protein X in the function of mammalian pyruvate dehydrogenase complex. *Biochem. Biophys. Res. Commun.* 160:715–721.

Graham, L.D., Packman, L.C., and Perham, R.N. 1989. Kinetics and specificity of reductive acylation of lipoyl domains from 2-oxo acid dehydrogenase multienzyme complexes. *Biochemistry* 28:1574–1581.

Gudi, R., Bowker-Kinley, M.M., Kedishvili, N.Y., Zhao, Y., and Popov, K.M. 1995. Diversity of the pyruvate dehydrogenase kinase gene family in humans. *J. Biol. Chem.* 270:28989–28994.

Hansford, R.G. 1994. Physiological role of mitochondrial Ca^{2+} transport. *J. Bioenerg. Biomembr.* 26:495–508.

Harris, R.A., Bowker-Kinley, M.M., Wu, P., Jeng, J., and Popov, K.M. 1997. Dihydrolipoamide dehydrogenase-binding protein of the human pyruvate dehydrogenase complex. DNA-derived amino acid sequence, expression, and reconstitution of the pyruvate dehydrogenase complex. *J. Biol. Chem.* 272:19746–19751.

Harris, R.A., Huang, B., and Wu, P. 2001. Control of pyruvate dehydrogenase kinase gene expression. *Advan. Enzyme Regul.* 41:269–288.

Hiromasa, Y. and Roche, T.E. 2003. Facilitated interaction between the pyruvate dehydrogenase kinase isoform 2 and the dihydrolipoyl acetyltransferase. *J. Biol. Chem.* 278: 33681–33693.

Hiromasa, Y., Fujisawa, T., Aso, Y., and Roche, T.E. 2004. Organization of the cores of the mammalian pyruvate dehydrogenqase complex formed by E2 and E2 plus the E3-binding protein and capacities to bind the E1 and E3 components. *J. Biol. Chem.* 279:6921–6933.

Hiromasa, Y., Hu, L., and Roche, T.E. 2006. Ligand-induced effects on pyruvate dehydrogenase kinase isoform 2. *J. Biol. Chem.* 281:12568–12579.

Howard, M.J., Fuller, C., Broadhurst, R.W., Perham, R.N., Tang, J.-G., Quinn, J., Diamond, A.G., and Yeaman, S.J. 1998. Three-dimensional structure of the major autoantigen in primary biliary cirrhosis. *Gastroenterology* 115:139–146.

Huang, B., Gudi, R., Wu, P., Harris, R.A., Hamilton, J., and Popov, K.M. 1998. Isozymes of pyruvate dehydrogenase phosphatase. DNA-derived amino acid sequences, expression, and regulation. *J. Biol. Chem.* 273:17680–17688.

Hucho, F., Randall, D.D., Roche, T.E., Burgett, M.W., Pelley, J.W., and Reed, L.J. 1972. α-keto acid dehydrogenase complexes. XVII. Kinetic and regulatory properties of pyruvate dehydrogenase kinase and pyruvate dehydrogenase phosphatase from bovine kidney and heart. *Arch. Biochem. Biophys.* 151:328–340.

Jilka, J.M., Rahmatullah, M., Kazemi, M., and Roche, T.E. 1986. Properties of a newly characterized protein of the bovine kidney pyruvate dehydrogenase complex. *J. Biol. Chem.* 261:1858–1867.

Karpova, T., Danchuk, S., Kolobova, E., and Popov, K.M. 2003. Characterization of the isozymes of pyruvate dehydrogenase phosphatase: Implications for regulation of pyruvate dehydrogenase activity. *Biochim. Biophys. Acta* 1652:126–135.

Kato, M., Chuang, J.L., Tso, S.-C., Wynn, R.M., and Chuang, D.T. 2005. Crystal structure of pyruvate dehydrogenase kinase 3 bound to lipoyl domain 2 of human pyruvate dehydrogenase complex. *EMBO J.* 24:1763–1774.

Klyuyeva, A., Tuganova, A., and Popov, K.M. 2005. The carboxy-terminal tail of pyruvate dehydrogenase kinase 2 is required for the kinase activity. *Biochemistry* 44:13573–13582.

Knoechel, T.R., Tucker, A.D., Robinson, C.M., Phillips, C., Taylor, W., Bungay, P.J., Kasten, S.A., Roche, T.E., and Brown, D.G. 2006. Regulatory roles of the N-terminal domain based on crystal structures of human pyruvate dehydrogenase kinase 2 containing physiological and synthetic ligands. *Biochemistry* 45:402–415.

Kong, Y., Ming, D., Wu, Y., Stoops, J.K., Zhou, Z.H., and Ma, J. 2003. Conformational flexibility of pyruvate dehydrogenase complexes: A computational analysis by quantized elastic deformational model. *J. Mol. Biol.* 330:129–135.

Lawson, J.E., Niu, X.-D., Browning, K.S., Trong, H.L., Yan, J., and Reed, L.J. 1993. Molecular cloning and expression of the catalytic subunit of bovine pyruvate dehydrogenase phosphatase and sequence similarity with protein phosphatase 2C. *Biochemistry* 32:8987–8993.

Lawson, J.E., Park, S.H., Mattison, A.R., Yan, J., and Reed, L.J. 1997. Cloning, expression and properties of the regulatory subunit of bovine pyruvate dehydrogenase phosphatase. *J. Biol. Chem.* 272:31625–31629.

Lee, H.H., Kim, D.J., Ahn, H.J., Ha, J.Y., and Suh, S.W. 2004. Crystal structure of T-protein of the glycine cleavage system. Cofactor binding, insights into H-protein recognition, and molecular basis for understanding nonketotic hyperglycinemia. *J. Biol. Chem.* 279:50514–50523.

Leys, D., Basran, J., and Scrutton, N.S. 2003. Channelling and formation of 'active' formaldehyde in dimethylglycine oxidase. *EMBO J.* 22:4038–4048.

Liu, S., Baker, J.C., and Roche, T.E. 1995a. Binding of the pyruvate dehydrogenase kinase to recombinant constructs containing the inner lipoyl domain of the dihydrolipoyl acetyltransferase component. *J. Biol. Chem.* 270:793–800.

Liu, S., Baker, J.C., Andrews, P.C., and Roche, T.E. 1995b. Recombinant expression and evaluation of the lipoyl domains of the dihydrolipoyl acetyltransferase component of human pyruvate dehydrogenase complex. *Arch. Biochem. Biophys.* 316:926–940.

Liu, S., Gong, X., Yan, X., Peng, T., Baker, J.C., Li, L., Robben, P.M., Ravindran, S., Andersson, L.A., Cole, A.B., and Roche, T.E. 2001. Reaction mechanism for mammalian pyruvate dehydrogenase using natural lipoyl domain substrates. *Arch. Biochem. Biophys.* 386:123–135.

Machius, M., Chuang, J.L., Wynn, R.M., Tomchick, D.R., and Chuang, D.T. 2001. Structure of rat BCK kinase: Nucleotide-induced domain communication in a mitochondrial protein kinase. *Proc. Natl. Acad. Sci. USA* 98:11218–11223.

Mande, S.S., Sarfaty, S., Allen, M.D., Perham, R.N., and Hol, W.G. 1996. Protein–protein interactions in the pyruvate dehydrogenase multienzyme complex: Dihydrolipoamide dehydrogenase complexed with the binding domain of dihydrolipoamide acetyltransferase. *Structure* 4:277–286.

Mattevi, A., Obmolova, G., Schulze, E., Kalk, K.H., Westpha, A.H., de Kok, A., and Hol, W.G.J. 1992. Atomic structure of the cubic core of the pyruvate dehydrogenase multienzyme complex. *Science* 225:1544–1550.

Morrell, J.A., Orme, J., Butlin, R.J., Roche, T.E., Mayers, R.M., and Kilgour, E. 2003. AZD7545 is a selective inhibitor of pyruvate dehydrogenase kinase 2. *Biochem. Soc. Trans.* 31:1168–1170.

Neagle, J., De Marcucci, O., Dunbar, B., and Lindsay, J.G. 1989. Component X of mammalian pyruvate dehydrogenase complex: Structural and functional relationship to the lipoate acetyltransferase E2 component. *FEBS Lett.* 353:11–15.

Oliver, R.M. and Reed, L.J. 1982. *Electron Microscopy of Proteins*. Ed. R. Harris, Vol. 2, pp. 1–48. London: Academic Press.

Ono, K., Radke, G.A., Roche, T.E., and Rahmatullah, M. 1993. Partial activation of the pyruvate dehydrogenase kinase by the lipoyl domain region of E2 and interchange of the kinase between lipoyl domain regions. *J. Biol. Chem.* 268:26135–26143.

Patel, M.S. and Korotchkina, L.G. 2006. Regulation of the pyruvate dehydrogenase complex. *Biochem. Soc. Trans.* 34:217–222.

Patel, M.S. and Roche, T.E. 1990. Molecular biology and biochemistry of pyruvate dehydrogenase complexes. *FASEB J.* 4:3224–3233.

Peng, T. 2001. Functional consequences of substitutions for the prosthetic group of the inner lipoyl domain of dihydrolipoyl acetyltransferase and the inner lipoyl domains mechanisms of activation of pyruvate dehydrogenase kinase isoform 3. PhD Thesis, Kansas State University.

Perham, R.N. 2000. Swinging arms and swinging domains in multifunctional enzymes: Catalytic machines for multistep reactions. *Ann. Rev. Biochem.* 69:961–1004.

Perham, R.N., Jones, D.D., Chauhan, H.J., and Howard, M.J. 2002. Substrate channeling in 2-osoacid dehydrogenase multienzyme complexes. *Biochem. Soc. Trans.* 30:47–51.

Pettit, F.H., Roche, T.E., and Reed, L.J. 1972. Function of calcium ions in pyruvate dehydrogenase phosphatase activity. *Biochem. Biophys. Res. Commun.* 49:563–571.

Pettit, F.H., Pelley, J.W., and Reed, L.J. 1975. Regulation of pyruvate dehydrogenase kinase and phosphatase by acetyl-CoA/CoA and NADH/NAD ratios. *Biochem. Biophys. Res. Commun.* 65:575–582.

Pick, U., Haramaki, N., Constantinescu, A., Handelman, G.J., Tritschier, H.J., and Packer, L. 1995. Glutathione reductase and lipoamide dehydrogenase have opposite stereospecificities for α-lipoic acid enantiomers. *Biochem. Biophys. Res. Commun.* 206:724–730.

Popov, K.M. 1997. Regulation of mammalian pyruvate dehydrogenase kinase. *FEBS Lett.* 419:197–200.

Powers-Greenwood, S.L., Rahmatullah, M., Radke, G.A., and Roche, T.E. 1989. Separation of protein X from the dihydrolipoyl transacetylase component of the mammalian pyruvate dehydrogenase complex and role of protein X. *J. Biol. Chem.* 264:3655–3657.

Pullen, K.E., Ng, H.L., Sung, P.Y., Good, M.C., Smith, S.M., and Alber, T. 2004. An alternate conformation and a third metal in PstP/Ppp, the M. tuberculosis PP2C-family Ser/Thr protein phosphatase. *Structure* 12:1947–1954.

Radke, G.A., Ono, K., Ravindran, S., and Roche, T.E. 1993. Critical role of a lipoyl cofactor of the dihydrolipoyl acetyltransferase in the binding and enhanced function of the pyruvate dehydrogenase kinase. *Biochem. Biophys. Res. Commun.* 190:982–991.

Rahmatullah, M. and Roche, T.E. 1985. Modification of bovine kidney pyruvate dehydrogenase kinase activity by CoA esters and their mechanism of action. *J. Biol. Chem.* 260:10146–10152.

Rahmatullah, M. and Roche, T.E. 1987. The catalytic requirements for reduction and acetylation of protein X and the related regulation of various forms of resolved pyruvate dehydrogenase kinase. *J. Biol. Chem.* 262:10265–10271.

Rahmatullah, M., Gopalakrishnan, S., Andrews, P.C., Chang, C.L., Radke, G.A., and Roche, T.E. 1989a. Subunit associations in the mammalian pyruvate dehydrogenase complex: Structure and role of protein X and the pyruvate dehydrogenase component binding domain of the dihydrolipoyl transacetylase component. *J. Biol. Chem.* 264:2221–2227.

Rahmatullah, M., Gopalakrishnan, S., Radke, G.A., and Roche, T.E. 1989b. Domain structures of the dehydrolipoyl transacetylase and the protein X components of mammalian pyruvate dehydrogenase complex–selective cleavage by protease Arg C. *J. Biol. Chem.* 264:1245–1251.

Rahmatullah, M., Radke, G.A., Andrews, P.C., and Roche, T.E. 1990. Changes in the core of the mammalian-pyruvate dehydrogenase complex upon selective removal of the lipoyl domain from the transacetylase component but not from the protein X component. *J. Biol. Chem.* 265:14512–14517.

Randle, P.J. 1986. Fuel selection in animals. *Biochem. Soc. Trans.* 14:1799–1806.

Randle, P.J. and Priestman, D.A. 1996. Shorter term and longer term regulation of pyruvate dehydrogenase kinases. In *Alpha-Keto Acid Dehydrogenase Complexes*. Ed. M.S. Patel, T.E. Roche, and R.A. Harris, pp. 151–161. Basel: Birkhauser Verlag.

Ravindran, S., Radke, G.A., Guest, J.R., and Roche, T.E. 1996. Lipoyl domain-based mechanism for integrated feedback control of pyruvate dehydrogenase complex by enhancement of pyruvate dehydrogenase kinase activity. *J. Biol. Chem.* 271: 653–662.

Reed, L.J. 1974. Multienzyme complexes. *Acc. Chem. Res.* 7:40–46.

Reed, L.J. and Hackert, M.L. 1990. Structure-function relationships in dihydrolipoamide acyltransferases. *J. Biol. Chem.* 265:8971–8974.

Roche, T.E. and Hiromasa, Y. 2007. Pyruvate dehydrogenase kinase regulatory mechanisms and inhibition in treating diabetes, heart ischemia, and cancer, *Cell. Mol. Life Sci.* 64:830–849.

Roche, T.E., Liu, J., Ravindran, S., Baker, J., and Wang, D. 1996. Role of the E2 core in the dominant mechanisms of regulatory control of mammalian pyruvate dehydrogenase complex. In *Alpha-Keto Acid Dehydrogenase Complexes*. Ed. M.S. Patel, T.E. Roche, and R.A. Harris, pp. 33–52. Basel: Birkhauser Verlag.

Roche, T.E., Baker, J.C., Yan, X., Hiromasa, Y., Gong, X., Peng, T., Dong, J., Turkan, A., and Kasten, S.A. 2001. Distinct regulatory properties of pyruvate dehydrogenase kinase and phosphatase isoforms. *Prog. Nucleic Acid Res. Mol. Biol.* 70:33–75.

Roche, T.E., Hiromasa, Y., Turkan, A., Gong, X., Peng, T., Dong, J., and Turkan, A. 2003a. Lipoyl domain-facilitated activated function and control of pyruvate dehydrogenase kinase and phosphatase isoform 1. *Eur. J. Biochem.* 270:1050–1056.

Roche, T.E., Hiromasa, A., Turkan, A., Gong, X., Peng, T., Yan, X., and Kasten, S.A. 2003b. Central organization of mammalian pyruvate dehydrogenase PD complex and lipoyl domain-mediated function and control of PD kinases and phosphatase 1. In *Thiamine: Catalytic Mechanisms and Role in Normal and Disease States*. Ed. F. Jordan and M.S. Patel, pp. 363–386. New York: Marcel Dekker.

Rowles, J., Scherer, S.W., Xi, T., Majer, M., Nickle, D.C., Rommens, J.M., Popov, K.M., Harris, R.A., Riebow, N.L., Xia, J., Tsui, L.-C., Bogardus, C., and Prochazka, M. 1996. Cloning and characterization of PDK4 on 7q21.3 encoding a fourth pyruvate dehydrogenase kinase isozyme in human. *J. Biol. Chem.* 271:22376–22382.

Sanderson, S.J., Miller, C.M., and Lindsay, G.J. 1996a. Stoichiometry, organisation and catalytic function of protein X of the pyruvate dehydrogenase complex from bovine heart. *Eur. J. Biochem.* 236:68–77.

Sanderson, S.J., Khan, S.S., McCartney, R.G., Miller, C.M., and Lindsay, G.J. 1996b. Reconstitution of mammalian pyruvate dehydrogenase and 2-oxoglutarate dehydrogenase complex: Analysis of protein X involvement and interaction of homologous and heterologous dihydrolipoamide dehydrogenase. *Biochem. J.* 319:109–116.

Smolle, M., Prior, A.E., Brown, A.E., Cooper, A., Byron, O., and Lindsay, G.J. 2006. A new level of architectural complexity in the human pyruvate dehydrogenase complex. *J. Biol. Chem.* 28:19772–19780.

Steussy, C.N., Popov, K.M., Bowker-Kinley, M.M., Sloan, R.B., Jr., Harris, R.A., and Hamilton, J.A. 2001. Structure of pyruvate dehydrogenase kinase. Novel folding pattern for a serine protein kinase. *J. Biol. Chem.* 276:37443–37450.

Sugden, M.C., Bulmer, K., and Holness, M.J. 2001. Fuel-sensing mechanisms integrating lipid and carbohydrate utilization. *Biochem. Soc. Trans.* 29:272–278.

Teague, W.M., Pettit, F.H., Wu, T.-L., Silberman, S.R., and Reed, L.J. 1982. Purification and properties of pyruvate dehydrogenase phosphatase from bovine heart and kidney. *Biochemistry* 21:5585–5592.

Thekkumkara, T.J., Ho, L., Wexler, I.D., Pons, G., Lui, T.-C., and Patel, M.S. 1988. Nucleotide sequence of a cDNA for the dihydrolipoamide acetyltransferase component of human pyruvate dehydrogenase complex. *FEBS Lett.* 240:45–48.

Thomas, A.P., Diggle, T.A., and Denton, R.M. 1986. Sensitivity of pyruvate dehydrogenase phosphatase to Mg^{2+} ions. *Biochem. J.* 238:83–91.

Tso, S.C., Kato, M., Chuang, J.L., and Chuang, D.T. 2006. Structural determinants for cross-talk between pyruvate dehydrogenase kinase 3 and lipoyl domain 2 of the human pyruvate dehydrogenase complex. *J. Biol. Chem.* 281:27197–28204.

Tuganova, A. and Popov, K.M. 2005. Role of protein–protein interactions in the regulation of pyruvate dehydrogenase kinase. *Biochem. J.* 387:147–153.

Tuganova, A., Boulantnikov, I. and Popov, K.M. 2002. Interaction between the individual isoenzymes of pyruvate dehydrogenase kinase and the inner lipoyl-bearing domain of transacetylase component of pyruvate dehydrogenase complex. *Biochem. J.* 366:129–136.

Turkan, A., Gong, X., Peng, T., and Roche, T.E. 2002. Structural requirements within the lipoyl domain for the Ca^{2+}-dependent binding and activation of pyruvate dehydrogenase phosphatase isoform 1 or its catalytic subunit. *J. Biol. Chem.* 277:14976–14985.

Turkan, A., Hiromasa, Y., and Roche, T.E. 2004. Formation of a complex of the catalytic subunit of pyruvate dehydrogenase phosphatase isoform 1 PDP1c and L2 domain forms a Ca^{2+}-binding site and captures PDP1c as a monomer. *Biochemistry* 43:15073–15085.

Wagner, M.A., Trickey, P., Chen, Z.W., Mathews, F.S., and Jorns, M.S. 2000. Monomeric sarcosine oxidase: 1. Flavin reactivity and active site binding determinants. *Biochemistry* 39:8813–8824.

Wallis, N.G., Allen, M.D., Broadhurst, R.W., Lessard, I.A.D., and Perham, R.N. 1996. Recognition of a surface loop of the lipoyl domain underlies substrate channelling in the pyruvate dehydrogenase multienzyme complex. *J. Mol. Biol.* 263:463–474.

Wu, P., Blair, P.V., Sato, J., Jaskiewicz, J., Popov, K.M., and Harris, R.A. 2000. Starvation increases the amount of pyruvate dehydrogenase kinase in several mammalian tissues. *Arch. Biochem. Biophys.* 381:1–7.

Wynn, R.M., Chaung, J.L., Cote, C.D., and Chuang, D.T. 2000. Tetrameric assembly and conservation in the ATP-binding domain of rat branched-chain alpha-ketoacid dehydrogenase kinase. *J. Biol. Chem.* 275:30512–30519.

Yan, J., Lawson, J.E., and Reed, L.J. 1996. Role of the regulatory subunit of bovine pyruvate dehydrogenase phosphatase. *Proc. Natl. Acad. Sci. USA* 93:4953–4956.

Yang, D., Song, J., Wagenknecht, T., and Roche, T.E. 1997. Assembly and full functionality of recombinantly expressed dihydrolipoyl acetyltransferase component of the human pyruvate dehydrogenase complex. *J. Biol. Chem.* 272:6361–6369.

Yang, D., Gong, X., Yakhnin, A., and Roche, T.E. 1998. Requirements for the adaptor protein role of dihydrolipoyl acetyltransferase in the upregulated function of the pyruvate dehydrogenase kinase and pyruvate dehydrogenase phosphatase. *J. Biol. Chem.* 273:14130–14137.

Yu, X., Hiromasa, Y., Tsen, H., Stoops, J.K., Roche, T.E., and Zhou, Z.H. 2007. Structures of the human pyruvate dehydrogenase complex cores: A highly conserved catalytic center with flexible N-terminal domains. *Structure*, in press.

Zhou, Z.H., Liao, W., Cheng, R.H., Lawson, J.E., McCarthy, D.B., Reed, L.J., and Stoops, J.K. 2001. Direct evidence for the size and conformational variability of the pyruvate dehydrogenase complex revealed by three-dimensional electron microscopy. *J. Biol. Chem.* 276:21704–21713.

8 Inactivation and Inhibition of Alpha-Ketoglutarate Dehydrogenase: Oxidative Modification of Lipoic Acid

Kenneth M. Humphries, Amy C. Nulton-Persson, and Luke I. Szweda

CONTENTS

In studies designed to establish molecular mechanisms by which free-radical species and derived products mediate mitochondrial dysfunction, lipoic acid presented itself as keenly susceptible to oxidative modification. Oxidative modification of this naturally occurring enzyme-bound cofactor results in profound deficits of specific enzymatic functions and, ultimately, deficits in oxidative phosphorylation. In the following chapter, we will discuss molecular mechanisms by which different forms of oxidative stress irreversibly inactivate and reversibly inhibit the multienzyme complex α-ketoglutarate dehydrogenase (KGDH) by virtue of this enzyme's reliance on the covalently attached lipoic acid for activity. A brief discussion of mitochondria as a source of oxygen derived free radicals and pro-oxidants and of oxidative modifications to macromolecules, particularly with respect to membrane lipids, is offered to provide relevance for these findings.

MITOCHONDRIA: A SOURCE OF REACTIVE OXYGEN SPECIES

The mitochondrial electron transport chain is responsible for the four-electron reduction of oxygen to water and occurs concurrently with the extrusion of protons from the mitochondrial matrix. By pumping protons out of the matrix, an electrochemical gradient is created which supplies the energy for the production of ATP [1]. Electrons are donated to the electron transport chain by NADH and FADH through complex I (NADH coenzyme Q reductase) and complex II (succinate dehydrogenase), respectively. Ubiquinone, which accepts electrons from complexes I and II, undergoes two sequential one-electron reductions to ubisemiquinone and ubiquinol (the Q cycle). Complex III (ubiquinol cytochrome *c* reductase) transfers the electrons from ubiquinol to cytochrome *c*. Reduced cytochrome *c* is then used by complex IV (cytochrome *c* oxidase) for the reduction of oxygen to water. Nevertheless, depending on the metabolic state of the mitochondria and, in particular, the rate of electron transport, one-electron reduction of oxygen can occur resulting in the formation of the free-radical species superoxide anion ($O_2^{\dot{-}}$) [2–10].

Molecular oxygen, as the terminal electron acceptor of the mitochondrial electron transport chain, is essential for aerobic energy production. Conversely, formation of reactive species derived from the incomplete reduction of oxygen can lead to oxidative damage and derangements in mitochondrial and cellular function. Under normal conditions, enzymatic and nonenzymatic antioxidant systems scavenge free radicals and preserve mitochondrial integrity [11]. However, during certain degenerative processes, such as cardiac ischemia/reperfusion [12–17], evidence suggests that mitochondrial generation of pro-oxidant species increases indicating that antioxidants become overwhelmed. Such conditions of oxidative stress are often associated with declines in mitochondrial respiratory activity [13,18–25]. Highly reactive and short-lived, oxygen radicals would be expected to cause damage at or near the site of their formation. Oxidative damage has therefore been speculated to play a role in the loss of mitochondrial function that occurs during certain pathophysiological events.

Oxygen radicals and pro-oxidants generated by respiring mitochondria include superoxide anion ($O_2^{\dot{-}}$), hydrogen peroxide (H_2O_2), and hydroxyl radical [3–10,26–29]. Production increases when the half-life of reduced components of the respiratory chain increases [3–10]. This would be expected during state 4 (ADP independent) relative to state 3 respiration (ADP dependent) and as a result of electron chain deficits distal to known sites of free-radical production.

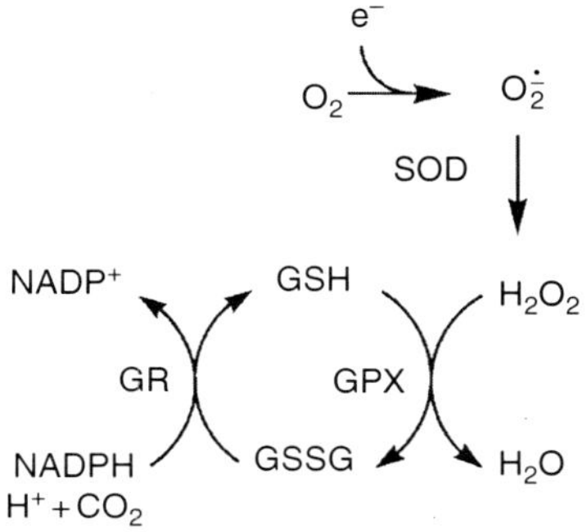

SCHEME 8.1 Enzymatic removal of superoxide anion. SOD, superoxide dismutase; GPX, glutathione peroxidase; GR, glutathione reductase.

$O_2^{\cdot-}$ produced by the electron transport chain is rapidly converted to H_2O_2 by superoxide dismutase [5,7,28,29]. Hydrogen peroxide can then be metabolized through the actions of glutathione peroxidase, catalase, peroxiredoxin, and thioredoxin/thioredoxin reductase [30–35], thereby minimizing the potential for oxidative damage to macromolecules (Scheme 8.1). If, however, the production of H_2O_2 is of sufficient magnitude or duration, oxidative modifications can occur. In the presence of metal ions, H_2O_2 can be converted to hydroxyl radical. Hydroxyl radical is capable of causing significant oxidative damage to DNA [2,26], membrane lipids containing polyunsaturated fatty acids [36,37], and protein [38,39].

LIPID PEROXIDATION AND FORMATION OF 4-HYDROXY-2-NONENAL

To determine how increased rates of mitochondrial free-radical production impact function, it is critical to consider likely mechanisms of oxidative damage. While protein, DNA, and lipids are all susceptible to oxidative damage, the oxidation of lipids is of interest for several reasons. Polyunsaturated fatty acids are thought to be the most susceptible cellular component to oxidative damage [11,36,37]. The susceptibility of polyunsaturated fatty acids arises from the conjugated diene system of the fatty acid backbone [11,36,37]. Fatty acids, especially within the mitochondrial membrane, would be expected to be particularly susceptible to oxidation because of their proximity to known sites of free-radical production. Lipid peroxidation is likely to impact mitochondrial function in multiple ways. Oxidation of fatty acids leads to the formation of aldehydes, alkenals, and hydroxyalkenals [36,37]. It has been demonstrated that these products of lipid peroxidation are highly reactive and cytotoxic [37]. It is therefore believed that lipid peroxidation and, in particular, the products of this process, mediate, in part, free-radical toxicity. Additionally, lipid peroxidation can alter membrane properties such as membrane fluidity [40]. This would in turn change the properties

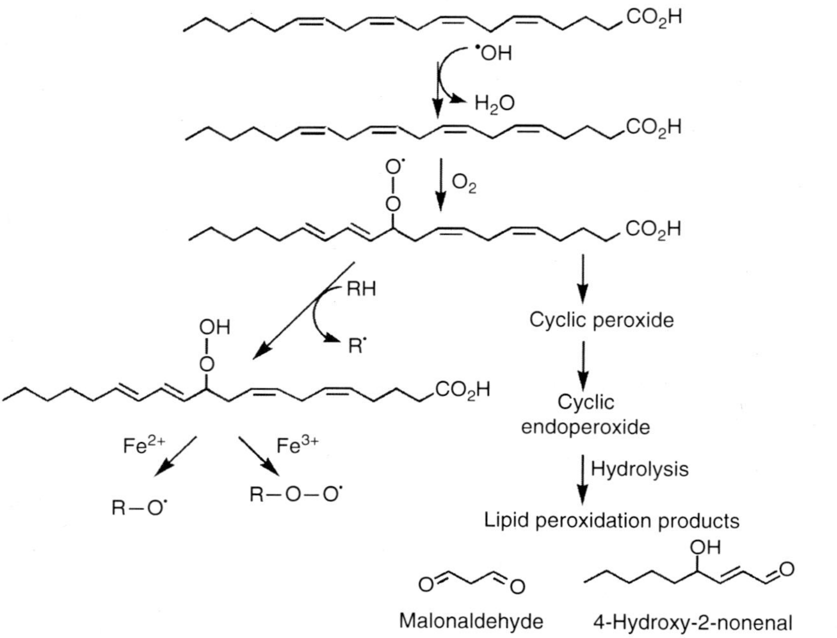

SCHEME 8.2 Lipid peroxidation and formation of lipid peroxidation products.

and functions of membrane proteins and could further enhance oxygen derived pro-oxidant production.

Peroxidation of membrane lipids occurs in a three-step process [11] (Scheme 8.2). In the first step, known as initiation, a hydrogen from a methylene group of a polyunsaturated fatty acid is abstracted by a free-radical species with sufficient oxidation potential, such as the hydroxyl radical. Polyunsaturated fatty acids are more susceptible to this type of reaction because once the hydrogen is abstracted, stabilization of the allylic radical is provided due to the conjugated diene system. Allylic radicals can then react with molecular oxygen to form peroxyl radicals. The formation of peroxyl radicals leads propagation of lipid peroxidation due to the ability of the peroxyl radical to abstract a hydrogen from another polyunsaturated fatty acid, thus forming a lipid hydroperoxide. It is this chain reaction that, if continued unabated, can damage the membrane integrity. Finally, the peroxyl radical may form a cyclic peroxide and/or a cyclic endoperoxide. These species can then undergo β-cleavage to form carbonyl containing compounds, including aldehydes, alkenals, and hydroxyalkenals [36,37]. As compared to certain free-radical species, which are diffusion limited, lipid peroxidation products are more stable and thus may more readily diffuse, allowing these products to cause damage distal to their initial site of production. Nevertheless, many of these products are hydrophobic and would be expected to be concentrated within the vicinity of the membrane.

REACTIVITY OF 4-HYDROXY-2-NONENAL WITH PROTEIN

In vitro experiments revealed that a variety of aldehydes, alkenals, and hydroxyalkenals are produced upon metal stimulated lipid peroxidation [36]. Malondialdehyde (MDA) is the major product formed during lipid peroxidation, but is not as reactive and cytotoxic at physiological pH as the hydroxyalkenals. 4-Hydroxy-2-nonenal (HNE) is the predominant hydroxyalkenal formed during in vitro microsomal peroxidation [36]. Therefore, HNE represents a likely mediator of free-radical damage. On the basis of the chemical properties of HNE, it is believed that the toxic effects of this lipid peroxidation product are mediated by its high reactivity with protein. HNE has three main functional groups that can participate alone or in sequence in chemical reactions with protein: the aldehyde group, the double bond, and the hydroxyl group [37]. Specifically, HNE can react with the sulfhydryl group of cysteine, the imidazole nitrogens of histidine, and the ε-amino group of lysine via 1,4 Michael addition to the double bond [37,41–49] (Scheme 8.3). The electron withdrawing hydroxyl group (C4) increases the

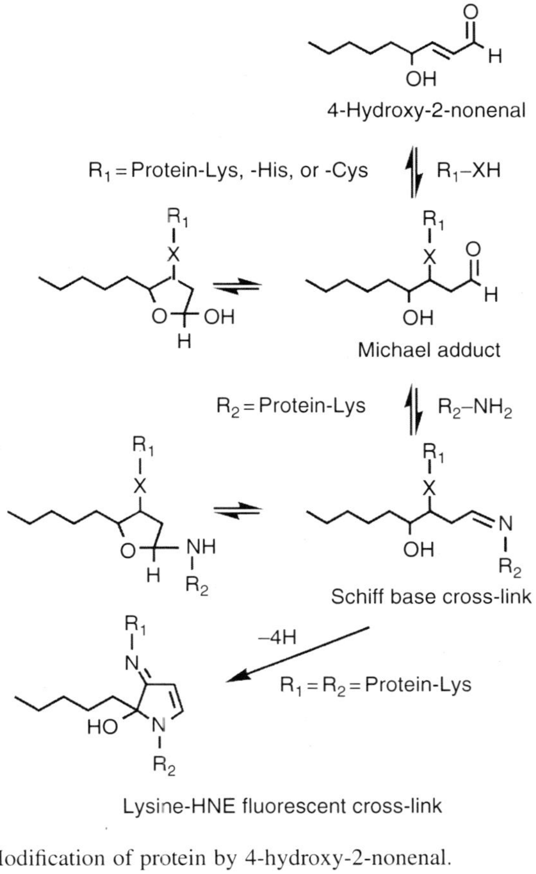

SCHEME 8.3 Modification of protein by 4-hydroxy-2-nonenal.

electrophilic properties of the double bond (C3), thereby increasing the capacity of HNE to undergo Michael addition. The implications of HNE modification of protein are many fold. HNE can rapidly inhibit enzymes by covalent modification of exposed histidine, cysteine, or lysine residues at an active site [37,41,46,48]. Alternatively, HNE modification may also cause enzyme inhibition by inducing conformational changes in protein structure [50]. Additionally, exposure of protein to HNE results in the formation of cross-linked protein and the appearance of HNE derived fluorescence products with properties reminiscent of lipofuscin [42,51–54]. A large number of pathophysiological conditions have been associated with an increase in HNE production in vivo. However, the consequences of HNE modification of protein on cellular function are difficult to determine in complex biological systems. In vitro studies examining the chemistry and mechanisms of HNE inactivation of enzymatic function therefore offer insight into the role and mechanism whereby lipid peroxidation leads to cellular damage during conditions of oxidative stress.

INHIBITION OF MITOCHONDRIAL RESPIRATION BY 4-HYDROXY-2-NONENAL

Mitochondria are a source of pro-oxidants and exhibit deficits in function under conditions of oxidative stress. As noted above, HNE is a potential mediator of oxidative damage by virtue of its reactivity with protein. It was therefore of interest to determine the effects of HNE on mitochondrial function and molecular mechanisms responsible for HNE-induced alterations in function. This information would be critical to assigning the relevance of such processes to the development of certain pathophysiological conditions associated with mitochondrial free-radical production and loss in function. Treatment of isolated cardiac mitochondria with HNE (micromolar concentrations) results in declines in NADH-linked ADP-dependent (state 3) respiration [55]. Inhibition is dependent upon both HNE concentration and time of incubation. Although HNE is an extremely reactive molecule, evidence suggested that the biochemical mechanism(s) by which HNE exerted effects on respiration were specific rather than global in nature. HNE did not inhibit ADP-independent (state 4) respiration or uncouple oxidative phosphorylation. In addition, HNE inhibited NADH-linked but not FADH-linked respiration. Overall, under conditions in which respiration was diminished, HNE did not appear to affect membrane integrity or electron transport chain activity [55].

The results of these experiments suggested that HNE inhibited mitochondrial respiration by reducing the availability of NADH as a source of electrons for electron transport. Subsequent studies confirmed that under conditions in which mitochondrial respiration is inhibited by HNE, there was a dramatic decrease in the production of NAD(P)H [55]. Thus, inhibition of mitochondrial respiration was due to the decrease in the activity of NADH-producing dehydrogenase(s). Following treatment of mitochondria with HNE, the effects on numerous dehydrogenases were assessed. Under the experimental conditions tested, the only dehydrogenase susceptible to HNE-mediated inhibition was KGDH [56]. Furthermore, a one-to-one relationship between the degree of KGDH inhibition and loss in the maximal rate of state 3 respiration was observed indicating that the effects of HNE on mitochondrial respiration were due, in large part, to inactivation of KGDH [55].

LIPOIC ACID AS A TARGET OF 4-HYDROXY-2-NONENAL MODIFICATION

α-Ketoglutarate dehydrogenase is a critical, rate-limiting enzyme of the Krebs cycle. Thus, inhibition of this enzyme would clearly have profound effects on the ability of the mitochondria to meet cellular energy requirements. It was therefore important to define the molecular mechanism of HNE-mediated inactivation of KGDH and to identify the feature(s) of the enzyme that makes it particularly susceptibility to inactivation.

α-Ketoglutarate dehydrogenase is a classically described multienzyme complex, structurally and catalytically similar to pyruvate and branched chain α-ketoacid dehydrogenase. These complexes are comprised of multiple copies of three enzymes, α-ketoacid decarboxylase (E1), dihydrolipoyl transacetylase (E2), and dihydrolipoamide dehydrogenase (E3) (Scheme 8.4). In addition, dehydrogenase activity requires three cofactors: thiamine pyrophosphate bound to E1, lipoic acid covalently bound to the E2 subunit, and FAD+ for E3 activity. Lipoic acid is covalently bound to these proteins by an amide linkage to the ε-amino group of specific lysine residues. Upon decarboxylation of substrate by E1, the lipoyllysine group functions as a carrier of reaction intermediates that can interact with both the E1 and E3 subunits of the ketoacid multienzyme complexes [57].

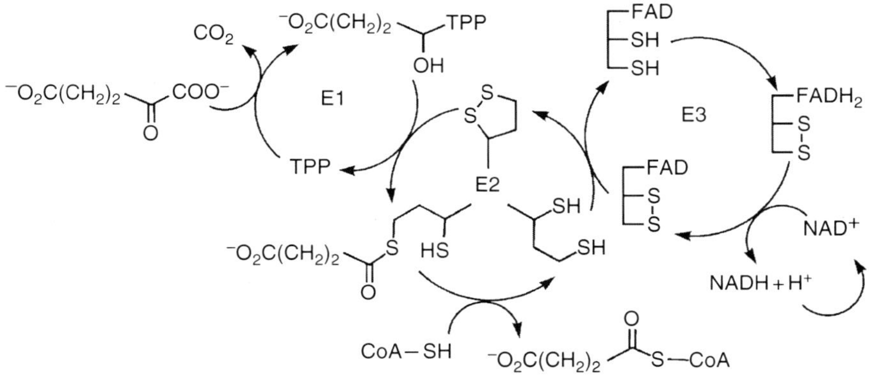

SCHEME 8.4 α-Ketoglutarate dehydrogenase. E1, α-ketoacid decarboxylase; E2, dihydrolipoyl transacetylase; and E3, dihydrolipoamide dehydrogenase.

Lipoic acid, first isolated and identified by Lester Reed in the 1950s as a microbial growth factor, is a sulfur containing coenzyme [58]. This cofactor is unique to five known proteins, all located within the mitochondria: the dihydrolipoamide acyltransferase (E2) subunits of pyruvate, α-ketoglutarate, and branched-chain α-ketoacid complexes, E3-binding protein (protein X) of the pyruvate dehydrogenase complex, and the H-protein of the glycine cleavage system [57,59–61]. The fact that lipoic acid is a strong nucleophile by virtue of its vicinal thiols was particularly interesting. As an electrophilic molecule, HNE would be expected to react readily with lipoic acid under physiological conditions.

Numerous lines of evidence indicate that indeed, the unique sensitivity of KGDH to HNE inactivation is conferred by the enzyme's required cofactor, lipoic acid (Scheme 8.5). First, it was found that HNE had minimal effects on the activity of purified KGDH. However, if the enzyme was incubated with HNE in the presence of α-ketoglutarate and CoASH or NADH, the enzyme could be readily inactivated. It was also noted that while PDH is resistant to HNE mediated inactivation in mitochondria, purified PDH could be inactivated. Furthermore, purified PDH was inhibited to a much larger extent in the presence of substrate (CoASH and pyruvate, or NADH). In the presence of NADH or CoASH and α-ketoglutarate, lipoic acids on the E2 subunit of KGDH are maintained in the reduced state [56] (Scheme 8.4).

The site of HNE modification was further probed by enzymatically measuring the transfer of an artificial substrate from the E2 to E3 subunits. In the presence of NADH, dithiolnitrobenzoic acid (DTNB/Ellman's reagent) binds to reduced lipoic acid on the E2 subunit and is then catalytically cleaved by the E3 subunit of the multienzyme complex [62,63]. The rate of this activity was measured in the presence and absence of HNE. The rate of DTNB reduction decreased with increasing concentrations of HNE, correlating closely with a loss in KGDH activity. Furthermore, under the same conditions, HNE had no effect on E3 activity as measured by the rate of reduction of NAD^+ in the presence of dihydrolipoamide [56].

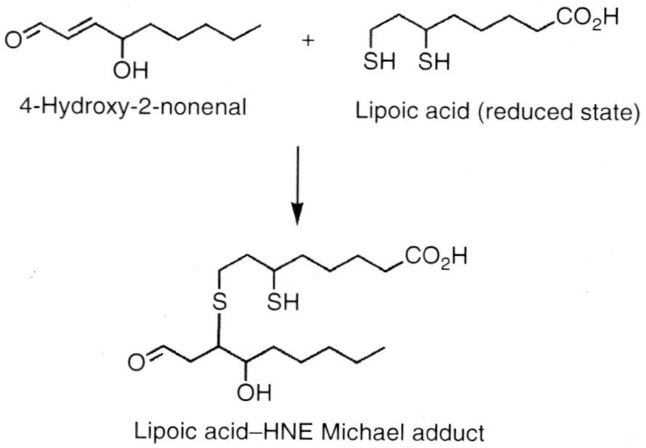

SCHEME 8.5 Reaction of 4-hydroxy-2-nonenal with lipoic acid.

The last evidence that HNE modifies lipoic acid was obtained immunochemically. Antibodies were raised against lipoic acid that recognize the native, protein-bound molecule. These antibodies do not recognize thiol-modified lipoic acid. Western blot analyses using these antibodies revealed that when KGDH (or PDH) is inhibited by HNE, there is a loss of antibody binding that is directly proportional to the extent of enzyme inactivation [56]. More recently it has been shown that treatment of purified KGDH, PDH, or HepG2 cells with HNE results in the inactivation of KGDH and PDH. Enzyme inactivation could be largely prevented by inclusion of reduced sulfhydryl-containing compounds including free lipoic acid [64]. These observations cumulatively and unequivocally show that, in the presence of substrate, lipoic acid is modified by HNE resulting in inactivation of KGDH. In contrast, the absence of substrate allows lipoic acid to remain predominantly in the oxidized (disulfide) state, protecting the moiety from HNE modification.

The susceptibility of lipoic acid to HNE modification is likely predicated by both molecules' chemical and structural properties. It is well established that HNE, an α,β-unsaturated aldehyde, can react with sulfhydryls [37]. Reduced lipoic acid, with its two sulfhydryl moieties, is a strong nucleophile at physiological pH and is located on the surface of the E2 subunit of KGDH and PDH [57,59]. Hydrophobic interactions are also likely to occur between the hydrocarbon chains of lipoic acid and HNE. Lipoic acid therefore represents a prime target for Michael addition to HNE (Scheme 8.5). Because many lipid peroxidation products are lipophilic electrophiles, this chemistry is not likely to be limited to HNE and may represent a common mechanism of cytotoxicity.

With KGDH and PDH so similar structurally and mechanistically, it was somewhat surprising that KGDH was much more readily inhibited by HNE in intact cardiac mitochondria. There are at least three potential explanations. First, HNE is a hydrophobic compound that would be at greatest concentration in the lipid bilayer. α-Ketoglutarate dehydrogenase is reported to be associated with the inner mitochondrial membrane and would therefore be particularly susceptible to modification [65–68]. Second, the E2 subunit of PDH contains two lipoic acids per molecule whereas the E2 subunit of KGDH contains only one. Studies have shown that removal of lipoic acid residues on the E2 subunit of PDH below a certain threshold have minimal effect on activity [69,70]. PDH may therefore have a higher capacity for HNE modification before losing activity. Third, the E3BP subunit of the PDH complex may confer a certain amount of protection to the enzyme complex that is not afforded to KGDH [64].

DETECTION OF 4-HYDROXY-2-NONENAL-MODIFIED LIPOIC ACID

The cellular environment created during the progression of a number of degenerative conditions suggests that HNE inactivation of KGDH and PDH is likely to play a significant role in disease-related losses of mitochondrial function. Upon cardiac ischemia/reperfusion, mitochondria exhibit declines in NADH-linked respiration [13,18–25] and increased rates of free-radical production [12–17]. This is accompanied by a rise in tissue and perfusate levels of HNE [71,72] and the appearance of HNE-modified protein [20,73,74]. In addition, the preceding ischemic event induces a buildup of reducing equivalents (NADH) and a fall in the level of ADP, conditions which would favor the reduced state of lipoic acid on KGDH and PDH, thereby priming these enzymes for modification by HNE during reperfusion. An age-dependent decline in KGDH activity occurs during cardiac ischemia/reperfusion [21]. As determined by Western blot analysis using anti-lipoic acid antibodies, this drop in activity is associated with a decrease in native lipoic acid residues on the E2 subunit of KGDH [22]. This result strongly suggests a role for oxidation of lipoic acid residues as a mediator of mitochondrial deficits following myocardial ischemia/reperfusion.

Although antibodies recognizing lipoic acid are a useful analytical tool, they are qualitative in that they cannot distinguish actual loss/degradation of lipoic acid versus modification. Therefore, techniques must be developed to directly measure HNE/lipoic acid conjugates under conditions of increased lipid peroxidation. Only then we will be able to directly assess the contribution of this mechanism to declines in KGDH activity and mitochondrial function.

FATE OF LIPOIC ACID 4-HYDROXY-2-NONENAL CONJUGATE

α-Ketoglutarate dehydrogenase and other lipoic acid containing proteins are large multienzyme complexes consisting of multiple copies of three enzymes [57,59,60]. It may therefore be energetically favorable to repair and restore enzyme activity rather than following a pathway of protein degradation and synthesis. Lipoamidase, an enzyme that catalyzes the hydrolysis of lipoic acid from lysine residues [75–78], may be critical in this repair process. Lipoamidase has a wide range of substrate specificity with a relatively higher affinity for hydrophobic compounds [75]. Thus HNE-modified lipoic acid, a very hydrophobic adduct, may be a potential substrate for lipoamidase and a signal for removal. If HNE were indeed a substrate for lipoamidase, it is also likely that mitochondria would contain the enzymatic machinery to replace lipoic acid once it is removed. The metabolic pathways of lipoic acid production and degradation are not well understood [79]. However, it has recently been reported that mitochondria contain the enzymatic machinery to produce lipoic acid de novo [80–86] and that lipoic acid obtained nutritionally is unable to be incorporated into enzyme complexes [87]. This suggests that lipoic acid is routinely turned over within the mitochondria.

Alternative to a repair mechanism, lipoic acid modification may be a signal for targeted degradation. To this extent, TFEC, a renal toxicant, selectively inactivates KGDH by covalently modifying the E2 and E3 subunits [88]. Modification leads to breakdown of the structural integrity of the multienzyme complex and association of the E2 and E3 subunits with the mitochondrial chaperone proteins HSP60 and HSP70 [88]. However, it is not known whether association of the modified proteins with chaperones leads to protein degradation, repair, or reassembly. More recently, it has been shown that KGDH stability can be regulated by the RING finger ubiquitin ligase, Siah [89]. While it is unclear how KGDH would be targeted by this cytosolic degradation system, it is interesting to hypothesize that lipoic acid modification plays a role in this process. Future studies must be performed to determine the possible association of HNE-modified KGDH with chaperone proteins and identify the mechanism(s) of enzyme regeneration or degradation that follow.

REVERSIBLE OXIDATIVE INHIBITION OF α-KETOGLUTARATE DEHYDROGENASE

Pro-oxidants with sufficient oxidizing potential to abstract an electron from polyunsaturated fatty acids must be produced for lipid peroxidation to occur. On its own, H_2O_2 is not capable of catalyzing lipid peroxidation. This pro-oxidant can, however, react with and oxidize sulfhydryl groups. H_2O_2-dependent sulfhydryl oxidation can progress from reversible to irreversible forms of modification [90–97]. By virtue of its susceptibility to reaction with the electrophilic lipid peroxidation product HNE, it stood to reason that lipoic acid bound to the E2 subunit of KGDH may be susceptible to other forms of oxidative modification. This is exemplified by observations that treatment of isolated synaptosomes with H_2O_2 and liver mitochondria with *t*-butyl hydroperoxide induces loss of KGDH activity [98,99].

Respiring cardiac mitochondria, challenged with micromolar concentrations of H_2O_2, exhibit declines in the maximal rate of NADH-dependent respiration and KGDH activity [100]. Following consumption of H_2O_2, however, respiratory activity is restored. Reversible alterations in respiratory activity are accompanied by the loss and regain in KGDH activity [100,101]. Due to the reversible nature of H_2O_2-induced inhibition, it is not likely that lipid peroxidation products are involved. Treatment of non-respiring, solubilized mitochondria, or purified KGDH with similar concentrations of H_2O_2 does not result in inhibition or inactivation of KGDH [100,101]. Thus, processes unique to the microenvironment of respiring mitochondria are required for the reversible inhibition of KGDH induced by H_2O_2.

GLUTATHIONYLATION OF α-KETOGLUTARATE DEHYDROGENASE

When mitochondria are challenged with H_2O_2, the content of NAD(P)H and reduced glutathione (GSH) decline [101]. This is consistent with the consumption of H_2O_2 by the glutathione peroxidase and glutathione reductase system. Following consumption of H_2O_2, the levels of NAD(P)H and GSH are rapidly restored [101]. In various systems, GSH is not quantitatively converted to GSSG in response to pro-oxidant stress. A considerable fraction of the glutathione

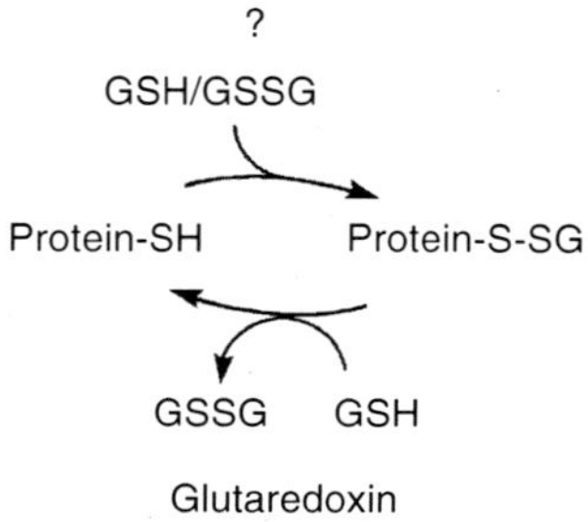

SCHEME 8.6 Reversible glutathionylation of protein.

pool appears to be involved in protein glutathionylation. This represents a reversible process given that the enzyme glutaredoxin can rapidly de-glutathionylate protein resulting in recovery of reduced sulfhydryl residues and protein activity [31,90,101–105] (Scheme 8.6). In this manner, critical and reactive sulfhydryl residues may be protected from irreversible oxidative modifications [102, 106,107]. Following treatment of respiring mitochondria with H_2O_2 and inhibition of KGDH, treatment of solubilized preparations with glutaredoxin results in full reactivation of KGDH. Thioredoxin/thioredoxin reductase fail to restore enzyme activity [101]. These findings offer support for the conclusion that KGDH is reversibly glutathionylated and inhibited in response to H_2O_2. This is perhaps not surprising given the requirement of redox cycling of vicinal sulfhydryl residues on both the E2 and E3 subunits of KGDH for enzyme activity. Two major questions that remain are the molecular mechanism by which KGDH becomes glutathionylated and the physiological implications of reversible redox-dependent inhibition of KGDH.

Attempts to reconstitute reversible glutathionylation and inhibition of KGDH have met with failure. α-Ketoglutarate dehydrogenase is resistant to inhibition upon direct treatment of the enzyme with various combinations of H_2O_2, GSH, GSSG, and enzyme substrates and cofactors [101]. In addition, while treatment of the enzyme with the oxidant diamide and GSH results in protein glutathionylation and inactivation, subsequent treatment with glutaredoxin fails to restore enzyme activity [101]. These results indicate that reversible glutathionylation and inhibition of KGDH in intact mitochondria in response to H_2O_2 is a highly regulated process that either involves factors not yet anticipated or enzymatic glutathionylation and deglutathionylation akin to protein phosphorylation and dephosphorylation by kinases and phosphatases. Mechanisms responsible for protein glutathionylation are an emerging area of investigation particularly with respect to redox regulation. Given the integral role of KGDH in controlling the rate NADH production and, thus, electron transport and ATP synthesis as well as the integrated and reversible manner in which KGDH responds to H_2O_2, physiological consequences of this process are important to elucidate.

GLUTATHIONYLATION OF α-KETOGLUTARATE DEHYDROGENASE AS AN ANTIOXIDANT RESPONSE

The most immediate and recognizable purpose of redox-dependent inhibition of KGDH would be to protect essential sulfhydryl residues from irreversible oxidative damage. This is perhaps most evident when considering HNE-induced inactivation of KGDH. As previously discussed, reaction of electrophilic HNE with the nucleophilic lipoic acid covalently attached to the E2 subunit of KGDH results in enzyme inactivation [55,56]. Glutathionylation of this residue would be expected to protect from HNE-induced inactivation (Scheme 8.7). Glutathionylation would also be expected to prevent the conversion of key enzyme sulfhydryl residues to sulfenic, sulfinic, and finally sulfonic acid derivatives. While sulfenic and sulfinic acids are reversible oxidation states, to date, sulfonic acid appears to represent irreversible oxidative modification [90,96,97,102,108,109]. Under certain conditions, KGDH is itself a source of free radicals that can lead to self-inactivation of the enzyme. It will be important to determine whether gluathionylation of KGDH diminishes enzyme-catalyzed production of free radicals and/or protects from self-inactivation [110–115]. Clearly, future elucidation of the site of glutathionylation and the mechanism (enzymatic versus pro-oxidant priming of GSH and/or protein-SH) will help establish the manner in which glutathionylation protects the enzyme from irreversible forms of oxidative modification.

Given that KGDH catalyzes a rate-limiting step in NADH production, electron transport, and ATP synthesis, one can envision other potential roles of redox-dependent inhibition of KGDH. A reduction in KGDH activity would limit

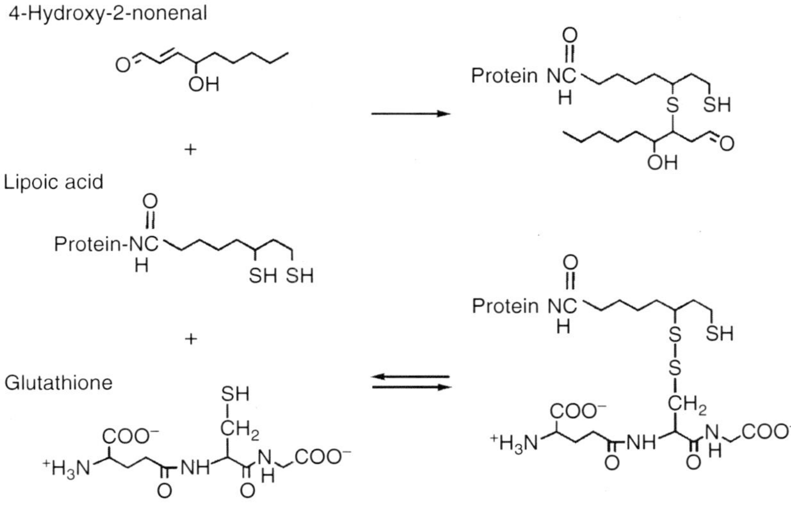

SCHEME 8.7 Protection of lipoic acid from 4-hydroxy-2-nonenal modification by glutathionylation.

the supply of NADH for electron transport. This, in turn, would be expected to reduce the rate of $O_2^{\dot{-}}$ production by the electron transport chain. H_2O_2-induced inhibition of KGDH may therefore represent a feedback mechanism to reduce free-radical production by a compromised electron transport chain and the potential for irreversible loss of protein and mitochondrial function [99–101]. Alternatively, it has previously been shown that inhibition of KGDH facilitates the release of cytochrome *c* from the mitochondria [116], an initiating step in the apoptotic pathway [117,118]. Thus, if oxidative stress persists or is of high enough magnitude, the continued inhibition or inactivation of KGDH may represent a pro-apoptotic event. The activity of KGDH may therefore serve as a metabolic sensor of mitochondrial/cellular redox status and the likelihood of repair.

CONCLUSION

By virtue of the chemistry of lipoic acid, this molecule represents a potent antioxidant and an important cofactor for redox reactions. The bioactivity of lipoic acid is therefore central to controlling the redox balance within the mitochondria and cell. Nevertheless, for the same reasons that lipoic acid is capable of participating in beneficial reactions, it is also susceptible to damage. The balance between these processes represents the precarious relationship that exists between life and death. It is therefore not surprising that KGDH appears to be exquisitely regulated in response to changes in redox status. Which forms of modification persist under various physiological and pathophysiological conditions will ultimately influence the outcome.

REFERENCES

1. Mitchell, P. and Moyle, J. 1965. Evidence discriminating between the chemical and the chemiosmotic mechanisms of electron transport phosphorylation. *Nature* 208, 1205–1206.
2. Beckman, K.B. and Ames, B.N. 1998. The free radical theory of aging matures. *Physiol Rev* 78, 547–581.
3. Boveris, A. and Cadenas, E. 1975. Mitochondrial production of superoxide anions and its relationship to the antimycin insensitive respiration. *FEBS Lett* 54, 311–314.
4. Cadenas, E., Boveris, A., Ragan, C.I., and Stoppani, A.O. 1977. Production of superoxide radicals and hydrogen peroxide by NADH-ubiquinone reductase and ubiquinol-cytochrome c reductase from beef-heart mitochondria. *Arch Biochem Biophys* 180, 248–257.
5. Cadenas, E. and Davies, K.J. 2000. Mitochondrial free radical generation, oxidative stress, and aging. *Free Radic Biol Med* 29, 222–230.
6. Chance, B. and Williams, G.R. 1956. The respiratory chain and oxidative phosphorylation. *Adv Enzymol Relat Subj Biochem* 17, 65–134.
7. Loschen, G., Azzi, A., and Flohe, L. 1973. Mitochondrial H_2O_2 formation: Relationship with energy conservation. *FEBS Lett* 33, 84–87.
8. Nohl, H., Breuninger, V., and Hegner, D. 1978. Influence of mitochondrial radical formation on energy-linked respiration. *Eur J Biochem* 90, 385–390.

9. Nohl, H. and Hegner, D. 1978. Do mitochondria produce oxygen radicals in vivo? *Eur J Biochem* 82, 563–567.
10. Turrens, J.F. and Boveris, A. 1980. Generation of superoxide anion by the NADH dehydrogenase of bovine heart mitochondria. *Biochem J* 191, 421–427.
11. Halliwell, B. and Gutteridge, J.M.C. 1989. *Free Radic Biol Med*, Oxford University Press, New York.
12. Ambrosio, G., Zweier, J.L., Duilio, C., Kuppusamy, P., Santoro, G., Elia, P.P., Tritto, I., Cirillo, P., Condorelli, M., Chiariello, M., et al. 1993. Evidence that mitochondrial respiration is a source of potentially toxic oxygen free radicals in intact rabbit hearts subjected to ischemia and reflow. *J Biol Chem* 268, 18532–18541.
13. Churchill, E.N. and Szweda, L.I. 2005. Translocation of deltaPKC to mitochondria during cardiac reperfusion enhances superoxide anion production and induces loss in mitochondrial function. *Arch Biochem Biophys* 439, 194–199.
14. Das, D.K., George, A., Liu, X.K., and Rao, P.S. 1989. Detection of hydroxyl radical in the mitochondria of ischemic-reperfused myocardium by trapping with salicylate. *Biochem Biophys Res Commun* 165, 1004–1009.
15. Otani, H., Tanaka, H., Inoue, T., Umemoto, M., Omoto, K., Tanaka, K., Sato, T., Osako, T., Masuda, A., Nonoyama, A., et al. 1984. In vitro study on contribution of oxidative metabolism of isolated rabbit heart mitochondria to myocardial reperfusion injury. *Circ Res* 55, 168–175.
16. Ueta, H., Ogura, R., Sugiyama, M., Kagiyama, A., and Shin, G. 1990. O_2- spin trapping on cardiac submitochondrial particles isolated from ischemic and non-ischemic myocardium. *J Mol Cell Cardiol* 22, 893–899.
17. Ferrari, R. 1996. The role of mitochondria in ischemic heart disease. *J Cardiovasc Pharmacol* 28 Suppl 1, S1–S10.
18. Duan, J. and Karmazyn, M. 1989. Relationship between oxidative phosphorylation and adenine nucleotide translocase activity of two populations of cardiac mitochondria and mechanical recovery of ischemic hearts following reperfusion. *Can J Physiol Pharmacol* 67, 704–709.
19. Hardy, L., Clark, J.B., Darley-Usmar, V.M., Smith, D.R., and Stone, D. 1991. Reoxygenation-dependent decrease in mitochondrial NADH:CoQ reductase (Complex I) activity in the hypoxic/reoxygenated rat heart. *Biochem J* 274 (Pt 1), 133–137.
20. Lucas, D.T. and Szweda, L.I. 1998. Cardiac reperfusion injury: Aging, lipid peroxidation, and mitochondrial dysfunction. *Proc Natl Acad Sci U S A* 95, 510–514.
21. Lucas, D.T. and Szweda, L.I. 1999. Declines in mitochondrial respiration during cardiac reperfusion: Age-dependent inactivation of alpha-ketoglutarate dehydrogenase. *Proc Natl Acad Sci U S A* 96, 6689–6693.
22. Sadek, H.A., Humphries, K.M., Szweda, P.A., and Szweda, L.I. 2002. Selective inactivation of redox-sensitive mitochondrial enzymes during cardiac reperfusion. *Arch Biochem Biophys* 406, 222–228.
23. Sadek, H.A., Nulton-Persson, A.C., Szweda, P.A., and Szweda, L.I. 2003. Cardiac ischemia/reperfusion, aging, and redox-dependent alterations in mitochondrial function. *Arch Biochem Biophys* 420, 201–208.
24. Veitch, K., Hombroeckx, A., Caucheteux, D., Pouleur, H., and Hue, L. 1992. Global ischaemia induces a biphasic response of the mitochondrial respiratory chain. Anoxic pre-perfusion protects against ischaemic damage. *Biochem J* 281 (Pt 3), 709–715.
25. Paradies, G., Petrosillo, G., Pistolese, M., Di Venosa, N., Federici, A., and Ruggiero, F.M. 2004. Decrease in mitochondrial complex I activity in ischemic/reperfused rat heart: Involvement of reactive oxygen species and cardiolipin. *Circ Res* 94, 53–59.

26. Giulivi, C., Boveris, A., and Cadenas, E. 1995. Hydroxyl radical generation during mitochondrial electron transfer and the formation of 8-hydroxydesoxyguanosine in mitochondrial DNA. *Arch Biochem Biophys* 316, 909–916.
27. Hansford, R.G. 1983. Bioenergetics in aging. *Biochim Biophys Acta* 726, 41–80.
28. Loschen, G., Azzi, A., Richter, C., and Flohe, L. 1974. Superoxide radicals as precursors of mitochondrial hydrogen peroxide. *FEBS Lett* 42, 68–72.
29. Loschen, G., Flohe, L., and Chance, B. 1971. Respiratory chain linked H(2)O(2) production in pigeon heart mitochondria. *FEBS Lett* 18, 261–264.
30. Arner, E.S. and Holmgren, A. 2000. Physiological functions of thioredoxin and thioredoxin reductase. *Eur J Biochem* 267, 6102–6109.
31. Gravina, S.A. and Mieyal, J.J. 1993. Thioltransferase is a specific glutathionyl mixed disulfide oxidoreductase. *Biochemistry* 32, 3368–3376.
32. Nohl, H. and Hegner, D. 1978. Evidence for the existence of catalase in the matrix space of rat-heart mitochondria. *FEBS Lett* 89, 126–130.
33. Radi, R., Turrens, J.F., Chang, L.Y., Bush, K.M., Crapo, J.D., and Freeman, B.A. 1991. Detection of catalase in rat heart mitochondria. *J Biol Chem* 266, 22028–22034.
34. Seo, M.S., Kang, S.W., Kim, K., Baines, I.C., Lee, T.H., and Rhee, S.G. 2000. Identification of a new type of mammalian peroxiredoxin that forms an intramolecular disulfide as a reaction intermediate. *J Biol Chem* 275, 20346–20354.
35. Rhee, S.G., Chae, H.Z., and Kim, K. 2005. Peroxiredoxins: A historical overview and speculative preview of novel mechanisms and emerging concepts in cell signaling. *Free Radic Biol Med* 38, 1543–1552.
36. Benedetti, A., Comporti, M., and Esterbauer, H. 1980. Identification of 4-hydroxynonenal as a cytotoxic product originating from the peroxidation of liver microsomal lipids. *Biochim Biophys Acta* 620, 281–296.
37. Esterbauer, H., Schaur, R.J., and Zollner, H. 1991. Chemistry and biochemistry of 4-hydroxynonenal, malonaldehyde and related aldehydes. *Free Radic Biol Med* 11, 81–128.
38. Berlett, B.S. and Stadtman, E.R. 1997. Protein oxidation in aging, disease, and oxidative stress. *J Biol Chem* 272, 20313–20316.
39. Stadtman, E.R. 1992. Protein oxidation and aging. *Science* 257, 1220–1224.
40. Chen, J.J. and Yu, B.P. 1994. Alterations in mitochondrial membrane fluidity by lipid peroxidation products. *Free Radic Biol Med* 17, 411–418.
41. Chen, J.J., Bertrand, H., and Yu, B.P. 1995. Inhibition of adenine nucleotide translocator by lipid peroxidation products. *Free Radic Biol Med* 19, 583–590.
42. Cohn, J.A., Tsai, L., Friguet, B., and Szweda, L.I. 1996. Chemical characterization of a protein-4-hydroxy-2-nonenal cross-link: Immunochemical detection in mitochondria exposed to oxidative stress. *Arch Biochem Biophys* 328, 158–164.
43. Friguet, B., Stadtman, E.R., and Szweda, L.I. 1994. Modification of glucose-6-phosphate dehydrogenase by 4-hydroxy-2-nonenal. Formation of cross-linked protein that inhibits the multicatalytic protease. *J Biol Chem* 269, 21639–21643.
44. Nadkarni, D.V. and Sayre, L.M. 1995. Structural definition of early lysine and histidine adduction chemistry of 4-hydroxynonenal. *Chem Res Toxicol* 8, 284–291.
45. Siems, W.G., Hapner, S.J., and van Kuijk, F.J. 1996. 4-hydroxynonenal inhibits Na(+)-K(+)-ATPase. *Free Radic Biol Med* 20, 215–223.
46. Szweda, L.I., Uchida, K., Tsai, L., and Stadtman, E.R. 1993. Inactivation of glucose-6-phosphate dehydrogenase by 4-hydroxy-2-nonenal. Selective modification of an active-site lysine. *J Biol Chem* 268, 3342–3347.
47. Tsai, L. and Sokoloski, E.A. 1995. The reaction of 4-hydroxy-2-nonenal with N alpha-acetyl-L-histidine. *Free Radic Biol Med* 19, 39–44.

48. Uchida, K. and Stadtman, E.R. 1992. Modification of histidine residues in proteins by reaction with 4-hydroxynonenal. *Proc Natl Acad Sci U S A* 89, 4544–4548.
49. Uchida, K. and Stadtman, E.R. 1993. Covalent attachment of 4-hydroxynonenal to glyceraldehyde-3-phosphate dehydrogenase. A possible involvement of intra- and intermolecular cross-linking reaction. *J Biol Chem* 268, 6388–6393.
50. Friguet, B., Szweda, L.I., and Stadtman, E.R. 1994. Susceptibility of glucose-6-phosphate dehydrogenase modified by 4-hydroxy-2-nonenal and metal-catalyzed oxidation to proteolysis by the multicatalytic protease. *Arch Biochem Biophys* 311, 168–173.
51. Itakura, K., Osawa, T., and Uchida, K. 1998. Structure of a fluorescent compound formed from 4-hydroxy-2-nonenal and N-alpha-hippuryllysine: A model for fluorophores derived from protein modifications by lipid peroxidation. *J Org Chem* 63, 185–187.
52. Szweda, P.A., Tsai, L., and Szweda, L.I. 2002. Immunochemical detection of a fluorophore derived from the lipid peroxidation product 4-hydroxy-2-nonenal and lysine. *Methods Mol Biol* 196, 277–290.
53. Tsai, L., Szweda, P.A., Vinogradova, O., and Szweda, L.I. 1998. Structural characterization and immunochemical detection of a fluorophore derived from 4-hydroxy-2-nonenal and lysine. *Proc Natl Acad Sci U S A* 95, 7975–7980.
54. Xu, G. and Sayre, L.M. 1998. Structural characterization of a 4-hydroxy-2-alkenal-derived fluorophore that contributes to lipoperoxidation-dependent protein cross-linking in aging and degenerative disease. *Chem Res Toxicol* 11, 247–251.
55. Humphries, K.M., Yoo, Y., and Szweda, L.I. 1998. Inhibition of NADH-linked mitochondrial respiration by 4-hydroxy-2-nonenal. *Biochemistry* 37, 552–557.
56. Humphries, K.M. and Szweda, L.I. 1998. Selective inactivation of alpha-ketoglutarate dehydrogenase and pyruvate dehydrogenase: Reaction of lipoic acid with 4-hydroxy-2-nonenal. *Biochemistry* 37, 15835–15841.
57. Reed, L.J. 1974. Multienzyme complexes. *Acc Chem Res* 7, 40–46.
58. Reed, L.J. 1957. The chemistry and function of lipoic acid. *Adv Enzymol Relat Subj Biochem* 18, 319–347.
59. Perham, R.N. 1991. Domains, motifs, and linkers in 2-oxo acid dehydrogenase multienzyme complexes: A paradigm in the design of a multifunctional protein. *Biochemistry* 30, 8501–8512.
60. Yeaman, S.J. 1989. The 2-oxo acid dehydrogenase complexes: Recent advances. *Biochem J* 257, 625–632.
61. Smolle, M., Prior, A.E., Brown, A.E., Cooper, A., Byron, O., and Lindsay, J.G. 2006. A new level of architectural complexity in the human pyruvate dehydrogenase complex. *J Biol Chem* 281, 19772–19780.
62. Erfle, J.D. and Sauer, F. 1968. An NADH dependent cleavage of DTNB by the alpha-ketoglutarate dehydrogenase complex. *Biochem Biophys Res Commun* 32, 562–567.
63. Brown, J.P. and Perham, R.N. 1976. Selective inactivation of the transacylase components of the 2-oxo acid dehydrogenase multienzyme complexes of Escherichia coli. *Biochem J* 155, 419–427.
64. Korotchkina, L.G., Yang, H., Tirosh, O., Packer, L., and Patel, M.S. 2001. Protection by thiols of the mitochondrial complexes from 4-hydroxy-2-nonenal. *Free Radic Biol Med* 30, 992–999.
65. Maas, E. and Bisswanger, H. 1990. Localization of the alpha-oxoacid dehydrogenase multienzyme complexes within the mitochondrion. *FEBS Lett* 277, 189–190.

66. Sumegi, B. and Srere, P.A. 1984. Complex I binds several mitochondrial NAD-coupled dehydrogenases. *J Biol Chem* 259, 15040–15045.
67. Porpaczy, Z., Sumegi, B., and Alkonyi, I. 1987. Interaction between NAD-dependent isocitrate dehydrogenase, alpha-ketoglutarate dehydrogenase complex, and NADH:Ubiquinone oxidoreductase. *J Biol Chem* 262, 9509–9514.
68. Fukushima, T., Decker, R.V., Anderson, W.M., and Spivey, H.O. 1989. Substrate channeling of NADH and binding of dehydrogenases to complex I. *J Biol Chem* 264, 16483–16488.
69. Rahmatullah, M., Radke, G.A., Andrews, P.C., and Roche, T.E. 1990. Changes in the core of the mammalian-pyruvate dehydrogenase complex upon selective removal of the lipoyl domain from the transacetylase component but not from the protein X component. *J Biol Chem* 265, 14512–14517.
70. Berman, J.N., Chen, G.X., Hale, G., and Perham, R.N. 1981. Lipoic acid residues in a take-over mechanism for the pyruvate dehydrogenase multienzyme complex of Escherichia coli. *Biochem J* 199, 513–520.
71. Blasig, I.E., Grune, T., Schonheit, K., Rohde, E., Jakstadt, M., Haseloff, R.F., and Siems, W.G. 1995. 4-Hydroxynonenal, a novel indicator of lipid peroxidation for reperfusion injury of the myocardium. *Am J Physiol* 269, H14–H22.
72. Grune, T., Siems, W.G., Schonheit, K., and Blasig, I.E. 1993. Release of 4-hydroxynonenal, an aldehydic mediator of inflammation, during postischemic reperfusion of the myocardium. *Int J Tissue React* 15, 145–150.
73. Chen, J., Henderson, G.I., and Freeman, G.L. 2001. Role of 4-hydroxynonenal in modification of cytochrome c oxidase in ischemia/reperfused rat heart. *J Mol Cell Cardiol* 33, 1919–1927.
74. Eaton, P., Li, J.M., Hearse, D.J., and Shattock, M.J. 1999. Formation of 4-hydroxy-2-nonenal-modified proteins in ischemic rat heart. *Am J Physiol* 276, H935–H943.
75. Oizumi, J. and Hayakawa, K. 1990. Lipoamidase is a multiple hydrolase. *Biochem J* 271, 45–49.
76. Oizumi, J. and Hayakawa, K. 1990. Lipoamidase (lipoyl-X hydrolase) from pig brain. *Biochem J* 266, 427–434.
77. Oizumi, J. and Hayakawa, K. 1997. Purification and properties of brain lipoamidase. *Methods Enzymol* 279, 202–210.
78. Suzuki, K. and Reed, L.J. 1963. Lipoamidase. *J Biol Chem* 238, 4021–4025.
79. Booker, S.J. 2004. Unraveling the pathway of lipoic acid biosynthesis. *Chem Biol* 11, 10–12.
80. Brody, S., Oh, C., Hoja, U., and Schweizer, E. 1997. Mitochondrial acyl carrier protein is involved in lipoic acid synthesis in Saccharomyces cerevisiae. *FEBS Lett* 408, 217–220.
81. Jordan, S.W. and Cronan, J.E., Jr. 1997. Biosynthesis of lipoic acid and posttranslational modification with lipoic acid in Escherichia coli. *Methods Enzymol* 279, 176–183.
82. Jordan, S.W. and Cronan, J.E., Jr. 1997. A new metabolic link. The acyl carrier protein of lipid synthesis donates lipoic acid to the pyruvate dehydrogenase complex in Escherichia coli and mitochondria. *J Biol Chem* 272, 17903–17906.
83. Wada, M., Yasuno, R., Jordan, S.W., Cronan, J.E., Jr., and Wada, H. 2001. Lipoic acid metabolism in Arabidopsis thaliana: Cloning and characterization of a cDNA encoding lipoyltransferase. *Plant Cell Physiol* 42, 650–656.
84. Wada, H., Shintani, D., and Ohlrogge, J. 1997. Why do mitochondria synthesize fatty acids? Evidence for involvement in lipoic acid production. *Proc Natl Acad Sci U S A* 94, 1591–1596.

85. Zhao, X., Miller, J.R., Jiang, Y., Marletta, M.A., and Cronan, J.E. 2003. Assembly of the covalent linkage between lipoic acid and its cognate enzymes. *Chem Biol* 10, 1293–1302.
86. Zhao, X., Miller, J.R., and Cronan, J.E. 2005. The reaction of LipB, the octanoyl-[acyl carrier protein]: Protein N-octanoyltransferase of lipoic acid synthesis, proceeds through an acyl-enzyme intermediate. *Biochemistry* 44, 16737–16746.
87. Yi, X. and Maeda, N. 2005. Endogenous production of lipoic acid is essential for mouse development. *Mol Cell Biol* 25, 8387–8392.
88. Bruschi, S.A., Lindsay, J.G., and Crabb, J.W. 1998. Mitochondrial stress protein recognition of inactivated dehydrogenases during mammalian cell death. *Proc Natl Acad Sci U S A* 95, 13413–13418.
89. Habelhah, H., Laine, A., Erdjument-Bromage, H., Tempst, P., Gershwin, M.E., Bowtell, D.D., and Ronai, Z. 2004. Regulation of 2-oxoglutarate (alpha-ketoglutarate) dehydrogenase stability by the RING finger ubiquitin ligase Siah. *J Biol Chem* 279, 53782–53788.
90. Barrett, W.C., DeGnore, J.P., Keng, Y.F., Zhang, Z.Y., Yim, M.B., and Chock, P.B. 1999. Roles of superoxide radical anion in signal transduction mediated by reversible regulation of protein–tyrosine phosphatase 1B. *J Biol Chem* 274, 34543–34546.
91. Claiborne, A., Yeh, J.I., Mallett, T.C., Luba, J., Crane, E.J., IIId, Charrier, V., and Parsonage, D. 1999. Protein-sulfenic acids: Diverse roles for an unlikely player in enzyme catalysis and redox regulation. *Biochemistry* 38, 15407–15416.
92. Denu, J.M. and Tanner, K.G. 2002. Redox regulation of protein tyrosine phosphatases by hydrogen peroxide: Detecting sulfenic acid intermediates and examining reversible inactivation. *Methods Enzymol* 348, 297–305.
93. Poole, L.B., Karplus, P.A., and Claiborne, A. 2004. Protein sulfenic acids in redox signaling. *Annu Rev Pharmacol Toxicol* 44, 325–347.
94. Rhee, S.G., Kang, S.W., Jeong, W., Chang, T.S., Yang, K.S., and Woo, H.A. 2005. Intracellular messenger function of hydrogen peroxide and its regulation by peroxiredoxins. *Curr Opin Cell Biol* 17, 183–189.
95. Song, H., Bao, S., Ramanadham, S., and Turk, J. 2006. Effects of biological oxidants on the catalytic activity and structure of group VIA phospholipase A2. *Biochemistry* 45, 6392–6406.
96. Thomas, J.A. and Mallis, R.J. 2001. Aging and oxidation of reactive protein sulfhydryls. *Exp Gerontol* 36, 1519–1526.
97. Woo, H.A., Chae, H.Z., Hwang, S.C., Yang, K.S., Kang, S.W., Kim, K., and Rhee, S.G. 2003. Reversing the inactivation of peroxiredoxins caused by cysteine sulfinic acid formation. *Science* 300, 653–656.
98. Rokutan, K., Kawai, K., and Asada, K. 1987. Inactivation of 2-oxoglutarate dehydrogenase in rat liver mitochondria by its substrate and t-butyl hydroperoxide. *J Biochem (Tokyo)* 101, 415–422.
99. Tretter, L. and Adam-Vizi, V. 2000. Inhibition of Krebs cycle enzymes by hydrogen peroxide: A key role of [alpha]-ketoglutarate dehydrogenase in limiting NADH production under oxidative stress. *J Neurosci* 20, 8972–8979.
100. Nulton-Persson, A.C. and Szweda, L.I. 2001. Modulation of mitochondrial function by hydrogen peroxide. *J Biol Chem* 276, 23357–23361.
101. Nulton-Persson, A.C., Starke, D.W., Mieyal, J.J., and Szweda, L.I. 2003. Reversible inactivation of alpha-ketoglutarate dehydrogenase in response to alterations in the mitochondrial glutathione status. *Biochemistry* 42, 4235–4242.

102. Barrett, W.C., DeGnore, J.P., Konig, S., Fales, H.M., Keng, Y.F., Zhang, Z.Y., Yim, M.B., and Chock, P.B. 1999. Regulation of PTP1B via glutathionylation of the active site cysteine 215. *Biochemistry* 38, 6699–6705.
103. Beer, S.M., Taylor, E.R., Brown, S.E., Dahm, C.C., Costa, N.J., Runswick, M.J., and Murphy, M.P. 2004. Glutaredoxin 2 catalyzes the reversible oxidation and glutathionylation of mitochondrial membrane thiol proteins: Implications for mitochondrial redox regulation and antioxidant DEFENSE. *J Biol Chem* 279, 47939–47951.
104. Lillig, C.H., Berndt, C., Vergnolle, O., Lonn, M.E., Hudemann, C., Bill, E., and Holmgren, A. 2005. Characterization of human glutaredoxin 2 as iron-sulfur protein: A possible role as redox sensor. *Proc Natl Acad Sci U S A* 102, 8168–8173.
105. Vlamis-Gardikas, A., Aslund, F., Spyrou, G., Bergman, T., and Holmgren, A. 1997. Cloning, overexpression, and characterization of glutaredoxin 2, an atypical glutaredoxin from Escherichia coli. *J Biol Chem* 272, 11236–11243.
106. Mieyal, J.J., Gravina, S.A., Mieyal, P.A., Srinivasan, U., and Starke, D.W. 1995. Glutathionyl specificity of thioltransferases: Mechanistic and physiological implications. In *Biothiols in Health and Disease* (Packer, L. and Cadenas, E., Eds) pp. 305–372, Marcel Dekker, New York.
107. Thomas, J.A., Poland, B., and Honzatko, R. 1995. Protein sulfhydryls and their role in the antioxidant function of protein S-thiolation. *Arch Biochem Biophys* 319, 1–9.
108. Biteau, B., Labarre, J., and Toledano, M.B. 2003. ATP-dependent reduction of cysteine-sulphinic acid by S. cerevisiae sulphiredoxin. *Nature* 425, 980–984.
109. Woo, H.A., Jeong, W., Chang, T.S., Park, K.J., Park, S.J., Yang, J.S., and Rhee, S.G. 2005. Reduction of cysteine sulfinic acid by sulfiredoxin is specific to 2-cys peroxiredoxins. *J Biol Chem* 280, 3125–3128.
110. Bunik, V.I. 2003. 2-Oxo acid dehydrogenase complexes in redox regulation. *Eur J Biochem* 270, 1036–1042.
111. Bunik, V. and Follmann, H. 1993. Thioredoxin reduction dependent on alpha-ketoacid oxidation by alpha-ketoacid dehydrogenase complexes. *FEBS Lett* 336, 197–200.
112. Bunik, V.I. and Sievers, C. 2002. Inactivation of the 2-oxo acid dehydrogenase complexes upon generation of intrinsic radical species. *Eur J Biochem* 269, 5004–5015.
113. Starkov, A.A., Fiskum, G., Chinopoulos, C., Lorenzo, B.J., Browne, S.E., Patel, M.S., and Beal, M.F. 2004. Mitochondrial alpha-ketoglutarate dehydrogenase complex generates reactive oxygen species. *J Neurosci* 24, 7779–7788.
114. Tretter, L. and Adam-Vizi, V. 2004. Generation of reactive oxygen species in the reaction catalyzed by alpha-ketoglutarate dehydrogenase. *J Neurosci* 24, 7771–7778.
115. Tretter, L. and Adam-Vizi, V. 2005. Alpha-ketoglutarate dehydrogenase: A target and generator of oxidative stress. *Philos Trans R Soc Lond B Biol Sci* 360, 2335–2345.
116. Huang, H.M., Ou, H.C., Xu, H., Chen, H.L., Fowler, C., and Gibson, G.E. 2003. Inhibition of alpha-ketoglutarate dehydrogenase complex promotes cytochrome c release from mitochondria, caspase-3 activation, and necrotic cell death. *J Neurosci Res* 74, 309–317.
117. Green, D.R. and Reed, J.C. 1998. Mitochondria and apoptosis. *Science* 281, 1309–1312.
118. Orrenius, S. 2004. Mitochondrial regulation of apoptotic cell death. *Toxicol Lett* 149, 19–23.

9 Lipoate-Protein Ligase A: Structure and Function

Kazuko Fujiwara, Harumi Hosaka, Atsushi Nakagawa, and Yutaro Motokawa

CONTENTS

INTRODUCTION

R-(+)-lipoic acid is a disulfide-containing cofactor of the pyruvate, α-ketoglutarate, and branched chain α-ketoacid dehydrogenase complexes and the glycine cleavage system. Lipoic acid is covalently bound to a specific lysine residue of the E2 component of the α-ketoacid dehydrogenase complexes and of H-protein of the glycine cleavage system via an amide linkage between the carboxyl group of lipoic acid and the ε-amino group of the lysine residue of the lipoate-dependent proteins. The lipoyllysine arm plays a pivotal role in the reaction sequence, shuttling the reaction intermediate and the reducing equivalents between the active sites of the components of the complex (Fujiwara et al., 1992; Perham, 2000).

There are two pathways for the attachment of lipoic acid to the proteins in most of the organisms (Figure 9.1). One pathway is termed the salvage pathway. Lipoate-protein ligase A (LplA) is responsible for the pathway and catalyzes the protein lipoylation reaction using free lipoic acid and ATP as substrates, producing lipoylated proteins and AMP. Biochemical studies on protein lipoylation

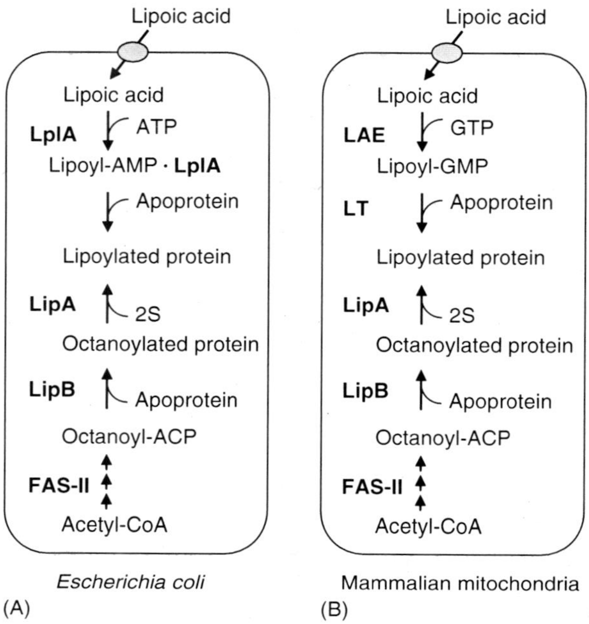

FIGURE 9.1 Two pathways of the protein lipoylation in *Escherichia coli* (A) and in mammalian mitochondria (B). Lipoate-protein ligase A (LplA) in bacteria is responsible for the salvage pathway of the protein lipoylation using free lipoic acid, whereas in mammalian mitochondria, both lipoate-activating enzyme (LAE) and lipoyltransferase (LT) are required for the pathway. Another lipoylation pathway is coupled with the de novo synthesis of lipoic acid. LipB transfers the octanoyl moiety from octanoyl-acyl carrier protein (octanoyl-ACP) to the lipoate-dependent protein, and then LipA catalyzes the insertion of sulfur atoms at the C6 and C8 positions of the octanoyl moiety. Enzymes are shown in boldface letters. FAS-II, type-II fatty acid synthetase complex.

were started about 50 years ago by Reed et al. (1958) using extracts from the lipoic acid-deficient *Streptococcus faecalis* and demonstrated the presence of the lipoate attachment system in the bacteria. Because lipoyl-AMP could replace lipoic acid and ATP, the following two-step mechanism was proposed for the covalent attachment of lipoic acid:

$$\text{Lipoic acid} + \text{ATP} \rightarrow \text{lipoyl-AMP} + \text{PP}_i \text{ (Reaction 1)}$$

$$\text{Lipoyl-AMP} + \text{apoprotein} \rightarrow \text{lipoylated protein} + \text{AMP (Reaction 2)}$$

The development of biotechnologies greatly advanced the research of the protein lipoylation, which enabled not only the isolation of genes responsible for the protein lipoylation but also the production of lipoate-acceptor apoproteins. The first gene which was cloned and over-expressed was the *Escherichia coli*

LplA (Morris et al., 1994). The *lplA* gene was mapped to min 99.6 of the *E. coli* chromosome. The purified recombinant LplA catalyzed the incorporation of [^{35}S] lipoic acid to the apoprotein substrate (an over-expressed lipoyl domain of E2 subunit of the pyruvate dehydrogenase complex fused to the acyl carrier protein). The reaction requires lipoic acid, ATP, $MgCl_2$, inorganic phosphate, and a lipoate-acceptor apoprotein.

Free lipoic acid is supplied exogenously from the environment. Alternatively, there must be a significant amount of the endogenous supply of lipoic acid by the recycling of the lipoylated proteins through breakdown and the following release of lipoic acid by the action of enzymes with amidase or lipoamidase-like activity.

In mammals, lipoyltransfersase, a homologue of lipoate-protein ligase, is responsible for the salvage pathway of the protein lipoylation. The lipoyltransferase was purified from bovine liver mitochondria (Fujiwara et al., 1994). The enzyme catalyzes the lipoylation reaction using lipoyl-AMP as a substrate, yielding lipoylated protein and AMP (Reaction 2). The human lipoyltransferase gene is localized to chromosome band 2q11.2, comprising four exons. The expression of mRNA of the lipoyltransferase is highly regulated and most abundant in skeletal muscle and heart. The expression pattern is corelated with those of lipoate-acceptor proteins, enabling the efficient lipoylation of the proteins in response to the requirement (Fujiwara et al., 1999). Because lipoyltransferase has no ability to activate lipoate to lipoyl-AMP (Reaction 1), a contribution of another enzyme, the lipoate-activating enzyme, is required for the protein lipoylation using free lipoic acid. The lipoate-activating enzyme was also purified from bovine liver mitochondria (Fujiwara et al., 2001). The amino acid sequence analysis showed that the enzyme was a mitochondrial medium-chain acyl-CoA synthetase. Lipoyl-AMP synthesized by the lipoate-activating enzyme was hardly released from the enzyme. Using GTP instead of ATP solved the problem because the lipoyl-GMP produced was easily released from the enzyme. The in vitro lipoylation reaction which was coupled with the lipoate-activating enzyme and the lipoyltransferase proceeded about 1000-fold faster with GTP than that with ATP. Since the de novo lipoic acid synthesis pathway (composed of LipA and LipB) and the lipoate-dependent multienzyme complexes are located in mitochondria, it is advantageous for the salvage pathway to work by the endogenous supply of the free lipoic acid. The free lipoic acid is also supplied exogenously through the sodium-dependent vitamin transporter, which uptakes pantothenate, biotin, and lipoate (Grassl, 1992; Prasad et al., 1998). Although the mammalian lipoyltransferase catalyzes only the second reaction, it would belong to the LplA family because it contains a unique RRXXGGGXVXHD motif conserved among LplAs, and the overall amino acid sequence shares about 30% identity with that of *E. coli* LplA (Figure 9.2) (Fujiwara et al., 1997; Fujiwara et al., 1999).

The presence of the second pathway for the protein lipoylation was first found in the analysis of *lplA* null mutants (Morris et al., 1995). In the experiment, *lplA* null mutants did not display growth defects unless combined with *lipA* or *lipB* null mutations, thus indicating the redundancy between the functions of LplA and LipA and/or LipB. The pathway uses endogenously synthesized lipoic acid. The

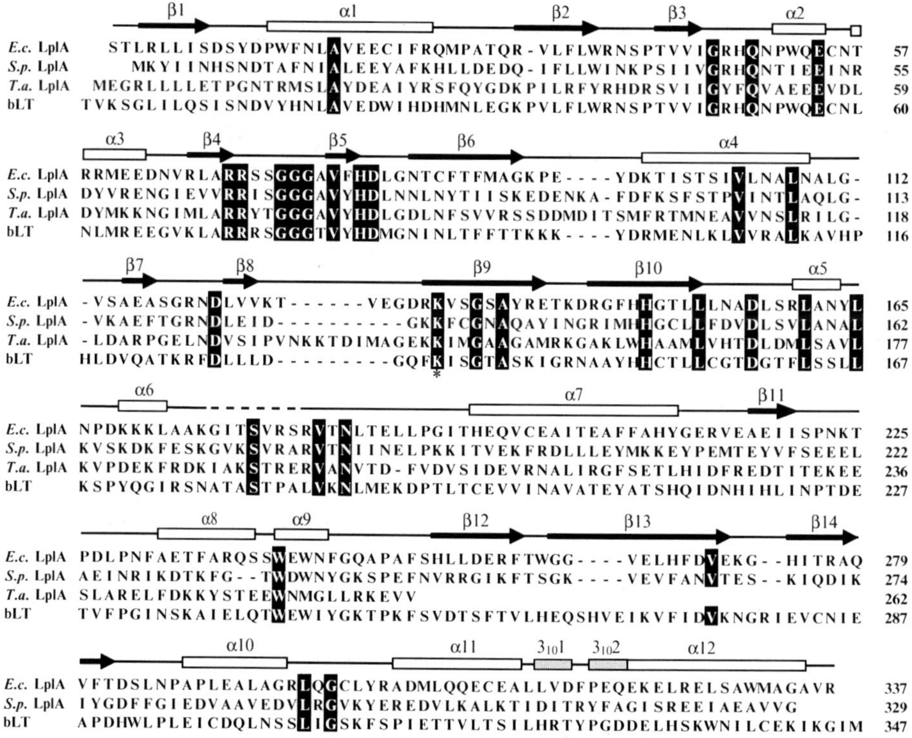

FIGURE 9.2 Amino acid sequence alignment with the secondary structural elements of *Escherichia coli* (*E.c.*) LplA. The amino acid sequence of *E. coli* LplA (Morris et al., 1994) is aligned with those of LplAs from *Streptococcus pneumoniae* (*S.p.*) and *Thermoplasma acidophilum* (*T.a.*) (Swiss-Prot accession code, Q9HKT1), and bovine lipoyltransferase (bLT) (Fujiwara et al., 1997). The residues identical among all proteins are shown on black backgrounds. The lysine residue strictly conserved among the LplA family is indicated with an asterisk under the sequence. The secondary structural elements of *E. coli* LplA (PDB code, 1X2H, MolA) assigned using the program DSSP (Kabsch and Sander, 1983) are shown above the sequence: α-helices, white rectangles; 3_{10} helices, gray rectangles; β-strands, arrows. A dashed line represents the disordered region. (From Morris, T.W., Reed, K.E., and Cronan, J.E., Jr., *J. Biol. Chem.*, 269, 16091, 1994; Fujiwara, K., Okamura-Ikeda, K., and Motokawa, Y., *J. Biol. Chem.*, 272, 31974, 1997; Kabsch, W. and Sander, C., *Biopolymers*, 22, 2577, 1983.)

de novo lipoic acid synthesis uses the octanoyl-acyl carrier protein as a precursor, which is provided by the type-II fatty acid synthesis pathway. The octanoyl moiety is first transferred to the lipoate-dependent proteins by the action of LipB, an octanoyl (lipoyl)-acyl carrier protein:protein *N*-lipoyltransferase (Jordan and Cronan, 2003; Zhao et al., 2005). Then the sulfur atom insertion occurs at the C6 and C8 positions of the octanoyl moiety by the function of LipA, an *S*-adenosylmethionine-dependent [Fe-S] cluster-containing enzyme (Miller et al., 2000; Cicchillo et al., 2004). The importance of the de novo lipoic acid synthesis in mammals has been demonstrated by the lipoic acid synthetase (Lias) gene-knockout experiment, in which homozygous mouse embryos lacking Lias die before 9.5 days postcoitum, and this lacking cannot be rescued by maternal lipoic acid or by supplementing the mother's diet with lipoic acid (Yi and Maeda, 2005).

Recently, the three-dimensional structures of LplA from *E. coli* (Fujiwara et al., 2005), *Thermoplasma acidophilum* (Kim et al., 2005; McManus et al., 2006), and *Streptococcus pneumoniae* have been determined. Although the *E. coli* LplA catalyzes both the lipoate-activating and lipoate-transfer reactions, the *T. acidophilum* LplA and the mammalian lipoyltransferase catalyze only a single reaction. The former catalyzes the first reaction and the latter catalyzes the second reaction. The structural comparison of these proteins should provide a clue to understand the reaction mechanism for the protein lipoylation. In the present review we describe the three-dimensional structures of LplA from *E. coli* and *T. acidophilum* and propose the reaction mechanism, which is deduced from the interactions between the LplA and the substrates and the reaction intermediate.

PROPERTIES AND STRUCTURE OF *E. coli* LIPOATE-PROTEIN LIGASE A

Properties of *E. coli* LplA

The *E. coli* LplA is composed of 337 amino acids with a molecular mass of 37,795 Da, excluding the initiating methionine residue which is cleaved off during the biosynthesis. (Morris et al., 1994). The K_m values of the recombinant LplA for *R*-lipoic acid, ATP, and apoprotein were determined to be 4.5, 15.8, and 1.2 μM, respectively, and an apparent V_{max} to be 258 nmol/min/mg protein using apoH-protein as a lipoate-acceptor protein (Fujiwara et al., 2005). The V_{max} value is comparable with that determined using a lipoate-acceptor fusion protein composed of the *E. coli* acyl carrier protein and the lipoyl domains of the E2 subunit of pyruvate dehydrogenase (Morris et al., 1994). Although *R*-lipoic acid is a naturally occurring enantiomer of lipoic acid, *E. coli* LplA can also use *S*-lipoic acid and lipoic acid analogues in the in vitro reaction with relative activities of 32% (*S*-lipoic acid), 70% (8-methyllipoic acid), and 13% (octanoic acid) to the activity with *R*-lipoic acid (Brookfield et al., 1991), and with a V_{max}/K_m value of 3.2% (selenolipoic acid) and 0.9% (6-thio-octanoic acid) relative to that with *R,S*-lipoic acid (Green et al., 1995). In an in vivo experiment by Loeffelhardt et al. (1996), *E. coli JRG26*, which had a defect in the lipoic acid biosynthesis, could

TABLE 9.1
Concentrations of Lipoic Acid Enantiomers and Their Analogues Required for the Growth of *Escherichia coli JRG26*

Compound	Concentration (M)	Cell Density (OD_{600})
R-Lipoic acid	4.85×10^{-10}	1.8
	0.49×10^{-10}	1.5
S-Lipoic acid	4.8×10^{-5}	1.5
	0.48×10^{-5}	0.16
R,S-Bisnorlipoic acid	5.6×10^{-4}	0.73
	0.56×10^{-4}	ND [a]
R,S-Dithiolane-3-caproic acid	4.5×10^{-6}	1.8
	0.45×10^{-6}	1.02

Source: From Loeffelhardt, S., Borbe, H.O., Locher, M., and Bisswanger, H., *Biochim. Biophys. Acta*, 1297, 90, 1996.

Note: *E. coli* strain *JRG26* was cultivated in a minimal medium, supplemented with 0.2% glucose and the lipoic acid analogues at the indicated concentrations. Cell densities were measured at the stationary phase.

[a] Not detectable.

grow using *R*-lipoic acid, *S*-lipoic acid, and their analogues although concentrations of *S*-lipoic acid and the analogues required for the growth were about four to six orders higher than that of *R*-lipoic acid (Table 9.1). The incorporations of *R,S*-lipoic acid and *R,S*-bisnorlipoic acid in the pyruvate dehydrogenase complex were confirmed using respective specific antibodies. It is interesting that the pyruvate dehydrogenase complexes which incorporated *R,S*-dithiolane-3-caproic acid and *R, S*-bisnorlipoic acid showed comparable activities (85% and 37%, respectively) to that incorporated *R*-lipoic acid.

Overall Structure of *E. coli* LplA

The selenomethionine-substituted *E. coli* LplA (Se-LplA) was produced in *E. coli* B834(DE3)pLysS in a culture medium supplemented with selenomethionine. The structure of Se-LplA was solved by the single-wavelength anomalous dispersion phasing method (Fujiwara et al., 2005). Although the *E. coli* LplA is a monomer in solution, there are three LplA molecules in an asymmetric unit (MolA, MolB, and MolC), comprising a dimer formation. All residues (1–337) of the MolC could be assigned, whereas residues 177–182 and 176–182 in MolA and MolB, respectively, were not determined because of the structural disorder.

The structure of the LplA comprises a large N-terminal domain (residues 1–244) and a small C-terminal domain (residues 253–337) connected by a single stretch of the polypeptide (residues 245–252) (Figure 9.3A). The N-terminal domain consists of two β-sheets, a large core β-sheet comprising six antiparallel β-strands (β1, β2, β6, β8–β10) and a small β-sheet comprising three antiparallel β-strands

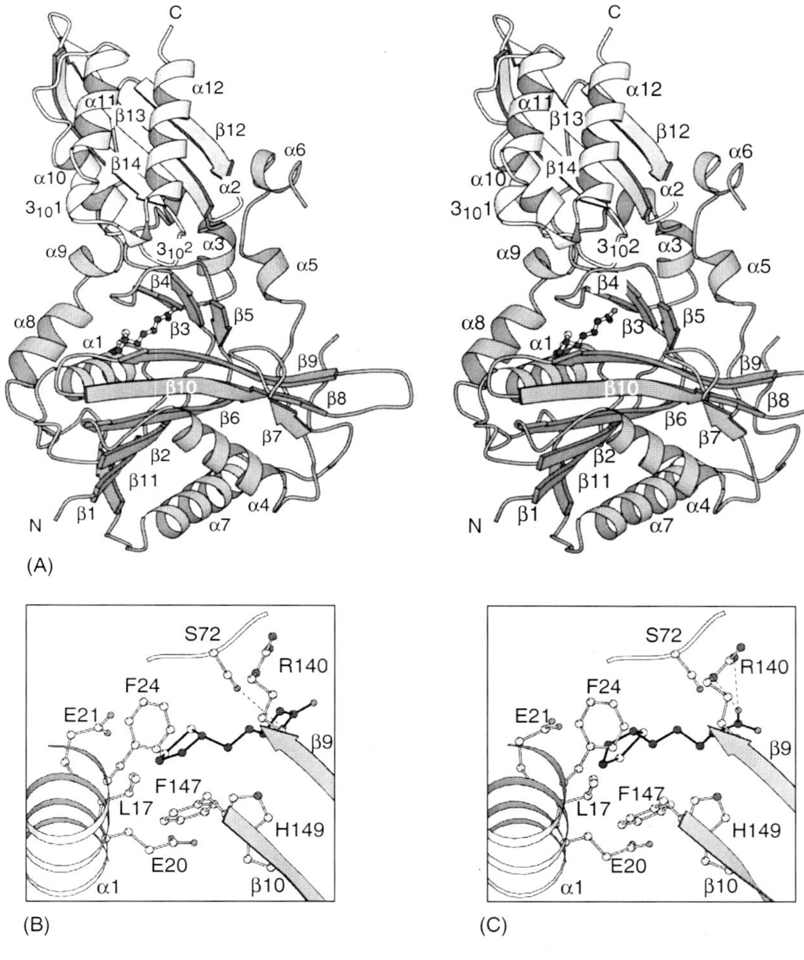

FIGURE 9.3 Structure of *Escherichia coli* LplA. (A) Stereo view of the LplA structure in complex with lipoic acid (1X2H, MolA). The protein is shown in a ribbon representation and lipoic acid is shown in a ball-and-stick mode. The secondary structural elements and the N- and C-termini are labeled. (B and C) Close-up views of the lipoic acid-binding site in MolA and MolB, respectively. Protein is shown in a ribbon representation, and lipoic acid (black) and interacting residues (gray) are shown in a ball-and-stick representation. Dashed lines show potential hydrogen bonds between atoms.

(β3–β5), and nine α-helices surrounding the β-sheets. The structure falls into α/β proteins (Brändén, 1980). The C-terminal domain consists of a β-sheet comprising three β-strands (β12–β14), three α-helices, and two 3_{10}-helices. The helices are located on the one side of the β-sheet.

A search using the DALI program (Holm and Sander, 1993) identified several proteins with the high structural similarity to the *E. coli* LplA. It includes the putative LplA from *S. pneumoniae* (1VQZ, Z-score, 28.9; root-mean-square deviation [rmsd], 3.1 Å; alignment, 297/326; sequence identity, 32%) and LplA from *T. acidophilum* (2ARS, Z-score, 20.6; rmsd, 2.9 Å; alignment, 200/232; sequence identity, 30%) (Kim et al., 2005), LipB from *Mycobacterium tuberculosis* (1W66, Z-score, 14.4; rmsd, 3.6 Å; alignment, 176/218; sequence identity, 18%) (Ma et al., 2006), BirA, bifunctional protein with biotin-protein ligase (BPL) activity from *E. coli* (1BIA, Z-score, 9.2; rmsd, 3.8 Å; alignment, 137/292; sequence identity, 12%) (Wilson et al., 1992), the BPL from *Pyrococcus horikoshi* OT3 (1WNL, Z-score, 9.0; rmsd, 3.5 Å; alignment, 133/235; sequence identity, 13%) (Bagautdinov et al., 2005), and phenylalanyl-tRNA synthetase from *Thermus thermophilus* (1PYS, Z-score, 5.7; rmsd, 8.4 Å; alignment, 121/785; sequence identity, 8%) (Mosyak et al., 1995). They all have functional similarities to the *E. coli* LplA with respect to the fact that LplAs, BPLs, and the catalytic core of phenylalanyl-tRNA synthetase catalyze the consecutive two-step reaction, implying a tightly bound acyl-adenylate intermediate, and that LipB catalyzes transfer of lipoyl (or octanoyl) moiety to the lipoate-dependent proteins. The superposition of the Cα trace of *E. coli* LplA onto that of BirA shows that several α-helices (α1, α4, and α5) and β-strands (β1, β2, β6, β7, β9, and β10) are well superimposed, and the substrates are situated in a similar position in the respective enzyme (Fujiwara et al., 2005). Although LplA, BPL, and the class II aminoacyl-tRNA synthetase share the structural similarity, residues contributing to the catalysis and kinetic features of the substrate binding order are different between LplA/BPL and class II aminoacyl-tRNA synthetase. Therefore, Wood et al. (2006) have proposed that LplA and BPL diverged early from a common ancestor that had been split from the class II aminoacyl-tRNA synthetase family, and after selection for cofactor (or amino acid) binding, the adenylation activity arose independently through functional convergence.

Lipoic Acid-Binding Site

The LplA crystal in complex with lipoic acid was produced by co-crystallization using a reservoir solution containing 3 mM *R*-lipoic acid (Fujiwara et al., 2005). Weak but significant electron density corresponding to lipoic acid was found in MolA and MolB. However, the positions of the lipoic acid in MolA and MolB were slightly different, presumably because of weak interactions between the LplA and lipoic acid (Figure 9.3B and C). Lipoic acid is bound in a hydrophobic pocket in the N-terminal domain formed by the core β-sheet, β4–β5 loop, and α1 helix. The dithiolane ring and the aliphatic tail of lipoic acid interact with Leu17, Phe24, Phe147, Ala138, His149, and aliphatic parts of Glu21 and Ser72 through the hydrophobic interactions. The carboxyl group of lipoic acid in MolA makes a hydrogen bond with the side-chain of Ser72, whereas in MolB, the carboxyl group hydrogen bonds with the side-chain of Arg140. Since van der Waals interactions are neither strong nor specific, and a weak hydrogen bond occurs at the carboxyl group, it may allow the LplA to bind not only *R*-lipoic acid but also *S*-lipoic acid, lipoic acid analogues, and octanoic acid. The binding mode properly explains why the LplA activates these carboxylic acids and transfers them to the lipoate-dependent proteins (Brookfield et al., 1991; Fujiwara et al., 1994; Green et al., 1995; Loeffelhardt et al., 1996).

As shown in Figure 9.4B, the lipoic acid-binding pocket of *E. coli* LplA which is sandwiched between the core β-sheet and the β4–β5 loop, is more open

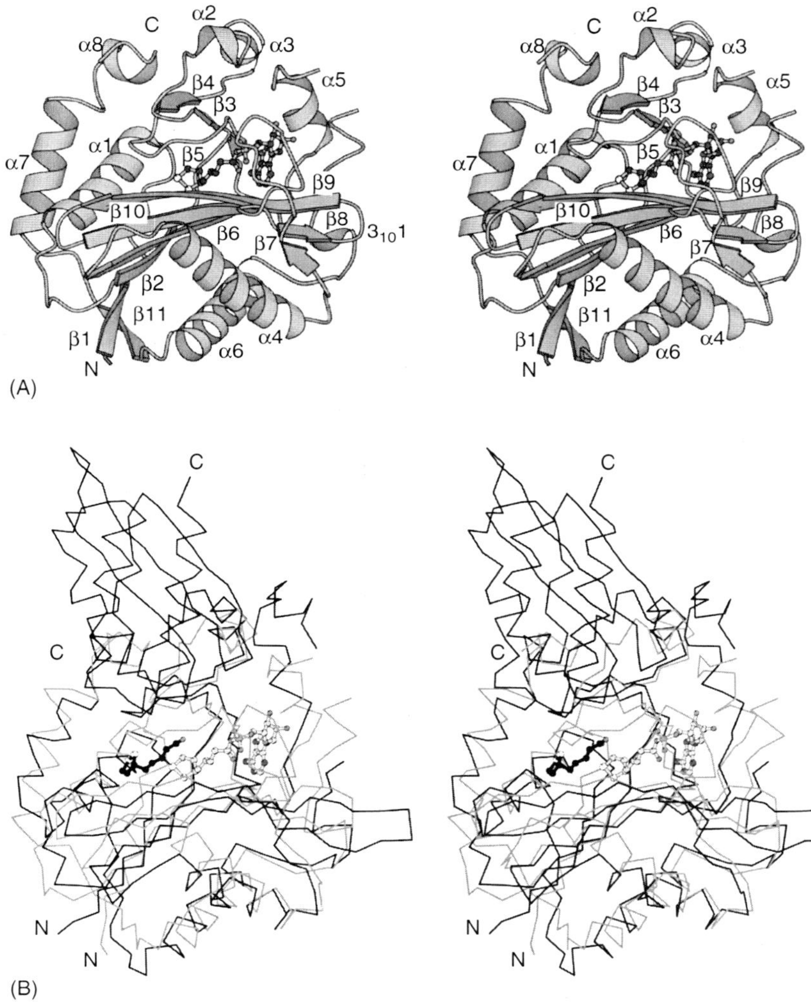

FIGURE 9.4 Structure of *Thermoplasma acidophilum* LplA in complex with lipoyl-AMP. (A) Stereo view of the LplA structure is drawn the same as in Figure 9.3A with the secondary structural elements (PDB code, 2ART). (B) Structural comparison between the *T. acidophilum* LplA·lipoyl-AMP complex (gray) and *Escherichia coli* LplA·lipoic acid complex (black). Ligands are shown in a ball-and-stick representation.

compared to that of *T. acidophilum* LplA, and the lipoic acid seems to have less hydrophobic interaction with the surrounding protein than the tightly bound lipoyl-AMP in the *T. acidophilum* LplA. The structure may enable lipoic acid to escape easily from the binding pocket. The significance of the open structure of the *E. coli* LplA active site remains unknown.

The LplA has a characteristic RRXXGGGXVXHD motif (residues 69–80), which is highly conserved among the LplA family. The motif is situated on the β4–β5 structure, constituting the upper surface of the lipoic acid-binding pocket. Arg70 forms a hydrogen bond network with the main-chain carbonyl group of Gly73 and Gly75 and the side-chain of Ser72, stabilizing the loop structure. An R72G point mutation of the human lipoyltransferase (equivalent to Arg-69 of the *E. coli* LplA) has been found in a hyperglycinemia patient, but no mutation has been found in the four genes comprising the glycine cleavage system: genes for P-protein, H-protein, T-protein, and L-protein (S. Kure, personal communication). The mutation may cause the reduction of lipoylated H-protein. The expression of the human recombinant R72G mutant failed because of the instability of the mutant protein. Thus, Arg69 and Arg70 in the *E. coli* LplA appear to play an important role in the stabilization of the three-dimensional structure.

Limited proteolysis of the *E. coli* LplA resulted in excision of a region (residues 171–181, a part of α5 helix and the following loop) accompanying an inactivation of the LplA, while a part of the region (residues 176–181) was protected from the excision by the addition of lipoic acid (McManus et al., 2006). The loop tends to be disordered in the *E. coli* and *T. acidophilum* LplA structures. It may contribute to holding of substrates or to the second lipoate-transfer reaction in conjunction with the C-terminal domain.

STRUCTURE OF *T. acidophilum* LIPOATE-PROTEIN LIGASE A

Overall Fold of the *T. acidophilum* LplA Molecule

The structure of the *T. acidophilum* LplA has been thoroughly determined by Kim et al. (2005) and McManus et al. (2006). The LplA consists of only one domain composed of 262 amino acid residues with a molecular mass of 29,871 Da. The protein lacks a C-terminal domain found in LplAs from *E. coli* and *S. pneumoniae*. Since a crystal of *T. acidophilum* LplA in complex with lipoyl-AMP is obtained by soaking a native crystal in the reservoir solution containing ATP and then by transferring the crystal to the reservoir solution containing lipoic acid, the LplA can catalyze the lipoate-adenylation reaction in the crystal. However, overall lipoylation of the lipoyl domain or E2 subunit of the pyruvate dehydrogenase complex did not occur, suggesting inability of the LplA to catalyze the second lipoate-transfer reaction. It has been suggested that a small protein, whose gene is located alongside the gene for the LplA, contributes to the lipoate-transfer reaction (McManus et al., 2006).

The structure of the *T. acidophilum* LplA consists of a large core β-sheet composed of six antiparallel β-strands (β2, β6–β10), a small β-sheet composed of three β-strands (β3–β5), and eight α-helices and a 3_{10} helix surrounding the β-sheets (Figure 9.4A). The structure falls into α/β proteins. The overall fold of the *T. acidophilum* LplA structure is similar to that of the N-terminal domain of *E. coli* LplA (Figure 9.4B). However, the narrow active site sandwiched between the core β-sheet and β4–β5 loop, the long β7–β8 loop structure, and the lipoic acid-binding site are different from those of *E. coli* LplA.

Substrate-Binding Site

Structures of the LplA in complex with lipoic acid (McManus et al., 2006), ATP, and lipoyl-AMP (Kim et al., 2005) have been determined. Lipoyl-AMP is bound deeply in the bifurcated pocket of the LplA adopting a U-shaped conformation (Figure 9.5). The lipoyl moiety of lipoyl-AMP is bound in a hydrophobic pocket formed with hydrophobic residues (Leu18, Ile46, Val79, Ala163), two histidine residues (His81, His161), and the aliphatic part of Arg72. The size and the hydrophobic environment of the pocket are not suitable to accommodate the biotin molecule. The carbonyl oxygen atom of the lipoyl moiety interacts with the invariant Lys145 through a hydrogen bond. The β4–β5 loop, which lines the upper side of the pocket, is proposed to be a discriminating loop against the entry of bulky biotin, interacting with lipoic acid, because a G75S mutation in the loop of the *E. coli* LplA (corresponding to Gly77 of the *T. acidophilum* LplA) makes the LplA resistant to the use of selenolipoic acid as a substrate, while it had

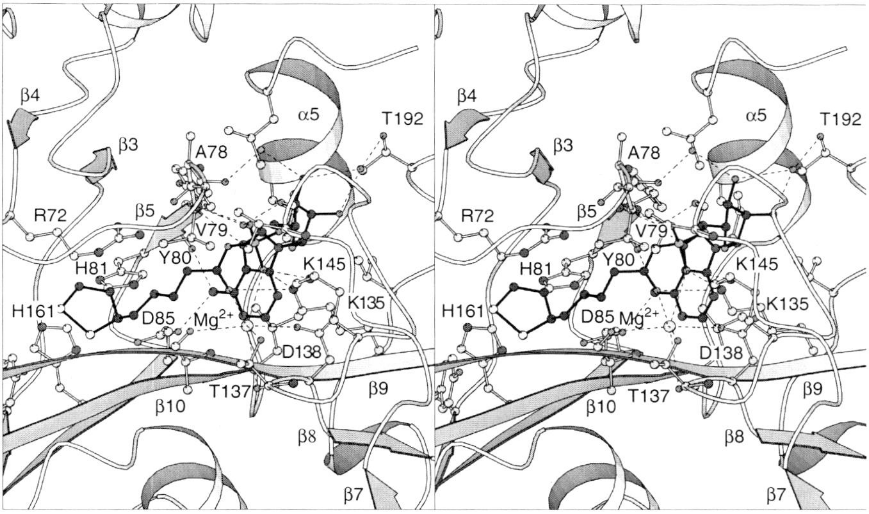

FIGURE 9.5 Close-up view of the active site of *Thermoplasma acidophilum* LplA in complex with lipoyl-AMP. The protein is shown in a ribbon representation. Lipoyl-AMP (black) and interacting residues (gray) are shown in a ball-and-stick representation. Dashed lines show potential hydrogen bonds and coordinations between atoms.

little effect on using lipoic acid which is smaller than selenolipoic acid (Morris et al., 1995).

The AMP moiety of lipoyl-AMP forms extensive hydrogen bonds with residues of the LplA. The phosphate moiety of lipoyl-AMP makes hydrogen bonds with the side-chain of Lys135, Lys145, and the main-chain nitrogen atom of Gly77.

O4* of the ribose moiety forms a hydrogen bond with the side-chain of Lys145. O3* is hydrogen-bonded with the main-chain oxygen atom of Ala78 and the side-chain of Thr192 through a respective water molecule. O2* forms a hydrogen bond with the side-chain of Thr192 through a water molecule.

N6 of the adenine ring makes hydrogen bonds with the main-chain oxygen atom of Tyr80 and the side-chain of Asp85. N1 and N7 form a hydrogen bond with the main-chain oxygen atom of Ala163 and the nitrogen atom of Tyr80, respectively. The hydrophobic parts of the adenine ring also make hydrophobic interactions with the hydrophobic residues (Val79, Ala150, Ala163, Leu165, Leu173, and Leu177) of the LplA. The residues, except Ala163, are conserved among proteins listed in Figure 9.2.

The LplA has a tightly bound magnesium ion. It forms a hexacoordinate complex with the phosphate oxygen atom and the carbonyl oxygen atom of lipoyl-AMP, the side-chain of Asp138, the side-chain and main-chain oxygen atoms of Thr137, and the main-chain oxygen atom of Ala-149.

The characteristic long β7–β8 loop (residues 124–138), which is short in the other LplAs, appears to function as a lid to the active site. Lys135, Val129 (through a water molecule), and Thr137 and Asp138 (both through the magnesium ion) in the loop interact with the phosphate group of lipoyl-AMP.

In the LplA·lipoic acid complex, lipoic acid is bound same as lipoyl moiety in the LplA·lipoyl-AMP complex. The carboxyl group interacts with the side-chain of Lys145. In the LplA·ATP complex, ATP interacts with the same sites as the AMP moiety in the LplA·lipoyl-AMP complex. Its β-phosphate group occupies the same position as the carbonyl group of lipoyl-AMP, and the γ-phosphate group interacts with the side-chain of Arg72, maintaining the bent conformation of triphosphate group. Kim et al. (2005) suggest that when lipoic acid enters into the pocket, the β- and γ-phosphate groups will be released from the binding pocket because of the hydrophobic environment in the area of the pocket.

PREDICTED CATALYTIC MECHANISM OF LPLA

Although the *E. coli* LplA is the only LplA to date which has been demonstrated to catalyze both the lipoate-activation reaction and the lipoate-transfer reaction in the protein lipoylation, its structure in complex with the substrates or the intermediate are not well characterized. On the other hand, the crystal structures of the *T. acidophilum* LplA in complex with lipoic acid, ATP, and the lipoyl-AMP intermediate in addition to the native structure are fully determined, although it catalyzes only the first lipoate-activation reaction. The structures of the various complexes provide insights into the active site corresponding to different steps of the lipoate-activation reaction. Recently we have determined the crystal structure of the bovine lipoyltransferase in complex with lipoyl-AMP (Fujiwara et al., 2007). The lipoyltransferase molecule consists of two domains, a large N-terminal domain and a small C-terminal domain, similar to that of the *E. coli* LplA. However, the conformation of the C-terminal domain is quite different from that of the *E. coli* LplA. Lipoyl-AMP is bound in the N-terminal domain. The overall fold surrounding the active site, the position of lipoyl-AMP, and interaction mode with lipoyl-AMP are similar to those of the *T. acidophilum* LplA although some noteworthy differences have been found. The carbonyl oxygen atom and the phosphate group of lipoyl-AMP are located within a hydrogen bonding distance from the conserved lysine residue of the lipoyltransferase (equivalent to Lys145 of the *T. acidophilum* LplA). The results suggest that all members of the LplA family follow a common reaction mechanism in the protein lipoylation.

The *E. coli* LplA has some analogies to the class II aminoacyl-tRNA synthetase in terms of the successive two-step reaction and the structural similarity as shown by the DALI search (Cavarelli et al., 1994; Arnez and Moras, 1997). In the first step, the amino acid is activated by the nucleophilic attack to the α-phosphate of ATP. The reaction proceeds via a putative penta-coordinate transition state, yielding a mixed anhydride intermediate, aminoacyl-adenylate, and inorganic pyrophosphate. In the second step, the aminoacyl moiety is transferred to the 3′-terminal ribose of a cognate tRNA, yielding an aminoacyl-tRNA and AMP. Their active sites are built on a six-stranded antiparallel β-sheet flanked by an additional parallel β-strand and surrounded by α-helices.

On the basis of the crystallographic results of the *T. acidophilum* LplA and the analogy of the reaction mechanism to the class II aminoacyl-tRNA synthetase, we propose a catalytic mechanism for the protein lipoylation as shown in Figure 9.6.

1. Prior to the initiation of the lipoate-activation reaction, ATP and lipoic acid bind to the respective pockets. ATP is bound adopting a bent conformation. The AMP moiety makes extensive hydrogen bonds and electrostatic interactions with the LplA, similar to the AMP moiety in the LplA·lipoyl-AMP complex as mentioned above. Lipoic acid is bound in the hydrophobic pocket, substituting for the β- and γ-phosphate moiety of ATP. The carboxyl group forms the hydrogen bond with Lys145. Thus, Lys145 plays an essential role in positioning lipoic acid near ATP. The α-phosphate group of ATP interacts with the positively charged Lys135, Lys145, and the magnesium ion. The bindings lower the activation energy barrier for the withdrawal of electrons from the α-phosphate group of ATP, stimulating the positive charge of the α-phosphorus atom, which facilitates the nucleophilic attack by lipoic acid.

   The nucleophilic attack by lipoic acid to the α-phosphorus atom leads to the formation of the lipoyl-AMP intermediate and the release of

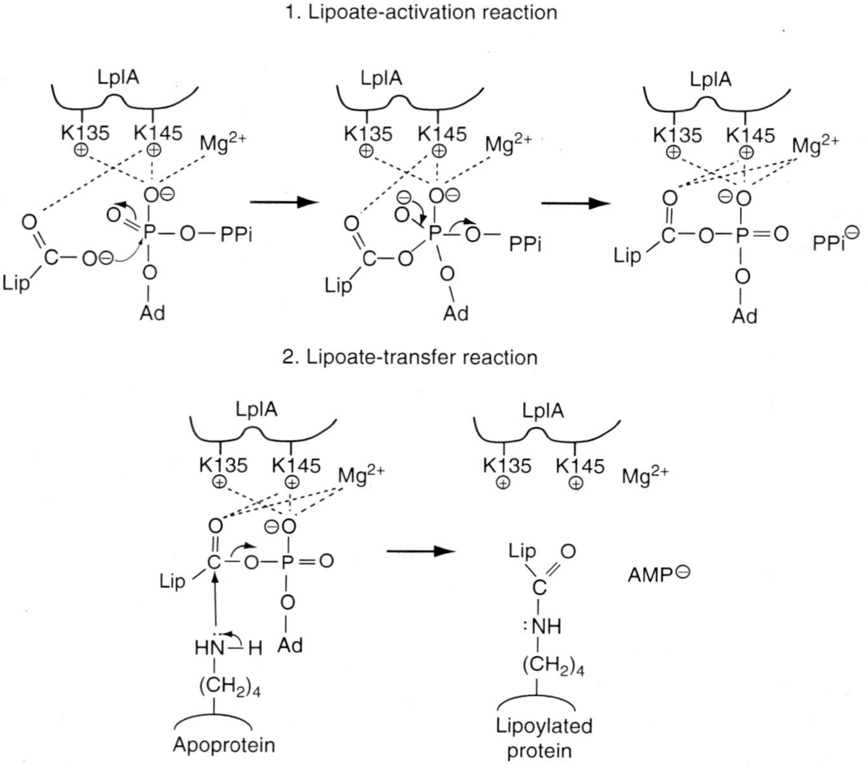

FIGURE 9.6 Predicted reaction mechanism of LplA. Schematic lipoate-activation reaction (1) and lipoate-transfer reaction (2) are drawn. K135 and K145 are lysine residues in the *Thermoplasma acidophilum* LplA. K145 is a strictly conserved lysine residue among the members of LplA family. Dashed lines represent potential hydrogen bonds, electrostatic interactions, and coordinations between atoms. Ad, adenosine moiety of ATP or lipoyl-AMP; Lip, dithiolane ring and the aliphatic chain of lipoic acid.

inorganic pyrophosphate via a putative penta-coordinate transition state, which is stabilized by the lysine residues and the magnesium ion.

2. In the second lipoate-transfer step, the ε-amino group of the lysine residue of the lipoate-acceptor apoprotein attacks the carbonyl carbon atom of lipoyl-AMP. The interaction of the carbonyl oxygen atom with Lys145 and the magnesium ion stimulates the positive charge of the carbonyl carbon atom and as a consequence, facilitates the nucleophilic attack by the ε-amino group, accompanying the cleavage of the bond between the phosphate group and the carbonyl group. The lipoyl moiety is transferred to the ε-amino group, giving rise to a lipoylated protein and AMP.

Thus the lysine residue, which is strictly conserved among the LplA family, and the magnesium ion appear to play important roles in the protein lipoylation reaction: first, in placing the substrates in juxtaposition; second, in the withdrawal of electrons from the α-phosphate group of ATP, facilitating the nucleophilic attack of lipoic acid on the α-phosphorus atom; third, in the stabilization of the penta-coordinate transition state by the neutralization of the negative charge of the state; and fourth, in the withdrawal of electrons from the carbonyl group of lipoyl-AMP, leading to the nucleophilic attack by the lysine residue of the lipoate-acceptor protein accompanying the bond cleavage between the carbonyl group and the phosphate group. The interaction between the LplA and the lipoate-acceptor protein has not been determined. The determination of the three-dimensional structure of the ternary complex between LplA, apoprotein, and lipoyl-AMP will be required for a deeper insight into the lipoate-transfer reaction.

REFERENCES

Arnez, J.G. and Moras, D. 1997. Structural and functional considerations of the aminoacylation reaction. *Trends Biochem. Sci.* 22: 211–216.

Bagautdinov, B., Kuroishi, C., Sugahara, M., and Kunishima, N. 2005. Crystal structures of biotin protein ligase from *Pyrococcus horikoshii* OT3 and its complexes: Structural basis of biotin activation. *J. Mol. Biol.* 353: 322–333.

Brändén, C.-I. 1980. Relation between structure and function of α/β-proteins. *Quart. Rev. Biophy.* 13: 317–338.

Brookfield, D.E., Green, J., Ali, S.T., Machado, R.S., and Guest, J.R. 1991. Evidence for two protein-lipoylation activities in *Escherichia coli. FEBS Lett.* 295: 13–16.

Cavarelli, J., Eriani, G., Rees, B., Ruff, M., Boeglin, M., Mitschler, A., Martin, F., Gangloff, J., Thierry, J.-C., and Moras, D. 1994. The active site of yeast aspartyl-tRNA synthetase: Structural and functional aspects of the aminoacylation reaction. *EMBO J.* 13: 327–337.

Cicchillo, R.M., Iwig, D.F., Jones, A.D., Nesbitt, N.M., Baleanu-Gogonea, C., Souder, M.G., Tu, L., and Booker, S.J. 2004. Lipoyl synthase requires two equivalents of *S*-adenosyl-L-methionine to synthesize one equivalent of lipoic acid. *Biochemistry* 43: 6378–6386.

Fujiwara, K., Okamura-Ikeda, K., and Motokawa, Y. 1992. Expression of mature bovine H-protein of the glycine cleavage system in *Escherichia coli* and in vitro lipoylation of the apoform. *J. Biol. Chem.* 267: 20011–20016.

Fujiwara, K., Okamura-Ikeda, K., and Motokawa, Y. 1994. Purification and characterization of lipoyl-AMP:N^{ε}-lysine lipoyltransferase from bovine liver mitochondria. *J. Biol. Chem.* 269: 16605–16609.

Fujiwara, K., Okamura-Ikeda, K., and Motokawa, Y. 1997. Cloning and expression of a cDNA encoding bovine lipoyltransferase. *J. Biol. Chem.* 272: 31974–31978.

Fujiwara, K., Suzuki, M., Okumachi, Y., Okamura-Ikeda, K., Fujiwara, T., Takahashi, E., and Motokawa, Y. 1999. Molecular cloning, structural characterization and chromosomal localization of human lipoyltransferase gene. *Eur. J. Biochem.* 260: 761–767.

Fujiwara, K., Takeuchi, S., Okamura-Ikeda, K., and Motokawa, Y. 2001. Purification, characterization, and cDNA cloning of lipoate-activating enzyme from bovine liver. *J. Biol. Chem.* 276: 28819–28823.

Fujiwara, K., Toma, S., Okamura-Ikeda, K., Motokawa, Y., Nakagawa, A., and Taniguchi, H. 2005. Crystal structure of lipoate-protein ligase A from *Escherichia coli*: Determination of the lipoic acid-binding site. *J. Biol. Chem.* 280: 33645–33651.

Fujiwara, K., Hosaka, H., Matsuda, M., Okamura-Ikeda, K., Motokawa, Y., Suzuki, M., Nakagawa, A., and Taniguchi, H. 2007. Crystal structure of bovine lipoyltransferase in complex with lipoyl-AMP. *J. Mol. Biol.* 371: 222–234.

Grassl, S.M. 1992. Human placental brush-border membrane Na^+-biotin cotransport. *J. Biol. Chem.* 267: 17760–17765.

Green, D.E., Morris, T.W., Green, J., Cronan, J.E., Jr., and Guest, J.R. 1995. Purification and properties of the lipoate protein ligase of *Escherichia coli. Biochem. J.* 309: 853–862.

Holm, L. and Sander, C. 1993. Protein structure comparison by alignment of distance matrices. *J. Mol. Biol.* 233: 123–138.

Jordan, S.W. and Cronan, J.E., Jr. 2003. The *Escherichia coli lipB* gene encodes lipoyl (octanoyl)-acyl carrier protein:protein transferase. *J. Bacteriol.* 185: 1582–1589.

Kabsch, W. and Sander, C. 1983. Dictionary of protein secondary structure: Pattern recognition of hydrogen-bonded and geometrical features. *Biopolymers* 22: 2577–2637.

Kim, D.J., Kim, K.H., Lee, H.H., Lee, S.J., Ha, J.Y., Yoon, H.J., and Suh, S.W. 2005. Crystal structure of lipoate-protein ligase A bound with the activated intermediate: Insights into interaction with lipoyl domains. *J. Biol. Chem.* 280: 38081–38089.

Loeffelhardt, S., Borbe, H.O., Locher, M., and Bisswanger, H. 1996. In vivo incorporation of lipoic acid enantiomers and homologues in the pyruvate dehydrogenase complex from *Escherichia coli. Biochim. Biophys. Acta* 1297: 90–98.

Ma, Q., Zhao, X., Eddine, A.N., Geerlof, A., Li, X., Cronan, J.E., Kaufmann, S.H.E., and Wilmanns, M. 2006. The *Mycobacterium tuberculosis* LipB enzyme functions as a cysteine/lysine dyad acyltransferase. *Proc. Natl. Acad. Sci. USA* 103: 8662–8667.

McManus, E., Luisi, B.F., and Perham, R.N. 2006. Structure of a putative lipoate protein ligase from *Thermoplasma acidophilum* and the mechanism of target selection for post-translational modification. *J. Mol. Biol.* 356: 625–637.

Miller, J.R., Busby, R.W., Jordan, S.W., Cheek, J., Henshaw, T.F., Ashley, G.W., Broderick, J.B., Cronan, J.E., Jr., and Marletta, M.A. 2000. *Escherichia coli* LipA is a lipoyl synthase: In vitro biosynthesis of lipoylated pyruvate dehydrogenase complex from octanoyl-acyl carrier protein. *Biochemistry* 39: 15166–15178.

Morris, T.W., Reed, K.E., and Cronan, J.E., Jr. 1994. Identification of the gene encoding lipoate-protein ligase A of *Escherichia coli*: Molecular cloning and characterization of the lplA gene and gene product. *J. Biol. Chem.* 269: 16091–16100.

Morris, T.W., Reed, K.E., and Cronan, J.E., Jr. 1995. Lipoic acid metabolism in *Escherichia coli*: The *lplA* and *lipB* genes define redundant pathways for ligation of lipoyl groups to apoprotein. *J. Bacteriol.* 177: 1–10.

Mosyak, L., Reshetnikova, L., Goldgur, Y., Delarue, M., and Safro, M.G. 1995. Structure of phenylalanyl-tRNA synthetase from *Thermus thermophilus. Nat. Struct. Biol.* 2: 537–547.

Perham, R.N. 2000. Swinging arms and swinging domains in multifunctional enzymes: Catalytic machines for multistep reactions. *Annu. Rev. Biochem.* 69: 961–1004.

Prasad, P.D., Wang, H., Kekuda, R., Fujita, T., Fei, Y.-J., Devoe, L.D., Leibach, F.H., and Ganapathy, V. 1998. Cloning and functional expression of a cDNA encoding a mammalian sodium-dependent vitamin transporter mediating the uptake of pantothenate, biotin, and lipoate. *J. Biol. Chem.* 273: 7501–7506.

Reed, L.J., Leach, F.R., and Koike, M. 1958. Studies on a lipoic acid-activating system. *J. Biol. Chem.* 232: 123–142.

Wilson, K.P., Shewchuk, L.M., Brennan, R.G., Otsuka, A.J., and Matthews, B.W. 1992. *Escherichia coli* biotin holoenzyme synthetase/bio repressor crystal structure delineates the biotin- and DNA-binding domains. *Proc. Natl. Acad. Sci. USA* 89: 9257–9261.

Wood, Z.A., Weaver, L.H., Brown, P.H., Beckett, D., and Matthews, B.W. 2006. Co-repressor induced order and biotin repressor dimerization: A case for divergent followed by convergent evolution. *J. Mol. Biol.* 357: 509–523.

Yi, X. and Maeda, N. 2005. Endogenous production of lipoic acid is essential for mouse development. *Mol. Cell. Biol.* 25: 8387–8392.

Zhao, X., Miller, J.R., and Cronan, J.E. 2005. The reaction of LipB, the octanoyl-[acyl carrier protein]:protein *N*-octanoyltransferase of lipoic acid synthesis, proceeds through an acyl-enzyme intermediate. *Biochemistry* 44: 16737–16746.

10 An Evaluation of the Stability and Pharmacokinetics of *R*-Lipoic Acid and *R*-Dihydrolipoic Acid Dosage Forms in Human Plasma from Healthy Subjects

David A. Carlson, Karyn L. Young, Sarah J. Fischer, and Heinz Ulrich

CONTENTS

INTRODUCTION: STRUCTURE, STEREOCHEMISTRY, AND MECHANISMS OF ACTION OF ALPHA LIPOIC ACID

Alpha lipoic acid (ALA) is a medium chain (C8) fatty acid with vicinal sulfur atoms at C6 and C8 existing either as free sulfhydryls (DHLA) or linked via an intramolecular disulfide (ALA). The C6 atom (alternatively designated as C3 of the dithiolane ring) of the octanoic acid chain is chiral and the molecules exist as four enantiomers (*R*-(+)-LA, *S*-(−)-LA, *R*-(−)-DHLA, and *S*-(+)-DHLA) (Figure 10.1) and two racemic mixtures (rac-ALA = +/−ALA = (*RS*)-ALA and rac-DHLA = (+/−)-ALA = *RS*-DHLA). These six forms of lipoic acid have

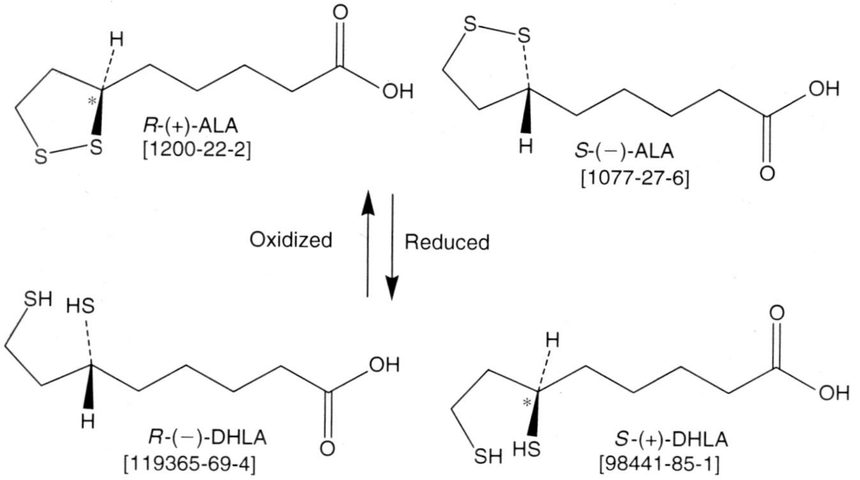

FIGURE 10.1 Three dimensional molecular diagrams depicting both the absolute configurations (*R* and *S*) which are mirror images and the change in specific rotations from (+) to (−) when RLA is reduced to *R*-DHLA of (−) to (+) when SLA is reduced to *S*-DHLA. The diagrams are useful in conceptualizing how one enantiomer can react differently than its mirror image with a receptor, transporter, signaling molecule, protein or enzyme in vivo. Only RLA is naturally occurring and proposed to be the eutomeric form of ALA.

biological similarities and differences in their mechanisms of action and should be considered pharmacologically distinct. RLA and *R*-DHLA are the naturally occurring enantiomers. The complete characterization of the pharmacological similarities and differences between the six forms is in its infancy but there are indications that the *R*-enantiomers are the eutomeric forms of ALA (Jacob et al. 1999; Packer et al. 2001). Stereochemical and non-stereochemical mechanisms of action of ALA have been identified and comprehensively reviewed (Carlson et al. 2007b). It is a well-known principle of enantioselective pharmacology that a given enantiomer may either be active (contributing to the PD), inactive (isomeric ballast), or detrimental (antimetabolite or competitive inhibitor, opposing the action of the preferred enantiomer) (Ariëns 1984; Jamali et al. 1989). Both in vitro and in vivo assays have revealed SLA may competitively inhibit the actions of RLA, which therefore becomes the limiting criteria therapeutically (Artwohl et al. 2000; Ulrich et al. 2001; May et al. 2006). Fundamental and yet currently unanswered questions are whether or not there is a single primary mechanism of action of ALA (or a summation of multiple factors) responsible for the therapeutic effects. It is unclear whether or not the therapeutic efficacy is primarily stereospecific or non-stereospecific or dose-concentration and time dependent.

The first description of the antioxidant properties of ALA was in 1959 (Rosenberg and Culik, 1959). In the late 1980s, rac-ALA and rac-DHLA were labeled as antioxidants due to the in vitro capacity to neutralize a variety of ROS, RNS, and RSS. By the early 1990s, ALA became available as a nutritional supplement (Packer et al. 1996) but the complex stereospecific and pleiotropic mechanisms of action in vivo have not been fully characterized (Pershadsingh 2007; Zhang et al. 2007). This is partially due to the fact that the bulk of 2320 papers published utilized only the racemic mixture and approximately 1% tested the enantiomers (Pub Med as of 3/21/07). Surprisingly, one study indicated that RLA was a better chelator than SLA and antioxidant (Ou et al. 1995). Recently, it was demonstrated that RLA was more effectively transported by the MCFA transporter in a human endothelial cell line than rac-ALA, indicating competitive inhibition by SLA (May et al. 2006). Recent studies attempting to elucidate the mechanisms of action have shifted primary emphasis away from direct scavenging of dangerous free radical species, single electron reductions of vitamin C, Vitamin E, and CoQ10 radicals, chelation of heavy and redox active transition metals due to the measurement of low micromolar concentrations in vivo (versus millimolar intracellular concentrations of GSH and ascorbate) and rapid disappearance from plasma and tissues subsequent to exogenous administration. These properties which undoubtedly have some effect in vivo must be reconciled with the short plasma half-lives of ALA, DHLA (all forms), and of the metabolites (Smith et al. 2004).

The new paradigm is founded on the ability of ALA to alter signal transduction and gene transcription by functioning as a metabolic and redox stressor and which may be mediated by either a pro-oxidant, reductive or redox effect via activation of early response genes (Packer et al. 1997; Dicter et al. 2002; Konrad 2005; Ogborne et al. 2005; Linnane et al. 2006). RLA has been shown to increase ATP production stereospecifically in the heart while SLA decreased it

(Hagen et al. 1999). Redox modulation may be both stereospecific and non-stereospecific and may result in increased expression of thioredoxin and other redox-sensitive genes (Lee et al. 2004; Larghero et al. 2006). RLA improved synthesis of GSH stereospecifically in the lens and may reverse age-related losses in GSH in the brain (Maitra et al. 1996; Suh et al. 2001; Bharat et al. 2002; Suh et al. 2004a). RLA binds to the IR and induces alterations in patterns of phosphorylation/dephosphorylation of various tyrosine kinases (Diesel et al. 2007) and phosphatases (Sommer et al. 2000; Cho et al. 2003), modulates ion transport (Sen 2000; Bishara et al. 2002), Na^+, K^+, ATPases (Arivazhagan and Panneerselvam 2004), plasma membrane redox system (Bera et al. 2005), the plasma redox status by lowering plasma L-Cys (Nolin et al. 2007), altering the free thiol/disulfide ratios (Gregus et al. 1992), signal transduction pathways and gene transcription associated with metabolic control genes (nutrigenes) associated with glucose and fat metabolism; peroxisome proliferator activated receptor-α (PPAR-α) and peroxisome proliferator activated receptor-γ (PPAR)-γ, (Pershadsingh et al. 2005; El Modaoui et al. 2006; Pershadsingh 2007), SREBP-1 (Moreau et al. 2007) leptin (Lee et al. 2006); downregulation of acute phase proteins such as CRP (Sola et al. 2005), IL-6 (Mantovani et al. 2003a; Sola et al. 2005), PAI (Sola et al. 2005), TNF-α, and IL-1β (Zhang and Frei 2001; Kiemer et al. 2002; Byun et al. 2005; Dulundu et al. 2007), transforming growth factor-β (TGF-β) (Oksala et al. 2007), adhesion molecules (ICAM-1,VCAM, MMP-9) (Marracci et al. 2004; Cantin et al. 2005; Yadav et al. 2005; Chaudhary et al. 2006; Kim et al. 2007b; Zhang et al. 2007), RAGE (receptor for advanced glycation end products) (Vincent et al. 2007), mediators of inflammation such as IFN-γ (Khanna et al. 1999), sPLA2 (Jameel et al. 2006), IL-18 (Lee et al. 2006), COX-2, PGE2, RANKL (receptor activator of NF-κB ligand), IL-1 (Ha et al. 2006), LOX-15 (Lapenna et al. 2003), iNOS (Demarco et al. 2004; Powell et al. 2004; Hurdag et al. 2005), induction or increased DNA binding of heat shock proteins (HSF-1, HSP-60, HSP-70) (Oksala et al. 2006; Oksala et al. 2007), regulation of apoptotic/antiapoptotic proteins such as caspase-3, BAX (Diesel et al. 2007), upregulation of the antiapoptotic protein Bcl_2 (Marsh et al. 2005) and phosphorylation of the apoptotic protein BAD (Kulhanek-Heinz 2004) inhibiting apoptosis, upregulation of eNOS (Visioli et al. 2002; Smith et al. 2003), and phase two detoxification enzymes via Nr-f2 and the ARE; SOD, (Balachandar et al. 2003), glutathione peroxidase (GPx), glutathione *S*-transferases [GST], NAD (P)H: quinone oxidoreductase 1 (Flier et al. 2002; Mantovani et al. 2003), γ-glutamylcysteine ligase (GCL) the rate limiting enzyme in the synthesis of GSH, hemi-oxygenase [HO-1] (Ogborne et al. 2005), increased B-lymphocytes (Wessner et al. 2006).

Many of these functions may be mediated via modulation of NF-κB, and activator protein-1 (AP-1), which was not stereospecific in inhibiting NF-κB in human aortic endothelial cells (HAEC) as SLA was equally effective (Packer 1998; Zhang and Frei 2001; Lee and Hughes 2002; Kim et al. 2007). The enantioselective preferences in modulation or expression of adenosine monophosphate protein kinase (AMPK) and PGC-1a (Kim et al. 2004; Lee et al. 2005b; Kim et al. 2007), Nr-f2 (Suh et al. 2001; Suh et al. 2004; Ogborne et al. 2005;

Suh et al. 2005) have not been tested but P38 MAPk and PI3K/Akt are critical proteins involved in the mechanisms of action of rac-ALA and there are indications of stereochemical preferences for RLA (Estrada et al. 1996). Recently, Zhang showed rac-ALA activated PI3K and P38 MAPk, which reduced the expression of LPS-induced inflammatory molecules but only rac-ALA was used (Abdul et al. 2007; Zhang et al. 2007). Some, but not all, of these proteins and signaling pathways have demonstrated stereospecific preferences for RLA but most of them have not yet been tested.

In a molecular modeling and energy-minimization study using lipoamide dehydrogenase, it was shown that only the active *R*-enantiomer is able to form direct contacts with the reactive thiol groups and imidazole base at the active site, whereas with the *S*-enantiomer the SH-group at C6 points away from the His450 base and functions as competitive inhibitor. This model may be useful to identify additional catalytic binding sites with enantioselective preferences for RLA or SLA in vivo (Raddatz et al. 1997).

STRUCTURE, PHYSICAL PROPERTIES, STABILITY, AND STEREOCHEMISTRY OF ALPHA LIPOIC ACID AND DIHYDROLIPOIC ACID

The first synthesis of rac-DHLA was reported by Stokstad's group at Lederle Laboratories and a number of successful synthetic procedures were developed and patented in the ensuing years (Bullock et al. 1957). Early investigators in the United States and Japan utilized rac-DHLA as an intermediate in the laboratory preparation of ALA, either by relying on spontaneous oxidation in solution, or more controllably with I_2, $FeCl_3$, or an alkyl nitrite to effect the desired transformation (Hornberger et al. 1952, 1953; Nakano and Sano 1956; Bullock et al. 1957; Deguchi 1960; Acker et al. 1963).

ALA exists as yellow crystalline needles or plates, depending on the solvents used and the temperature under which it forms but may also form polymeric chains of linear disulfides. The melting point of rac-ALA is 58°C–61°C and various melting points for RLA and SLA have been reported in the range of 46°C–51°C (Acker and Wayne 1957; Walton et al. 1995; Laban et al. 2007). The melting point of the enantiomers is lower, leading to reduced stability and increases the tendency to polymerize relative to rac-ALA. Interestingly, even RLA or SLA with high chemical and enantiomeric purities can either melt cleanly or sinter, which may be due to the presence of trace metals (Ames 2006).

The propensity of RLA to polymerize has a significant impact on the shelf life and bioavailability of derived dosage forms (Carlson 2007b). Subsequent to laboratory or industrial synthesis, the ALA polymers may be the predominant isolatable form. ALA can be separated from the polymer by repeated low temperature extraction with hydrocarbon solvents or by thermal depolymerization under high vacuum (Hornberger et al. 1953; Thomas and Reed, 1956). Zinc and HCl have been utilized to reduce the ALA polymer to DHLA with subsequent reoxidation in solution with I_2 to isolate crystalline ALA. Dilute (1 mM/ethanol) alkaline solutions (pH 12) of DHLA were used to depolymerize the linear disulfide polymer of ALA by acting as either a direct reducing agent or via thiolate anions and the thiol-disulfide exchange reaction (Thomas and Reed, 1956).

R-(+)-α lipoic acid has a large positive specific rotation due to the ring strain of the dithiolane ring and *R*-DHLA has a small negative specific rotation due to opening of the ring. The highest specific rotation reported for RLA is +120 ($[\alpha]_D^{23} = +120$ c = 1 EtOH) (Carlson et al. 2007b). DHLA (all forms) occurs physically as a clear to lightly colored oil with a boiling point of 154°C–156°C (0.3 mm) (Reed and Niu 1954). Pure *R*-DHLA has a negative specific rotation ($[\alpha]_D^{23} = -14$ to -15 EtOH) and was first prepared and reported by Gunsalus. This group also reported distillation of *R*-(−)-DHLA at <0.1 mm/Hg caused partial racemization, because of the observed increase in the specific rotation from −14.7 [ee = 98%] to −8.9 [ee = 59.3%] (Gunsalus et al. 1956, 1957). Contrary to the original report, it was reported that thin-film distillation of *R*-DHLA yielded a clear product with a 99.6% ee value with trace amounts of RLA formed due to oxidation (Bringmann et al. 1999).

STABILITY OF DIHYDROLIPOIC ACID IN VITRO AND DURING SAMPLE PREPARATION

Racemic dihydrolipoic acid has been utilized in various in vitro models. Calibration solutions of DHLA can be prepared in 50% water/methanol and stored at −80°C to prevent oxidation to rac-ALA (Constantinescu et al. 1995; Haramaki et al. 1997a). Another study by Packer's group claimed that rac-DHLA is unstable and rapidly oxidizes to rac-ALA at room temperature. To quantify rac-DHLA, it had to be dissolved in mobile phase at pH 2.9, and stored at −80°C. It was also necessary to continuously sparge the solutions with helium, which caused evaporation of the acetonitrile (ACN) from the mobile phase, leading to gradual increases in the retention times (Han et al. 1995). This could be obviated by sparging the buffer solution and ACN separately and allowing the gradient pump to mix them while continuously sparging with helium or using an in-line degasser. Rac-DHLA from a sealed ampoule under nitrogen was stable for less than 20 min in an electrophoretic assay before being oxidized to ALA (Panak et al. 1996). In studies testing the binding affinities of rac-ALA and rac-DHLA with BSA-1, Shepkin found that DHLA dissolved in buffer solution (20 mM deuterium phosphate buffer with 150 mM NaCl) was oxidized to rac-ALA to the extent of 50% (Schepkin et al. 1994).

BASELINE LEVELS OF *R*-DIHYDROLIPOIC ACID IN HUMAN PLASMA

Originally microbiological assays were used to quantify serum levels of RLA at baseline and total ALA after administration of rac-ALA. Although the POF assay was sensitive to RLA and *R*-DHLA (the *S*-enantiomers were inactive) the assay did not differentiate RLA from *R*-DHLA (which would not survive the harsh sample treatment, i.e., steam autoclaving at 120°C) utilized to remove baseline RLA from plasma proteins (Gunsalus et al. 1957; Guedes et al. 1965; Natraj et al. 1984). In more recent studies, baseline levels of RLA were detected. Teichert's group reported different values depending on the conditions of separating RLA from plasma proteins. Following acid hydrolysis, 12.3–43.1 ng/mL were detected. Enzymatic hydrolysis of plasma proteins yielded 1.4–11.6 ng/mL and <1–38.2 ng/mL using subtilisin and Alcalase, respectively. Baseline *R*-DHLA was detected in the range of 33–145 ng/mL after enzymatic hydrolysis with Alcalase (Teichert and Preiss 1992, 1995, 1997).

The Asta Medica group developed an enantioselective method utilizing a fluorescent adduct of *R*-DHLA or *S*-DHLA after $SnCl_2$ reduction of rac-ALA (Niebch et al. 1997). The method works equally well for derivatization in solvent, buffer, or plasma with either *R*-DHLA or *S*-DHLA subsequent to reduction of RLA and SLA standards, respectively, and can be used to determine the enantiomeric purities of the standards. No reports exist testing the reaction with preformed or baseline plasma *R*-DHLA. Since the authors claimed that DHLA was not stable in plasma, no attempt was made to split the sample in two portions to differentiate plasma RLA from *R*-DHLA and that formed during the ex vivo reduction step (Hermann and Niebch 1997). The method was utilized in several pharmacokinetic (PK) trials (Gleiter et al. 1999, 1996; Hermann et al. 1998; Breithaupt-Grogler et al. 1999; Krone 2002). The enantioselective method relies on reduction of ALA to DHLA and derivatization to an isoindole with two chiral centers by reaction with *o*-phthaldehyde (OPA) and D-phenylalanine (D-PA). The resulting diastereomers can be separated on a C_{18} column. Baseline levels of RLA were not detected (Niebch et al. 1997). Researchers at Degussa found endogenous levels of RLA to be 112 ± 67 ng/mL using essentially the same method, but neither group attempted to quantify baseline *R*-DHLA (which could be determined, if present by omitting the reduction step) due to the assumption that *R*-DHLA did not occur in plasma (Hermann et al. 1997; Bernkop-Schnurch et al. 2004).

In developing an high-performance liquid chromatography/electrochemical detection (HPLC/ECD) method for simultaneous detection of rac-ALA and rac-DHLA, it was impossible to prepare calibration solutions in human plasma due to rapid oxidation, but aqueous calibration solutions were used to quantify *R*-DHLA in human plasma. In some cases, when the entire workup was done under inert gas, DHLA could be quantified (Teichert and Preiss 1992, 1995, 1997). It was claimed that DHLA "as such" does not occur in plasma and therefore need not be considered as a possible interfering substance in the measurement of plasma ALA (Hermann and Niebch 1997). Later, Teichert and Preiss developed an HPLC method with integrated pulsed amperometric detection (IPAD) for simultaneous detection of ALA and its primary plasma and urinary metabolites and were unable to detect baseline levels of RLA, which they stated are in the range of 1 ng/mL (4.85 pmol/mL) with the LOQ = 0.022 nmol/mL [4.54 ng/mL] ALA. This is difficult to reconcile with the earlier range of measurements of baseline RLA (between 1 and 112 times greater, depending on the method). There was no report of attempts to quantify baseline levels of

R-DHLA (Teichert et al. 2002). Rac-DHLA could be recovered and detected immediately after being spiked into plasma (0.5 and 5.0 mg/mL; 70% recovery) but neither RLA nor *R*-DHLA was detected at baseline (Biewenga et al. 1999). *R*-DHLA could not be detected as a plasma metabolite after administration of 1 g of pure RLA utilizing an HPLC/ECD (coulometric) similar to the method reported by Sen (Sen et al. 1999). To test the stability in plasma, rac-DHLA was spiked into the sample (0.5 mg/mL) and stored overnight at −25°C. The following day, 94% had reverted to ALA. The authors commented that aqueous solutions of DHLA are more stable than DHLA in plasma since DHLA reduces plasma components faster than it is oxidized by (dissolved) air (Biewenga et al. 1999). If this were the case in vivo, a corresponding change or increases in RLA levels and higher concentrations of reduced amino thiols would be expected. High levels of hepatic and plasma thiols have been reported subsequent to 30 mg/kg intravenous (IV) injection (direct rather than infused) of rac-ALA into rats, suggesting reduction to rac-DHLA, which acts as the reducing agent (Gregus et al. 1992). In contrast, 100 mg/kg IP in rats oxidized plasma proteins but had no effect on plasma thiols (Cakatay and Kayali, 2005). Assuming linearity, extrapolation of Krone's data (from 20 mg/kg IV) indicates this dose would lead to a C_{max} of 103 μg/mL (0.5 mM) and an AUC of 11,676 ng × h/mL (Krone 2002). It has been shown that rac-DHLA increases plasma thiol concentrations and protects plasma proteins from oxidative stress (van der Vliet et al. 1995) but may be toxic in the range of 0.1–1 mM (20.83–208.33 μg/mL) (Kis et al. 1997; Kulhanek-Heinz 2004).

ENZYMATIC REDUCTION OF ALPHA LIPOIC ACID: IS DIHYDROLIPOIC ACID AN IN VIVO PLASMA METABOLITE OF ALPHA LIPOIC ACID?

Endogenous *R*-DHLA is bound to the E2 subunits of the mitochondrial 2-oxo acid dehydrogenase complexes where it stereospecifically functions as an acetyltransferase (Gunsalus et al. 1956). Watanabe proposed *R*-DHLA as the physiological substrate for $mPGES_2$, indicating a role in modulation of inflammation pathways and serving the first function for the endogenously produced dithiol outside the mitochondrial 2-oxo acid dehydrogenase complexes (Watanabe et al. 2003).

In the mid-1990s, several papers appeared reporting the cytosolic and mitochondrial enzymes responsible for reduction of rac-ALA, RLA, or SLA to the corresponding reduced forms (Bunik et al. 1995; Biewenga et al. 1996; Haramaki et al. 1997). Glutathione reductase was first believed to be the primary NADPH-dependent enzyme for cytosolic reduction, later it was determined that all or most of the reduction should be attributed to thioredoxin reductase [Trx-1] (Arner et al. 1996; Jones et al. 2002; Biaglow et al. 2003), which has a preference for RLA in mammalian mitochondria [Trx-2] (Bunik et al. 1995). Pig heart PDH also has a preference for RLA (Loffelhardt et al. 1995) and RLA increases ATP production in the heart whereas SLA diminishes it (Zimmer et al. 1995; Hagen et al. 1999).

It is widely believed that most of the antioxidant properties and therapeutic potential of rac-ALA and RLA are due to the in vivo reduction to rac-DHLA and *R*-DHLA (Packer et al. 1995) although recently it was determined that the disulfide radical cation should be considered as a candidate for the pharmacologically active species (Bucher et al. 2006). Most of the evidence to date that DHLA is the more physiologically active entity of the ALA/DHLA redox couple comes from extrapolation of high concentration (0.5–10 mM) in vitro assays using various cell lines where ALA and DHLA can be maintained for up to 72 h without significant metabolic transformation (except for redox cycling of the dithiolane ring and small amount of B-oxidized products) (Constantinescu et al. 1995; Roy et al. 1997; Han 2007). In vivo, DHLA is difficult to detect because of its rapid side chain oxidation and S,S-dimethylation. BMHA & BMBA (the 6 & 4 carbon metabolites) appear rapidly in plasma and have longer half-lives than the parent compound, rac-ALA (Schupke et al. 2001; Teichert et al. 2002).

Any attempt to characterize the in vivo mechanisms of action of ALA or DHLA must take into account the short plasma half-life of ALA, its rapid transformation into five primary metabolites, the lack of accumulation of the parent compound or its metabolites, and the rapid loss of PD activity by S-methylation of the dithiolane ring such that several sensitive assays have been unable to detect even low levels of DHLA in plasma.

Despite the pharmacological limitations of DHLA, rac-ALA and RLA possess therapeutic efficacy in humans (IV and PO), demonstrated by a reduction in the dose necessary to produce a similar benefit as determined by the glucose challenge and insulin clamp tests (DeFronzo et al. 1979; Jacob et al. 1999). Intravenous load doses of rac-ALA beginning at 600–1000 mg for 10 days could be reduced to 500 mg to achieve the same result (improved glucose metabolic clearance rate and improved insulin sensitivity index [ISI]) as per oral dosing of 600–1200 mg daily for 4 weeks. These treatments led to similar reductions in the effective doses. In two PK trials, one using 600 mg rac-ALA and the second 600 mg RLA-Tris salt (based on the weight of RLA), RLA-Tris produced average $C_{\max}$ plasma levels 2.3 times higher and the bioavailability (AUC) 1.9 times greater than rac-ALA (Krone 2002, pp. 163).

It is still unclear whether or not these effects are produced by an initial prooxidative or antioxidant signal, although evidence is mounting that the primary stimulus is oxidative which results in the induction of genes that improves the overall antioxidant status (Konrad 2005; Dicter et al. 2004; Cheng et al. 2006; Linnane et al. 2006). RLA is insulin mimetic and like insulin activates the insulin signaling pathway in 3T3-L1 adipocytes via hydrogen peroxide generation (Moini et al. 2002; Cho et al. 2003).

To fully define the in vivo mechanisms of action of rac-ALA and RLA it is essential to evaluate PK data in humans of RLA and its metabolites. It is an established principle in enantioselective PK that a racemic compound will not react in the same manner as a single enantiomer in a chiral environment due to differences in dissolution, disintegration, or any of the PK parameters such as absorption, distribution, metabolism, and elimination. Differences in these PK parameters will present different concentrations to any stereospecific target such as receptor sites, transporters, and enzymes and may differentially affect signal transduction and gene transcription.

Racemic dihydrolipoic acid was not detected even as a low-level plasma metabolite in four animal species (dog, mouse, rat, and human) after administration of rac-ALA, but was a likely, albeit transient intermediate in the formation of three ALA metabolites, BMOA, BMHA, and BMBA (Schupke et al. 2001; Teichert et al. 2002) (Figure 10.2). Krone discovered enantiose-

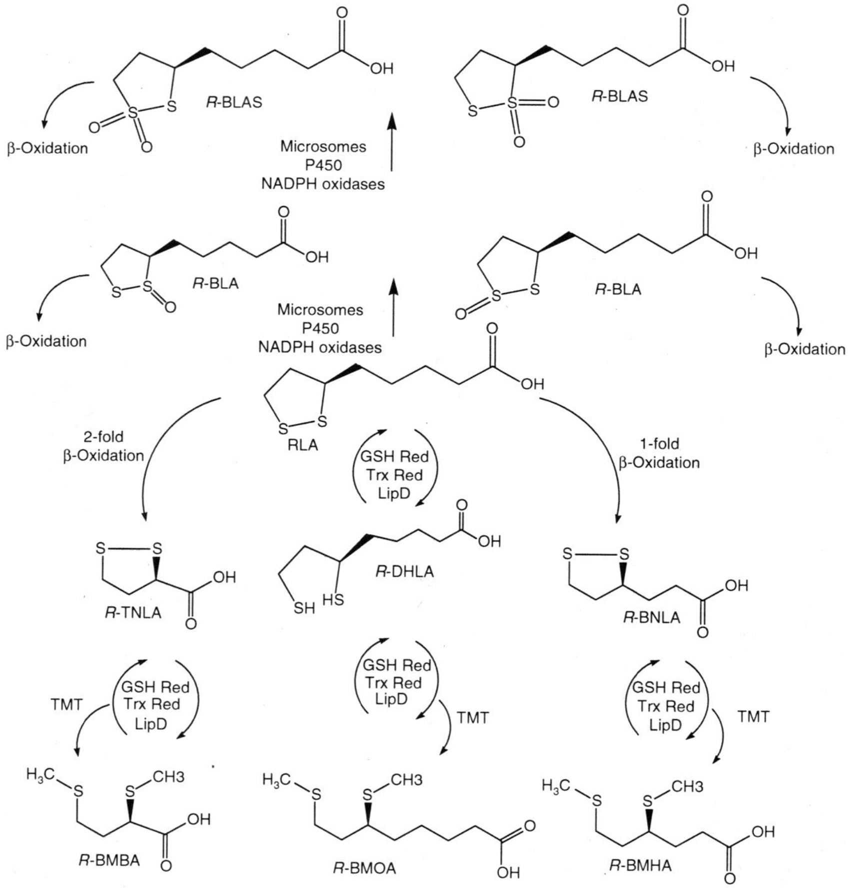

FIGURE 10.2 The metabolites of *R*-lipoic acid, *R*-tetranorlipoic acid (R-TNLA), *R*-bisnorlipoic acid (*R*-BNLA), and *R*-β-lipoic acid, most likely contribute to the pharmacodynamics of *R*-lipoic acid. The presence of *R*-BNLA and *R*-TNLA in plasma in higher concentrations than rac-BNLA and rac-TNLA indicates concentration effects or enantioselective metabolism and possible in vivo differences between *R*-lipoic acid and rac-α lipoic acid (Krone 2002). Methylation and oxidation of the sulfhydryls presumably causes a loss of activity and enhances renal elimination. (Lang 1992, Schupke et al. 2001, Krone 2002, Teichert et al. 2002).

lective differences in the plasma PK and metabolism between rac-ALA and RLA in humans but did not identify *R*-DHLA after the administration of 600 mg of RLA-Tris salt. The metabolite profile of RLA revealed substantially higher levels of *R*-bisnorlipoic acid (*R*-BNLA) whereas this metabolite was found in low levels or was absent after administration of rac-ALA to healthy volunteers (Schupke et al. 2001; Krone 2002; Teichert et al. 2002).There are three possible metabolic pathways for the formation of these primary metabolites from the parent compound (Krone 2002). The first pathway involves the metabolism of RLA to the metabolite BMOA by opening of the dithiolane ring to *R*-DHLA, followed by S,S-dimethylation and onefold or twofold β-oxidation reactions, yielding the metabolites BMHA and BMBA. The second possible pathway involves the metabolism of RLA via β-oxidation to the metabolite BNLA, followed by S,S-dimethylation of the dithiolane ring, which precludes the presence of *R*-DHLA as an intermediate metabolite. The third and last possible pathway for the formation of the metabolite *R*-tetranorlipoic acid (*R*-TNLA) from RLA is by a twofold β-oxidation, followed by S,S-dimethylation, which will produce the metabolite *R*-BMBA. This potential pathway also severely limits the mean residence time of *R*-DHLA in plasma, as a potential metabolite and as the agent responsible for the PD effects of rac-ALA or RLA. The formation-rate of the metabolite *R*-TNLA is not dependent on the concentrations of the metabolite *R*-BNLA in the plasma (Krone 2002). The enzymes responsible for S-methylation, the *S*-methyl-transferases are found in the intestinal mucosa, especially in the cecum and the colon, liver, kidneys, erythrocytes, and the lymphocytes (Stevens et al. 1990; Creveling 2002).

On the basis of the rapid appearance of *R*-BMHA in plasma after consumption of Na-RALA, K-RALA (5 min), or RLA-Tris (30–45 min), it is probable that a significant portion of the reduction and S-methylation occurring in the intestines is released into the mesenteric circulation and transported to the liver for uptake and further transformation. The metabolite *R*-TNLA was measured in the plasma in higher concentrations than *R*-BNLA after administration of 600 mg RLA-Tris salt and in higher concentrations in muscle than in plasma, indicating tissue-specific metabolism. It can be assumed that in humans the twofold β-oxidation (loss of four carbons from the side chain) is the favored pathway. Interestingly, *R*-TNLA was not identified after administration of 1 g RLA (yielding a C_{max} of one-seventh to one-tenth that of 600 mg RLA-Tris) indicating the possibility of dose-concentration dependent differences in the metabolite profiles of RLA (Biewenga et al. 1999; Krone 2002).

The results of ALA PK studies suggest that if untransformed *R*-DHLA can be detected as a plasma component, it will be found in low concentrations by escaping the highly efficient *S*-methyl transferases and by hydrophobic or covalent binding (SH/S–S) to plasma and tissue proteins. Additionally, any potential antioxidant property or therapeutic effect of *R*-DHLA or rac-DHLA in vivo is likely via a rapid, catalytic activation or modulation of specific receptors or signaling pathways since it is rapidly metabolized. The pharmacodynamic effect of RLA/*R*-DHLA is correlated to the redox properties of the dithiolane ring, but this is lost upon S-methylation and subsequent S-oxidation (Krone 2002). Alternatively, whereas DHLA in vitro is rapidly exported from the cell it is possible that DHLA like other thiols binds in vivo via thiol-disulfide exchange reactions (Smith et al. 2004). This may account for the low (12.4%) 24 h urinary excretion of RLA and its primary metabolites in humans (Teichert et al. 2003). Interestingly, even though baseline levels of RLA have been reported in the 80 nM–545 nM (16.5–112 ng/mL) range, it may have a significant role in maintaining the plasma redox status as it constitutes the second anodic wave as determined by cyclic voltammetry (Chevion et al. 1997; Takenouchi et al. 1960; Bernkop-Schnurch et al. 2004).

ANIMAL AND HUMAN PLASMA PHARMACOKINETICS OF RAC-DIHYDROLIPOIC ACID

Donatelli provided the first report of rac-ALA PK (as Na-ALA). Rabbits administered a 50 mg/kg IV dose of rac-ALA reached C_{max} in plasma of 76 μg/mL within 5 min and IM injections of 100 mg/kg gave C_{max} of 16 μg/mL within 20–30 min, which were completely in the disulfide form. There was no evidence of circulating rac-DHLA from either route of administration (Donatelli 1955).

In contrast, Riedel found rac-DHLA present in febrile and afebrile rabbit plasma after injection of 80 mg/kg rac-ALA, utilizing a modification of Haj-Yehia's assay (the sulfhydryls were blocked as the *S,S*-DEOC derivatives, and the fluorescent amide was made with panacyl bromide rather than APMB) by exposing rabbits to LPS to induce fever (Riedel et al. 2003). In afebrile rabbits, this IV dose yielded maximum plasma concentrations of 96.5 +/− 10.7 μg/mL (467 +/− 51.86 μM) rac-ALA and 2.2 +/− 0.26 μg/mL (10.68 +/− 1.3 μM) DHLA increasing to 8.6 +/− 2.28 μg/mL (41.7 +/− 1.1 μM) within 30 min of injection. In febrile rabbits, injection of 80 mg/kg rac-ALA yielded similar plasma concentrations of rac-ALA but DHLA concentrations decreased from 2.12 +/− 0.3 μg/mL (10.29 +/− 1.45 μM) to 0.84 +/− 0.22 μg/mL (4.07 +/− 1.07 μM) within 45 min, following the injection of LPS. The important findings of this study were that rac-DHLA could be measured in a quantity 2%–11% the amount of administered rac-ALA, was sufficiently stable to measure over time, and was found in a lower quantity (0.8% the amount of ALA) when the animal was subjected to oxidative stress subsequent to injection of LPS due to oxidation to rac-ALA. Riedel demonstrated that LPS-induced fever was dependent on the induction of systemic oxidative stress, which activated the NMDA receptor via oxidation of receptor thiols.

Hill measured rac-ALA and rac-DHLA (but no other metabolites) in dogs and cats using a modification (sample prep) of an HPLC/ECD procedure (Sen et al. 1999). The studies showed significant differences in metabolism of rac-ALA between the two species but both demonstrated cyclic patterns of excretion of both rac-ALA and rac-DHLA in urine, detectable up to 10 days after administration, suggestive of tight plasma protein binding (Hill et al. 2002, 2004, 2005), and inefficient *S*-methyltransferases. No rac-DHLA was detected in dog plasma after PO or IV administration but was found in urine in equal or larger amounts than rac-ALA after IV injection of 25 mg/kg rac-ALA; suggesting reduction in the periphery. Rac-DHLA could be detected in dog urine on day 1; reached highest levels on day 3; and was still detectable on days 4, 7, and 10 after administration. Cats fed 15 mg/kg had equal or greater concentrations of rac-DHLA in plasma and urine than rac-ALA; reaching levels of 1.7 μM (0.35 μg/mL), at $T = 10$ h, whereas rac-ALA reached levels of 0.3 μM (0.062 μg/mL). Rac-DHLA could be detected up to 10 days post-administration, but reached peak levels on days 2 and 5. These findings are not consistent with those of Schupke who was unable to detect [7,8-^{14}C] rac-DHLA as a metabolite in dog urine up to 72 h post-dosing (Schupke et al. 2001).

Racemic dihydrolipoic acid was first identified in human plasma and urine (HPLC with fluorescence detection) after a volunteer consumed an undisclosed amount of commercial rac-ALA but the concentrations were not reported and the chromatograms were apparently not drawn to scale (Haj-Yehia et al. 2000). As reported above, *R*-DHLA was not detected after administration of 1 g RLA in a human volunteer although the LOD of rac-DHLA in plasma was reported to be sufficient for low level detection (12.9 ng/mL) (Biewenga et al. 1999).

Taken together, this suggested the possibility of reduction of rac-ALA/RLA to rac-DHLA/*R*-DHLA in the complex milieu of the GI tract or in the blood and becoming tightly bound to plasma proteins. An alternative explanation for detectable DHLA would be that rac-ALA/RLA was transported into peripheral cells after escape of the hepatic first pass effect, reduced and a small portion of the effluxed DHLA escaped S-methylation and became bound to plasma proteins. It was also possible that several studies could not detect DHLA because the methods used to remove rac-ALA or RLA from plasma proteins were inadequate for removal of rac-DHLA or *R*-DHLA, due to possible covalent binding to Cys-34 of albumin or other free plasma thiols (Hortin et al. 2006; Gregus et al. 1992). Regardless of the mechanisms, concentrations were expected to be low due to which may be some researchers identified DHLA as a plasma component and others did not.

Early Japanese studies revealed disulfide reductase activity in the gut since the mixed disulfide bond between thiamine and 6-acetyl DHLA (thiamine-8-(methyl-6-acetyl-dihydrothioctate)-disulfide) (TATD) was reduced during GI absorption so that the each of the free thiols could be measured in plasma (Sugiyama and Yoshiga 1963; Matsuda 1968) (Figure 10.3). This suggests the intramolecular disulfide of ALA could be similarly reduced.

In an often cited paper concerning the extent of ALA absorption from the gut there is no mention of rac-DHLA as a metabolite of rac-ALA or [7,8-^{14}C] rac-α-lipoic acid (Peter and Borbe 1995).

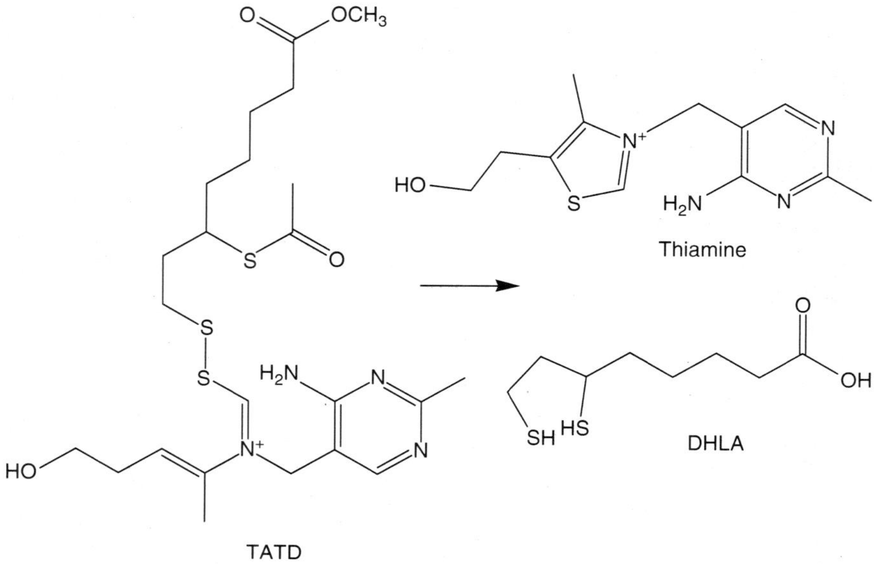

FIGURE 10.3 Thiamine-8-(methyl-6-acetyl-dihydrothioctate)-disulfide (TATD) is reduced by disulfide reductases releasing the free thiols into mesenteric venous circulation.

PLASMA PHARMACOKINETICS OF *R*-(+)-α LIPOIC ACID IN HUMANS

Reports of human plasma levels of RLA (after administration of the single enantiomer) are scarce but were mentioned in a book chapter, a paper, and a patent by a Dutch group. A 1 g dose of RLA generated a $C_{max} = 400$ ng/mL and later 1154 ng/mL but the AUC and other PK parameters were not reported (Biewenga et al. 1999; Biewenga, 1997; Biewenga et al. 1997). Adjusted for weight, and assuming linearity (Hermann et al. 1997), rac-ALA reaches peak plasma concentrations 16.45–31.17 times higher than pure RLA (based on the lower value), and 5.71–10.8 (based on the higher value) since 600 mg of rac-ALA has been reported to give average C_{max} 3949 ng/mL (Krone 2002) up to 7467 ng/mL (Teichert et al. 2003).

Owing to the unequal distribution of the enantiomers in plasma [RLA: SLA = 1.6–2.0:1] (Hermann et al. 1997; Niebch et al. 1997), the RLA content of 1 g rac-ALA would give a plasma C_{max} of 4114 ng/mL, 3.56–10-fold higher than the level achieved by administration of pure RLA. This limits the therapeutic potential of RLA since 4–10 g would have to be consumed to achieve therapeutic concentrations. Also reported was the C_{max} of two RLA metabolites; 3-keto-RLA (2092 ng/mL) and *R*-BNLA (704 ng/mL), and a partial plasma concentration versus time curve for RLA and the two metabolites was presented (Biewenga 1997b, Biewenga et al. 1999). In a recent clinical trial with rac-ALA in MS patients, the greatest therapeutic benefit was correlated to C_{max} (Yadav et al. 2005). The highest plasma concentration in this study (18 μg/mL) is unobtainable with pure RLA. On the basis of the work of Krone with RLA-Tris salt and the comments of Locher (who claimed that concentrations of 50 μM, 10.3 μg/mL could be reached with 600 mg RLA-Tris), the only way to achieve therapeutic levels (10 μg/mL) of RLA was to use a stable, salt form (Walgren et al. 2004) that is water soluble and not as prone to polymerization as pure RLA. Doses of RLA salts reached levels higher than an equal weight of rac-ALA, which was significantly higher than pure RLA (Krone 2002). For example, 600 mg PO rac-ALA gave an average ($n=9$; 6 males and 3 females) C_{max} of 3949 ng/mL (range 3065–5087 ng/mL) and an AUC of 3098 ng × h/mL (range 2513–3817 ng × h/mL) whereas 600 mg RLA-Tris gave an average ($n = 12$ males) C_{max} of 7235 ng/mL (range 5,099–10,267 ng/mL) and ($n = 12$ females) 11,357 ng/mL (range 7,331–11,209 ng/mL] and an AUC of 4690 males ng × h/mL (range 3956–5560) and 7304 females ng × h/mL (range 6146–8680) (Krone 2002).

MATERIALS

All of the test compounds and reference standards were synthesized, purified, identified, characterized, and validated by GeroNova Research, Inc. using NMR, GC–MS, polarimetry and a validated chiral HPLC assay (Carlson et al. 2007b). Rac-ALA, *R*-(+)-LA, rac-DHLA, *R*-(−)-DHLA had chemical purities >99% and the chiral compounds had enantiomeric excesses >99%. *R*-(+)-LA $[\alpha]_D{}^{23} = +120°$; *R*-(−)-DHLA $[\alpha]_D{}^{23} = -14.8°$. The RLA and *R*-DHLA mixture containing 30%–40% of the dimeric compound had an $[\alpha]_D{}^{23} = +41.4°–52°$, depending on the dimer content and the conditions under which it was formed. Higher temperatures and longer reaction times generally reduced the specific rotation. *R*-DHLA (as well as rac-DHLA and *S*-DHLA) can be readily identified and differentiated from the corresponding oxidized forms using the USP method for rac-ALA. The method is non-enantioselective but works equally well for all forms and can be quantified as total ALA and total DHLA. Standards must be made separately to prevent reaction between ALA and DHLA.

High-performance liquid chromatography grade solvents and reagent grade chemicals were purchased from GFS, Alfa Aesar, and TCI. APMB was purchased from City Chemical and purified by base-acid, 2 × recrystallizations from $CHCl_3$, and column chromatography.

ANALYTICAL EQUIPMENT AND METHODS

Specific rotations of the test compounds were measured on a Rudolph Research Autopol III Polarimeter. All specific rotations were run at room temperature (23°C) with $c = 1$ in EtOH. Comparative tests in benzene, acetonitrile, and methanol gave essentially the same results as long as the RLA did not polymerize upon dissolution and the resulting solutions remained clear.

High-performance liquid chromatography equipment: For comparison of published methods, two HPLC systems and three methods were used for plasma analysis of RLA and *R*-DHLA.

Method 1, System 1 was comprised of a Hewlett-Packard (HP) 1090 HPLC with an HP Model 1046A Programmable Fluorescence Detector (Haj-Yehia et al. 2000). A 250 μL injection loop was used. Separation of the fluorescent amides of *S,S*-DEOC-(*R*)-DHLA-APMB and RLA-APMB was attempted on an Agilent Zorbax Eclipse × DB-C8 column (4.6 × 15 mm, 5 μm) and detected at excitation 343 nm and emission 423 nm.

Method 2, System 1 was comprised of a Hewlett-Packard (HP) 1090 HPLC with an HP Model 1046A Programmable Fluorescence Detector. A 250 μL injection loop was used. Separation of the diastereomers formed by reaction of *R*-DHLA with OPA and DPA is accomplished by reverse phase HPLC on a Merck, LiChrospher 60 SelectB or Phenomenex Gemini column (25 cm × 4 mm I.D.: 5 μm particle size at 35°C) utilizing fluorescence detection with an excitation wavelength of 230 nm and an emission wavelength of 418 nm (Niebch et al. 1997; Bernkop-Schnurch et al. 2004). The standards and samples are eluted with a mixture of 55% 0.2 M K_2HPO_4 or Na_2HPO_4 (pH 5.8) and 45% acetonitrile/methanol (1:1) as the mobile phase (1.7 mL/min). Identification and quantification are performed by both the external standard and internal standard methods.

The original methods were modified to differentiate baseline or plasma *R*-DHLA from baseline or plasma RLA by splitting the plasma sample in two portions (Niebch et al. 1997; Bernkop-Schnurch et al. 2004). The first portion is reacted directly with the derivatizing reagents to detect *R*-DHLA. The second portion containing RLA and *R*-DHLA was reduced with $SnCl_2$ and derivatized in an identical manner as the first. In this way it was possible to quantify RLA and *R*-DHLA by difference.

Method 3, System 2 was comprised of a Hewlett-Packard (HP) 1050 HPLC and an ESA Coulochem II multielectrode detector, fitted with ESA 5010 Analytical Cell (electrode 1), a 5011A high sensitivity analytical cell (electrode 2), and an ESA 5020 guard cell (Sen et al. 1999). The electrodes were set at 0.4, 0.85, and 0.9 V, respectively. Separation was achieved with a Phenomenex C18 Gemini Column (250 × 4.6 mm; 5 μm). The mobile phase consisted of 50% of 50 mM NaH2PO4 (pH 2.7), 30% ACN, and 20% MeOH, which were filtered separately through a 0.45 μm filter and mixed with the gradient pump. The system was continuously sparged with helium. The flow rate was set at 1 mL/min.

PLASMA COLLECTION AND COMPARISON OF ANTICOAGULANTS

Each participant gave their informed and written consent before their inclusion in the study. Neither volunteer were taking prescription drugs or being treated for any known medical conditions. Pretrial screening for blood pressure, blood lipid profiles, homocysteine, and CRP were all within the reference ranges. The volunteers were regular users of RLA and *R*-DHLA, as well as followers of complex daily vitamin regimens. On the basis of a previous trial showing that rac-ALA does not accumulate or modify subsequent PK profiles, a 3 day wash out for RLA/*R*-DHLA was considered adequate (Teichert et al. 2003). Volunteers were instructed to avoid consumption of any alcohol or nutritional supplements for at least 3 days before the trial. Blood glucose was checked at each time point in the Na-RALA trials using an Onetouch Ultra monitor.

Whole blood was drawn from fasted subjects (10–12 h) by our in-house licensed phlebotomist (S.J. Fischer) into 4.5–9 mL evacuated tubes and chilled for 5 min in an ice–water bath, which reduced hemolysis. Blood was collected from the medial antibrachial or medial cubital veins (with applied tourniquet) and an indwelling 22 gauge catheter at $T = 0, 5, 10, 15, 20, 25, 30, 35, 40, 45, 50, 55$, and 60 min for RLA salts and $T = 0, 30, 60, 90, 120, 150, 180, 210$, and 240 min for RLA and *R*-DHLA into prechilled, acidic citrate (Biopool Stabilyte pH 4.5) or Li Heparin tubes, with modifications of previously published procedures. All of the common anticoagulants, Na, Li, NH_4 heparin, EDTA, and citrate, have been utilized in previous PK trials of ALA. Since one of our primary objectives was to attempt the detection of baseline as well as post-dosing levels of RLA and *R*-DHLA, we faced the same challenges as investigators attempting to measure the true in vivo plasma thiol redox status (Kleinman et al. 2000). DHLA, like plasma amino thiols, is prone to rapid oxidation so we found it necessary to test the stability of *R*-DHLA in the various anticoagulants. We selected acidic citrate (Biopool, Stabilyte) based on the report that plasma amino thiols were more stable in this anticoagulant than in heparin or EDTA (Williams et al. 2001). Additionally, we were able unable to detect *R*-DHLA in our preliminary PK measurements on the mixture of RLA and *R*-DHLA using EDTA or heparin tubes. Two tubes were used for the baseline blood draws to have sufficient blank plasma for calibration curves. RLA and *R*-DHLA were separated from plasma proteins using the method of Chen (Chen et al. 2005). Although recoveries of >90% were reported, 52%–63% of recoveries were in our hands and the calculations of plasma concentrations were adjusted based on the percentage recoveries of RLA or *R*-DHLA from each volunteer's plasma, spiked at different concentrations above and below C_{max}. This was done to adjust for interindividual differences in plasma protein binding. In a later study involving 12 healthy subjects we were able to improve the percent recoveries from 84% to 93% (Carlson et al. 2007a).

COMPARISONS OF HIGH-PERFORMANCE LIQUID CHROMATOGRAPHY METHODS OF ANALYSIS

The literature was searched for methods reported to allow either simultaneous quantification of RLA and *R*-DHLA (with or without enantioselective considerations) or that could be differentiated by splitting the samples into two fractions and measuring the RLA and *R*-DHLA separately (Teichert and Preiss 1992, 1995, 2002; Niebch et al. 1997; Witt and Rustow 1998; Biewenga et al. 1999; Sen et al. 1999; Haj-Yehia et al. 2000; Satoh et al. 2007). Interestingly, of all the methods allowing simultaneous detection of ALA and DHLA, only Teichert's and

Bernkop-Schnurch's groups detected baseline RLA, and only Teichert detected baseline *R*-DHLA. Although Haj-Yehia's group claimed the method was used in a PK trial, no data was reported and Witt's group claimed the method could be used for quantifying DHLA in plasma, but reported values only for spiked plasma and no PK data.

The first method under consideration was from the group first identifying rac-DHLA in human plasma and urine subsequent to consumption of rac-ALA (Haj-Yehia et al. 2000). Blocking the free thiols was readily accomplished as long as fresh bottles of ECF were used. Old reagents gave incomplete reactions due to hydrolysis of the ECF. The biggest problem encountered was the coupling of APMB with RLA and *S*,*S*-DEOC-(*R*)-DHLA even with a variety of modifications of the original procedure (Assaf 2005). At times, the fluorophore of *S*,*S*-DEOC-(*R*)-DHLA-APMB amide was detected in solvent and at other times it would completely disappear. Attempts at detection in plasma using heparin or EDTA gave poor results despite the fact that the original procedure utilized EDTA. The RLA-APMB amide was more readily detected by UV detection. Unfortunately, after considerable effort to perfect and optimize the procedure, we were unable to obtain consistent results and it was abandoned. Even if we were able to ultimately perfect the method, it was believed to be too time consuming and labor intensive for routine high-throughput analysis of ALA and DHLA in human PK trials with multiple volunteers.

A more rapid procedure was desired and the coulometric methods developed by Sen and Biewenga (apparently independently of one another), which Hill utilized for detection of rac-ALA and rac-DHLA in dog and cat plasma, urine, and feces, was selected. Although the method has been reported to have a sufficiently low LOD for DHLA (1–5 pmol/0.206–1.03 ng) in plasma, in preparation, we were unable to use it for detecting baseline or low levels of *R*-DHLA. The method proved to be the method of choice for rapid quantification of RLA in PK trials of RLA dosage forms. In some samples, detection of baseline levels of RLA was possible, including those from purchased and pooled plasma.

RESULTS AND DISCUSSION: A REEVALUATION OF THE STABILITY OF *R*-DIHYDROLIPOIC ACID

We have worked extensively with *R*-DHLA for several years and have tested its stability, neat, in solution, and, recently, in plasma. We herein report our findings regarding the stability of *R*-DHLA under various conditions and its contribution to the stability and bioavailability of RLA. While developing industrial-scale preparations of RLA and *R*-DHLA, we discovered that *R*-DHLA is quite stable in air when dry and free of trace metals. A vial of pure, neat *R*-DHLA open to the air had oxidized (to RLA) by ~14% (USP method for rac-ALA, HPLC/UV) in one year, although a solution left drying overnight over molecular sieves had mostly reverted to RLA by the next morning.

In agreement with Bringmann, we discovered that *R*-DHLA can be distilled under inert gas with formation of 1%–3% RLA and with no evidence of racemization as had been previously reported by Gunsalus. If *R*-DHLA is exposed to high temperatures and strong reducing conditions or is improperly distilled, several impurities form with extremely obnoxious odors. We have identified these impurities by LC/MS. The major impurity is the seven-membered thiolactone but reductive desulfurization also occurs, generating small amounts of alkenes, thietanes, and thiophenes (Figure 10.4).

Our initial interest in *R*-DHLA was not as a dosage form but to see if we could expand on Thomas and Reed's initial report and use it to depolymerize RLA polysulfides, neat and on a preparative scale. RLA is more prone to polymerization (which may occur instantly at its sintering point [46°C –47°C]) than rac-ALA

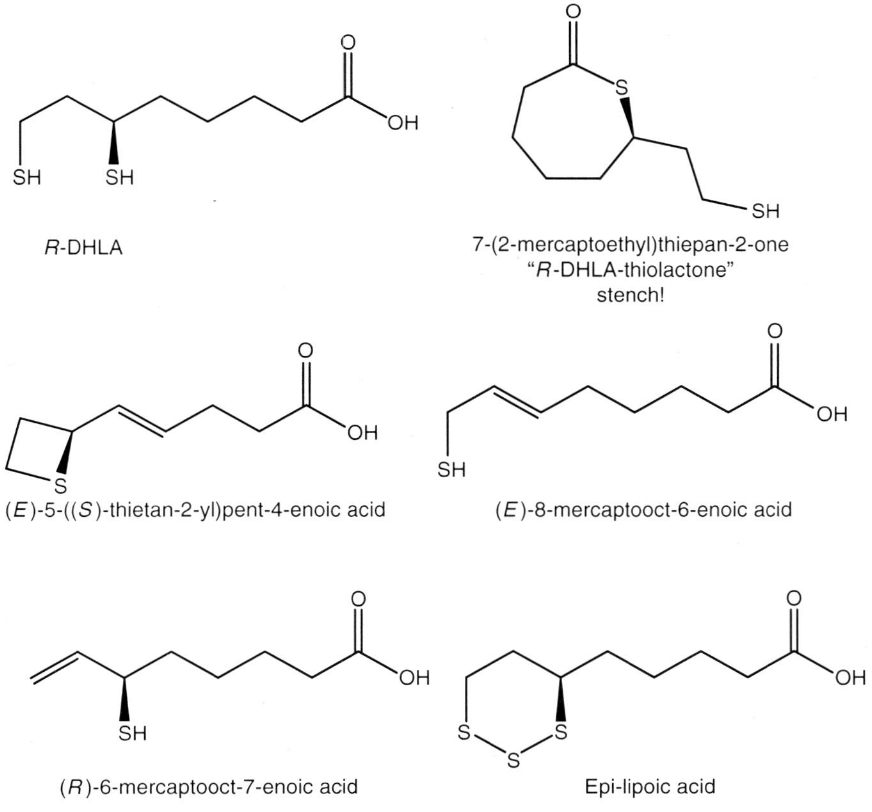

FIGURE 10.4 Impurities identified by LC/MS in commercial *R*-dihydrolipoic acid (*R*-DHLA). The primary impurity is *R*-DHLA-thiolactone. Epi-lipoic acid (epi-LA) is a trisulfide impurity found in commercial rac-lipoic acid (0.1%–3%) and in lower quantities in *R*-lipoic acid (<0.1%).

(Thomas and Reed 1956). To our surprise, we discovered that equal weights of RLA and *R*-DHLA can be mixed at room temperature or heated together on a cover slip using a digital micro-block apparatus, with no evidence of the sticky disulfide polymer or oxidation. The mixture was soluble in organic solvents or dilute base and both compounds could still be separated by HPLC and detected by UV (215 nm). We tested the stability of *R*-DHLA by heating it together with RLA at temperatures up to 160°C for 4 h with both compounds still detectable by UV and no evidence of decomposition or racemization.

At first, we were unaware of the formation of a new stable dimeric compound due to the observation of the chromatogram that displayed RLA and *R*-DHLA still present in the same ratios with the same peak areas as the standard solution, which had been made by mixing RLA and *R*-DHLA together in solvent. It was only when we ran the standards separately that we observed a 30%–40% reduction in the peak areas of both RLA and *R*-DHLA. When the mixture was reduced with alkaline $NaBH_4$, the theoretical amount of *R*-DHLA was recovered. The mixture was titrated with 0.1 N NaOH to 98% of the theoretical amount indicating the carboxyl groups were both free. This led us to believe that we were looking at a stable thiol–disulfide intermediate (Figure 10.5). The unexpected air and heat stability and the formation of the newly identified dimer led us to speculate that *R*-DHLA alone or in the presence of RLA might also be more stable in plasma than previously suggested (at a particular concentration and if absorption kinetics were the same for RLA and *R*-DHLA). Alternatively, we thought *R*-DHLA or the dimer could serve as a pro-drug or delivery system to improve the bioavailability of RLA and extend its residence time in plasma or tissues. Investigations of the effect of the dimer on the PK of RLA and *R*-DHLA are underway.

Even transient stability in plasma could have a significant biological or therapeutic impact that may be different than RLA itself, since rac-DHLA effluxed from the cell in vitro was rapidly converted to rac-ALA but was stable long enough to be able to significantly increase the reduction of cystine and increase intracellular uptake of cysteine (Han et al. 1997; Kis et al. 1997).

R-(+)-α lipoic acid is a difficult compound to manufacture, isolate (on an industrial scale), handle, store, and convert into dosage forms because of its propensity to polymerize. Commercial dosage forms of RLA have extremely poor shelf life, low GI absorption, and bioavailability. It was difficult to reconcile some of the existing reports of the alleged instability of DHLA in vitro with therapeutic claims. Despite many claims to the inherent instability of DHLA, it has been utilized in a number of experimental models and has been suggested for therapeutic applications (Lee et al. 2005; Ho et al. 2007; Holmquist et al. 2007).

PHARMACOKINETICS OF Na-RLA VERSUS *R*-(+)-α LIPOIC ACID AND *R*-DIHYDROLIPOIC ACID

Pure RLA has poor stability and bioavailability due to the propensity to polymerize. Plasma levels of from 1 g RLA have been reported to reach a C_{max} of

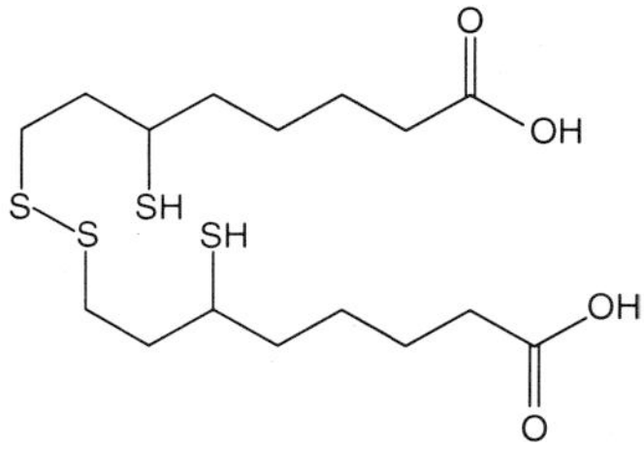

1

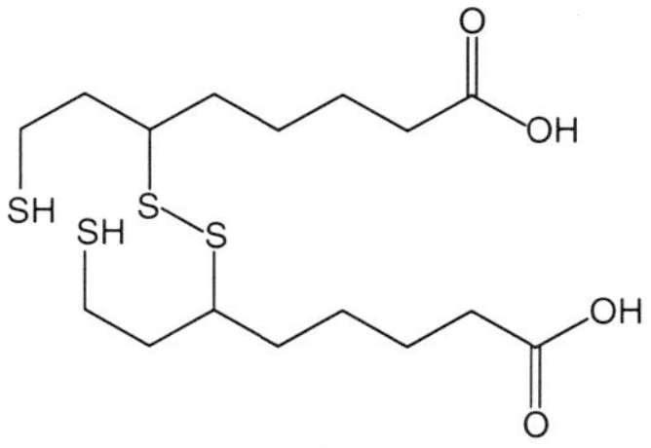

2

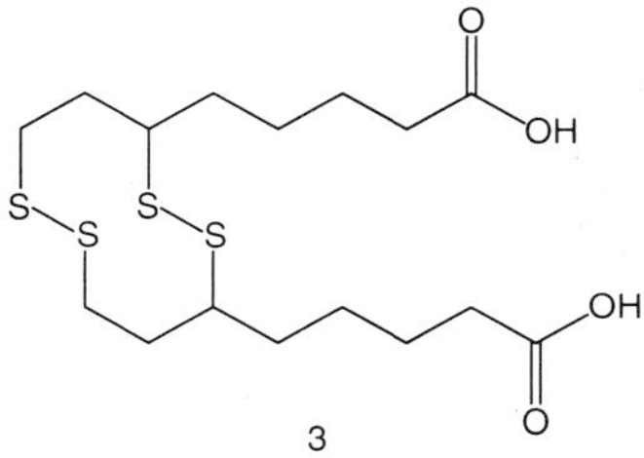

3

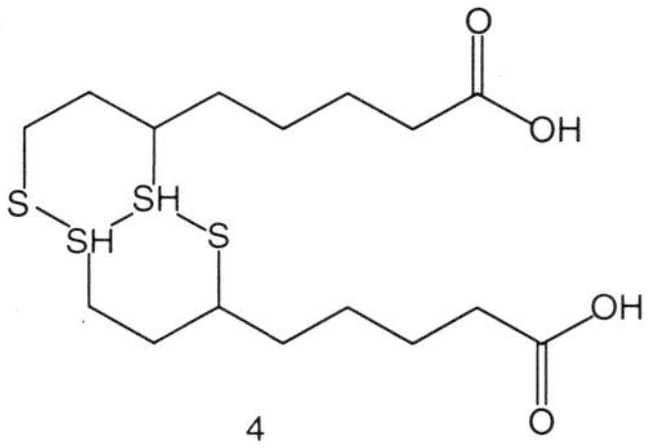

4

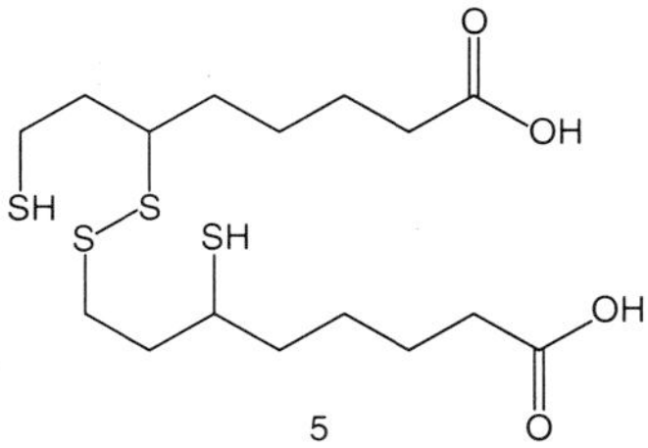

5

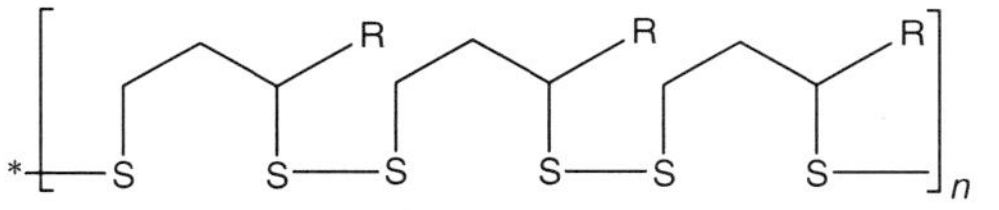

6

FIGURE 10.5 Racemic α-lipoic acid and to a greater extent *R*-lipoic acid form insoluble chains of linear disulfides (structure 6), which occur physically as sticky yellow masses (Thomas and Reed 1956). *R*-dihydrolipoic acid (*R*-DHLA) reduces the polymer to oil and stabilizes the structure to heat and further polymerization. Preliminary pharmacokinetic studies indicate the presence of *R*-DHLA increases the bioavailability of *R*-lipoic acid above (~2 fold) that of *R*-lipoic acid alone. Attempts to elucidate the structure are underway. Structures 1 and 3 are the most probable structures.

1.154 μg/mL, which argues against a therapeutic effect from low supplemental doses of RLA that yields concentrations barely distinguishable from baseline. This limitation is overcome by using 600–800 mg (based on RLA content) of a salt form of RLA; Na-RLA, K-RLA, and RLA-Tris that can reach plasma levels from 8–18 μg/mL. This falls into the proposed therapeutic range of 10–20 μg/mL (50–100 μM) (Anderwald et al. 2002; Krone 2002). A dose of 1200 mg Na-RLA (based on RLA content) reached $C_{max} = 45.1$ μg/mL (225 μM), which rivals the C_{max} of a 1200 mg infusion of rac-ALA (Hermann et al. 1997) and is suggested to be the upper limit for potential therapeutic action. Therefore, millimolar concentrations used in vitro or high-dose animal experiments have little physiological relevance for humans. It is relevant that load oral doses in rats led to C_{max} in tissues of 150 μM (Smith et al. 2004; Hagen 2007).

The nonlinear response between the 600 and 1200 mg dose of Na-RALA in Subject 1 suggests the possibility of saturation of the first pass mechanisms thus allowing more RLA into circulation and the periphery. More volunteers are needed to see if this trend is maintained. The dose, 45.1 μg/mL (225 μM), is close to the upper limit of tolerability in humans since it produced transient nausea at $T = 20$ min (10 min after C_{max}) for 10 min duration and was not associated with a sudden drop in blood sugar. Subject 2 also experienced brief nausea at $T = 20$ min, 10 min after $C_{max} = 18$ μg/mL. This also was not associated with any changes in blood sugar. The IV infusion of 1200 mg rac-ALA over 35 min produced a similar concentration time curve and induced nausea in 33% of the volunteers (28 of 86 people; assuming 50–70 kg/volunteer = 17–24 mg/kg) (Biewenga 1997B) (Table 10.1).

R-(−)-dihydrolipoic acid was sufficiently stable in plasma to be quantified (266 ng/mL) at baseline in Subject 2 but was undetected in Subject 1. It is unknown whether the 3 day wash out period was sufficient to clear *R*-DHLA from plasma (from daily consumption of Na-RALA) or whether it was from endogenous sources. Either way, it was sufficiently stable to be measured. Like the measurements in rabbit, it was possible to measure *R*-DHLA in human plasma at C_{max} of RLA in quantities 68 times lower. Our preliminary results show that 600 mg *R*-DHLA is absorbed slowly from the gut and reaches plasma levels 51 times lower than an equal weight of Na-RLA. Two unidentified metabolites appear in plasma and return to baseline within 4 h while *R*-DHLA is still quantified (223 ng/mL at $T = 240$ min), with no sign of decline. This suggests that *R*-DHLA may undergo slow thiol–disulfide reactions with gut and

TABLE 10.1

Comparisons of Pharmacokinetic Values between 600 mg *R*-Lipoic Acid and 600 mg *R*-Lipoic Acid (as Sodium *R*-Lipoate) in a Single Male (Subject 1) and Female Subject (Subject 2) Reveal Significant Increases in the C_{max} and AUC Values When *R*-Lipoic Acid Is Administered in the Form of Sodium *R*-Lipoate. An Unsuccessful Attempt Was Made to Determine the Pharmacokinetic Values for 600 mg *R*-Dihydrolipoic Acid (Subject 1)

Subject	Dosage (mg)	C_{max} (μg/mL)	T_{max} (min)	$T_{\frac{1}{2}}$ (min)	AUC 0–*t* (μg min/mL)	AUC 0–*t* (μg h/mL)
1	600 R-DHLA	0.26	120	ND	ND	ND
1	600 RLA	0.70	120.0	24.36	74.5	1.24
1	600 mg Na-RALA	14.10	10.0	5.34	311.1	5.185
2	600 RLA	1.01	70.0	31.67	130.0	2.17
2	600 Na-RALA	18.10	15.0	7.06	342.8	5.71

ND = Not determined.

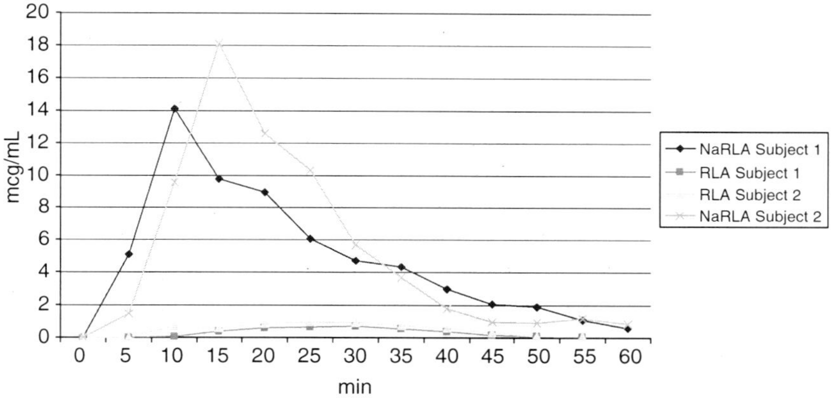

NaRLA Subject 1; C_{max} = 14.1 μg/mL; AUC 5.19 μg h/mL (Cmax = 20.1x, AUC = 4.2)
RLA; Subject 1; C_{max} = 0.7 μg/mL; AUC 1.24 μg h/mL
RLA; Subject 2; C_{max} = 1.01 μg/mL; AUC 2.17 μg h/mL
NaRLA Subject 2; C_{max} = 18.1 μg/mL; AUC 5.71 μg h/mL (Cmax = 17.92, AUC = 2.6)

CHART 10.1 Comparisons of plasma concentration versus time curves of 600 mg R-lipoic acid administered as the free acid or an equivalent amount of sodium R-lipoate in subjects one and two.

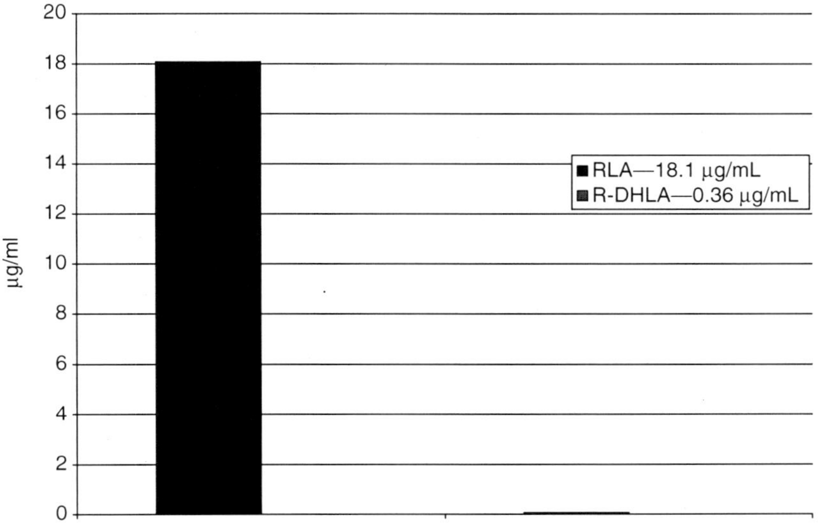

CHART 10.2 *R*-Dihydrolipoic acid at C_{max} 600 mg Na-*R*-lipoate Subject 2.

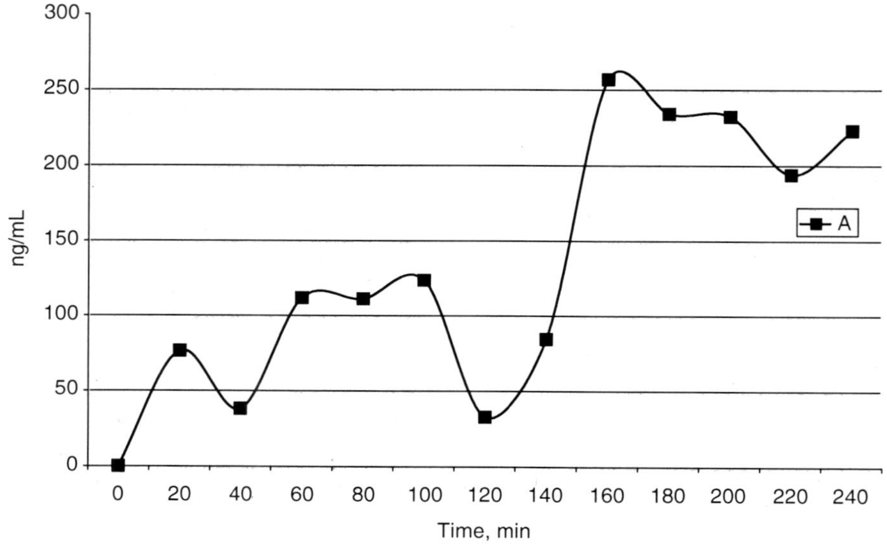

CHART 10.3 *R*-dihydrolipoic acid (*R*-DHLA) 600 mg concentration versus time curves for Subject 1.

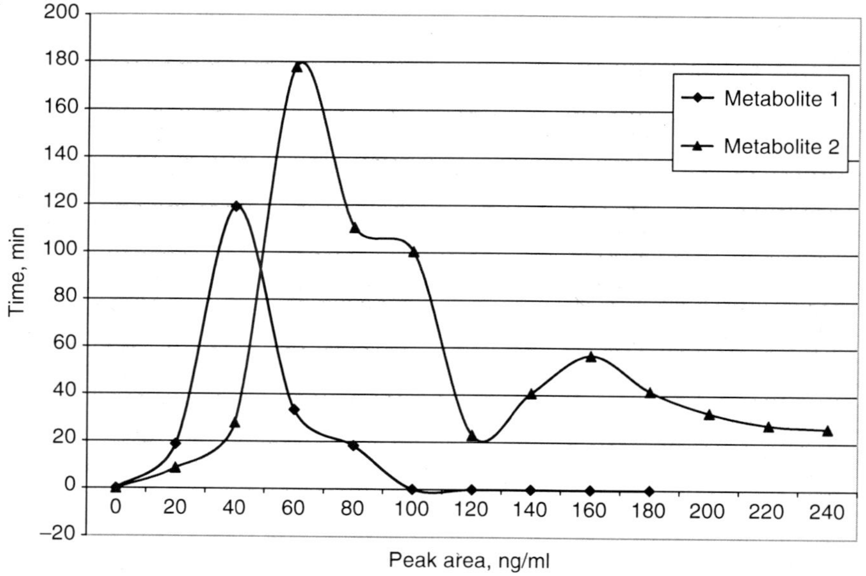

CHART 10.4 Unknown human plasma metabolites identified by derivatization and flourescence detection after consumption of 600 mg *R*-DHLA.

plasma thiols or bis-methylated. It was impossible to calculate the elimination half-life or the AUC. Blood should be collected in future studies out to at least 480 min (Charts 10.1 through 10.4).

Studies are in progress to systematically explore individual differences in protein binding, and expand the number of volunteers with Na-RALA, K-RALA, *R*-DHLA and the mixed dosage form containing RLA and *R*-DHLA and the new dimer.

ABBREVIATIONS

ACN = acetonitrile
AMPK = adenosine monophosphate protein kinase
AP-1 = activator protein-1
APMB = 2-(4-aminophenyl)-6-methylbenzothiazole
ARE = antioxidant response element
ATP = adenosine triphosphate
BAD = Bcl-xL/Bcl-2 associated death promoter
BAX = Bcl-2 associated x protein
Bcl2 = B-cell lymphoma 2
BMBA = bis(methylthio)butanoic acid
BMHA = bis(methylthio)hexanoic acid
BMOA = bis(methylthio)octanoic acid
BNLA = bisnorlipoic acid
BSA-1 = Bovine serum albumin-1
COX-2 = cyclooxygenase-2
CRP = C-Reactive Protein
D-PA = D-phenylalanine
ECF = ethyl chloroformate
ee = enantiomeric excess
eNOS = endothelial nitric oxide synthase
GCL = γ-glutamylcysteine ligase
GPx = glutathione peroxidase
GSH = reduced glutathione
GSH Red = glutathione reductase
GST = glutathione *S*-transferases
HAEC = human aortic endothelial cells
HO-1 = hemi-oxygenase-1
HSF-1 = heat shock factor-1
HSP-60 = heat shock protein-60
HSP-70 = heat shock protein-70
ICAM-1 = intercellular adhesion molecule-1
IFN-γ = interferon-γ
IL-1 = interleukin-1
IL-6 = interleukin-6
IL-1β = interleukin-1β
IL-18 = interleukin-18

iNOS	=	inducible nitric oxide synthase
IPAD	=	integrated pulsed amperometric detection
IR	=	insulin receptor
L-Cys	=	L-cysteine
LipD	=	lipoamide dehydrogenase
LOD	=	limit of detection
LOQ	=	limit of quantification
LOX-15	=	lipoxygenase-15
MCFA	=	medium chain fatty acid
MMP-9	=	matrix metalloproteinase-9
mPGES2	=	membrane-associated prostaglandin E2 synthase
NF-κB	=	nuclear factor κβ
NQOR	=	NAD(P)H: quinone oxidoreductase-1
Nr-f2	=	nuclear factor E2 p45-related factor-2
OPA	=	*o*-phthaldehyde
PI3K	=	phosphatidylinositol-3-kinase
P38 MAPk	=	P38 mitogen-activated protein kinase
P450	=	cytochrome P450
PAI	=	plasminogen activator inhibitor-1
PD	=	pharmacodynamic
PGC-1α	=	peroxisome proliferator activator receptor gamma-coactivator 1α
PGE2	=	prostaglandin E2
PK	=	pharmacokinetic
POF	=	Pyruvate oxidation factor
PPAR-α	=	peroxisome proliferator activated receptor-α
PPAR-γ	=	peroxisome proliferator activated receptor-γ
rac-ALA	=	racemic α-lipoic acid
rac-DHLA	=	racemic dihydrolipoic acid
RAGE	=	receptor for advanced glycation end products
RANKL	=	receptor activator of NF-κB ligand
R-BLA	=	*R*-β-lipoic acid
R-BLAS	=	*R*-β-lipoic acid sulfoxide
R-DHLA	=	*R*-(−)-dihydrolipoic acid
RLA	=	*R*-(+)-α lipoic acid
S-DHLA	=	*S*-(+)-dihydrolipoic acid
SLA	=	*S*-(−)-α lipoic acid
SOD	=	superoxide dismutase
sPLA2	=	phospholipase A-2
SREBP-1	=	sterol-regulatory element binding protein-1
TGF-β	=	transforming growth factor-β
TMT	=	thiol methyltransferase
TNF-α	=	tumor necrosis factor-α
TNLA	=	tetranorlipoic acid
Trx Red	=	thioredoxin reductase
VCAM	=	vascular cell adhesion molecule

REFERENCES

Abdul HM and Butterfield DA. Involvement of PI3K/PKG/ERK1/2 signaling pathways in cortical neurons to trigger protection by cotreatment of acetyl-L-carnitine and alpha-lipoic acid against HNE-mediated oxidative stress and neurotoxicity: Implications for Alzheimer's disease. *Free Rad. Biol. Med.* 2007 Feb 1; 42(3): 371–384.

Acker DS. Syntheses of reduced lipoic acid and analogs of lipoic acid. *J.Org. Chem.* 1963 28(10):2533–2536.

Acker DS and Wayne WJ. Optically active and radioactive α-lipoic acids. *J. Am. Chem. Soc.* 1957 79: 6483.

Ames B; Personal communication; Dr Bruce Ames 2006.

Anderwald C, Koca G, Furnsinn C, Waldhausl W, and Roden M. Inhibition of glucose production and stimulation of bile flow by *R* (+)-alpha-lipoic acid enantiomer in rat liver. *Liver.* 2002 Aug; 22(4):355–362.

Ariëns EJ. Stereochemistry, a basis for sophisticated nonsense in pharmacokinetics and clinical pharmacology. *Eur. J. Clin. Pharmacol.* 1984 26(6):663–668.

Arivazhagan P and Panneerselvam C. Alpha-lipoic acid increases Na^+ K^+ ATPase activity and reduces lipofuscin accumulation in discrete brain regions of aged rats. *Ann. N. Y. Acad. Sci.* 2004 Jun; 1019:350–354.

Arner ES, Nordberg J, and Holmgren A. Efficient reduction of lipoamide and lipoic acid by mammalian thioredoxin reductase. *Biochem. Biophys. Res. Commun.* 1996 Aug 5; 225(1):268–274.

Artwohl M, Schmetterer L, Rainer G, Waldhausl W, and Baumgartner-Parzer S. Abstract 274. Modulation by antioxidants of endothelial apoptosis, proliferation and associated gene/protein expression; European Association for the Study of Diabetes Program 36, Jerusalem, Israel 2000.

Assaf P. Personal communication Dr Peter Assaf 2005.

Balachandar AV, Malarkodi KP, and Varalakshmi P. Protective role of DL-alpha-lipoic acid against adriamycin-induced cardiac lipid peroxidation. *Hum. Exp. Toxicol.* 2003 May; 22(5):249–254.

Bernkop-Schnurch A, Reich-Rohrwig E, Marschutz M, Schuhbauer H, and Kratzel M. Development of a sustained release dosage form for alpha-lipoic acid. II. Evaluation in human volunteers. *Drug. Dev. Ind. Pharm.* 2004 Jan 30; (1):35–42.

Bera T, Lakshman K, Ghanteswari D, Pal S, Sudhahar D, Islam MN, Bhuyan NR, and Das P. Characterization of the redox components of transplasma membrane electron transport system from *Leishmania donovani* promastigotes. *Biochim. Biophys. Acta.* 2005 Oct 10; 1725(3):314–326.

Bharat S, Cochran BC, Hsu M, Liu J, Ames BN, and Andersen JK. Pre-treatment with *R*-lipoic acid alleviates the effects of GSH depletion in PC12 cells: Implications for Parkinson's disease therapy. *Neurotoxicology* 2002 Oct; 23(4–5):479–486.

Biaglow JE, Ayene IS, Koch CJ, Donahue J, Stamato TD, and Mieyal JJ, Tuttle SW. Radiation response of cells during altered protein thiol redox. *Radiat. Res.* 2003 Apr; 159(4):484–494.

Biewenga G, Haenen GR, and Bast A. Thioctic metabolites and methods of use thereof. United States Patent 5,925,668. 1999 July 20.

Biewenga G. An overview of lipoate chemistry, in: *Lipoic Acid in Health and Disease.* Eds: Fuchs J, Packer L, and Zimmer G, Marcel Dekker, Inc. New York, Basel. 1997, pp. 7–8.

Biewenga GP, Haenen GR, and Bast A. The pharmacology of the antioxidant lipoic acid. *Gen. Pharmacol.* 1997 Sep; 29(3):315–331.

Biewenga GP, Dorstijn MA, Verhagen JV, Haenen GR, and Bast A. Reduction of lipoic acid by lipoamide dehydrogenase. *Biochem. Pharmacol.* 1996 Feb 9; 51(3):233–238.

Bishara NB, Dunlop ME, Murphy TV, and Darby IA, Sharmini Rajanayagam MA, Hill MA. Matrix protein glycation impairs agonist-induced intracellular Ca^{2+} signaling in endothelial cells. *J. Cell. Physiol.* 2002 Oct; 193(1):80–92.

Bringmann G, Herzberg D, Adam G, Balkenhohl F, and Paust J; A Short and productive synthesis of (*R*)-α-lipoic acid. Z. *Naturforch.* 1999; 54b:655–661.

Breithaupt-Grogler K, Niebch G, Schneider E, Erb K, Hermann R, Blume HH, Schug BS, and Belz GG. Dose-proportionality of oral thioctic acid—coincidence of assessments via pooled plasma and individual data. *Eur. J. Pharm. Sci.* 1999 Apr 8; 8(1):57–65.

Bucher G, Lu C, and Sander W. The photochemistry of lipoic acid: Photoionization and observation of a triplet excited state of a disulfide. *Chem. Physchem.* 2006 Jan 16; 7(1):14.

Bullock MW, Hand JJ, and Stokstad ELR. Convenient synthesis of thioctic acid. *J. Am. Chem. Soc.* 1957; 79(8): 1978–1982.

Bunik V, Shoubnikova A, Loeffelhardt S, Bisswanger H, Borbe HO, and Follmann H. Using lipoate enantiomers and thioredoxin to study the mechanism of the 2-oxoacid-dependent dihydrolipoate production by the 2-oxoacid dehydrogenase complexes. *FEBS Lett.* 1995 Sep 4; 371(2):167–170.

Byun CH, Koh JM, Kim DK, Park SI, Lee KU, and Kim GS. Alpha-lipoic acid inhibits TNF-alpha-induced apoptosis in human bone marrow stromal cells. *J. Bone. Miner. Res.* 2005 July; 20(7):1125–1135.

Cakatay U and Kayali R. Plasma protein oxidation in aging rats after alpha-lipoic acid administration. *Biogerontology* 2005; 6(2):87–93.

Cantin AM, Martel M, Drouin G, and Paquette B. Inhibition of gelatinase B (matrix metalloproteinase-9) by dihydrolipoic acid. *Can. J. Physiol. Pharmacol.* 2005 Mar; 83(3):301–308.

Carlson DA, Smith AR, Fischer SJ, Young KL, and Packer L. The plasma pharmacokinetics of R-(+)-lipoic acid administered as sodium R-(+)-lipoate to healthy human subjects. *Alterm. Med. Rev.* 2007 Dec; 12(4): 343–351.

Carlson DA, Smith AR, Fischer SJ, Young KL, and Packer L. Stereochemical and non-stereochemical mechanisms in the therapeutic action of lipoic acid. 2007b; manuscript in preparation.

Chaudhary P, Marracci GH, and Bourdette DN. Lipoic acid inhibits expression of ICAM-1 and VCAM-1 by CNS endothelial cells and T cell migration into the spinal cord in experimental autoimmune encephalomyelitis. *J. Neuroimmunol.* 2006 Jun; 175 (1–2):87–96.

Chen J, Jiang W, Cai J, Tao W, Gao X, and Jiang X. Quantification of lipoic acid in plasma by high-performance liquid chromatography–electrospray ionization mass spectrometry. *J. Chromatogr. B. Analyt. Technol. Biomed. Life Sci.* 2005 Sep 25; 824 (1–2):249–257.

Cheng PY, Lee YM, Shih NL, Chen YC, and Yen MH. Heme oxygenase-1 contributes to the cytoprotection of alpha-lipoic acid via activation of p44/42 mitogen-activated protein kinase in vascular smooth muscle cells. *Free Radic. Biol. Med.* 2006 Apr 15; 40(8):1313–1322.

Chevion S, Hofmann M, Ziegler R, Chevion M, and Nawroth PP. The antioxidant properties of thioctic acid: Characterization by cyclic voltammetry. *Biochem. Mol. Biol. Int.* 1997 Feb; 41(2):317–327.

Cho KJ, Moini H, Shon HK, Chung AS, and Packer L. Alpha-lipoic acid decreases thiol reactivity of the insulin receptor and protein tyrosine phosphatase 1B in 3T3-L1 adipocytes. *Biochem Pharmacol.* 2003 Sep 1; 66(5):849–858.

Constantinescu A, Pick U, Handelman GJ, Haramaki N, Han D, Podda M, Tritschler HJ, and Packer L. Reduction and transport of lipoic acid by human erythrocytes. *Biochem. Pharmacol.* 1995 July 17; 50(2):253–261.

Creveling CR. *Methyltransferase in Enzyme Systems that Metabolise Drugs and Other Xenobiotics.* Current toxicology series. Ed. Ioannides C. John Wiley & Sons, Ltd. 2002.

DeFronzo RA, Tobin JD, and Andres R. Glucose clamp technique: A method for quantifying insulin secretion and resistance. *Am. J. Physiol. Endocrinol. Metab.* 237: E214-E223, 1979.

Deguchi Y. Synthesis of thioctic acid and its related compounds. (V) *Yakugaku Zasshi.* 1960 80:937–941.

Demarco VG, Scumpia PO, Bosanquet JP, and Skimming JW. Alpha-lipoic acid inhibits endotoxin-stimulated expression of iNOS and nitric oxide independent of the heat shock response in RAW 264.7 cells. *Free Radic. Res.* 2004 July 38(7): 675–682.

Dicter N, Madar Z, and Tirosh O. Alpha-lipoic acid inhibits glycogen synthesis in rat soleus muscle via its oxidative activity and the uncoupling of mitochondria. *J. Nutr.* 2002 Oct;132 (10):3001–3006.

Diesel B, Kulhanek-Heinze S, Holtje M, Brandt B, Holtje HD, Vollmar AM, and Kiemer AK. Alpha-lipoic acid as a directly binding activator of the insulin receptor: Protection from hepatocyte apoptosis. *Biochemistry* 2007 Feb 27; 46 (8):2146–2155.

Donatelli L. Pharmacology of thioctic acid, pp. 45–143: International Symposium on Thioctic Acid. University of Naples. 1955 November 28–29, Orientamenti Sulla Farmacologia dell' acido tioctico; Atti del Simposito Internazionale su L' acido tioctico, Universita di Napoli.

Dulundu E, Ozel Y, Topaloglu U, Sehirli O, Ercan F, Gedik N, and Sener G. Alpha-lipoic acid protects against hepatic ischemia-reperfusion injury in rats. *Pharmacology* 2007 Jan 24; 79(3):163–170.

El Midaoui A, Wu L, Wang R, and de Champlain J. Modulation of cardiac and aortic peroxisome proliferator-activated receptor-gamma expression by oxidative stress in chronically glucose-fed rats. *Am. J. Hypertens.* 2006 Apr; 19(4):407–412.

Estrada DE, Ewart HS, Tsakiridis T, Volchuk A, Ramlal T, Tritschler H, and Klip A. Stimulation of glucose uptake by the natural coenzyme alpha-lipoic acid/thioctic acid: Participation of elements of the insulin signaling pathway. *Diabetes* 1996 Dec; 45(12):1798–1804.

Flier J, Van Muiswinkel FL, Jongenelen CA, and Drukarch B; The neuroprotective antioxidant alpha-lipoic acid induces detoxication enzymes in cultured astroglial cells. *Free Radic. Res.* 2002 Jun; 36(6):695–699.

Gleiter CH, Schreeb KH, Freudenthaler S, Thomas M, Elze M, Fieger-Buschges H, Potthast H, Schneider E, Schug BS, Blume HH, and Hermann R. Lack of interaction between thioctic acid, glibenclamide and acarbose. *Br. J. Clin. Pharmacol.* 1999 Dec; 48(6):819–825.

Gleiter CH, Schug BS, Hermann R, Elze M, Blume HH, and Gundert-Remy U. Influence of food intake on the bioavailability of thioctic acid enantiomers. *Eur. J. Clin. Pharmacol.* 1996; 50(6):513–514.

Gregus Z, Stein AF, Varga F, and Klaassen CD. Effect of lipoic acid on biliary excretion of glutathione and metals. *Toxicol. Appl. Pharmacol.* 1992 May; 114(1):88–96. Mar; 35(3):369–375.

Guedes MF, Santos Mota JM, and Abreu ML. Some technical problems of the thioctic acid microbiological assay in human serum. *Med. Pharmacol. Exp. Int. J. Exp. Med.* 1965; 13(1):1–6.

Gunsalus IC and Razzell WE. Preparation and assay of lipoic acid derivatives. *Methods Enzymol.* 1957; 3:941–946.

Gunsalus IC, Barton LS, and Gruber W. Biosynthesis and structure of lipoic acid derivatives. *J. Am. Chem. Soc.* 1956; 78(8):1763–1766.

Ha H, Lee JH, Kim HN, Kim HM, Kwak HB, Lee S, Kim HH, and Lee ZH. Alpha-lipoic acid inhibits inflammatory bone resorption by suppressing prostaglandin E2 synthesis. *J. Immunol.* 2006 Jan 1; 176(1):111–117.

Hagen TM. Personal Communication; Dr Tory Hagen 2007.

Hagen TM, Ingersoll RT, Lykkesfeldt J, Liu J, Wehr CM, Vinarsky V, Bartholomew JC, and Ames AB; (*R*)-alpha-lipoic acid-supplemented old rats have improved mitochondrial function, decreased oxidative damage, and increased metabolic rate. *FASEB J.* 1999 Feb; 13(2):411–418.

Haj-Yehia AI, Assaf P, Nassar T, and Katzhendler J. Determination of lipoic acid and dihydrolipoic acid in human plasma and urine by high-performance liquid chromatography with fluorimetric detection. *J. Chromatogr. A* 2000 Feb 18; 870 (1–2):381–388.

Han D. Personal Communication. 2007.

Han D, Handelman GJ, and Packer L. Analysis of reduced and oxidized lipoic acid in biological systems by high performance liquid chromatography. *Methods Enzymol.* 1995; Vol. 251:315–325.

Han D, Handelman G, Marcocci L, Sen CK, Roy S, Kobuchi H, Tritschler HJ, Flohe L, and Packer L. Lipoic acid increases de novo synthesis of cellular glutathione by improving cystine utilization. *Biofactors.* 1997; 6(3): 321–338.

Haramaki N and Handelman GJ. Tissue-specific pathways of α-lipoate reduction in mammalian systems, in *Lipoic Acid in Health and Disease*. Eds. Fuchs J, Packer L, and Zimmer G. Marcel Dekker, Inc. New York, Basel. 1997a, pp. 145–162.

Haramaki N, Han D, Handelman GJ, Tritschler HJ, and Packer L. Cytosolic and mitochondrial systems for NADH- and NADPH-dependent reduction of alpha-lipoic acid. *Free Radic. Biol. Med.* 1997b; 22(3):535–542.

Hermann R, Wildgrube HJ, Ruus P, Niebch G, Nowak H, and Gleiter CH. Gastric emptying in patients with insulin dependent diabetes mellitus and bioavailability of thioctic acid-enantiomers. *Eur. J. Pharm. Sci.* 1998 Jan; 6(1):27–37.

Hermann R and Niebch G. Human pharmacokinetics of alpha lipoic acid, Chapter 22 in *Lipoic Acid in Health and Disease*. Eds. Fuchs J, Packer L, and Zimmer G. Marcel Dekker, Inc. Hermann & Niebch; Basel and New York. 1997, pp. 337–360.

Hill AS, Rogers QR, O'Neill SL, and Christopher MM. Effects of dietary antioxidant supplementation before and after oral acetaminophen challenge in cats. *Am. J. Vet. Res.* 2005 Feb; 66(2):196–204.

Hill AS, Werner JA, Rogers QR, O'Neill SL, and Christopher MM. Lipoic acid is 10 times more toxic in cats than reported in humans, dogs or rats. *J. Animal Physiol. Anim. Nutr. (Berl).* 2004 Apr; 88(3–4):150–156.

Hill AS. Pharmacokinetics of alpha lipoic acid in adult cats and dogs. Ph.D. Thesis UC Davis 2002.

Ho YS, Lai CS, Liu HI, Ho SY, Tai C, Pan MH, and Wang YJ. Dihydrolipoic acid inhibits skin tumor promotion through anti-inflammation and anti-oxidation. *Biochem. Pharmacol.* 2007 Jun 1; 73(11):1786–1795.

Holmquist L, Stuchbury G, Berbaum K, Muscat S, Young S, Hager K, Engel J, and Munch G. Lipoic acid as a novel treatment for Alzheimer's disease and related dementias. *Pharmacol. Ther.* 2007 Jan; 113(1):154–164.

Hornberger CS, Heitmiller RF, Gunsalus IC, Schnakenberg GHF, and Reed LJ. Synthesis of DL-lipoic acid. *J. Am. Chem. Soc.* 1953; 75(6):1273–1277.

Hornberger CS, Heitmiller RF, Gunsalus IC, Schnakenberg GHF, and Reed LJ. Synthetic preparation of lipoic acid. *J. Am. Chem. Soc.* 1952; 74(9):2382–2382.

Hortin GL, Seam N, and Hoehn GT. Bound homocysteine, cysteine, and cysteinylglycine distribution between albumin and globulins. *Clin. Chem.* 2006 Dec; 52(12): 2258–2264.

Hurdag C, Ozkara H, Citci S, Uyaner I, and Demirci C. The effects of alpha-lipoic acid on nitric oxide synthetase dispersion in penile function in streptozotocin-induced diabetic rats. *Int. J. Tissue React.* 2005; 27(3):145–150.

Jacob S, Ruus P, Hermann R, Tritschler HJ, Maerker E, Renn W, Augustin HJ, Dietze GJ, and Rett K. Oral administration of RAC-alpha-lipoic acid modulates insulin sensitivity in patients with type-2 diabetes mellitus: A placebo-controlled pilot trial. *Free Radic. Biol. Med.* 1999 Aug; 27(3–4):309–314.

Jamali F, Mehvar R, and Pasutto FM. Enantioselective aspects of drug action and disposition: Therapeutic pitfalls. *J. Pharm. Sci.* 1989 Sep; 78(9):695–715.

Jameel NM, Shekhar MA, and Vishwanath BS. Alpha-lipoic acid: An inhibitor of secretory phospholipase A2 with anti-inflammatory activity. *Life Sci.* 2006 Dec 14; 80(2): 146–153.

Jones W, Li X, Qu ZC, Perriott L, Whitesell RR, and May JM. Uptake, recycling, and antioxidant actions of alpha-lipoic acid in endothelial cells. *Free Radic. Biol. Med.* 2002 Jul 1; 33(1):83–93.

Khanna S, Roy S, Packer L, and Sen CK. Cytokine-induced glucose uptake in skeletal muscle: Redox regulation and the role of alpha-lipoic acid. *Am. J. Physiol.* 1999 May; 276(5 Pt 2):R1327–R1333.

Kiemer AK, Muller C, and Vollmar AM. Inhibition of LPS-induced nitric oxide and TNF-alpha production by alpha-lipoic acid in rat Kupffer cells and in RAW 264.7 murine macrophages. *Immunol. Cell Biol.* 2002 Dec; 80(6):550–557.

Kim HJ, Park KG, Yoo EK, Kim YH, Kim YN, Kim HS, Kim HT, Park JY, Lee KU, Jang WG, Kim JG, Kim BW, and Lee IK. Effects of PGC-1alpha on TNF-alpha-induced MCP-1 and VCAM-1 expression and NF-kappa B activation in human aortic smooth muscle and endothelial cells. *Antioxid. Redox Signal.* 2007a Jan 1.

Kim HS, Kim HJ, Park KG, Kim YN, Kwon TK, Park JY, Lee KU, Kim JG, and Lee IK. Alpha-lipoic acid inhibits matrix metalloproteinase-9 expression by inhibiting NF-kappa B transcriptional activity. *Exp. Mol. Med.* 2007b Feb 28; 39(1): 106–113.

Kim MS, Park JY, Namkoong C, Jang PG, Ryu JW, Song HS, Yun JY, Namgoong IS, Ha J, Park IS, Lee IK, Viollet B, Youn JH, Lee HK, and Lee KU. Anti-obesity effects of alpha-lipoic acid mediated by suppression of hypothalamic AMP-activated protein kinase. *Nat. Med.* 2004 Jul; 10(7):727–733.

Kis K, Meier T, Multhoff G, and Issels RD. Lipoate modulation of lymphocyte cysteine uptake, in *Lipoic Acid in Health and Disease*. Eds. Fuchs J, Packer L, and Zimmer G. Marcel Dekker, Inc. Basel, New York. 1997, pp. 113–129.

Kleinman WA and Richie JP Jr. Status of glutathione and other thiols and disulfides in human plasma. *Biochem. Pharmacol.* 2000 July 1; 60(1):19–29.

Konrad D. Utilization of the insulin-signaling network in the metabolic actions of alpha-lipoic acid-reduction or oxidation? *Antioxid. Redox Signal.* 2005 July–Aug; 7(7–8):1032–1039.

Krone D. The pharmacokinetics and pharmacodynamics of *R*-(+)-alpha lipoic acid. Ph.D. Thesis Johann Wolfgang Goethe University Frankfurt am Main, Germany 2002.

Kulhanek-Heinz S. Characterization of anti-apoptotic signaling pathways in hepatocytes activated by *R*-alpha lipoic acid and atrial natriuretic peptide, Ph.D. Thesis Ludwig-Maximillians-University of Munich 2004.

Laban G, Meisel P, and Muller G. Method for producing thioctic acid. US Patent 7,208,609. 2007 Apr; 24.

Lang G. In-vitro Metabolisierung von a-liponsaure unter besonderer Berucksichtigung enantioselektiver Biotranformationen. Inaugural-Dissertation zur Erlangung des Doktorgrades der Naturwissenschaften im Fachbereich Chemie der Wilhelms-Universitat Munster, 1992.

Lapenna D, Ciofani G, Pierdomenico SD, Giamberardino MA, and Cuccurullo F. Dihydrolipoic acid inhibits 15-lipoxygenase-dependent lipid peroxidation. *Free Radic. Biol. Med.* 2003 Nov 15; 35(10):1203–1209.

Larghero P, Vene R, Minghelli S, Travaini G, Morini M, Ferrari N, Pfeffer U, Noonan D, Albini A, and Benelli R. Biological assays and genomic analysis reveal lipoic acid modulation of endothelial cell behavior and gene expression. *Carcinogenesis* 2006 May; 28(5):1008–1020.

Lee KS, Kim SR, Park SJ, Min KH, Lee KY, Jin SM, Yoo WH, and Lee YC. Antioxidant down-regulates interleukin-18 expression in asthma. *Mol. Pharmacol.* 2006a Oct; 70 (4):1184–1193.

Lee Y, Naseem RH, Park BH, Garry DJ, Richardson JA, Schaffer JE, and Unger RH. Alpha-lipoic acid prevents lipotoxic cardiomyopathy in acyl CoA-synthase transgenic mice. *Biochem. Biophys. Res. Commun.* 2006b May 26; 344(1):446–452.

Lee RL, Rancourt RC, del Val G, Pack K, Pardee C, Accurso FJ, and White CW. Thioredoxin and dihydrolipoic acid inhibit elastase activity in cystic fibrosis sputum. *Am. J. Physiol. Lung Cell Mol. Physiol.* 2005a Nov; 289(5):L875–L882.

Lee WJ, Song KH, Koh EH, Won JC, Kim HS, Park HS, Kim MS, Kim SW, Lee KU, and Park JY. Alpha-lipoic acid increases insulin sensitivity by activating AMPK in skeletal muscle. *Biochem. Biophys. Res. Commun.* 2005b Jul 8; 332(3):885–891.

Lee CK, Pugh TD, Klopp RG, Edwards J, Allison DB, Weindruch R, and Prolla TA. The impact of alpha-lipoic acid, coenzyme Q10 and caloric restriction on life span and gene expression patterns in mice. *Free Radic. Biol. Med.* 2004 Apr 15; 36(8): 1043–1057.

Lee HA and Hughes DA. Alpha-lipoic acid modulates NF-kappa B activity in human monocytic cells by direct interaction with DNA. *Exp. Gerontol.* 2002 Jan–Mar; 37 (2–3):401–410.

Linnane AW and Eastwood H. Cellular redox regulation and prooxidant signaling systems: A new perspective on the free radical theory of aging. *Ann. N.Y. Acad. Sci.* 2006; 1067: 47–55.

Loffelhardt S, Bonaventura C, Locher M, Borbe HO, and Bisswanger H. Interaction of alpha-lipoic acid enantiomers and homologues with the enzyme components of the mammalian pyruvate dehydrogenase complex. *Biochem. Pharmacol.* 1995 Aug 25; 50(5):637–646.

Maitra I, Serbinova E, Tritschler HJ, and Packer L. Stereospecific effects of *R*-lipoic acid on buthionine sulfoximine-induced cataract formation in newborn rats. *Biochem. Biophys. Res. Commun.* 1996 Apr 16; 221(2):422–429.

Mantovani G, Maccio A, Madeddu C, Mura L, Massa E, Gramignano G, Lusso MR, Murgia V, Camboni P, and Ferreli L. Reactive oxygen species, antioxidant mechanisms, and serum cytokine levels in cancer patients: Impact of an antioxidant treatment. *J. Environ. Pathol. Toxicol. Oncol.* 2003a; 22(1):17–28.

Mantovani G, Maccio A, Madeddu C, Mura L, Gramignano G, Lusso MR, Murgia V, Camboni P, Ferreli L, Mocci M, and Massa E. The impact of different antioxidant agents alone or in combination on reactive oxygen species, antioxidant enzymes and cytokines in a series of advanced cancer patients at different sites: Correlation with disease progression. *Free Radic. Res.* 2003b Feb; 37(2):213–223.

Marracci GH, McKeon GP, Marquardt WE, Winter RW, Riscoe MK, and Bourdette DN. Alpha lipoic acid inhibits human T-cell migration: Implications for multiple sclerosis. *J. Neurosci. Res.* 2004 Nov 1; 78(3):362–370.

Marsh SA, Laursen PB, Pat BK, Gobe GC, and Coombes JS. Bcl-2 in endothelial cells is increased by vitamin E and alpha-lipoic acid supplementation but not exercise training. *J. Mol. Cell Cardiol.* 2005 Mar; 38(3):445–451.

Matsuda T. Intestinal absorption of lipoic acid and its derivative: in vivo study on absorption of Thiamine-8-(methyl-6-acetyldihydrothioctate)-disulfide using ligated dog intestine. *Vitamins* 1968; 38, 68.

May JM, Qu ZC, and Nelson DJ. Cellular disulfide-reducing capacity: An integrated measure of cell redox capacity. *Biochem. Biophys. Res. Commun.* 2006 Jun 16; 344(4):1352–1359.

Moreau R and Hagen TM. *R*-lipoic acid down-regulates lipogenic gene expression, diet and optimum nutrition; Linus Pauling Institute; Poster Presentation May 2007; 16–19.

Moini H, Tirosh O, Park YC, Cho KJ, and Packer L. *R*-alpha-lipoic acid action on cell redox status, the insulin receptor, and glucose uptake in 3T3-L1 adipocytes. *Arch. Biochem. Biophys.* 2002 Jan 15; 397(2):384–391.

Nakano I and Sano M. Alpha lipoic acid and its related compounds. II. Synthesis of S-substituted DL-dihydrolipoic acid. *Yakugaku Zasshi.* 1956; 76:943–947.

Natraj CV, Gandhi VM, and Menon KKG. Lipoic acid and diabetes: Effect of dihydrolipoic acid administration in diabetic rats and rabbits. *J. Biosci.* 1984; 6(1):1, 37–46.

Niebch G, Buchele B, Blome J, Grieb S, Brandt G, Kampa P, Raffel HH, Locher M, Borbe HO, Nubert I, and Fleischhauer I. Enantioselective high-performance liquid chromatography assay of (+)*R*- and (−)*S*-alpha-lipoic acid in human plasma. *Chirality* 1997; 9(1):32–36.

Nolin TD, McMenamin ME, and Himmelfarb J. Simultaneous determination of total homocysteine, cysteine, cysteinylglycine, and glutathione in human plasma by high-performance liquid chromatography: Application to studies of oxidative stress. *J. Chromatogr. B. Analyt. Technol. Biomed. Life Sci.* 2007 Jun 1; 852(1–2):554–561.

Ogborne RM, Rushworth SA, and O'Connell MA. Alpha-lipoic acid-induced heme oxygenase-1 expression is mediated by nuclear factor erythroid 2-related factor 2 and p38 mitogen-activated protein kinase in human monocytic cells. *Arterioscler. Thromb. Vasc. Biol.* 2005 Oct; 25(10):2100–2105.

Oksala NK, Lappalainen J, Laaksonen DE, Khanna S, Kaarniranta K, Sen CK, and Atalay M. Alpha-lipoic acid modulates heat shock factor-1 expression in streptozotocin-induced diabetic rat kidney. *Antioxid. Redox Signal.* 2007 Apr; 9(4):497–506.

Oksala NK, Laaksonen DE, Lappalainen J, Khanna S, Nakao C, Hanninen O, Sen CK, and Atalay M. Heat shock protein 60 response to exercise in diabetes: Effects of alpha-lipoic acid supplementation. *J. Diabetes Complications* 2006 July–Aug; 20(4): 257–261.

Ou P, Tritschler HJ, and Wolff SP. Thioctic (lipoic) acid: A therapeutic metal-chelating antioxidant? *Biochem Pharmacol.* 1995 Jun 29; 50(1):123–126.

Packer L, Kraemer K, and Rimbach G. Molecular aspects of lipoic acid in the prevention of diabetes complications. *Nutrition* 2001 Oct; 17(10):888–895.

Packer L. Alpha-lipoic acid: A metabolic antioxidant which regulates NF-kappa B signal transduction and protects against oxidative injury. *Drug Metab. Rev.* 1998 May; 30 (2):245–275. Review.

Packer L, Roy S, and Sen CK. Alpha-lipoic acid: A metabolic antioxidant and potential redox modulator of transcription. *Adv. Pharmacol.* 1997 38:79–101. Review.

Packer L and Tritschler HJ. Alpha-lipoic acid: The metabolic antioxidant. *Free Radic. Biol. Med.* 1996; 20(4):625–626.

Packer L, Witt EH, Tritschler HJ, Wessel K, and Ulrich H. Antioxidant properties and clinical implications of α-lipoic acid, Chapter 22 in *Biothiols in Health and Disease*. Eds. Packer L and Cadenas E. Marcel Dekker, Inc. New York, Basel, Hong Kong. 1995 p. 479.

Panak KC, Ruiz OA, Giorgieri SA, and Diaz LE. Direct determination of glutathione in human blood by micellar electrokinetic chromatography: Simultaneous determination of lipoamide and lipoic acid. *Electrophoresis.* 1996 Oct; 17(10): 1613–1616.

Pershadsingh HA. Alpha-lipoic acid: Physiologic mechanisms and indications for the treatment of metabolic syndrome. *Expert Opin. Investig. Drugs* 2007 Mar; 16(3):291–302.

Pershadsingh HA, Ho CI, Rajaman J, Lee C, Chittiboyina AG, Deshpande R, Kurtz TW, Chan JY, Avery MA, and Benson SC. α-Lipoic acid is a weak dual PPARa/g agonist: An ester derivative with increased PPARa/g efficacy and antioxidant activity. *J. Applied Res.* 2005 5:510–523.

Peter G and Borbe HO. Absorption of [7, 8-14C]rac-α-lipoic acid from in situ ligated segments of the gastrointestinal tract of the rat. *Arzneimittelforschung* 1995 Mar; 45 (3):293–299.

Powell LA, Warpeha KM, Xu W, Walker B, and Trimble ER. High glucose decreases intracellular glutathione concentrations and upregulates inducible nitric oxide synthase gene expression in intestinal epithelial cells. *J. Mol. Endocrinol.* 2004 Dec; 33(3):797–803.

Raddatz G and Bisswanger H. Receptor site and stereospecificity of dihydrolipoamide dehydrogenase for *R*- and *S*-lipoamide: A molecular modeling study. *J. Biotechnol.* 1997 Oct 17; 58(2):89–100.

Reed LJ and Niu C. Synthesis of DL-α-lipoic acid. *J. Am. Chem. Soc.* 1954 77:416.

Riedel W, Lang U, Oetjen U, Schlapp U, and Shibata M. Inhibition of oxygen radicul formation by methylene blue, aspirin, or alpha-lipoic acid, prevents bacterial-lipopolysaccharide-induced fever. *Mol. Cell. Biochem* 2003 May; 247(1–2):83–94.

Rosenberg HR and Culik R. Effect of α-lipoic acid supplementation on vitamin C and vitamin E deficiencies. *Arch. Biochem. Biophys.* 1959 8:86–93.

Roy S, Sen CK, Tritschler HJ, and Packer L. Modulation of cellular reducing equivalent homeostasis by alpha-lipoic acid. Mechanisms and implications for diabetes and ischemic injury. *Biochem. Pharmacol.* 1997 Feb 7; 53(3):393–399.

Satoh S, Toyo'oka T, Fukushima T, and Inagaki S. Simultaneous determination of alpha-lipoic acid and its reduced form by high-performance liquid chromatography with fluorescence detection. *J. Chromatogr. B Analyt. Technol. Biomed. Life Sci.* 2007, Jul 1; 854(1–2):109–115.

Sen CK. Cellular thiols and redox-regulated signal transduction. *Curr. Top Cell Regul.* 2000; 36:1–30. Review.

Sen CK, Roy S, Khanna S, and Packer L. Determination of oxidized and reduced lipoic acid using high-performance liquid chromatography and coulometric detection. *Methods Enzymol.* 1999; 299:239–246.

Schepkin V, Kawabata T, and Packer L. NMR study of lipoic acid binding to bovine serum albumin. *Biochem. Mol. Biol. Int.* 1994 Aug; 33(5):879–886.

Schupke H, Hempel R, Peter G, Hermann R, Wessel K, Engel J, and Kronbach T. New metabolic pathways of alpha-lipoic acid. *Drug Metab. Dispos.* 2001 Jun; 29(6):855–862.

Smith AR, Shenvi SV, Widlansky M, Suh JH, and Hagen TM. Lipoic acid as a potential therapy for chronic diseases associated with oxidative stress. *Curr. Med. Chem.* 2004 May; 11(9):1135–1146.

Smith AR and Hagen TM. Vascular endothelial dysfunction in aging: Loss of Akt-dependent endothelial nitric oxide synthase phosphorylation and partial restoration by (*R*)-alpha-lipoic acid. *Biochem. Soc. Trans.* 2003 Dec; 31(Pt 6):1447–1449.

Sola S, Mir MQ, Cheema FA, Khan-Merchant N, Menon RG, Parthasarathy S, and Khan BV. Irbesartan and lipoic acid improve endothelial function and reduce markers of inflammation in the metabolic syndrome: Results of the Irbesartan and lipoic acid in endothelial dysfunction (ISLAND) study. *Circulation* 2005 Jan 25; 111(3):343–348.

Sommer D, Fakata KL, Swanson SA, and Stemmer PM. Modulation of the phosphatase activity of calcineurin by oxidants and antioxidants in vitro. *Eur. J. Biochem.* 2000 Apr; 267(8):2312–2322.

Sugiyama N and Yoshiga S. Basic experiments with massive doses of tatd and clinical trial in therapy. *Chiryo.* 1963 Sep; 45:1722–1725.

Suh JH, Moreau R, Heath SH, and Hagen TM. Dietary supplementation with (*R*)-alpha-lipoic acid reverses the age-related accumulation of iron and depletion of antioxidants in the rat cerebral cortex. *Redox Rep.* 2005; 10(1):52–60.

Suh JH, Wang H, Liu RM, Liu J, and Hagen TM. (*R*)-alpha-lipoic acid reverses the age-related loss in GSH redox status in post-mitotic tissues: Evidence for increased cysteine requirement for GSH synthesis. *Arch. Biochem. Biophys.* 2004a Mar 1; 423(1):126–135.

Suh JH, Shenvi SV, Dixon BM, Liu H, Jaiswal AK, Liu RM, and Hagen TM. Decline in transcriptional activity of Nrf2 causes age-related loss of glutathione synthesis, which is reversible with lipoic acid. *Proc. Natl. Acad. Sci. U S A* 2004b Mar 9; 101 (10):3381–3386.

Suh JH, Shigeno ET, Morrow JD, Cox B, Rocha AE, Frei B, and Hagen TM. Oxidative stress in the aging rat heart is reversed by dietary supplementation with (*R*)-(alpha)-lipoic acid. *FASEB J* 2001 Mar; 15(3):700–706.

Takenouchi K, Aso K, and Namiki T. Alpha lipoic acid metabolism in various diseases. *I. J. Jap. Derm. Society.* 1960; 70:11.

Teichert J, Hermann R, Ruus P, and Preiss R. Plasma kinetics, metabolism, and urinary excretion of alpha-lipoic acid following oral administration in healthy volunteers. *J. Clin. Pharmacol.* 2003 Nov; 43(11):1257–1267.

Teichert J and Preiss R. High-performance liquid chromatographic assay for alpha-lipoic acid and five of its metabolites in human plasma and urine. *J. Chromatogr. B Analyt. Technol. Biomed. Life Sci.* 2002 Apr 5; 769(2):269–281.

Teichert J and Preiss R. High-performance liquid chromatography methods for determination of lipoic and dihydrolipoic acid in human plasma. *Methods Enzymol.* 1997; 279:159–166. Review.

Teichert J and Preiss R. Determination of lipoic acid in human plasma by high-performance liquid chromatography with electrochemical detection. *J. Chromatogr. B Biomed. Appl.* 1995 Oct 20; 672(2):277–281.

Teichert J and Preiss R. HPLC-methods for determination of lipoic acid and its reduced form in human plasma. *Int. J. Clin. Pharmacol. Ther. Toxicol.* 1992 Nov; 30(11): 511–512.

Thomas RC and Reed LJ. Disulfide polymers of DL-lipoic acid. *J. Am. Chem. Soc.* 1956; 78 (23):6148–6149.

Ulrich H, Weischer CH, Engel J, and Hettche H. Pharmaceutical compositions containing *R*-alpha-lipoic acid or *S*-alpha-lipoic acid as active ingredient United States Patent 6,271,254 (August 7, 2001).

van der Vliet A, Cross CE, Halliwell B, and O'Neill CA. Plasma protein sulfhydryl oxidation: Effect of low molecular weight thiols. *Methods Enzymol.* 1995; 251:448–455.

Vincent AM, Perrone L, Sullivan KA, Backus C, Sastry AM, Lastoskie C, and Feldman EL. Receptor for advanced glycation end products activation injures primary sensory neurons via oxidative stress. *Endocrinology* 2007 Feb; 148(2):548–558.

Visioli F, Smith A, Zhang W, Keaney JF Jr, Hagen T, and Frei B. Lipoic acid and vitamin C potentiate nitric oxide synthesis in human aortic endothelial cells independently of cellular glutathione status. *Redox Rep.* 2002; 7(4):223–227.

Walgren JL, Amani Z, McMillan JM, Locher M, and Buse MG. Effect of *R*(+)alpha-lipoic acid on pyruvate metabolism and fatty acid oxidation in rat hepatocytes. *Metabolism* 2004 Feb; 53(2):165–173.

Walton W, Wagner A, Bachelor F, Peterson L, et al. Synthesis of (+)-α-lipoic acid and its optical antipode. *J. Am. Chem. Soc.* 1955 77:5144.

Watanabe K, Ohkubo H, Niwa H, Tanikawa N, Koda N, Ito S, and Ohmiya Y. Essential 110Cys in active site of membrane-associated prostaglandin E synthase-2. *Biochem. Biophys. Res. Commun.* 2003 Jun 27; 306(2):577–581.

Wessner B, Strasser EM, Manhart N, and Roth E. Supply of *R*-alpha-lipoic acid and glutamine to casein-fed mice influences the number of B lymphocytes and tissue glutathione levels during endotoxemia. *Wien. Klin. Wochenschr.* 2006 Mar; 118(3–4):100–107.

Williams RH, Maggiore JA, Reynolds RD, and Helgason CM. Novel approach for the determination of the redox status of homocysteine and other aminothiols in plasma from healthy subjects and patients with ischemic stroke. *Clin. Chem.* 2001 Jun; 47 (6):1031–1039.

Witt W and Rustow B. Determination of lipoic acid by precolumn derivatization with monobromobimane and reversed-phase high-performance liquid chromatography. *J. Chromatogr. B Biomed. Sci. Appl.* 1998 Jan 23; 705(1):127–131.

Yadav V, Marracci G, Lovera J, Woodward W, Bogardus K, Marquardt W, Shinto L, Morris C, and Bourdette D. Lipoic acid in multiple sclerosis: A pilot study. *Mult. Scler.* 2005 Apr; 11(2):159–165.

Zhang WJ, Wei H, Hagen T, and Frei B. Alpha-lipoic acid attenuates LPS-induced inflammatory responses by activating the phosphoinositide 3-kinase/Akt signaling pathway. *Proc. Natl. Acad. Sci. U S A* 2007 Mar 6; 104(10):4077–4082.

Zhang WJ and Frei B. Alpha-lipoic acid inhibits TNF-alpha-induced NF-kappa B activation and adhesion molecule expression in human aortic endothelial cells. *FASEB J.* 2001 Nov; 15(13):2423–2432.

Zimmer G, Mainka L, and Ulrich H. ATP synthesis and ATPase activities in heart mitoplasts under influence of *R*- and *S*-enantiomers of lipoic acid. *Methods Enzymol.* 1995; 251:332–340.

11 Pharmacokinetics, Metabolism, and Renal Excretion of Alpha-Lipoic Acid and Its Metabolites in Humans

Jens Teichert and Rainer Preiss

CONTENTS

INTRODUCTION

Alpha-lipoic acid (1,2-dithiolane-3-pentanoic acid), also known as thioctic acid, contains an intramolecular disulfide bond that can be enzymatically reduced in vivo resulting in the formation of two highly reactive vicinal sulfhydryl groups. Due to the low negative redox potential (−0.32 V) the reduced form of dihydrolipoic acid (6,8-dimercapto-octanoic acid) can regenerate ascorbate directly and is capable of regenerating endogenous thiols involved in physiological antioxidant redox systems such as cysteine and glutathione. The naturally occurring redox couple containing the *R*-enantiomer of α-lipoic/dihydrolipoic acid is covalently bound to a lysine residue forming an essential lipoamide that functions as a coenzyme of E2 subunit of four mitochondrial multienzyme complexes, e.g., the pyruvate dehydrogenase. Exogenously administered α-lipoic acid is reduced intracellularly by several enzymes and released as dihydrolipoic acid into the extracellular milieu. Racemic α-lipoic acid has been used in Germany in the treatment of diabetic polyneuropathy for many years. Because α-lipoic acid has been reported to have a number of potentially beneficial effects in both prevention and treatment of free-radical mediated diseases, numerous preclinical and clinical trials especially for the therapy of type-2 diabetes-induced diabetic polyneuropathy have been conducted. The efficacy of α-lipoic acid treatment on neuropathic disorders has been contradictorily discussed because the results of the individual trials varied and a conclusive interpretation is difficult.

Only few data have been published concerning the pharmacokinetics of exogenously administered α-lipoic acid. α-Lipoic acid is structurally related to medium-chained fatty acids, which are being readily and completely absorbed from the gastrointestinal tract. As indicated by the results of a preclinical study, α-lipoic acid is easily absorbed from the gastrointestinal tract including the stomach (Peter and Borbe, 1995). Absorption was markedly lower from stomach than from intestine. The pharmacokinetics of α-lipoic acid is characterized by rapid absorption, short plasma half-lives, extensive hepatic metabolism, and low and varying bioavailability. Dose linearity was demonstrated after oral doses of 50–600 mg (Breithaupt-Grogler et al., 1999). The bioavailability, C_{max}, and AUC values were decreased by the coingestion of food (Gleiter et al., 1996). Delayed gastric emptying reduced absolute bioavailability, C_{max}, and AUC but did not substantially affect the rate of absorption (Hermann et al., 1998).

METABOLIC PATHWAYS OF α-LIPOIC ACID

McCormick et al. revealed that α-lipoic acid undergoes predominantly β-oxidation as indicated by the detection of bisnorlipoic acid, β-hydroxy-bisnorlipoic acid and tetranorlipoic acid in the urine of rats after treatment of α-lipoic acid and exhalation of 30% of the administered dose as carbon dioxide (Harrison and McCormick, 1974; Spence and McCormick, 1976). This finding was further supported by data from preclinical studies in dogs, mice, and rats where several metabolites formed by β-oxidation and further metabolization were detected by LC–MS/MS in plasma and urine. In addition, in human urine and plasma samples taken from a clinical study in healthy volunteers after oral administration of 600 mg α-lipoic acid, six derivatives of bisnorlipoic acid and tetranorlipoic acid were identified (Schupke et al., 2001). β-Ketolipoic acid, which is formed by conversion of the 3-hydroxyl group into a keto group during step 3 of the β-oxidation, was detected in the plasma samples from humans and rats (Schupke et al., 2001). The product of thiolysis, the two carbons shorter analogue bisnorlipoic acid, was identified in human plasma and urine by HPLC and electrochemical detection

(Teichert et al., 2003). The final product of β-oxidation, tetranorlipoic acid, was detected in dog plasma and human plasma (Schupke et al., 2001) as well as in human urine (Teichert et al., 2003). After enzymatic reduction of the α-lipoic acid, bisnorlipoic acid, and tetranorlipoic acid, the resulting dimercapto-carboxylic acids can be further metabolized to the corresponding S-bismethylated carboxylic acids catalyzed by *S*-methyl transferases. These bismethylthio-carboxylic acids can undergo further degradation by β-oxidation except the S-methylated tetranorlipoic acid, namely 2,4-bis(methylthio)butanoic acid. The metabolic pathways of α-lipoic acid in humans as indicated by the detection of five metabolites in plasma and urine of 24 subjects are shown in Figure 11.1. Some authors have described the appearance of β-ketolipoic acid in human plasma (Biewenga et al., 1999; Schupke et al., 2001). We failed to identify this metabolite caused by lack of availability of the corresponding reference compound necessary for electrochemical detection. However, some unmatched peaks in the chromatogram could not be assigned to any available reference compound. These peaks may represent unknown metabolites such as the above-mentioned ketolipoic acid as well as various sulfoxides. Furthermore, there is evidence of glucuronic acid conjugation

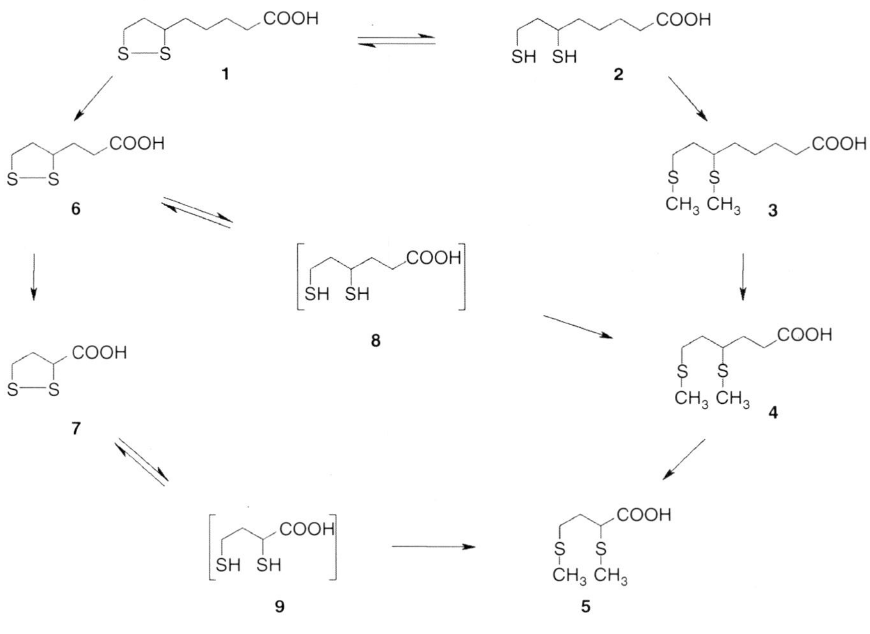

FIGURE 11.1 The metabolic pathways of α-lipoic acid supported by quantitation of the compounds **1** as well as **3–7** in human plasma and urine. Postulated compounds **8** and **9** are depicted in square brackets. **1**, α-lipoic acid; **2**, dihydrolipoic acid; **3**, 6,8-bis(methylthio) octanoic acid; **4**, 4,6-bis(methylthio)hexanoic acid; **5**, 2,4-bis(methylthio)butanoic acid; **6**, bisnorlipoic acid; **7**, tetranorlipoic acid; **8**, 4,6-dimercapto-hexanoic acid; **9**, 2,4-dimercapto-butanoic acid.

for α-lipoic acid and some metabolites as indicated by elevated concentrations of these compounds after treatment of human urine samples with β-glucuronidase/arylsulfatase (Teichert et al., 2003). In contrast to all other species investigated, a glycine conjugate was detected in the urine of mice (Schupke et al., 2001).

PHARMACOKINETIC PROPERTIES OF α-LIPOIC ACID AND ITS METABOLITES IN HEALTHY VOLUNTEERS

Single-Dose Trials

Enantioselective pharmacokinetic data from a previous study revealed short time to reach plasma peak levels of 54 ± 44 min following administration of a single tablet and 12.6 ± 4.2 min after administration of an oral solution, both containing 200 mg of racemic α-lipoic acid (Hermann et al., 1996). The corresponding mean peak levels were 0.49 ± 0.27 and 0.31 ± 0.16 μg/mL for *R*(+)-α-lipoic acid and *S*(−)-α-lipoic acid, respectively for the tablet and 2.24 ± 1.21 and 1.32 ± 0.69 μg/mL, respectively for the oral solution. Similar differences between both enantiomers were found for mean AUC data (24.6 ± 7.8 versus 15.0 ± 5.4 μg·min/mL for the tablet and 40.8 ± 14.4 versus 23.4 ± 8.4 μg·min/mL for the oral solution). The absolute bioavailability of the *R*(+)-enantiomer was significantly higher than for the *S*(−)-enantiomer 24.1% ± 12.7% versus 19.1% ± 12.8%. The absolute bioavailability of the oral solution was 38 ± 15 and 28% ± 14% for *R*(+)- and *S*(−)-α-lipoic acid, respectively. The terminal elimination half-life after administration of a 200 mg tablet was 19.8 min for both enantiomers and was comparable to those obtained for the *R*(+)- (22.2 min) and *S*(−)-enantiomer (19.2 min) after intravenous administration (Hermann et al., 1996). It should be noted that both intravenous and oral solution contained the trometamole salt of α-lipoic acid, whereas the tablets contained the undissociated acid. Reduced solubility of the acid compared to its salts in aqueous solutions can influence C_{max} and AUC and may cause prolonged time to reach plasma peak level. The findings of this study were supported by own results of a pharmacokinetic trial in which an absolute bioavailability of 29.1% ± 10.3% was determined for racemic α-lipoic acid after oral single dose of 200 mg in 12 healthy volunteers (Teichert et al., 1998). Figure 11.2 shows the mean plasma concentration–time curves after administration of 200 and 600 mg racemic α-lipoic acid. Dose linearity was observed after threefold oral dose as indicated by the triplicate AUC values. The mean AUC increased 4.6-fold when the 600 mg dose was given intravenously. These nonlinearly increased plasma AUC values at higher dose levels may be caused by a transient saturation of the first-pass metabolism. The pharmacokinetic parameters of α-lipoic acid are summarized in Table 11.1. The mean total plasma clearance was 18.7 ± 3.4 and 11.1 ± 3.2 mL/min·kg for the 200 and 600 mg IV, respectively, and is in good accordance with the values for the *R*(+)- and *S*(−)-α-lipoic acid (12.2 and 15.6 mL/min·kg) (Hermann et al., 1996). These values are closely related to the physiological plasma flow to the liver. For highly extractable drugs, the extraction ratio is dependent on hepatic blood flow. Hence, we assume

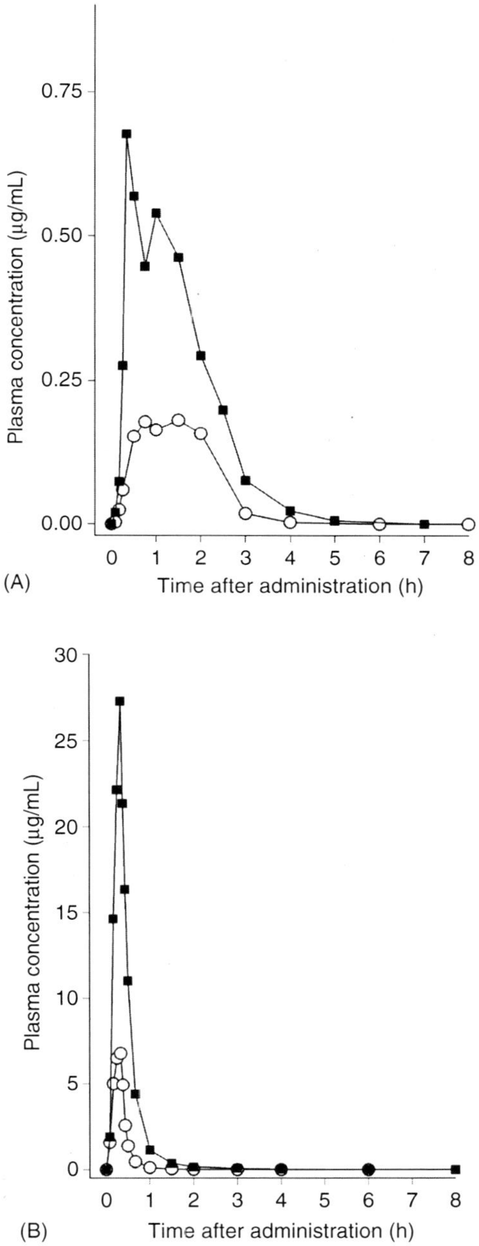

FIGURE 11.2 Mean plasma concentration–time curves of α-lipoic acid in 12 subjects after single oral doses (A) and intravenous doses (B) of 200 mg (white circles) and 600 mg (black squares) racemic α-lipoic acid.

TABLE 11.1
Pharmacokinetic Parameters of α-Lipoic Acid after Oral Single Dose and Constant-Rate Intravenous Infusion (20 min) of 200 and 600 mg Racemic α-Lipoic Acid in Healthy Volunteers

Dose	*n*	t_{max} (min)	C_{max} (μg/mL)	AUC (μg·min/mL)	$t_{1/2}$(min)
200 mg PO	12	80.0 ± 50.5	0.66 ± 0.33	46.82 ± 21.46	32.7 ± 15.2
600 mg PO	12	87.9 ± 75.4	1.96 ± 1.23	148.08 ± 58.67	25.8 ± 6.6
200 mg IV	12	19.1 ± 3.9	8.32 ± 2.35	157.97 ± 35.05	23.4 ± 10.8
600 mg IV	12	18.6 ± 3.5	28.57 ± 5.78	735.44 ± 225.67	32.8 ± 9.4

Note: Values are means ± standard deviations.

flow limited hepatic clearance for α-lipoic acid. Therefore, the pharmacokinetics of α-lipoic acid is more likely to be affected by alteration of liver blood flow.

Multiple-Dose Trials

α-Lipoic acid used for the management of diabetic neuropathy will likely be administered on a long-term basis. It is thus important to determine whether the pharmacokinetics is significantly altered with repeated administration. In this study, we administered film tablets containing 600 mg of racemic α-lipoic acid with 150 mL tap water following a 12 h overnight fasting period to nine healthy volunteers (Teichert et al., 2003). Blood samples were collected at 18 time points including predose on two profile days. On days 2 and 3, predose samples were collected. Time versus plasma concentration curves recorded on profile days 1 and 4 and summaries of pharmacokinetic parameters for α-lipoic acid and all metabolites measured in this study are given in Figure 11.3A–F and Table 11.2, respectively. Bisnorlipoic acid appeared in plasma for a short period and the concentrations were consistently low. The apparent oral systemic clearance of α-lipoic acid uncorrected for bioavailability (CL/F) was 43.0 ± 12.2 mL/min·kg after single oral dose on day 1. The apparent oral steady-state clearance (CL_{ss}/F) was 38.4 ± 7.8 mL/min·kg and was calculated using the area under the curve from 0 to 24 h ($AUC_{0-\tau}$) from day 4 data. The apparent oral volume of distribution during the terminal phase (V_Z/F) and at steady state was 169.8 ± 70.3 and 136.1 ± 37.9 L, respectively. There were no statistically significant changes in the pharmacokinetic parameters for α-lipoic acid as well its metabolites over four days of 600 mg once-daily oral dosing. In the predose samples of days 2, 3, and 4, mean plasma concentrations of 25, 32, and 27 ng/mL, respectively, were measured for the metabolite 2,4-bis(methylthio)butanoic acid. Neither α-lipoic acid nor any other metabolite was detected in these samples. The accumulation factor calculated by the computer program WinNonlin 4.0 was 1.0000 for each compound.

In addition, to evaluate the cumulative excretion of α-lipoic acid and its metabolites over time, eight urine samples were collected up to 24 h post-dose.

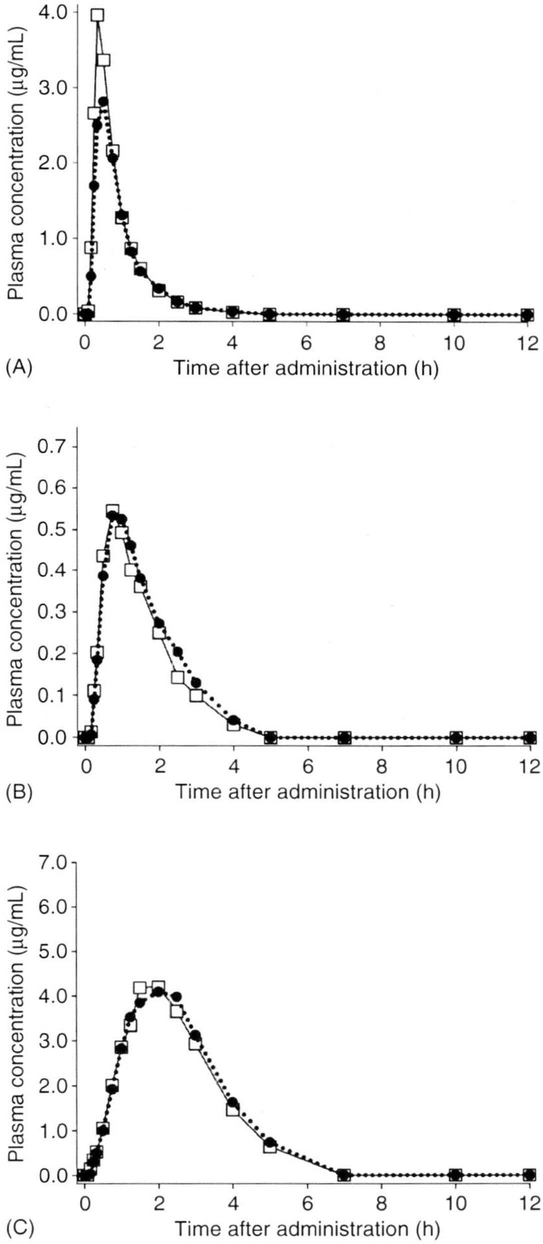

FIGURE 11.3 Geometric mean plasma concentration–time curves of (A) α-lipoic acid, (B) BMOA (6,8-bis(methylthio)octanoic acid), (C) BMHA (4,6-bis(methylthio)hexanoic acid).

(continued)

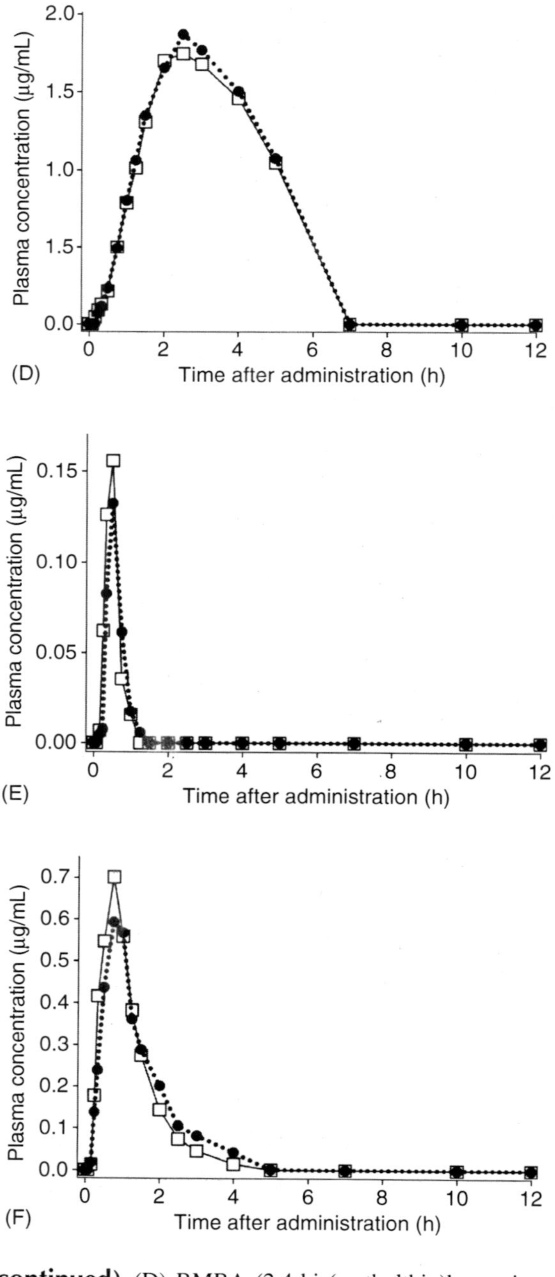

FIGURE 11.3 (continued) (D) BMBA (2,4-bis(methylthio)butanoic acid), (E) BNLA (bisnorlipoic acid), and (F) TNLA (tetranorlipoic acid) in nine healthy volunteers recorded on day 1 (circles, dotted line) and day 4 (squares, solid line) after once-daily oral administration of 600 mg α-lipoic acid for four days.

TABLE 11.2
Pharmacokinetic Parameters of α-Lipoic Acid and Its Metabolites after Once-Daily Oral Doses of 600 mg for Four Days in Healthy Volunteers Measured on Days 1 and 4

Substance	Day	*n*	t_{max} (min)	C_{max} (μg/mL)	AUC (μg·min/mL)	$t_{1/2}$ (min)
α-LA	1	9	30.6 ± 12.6	7.47 ± 2.14	190.54 ± 44.75	34.8 ± 8.7
	4	9	26.1 ± 8.6	6.49 ± 1.69	206.10 ± 30.08	31.9 ± 7.5
BMOA	1	9	51.7 ± 15.2	0.68 ± 0.33	67.49 ± 32.34	51.1 ± 12.8
	4	9	45.0 ± 26.0	0.64 ± 0.27	59.33 ± 22.23	44.8 ± 8.8
BMHA	1	9	113.3 ± 25.0	4.52 ± 1.18	851.87 ± 214.05	118.1 ± 31.9
	4	9	110.0 ± 21.2	4.61 ± 1.02	839.10 ± 276.62	112.6 ± 35.5
BMBA	1	9	160.0 ± 42.4	2.04 ± 0.57	567.99 ± 112.47	124.3 ± 28.3
	4	9	160.0 ± 36.7	1.91 ± 0.23	543.92 ± 83.44	124.2 ± 24.2
BNLA	1	9	33.3 ± 13.9	0.21 ± 0.15	6.00 ± 4.17	13.1 ± 6.5
	4	8	26.9 ± 8.8	0.23 ± 0.12	7.69 ± 4.03	12.5[a]
TNLA	1	9	46.1 ± 16.4	0.89 ± 0.71	62.79 ± 37.42	43.9 ± 12.2
	4	9	47.2 ± 20.9	1.01 ± 0.53	65.21 ± 29.72	38.5 ± 10.9

Note: Values are means ± standard deviations. AUC, area under the curve from the time of dosing to the last measurable concentration; α-LA, α-lipoic acid; BMOA, 6,8-bis(methylthio)octanoic acid; BMHA, 4,6-bis(methylthio)hexanoic acid; BMBA, 2,4-bis(methylthio)butanoic acid; BNLA, bisnorlipoic acid; TNLA, tetranorlipoic acid.

[a] Represents an individual value.

Bisnorlipoic acid was not detectable in the urine samples. The summarized urine pharmacokinetic data are depicted in Table 11.3.

Summary

As is evident from Table 11.2, the times of maximum plasma concentration of the metabolites correspond with the chronological order of their formation. Compared to *S*-methyl transferase-mediated alkylation and subsequent β-oxidation, the T_{max} values of 6,8-bis(methylthio)octanoic acid, bisnorlipoic acid, and tetranorlipoic acid provide evidence that β-oxidation of α-lipoic acid is the faster metabolic pathway. We observed a wide individual variation in the pharmacokinetics for α-lipoic acid and its metabolites after once-daily administration of α-lipoic acid for four days. However, the pharmacokinetic parameters of α-lipoic acid and the metabolites are not significantly altered during repeated oral dosing. As indicated by the accumulation factor, α-lipoic acid as well as its metabolites did not accumulate in plasma, albeit nonsignificantly elevated trough levels of 2,4-bis(methylthio)butanoic acid ranging near the lower limit of quantitation were found.

A total of 12.8% and 12.0% of the administered doses were recovered in urine as unchanged parent and four metabolites over 24 h post-dose on days 1 and 4,

TABLE 11.3
Urinary Excreted Amounts of α-Lipoic Acid and Its Metabolites after Once-Daily Oral Doses of 600 mg for Four Days in Healthy Volunteers Measured on Days 1 and 4

Substance	Day	*n*	Ae (mg)	Ae (% dose)	CL_R (mL/min)
α-LA	1	9	1.149 ± 0.332	0.19 ± 0.06	6.37 ± 2.30
	4	9	1.288 ± 0.408	0.21 ± 0.07	6.50 ± 2.35
BMOA	1	9	5.193 ± 2.516	0.76 ± 0.37	71.99 ± 16.55
	4	9	4.693 ± 2.299	0.68 ± 0.33	73.71 ± 24.74
BMHA	1	9	48.369 ± 18.381	7.98 ± 3.03	56.67 ± 18.47
	4	9	42.424 ± 16.280	7.00 ± 2.69	50.29 ± 14.32
BMBA	1	9	19.432 ± 3.689	3.71 ± 0.70	34.07 ± 9.53
	4	9	20.591 ± 4.318	3.93 ± 0.82	37.62 ± 10.36
TNLA	1	9	0.739 ± 0.318	0.17 ± 0.07	12.53 ± 4.03
	4	8	0.916 ± 0.457	0.21 ± 0.10	14.02 ± 5.15
Total	1		74.882	12.81	
	4		69.912	12.03	

Note: Values are means ± standard deviations. Ae, Amount excreted in urine; CL_R, renal clearance; α-LA, α-lipoic acid; BMOA, 6,8-bis(methylthio)octanoic acid; BMHA, 4,6-bis(methylthio) hexanoic acid; BMBA, 2,4-bis(methylthio)butanoic acid; TNLA, tetranorlipoic acid.

respectively. The major urinary metabolites recovered from human urine over 24 h were the conjugated 4,6-bis(methylthio)hexanoic acid and 2,4-bis (methylthio)butanoic acid. Conjugation was found to a lesser extent for the more polar butanoic acid derivative, whereas no non-conjugated 6,8-bis (methylthio)octanoic acid was found in urine. All urine samples were treated with β-glucuronidase/arylsulfatase during sample preparation throughout this study. With quantitative gastrointestinal absorption provided, these results reveal that renal elimination of α-lipoic acid and at least four of its metabolites is not a predominant route of α-lipoic acid clearance. This finding contrasts with the urinary recovery of more than 80% of the administered radioactive-labeled dose in animal studies. Either further metabolites such as sulfoxides not detected in this study or additional routes of elimination should therefore be considered.

PHARMACOKINETIC PROPERTIES OF α-LIPOIC ACID AND ITS METABOLITES IN PATIENTS WITH RENAL IMPAIRMENT

Patients with Severe Renal Dysfunction

The effect of renal function on the pharmacokinetics of α-lipoic acid and its metabolites was investigated in eight patients with severe kidney damage as indicated by a creatinine clearance in the range from 12 to 29 mL/min (Teichert et al., 2005). For the renally impaired subjects, an identical study design as described above for the healthy volunteers was chosen. Therefore, they received oral single doses of 600 mg racemic α-lipoic acid daily for four days. Predose blood samples were drawn daily throughout the four study days. Seventeen blood samples were collected post-dose on days 1 and 4. Urine was collected at predose and during eight intervals up to 24 h post-dose on days 1 and 4. Two additional urine samples were collected at 24–36 and 36–48 h following the last dosing on day 4. The 24 h creatinine clearance was determined before dosing on day 1 of drug administration and during poststudy laboratory analyses to confirm renal status. Liver function was assessed by standard laboratory parameters, i.e., alanine and aspartate aminotransferase, alkaline phosphatase, bilirubin, and γ-glutamyl transferase before and after drug administration. The subjects did not exhibit abnormal liver function and the treatment with α-lipoic acid did not alter these parameters. The pharmacokinetic parameters of α-lipoic acid and the metabolites are summarized in Table 11.4. We failed to detect tetranorlipoic acid in the plasma samples from the renally impaired patients. Bisnorlipoic acid was quantifiable only in few samples. Pharmacokinetic parameters of bisnorlipoic acid could therefore not be completely calculated and should be seen as orientating. Area under the curve values of α-lipoic acid and its metabolites was not significantly different compared to healthy subjects except for 4,6-bis(methylthio) hexanoic acid, which showed significantly increased values. Significantly

TABLE 11.4
Pharmacokinetic Parameters of α-Lipoic Acid and Its Metabolites after Once-Daily Oral Doses of 600 mg for Four Days in Patients with Severe Kidney Damage Measured on Days 1 and 4

Substance	Day	n	t_{max}(min)	C_{max}(μg/mL)	AUC (μg·min/mL)	$t_{1/2}$ (min)
α-LA	1	9	76.1 ± 65.4	3.57 ± 2.91	227.00 ± 159.80	21.9 ± 4.8
	4	9	35.0 ± 23.7	4.32 ± 3.70	225.70 ± 158.68	22.3 ± 5.3
BMOA	1	9	115.0 ± 68.7	1.01 ± 0.63	131.96 ± 85.36	53.8 ± 9.7
	4	9	61.7 ± 32.2	0.96 ± 0.68	129.74 ± 105.88	54.4 ± 16.3
BMHA	1	9	170.0 ± 60.0	6.42 ± 2.18	1808.39 ± 853.93	121.3 ± 36.1
	4	9	141.7 ± 46.9	7.53 ± 2.16	1808.33 ± 669.72	138.4 ± 32.3
BMBA	1	9	203.3 ± 76.2	1.26 ± 0.94	595.11 ± 536.79	263.6 ± 114.5
	4	9	173.3 ± 61.4	1.35 ± 0.83	506.13 ± 323.68	230.1 ± 67.0
BNLA	1	4	20.0 ± 8.2	0.18 ± 0.10	4.02 ± 2.07	27.7 ± 5.6
	4	2	25.0 ± 7.1	0.24 ± 0.09	[a]	21.7 ± 13.9

Note: Values are means ± standard deviations. AUC, area under the curve from the time of dosing to the last measurable concentration; α-LA, α-lipoic acid; BMOA, 6,8-bis(methylthio)octanoic acid; BMHA, 4,6-bis(methylthio)hexanoic acid; BMBA, 2,4-bis(methylthio)butanoic acid; BNLA, bisnorlipoic acid.

[a] Not calculable due to missing data.

elevated trough levels of both metabolites 4,6-bis(methylthio)hexanoic acid and 2,4-bis(methylthio)butanoic acid were measured before dosing on days 2, 3, and 4 compared to healthy volunteers. However, the accumulation factor was 1.0000 for α-lipoic acid and the metabolites. There were no statistically significant differences between the pharmacokinetic parameters AUC, C_{max}, and $t_{1/2}$ for α-lipoic acid and the metabolites on days 1 and 4. The geometric mean plasma concentration–time curves for α-lipoic acid and the metabolites are depicted in Figure 11.4A–D.

The amounts of α-lipoic acid and the metabolites recovered in the urine from subjects with severely reduced kidney function are detailed in Table 11.5. No quantifiable concentrations of bisnorlipoic and tetranorlipoic acids were found in the urine samples of these subjects. Compared to healthy subjects, we detected a significant reduction in renal clearance of α-lipoic acid, 6,8-bis(methylthio)octanoic

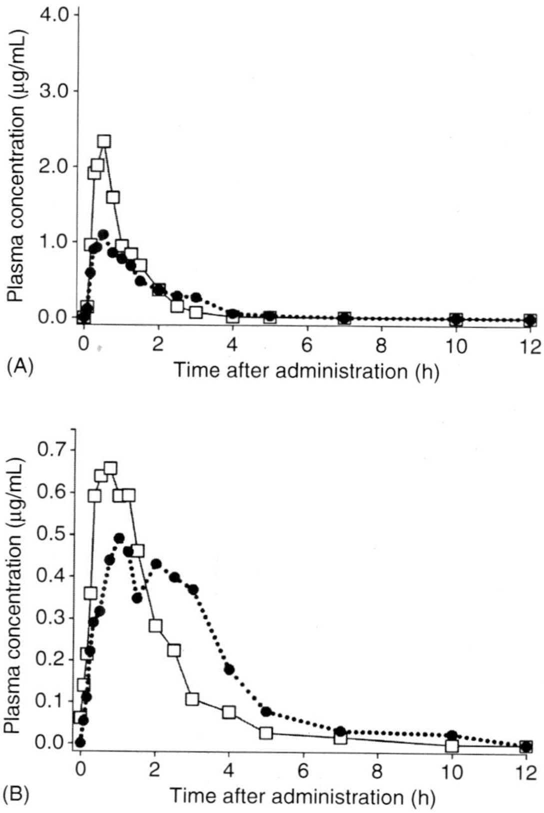

FIGURE 11.4 Geometric mean plasma concentration–time curves of (A) α-lipoic acid, (B) BMOA (6,8-bis(methylthio)octanoic acid).

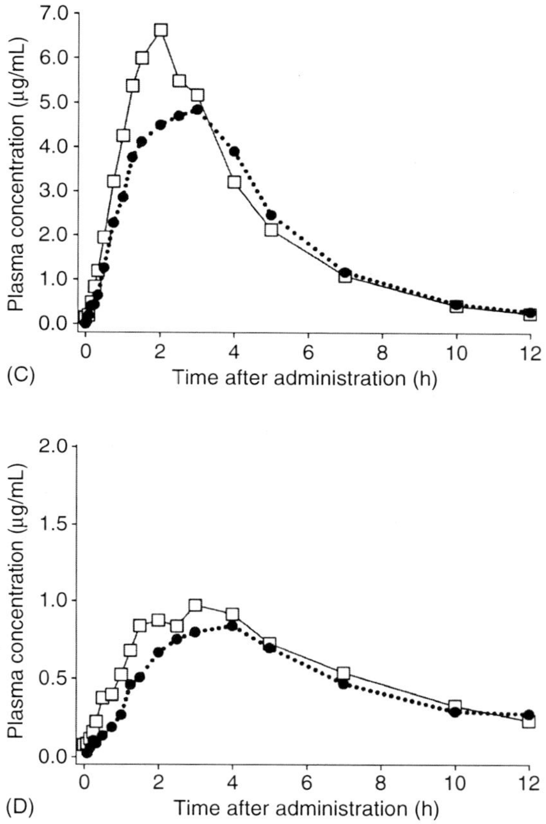

FIGURE 11.4 (continued) (C) BMHA (4,6-bis(methylthio)hexanoic acid), and (D) BMBA (2,4-bis(methylthio)butanoic acid) in nine subjects with severe kidney damage recorded on day 1 (circles, dotted line) and day 4 (squares, solid line) after once-daily oral administration of 600 mg α-lipoic acid for four days.

acid, 4,6-bis(methylthio)hexanoic acid, and 2,4-bis(methylthio)butanoic acid as well as a significant reduction of urinary excreted amounts of α-lipoic acid and 2,4-bis(methylthio)butanoic acid accompanied by a significantly prolonged half-life for 2,4-bis(methylthio)butanoic acid and a significantly decreased half-life for α-lipoic acid. The apparent oral systemic clearance of α-lipoic acid (CL/F) was 53.5 ± 37.4 mL/min·kg after single oral dose on day 1. The apparent oral steady-state clearance (CL_{ss}/F) was 55.9 ± 45.4 mL/min·kg calculated from day 4 data. The apparent oral volume of distribution during the terminal phase (V_Z/F) and at steady state was 127.3 ± 90.2 and 131.6 ± 108.3 L, respectively. Statistical comparisons were performed using ANOVA. A level of $P < .05$ was considered statistically significant.

TABLE 11.5
Urinary Excreted Amounts of α-Lipoic Acid and Its Metabolites after Once-Daily Oral Doses of 600 mg for Four Days in Patients with Severe Kidney Damage Measured on Days 1 and 4

Substance	Day	*n*	Ae (mg)	Ae (% dose)	CL_R (mL/min)
α-LA	1	9	0.311 ± 0.134	0.05 ± 0.02	1.97 ± 1.22
	4	9	0.283 ± 0.144	0.04 ± 0.02	1.53 ± 0.87
BMOA	1	9	3.837 ± 1.731	0.64 ± 0.29	44.02 ± 31.82
	4	9	3.633 ± 2.756	0.61 ± 0.46	35.15 ± 23.23
BMHA	1	9	27.346 ± 14.530	4.56 ± 2.42	18.01 ± 12.61
	4	9	31.635 ± 19.456	5.27 ± 3.24	20.00 ± 15.41
BMBA	1	9	4.174 ± 3.656	0.70 ± 0.61	7.25 ± 3.78
	4	9	6.303 ± 3.435	1.05 ± 0.57	11.95 ± 5.13
Total	1		35.668	5.94	
	4		41.854	6.97	

Note: Values are means ± standard deviations. Ae, amount excreted in urine; CL_R, renal clearance; α-LA, α-lipoic acid; BMOA, 6,8-bis(methylthio)octanoic acid; BMHA, 4,6-bis(methylthio) hexanoic acid; BMBA, 2,4-bis(methylthio)butanoic acid.

Patients with End-Stage Renal Disease

To evaluate the pharmacokinetics of α-lipoic acid and its metabolites in patients with end-stage renal failure and the effect of hemodialysis on pharmacokinetics, blood samples were collected from eight patients with end-stage renal disease (ESRD) requiring hemodialysis during dialysis on study day 1 and in the dialysis-free interval on day 4. In contrast to renally impaired as well as healthy subjects, ESRD patients did not receive oral single doses of α-lipoic acid on days 2 and 3 and therefore, no predose samples were collected on those days.

To evaluate the hemodialysability of α-lipoic acid and its metabolites, two 10 mL aliquots of dialysate were collected before and at 15 min intervals during hemodialysis. End-stage renal disease patients received the single oral doses of 600 mg α-lipoic acid when the 5 h lasting dialysis sessions were started. The pharmacokinetic parameters calculated for ESRD patients during hemodialysis and in the dialysis-free interval are shown in Table 11.6. Plots of geometric mean plasma concentrations of α-lipoic acid and three metabolites in ESRD patients over a 12 h period recorded during hemodialysis and in the dialysis-free interval are given in Figure 11.5A–D. No pharmacokinetic parameters were calculable for bisnorlipoic and tetranorlipoic acids due to intermittent plasma levels. Hence, no plasma concentration versus time curve is depicted for these metabolites. Amounts of α-lipoic acid and the metabolites removed by hemodialysis are listed in Table 11.7. The total volume of processed dialysate was calculated

TABLE 11.6
Pharmacokinetic Parameters of α-Lipoic Acid and Its Metabolites as Observed after a Single Oral Dose of 600 mg α-Lipoic Acid on Day 1 (during Hemodialysis) and Day 4 (Dialysis-Free Interval) in Patients with End-Stage Renal Disease

Substance	Day	n	t_{max} (min)	C_{max} (μg/mL)	AUC (μg·min/mL)	$t_{1/2}$ (min)
α-LA	1	8	44.4 ± 29.2	4.25 ± 1.64	247.53 ± 111.53	21.8 ± 5.8
	4	7	57.1 ± 46.2	4.55 ± 3.25	221.42 ± 135.45	23.6 ± 4.7
BMOA	1	8	82.5 ± 46.8	0.80 ± 0.34	89.77 ± 48.66	36.7 ± 6.6
	4	7	71.4 ± 42.9	0.58 ± 0.33	58.36 ± 33.37	46.2 ± 8.8
BMHA	1	8	136.9 ± 50.4	5.73 ± 1.92	1147.11 ± 431.08	93.9 ± 22.3
	4	7	150.0 ± 45.8	7.01 ± 2.77	1487.26 ± 591.94	123.9 ± 28.5
BMBA	1	8	152.1 ± 81.4	0.85 ± 0.37	260.17 ± 151.68	246.5 ± 107.5
	4	7	167.1 ± 41.9	1.27 ± 0.67	392.92 ± 225.78	236.5 ± 69.8
BNLA	1	3	55.0 ± 31.2	0.40 ± 0.18	19.22 ± 16.08	50.4 ± 39.1
	4		[a]	[a]	[a]	[a]

Note: Values are means ± standard deviations. AUC, area under the curve from the time of dosing to the last measurable concentration; α-LA, α-lipoic acid; BMOA, 6,8-bis(methylthio)octanoic acid; BMHA, 4,6-bis(methylthio)hexanoic acid; BMBA, 2,4-bis(methylthio)butanoic acid; BNLA, bisnorlipoic acid.

[a] Not calculable due to missing data.

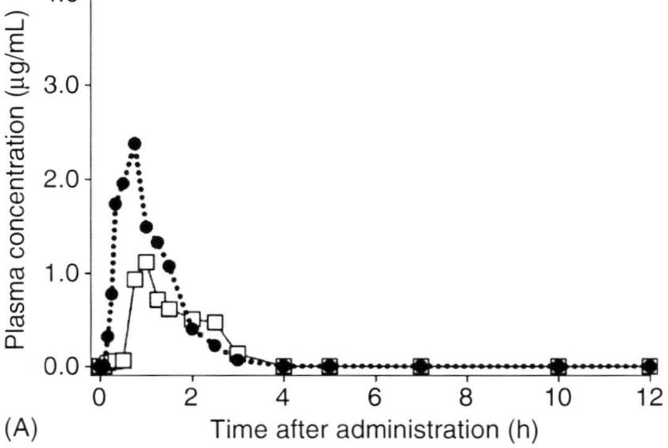

FIGURE 11.5 Geometric mean plasma concentration–time curves of (A) α-lipoic acid.

(continued)

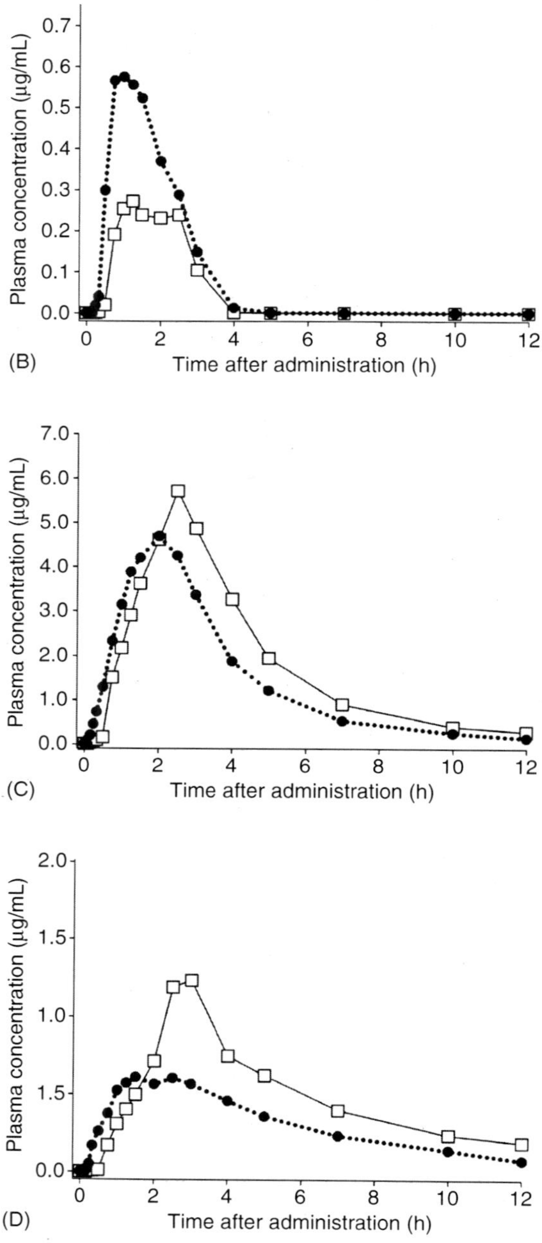

FIGURE 11.5 (continued) (B) BMOA (6,8-bis(methylthio)octanoic acid), (C) BMHA (4,6-bis(methylthio)hexanoic acid), and (D) BMBA (2,4-bis(methylthio)butanoic acid) in eight patients with end-stage renal disease (ESRD) recorded on day 1 (during hemodialysis, circles, dotted line) and day 4 (dialysis-free interval, squares, solid line) each after a single oral dose of 600 mg α-lipoic acid.

TABLE 11.7
Amounts of α-Lipoic Acid and Its Metabolites Removed by Hemodialysis after a Single Oral Dose of 600 mg α-Lipoic Acid on Day 1 (during Hemodialysis) and Day 4 (Dialysis-Free Interval) in Patients with End-Stage Renal Disease

Substance	Day	*n*	Ae (mg)	Ae (% dose)	CL_{HD} (mL/min)
α-LA	1	8	1.054 ± 0.539	0.18 ± 0.09	4.71 ± 1.98
	4		—	—	—
BMOA	1	8	0.583 ± 0.363	0.08 ± 0.05	6.76 ± 2.43
	4		—	—	—
BMHA	1	8	14.292 ± 6.712	2.36 ± 1.11	12.53 ± 3.40
	4		—	—	—
BMBA	1	8	4.739 ± 2.413	0.90 ± 0.46	23.12 ± 9.82
	4		—	—	—
BNLA	1	6	0.672 ± 0.499	0.36 ± 0.30	[a]
	4		—	—	—
TNLA	1	8	1.608 ± 1.326	0.13 ± 0.10	[a]
	4		—	—	—
Total	1		22.949	4.01	
	4		—	—	

Note: Values are means ± standard deviations. Ae, amount removed by hemodialysis; CL_{HD}, hemodialysis clearance; α-LA, α-lipoic acid; BMOA, 6,8-bis(methylthio)octanoic acid; BMHA, 4,6-bis(methylthio)hexanoic acid; BMBA, 2,4-bis(methylthio)butanoic acid.

[a] Not calculable due to missing data.

by the dialysate flow rate (500 mL/min) and the length of the procedure. The significantly decreased terminal half-life for α-lipoic acid and the significantly increased terminal half-life for 2,4-bis(methylthio)butanoic acid found in ESRD patients are in good accordance with the findings in subjects with severe kidney dysfunction. In the same manner, area under the curve values of 4,6-bis(methylthio)hexanoic acid was significantly increased. In accordance to the results obtained from subjects with normal as well as severely impaired kidney function, no statistically significant difference for α-lipoic acid and the metabolites 6,8-bis(methylthio)octanoic acid, 4,6-bis(methylthio)hexanoic acid, and 2,4-bis(methylthio)butanoic acid was detectable in ESRD patients between day 1 (during hemodialysis) and day 4 (dialysis-free interval).

The apparent oral systemic clearance of α-lipoic acid (CL/F) was 50.6 ± 34.5 and 59.9 ± 39.6 mL/min·kg after single oral doses on days 1 and 4, respectively. The apparent oral volume of distribution during the terminal phase (V_Z/F) on day 1 (hemodialysis) and day 4 (dialysis-free interval) was 105.1 ± 87.9 and 126.7 ± 80.6 L, respectively.

Summary

Intersubject variability in individual plasma concentration–time curves of α-lipoic acid and its metabolites was strongly increased in patients with severe renal impairment and ESRD compared to healthy subjects. The individual plasma concentration–time curves of α-lipoic acid in renally impaired subjects exhibit large intersubject variability in particular with respect to the time to peak (t_{max}). Hence, geometric mean plasma concentration–time curves do not represent real circumstances. As detailed in Table 11.6, we detected no statistically significant differences between days 1 and 4 for AUC values.

There was a significant, positive correlation between creatinine clearance and renal clearance for α-lipoic acid ($R = .8565$, $P < .0001$), 6,8-bis(methylthio)octanoic acid ($R = .7926$, $P < .0001$), 4,6-bis(methylthio)hexanoic acid ($R = .8953$, $P < .0001$), and 2,4-bis(methylthio)butanoic acid ($R = .8852$, $P < .0001$). On the other side, there was no significant difference in CL/F as well most pharmacokinetic parameters between subjects with severely reduced renal function and subjects with normal renal function. Hence, apparent total clearance (CL/F) does not significantly depend on creatinine clearance. Although the mean values were not significantly different compared to healthy subjects, increased intersubject variability was detected for CL/F and V_Z/F in subjects with severely reduced kidney function as well as ESRD patients as indicated by a significantly raised coefficient of variation. Furthermore, the total amount of α-lipoic acid and the metabolites removed by hemodialysis was comparable to the total urinary excreted amount from patients with severely reduced kidney function. In conclusion, hemodialysis has no significant effect on the elimination of α-lipoic acid and its metabolites.

CONCLUSIONS

As a highly permeable substance, orally administered α-lipoic acid should be nearly completely absorbed when the active agent is quantitatively dissolved. Although anionic lipoate is very soluble, undissociated α-lipoic acid has limited water solubility. As an acidic drug, undissociated α-lipoic acid shows reduced absorption due to poor water solubility at the low pH in stomach followed by strongly increased absorption in intestine. Therefore, varying absorption of α-lipoic acid due to limited dissolution of drug particles after disintegration of the solid dosage form can cause high intrasubject and intersubject pharmacokinetic variability. The influence of drug release properties of the formulation on absolute bioavailability and variability of pharmacokinetic parameters is illustrated by the pharmacokinetics obtained after administration of an oral solution containing the dissolved trometamole salt of α-lipoic acid. Thus, modified drug-delivery technologies are capable of enhanced drug dissolution and absorption, which can alter pharmacokinetics and diminish variability. In the meantime, the available so-called HR film tablets provide a high absorption of α-lipoic acid comparable to the absorption from the oral solution. The pharmacokinetics of α-lipoic acid following administration of conventional quick release tablets, which is characterized by limited bioavailability due to extensive hepatic presystemic elimination as well as extremely short half-life, provides limited clinical effects because of quickly declining plasma levels. To maintain a sufficient and constant concentration of α-lipoic acid in the systemic circulation, which is clinically efficient, multiple oral daily doses or intravenous administration are needed. In order to overcome the disadvantage of multiple daily dosing, a sustained release delivery system for α-lipoic acid has been reported (Bernkop-Schnurch et al., 2004). We used conventional quick release film tablets containing 600 mg of α-lipoic acid. The results presented herein and the drawn conclusions are therefore representative only for this pharmaceutical formulation.

Assuming an absolute bioavailability of about 30%, the mean total plasma clearance would be in the range from 11.4 to 13.1 mL/min·kg for all subjects unaffected by the kidney function. These values are closely related to total hepatic blood flow. Renal impairment should not affect total systemic clearance of α-lipoic acid because the hepatic blood flow is not altered in renal failure and the hepatic clearance of highly extracted drugs depends primarily on the hepatic blood flow rather than on intrinsic clearance or protein binding.

The pharmacokinetic profile of α-lipoic acid after daily oral doses of 600 mg makes the drug suitable for a wide range of patients from varying degrees of renal dysfunction to ESRD. Despite the fact that the renal clearance of α-lipoic acid and its major circulating metabolites is decreased in subjects with severe kidney damage, no significant differences in the pharmacokinetics of α-lipoic acid were found compared to subjects with normal renal function. We assume that most of the apparent total clearance of α-lipoic acid is attributed to nonrenal clearance. On the other hand, the contribution of unidentified metabolites to renal clearance should be considered. Significantly increased trough levels of 4,6-bis (methylthio)hexanoic acid and 2,4-bis(methylthio)butanoic acid were detected in subjects with severe kidney damage. The latter cannot be further degraded by the β-oxidation pathway. Nevertheless, neither α-lipoic acid nor any metabolite does accumulate in plasma after once-daily single oral doses of 600 mg α-lipoic acid for four days. It has been reported that chronic renal failure can significantly reduce nonrenal clearance of drugs that are cleared predominantly hepatically (Korashy et al., 2004). Alterations in the pharmacokinetics of several metabolites in renally impaired patients are therefore, maybe, caused by reduced hepatic elimination (e.g., hepatic conjugation such as glucuronidation). Chronic renal failure can cause reduced as well as increased glucuronidation as observed in animal experiments and clinical studies (Dreisbach and Lertora, 2003). The question whether the conjugation of the α-lipoic acid and the metabolites is affected by chronic renal failure remains unanswered because we treated all urine samples with β-glucuronidase prior to analysis. In summary, severely diminished renal clearance did not affect the overall pharmacokinetic profile of α-lipoic acid following once-daily oral dose of 600 mg. Thus, dosing changes are not required in patients with kidney disease.

Likewise, the pharmacokinetics of α-lipoic acid and its metabolites are not altered in ESRD patients needing hemodialysis at blood and dialysate flow rates commonly used in most dialysis facilities in Germany. At higher flow rates, an expected slightly increased dialyzer clearance would not change the overall pharmacokinetics. Apparent total clearance of α-lipoic acid was not significantly altered even in the dialysis-free interval after oral administration of 600 mg α-lipoic acid. In conclusion, α-lipoic acid and five of its known metabolites are not significantly eliminated by hemodialysis in patients with end-stage renal disease.

REFERENCES

Bernkop-Schnurch, A., Reich-Rohrwig, E., Marschutz, M., Schuhbauer, H., and Kratzel, M. Development of a sustained release dosage form for alpha-lipoic acid. II. Evaluation in human volunteers. *Drug Dev Ind Pharm* **30** (1): 35–42, 2004.

Biewenga, G., Haenen, G.R.M.M., and Bast, A. Thioctic metabolites and methods of use thereof. United States Patent 5,925,668 (July 20), 1999.

Breithaupt-Grogler, K., Niebch, G., Schneider, E., Erb, K., Hermann, R., Blume, H.H., Schug, B.S., and Belz, G.G. Dose-proportionality of oral thioctic acid—coincidence of assessments via pooled plasma and individual data. *Eur J Pharm Sci* **8** (1): 57–65, 1999.

Dreisbach, A.W. and Lertora, J.J.L. The effect of chronic renal failure on hepatic drug metabolism and drug disposition. *Semin Dial* **16**: 45–50, 2003.

Gleiter, C.H., Schug, B.S., Hermann, R., Elze, M., Blume, H.H., and Gundert-Remy, U. Influence of food intake on the bioavailability of thioctic acid enantiomers. *Eur J Clin Pharmacol* **50** (6): 513–514, 1996.

Harrison, E.H. and McCormick, D.B. The metabolism of dl-(1,6-14C)lipoic acid in the rat. *Arch Biochem Biophys* **160** (2): 514–522, 1974.

Hermann, R., Niebch, G., Borbe, H.O., FiegerBuschges, H., Ruus, P., Nowak, H., RiethmullerWinzen, H., Peukert, M., and Blume, H. Enantioselective pharmacokinetics and bioavailability of different racemic alpha-lipoic acid formulations in healthy volunteers. *Eur J Pharm Sci* **4** (3): 167–174, 1996.

Hermann, R., Wildgrube, H.J., Ruus, P., Niebch, G., Nowak, H., and Gleiter, C.H. Gastric emptying in patients with insulin dependent diabetes mellitus and bioavailability of thioctic acid-enantiomers. *Eur J Pharm Sci* **6** (1): 27–37, 1998.

Korashy, H.M., Elbekai, R.H., and El-Kadi, A.O.S. Effects of renal diseases on the regulation and expression of renal and hepatic drug-metabolizing enzymes: A review. *Xenobiotica* **34**: 1–29, 2004.

Peter, G. and Borbe, H.O. Absorption of [7,8-14C]rac-a-lipoic acid from in situ ligated segments of the gastrointestinal tract of the rat. *Arzneimittelforschung* **45** (3): 293–299, 1995.

Schupke, H., Hempel, R., Peter, G., Hermann, R., Wessel, K., Engel, J., and Kronbach, T. New metabolic pathways of alpha-lipoic acid. *Drug Metab Dispos* **29** (6): 855–862, 2001.

Spence, J.T. and McCormick, D.B. Lipoic acid metabolism in the rat. *Arch Biochem Biophys* **174** (1): 13–19, 1976.

Teichert, J., Hermann, R., Ruus, P., and Preiss, R. Plasma kinetics, metabolism, and urinary excretion of alpha-lipoic acid following oral administration in healthy volunteers. *J Clin Pharmacol* **43** (11): 1257–1267, 2003.

Teichert, J., Kern, J., Tritschler, H.J., Ulrich, H., and Preiss, R. Investigations on the pharmacokinetics of alpha-lipoic acid in healthy volunteers. *Int J Clin Pharmacol Ther* **36** (12): 625–628, 1998.

Teichert, J., Tuemmers, T., Achenbach, H., Preiss, C., Hermann, R., Ruus, P., and Preiss, R. Pharmacokinetics of alpha-lipoic acid in subjects with severe kidney damage and end-stage renal disease. *J Clin Pharmacol* **45** (3): 313–328, 2005.

12 Modulation of Cellular Redox and Metabolic Status by Lipoic Acid

Derick Han, Ryan T. Hamilton, Philip Y. Lam, and Lester Packer

CONTENTS

INTRODUCTION

Lipoic acid (LA) has recently gained substantial attention as a therapeutic agent and nutritional supplement (Packer et al., 2001; Ames and Liu, 2004; Smith et al., 2004). Exogenous LA supplementation has been reported to decrease ischemia–reperfusion injury in peripheral nerves (Mitsui et al., 1999), brain (Cao and Phillis, 1995), and liver (Muller et al., 2003; Dulundu et al., 2007), and has been shown to be beneficial in the treatment of disorders such as polyneuropathy and diabetes (Bustamante et al., 1998; Packer et al., 2001). In cell culture studies,

LA supplementation has been demonstrated to modulate NF-κB signaling (Suzuki et al., 1992), MAPK signaling (Cho et al., 2003b), phosphoinositide 3-kinase (PI3K)/protein kinase B (Akt) signaling (Muller et al., 2003; Zhang et al., 2007), inhibit tyrosine phosphatase activity (Cho et al., 2003a), inhibit HIV replication (Baur et al., 1991), increase cellular glutathione (GSH) levels (Han et al., 1995b; Han et al., 1997a) as well as promote glucose uptake (Roy et al., 1997; Khanna et al., 1999). The mechanism by which LA can modulate so many activities in cells has not been completely characterized due to the complex chemistry of LA. Structurally, LA is a very unique compound containing (1) a medium length fatty acid tail and (2) a thiolane ring that provides antioxidant functions as well as undergoes redox changes similar to GSH (Figure 12.1). The thiolane ring of LA can be reduced to vicinal thiols (dihydrolipoic acid [DHLA]) in cells utilizing NADH and NADPH reducing equivalents (Handelman et al., 1994; Haramaki et al., 1997). The biological effects of LA may be mainly attributed to LA's (1) antioxidant capacity, (2) fatty acid properties, and (3) redox chemistry.

Antioxidant capacity: Both LA and DHLA have been reported to scavenge numerous reactive oxygen and nitrogen species such as H_2O_2, hydroxyl radical ($HO^•$), hypochlorous acid (HOCl), and peroxynitrite ($ONOO^-$) (Scott et al., 1994; Packer et al., 2001; Moini et al., 2002). Dihydrolipoic acid can also recycle antioxidants such as ascorbate, vitamin E, and ubiquinol, and thus bolster the antioxidant capacity of cells (Packer et al., 1995). In addition, LA supplement has been shown to reduce cellular injury caused by antioxidant deficiency. Lipoic acid

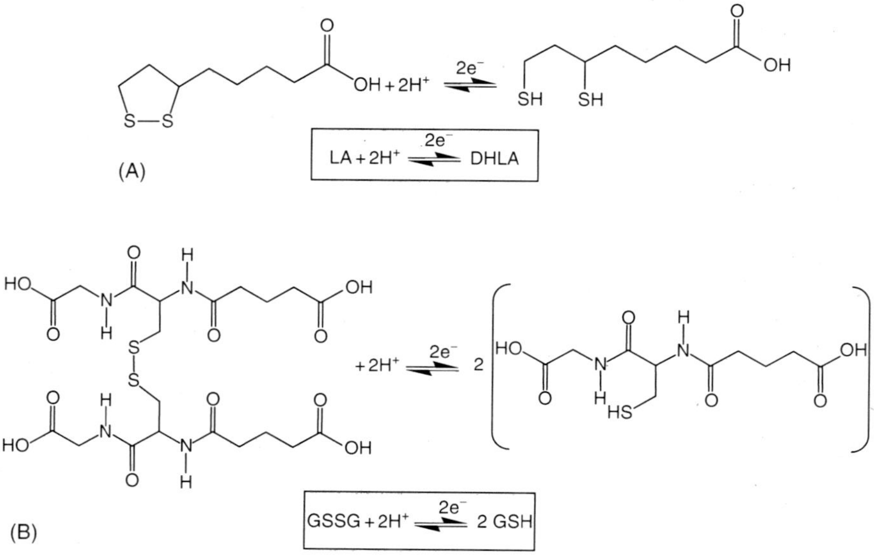

FIGURE 12.1 Structure of lipoic acid and glutathione. Both lipoic acid and glutathione exist in reduced and oxidized forms.

supplementation has been demonstrated to prevent symptoms of vitamin E deficiency in mice (Podda et al., 1994) as well as inhibit neurotoxicity caused by GSH depletion in neuronal cells (Han et al., 1997b). Consequently many biological effects of LA supplementation can be attributed to the antioxidant properties of LA and DHLA. The antioxidant capacity of LA and DHLA has been previously reviewed and will not be extensively discussed in this chapter (Packer et al., 1995; Packer et al., 2001; Moini et al., 2002).

Fatty acid properties: The fatty acid tail of lipoic acid is structurally similar to the medium-chain fatty acid, octanoic acid (Figure 12.2). Two important activities can be ascribed to the fatty acid chain of LA: (1) transport of LA and DHLA in and out of both cells and mitochondria and (2) mitochondrial β-oxidation. Lipoic acid is rapidly transported into cells and mitochondria upon treatment (Handelman et al., 1994; Han et al., 1995a; Han et al., 1997a). Although the membrane transporters have not been extensively characterized, LA transport is partially inhibited by octanoic acid suggesting that a medium-length fatty acid transporter is responsible, at least in part, for LA incorporation into cells and mitochondria (May et al., 2006). Although often overlooked, the efficient transport of LA into cells is largely the result of its fatty acid tail. Thus, the tail of LA plays a significant role in potentiating the biological effects of LA by increasing its uptake and therefore its bioavailability to cells. Most disulfides are not easily transported into cells (Bannai, 1984). Cystine (the disulfide form of cysteine) is important in modulating GSH levels in cells, but is poorly transported into cells through the x_c^- system (Bannai and Tateishi, 1986; Deneke and Fanburg, 1989; Droge and Breitkreutz, 2000). In addition, the tail of LA likely plays an important role in transport of LA across the blood–brain barrier. Other thiol therapeutic agents such as GSH esters appear to poorly cross the blood–brain barrier (Anderson et al., 1985). The ability of LA to cross the blood–brain barrier makes LA a potent brain antioxidant and therapeutic agent (Packer et al., 1997).

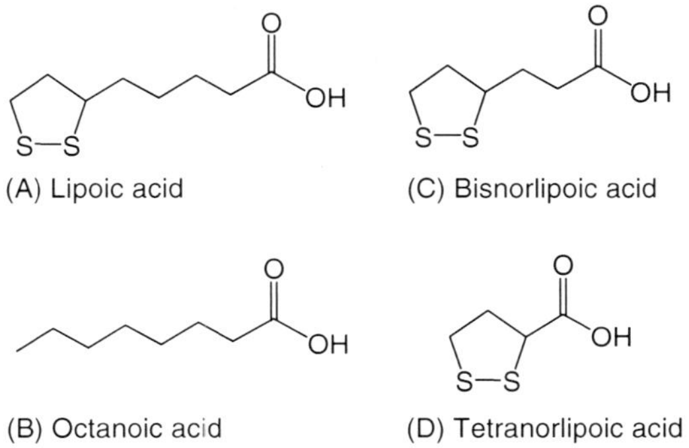

FIGURE 12.2 Structural analogs of lipoic acid. The fatty acid tail of (A) lipoic acid is structurally similar to (B) octanoic acid. Lipoic acid is metabolized by β-oxidation in mitochondria to form (C) bisnorlipoic acid and (D) tetranorlipoic acid.

The fatty acid tail of lipoic acid is also important in metabolism; LA can undergo β-oxidation in mitochondria to generate bisnorlipoic acid and tetranorlipoic acid (Figure 12.2). Bisnorlipoic acid, tetranorlipoic acid, β-hydroxybisnorlipoic acid as well as other metabolic products were formed from the β-oxidation of lipoic acid and were observed in the urine of rats and humans following LA supplementation (Harrison and McCormick, 1974; Schupke et al., 2001). Whether the β-oxidation products of lipoic acid were in the reduced or oxidized form was not investigated in this study. However, we observed that lipoic acid treatment to cells resulted in the formation of both the reduced and oxidized forms of tetranorlipoic acid and bisnorlipoic acid (unpublished results). This suggests that the biological effects of LA supplementation cannot be solely attributed to LA but the biological effects of DHLA, reduced and oxidized tetranorlipoic acid, as well as reduced and oxidized bisnorlipoic acid must be taken into account. Consequently, the metabolites of LA must be taken into consideration when analyzing the effects of LA treatment.

Redox chemistry: The redox reactions of LA are dynamic following its treatment to cells (Handelman et al., 1994; Han et al., 1997a). Lipoic acid is rapidly reduced to DHLA through NADH- and NADPH-dependent pathways including the mitochondrial electron transport chain, thioredoxin and thioredoxin reductase, lipoamide dehydrogenase, and oxidized glutathione (GSSG) reductase (Pick et al., 1995; Arner et al., 1996; Haramaki et al., 1997). Dihydrolipoic acid is a strong reducing agent (−320 mV), and consequently can affect other redox couples in cells, tissues, and extracellular fluids. The LA/DHLA redox potential will strongly influence other thiol-disulfide couples such as cystine–cysteine (−220 mV), GSH–GSSG (−240 mV), and protein thiol-disulfides (i.e., thioredoxin = −200 mV) (Moore et al., 1964; Jocelyn, 1967). Brown et al. using an LA derivative (triphenylphosphonium-conjugated α-lipoyl derivative) suggest that reduction of LA to DHLA is a key feature in protecting cells and mitochondria (Brown et al., 2007). The LA derivative was not observed to be significantly reduced (primarily by thioredoxin) and consequently was less effective than LA in protecting cells and mitochondria from oxidative stress. Thus, many biological effects that lipoic acid induces are mediated through reduction of LA to DHLA and consequent changes in redox status. In addition, redox changes that LA supplementation induces in cells may be important in modulating signaling pathways in cells as well as explain many biological actions of LA supplementation such as promoting glucose uptake in cells. This chapter will focus on the redox modulation of intra- and extracellular environment by LA treatment. In addition, the importance of redox changes elicited by LA in modulating signaling and metabolic pathways in cells will be discussed.

REDOX STATUS OF CELLS AND THE EXTRACELLULAR ENVIRONMENT

Before discussing LA redox chemistry, a brief review of cellular and extracellular redox status will be presented. Cellular or extracellular redox status is estimated by measuring the ratio of interconvertible reduced/oxidized forms of abundant redox molecules such as NAD^+/NADH, $NADP^+$/NADPH, cystine/cysteine, and GSH/GSSG (Schafer and Buettner, 2001; Han et al., 2006b). The GSH/GSSG ratio is the most commonly measured redox couple used to obtain an estimate of redox state in cells. The GSH/GSSG is frequently measured because it is found at high levels in cells (100–10,000 times greater than other redox couples) and because the GSH/GSSG ratio is important in determining redox status of proteins (thiol-disulfide redox status of cysteine residues), which influences protein function and activity. The GSH/GSSG redox is also used to estimate extracellular redox status, however in plasma the cysteine–cystine levels are at much higher levels than GSH/GSSG (~2–5 μM GSH–GSSG versus ~80–120 μM for cysteine–cystine) and thus may be a better estimate of extracellular redox status.

Cellular Redox Status

The cytoplasm of cells is highly reduced under normal conditions with intracellular GSH/GSSG ratios being greater than 100:1. Within the cells, various organelles have different redox potentials. Using the Nernst equation, it has been estimated that the cytosol has a redox potential of −220 to −260 mV ($[GSH]^2/[GSSG]$), whereas mitochondria are more reduced having a potential of −270 mV (Kirlin et al., 1999; Schafer and Buettner, 2001). On the other hand the endoplasmic reticulum has a more oxidized potential of −150 mV (Hansen et al., 2006). Due to the high reducing GSH/GSSG ratio in cytoplasm and mitochondria, cysteine residues in proteins in the cytoplasm and mitochondria generally exists as thiols (Figure 12.3). Protein disulfides may exist intracellularly within proteins

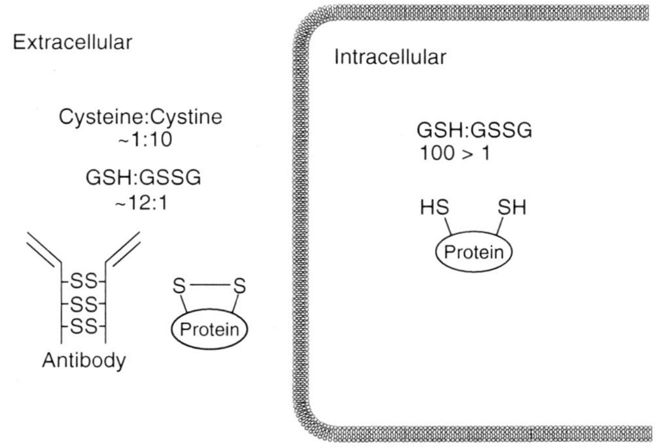

FIGURE 12.3 Redox status of the intra- and extracellular environment. Because of the high GSH/GSSG ratio the intracellular environment is very reduced, with cysteine in proteins being in the reduced thiol state. In contrast, the extracellular environment is oxidized with the majority of free cysteine and cysteine in proteins being in the disulfide form.

if they are inaccessible to GSH and thioredoxin (the major protein reductant in cells). However, almost all exposed cysteine residues in cytoplasmic and mitochondrial proteins are believed to exist in the thiol form, which is important for protein function. Alterations of protein thiols to disulfides or other cysteine oxidation products (sulfenic acid, sulfonic acids, etc) have been found to alter protein function and activity (Klatt and Lamas, 2000; Hansen et al., 2006; Han et al., 2006b).

Reversible thiol modifications in proteins have been suggested to be nanoswitches, to turn on and off proteins, in cells similar to phosphorylation (Schafer and Buettner, 2001; Hansen et al., 2006). Numerous proteins contain critical thiols that undergo a loss or decline in activity if oxidized. For example, redox alterations to the thiols of transcriptional factors such as NF-κB will alter binding of the transcription factor to DNA and consequently transcriptional activity (Matthews et al., 1992; Klatt and Lamas, 2000). Similarly, many kinases (i.e., PKC) and phosphatases are believed to contain redox-sensitive thiol residues that inhibit/modulate activity and thus cell signaling (Kwon et al., 2003; Rhee et al., 2005; Hansen et al., 2006; Han et al., 2006b).

Thiol modifications may also be a mechanism by which cells sense and regulate cellular redox status. The transcription factor, NF-E2 related factor 2 (Nrf2); plays an essential role in response to oxidative stress through the induction of antioxidant genes, including GCL which upregulate GSH levels (Motohashi and Yamamoto, 2004; Zhang, 2006). Changes in cellular redox status caused by GSH depletion by oxidants such as H_2O_2 will cause Nrf2 translocation to the nucleus, transcription of GCL, and increase GSH synthesis to help restore the cellular redox potential. The Nrf2 is normally found in the cytoplasm bound to the inhibitory protein, Kelch-like ECH-associated protein 1 (Keap 1). Thiols in Keap 1 act as redox sensors and oxidation of thiols in Keap 1 thiols will cause Nrf2 dissociation and translocation to the nucleus (Dinkova-Kostova et al., 2002; Wakabayashi et al., 2004). Thus, Keap 1 and Nfr2 act together to sense and respond to changes in cellular redox status by upregulating genes necessary to restore cellular redox status.

Cells expend energy in the form of NADPH to maintain the intracellular environment in a reduced state. The GSH/GSSG ratio is maintained by the enzyme GSSG reductase, which utilizes the reducing power of NADPH to convert GSSG to 2GSH (Kaplowitz et al., 1985). Glutathione can reduce intra- and intermolecular protein disulfides, crosslinks of proteins, which can form during oxidative stress. However, GSH reduction of some disulfide bonds, particularly vicinal disulfides, is relatively slow and is generally catalyzed by thioredoxin in cells. Thioredoxin contains vicinal thiols that react with and reduce disulfide crosslinks (Reactions 12.1 and 12.2).

$$\text{Trx-S}_2\text{H}_2 + \text{Protein-S-S-Protein} \rightarrow \text{Trx-S}_2 + 2\ \text{Protein-SH} \qquad (12.1)$$

$$\text{Trx-S}_2 + \text{NADPH} \rightarrow \text{Trx-S}_2\text{H}_2 + \text{NADP}^+ \qquad (12.2)$$

The selenium enzyme Trx reductase uses the reducing power of NADPH to regenerate Trx-SH_2 from Trx-S_2 (Nakamura et al., 1997; Holmgren et al., 2005).

The GSH/GSSG ratio and protein redox status is thus ultimately tied to NADPH levels, which is determined by energy status of the cell. The thioredoxin redox potential has been estimated to be −280 mV in the cytoplasm, while in mitochondria and the nucleus thioredoxin has been estimated to be more reduced with potentials of −340 and −300 mV, respectively (Hansen et al., 2006).

Glutathione and thioredoxin help to maintain protein thiols in a reduced state even in the presence of steady-state levels of reactive oxygen species generated from mitochondria, the major sources of reactive oxygen species in cells. Oxidative stress has been traditionally defined as "a disturbance in the prooxidant-antioxidant balance in favor of the former" (Sies, 1985). However, given the importance of redox changes in cellular processes, oxidative stress has recently been redefined as "an imbalance between oxidants and antioxidants in favor of the oxidants, leading to a disruption of redox signaling and control and/or molecular damage" (Jones, 2006b), as recently reviewed (Packer and Cadenas, 2007). During oxidative stress the GSH/GSSG ratio can fall (as low as 4:1) to perturb protein redox status as well as cell signaling. Changes in the GSH/GSSG redox potential are believed to regulate proliferation, differentiation, apoptosis, and necrosis in cells through redox regulation of key-signaling proteins (Kirlin et al., 1999; Kwon et al., 2003; Han et al., 2006a). In addition, bursts of hydrogen peroxide from the mitochondrial electron transport chain or NADPH oxidase or both have been shown to occur following treatment with ligand-signaling molecules such as TNF-α. These bursts of hydrogen peroxide have been suggested to be second messengers that modulate protein thiol-disulfide redox status and cell-signaling pathways (Suzuki et al., 1997; Rhee et al., 2005). Consequently, fluxes of reactive oxygen and nitrogen species are important in cell signaling through modulation of redox status in key proteins (Han et al., 2006b).

Extracellular Redox Status

In contrast to the reduced intracellular environment, the extracellular environment is very oxidized. In plasma, glutathione is found at very low levels (~2–5 μM for plasma versus 2–14 mM for cells), and the GSH/GSSG ratio has been estimated to be around ~12:1 (Jones et al., 2000). Cysteine–cystine is found at much higher levels (~100 μM), but the cysteine/cystine redox ratio remains very low (~1:10) (Jones et al., 2002). Proteins in the extracellular environment generally have a higher percentage of disulfide bonds that are important for structural integrity. A classic example is antibodies which contain disulfide bonds that bind light- and heavy-chain components of the immunoglobins. Disulfide bonds are formed in the ER before secretion into the oxidized extracellular environment, which as previously mentioned contain a more oxidized environment due to a lower GSH/GSSG ratio (Hwang et al., 1992; Frand et al., 2000). The extracellular medium of cells in culture is even more oxidized containing only cystine and GSSG (from GSH that is placed in some media formulas). Any thiols placed in media such as cysteine and GSH will eventually become oxidized in the air (21% oxygen) and in a neutral pH environment. Like most thiols, the reactivity of

GSH increases when it becomes deprotonated to form a nucleophilic sulfur ion. Consequently, stabilization of thiols such as GSH during cell or tissue extraction generally requires treating cells or tissues with acidic solutions (i.e., metaphosphoric acid) in order to keep GSH protonated and stable during sample preparation (Han et al., 2000). However, cells in culture generally release thiols such as GSH to slightly reduce the extracellular environment (Bannai and Ishii, 1980; Aw et al., 1986). The redox status of culture media also depends on the density and duration of incubation with cells.

EFFECT OF LIPOIC ACID ON CELLULAR AND EXTRACELLULAR REDOX STATUS

Since the posttranslational modification of thiols on cysteine residues are an important mechanism in modulating cell signaling, LA, a redox modulator, is likely to affect many signaling pathways by altering the redox status of proteins in cells. The disulfide bonds of LA can potentially perturb the cellular redox environment by oxidizing protein thiols and altering signaling pathways. However, LA chemistry is complex, since LA is rapidly converted to DHLA, a strong reductant (Handelman et al., 1994; Han et al., 1997a). Lipoic acid treatment will affect the cellular and extracellular redox status in several sequential stages: (1) LA transport into cells and perturbation of the cellular redox status, (2) LA reduction to DHLA and redox changes caused by DHLA, and (3) DHLA release into the extracellular environment and consequent reduction of the extracellular environment. Protein redox status may be altered by LA/DHLA in each of these stages to modulate cell-signaling pathways.

Lipoic Acid Transport into Cells and Perturbation of the Cellular Redox Status

As previously mentioned, the fatty acid tail of LA allows its rapid transport into cells following supplementation (Handelman et al., 1994; Han et al., 1995a; Han et al., 1997a). Disulfides in the intracellular environment are potentially disruptive due to their ability to undergo thiol-disulfide exchange to alter protein redox status and function (Han et al., 2006b). In this sense, LA can be considered an oxidant (albeit a weak one), since it can potentially oxidize protein disulfides. However, the vicinal disulfide of lipoic acid acts very different than other disulfides such as GSSG. When GSSG levels increase during oxidative stress, protein disulfides are formed through a process known as glutathionylation (Figure 12.4A) (Hurd et al., 2005). Consequently, to avoid glutathionylation cells attempt to keep the steady-state GSSG levels low by reducing GSSG through GSSG reductase or by pumping GSSG out of cells. Glutathionylation of many signaling proteins has been shown to alter protein activity and affect cell signaling (Anselmo and Cobb, 2004; Han et al., 2005; Hurd et al., 2005). Lipoic acid having a vicinal disulfide group can interact with protein thiols, however, is less likely to crosslink with proteins due to the reactivity of the newly generated

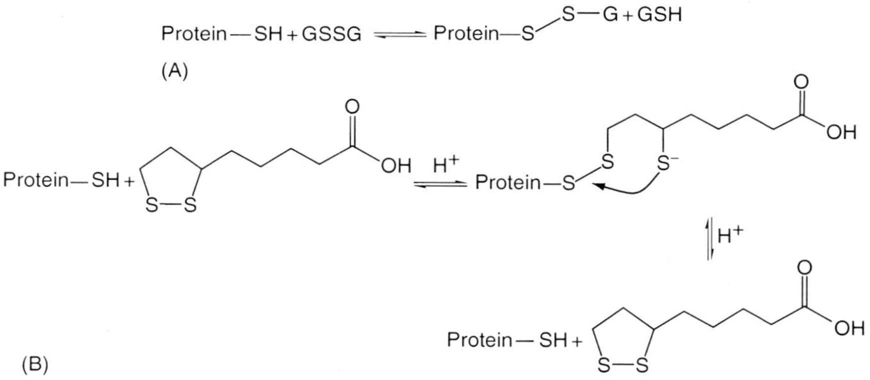

FIGURE 12.4 Reaction of LA and GSSG with protein thiols. (A) GSSG can undergo mixed disulfide linkages with protein thiols by glutathionylation. (B) The vicinal sulfide group of LA makes it less likely to crosslink with protein thiols. The free thiol group generated from the reaction of LA and proteins is likely to reverse any crosslinks formed.

free thiol of LA following disulfide-thiol exchange (Figure 12.4B). Lipoic acid is therefore a less disruptive disulfide than GSSG and is less likely to alter protein redox status. However, if LA concentrations reach high levels in cells, proteins with vicinal thiols and a strong reduction potential such as thioredoxin (−200 mV) could possibly reduce LA to DHLA when driven by a concentration gradient. Thus, LA treatment could initially induce a slightly oxidized intracellular environment and alter signaling pathways through protein disulfide bond formation.

One of the proteins that may be oxidized by LA is Keap 1 of the Nrf2–Keap 1 complex. Lipoic acid administration has been shown to induce Nrf2 translocation to the nucleus and improve GSH synthesis in rats (Suh et al., 2004). As previously discussed, Nrf2 remains bound to Keap 1 in the cytoplasm (Motohashi and Yamamoto, 2004; Zhang, 2006). Oxidation of critical thiols in Keap 1 can cause Nrf2 dissociation and translocation to the nucleus (Dinkova-Kostova et al., 2002; Wakabayashi et al., 2004). Following oxidation of thiols, Keap 1 is ubiquitinated and degraded by proteasomes. The induction of Nrf2 translocation by LA treatment suggests that LA is likely oxidizing thiols in Keap 1 to liberate Nrf2 for translocation to the nucleus. Other redox sensors with critical thiols such as Keap 1 may be potential targets of LA oxidation.

Lipoic Acid Reduction to DHLA and Redox Changes Caused by DHLA

Lipoic acid was rapidly converted to DHLA in all primary cells (e.g., lymphocytes, red blood cells, fibroblasts) and cell lines examined (Handelman et al., 1994; Constantinescu et al., 1995; Han et al., 1995a). In Jurkat cells treated with LA

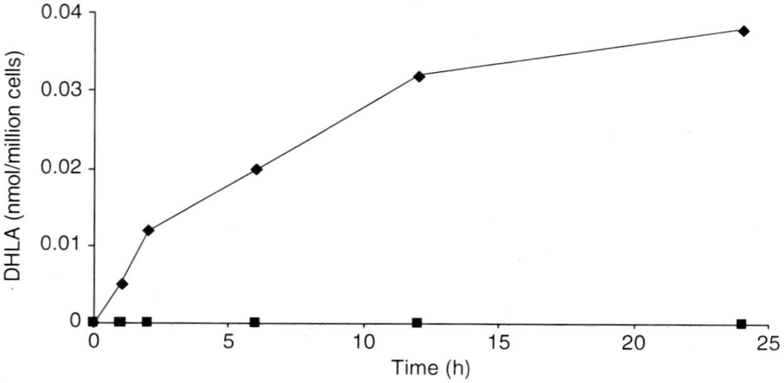

FIGURE 12.5 Accumulation of DHLA in cells following LA treatment. Jurkat cells were incubated with 100 μM LA (◆) or left untreated (control (■)) in fresh RPMI 1640 with 10% serum for 24 h. At various times, cells were harvested and treated with 5% metaphosphoric acid to stabilize DHLA. Dihydrolipoic acid levels were determined by HPLC as previously described.

(100 μM) there was a steady buildup of DHLA in cells (Figure 12.5). The major disulfide reductases in cells, GSSG reductase and thioredoxin reductase, have been demonstrated to convert LA to DHLA by utilizing the reducing power of NADPH (Pick et al., 1995; Arner et al., 1996; Haramaki et al., 1997). Similarly, mitochondria were also shown to convert LA to DHLA utilizing NADH (Haramaki et al., 1997). Cells appear to use a large amount of energy in the form of NADH or NADPH to convert LA to DHLA. Because of its vicinal thiols, DHLA is a strong reductant that shares more characteristics with the dithiol thioredoxin than the monothiol GSH. For example, the rate of GSH reducing proteins with vicinal disulfides is relatively slow because of the reversibility of the reaction (Figure 12.6A). Dihydrolipoic acid, on the other hand, may act like thioredoxin and reduce vicinal disulfides of proteins to their thiol constituents through a nonreversible intermediate (Figure 12.6B). This suggests that the intracellular environment, which may have become slightly oxidized by LA, will become more reduced as DHLA levels accumulate in the cells (Han et al., 1997a).

DHLA Release into the Extracellular Cellular Environment and Consequent Redox Changes

Once DHLA is formed in the cells, a significant portion appears to be released into the medium following its concentration gradient in cells (Handelman et al., 1994; Han et al., 1995a; Han et al., 1995b; Han et al., 1997a). In Jurkat cells, a steady increase of DHLA in the medium was observed following LA treatment (Figure 12.7A). Dihydrolipoic acid accumulates in media, although a significant

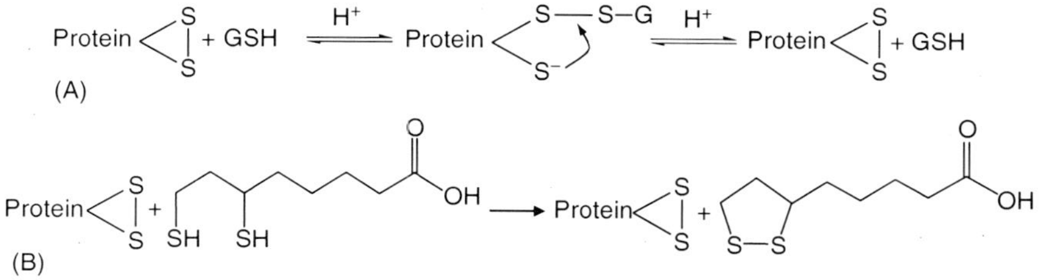

FIGURE 12.6 Reaction of GSH and DHLA with vicinal disulfides in proteins. (A) Glutathione can reduce vicinal disulfides in proteins but the reaction is relatively slow because of the reversibility of the reaction. Protein disulfide reduction requires two GSH molecules. However, the free thiol in proteins can reverse crosslinks between GSH and protein before a second GSH can reduce the bond. (B) Dihydrolipoic acid, on the other hand, being a dithiol can directly reduce protein disulfides simultaneously in a mechanism similar to thioredoxin.

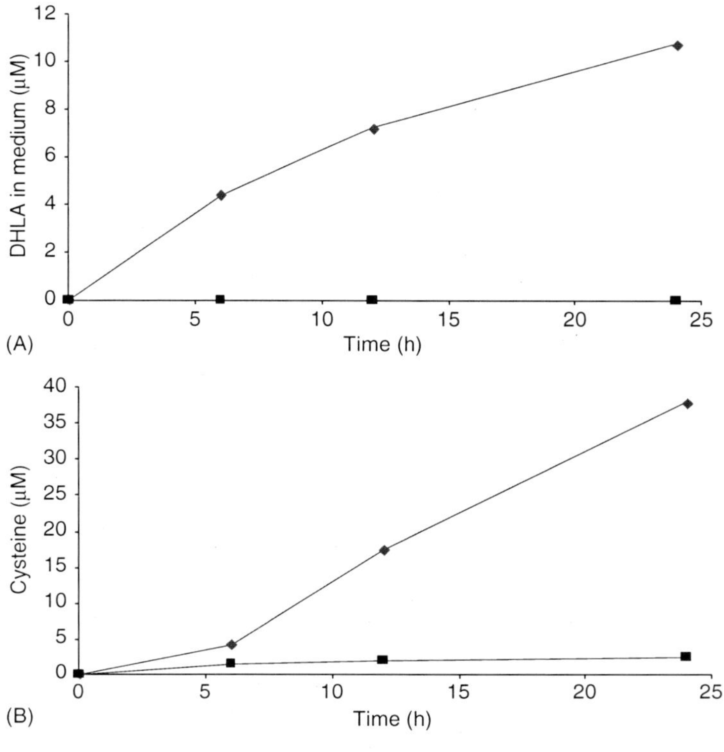

FIGURE 12.7 Accumulation of DHLA and cysteine in media following LA treatment to cells. Jurkat cells were incubated with 100 μM LA (◆) or left untreated (control [■]) in fresh RPMI 1640 with 10% serum for 24 h. At various times, cells were harvested and the media was analyzed by HPLC for DHLA and cysteine.

portion of DHLA will be reoxidized to LA through the following reactions: (1) reduction of disulfides in culture media, especially cystine; (2) auto-oxidation to LA; and (3) reduction of metals and consequent ROS generation.

(1) *Reduction of disulfides in cells, particularly cystine, by DHLA*. DHLA released by cells rapidly reduces disulfides in the extracellular environment, particularly cystine the major disulfide in culture media and plasma (Han et al., 1997a). The reduction of extracellular disulfides by DHLA may affect signaling pathways through several mechanisms: (1) increase cysteine transport and GSH levels in the cells and (2) reduction of disulfide bonds of membrane-bound surface proteins, particularly receptors.

Since cysteine is nonexistent in culture media and found at low levels in plasma, cystine is an extremely important source of cysteine needed for GSH synthesis (Bannai and Tateishi, 1986; Deneke and Fanburg, 1989; Droge and Breitkreutz, 2000). However, cystine needed for GSH synthesis is weakly transported into cells by the x_{c-} transporter in most cells (i.e., lymphocytes). The synthesis of GSH in cells, particularly in culture, is limited by cystine transport through the x_{c-} system. Dihydrolipoic acid released into the medium by cells following LA treatment will reduce cystine to cysteine (Figure 12.7b), which will facilitate the rapid transport of cysteine into cells for GSH synthesis through the efficient ASC system. Thus, in primary lymphocytes and many cell lines, LA treatment was observed to increase cellular GSH levels (>50%) due to extracellular DHLA release, reduction of cystine to cysteine, and the increased transport of cysteine into cells by the ASC system (Figure 12.8) (Han et al., 1997a). This pathway bypasses the weak x_{c-} cystine transporter to increase bioavailability of cysteine for GSH synthesis in cells. Addition of reducing agents such as 2-mercapoethanol, dithiothreitol (DTT), and even GSH to culture media is believed to increase cellular GSH levels through the reduction of cystine to cysteine which is readily transported into cells through the ASC system (Bannai and Tateishi, 1986; Deneke and Fanburg, 1989). Glutathione is not transported into cells, and when added to culture media increases intracellular GSH levels through the reduction of cystine to cysteine. Although LA is not a reducing agent, LA treatment generates a more reduced extracellular medium through the cellular reduction of LA to DHLA, and the release of DHLA into the extracellular environment (Han et al., 1997a).

Many membrane-bound receptors contain disulfide bonds that could be reduced by DHLA, which may alter receptor function and signaling events (Hogg, 2002). A secreted form of thioredoxin has been characterized and suggested to play a role in cell signaling by the reduction of disulfide bonds of extracellular proteins (Pekkari and Holmgren, 2004). Therefore, DHLA released into the extracellular environment may act similarly to secreted thioredoxin and reduce disulfide bonds on membrane-bound proteins and receptors to alter cell signaling.

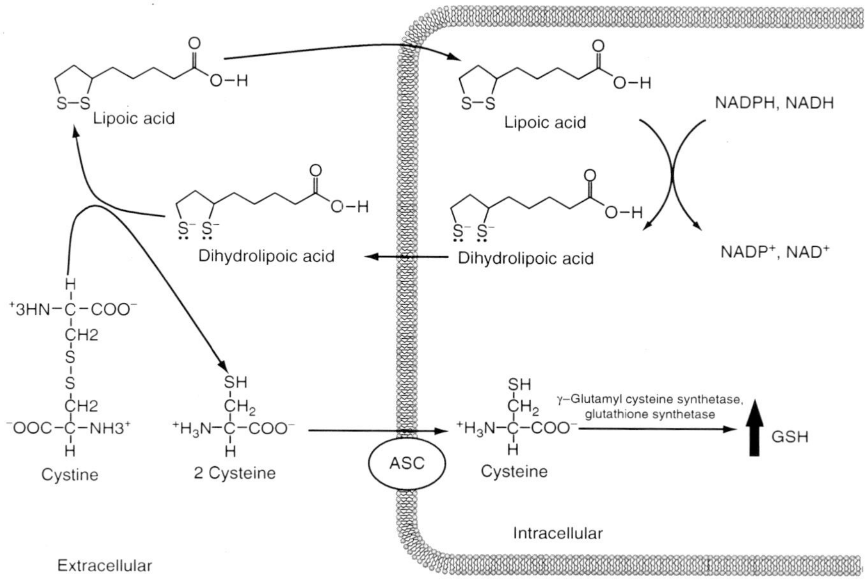

FIGURE 12.8 Mechanism by which lipoic acid treatment increases GSH levels in cells. Upon treatment, lipoic acid is rapidly transported into cells, where it is reduced to DHLA by NADPH or NADH dependent pathways. Dihydrolipoic acid is subsequently released into the media where it reduces cystine to cysteine. Cysteine is efficiently transported into cells by the ASC transporter, while cystine is weakly transported into cells by the x_{c-} system. Since cysteine is the rate-limiting amino acid for GSH synthesis, increased cysteine transport into cells, through the ASC system, increases GSH synthesis.

(2) *Auto-oxidation of DHLA*. Like most thiols, DHLA is not stable in culture medium (pH 7.4) and will rapidly auto-oxidize to form LA. The vicinal thiols of DHLA make it especially prone to auto-oxidation, by an intramolecular disulfide attack. We observed long-term storage of DHLA to be difficult and that some LA formation occurred over time even in acidic conditions and at low temperatures.

(3) *Dihydrolipoic acid reduction of metals and consequently ROS generation*. Dihydrolipoic acid, like many reductants (i.e., cysteine, ascorbate), can potentially reduce metals in media which can react with oxygen to generate superoxide (Moini et al., 2002). Reductants under certain conditions may act as pro-oxidants and promote superoxide and hydrogen peroxide formation (James et al., 2004). Dihydrolipoic acid has been shown to induce reactive oxygen species formation in the presence of transition metals. Culture media and plasma contain various amounts of transition metals, making it likely that DHLA released in the media will promote the formation of reactive oxygen species. Since hydrogen peroxide can activate various signaling pathways in cells, LA may

promote activation or deactivation of cell-signaling pathways through hydrogen peroxide formation in media. However, since DHLA and LA are antioxidants, their ability to scavenge reactive oxygen species such as superoxide and hydrogen peroxide should limit the formation of reactive oxygen species generated from the reaction of metals with DHLA.

OVERVIEW OF LIPOIC ACID REDOX CYCLING IN CELLS

Treatment of cells with LA results in a redox cycle in which LA is reduced to DHLA in cells, released into the medium and subsequently reoxidized to LA, which will reenter cells to repeat the redox cycle (Han et al., 1997a). This LA redox cycle can almost be seen as a "futile cycle" since LA is constantly being regenerated in cells. It remains to be determined whether DHLA levels following LA treatment ever reaches a steady state. The dose of LA administered is likely to be an important factor in determining if steady-state levels of DHLA are achieved in cells. When 100 μM of LA was added to Jurkat cells, DHLA continued to accumulate in the medium and in cells (up to 24 h when the experiments were terminated), suggesting that the rate of LA reduction was faster than DHLA release and oxidation in the media (Han et al., 1995b; Han et al., 1997a). With lower LA doses (<50 μM), it is possible that a DHLA steady-state may be achieved, with most being in the reduced form (DHLA). Regardless of whether a steady state is achieved, redox cycling of LA to DHLA and its auto-oxidation will ultimately decrease upon β-oxidation of LA and possibly DHLA, to bisnorlipoic acid and tetranorlipoic acid (Harrison and McCormick, 1974; Schupke et al., 2001). It remains to be determined if bisnorlipoic acid or tetranorlipoic acid, which lacks the fatty acid tail, is transported into cells and redox cycles. The reduction of LA to DHLA was not cell specific and was observed in all cells including red blood cells, primary lymphocytes, primary fibroblasts as well as in tissue extracts examined (Handelman et al., 1994; Constantinescu et al., 1995; Han et al., 1995a; Haramaki et al., 1997).

Following LA treatment, there are many points in which cellular and extracellular redox may be affected to alter signaling pathways. Initially, LA treatment may result in a more oxidized intracellular environment due to large amounts of LA being transported into cells. Although a weak oxidant, LA could potentially oxidize protein thiols especially those with a low redox potential (i.e., thioredoxin). However, LA treatment does not appear to induce a very oxidized environment since cellular GSH levels steadily increases following LA treatment (Han et al., 1995b; Han et al., 1997a). Accumulation of DHLA and increased GSH levels in cells suggests that a highly reduced intracellular environment occurs with time following LA treatment. The Nernst equation predicts that increases in GSH levels without changes in the GSH/GSSG ratio will increase the redox potential of cells. The increased DHLA and GSH levels could potentially decrease levels of intracellular levels of reactive oxygen species or prevent disulfide bond formation essential for cell signaling. Finally, DHLA in the extracellular environment will reduce disulfide bonds of proteins attached to the membrane, particularly receptors, to modulate signal transduction pathways in cells. Further studies with LA are needed to better characterize the protein redox changes that LA elicits in cells to better understand its mechanism of action in modulating signaling pathways.

MODULATION OF CELLULAR ENERGY STATUS BY LIPOIC ACID

The reduction of LA to DHLA in cells has been shown to occur through both NADPH (GSSG reductase, thioredoxin reductase) and NADH-dependent (mitochondria) pathways (Pick et al., 1995; Arner et al., 1996; Haramaki et al., 1997). This suggests that redox cycling of LA occurs at the expense of cellular energy in the form of NADH and NADPH. Normally cells spend energy in the form of NADPH to maintain a reduced intracellular environment, but LA treatment ultimately causes both the intracellular (increase of GSH and DHLA) and extracellular environment (increase of DHLA and cysteine) to become more reduced. The continuous reduction of LA to DHLA may stress the energy generation capacity of cells particularly at high doses of LA. In Wurzburg cells addition of 0.5 mM LA caused a 30% decrease in NADH levels and 20% decrease in NADPH after 24 h, suggesting an increase in energy expenditure due to the conversion of LA to DHLA (Roy et al., 1997). Associated with the drop in NADH and NADPH levels following LA treatment, was an increase in glucose uptake (Roy et al., 1997). Indeed glucose uptake appears to be a major consequence of LA treatment in most cells (Khanna et al., 1999; Rudich et al., 1999). These findings suggest a possible link between energy used in LA redox cycling and glucose uptake. Decreases in NADH and NADPH levels caused by the reduction of LA to DHLA are likely to be sensed by cells and likely to induce activation of signals that upregulate glucose transport to compensate for the increased demand of NADH and NADPH reducing equivalents. Increases in glucose uptake may be an important compensatory mechanism to maintain NADH and NADPH levels depleted by the continuous reduction of LA to DHLA. NADH is required for ATP production and therefore leaves cells in an energy crisis when NADH and NADPH levels are decreased below a certain threshold.

The decrease in NADH and NADPH levels following LA treatment suggests that β-oxidation of LA is insufficient to replenish energy consumed in the reduction of LA to DHLA. We observed that the reduction of LA to DHLA occurred some ~20 times faster than the β-oxidation of LA in Jurkat cells (unpublished results). Detailed studies examining β-oxidation of LA in cells are warranted to understand its importance in LA metabolism. However, it is clear that β-oxidation is the ultimate fate of LA in vivo and in cultured cells.

LIPOIC ACID AND MITOCHONDRIAL FUNCTIONALITY

Extensive studies conducted in the laboratory of Bruce Ames and colleagues have shown beneficial effects of LA supplementation on mitochondrial function in aging animals. Lipoic acid feeding was shown to increase ambulant activity, mitochondrial respiration, and carnitine acetyltransferase activity as well as decrease oxidative stress in old rats (Lykkesfeldt et al., 1998; Hagen et al., 1999; Hagen et al., 2000). Aging is associated with a decline in redox potential in both cytoplasm and mitochondria, which may modulate protein activity (Cadenas and Davies, 2000; Jones, 2006a). Lipoic acid supplementation has been found to increase GSH levels in aged animals suggesting that LA feeding increases the redox potential of cells and mitochondria (Hagen et al., 1999). Consequently, LA feeding may restore the decline in redox potential that occurs during aging in cytoplasm and mitochondria.

Interestingly, supplementation of rats with both LA and acetyl-carnitine improved mitochondrial function more than supplementation of rats with either compound alone (Hagen et al., 2002; Liu et al., 2002; Ames and Liu, 2004). Carnitine is an important cofactor for the transport of fatty acids into mitochondria for β-oxidation. It is therefore possible that carnitine supplementation helps LA transport into mitochondria for reduction and β-oxidation, or carnitine supplementation may increase the efficiency of β-oxidation of fatty acids to provide reducing equivalents needed for the reduction of LA to DHLA. Whether LA and carnitine work synergistically to increase the redox potential of cells and mitochondria remains to be determined and merits further investigation.

PERSPECTIVE

Although LA is a naturally occurring compound that is synthesized in cells (hence not classified a vitamin), it is only present in trace amounts as a cofactor of the E_2 subunit of the pyruvate dehydrogenase complex and other mitochondrial α-keto acid dehydrogenases (Koike and Reed, 1960). Consequently, cells never experience large amounts of LA unless it is supplemented or provided exogenously. Because of its unique chemical structure, LA will greatly modify the redox and energy status of cells. Lipoic acid is a unique disulfide because its fatty acid tail allows for rapid transport into cells. Once inside cells, the vicinal disulfide of LA readily undergoes redox cycling and modulates the intra- and extracellular redox environment. There has been a debate as to whether LA is an oxidant, a reductant, or an antioxidant in cells. Our work suggests that timing is an important factor. Immediately following LA treatment, LA enters cells and because of its disulfide bonds can potentially oxidize some protein thiols to act as a pro-oxidant. Lipoic acid is rapidly converted to DHLA, a strong reductant, and thus will reduce protein disulfide bonds. Dihydrolipoic acid is released from cells by its concentration gradient and will reduce disulfides in the extracellular environment, particularly cystine. The extensive reduction of LA to DHLA occurs at the expense of NADH or NADPH. Lipoic acid will stress the energy generative capacity of cells and in this sense can be considered a stressor to cells. Although a certain amount of energy is generated from the β-oxidation of LA, constant redox cycling of LA is likely to drain NADH or NADPH energy equivalents of cells. The depletion of NADH or NADPH equivalents caused by redox cycling of LA may be an important factor in the activation of energy generating pathways such as glucose transport into cells. In addition, protein redox alterations induced by LA and DHLA may be an important factor in altering cell-signaling pathways. Consequently, redox cycling of LA to DHLA appears to be a key factor in explaining the multiple effects of LA treatment.

ABBREVIATIONS

DHLA,	dihydrolipoic acid
ER,	endoplasmic reticulum
GSH,	glutathione
GSSG,	oxidized glutathione
Keap 1,	Kelch-like ECH-associated protein 1
LA,	lipoic acid
Nrf2,	NF-E2 related factor 2
TRx,	thioredoxin

REFERENCES

Ames, B.N. and Liu, J. 2004. Delaying the mitochondrial decay of aging with acetylcarnitine. *Ann N Y Acad Sci 1033*, 108–116.

Anderson, M.E., Powrie, F., Puri, R.N., and Meister, A. 1985. Glutathione monoethyl ester: Preparation, uptake by tissues, and conversion to glutathione. *Arch Biochem Biophys 239*, 538–548.

Anselmo, A.N. and Cobb, M.H. 2004. Protein kinase function and glutathionylation. *Biochem J 381*, e1–e2.

Arner, E.S., Nordberg, J., and Holmgren, A. 1996. Efficient reduction of lipoamide and lipoic acid by mammalian thioredoxin reductase. *Biochem Biophys Res Commun 225*, 268–274.

Aw, T.Y., Ookhtens, M., Ren, C., and Kaplowitz, N. 1986. Kinetics of glutathione efflux from isolated rat hepatocytes. *Am J Physiol 250*, G236–G243.

Bannai, S. 1984. Transport of cystine and cysteine in mammalian cells. *Biochim Biophys Acta 779*, 289–306.

Bannai, S. and Ishii, T. 1980. Formation of sulfhydryl groups in the culture medium by human diploid fibroblasts. *J Cell Physiol 104*, 215–223.

Bannai, S. and Tateishi, N. 1986. Role of membrane transport in metabolism and function of glutathione in mammals. *J Membr Biol 89*, 1–8.

Baur, A., Harrer, T., Peukert, M., Jahn, G., Kalden, J.R., and Fleckenstein, B. 1991. Alpha-lipoic acid is an effective inhibitor of human immunodeficiency virus (HIV-1) replication. *Klin Wochenschr 69*, 722–724.

Brown, S.E., Ross, M.F., Sanjuan-Pla, A., Manas, A.R., Smith, R.A., and Murphy, M.P. 2007. Targeting lipoic acid to mitochondria: Synthesis and characterization of a triphenylphosphonium-conjugated alpha-lipoyl derivative. *Free Radic Biol Med 42*, 1766–1780.

Bustamante, J., Lodge, J.K., Marcocci, L., Tritschler, H.J., Packer, L., and Rihn, B.H. 1998. Alpha-lipoic acid in liver metabolism and disease. *Free Radic Biol Med 24*, 1023–1039.

Cadenas, E. and Davies, K.J. 2000. Mitochondrial free radical generation, oxidative stress, and aging. *Free Radic Biol Med 29*, 222–230.

Cao, X. and Phillis, J.W. 1995. The free radical scavenger, alpha-lipoic acid, protects against cerebral ischemia–reperfusion injury in gerbils. *Free Radic Res 23*, 365–370.

Cho, K.J., Moini, H., Shon, H.K., Chung, A.S., and Packer, L. 2003a. Alpha-lipoic acid decreases thiol reactivity of the insulin receptor and protein tyrosine phosphatase 1B in 3T3-L1 adipocytes. *Biochem Pharmacol 66*, 849–858.

Cho, K.J., Moon, H.E., Moini, H., Packer, L., Yoon, D.Y., and Chung, A.S. 2003b. Alpha-lipoic acid inhibits adipocyte differentiation by regulating pro-adipogenic transcription factors via mitogen-activated protein kinase pathways. *J Biol Chem 278*, 34823–34833.

Constantinescu, A., Pick, U., Handelman, G.J., Haramaki, N., Han, D., Podda, M., Tritschler, H.J., and Packer, L. 1995. Reduction and transport of lipoic acid by human erythrocytes. *Biochem Pharmacol 50*, 253–261.

Deneke, S.M. and Fanburg, B.L. 1989. Regulation of cellular glutathione. *Am J Physiol 257*, L163–L173.

Dinkova-Kostova, A.T., Holtzclaw, W.D., Cole, R.N., Itoh, K., Wakabayashi, N., Katoh, Y., Yamamoto, M., and Talalay, P. 2002. Direct evidence that sulfhydryl groups of Keap1 are the sensors regulating induction of phase 2 enzymes that protect against carcinogens and oxidants. *Proc Natl Acad Sci U S A 99*, 11908–11913.

Droge, W. and Breitkreutz, R. 2000. Glutathione and immune function. *Proc Nutr Soc 59*, 595–600.

Dulundu, E., Ozel, Y., Topaloglu, U., Sehirli, O., Ercan, F., Gedik, N., and Sener, G. 2007. Alpha-lipoic acid protects against hepatic ischemia–reperfusion injury in rats. *Pharmacology 79*, 163–170.

Frand, A.R., Cuozzo, J.W., and Kaiser, C.A. 2000. Pathways for protein disulphide bond formation. *Trends Cell Biol 10*, 203–210.

Hagen, T.M., Ingersoll, R.T., Lykkesfeldt, J., Liu, J., Wehr, C.M., Vinarsky, V., Bartholomew, J. C., and Ames, A.B. 1999. (*R*)-Alpha-lipoic acid-supplemented old rats have improved mitochondrial function, decreased oxidative damage, and increased metabolic rate. *FASEB J 13*, 411–418.

Hagen, T.M., Liu, J., Lykkesfeldt, J., Wehr, C.M., Ingersoll, R.T., Vinarsky, V., Bartholomew, J.C., and Ames, B.N. 2002. Feeding acetyl-L-carnitine and lipoic acid to old rats significantly improves metabolic function while decreasing oxidative stress. *Proc Natl Acad Sci U S A 99*, 1870–1875.

Hagen, T.M., Vinarsky, V., Wehr, C.M., and Ames, B.N. 2000. (*R*)-Alpha-lipoic acid reverses the age-associated increase in susceptibility of hepatocytes to *tert*-butylhydroperoxide both in vitro and in vivo. *Antioxid Redox Signal 2*, 473–483.

Han, D., Canali, R., Garcia, J., Aguilera, R., Gallaher, T.K., and Cadenas, E. 2005. Sites and mechanisms of aconitase inactivation by peroxynitrite: Modulation by citrate and glutathione. *Biochemistry 44*, 11986–11996.

Han, D., Hanawa, N., Saberi, B., and Kaplowitz, N. 2006a. Hydrogen peroxide and redox modulation sensitize primary mouse hepatocytes to TNF-induced apoptosis. *Free Radic Biol Med 41*, 627–639.

Han, D., Hanawa, N., Saberi, B., and Kaplowitz, N. 2006b. Mechanisms of liver injury. III. Role of glutathione redox status in liver injury. *Am J Physiol Gastrointest Liver Physiol 291*, G1–G7.

Han, D., Handelman, G., Marcocci, L., Sen, C.K., Roy, S., Kobuchi, H., Tritschler, H.J., Flohe, L., and Packer, L. 1997a. Lipoic acid increases de novo synthesis of cellular glutathione by improving cystine utilization. *Biofactors 6*, 321–338.

Han, D., Handelman, G.J., and Packer, L. 1995a. Analysis of reduced and oxidized lipoic acid in biological samples by high-performance liquid chromatography. *Methods Enzymol 251*, 315–325.

Han, D., Loukianoff, S., and McLaughlin, L. 2000. Oxidative stress indices: Analytical aspects and significance, In *Handbook of Oxidants and Antioxidants in Exercise*, C.K. Sen, L. Packer, and O. Hanninen (Eds.), New York: Elsevier, pp. 433–484.

Han, D., Sen, C.K., Roy, S., Kobayashi, M.S., Tritschler, H.J., and Packer, L. 1997b. Protection against glutamate-induced cytotoxicity in C6 glial cells by thiol antioxidants. *Am J Physiol 273*, R1771–R1778.

Han, D., Tritschler, H.J., and Packer, L. 1995b. Alpha-lipoic acid increases intracellular glutathione in a human T-lymphocyte Jurkat cell line. *Biochem Biophys Res Commun 207*, 258–264.

Handelman, G.J., Han, D., Tritschler, H., and Packer, L. 1994. Alpha-lipoic acid reduction by mammalian cells to the dithiol form, and release into the culture medium. *Biochem Pharmacol 47*, 1725–1730.

Hansen, J.M., Go, Y.M., and Jones, D.P. 2006. Nuclear and mitochondrial compartmentation of oxidative stress and redox signaling. *Annu Rev Pharmacol Toxicol 46*, 215–234.

Haramaki, N., Han, D., Handelman, G.J., Tritschler, H.J., and Packer, L. 1997. Cytosolic and mitochondrial systems for NADH- and NADPH-dependent reduction of alpha-lipoic acid. *Free Radic Biol Med 22*, 535–542.

Harrison, E.H. and McCormick, D.B. 1974. The metabolism of dl-(1,6-14C)lipoic acid in the rat. *Arch Biochem Biophys 160*, 514–522.

Hogg, P.J. 2002. Biological regulation through protein disulfide bond cleavage. *Redox Rep 7*, 71–77.

Holmgren, A., Johansson, C., Berndt, C., Lonn, M.E., Hudemann, C., and Lillig, C.H. 2005. Thiol redox control via thioredoxin and glutaredoxin systems. *Biochem Soc Trans 33*, 1375–1377.

Hurd, T.R., Filipovska, A., Costa, N.J., Dahm, C.C., and Murphy, M.P. 2005. Disulphide formation on mitochondrial protein thiols. *Biochem Soc Trans 33*, 1390–1393.

Hwang, C., Sinskey, A.J., and Lodish, H.F. 1992. Oxidized redox state of glutathione in the endoplasmic reticulum. *Science 257*, 1496–1502.

James, A.M., Smith, R.A., and Murphy, M.P. 2004. Antioxidant and prooxidant properties of mitochondrial coenzyme Q. *Arch Biochem Biophys 423*, 47–56.

Jocelyn, P.C. 1967. The standard redox potential of cysteine–cystine from the thiol-disulphide exchange reaction with glutathione and lipoic acid. *Eur J Biochem 2*, 327–331.

Jones, D.P. 2006a. Extracellular redox state: Refining the definition of oxidative stress in aging. *Rejuvenation Res 9*, 169–181.

Jones, D.P. 2006b. Redefining oxidative stress. *Antioxid Redox Signal 8*, 1865–1879.

Jones, D.P., Carlson, J.L., Mody, V.C., Cai, J., Lynn, M.J., and Sternberg, P. 2000. Redox state of glutathione in human plasma. *Free Radic Biol Med 28*, 625–635.

Jones, D.P., Mody, V.C., Jr., Carlson, J.L., Lynn, M.J., and Sternberg, P., Jr. 2002. Redox analysis of human plasma allows separation of pro-oxidant events of aging from decline in antioxidant defenses. *Free Radic Biol Med 33*, 1290–1300.

Kaplowitz, N., Aw, T.Y., and Ookhtens, M. 1985. The regulation of hepatic glutathione. *Annu Rev Pharmacol Toxicol 25*, 715–744.

Khanna, S., Roy, S., Packer, L., and Sen, C.K. 1999. Cytokine-induced glucose uptake in skeletal muscle: Redox regulation and the role of alpha-lipoic acid. *Am J Physiol 276*, R1327–R1333.

Kirlin, W.G., Cai, J., Thompson, S.A., Diaz, D., Kavanagh, T.J., and Jones, D.P. 1999. Glutathione redox potential in response to differentiation and enzyme inducers. *Free Radic Biol Med 27*, 1208–1218.

Klatt, P. and Lamas, S. 2000. Regulation of protein function by S-glutathiolation in response to oxidative and nitrosative stress. *Eur J Biochem 267*, 4928–4944.

Koike, M. and Reed, L.J. 1960. Alpha-Keto acid dehydrogenation complexes. II. The role of protein-bound lipoic acid and flavin adenine dinucleotide. *J Biol Chem 235*, 1931–1938.

Kwon, Y.W., Masutani, H., Nakamura, H., Ishii, Y., and Yodoi, J. 2003. Redox regulation of cell growth and cell death. *J Biol Chem 384*, 991–996.

Liu, J., Killilea, D.W., and Ames, B.N. 2002. Age-associated mitochondrial oxidative decay: Improvement of carnitine acetyltransferase substrate-binding affinity and activity in brain by feeding old rats acetyl-L-carnitine and/or *R*-alpha-lipoic acid. *Proc Natl Acad Sci U S A 99*, 1876–1881.

Lykkesfeldt, J., Hagen, T.M., Vinarsky, V., and Ames, B.N. 1998. Age-associated decline in ascorbic acid concentration, recycling, and biosynthesis in rat hepatocytes–reversal with (*R*)-alpha-lipoic acid supplementation. *FASEB J 12*, 1183–1189.

Matthews, J.R., Wakasugi, N., Virelizier, J.L., Yodoi, J., and Hay, R.T. 1992. Thioredoxin regulates the DNA binding activity of NF-kappa B by reduction of a disulphide bond involving cysteine 62. *Nucleic Acids Res 20*, 3821–3830.

May, J.M., Qu, Z.C., and Nelson, D.J. 2006. Cellular disulfide-reducing capacity: An integrated measure of cell redox capacity. *Biochem Biophys Res Commun 344*, 1352–1359.

Mitsui, Y., Schmelzer, J.D., Zollman, P.J., Mitsui, M., Tritschler, H.J., and Low, P.A. 1999. Alpha-lipoic acid provides neuroprotection from ischemia–reperfusion injury of peripheral nerve. *J Neurol Sci 163*, 11–16.

Moini, H., Packer, L., and Saris, N.E. 2002. Antioxidant and prooxidant activities of alpha-lipoic acid and dihydrolipoic acid. *Toxicol Appl Pharmacol 182*, 84–90.

Moore, E.C., Reichard, P., and Thelander, L. 1964. Enzymatic synthesis of deoxyribonucleotides.V. Purification and properties of thioredoxin reductase from *Escherichia coli* B. *J Biol Chem 239*, 3445–3452.

Motohashi, H. and Yamamoto, M. 2004. Nrf2-Keap1 defines a physiologically important stress response mechanism. *Trends Mol Med 10*, 549–557.

Muller, C., Dunschede, F., Koch, E., Vollmar, A.M., and Kiemer, A.K. 2003. Alpha-lipoic acid preconditioning reduces ischemia–reperfusion injury of the rat liver via the PI3-kinase/Akt pathway. *Am J Physiol Gastrointest Liver Physiol 285*, G769–G778.

Nakamura, H., Nakamura, K., and Yodoi, J. 1997. Redox regulation of cellular activation. *Annu Rev Immunol 15*, 351–369.

Packer, L. and Cadenas, E. 2007. Oxidants and antioxidants revisited. New concepts of oxidative stress. *Free Radic Res* 41, 951–952.

Packer, L., Kraemer, K., and Rimbach, G. 2001. Molecular aspects of lipoic acid in the prevention of diabetes complications. *Nutrition 17*, 888–895.

Packer, L., Tritschler, H.J., and Wessel, K. 1997. Neuroprotection by the metabolic antioxidant alpha-lipoic acid. *Free Radic Biol Med 22*, 359–378.

Packer, L., Witt, E.H., and Tritschler, H.J. 1995. Alpha-lipoic acid as a biological antioxidant. *Free Radic Biol Med 19*, 227–250.

Pekkari, K. and Holmgren, A. 2004. Truncated thioredoxin: Physiological functions and mechanism. *Antioxid Redox Signal 6*, 53–61.

Pick, U., Haramaki, N., Constantinescu, A., Handelman, G.J., Tritschler, H.J., and Packer, L. 1995. Glutathione reductase and lipoamide dehydrogenase have opposite stereospecificities for alpha-lipoic acid enantiomers. *Biochem Biophys Res Commun 206*, 724–730.

Podda, M., Tritschler, H.J., Ulrich, H., and Packer, L. 1994. Alpha-lipoic acid supplementation prevents symptoms of vitamin E deficiency. *Biochem Biophys Res Commun 204*, 98–104.

Rhee, S.G., Kang, S.W., Jeong, W., Chang, T.S., Yang, K.S., and Woo, H.A. 2005. Intracellular messenger function of hydrogen peroxide and its regulation by peroxiredoxins. *Curr Opin Cell Biol 17*, 183–189.

Roy, S., Sen, C.K., Tritschler, H.J., and Packer, L. 1997. Modulation of cellular reducing equivalent homeostasis by alpha-lipoic acid. Mechanisms and implications for diabetes and ischemic injury. *Biochem Pharmacol 53*, 393–399.

Rudich, A., Tirosh, A., Potashnik, R., Khamaisi, M., and Bashan, N. 1999. Lipoic acid protects against oxidative stress induced impairment in insulin stimulation of protein kinase B and glucose transport in 3T3-L1 adipocytes. *Diabetologia 42*, 949–957.

Schafer, F.Q. and Buettner, G.R. 2001. Redox environment of the cell as viewed through the redox state of the glutathione disulfide/glutathione couple. *Free Radic Biol Med 30*, 1191–1212.

Schupke, H., Hempel, R., Peter, G., Hermann, R., Wessel, K., Engel, J., and Kronbach, T. 2001. New metabolic pathways of alpha-lipoic acid. *Drug Metab Dispos 29*, 855–862.

Scott, B.C., Aruoma, O.I., Evans, P.J., O'Neill, C., Van der Vliet, A., Cross, C.E., Tritschler, H., and Halliwell, B. 1994. Lipoic and dihydrolipoic acids as antioxidants. A critical evaluation. *Free Radic Res 20*, 119–133.

Sies, H. 1985. *Oxidative Stress*, New York: Academic Press.

Smith, A.R., Shenvi, S.V., Widlansky, M., Suh, J.H., and Hagen, T.M. 2004. Lipoic acid as a potential therapy for chronic diseases associated with oxidative stress. *Curr Med Chem 11*, 1135–1146.

Suh, J.H., Shenvi, S.V., Dixon, B.M., Liu, H., Jaiswal, A.K., Liu, R.M., and Hagen, T.M. 2004. Decline in transcriptional activity of Nrf2 causes age-related loss of glutathione synthesis, which is reversible with lipoic acid. *Proc Natl Acad Sci USA 101*, 3381–3386.

Suzuki, Y.J., Aggarwal, B.B., and Packer, L. 1992. Alpha-lipoic acid is a potent inhibitor of NF-kappa B activation in human T cells. *Biochem Biophys Res Commun 189*, 1709–1715.

Suzuki, Y.J., Forman, H.J., and Sevanian, A. 1997. Oxidants as stimulators of signal transduction. *Free Radic Biol Med 22*, 269–285.

Wakabayashi, N., Dinkova-Kostova, A.T., Holtzclaw, W.D., Kang, M.I., Kobayashi, A., Yamamoto, M., Kensler, T.W., and Talalay, P. 2004. Protection against electrophile and oxidant stress by induction of the phase 2 response: Fate of cysteines of the Keap1 sensor modified by inducers. *Proc Natl Acad Sci U S A 101*, 2040–2045.

Zhang, D.D. 2006. Mechanistic studies of the Nrf2-Keap1 signaling pathway. *Drug Metab Rev 38*, 769–789.

Zhang, W.J., Wei, H., Hagen, T., and Frei, B. 2007. Alpha-lipoic acid attenuates LPS-induced inflammatory responses by activating the phosphoinositide 3-kinase/Akt signaling pathway. *Proc Natl Acad Sci U S A 104*, 4077–4082.

13 Redoxin Connection of Lipoic Acid

José Antonio Bárcena, Pablo Porras, Carmen Alicia Padilla, José Peinado, José Rafael Pedrajas, Emilia Martínez-Galisteo, and Raquel Requejo

CONTENTS

INTRODUCTION

Lipoic acid in the form of enzyme-bound lipoamide plays an important role in the mitochondria as a thiol–disulfide redox couple that captures two energetic electrons produced during the highly exergonic decarboxylation of 2-oxoacids or glycine and transfers them to NAD^+, thus contributing to the obtention of metabolically useful reducing power. The reduced form of this redox couple, dihydrolipoamide, is the strongest cellular thiol reductant ($E_o{}' = -0.320$ V).

One undesired consequence for the cell of the thermodynamically favored oxidation of lipoic acid could be its accidental oxidation by any of the oxidants that abound in the mitochondria, including reactive oxygen species (ROS), what could result in the inactivation of the enzymes it is a cofactor of. This adverse effect could be prevented or repaired if any of the mitochondrial thiolic defensive systems could gain access to and reduce the oxidized lipoyl moiety.

On the other hand, the involvement of dihydrolipoamide in redox exchange reactions with other thiol–disulfide couples could be seized by the mitochondria to divert reducing equivalents for thiol–disulfide redox regulatory or defense purposes, provided that a specific mechanism makes it kinetically feasible.

This chapter compiles data demonstrating that lipoamide actually interacts with the glutathione system and with thiol–disulfide redoxins, namely glutaredoxin (Grx), peroxiredoxin (Prx), and thioredoxin (Trx), either to protect itself from unproductive oxidation or to channel part of the energy of central catabolic pathways toward antioxidant defensive purposes.

OXIDATIVE STRESS AND THIOL–DISULFIDE HOMEOSTASIS

Molecular oxygen plays a vital role for the development of cellular functions, mainly the obtaining of energy through respiration, but together with its beneficial role, it bears the unavoidable inconvenience of the formation of ROS. These are superoxide ($O_2^{\bullet-}$), peroxide (H_2O_2), and hydroxyl (•OH). Nitric oxide (•NO) is a gaseous free radical, which is produced inside the cells in response to a great number of stimuli, depending on the cell type. When NO reacts with oxygen, reactive nitrogen species (RNS) like N_2O_3 are formed, which can easily transfer one nitroxonium radical (NO^+) to a nucleophilic center like a sulfur atom (Davis et al., 2001).

The production of ROS is particularly intense in mitochondria where several defensive systems are present either to directly eliminate ROS like superoxide dismutase and peroxidases or to counteract their effects on proteins, like the redoxin systems.

The ROS and RNS do effectively damage the macromolecules and it has been argued that the detrimental effects of ROS on proteins are nonspecific, which is obviously true for •OH, a species that reacts with almost any molecule at very high rates leading to cell injury. It may also be true for other ROS species, particularly, when produced at high concentrations in a global cellular scale, but if confined and at low levels, some ROS like H_2O_2 and $O_2^{\bullet-}$, are known to play a regulatory role, comparable to the well-documented role of •NO. Actually, their reactivity is not as high as their names imply, can be produced endogenously after receptor activation, and are metabolized by specific enzymes. The concept of their acting as second messengers in cell signaling has gained acceptance and is being established in the light of many evidences (Rhee, 1999; Forman et al., 2004; Jones, 2006a).

The action of ROS and RNS on proteins is primarily targeted to cysteines, although other amino acids can also be oxidized. Cysteine is one of the less-frequent amino acids present in the proteome of eukaryotic organisms whose genomes have been sequenced so far (Pe'er et al., 2004). The thiolate anion ($R\text{-}S^-$) is the most powerful nucleophile available among the amino acids present in proteins, what makes it particularly suited for catalytic roles. This is the reason it is commonly found in enzyme active sites and is also well designed for structural and regulatory roles. But on the other hand, its high reactivity toward electrophiles makes it vulnerable to accidental oxidation. Hence, thiols of proteins are especially sensitive to oxidation by ROS entailing undesired loss of functionality. So, it is very important for the cell to maintain the thiol redox homeostasis.

The reactivity of the thiolate anion depends on the pK_a of the Cys residue and is higher at alkaline pH. Concerning the first point, the pK_a of a Cys residue can vary dramatically as a function of its local environment and concerning the second, it must be taken into account that in the mitochondrial matrix the pH value is typically around 8, i.e., more alkaline than in the cytosol.

Reaction of cysteinic thiolates with ROS or RNS may result in thiol modifications with a varied degree of oxidation ranging from sulfenic (-SOH), sulfinic ($-SO_2H$), and sulfonic ($-SO_3H$), to disulfide bridge formation (-S–S-), nitrosylation (-SNO), and conjugation with glutathione (-S–SG). Except for oxidations to sulfinate and sulfonate, the rest of the modifications are easily reverted to recover protein functionality and to keep the thiol redox state at normal values (Figure 13.1). With low molecular weight thiols, sulfenates can form mixed disulfides in a process known as S-thiolation or more specifically, S-glutathionylation when the thiol is

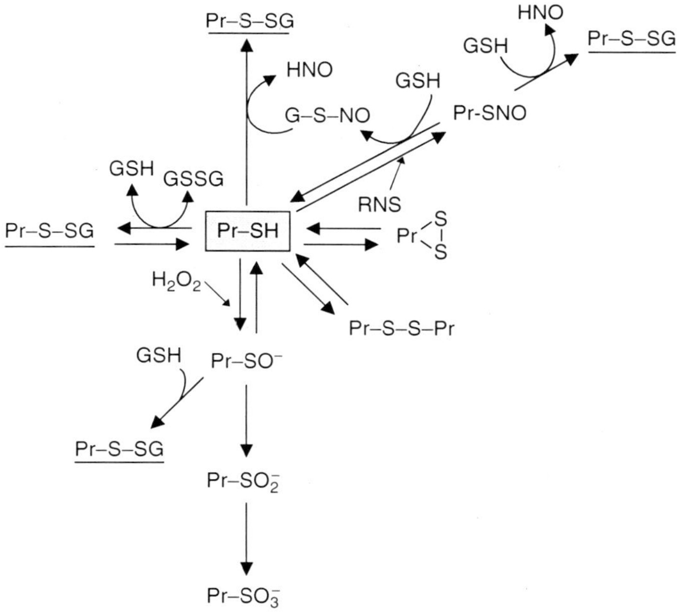

FIGURE 13.1 Redox modifications of protein thiols by reactive oxygen species (ROS) and reactive nitrogen species (RNS). One electron oxidation to thiyl radical has not been included. Mixed disulfide of the protein with glutathione or glutathionylated species are underlined. See main text for further descriptions.

glutathione. S-glutathionylation of proteins can also originate through other mechanisms, but we will come back to this topic later. This process alleviates the effects of oxidative stress, protects thiols against irreversible oxidation, and can be used to regulate the activity of the affected proteins (Cotgreave and Gerdes, 1998; Hurd et al., 2005).

The important task of maintaining the thiol redox homeostasis relies mainly on the catalytic action of redoxins, i.e., the Trx and the glutathione/Grx systems. Both require reducing power, which is provided by nicotinamide adenine dinucleotide phosphate (NADPH) in the first instance. The former is composed of NADPH (also ferredoxin in photosynthetic organisms), and the proteins thioredoxin reductase (TR) and Trx, while the latter is composed of NADPH, glutathione (GSH), and the proteins glutathione reductase (GR) and Grx. Thioredoxin and Grx are oxidoreductases with structural similarities highly conserved through evolution specialized on the reduction of disulfide bonds and mixed disulfides with glutathione, respectively. Another group of redoxins, peroxiredoxins (Prxs), can reduce peroxides in connection with either the Trx or the Grx systems as suppliers of reducing power. All these systems are revised briefly in another section below.

The essentially reversible nature of cysteine oxidative modifications allows them for a role in the modulation of protein function as a redox switch involved in cell signaling. Reversible S-nitrosylation of many proteins is already an established mechanism for protein functionality modulation analogous to phosphorylation (Hess et al., 2001; Mannick and Schonhoff, 2002) and there is a growing list of proteins whose function is regulated through reactive Cys residue modification by H_2O_2 or simply by thiol–disulfide exchange (Nordberg and Arner, 2001; Droge, 2002). This includes transcription factors, enzyme activities, and components of signal transduction cascades (Ghezzi et al., 2005). The regulation of the activity of several photosynthetic enzymes by reversible modification through a thiol–disulfide exchange mechanism specifically mediated by Trx is worth to mention for its pioneering role in this context (Buchanan and Balmer, 2005).

It must be stressed that for specific cysteines to suffer redox modification, the surrounding milieu does not have to experiment a global oxidative stress, that is, the whole cellular redox equilibrium does not have to be altered. According to this concept, that has gained strength in recent years, individual signaling and control events occur through discrete redox pathways rather than through mechanisms that are directly responsive to a global thiol/disulfide balance. Consequently oxidative stress may be better defined as a disruption of redox signaling and control (Jones, 2006b).

SOURCES OF REDUCING POWER FOR DEFENSIVE PURPOSES IN MITOCHONDRIA

The electron transport chain in mitochondria is quantitatively the most significant redox system in aerobic organisms and accounts for perhaps 98%–99% of all O_2 consumption (Jones, 2006a). Hence, mitochondria are the main source of endogenous ROS so that protective mechanisms in these organelles are a must.

It has been stated that free protein thiols are present at higher concentration than GSH in mitochondria and as a consequence, they are prone to act as scavengers of ROS, particularly those that bear a stabilized thiolate (Hurd et al., 2005). Proteins that are oxidatively damaged in that way have to be restored to their functional free thiol state by ad hoc mechanisms. Trx and Grx/GSH systems play a prominent role here as well, but again a source of reducing power is necessary for their action that is canonically provided by NADPH in conjunction with TR and GR, which are strictly NADPH dependent.

The availability of NADPH in the mitochondria is not as obvious as it is in the cytosol, where the pentose phosphate pathway is operative. On the contrary, the other pyridine nucleotide, NADH, is constantly produced at high rate in active mitochondria. Since this form of reducing power is not useful for NADPH-specific antioxidant defensive purposes, the mitochondria has evolved other mechanisms to transfer the reducing capacity of NADH to NADPH, i.e., isocitrate dehydrogenase (Jo et al., 2001), pyridine nucleotide transhydrogenase (Hatefi and Yamaguchi, 1996), and NADH kinase (Outten and Culotta, 2003).

The energy-transducing nicotinamide nucleotide transhydrogenases of mammalian mitochondria and bacteria are structurally related membrane-bound enzymes that catalyze the direct transfer of a hydride ion between NAD(H) and NADP(H) in a reaction that is coupled to transmembrane proton translocation. Hence, $NADP^+$ reduction by NADH is coupled to energy production in mitochondria. In the presence of uncoupler reagents, reduction of $NADP^+$ is clearly unfavored (Hatefi and Yamaguchi, 1996).

Since the redox potential, E_o', of both redox pairs is the same (−320 mV), this reaction has an equilibrium constant, $K_{eq} = 1$ (Scheme 13.1a). With these thermodynamic parameters, the reaction will transfer electrons in either direction, depending on the current $NAD(P)^+/NAD(P)H$ ratios of both pairs. However, equilibration between NADH and NADPH does not and must not occur in mitochondria neither under normal respiring conditions not, even, under oxidative stress (Tribble and Jones, 1990). Moreover, their respective total levels and their redox state are far from similar (Sies, 1982), which is difficult to reconcile with the operating of this system.

The product of the *POS5* gene in *Saccharomyces cerevisiae* has been shown to code for an NAD(H) kinase that localizes to the mitochondrial matrix where it provides NADPH and contributes to supply this reductant for defensive and other metabolic purposes (Outten and Culotta, 2003). Mutants lacking the gene are strongly sensitive to oxidative stress conditions and suffer from iron homeostasis imbalance. This system for NADPH production in mitochondria will work in the direction of NADPH formation when the concentration of NADH is higher than that of NADPH, but independent of the redox potentials, since no redox reaction is involved (Scheme 13.1b). However, it was proposed that NADPH synthesis in the mitochondria through this system would require the concurrent activity of $NADP^+$ phosphatase so that NAD^+ is produced and introduced back into the NADH-generating pathways, thus avoiding depletion of the NADH pool as determined by morphological observations (Outten and Culotta, 2003).

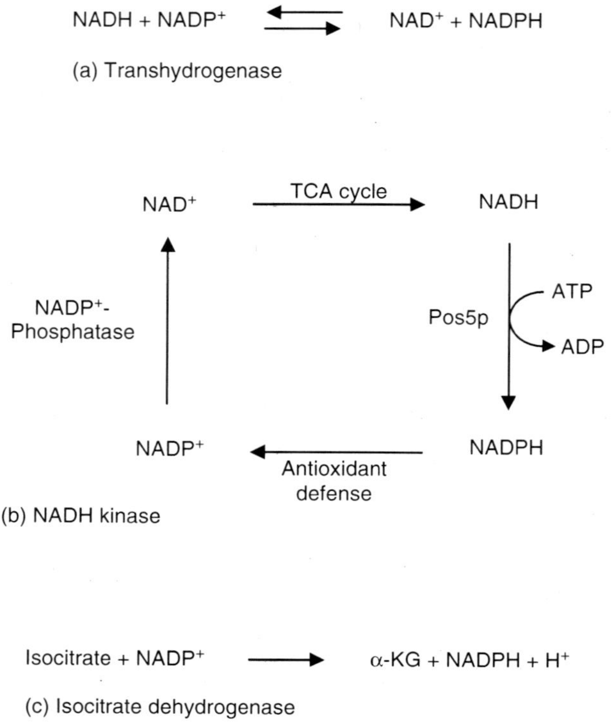

SCHEME 13.1 Sources of NADPH in the mitochondria. See main text for description.

In mammalian mitochondria, isocitrate dehydrogenase (mICD) has been reported to play a role in the defense against oxidative damage to cellular macromolecules and to mitochondrial injury in particular (Jo et al., 2001) (Scheme 13.1c). The enzyme is induced by ROS and has an influence on NADPH and GSH levels in the mitochondria. However, the role of this enzyme in the yeast *S. cerevisiae* may be different since suppression of mIDH by deletion of the corresponding gene had no effect on oxidative stress sensitivity (Minard et al., 1998).

ENZYME-BOUND LIPOIC ACID AS AN ACTOR IN THIOL REDOX HOMEOSTASIS SCENE

Alternative or complementary mechanisms as sources of reducing power cannot be discarded if we take into account the tremendous task of maintaining redox homeostasis in the mitochondria. It has been calculated that electron flow through NADPH pathways in hepatocyte mitochondria is only 12% that of NADH

(Tribble and Jones, 1990). It will be advantageous for the cell if a mechanism exist that could allow to directly tap reducing power from the catabolic mainstream instead of relying on the availability of the final product, NADH and its conversion to NADPH. A good candidate to this end would be lipoic acid or more precisely, its physiological form, enzyme-bound lipoamide.

Lipoic acid, in the form of lipoamide, is covalently bound to the E2 subunit of 2-oxoacid dehydrogenase complexes (ODCs) and to protein H of the glycine cleavage system (GCS). It plays important functions in the energetic and biosynthetic mitochondrial metabolism. In both types of reactions, lipoic acid is reduced to partially capture the energy produced during the decarboxylation of 2-oxoacids or glycine so as to channel it toward the production of useful reducing power in the form of NADH. The very same flavoenzyme, lipoamide dehydrogenase (LPD), catalyzes this reaction in both systems (Perham, 2000; Douce et al., 2001).

Here, we will center our attention on the thiol–disulfide nature of this important cofactor, considering that the very low redox potential of the lipoyl redox pair, dihydrolipoic acid—or its functional form, dihydrolipoamide—is the most powerful thiolic cellular reductant and could be very useful in other roles if its reducing power could be transferred to the thiol redox pool. Several reasons support this hypothesis.

1. The midpoint redox potential of the lipoamide/dihydrolipoamide couple calculated from equilibrium with NAD^+/NADH in the presence of LPD is $E_o' = -290$ mV (Sanadi et al., 1959) but $E_o' = -320$ mV as determined by polarography (Ke, 1957), a value which is more compatible with later experimental results (Jocelyn, 1967). This value is as high as that of the pyridine nucleotide pairs $NAD(P)^+$/NAD(P)H and likely higher than that of NAD^+/NADH under in vivo conditions, where the NAD^+/NADH ratio is above 10. Dihydrolipoamide is then a reductant strong enough to fulfill protective roles in the mitochondria provided there are appropriate connecting mechanisms with the thiol redox pool.
2. The thiolic nature of lipoamide makes it particularly suitable for thiol-disulfide redox exchange with the Trx and Grx/GSH systems. For instance, dihydrolipoamide can reduce Trx with fast kinetics (Holmgren, 1979). That disulfides might be reduced by exchange with lipoic acid was suggested many years ago based on the fact that a number of disulfides were reduced in mitochondria by exchange with reduced lipoic acid, which is generated by the oxidation of 2-oxoacids or by reduction of lipoic acid by NADH in the presence of LPD (Skrede, 1968).

However, these results could not be extrapolated to GSSG reduction since working with mitochondrial suspensions, addition of Krebs' cycle acids restored the levels of GSH from GSSG but precisely the oxoacids oxaloacetate, pyruvate, and 2-oxoglutarate did not (Jocelyn, 1975). The experimental approach included pretreatment of the isolated mitochondrial preparations with *t*-butylhydroperoxide that could have led to unnoticed inactivation of the 2-oxoacid complexes by ROS

(Bunik and Sievers, 2002), an undocumented fact by that time (Jocelyn, 1975). Another likely explanation already given by the author was that GSSG could have an inhibitory effect by the formation of mixed disulfides, nowadays named glutathionylation, with the enzymes –SH groups (Jocelyn, 1975).

On the other hand, it is known that GSH and lipoic acid do not show particularly rapid thiol–disulfide exchange rates although it could be amenable to catalysis (Forman et al., 2004). Other unexplained results have been reported that show up differences between the behavior of KGDH and PDH assayed in situ as compared to the purified complexes (Nulton-Persson et al., 2003). However, more recent experiments in other laboratories have shown pyruvate (J. Peinado, unpublished observations) and NADH (Haramaki et al., 1997) dependent GSSG reduction in isolated mitochondrial preparations.

Whatever is the case, the unresolved discrepancies justify the search for definitive evidences to shed light on this issue.

3. The lipoyl group of E2 in 2-oxoacid dehydrogenases is bound to a lysine residue in the protein thus forming a long flexible arm, which in its reducing form travels a long distance, estimated ≈200 Å, between E1 and E3 during the catalytic cycle (Perham, 2000) (Figure 13.2).

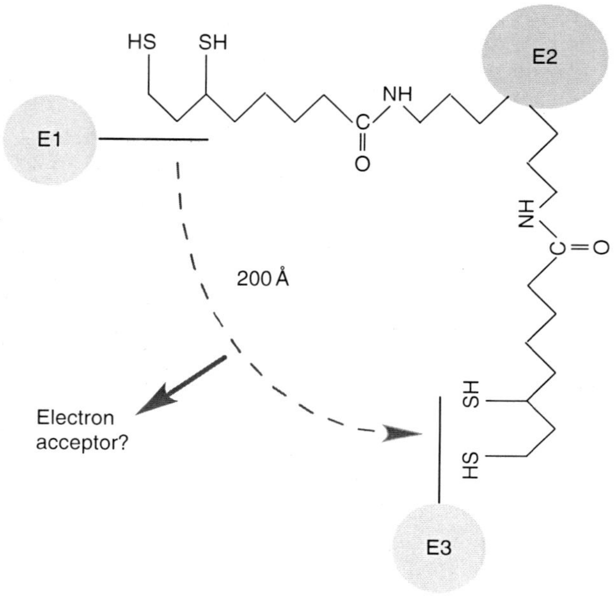

FIGURE 13.2 Oscillation of the lipoyl arm between subunits E1 and E3 in the oxoacid dehydrogenase complex (ODC). The possible transfer of reducing equivalents to a suitable electron acceptor has been enhanced.

Moreover, the binding lysine residue lays itself in a protruding beta turn region on the surface of the lipoyl domain of E2. To add up to this argument, the lipoyl domains are present in tandems of several units linked by flexible regions, so that the outermost domain can reach the surface of the complex despite the fact that its apoenzyme E2 constitutes the very core of the large structure (Guest et al., 1997). All these characteristics might render the lipoyl unit accessible to interacting protein partners like specific or housekeeping redoxins. This point is further discussed below.

4. The lipoyl groups in these enzymes are thought to be present in an unnecessary excess relative to the needs for optimal electron transfer (Reed and Hackert, 1990). Abnormal acyl group transfer reactions between neighboring lipoyl residues are also known to occur (Danson et al., 1978), so that one can envisage a similar transfer of electrons from these lipoyl residues to appropriate electron acceptors to download its reducing power during transit.
5. Under conditions of low NAD^+, reduced lipoamide would not be oxidized and could accumulate; then it could get oxidized in the wrong direction toward irreversible or ROS generating redox states, thus posing the risk of damage to the enzyme and neutralization of the whole complex activity (Bunik and Sievers, 2002). A controlled secure way to alleviate the electrons surcharge of dihydrolipoyl residues toward useful commitments would help in preserving the integrity of the complex.

Before proceeding to describe and discuss the exact connections between lipoic acid/lipoamide and the redoxin proteins, let us make a brief introduction to this family of proteins and, particularly, to their mitochondrial counterparts together with other thiolic systems also present in mitochondria.

MITOCHONDRIAL REDOXIN PROTEINS AND OTHER THIOLIC SYSTEMS

Specific mitochondrial Trx, Prx, and Grx systems have been characterized recently in every type of cell, including yeast (Pedrajas et al., 1999). The possible role of these oxidoreductases as cellular redox homeostasis thiolic systems against oxidative stress has attracted much attention, but they are also involved in regulatory processes affecting mitochondrial function (Hurd et al., 2005). Oxidative processes in mitochondria initiate or are concomitant with a series of phenomena that may lead to activation of the cellular apoptotic program (Cai and Jones, 1998). In general, oxidative processes favor the onset of neurodegenerative diseases and aging (Beal, 1995).

Grx and Trx are small proteins, 9–16 kDa, belonging to the "Trx/Grx fold proteins" superfamily (Martin, 1995), ubiquitous in all living beings with thiol–disulfide oxidoreductase activity and an active site with the CXXC motif. The pK_a and the nature of the central dipeptide $-X_1-X_2-$, which is conserved in each subfamily, are the determining factor for the redox properties of theses proteins, although other structural elements also have an influence (Holmgren, 1989; Nordberg and Arner, 2001; Fernandes and Holmgren, 2004).

Glutaredoxins

Glutaredoxins have arisen recently as important mitochondrial oxidoreductases, related to different metabolic processes in this organella. These small oxidoreductases receive reducing power from glutathione and can catalyze several thiol reactions through both monothiol and dithiol mechanisms (Holmgren, 1989) (Scheme 13.2).

Through these mechanisms, Grx can act on a number of different substrates, having a role in various cellular processes, such as electron donor for ribonucleotide reductase, dehydroascorbate reduction, regulation of several enzymatic activities and DNA binding of transcription factors, taking part of different redox signaling mechanisms, or assembly of Fe–S clusters (Fernandes and Holmgren, 2004). One of their most prominent functions is glutathionylation/ deglutathionylation of proteins, an antioxidant and regulatory process that is

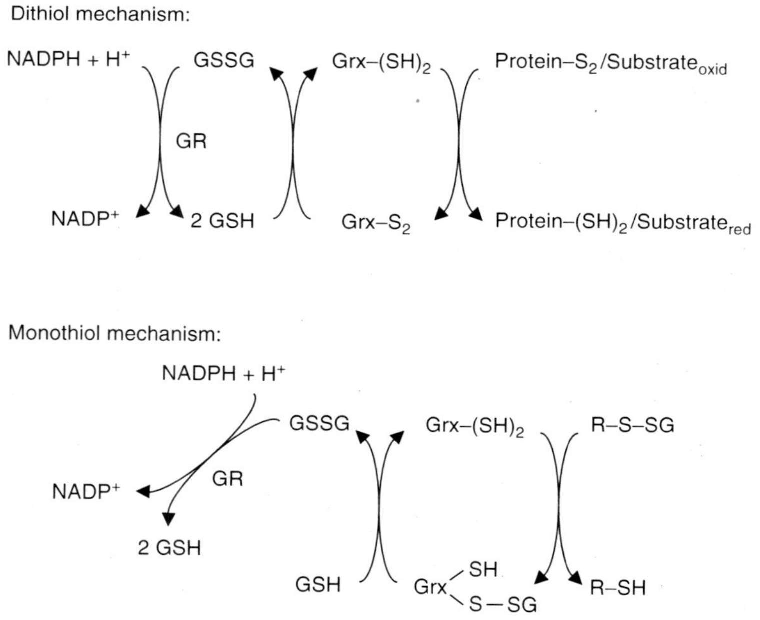

SCHEME 13.2 Reaction mechanisms of Grx. In the monothiol mechanism, one of the Cys at the active site is dispensable, whereas in the dithiol mechanism, an internal disulfide is formed at the active site during the catalytic cycle.

catalyzed by Grxs through the monothiol mechanism (Chrestensen et al., 2000; Murata et al., 2003).

Glutaredoxins are classified into three groups (Fernandes and Holmgren, 2004):

- Canonical dithiolic Grxs: the most abundant group is formed by proteins 10–12 kDa, with typical active site CPY(F)C and the typical three-dimensional structure of redoxins, called the redoxin fold.
- Monothiolic Grxs: the monothiolic Grxs lack the C-terminal cysteine of the active site. They are larger proteins (from 15 to 30 kDa) without oxidoreductase activity.
- Glutaredoxins structurally related to GR: they have both Grx and glutathione transferase activities and the main example is Grx2 from *E. coli*, a protein with two different structural domains, one with the typical redoxin fold and another one with α-helix structure.

Mitochondrial Grxs have been best studied in *S. cerevisiae* and human. Grx2 and Grx5 of the former organism share mitochondrial localization (Pedrajas et al., 2002; Rodriguez-Manzaneque et al., 2002). Two isoforms of Grx2 have been found in the mitochondria: one in the matrix and another, slightly larger, anchored to the outer membrane. There are also cytosolic and endoplasmic reticulum isoforms of this protein; the mechanism responsible for this diversity of localizations of the product of one single transcript is due to the conjunction of two phenomena: translation initiation at two different start sites and inefficient mitochondrial translocation (Porras et al., 2006). The functional meaning of such a complex distribution is still unclear, though it is tempting to assign a potential role on yeast respiratory physiology to the membrane form. Grx5 is a mitochondrial matrix protein that has a seminal role in iron–sulfur cluster assembly and has been found also in humans and other organisms such as the zebra fish (Wingert et al., 2005). The double mutant grx2 grx5 is nonviable in *S. cerevisiae*, showing that the presence of at least one mitochondrial Grx is essential in this organism (Rodriguez-Manzaneque et al., 1999).

Human Grx2 is a canonical Grx with a peculiar active site CSYC and also has both cytosolic and mitochondrial localization, though its distribution on two different compartments comes from alternative splicing (Gladyshev et al., 2001; Lundberg et al., 2001). The active site of Grx2 confers it an unusually high deglutathionylation capacity and the ability to receive reducing power from TR (Johansson et al., 2004). Recently, it has been found that this protein, under normal reducing conditions, forms a non-active dimer that is monomerized under oxidant conditions. The analysis of the dimer revealed the presence of a four cysteine-coordinated, non-oxidizable $[2Fe–2S]^{2+}$ cluster; the human Grx2 being the first redoxin to show such kind of structure and cofactor (Lillig et al., 2005; Johansson et al., 2006). These characteristics along with its mitochondrial localization confer this protein an important implication on defense against apoptosis induced by ROS action.

Thioredoxins

Thioredoxins, which have a low redox potential, reduce protein disulfides utilizing two cysteine residues in their Cys-Gly-Pro-Cys active site. The resulting active site disulfide in Trx is reduced by TR and not by GSH, at difference with Grxs.

The Trx system, consisting of Trx, TR, and NADPH (Scheme 13.3), plays an important role in maintaining the redox state of the cell (Holmgren, 1985). The Trx system is widely distributed in prokaryotes an eukaryotes and is involved in many cellular functions including synthesis of deoxyribonucleotides (Laurent et al., 1964), redox control of numerous transcription factors, protection against oxidative stress (Nakamura et al., 1992), and cell growth and cancer (Zhao et al., 2006). There are Trx and TR isoforms differing in their molecular properties and subcellular or tissular localization. Reversible redox regulation of the activity of several enzymes of the carbon assimilatory metabolism in plants by Trx is a well-established topic (Buchanan and Balmer, 2005). The Trx system plays an important antioxidant role in the mitochondria either as reductant of disulfide bridges, whose substrates include proteins affected by oxidative damage or proteins suffering thiol–disulfide changes as part of a regulatory mechanism. The Trx system also plays an important antioxidant role by providing reducing power to Prxs, what has been demonstrated to occur in mitochondria (Pedrajas et al., 1999; Pedrajas et al., 2000) and may suggest a defense system against H_2O_2 produced by the mitochondrial respiratory chain (Lee et al., 1999). Surprisingly, recent studies have evidenced the role of the Trx system in the protection against reductive stress that results in ribosome aggregation. This function is dependent on its activity as a cofactor for the Tsa peroxiredoxin (Rand and Grant, 2006).

Three mammalian TR isoenzymes have been described: cytosolic TR1 (Gasdaska et al., 1995; Holmgren, 1977), mitochondrial TR2 (Lee et al., 1999; Miranda-Vizuete et al., 1999a; Miranda-Vizuete et al., 1999b), and the testis-specific TGR (Trx GSH reductase) (Sun et al., 2001). Both TR1 and TR2 are also expressed as different isoforms derived from alternative splicing, resulting in a highly complex and cell type-specific expression pattern. They are flavin homodimeric oxidoreductases with an essential redox-active cysteine-selenocysteine (SeCys) conserved at their C-terminal part (Mustacich and Powis, 2000; Zhong et al., 2000), which is the

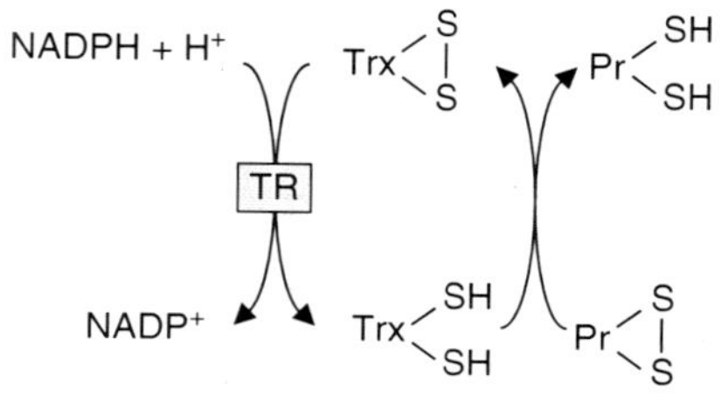

SCHEME 13.3 Electron transfer chain at the Trx system. See main text for further details.

reason for TRs' wide substrate specificity, including non-disulfide compounds as selenite and peroxides (Zhong and Holmgren, 2000).

Mammalian cells contain two distinct Trxs. Trx1 is a cytosolic or nuclear protein, which can be exported and can act as cytokine or chemokine (Pekkari and Holmgren, 2004), whereas Trx2 is targeted to mitochondria (Spyrou et al., 1997). Recent studies show that Trx2 is essential not only for mitochondrial-dependent apoptosis (Tanaka et al., 2002) but also for cell growth and mammalian development (Nonn et al., 2003).

Peroxiredoxins

Peroxiredoxins are ubiquitous antioxidant proteins that were evidenced in the late 1980s (Kim et al., 1988). They use redox-active cysteines for reducing a broad range of peroxides including peroxynitrite, obtaining reducing equivalents from a variety of thiols. Each Prx contains a conserved cysteine residue near the N-terminal that exerts a nucleophilic attack on a peroxide molecule, leading to a sulfenic group (Scheme 13.4). Later, the sulfenic group reacts with other thiol, generally from other cysteine belonging to the very same Prx molecule to form a disulfide bridge. Finally, the disulfide is reverted to sulfhydryl by the acceptance of a pair of electrons from an exogenous thiol compound. Most Prxs receive reducing equivalents from NADPH

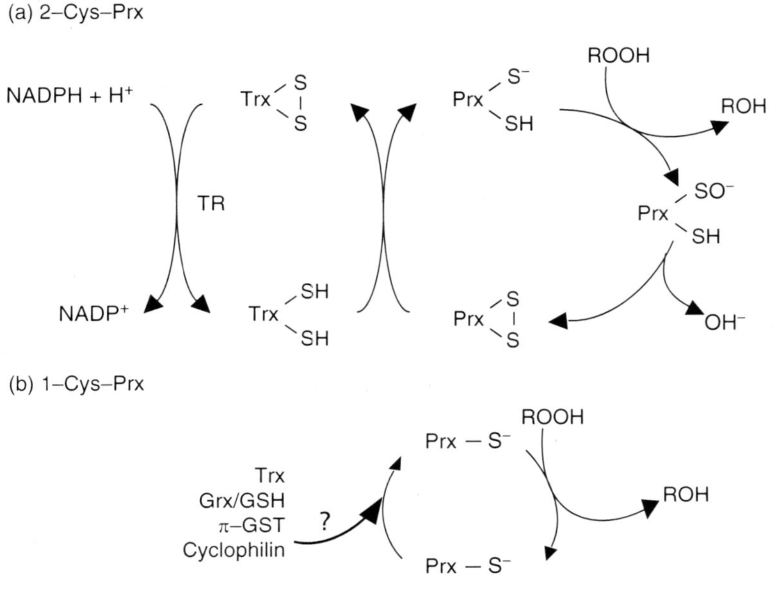

SCHEME 13.4 Reaction mechanism at the active site and electron transfer chain of Thioredoxin Peroxidases. (a) Mechanism of 2-Cys Prx belonging to groups I, II and III. (b) Simplified scheme to indicate that electron donors are diverse and controversial in the cases of various 1-Cys Prx. See text for further discussion.

through the Trx system and, therefore, Prxs have also been termed thioredoxin peroxidases (TPxs) (Chae et al., 1994a; Chae et al., 1994b).

Peroxiredoxins have been classified into four groups according to both structural and mechanistic differences (Verdoucq et al., 1999; Dietz, 2003; Wood et al., 2003b). Type-I TPxs or typical 2-Cys-Prxs are denominations for a group of Prxs that require a homodimeric structure for their functionality. Type-II TPxs, also termed atypical 2-Cys-Prxs, are another group of Prxs whose members can form homodimers but each monomer has an independent catalytic site. Another group of Prxs, named "homologous to BCP" or PrxQs, has been characterized in microorganisms and plants and is structurally and functionally monomeric. The Prxs from all three groups accept reducing equivalents from the Trx system to reduce hydroperoxides, and each monomer contains two catalytic cysteines necessary for TPx activity (Jeong et al., 1999; Jeong et al., 2000; Park et al., 2000).

The 1-Cys-Prxs constitute the fourth group of Prxs, homodimers that contain only just one conserved catalytic cysteine per monomer. The nature of their electron donor is not yet clear, but it is generally accepted that these peroxidases are not reduced by Trxs. This holds true for mammalian 1-Cys-Prxs, but surprisingly, the 1-Cys-Prx from *S. cerevisiae* showed TPrx activity (Kang et al., 1998; Pedrajas et al., 2000; Peshenko and Shichi, 2001). Other electron donors, like glutathione or cyclophilin A, have been proposed for mammalian 1-Cys-Prxs (Chen et al., 2000; Lee et al., 2001).

Eukaryotic cells usually contain several isoforms of Prxs, differentially distributed in various subcellular compartments from nucleus and cytosol, to peroxisomes, chloroplasts, and mitochondria. The type of peroxiredoxin and the organelle it is localized in is different depending on the organism (Kang et al., 1998; Jeong et al., 1999; Dietz, 2003). For instance, mitochondria from *S. cerevisiae* contain a 1-Cys-Prx (Prx1p), which is the only 1-Cys-Prx characterized so far that consumes reducing equivalents from NADPH through the Trx system to carry out the reduction of hydroperoxides (Pedrajas et al., 2000). However, in mammals, one typical and another atypical 2-cys-Prx, named PRX III and PRX V, respectively, are localized in mitochondria and both are reduced by Trx (Watabe et al., 1994; Knoops et al., 1999).

Peroxiredoxins have been considered as simple peroxidases involved in the defense against oxidative stress, but their kinetic characteristics have raised some questions about their specific biological function. Prxs have lower catalytic efficiencies ($\sim 10^{-5}$ M^{-1} s^{-1}) compared to glutathione peroxidases ($\sim 10^{-8}$ M^{-1} s^{-1}) and catalases ($\sim 10^{-6}$ M^{-1} s^{-1}); however, their affinity to peroxides (for example, H_2O_2 K_m < 20 mM) is markedly higher than that from other peroxidases (Jeong et al., 1999; Pedrajas et al., 2000; Hofmann et al., 2002). In addition, peroxiredoxins are particularly sensitive to relatively high concentrations of peroxides, losing the peroxidatic activity when the catalytic sulfhydryl group is overoxidized to sulfinic or sulfonic (Chae et al., 1994a; Yang et al., 2002). In this sense, a new family of antioxidant proteins named sulfiredoxins has been recently characterized that specifically reduce the catalytic overoxidized thiols of some peroxiredoxins (Biteau et al., 2003; Chang et al., 2004). Therefore, the role of peroxiredoxins as mere defensive peroxidases may be questioned and, because these are among the most sensitive proteins for the detection of peroxides, it has been hypothesized that they play a more significant role in the modulation of cellular signaling through peroxides (Jin et al., 1997; Wood et al., 2003a; Toledano et al., 2004).

On the other hand, some peroxiredoxins are able to form toroidal structures by oligomerization (Wood et al., 2003b) thus acquiring a new function with the peroxiredoxin acting as chaperones that protects sensitive proteins from oxidative damage (Jang et al., 2004).

As mentioned above, most peroxiredoxins obtain the reducing power from Trxs. Nevertheless, other biological electron donors have been recently characterized for some peroxiredoxins. A poplar peroxiredoxin, homologous to mammalian PRX V, accepts both Grx and Trx as electron donors (Rouhier et al., 2002; Noguera-Mazon et al., 2006). Recently, it has been described that reduction of the mammalian PRX VI, a 1-Cys-Prx, occurs by disulfide linking of the peroxiredoxin to the π-glutathione-*S*-transferase followed by glutathionylation (Manevich et al., 2004). Initial studies, which were carried out with a 2-Cys-Prx, showed that both glutathione and lipoic are poor electron donors and hence the impression was that they cannot be principal donors (Kim et al., 1988). However, peroxiredoxins constitute a very heterogeneous family of proteins so that the possibility of new connections with other redoxins and with lipoic acid is yet to be discovered. Actually, one peroxiredoxin that uses an intermediary specific protein to connect with lipoamide has been thoroughly studied in *Mycobacterium tuberculosis* (Bryk et al., 2002). But this is described later.

Glutathione

The reduced form of glutathione (GSH) is the main low molecular weight water soluble antioxidant in the cells and its concentration is usually two orders of magnitude higher than that of its oxidized disulfide form (GSSG) under normal conditions (Meister and Anderson, 1983; Reed and Hackert, 1990). However, this may bė much lower under oxidative stress (Gilbert, 1995). The total GSH concentration in the cell is estimated in 5–10 mM (Meredith and Reed, 1982; Griffith and Meister, 1985b) although it was earlier noticed that the mitochondrial pool is separated from that of the cytosol (Jocelyn, 1975). A specific transport system is now known (Griffith and Meister, 1985b) that allows for the maintenance of mitochondrial glutathione concentration at higher levels that in the cytosol (Soderdahl et al., 2003).

The maintenance of the glutathione system depends on the capacity to recycle GSSG to GSH. This recycling is particularly important in mitochondria where de novo GSH synthesis does not take place (Griffith and Meister, 1985a) and GSSG cannot be exported (Jocelyn, 1975) since unlike the plasma membrane, the mitochondrial inner membrane does not have the ability to export GSSG, at least under oxidative conditions (Olafsdottir and Reed, 1988). The role of glutathione recycling is played by GR (Outten and Culotta, 2004) that is NADPH specific. As a consequence, the reduction of mitochondrial GSSG requires a source of NADP(H) through any of the mechanisms described above (Jo et al., 2001; Outten and Culotta, 2003).

As has already been mentioned above, reaction of glutathione with protein thiols leads to its glutathionylation thus preventing them from deeper oxidation to irreversible species (see Figure 13.1). The way protein thiols react with glutathione can follow different mechanisms, depending on the circumstances (Schafer and Buettner, 2001). One way is through thiol–disulfide exchange between the protein thiol and the oxidized form of glutathione, GSSG. This reaction would only be thermodynamically favored if the GSH/GSSG ratio diminishes markedly, as is the case at the onset of oxidative stress conditions, when GSH is being used enzymatically as a counteracting antioxidant (Schafer and Buettner, 2001). In the mitochondrial matrix there is an additional condition in favor of this mechanism, that is, the high pH around 8 that would facilitate the required ionization of the thiol to thiolate thus enhancing its reactivity. The final outcome will depend on the displacement of the glutathione redox pair toward the oxidized GSSG, or the reduced GSH, forms (Schafer and Buettner, 2001).

Glutathionylation can also take place by reaction of the reduced form of the redox couple, GSH, either with a thiyl radical (Pr-S$^{\bullet}$) or with a protein sulfenate (Pr-SO^{-}). The former originates when a protein thiol suffers 1 electron oxidation and after mixed disulfide formation, superoxide anion ($O_2^{\bullet-}$) results as a side product. The latter occurs when the protein thiol is 2-electron oxidized as is the case by reaction with peroxides or peroxynitrite that will then act as signals altering the protein glutathionylation status (Hurd et al., 2005).

CONNECTION OF LIPOAMIDE WITH GLUTATHIONE VIA GLUTAREDOXIN

According to the hypothesis stated above, we have addressed the possible connection between lipoic acid and the glutathione/Grx system through an in vivo approach with purified proteins. Grx activity is assayed routinely by means of a standard protocol based on the ability of these proteins to efficiently accelerate the transfer of reducing equivalents from reduced glutathione (GSH) to the nonphysiological compound, hydroxyethyldisulfide (HED), the symmetrical disulfide of β-mercaptoethanol (Luthman and Holmgren, 1982) (Scheme 13.5). Grxs are known for having a binding site with high affinity for the glutathione moiety (-SG) that allows them to recognize any -SG label, no matter whether it belongs to reduced GSH, to oxidized GSSG, or to any other mixed disulfide of glutathione provided there are no steric constraints (Bushweller et al., 1992).

On these grounds, we designed a protocol to test whether Grx could catalyze the reduction of GSSG by dihydrolipoamide (Porras et al., 2002). In this assay, 0.75 mM lipoamide ($lipS_2$) is first reduced in the assay cuvette with 1 mM NADH and purified commercial LPD until equilibrium is reached. At this point, the ratio of $[lip(SH)_2]/[lipS_2] = 1.6$. Then 0.5 mM GSSG and an appropriate amount of

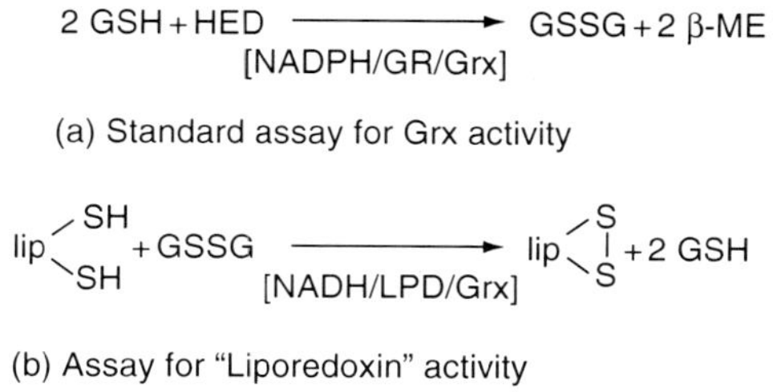

$$2\,GSH + HED \xrightarrow{[NADPH/GR/Grx]} GSSG + 2\,\beta\text{-ME}$$

(a) Standard assay for Grx activity

$$lip\!\!\begin{array}{c}\diagup SH\\ \diagdown SH\end{array}\!\! + GSSG \xrightarrow{[NADH/LPD/Grx]} lip\!\!\begin{array}{c}\diagup S\\ \diagdown S\end{array}\!\!\!\!| + 2\,GSH$$

(b) Assay for "Liporedoxin" activity

SCHEME 13.5 Schematic description of the standard glutaredoxin (Grx) and liporedoxin (Lpx) assays. See text for detailed descriptions

pure Grx are added, and the reduction of GSSG coupled to the oxidation of NADH is followed spectrophotometrically at 340 nm. In addition, the levels of the thiol reagents involved in the assay, GSSG, GSH, lipoamide, and dihydrolipoamide, have been quantified after derivatization with monobromobimane and further separation by HPLC (J. Peinado, unpublished results). We have determined that Grx catalyzes the reduction of one mol GSSG by 1 mol of dihydrolipoamide. The reaction is specific for Grxs with k_{cat} ranging between 200 min^{-1} and 3200 min^{-1} and has been characterized thoroughly regarding its kinetic and chemicophysical parameters. We call this activity "liporedoxin" (Lpx).

The ratio of Lpx to standard HED activities (Lpx/HED) was calculated for eight representative Grx isoforms from *E. coli*, yeast, and human (Table 13.1). There was a broad range of values but interestingly, it was the mitochondrial soluble isoform from yeast Grx2 the one with the highest ratio indicating a favorable relationship between lipoamide and Grx2 (Porras et al., 2002). This result is coherent with the mitochondrial localization of both components in vivo thus reinforcing the likelihood of its physiological meaning.

This was the first demonstration that Grxs catalyze the transfer of reduction equivalents from dihydrolipoamide to GSSG. We started from the fact that a possible reaction between lipoic acid and glutathione is thermodynamically highly favored in the direction of GSSG reduction and not in the opposite direction under most circumstances in the mitochondria (Table 13.2).

The two striking points are that this activity constitutes a new GR activity that depends on NADH instead of NADPH and that Grx can be reduced by dihydrolipoamide in the absence of GSH and NADPH.

As a whole, this activity introduces a drastic change in the history of Grx as it implies an impressive function acting on the opposite direction as the one accepted so far, which is, capturing reducing power from dihydrolipoamide to keep the glutathione system in the reduced state, instead of using the reducing power of glutathione to reduce other disulfides.

In doing so, one can hypothesize that Grx would capture reducing power from the lipoyl residue of protein H or from E2 of the 2-oxoacid dehydrogenase complexes, previously reduced by LPD. Actually, the latter reaction is known

TABLE 13.1
Lpx/HED Activity Ratio for Eight Glutaredoxins

Type of Grx	Lpx/HED
E. coli	
Grx1	0,13
Grx2	0,59
Grx3C65Y	0,30
Yeast	
Grx2	1,20
Rat	
Grx1	0,35
Human	
Grx1	0,36
Grx2Δ41-164	0,23
Grx2a	0,33

Source: Data extracted from Porras, P., Pedrajas, J.R., Martínez-Galisteo, E., Johansson, C., Holmgren, A., Padilla, C.A., and Bárcena, J.A., *Biochem. Biophys. Res. Commun.*, 295, 1046, 2002.

Note: All eight Grxs were active with both assays, but there were marked differences in their efficiencies.

to occur. This setting would constitute an NADH-dependent reduction of Grx that could result in the reduction of GSSG, so to say, a kind of NADPH-independent GR. In the majority of organisms, this activity could afford the maintenance of thiol redox homeostasis straightforward via GSH without the need for NADH conversion to NADPH. On the other hand, it could be the solution for organisms that synthesize and use GSH but whose proteome lacks GR, as is the case with the cyanobacterium *Synechocystis*, p.e. (Florencio et al., 2006).

TABLE 13.2
$\Delta G'$ (kJ/mol) for the Reduction of GSSG by $Lip(SH)_2$

GSH/GSSG	GSH + GSSG (mM)		
	1	5	10
100	−7.85	−3.92	−2.23
10	−12.26	−9.72	−8.03
1	−20.75	−16.83	−15.15

Note: Calculated at pH 7; 25°C, for $lip(SH)_2/lipS_2 = 2$ and E_o' for lipoamide and glutathione −0.32 and −0.23 V, respectively.

Another role that can be envisaged for this connection between lipoamide and glutathione is the reduction of GSSG via Grx without the involvement of LPD, that is, by directly "tapping" reducing equivalents from E2-lipoamide during the normal energy flow initiated by the highly exergonic oxidative decarboxylation of pyruvate (or other 2-oxoacids) through the PDH complex (Figure 13.3). This

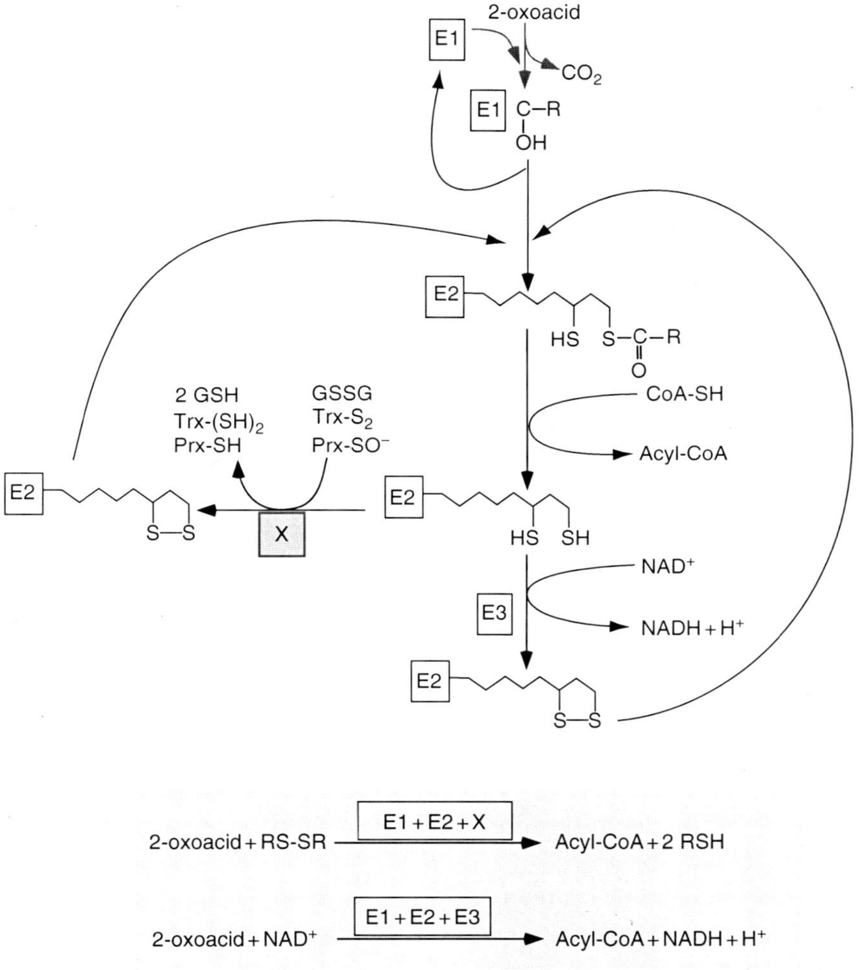

FIGURE 13.3 Schematic depiction of oxoacid dehydrogenase complex (ODC) catalytic cycle centered around the lipoamide cofactor. The redox cycling of lipoamide is highlighted. The normal catalytic reaction is shown at the right-hand side; shunting of reducing power from dihydrolipoamide toward the mitochondrial thiolic system and peroxides is shown in left. X denotes a protein, specific or not, that interacts with the lipoyl domain to catalyze the capturing of electrons. The global reactions for both processes are summarized at the bottom. See the main text for in-depth description.

scheme would go on continuously with the production of acetyl-CoA. However, the fate of the electrons captured by lipoamide could be either NAD^+ or Grx/GSSG, depending both on the thermodynamics and the kinetics of the situation.

The redox midpoint potential of lipoamide and NAD pairs is equal (−0.320 V) and, consequently, they react with a $\Delta G_o' = 0.0$ kJ/mol; $K_{eq}' = 1$. These thermodynamic parameters indicate that both ratios will be reciprocal: for $NAD^+/NADH = 10$, $lip(SH)_2/lipS_2 = 0.1$, but when the need for NADH is not highly demanding and its production at other points downstream at the Krebs cycle is sufficient, the NAD and lipoamide redox pairs will equilibrate and the transfer of electrons from reduced lipoamide to oxidized NAD^+ could become thermodynamically forbidden. Under these conditions, accumulation of reduced lipoamide would favor its undesired oxidation by ROS. This would be detrimental for the PDH activity unless the electron surcharge of dihydrolipoamide is alleviated by diversion to another type of useful intended oxidation, that is, through thiol–disulfide exchange with GSSG catalyzed by Grx. Actually, studies on the coupling of acetyl groups and electrons in the PDH complex of *E. coli* indicated that only half of the lipoyl residues are coupled to NADH formation (Frey et al., 1978). So, besides serving as a useful source of reducing power, diversion of electrons to GSSG would help protecting the whole complex from inactivation.

On the other hand, the transfer of electrons from reduced lipoamide to GSSG is thermodynamically highly favored even for GSH/GSSG ratios as high as 100 and a LipSH2/LipS2 ratio of 1 (Table 13.2).

A crucial question for all these hypothesis to be true is the requirement for physical interaction between Grx and either the lipoyl domain of E2 or protein H. Interaction with protein H would initially pose less restrictions than the 2-oxoacid dehydrogenase complexes, since all four components of the glycine cleavage system do not form a stable structure. It is thought that LPD and protein H are free in the mitochondrial matrix (Douce et al., 2001). However, the lipoyl moiety in this protein does not seem to suffer marked oscillations during its catalytic cycle at difference with the other lipoate-containing proteins (Guilhaudis et al., 1999). The validity of the hypothesis has to be tested experimentally.

Concerning the possible interaction of Grx with the 2-oxoacid dehydrogenase complexes, several favorable factors can be considered a priori. The abundance of flexible elements around the lipoyl domain and its projection outward the complex core by means of extended parts of the polypeptide chain (Perham, 2000) is relevant in this respect. Moreover, the number of lipoyl domains per E2 chain can be as high as three (Neveling et al., 1998) but the excess of lipoyl groups is apparently superfluous, as indicated by the ability to genetically or chemically remove lipoyl groups, without a matching decrease in overall complex activity (Reed and Hackert, 1990). Three tandem lipoyl groups confer survival advantages in vivo (Dave et al., 1995) but one single lipoyl group suffices for full activity in vitro (Guest et al., 1997). Thus, the advantage of three lipoyl domains for the microorganism does not reside in their catalytic role, but on the capacity to extend the reach of the outmost lipoyl cofactor (Guest et al., 1997).

Meanwhile, although interactions of the lipoyl domains in ODC with other proteins of a larger size would seem more difficult at first sight, they have been actually documented at atomic resolution. This is the case for the interaction of the human regulatory enzyme pyruvate dehydrogenase kinase 3 (PDHK) with the lipoyl domain, which has been nicely elucidated by x-ray crystallography (Kato et al., 2005). This enzyme binds to the pyruvate dehydrogenase complex (PDC) by strong specific interaction with one inner lipoyl domain of the E2 core, showing that, although buried in the core of the PDH complex, the lipoyl domain is accessible to the large PDHK (Liu et al., 1995).

Interestingly, reduction of the lipoyl group significantly enhances the affinity of the lipoyl domain toward PDHK compared to the domain containing the coenzyme in the oxidized state (Roche et al., 2003). Although these data are promising, evidences for the interaction of Grx with the lipoyl domains are still awaiting. However, one relevant result has been reported recently demonstrating the existence of a Trx-like protein in *M. tuberculosis* that specifically taps reducing power from lipoamide in 2-oxoacid complexes for antioxidant purposes (Bryk et al., 2002). This system is described in depth in another section below.

Glutaredoxins coupled to GR are key players in antioxidant cellular processes based on the transfer of reducing equivalents from NADPH to disulfides via GSH. Now, a new important function can be envisaged for Grxs working the opposite way, i.e., tapping reducing power from the main stream of catabolic pathways to maintain glutathione in the reduced state. This would be another supplementary way to serve processes in which reduced glutathione is involved but without NAD(P)H consumption. This in turn, would have the additional bonus of protecting the complex from inactivation by accumulation of reduced lipoamide.

In line with the reasoning developed herein, we are working on the search for additional experimental evidences to support it. For instance, pyruvate-dependent reduction of GSSG in isolated mitochondria has been observed. Moreover, yeast mitochondria lacking Grx2 have a lower PDH activity compared to wild-type mitochondria (R. Requejo, unpublished observations).

More evidences of a connection between Grx and ODC have been reported. PDH and KGDH are preferred targets of ROS attack in the mitochondria, both by direct oxidation of key cysteines and by formation of carbonyl residues (Cabiscol et al., 2000). Numerous degenerative pathologies associated to oxidative stress show a concurrent loss of KGDH activity in neuronal mitochondria (Beal, 1995). Using experimental models, it has been shown that H_2O_2 treatment at μM concentrations transiently affects mitochondrial performance with parallel reversible loss of KGDH activity (Nulton-Persson and Szweda, 2001). The underlying mechanism is based on modification of thiol groups on KGDH by H_2O_2 under conditions that decrease mitochondrial GSH levels. Since the process was reverted by Grx, with or without GSH, but not by Trx, it was suggested that the inactivation consisted of a glutathionylation process (Nulton-Persson et al., 2003). Curiously, these phenomena were only observed when analyzed in isolated mitochondrial preparations, but do not take place in vitro with the purified PDH, indicating the need for the intrinsic structure of PDH within the mitochondria and the coordinated response of multiple mitochondrial components. We and other researchers familiar with in vitro assay of ODC have experienced some kind anomalous results related to thiols, indicating so far unforeseen intricacies of the complex behavior.

Lipoyl–lysine of the E2 subunit of PDH acts as powerful autoantigen due to inappropriate release during apoptosis leading to primary biliary cirrhosis (Mato et al., 2004). The autoantibodies preferentially recognize PDC-E2 with reduced sulfhydryl groups. However, apoptosis also initiates a special sequence of events such that cells mask their E2 lysine–lipoyl group to prevent the accumulation and subsequent exposure of these potentially self-reactive antigens. But among several xenobiotic derivatizations tested, the only regulatory mechanism that prevents the accumulation of autoantigenic peptides is glutathionylation (Mato et al., 2004), which is known to take place during apoptosis. Therefore, it appears that proteins containing lysine-lipoate must be highly regulated to prevent the inappropriate release of these proteins.

The interesting point for the purpose of this review is that the involvement of a glutathionylation mechanism immediately brings Grx at close up as a protective actor in this pathological scenario.

CONNECTION OF LIPOAMIDE WITH THIOREDOXIN

Thioredoxin also catalyzed lipoamide-dependent reduction of the insulin disulfides in a coupled system with NADH, lipoamide, and LPD. It was suggested that the fast spontaneous reaction between dihydrolipoamide and Trx-S_2 would provide a mechanism for NADH or pyruvate-dependent disulfide reduction (Holmgren, 1979). Although no other results were obtained in this line of research that could confirm this possibility, another connection of the Trx system with lipoic acid was found (Arner et al., 1996a). TR from calf thymus, calf liver, human placenta, and rat liver efficiently catalyzed reduction of both lipoic acid and lipoamide in NADPH-dependent reactions in vitro. Lipoic acid was reduced almost as efficiently as lipoamide at difference with LPD. Under equivalent conditions, mammalian TR reduced lipoic acid by NADPH 15 times more efficiently than the corresponding NADH-dependent reduction catalyzed by LPD. It was then suggested a role for TR in the reduction of lipoic acid when exogenously administered for therapeutical purposes (Arner et al., 1996b).

Exogenous lipoic acid is up taken by mammalian cells and is rapidly reduced to dihydrolipoic inside the cell, presumably by LPD, where it can exert antioxidant protection for clinical use (Packer et al., 2001). However, LPD cannot be the sole enzyme catalyzing reduction of lipS_2, since cells lacking LPD, like the erythrocytes, still reduce lipS_2 acid to lip$(SH)_2$ in the presence of glucose by an NADPH-dependent reaction (Constantinescu et al., 1995). Another candidate for this reaction could be GR but it has been argued that its efficiency is too low compared with the catalytic power of this enzyme as is evident from its activity with GSSG (Pick et al., 1995). On the other hand, the efficient reduction of lipoic acid by TR is not surprising given its well-known wide substrate specificity (Holmgren, 1989).

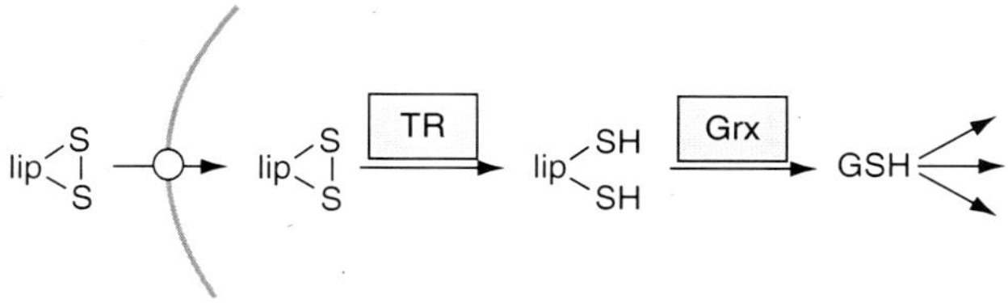

FIGURE 13.4 Proposed sequence of events for the reduction of exogenous lipoic acid. Administered lipoic acid would be reduced in the cell by the tandem action of TR and Grx and will help maintain the thiol redox homeostasis through the glutathione system. More explanations are provided in the main text.

These results together with the results described in the preceding section (Porras et al., 2002) suggest that in mammalian cells, administered lipoic acid could be reduced by TR and its redox equivalents passed on to the more versatile GSH through the action of Grx (Figure 13.4). This would be an interesting connection between members of the glutathione and Trx systems with separated TR and Grx acting in tandem. Other tight connection between these two redoxin systems exists in Drosophila in the form of a unique molecular entity—a Grx module fused to TR with the capacity to catalyze the reduction of GSSG by NADPH (Sun et al., 2001). Moreover, in a recent report, it has been shown that TR catalyzes NADPH reduction of human mitochondrial Grx2 working in tandem in a glutathionylation pathway (Johansson et al., 2004).

Thioredoxin connection with the lipoyl system has also been reported from studies with KGDH. Accumulation of the dihydrolipoate intermediate during the catalytic cycle of the enzymatic complex causes inactivation of the first enzyme of the complexes, E1. With the mammalian pyruvate dehydrogenase (PDH), the phosphorylation system is involved in the lipoate-dependent regulation, whereas mammalian 2-oxoglutarate dehydrogenase exhibits a higher sensitivity to direct regulation by the complex-bound dihydrolipoate/lipoate and external SH/S–S, including mitochondrial Trx. Trx efficiently protects the complexes from self-inactivation during catalysis at low NAD^+. (Bunik and Sievers, 2002; Bunik, 2003). Under conditions of low NAD^+ availability, E2-bound reduced lipoamide cannot be recycled and, thus, accumulates, resulting in the transient formation of thiyl radical (E2-($S^\bullet$)-SH), both anaerobically and in the presence of oxygen (Bunik and Sievers, 2002). The reactive thiyl radical produces superoxide anion ($O_2^{\bullet-}$) upon 1 electron oxidation by oxygen, but also inactivates E1 upon 1 electron reduction by the E1 catalytic intermediate (Figure 13.5). Trx but not GSH protects the complex from this inactivation process. It has been proposed that Trx performs this protecting action by dismutation of the E2-lipoamide thiyl radical. Support for this hypothesis comes from the determined high stability of the Trx thiyl radical (Lmoumene et al., 2000).

Overall, these redox side reactions of the lipoamide residue create a link between the ODC and the surrounding medium, that if coupled to mitochondrial

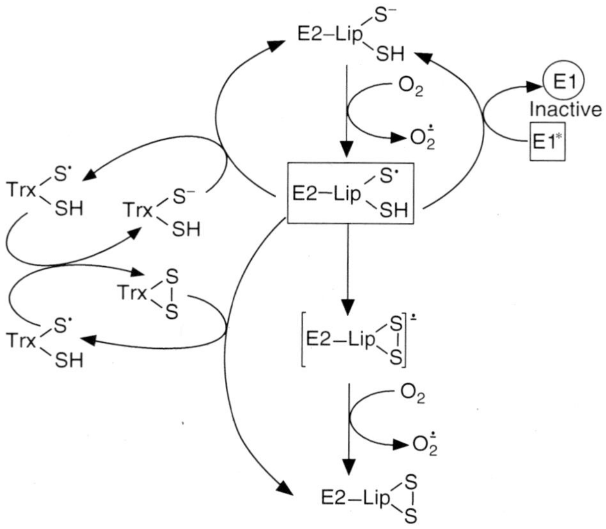

FIGURE 13.5 Summary of the interaction between lipoic acid and Trx at the KGDH complex. The scheme depicted is a compilation of the studies and discussions by Bunik and coworkers. See main text for detailed description.

Trx-dependent peroxidases (Watabe et al., 1994; Knoops et al., 1999; Pedrajas et al., 2000) could cooperate in the scavenging of ROS.

INTERACTION OF LIPOAMIDE WITH PEROXIREDOXIN

But the definitive data demonstrating the likelihood of lipoamide from the ODC acting as an electronic shunt between main catabolic pathways and the antioxidant defense pathways come from the studies by C. Nathan and coworkers with the infectious bacterium *Mycobacterium tuberculosis* that we describe in the next section.

Mycobacterium tuberculosis, the causative agent of tuberculosis, survives inside macrophages in spite of an aggressive oxidative and nitrosative attack displayed by the host cell. The persistence of the bacteria is due to their possessing several antioxidant systems. One of them is the NADH-driven hydroperoxide and peroxynitrite reduction by the peroxiredoxin alkyl hydroperoxide reductase subunit C (AhpC) (Bryk et al., 2002). In a search for the physiological reductase of AhpC, it was observed that the Trx system from the same organism was ineffective as reductant. However, a protein, AhpD coded by a gene located next to the *aphC* gene, could promote very efficient NADH-dependent peroxidase activity by AhpC when added to an assay mixture containing a whole bacterium lysate. Further purification from the lysate led to the identification of Lpd, the sole dihydrolipoamide dehydrogenase from *M. tuberculosis*. Further screening ended

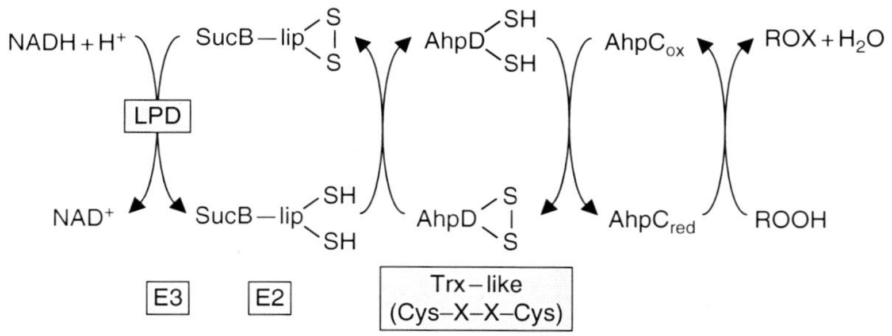

FIGURE 13.6 Electron transport chain from NADH to peroxides in *Mycobacterium tuberculosis*. The bacterium has no Trx system capable to provide reducing power for the reduction of peroxides but bears a specifically designed protein to capture electrons from the KGDH complex and to transfer them to a canonical Prx (on the basis of the work by Nathan and coworkers).

up with the discovery of a need for lipoamide to sustain the peroxidase activity. Since there is not free lipoamide in cells, a lipoamide containing protein should be the provider for lipoamide in vivo and it happened to be a 85 kDa protein identified as the product of gene *sucB*, a homologue of the E2 component of KGDH complex. Finally, a system was reconstituted in vitro with the recombinant proteins establishing the electron transfer chain from NADH to peroxides as NADH → Lpd → SucB → AhpD → AhpC → ROOH (Bryk et al., 2002) (Figure 13.6).

AhpD was characterized showing that it catalyzes Trx-like thiol–disulfide oxidoreductase activity by reducing DTNB when provided electrons from NADH via Lpd + SucB. Moreover, AhpD possesses an active site dithiol with the Trx signature Cys–X–X–Cys, likely to oscillate between the dithiol/disulfide states during the catalytic cycle. Apart from these striking similarities with Trxs, the tertiary structure of AhpD is nearly all helical and does not resemble any structure previously reported including those of the Trx-fold family; moreover, it is assembled as compact trimers (Bryk et al., 2002).

It is the demonstration of this electron capturing system through one ad hoc protein specifically interacting with the lipoyl domains that makes plausible the hypothesis developed along this review for shunting reducing power from main catabolic pathways toward defensive tasks in the mitochondria, and that keeps the search for definitive evidences alive.

CONCLUDING REMARKS

The energetic carbon oxidative flow through the main catabolic pathway, that is, the Krebs cycle, can be diverted from its primary aim of producing ATP, to provide reducing equivalents for other purposes like antioxidative defense.

This can be specific, like the elimination of hydroperoxides involving a dedicated Trx-like protein or more general as is the maintenance of the housekeeping thiol–disulfide pair GSH/GSSG in its reduced state involving

Grx and the spreading of a thiolic state to sensitive protein disulfides through the Trx regulatory network.

One additional advantage of this escaping of reducing equivalents is that accumulation of dihydrolipoamide and the concomitant production of thiyl and superoxide radicals is avoided, thus protecting the 2-oxoacid dehydrogenase complex from irreversible inactivation.

This type of reducing power shunt can be of utmost importance in organisms lacking the canonical NADPH-dependent reductases, like GR or TR. This has already been demonstrated for *M. tuberculosis* where TR is not operative for hydroperoxides reduction and could also explain the mechanism of GSSG reduction in *Synechocystis* where GR is absent.

Apart from the above-mentioned Trx-like protein of *M. tuberculosis*, other redoxin connections of lipoic acid have only been detected indirectly and the conclusive demonstration of a direct physical interaction between the proteins involved is still awaiting.

ABBREVIATIONS

GR,	glutathione reductase
Grx,	glutaredoxin
KGDC,	α-ketoglutarate dehydrogenase complex
KGDH,	α-ketoglutarate dehydrogenase
LPD,	lipoamide dehydrogenase
Lpx,	liporedoxin activity
ODC,	2-oxoacid dehydrogenase complex
PDC,	pyruvate dehydrogenase complex
PDH,	pyruvate dehydrogenase
Prx,	peroxiredoxin
TR,	thioredoxin reductase
Trx,	thioredoxin

REFERENCES

Arner, E.S., Nordberg, J., and Holmgren, A. 1996a. Efficient reduction of lipoamide and lipoic acid by mammalian thioredoxin reductase. *Biochem. Biophys. Res. Commun.*, **225**, 268–274.

Arner, E.S.J., Nordberg, J., and Holmgren, A. 1996b. Efficient reduction of lipoamide and lipoic acid by mammalian thioredoxin reductase. *Biochem. Biophys. Res. Commun.*, **225**, 268–274.

Beal, M.F. 1995. Aging, energy, and oxidative stress in neurodegenerative diseases. *Ann. Neurol.*, **38**, 357–366.

Biteau, B., Labarre, J., and Toledano, M.B. 2003. ATP-dependent reduction of cysteine-sulphinic acid by *S. cerevisiae* sulphiredoxin. *Nature*, **425**, 980–984.

Bryk, R., Lima, C.D., Erdjument-Bromage, H., Tempst, P., and Nathan, C. 2002. Metabolic enzymes of mycobacteria linked to antioxidant defense by a thioredoxin-like protein. *Science*, **295**, 1073–1077.

Buchanan, B.B. and Balmer, Y. 2005. Redox regulation: A broadening horizon. *Annu. Rev. Plant Biol.*, **56**, 187–220.

Bunik, V.I. 2003. 2-Oxo acid dehydrogenase complexes in redox regulation. *Eur. J. Biochem.*, **270**, 1036–1042.

Bunik, V.I. and Sievers, C. 2002. Inactivation of the 2-oxo acid dehydrogenase complexes upon generation of intrinsic radical species. *Eur. J. Biochem.*, **269**, 5004–5015.

Bushweller, J.H., Aslund, F., Wuthrich, K., and Holmgren, A. 1992. Structural and functional characterization of the mutant *Escherichia coli* glutaredoxin (C14-S) and its mixed disulfide with glutathione. *Biochemistry*, **31**, 9288–9293.

Cabiscol, E., Piulats, E., Echave, P., Herrero, E., and Ros, J. 2000. Oxidative stress promotes specific protein damage in *Saccharomyces cerevisiae*. *J. Biol. Chem.*, **275**, 27393–27398.

Cai, J. and Jones, D.P. 1998. Superoxide in apoptosis. Mitochondrial generation triggered by cytochrome c loss. *J. Biol. Chem.*, **273**, 11401–11404.

Chae, H.Z., Chung, S.J., and Rhee, S.G. 1994a. Thioredoxin-dependent peroxide reductase from yeast. *J. Biol. Chem.*, **269**, 27670–27678.

Chae, H.Z., Robison, K., Poole, L.B., Church, G., Storz, G., and Rhee, S.G. 1994b. Cloning and sequencing of thiol-specific antioxidant from mammalian brain: Alkyl hydroperoxide reductase and thiol-specific antioxidant define a large family of antioxidant enzymes. *Proc. Natl. Acad. Sci. U S A*, **91**, 7017–7021.

Chang, T.S., Jeong, W., Woo, H.A., Lee, S.M., Park, S., and Rhee, S.G. 2004. Characterization of mammalian sulfiredoxin and its reactivation of hyperoxidized peroxiredoxin through reduction of cysteine sulfinic acid in the active site to cysteine. *J. Biol. Chem.*, **279**, 50994–51001.

Chen, J.W., Dodia, C., Feinstein, S.I., Jain, M.K., and Fisher, A.B. 2000. 1-Cys peroxiredoxin, a bifunctional enzyme with glutathione peroxidase and phospholipase A2 activities. *J. Biol. Chem.*, **275**, 28421–28427.

Chrestensen, C.A., Starke, D.W., and Mieyal, J.J. 2000. Acute cadmium exposure inactivates thioltransferase (Glutaredoxin), inhibits intracellular reduction of protein-glutathionyl-mixed disulfides, and initiates apoptosis. *J. Biol. Chem.*, **275**, 26556–26565.

Constantinescu, A., Pick, U., Handelman, G.J., Haramaki, N., Han, D., Podda, M., Tritschler, H.J., and Packer, L. 1995. Reduction and transport of lipoic acid by human erythrocytes. *Biochem. Pharmacol.*, **50**, 253–261.

Cotgreave, I.A. and Gerdes, R.G. 1998. Recent trends in glutathione biochemistry—glutathione-protein interactions: A molecular link between oxidative stress and cell proliferation? *Biochem. Biophys. Res. Commun.*, **242**, 1–9.

Danson, M.J., Fersht, A.R., and Perham, R.N. 1978. Rapid intramolecular coupling of active sites in the pyruvate dehydrogenase complex of *Escherichia coli*: Mechanism for rate enhancement in a multimeric structure. *Proc. Natl. Acad. Sci. U S A*, **75**, 5386–5390.

Dave, E., Guest, J.R., and Attwood, M.M. 1995. Metabolic engineering in *Escherichia coli*: Lowering the lipoyl domain content of the pyruvate dehydrogenase complex adversely affects the growth rate and yield. *Microbiology*, **141 (Pt 8)**, 1839–1849.

Davis, K.L., Martin, E., Turko, I.V., and Murad, F. 2001. Novel effects of nitric oxide. *Annu. Rev. Pharmacol. Toxicol.*, **41**, 203–236.

Dietz, K.J. 2003. Plant peroxiredoxins. *Annu. Rev. Plant Biol.*, **54**, 93–107.

Douce, R., Bourguignon, J., Neuburger, M., and Rebeille, F. 2001. The glycine decarboxylase system: A fascinating complex. *Trends Plant Sci.*, **6**, 167–176.

Droge, W. 2002. Aging-related changes in the thiol/disulfide redox state: Implications for the use of thiol antioxidants. *Exp. Gerontol.*, **37**, 1333–1345.

Fernandes, A.P. and Holmgren, A. 2004. Glutaredoxins: Glutathione-dependent redox enzymes with functions far beyond a simple thioredoxin backup system. *Antioxid. Redox Signal.*, **6**, 63–74.

Florencio, F.J., Perez-Perez, M.E., Lopez-Maury, L., Mata-Cabana, A., and Lindahl, M. 2006. The diversity and complexity of the cyanobacterial thioredoxin systems. *Photosynth. Res.*, **89**, 157–171.

Forman, H.J., Fukuto, J.M., and Torres, M. 2004. Redox signaling: Thiol chemistry defines which reactive oxygen and nitrogen species can act as second messengers. *Am. J. Physiol. Cell Physiol.*, **287**, C246–C256.

Frey, P.A., Ikeda, B.H., Gavino, G.R., Speckhard, D.C., and Wong, S.S. 1978. *Escherichia coli* pyruvate dehydrogenase complex. Site coupling in electron and acetyl group transfer pathways. *J. Biol. Chem.*, **253**, 7234–7241.

Gasdaska, P.Y., Gasdaska, J.R., Cochran, S., and Powis, G. 1995. Cloning and sequencing of a human thioredoxin reductase. *FEBS Lett.*, **373**, 5–9.

Ghezzi, P., Bonetto, V., and Fratelli, M. 2005. Thiol–disulfide balance: From the concept of oxidative stress to that of redox regulation. *Antioxid. Redox Signal.*, **7**, 964–972.

Gilbert, H.F. 1995. Thiol/disulfide exchange equilibria and disulfide bond stability. *Methods Enzymol.*, **251**, 8–28.

Gladyshev, V.N., Liu, A., Novoselov, S.V., Krysan, K., Sun, Q.A., Kryukov, V.M., Kryukov, G.V., and Lou, M.F. 2001. Identification and characterization of a new mammalian glutaredoxin (thioltransferase) Grx2. *J. Biol. Chem.*, **276**, 30374–30380.

Griffith, O.W. and Meister, A. 1985a. Origin and turnover of mitochondrial glutathione. *Proc. Natl. Acad. Sci. U S A*, **82**, 4668–4672.

Griffith, O.W. and Meister, A. 1985b. Origin and turnover of mitochondrial glutathione. *Proc. Natl. Acad. Sci. U S A*, **82**, 4668–4672.

Guest, J.R., Attwood, M.M., Machado, R.S., Matqi, K.Y., Shaw, J.E., and Turner, S.L. 1997. Enzymological and physiological consequences of restructuring the lipoyl domain content of the pyruvate dehydrogenase complex of *Escherichia coli. Microbiology*, **143 (Pt 2)**, 457–466.

Guilhaudis, L., Simorre, J.P., Blackledge, M., Neuburger, M., Bourguignon, J., Douce, R., Marion, D., and Gans, P. 1999. Investigation of the local structure and dynamics of the H subunit of the mitochondrial glycine decarboxylase using heteronuclear NMR spectroscopy. *Biochemistry*, **38**, 8334–8346.

Haramaki, N., Han, D., Handelman, G.J., Tritschler, H.J., and Packer, L. 1997. Cytosolic and mitochondrial systems for NADH- and NADPH-dependent reduction of alpha-lipoic acid. *Free Radic. Biol. Med.*, **22**, 535–542.

Hatefi, Y. and Yamaguchi, M. 1996. Nicotinamide nucleotide transhydrogenase: A model for utilization of substrate binding energy for proton translocation. *FASEB J.*, **10**, 444–452.

Hess, D.T., Matsumoto, A., Nudelman, R., and Stamler, J.S. 2001. S-nitrosylation: Spectrum and specificity. *Nature Cell Biol*, **3**, E46–E49.

Hofmann, B., Hecht, H.J., and Flohe, L. 2002. Peroxiredoxins. *Biol. Chem.*, **383**, 347–364.

Holmgren, A. 1977. Bovine thioredoxin system. Purification of thioredoxin reductase from calf liver and thymus and studies of its function in disulfide reduction. *J. Biol. Chem.*, **252**, 4600–4606.

Holmgren, A. 1979. Thioredoxin catalyzes the reduction of insulin disulfides by dithiothreitol and dihydrolipoamide. *J. Biol. Chem.*, **254**, 9627–9632.

Holmgren, A. 1985. Thioredoxin. *Annu. Rev. Biochem.*, **54**, 237–271.

Holmgren, A. 1989. Thioredoxin and glutaredoxin systems. *J. Biol. Chem.*, **264**, 13963–13966.

Hurd, T.R., Costa, N.J., Dahm, C.C., Beer, S.M., Brown, S.E., Filipovska, A., and Murphy, M.P. 2005. Glutathionylation of mitochondrial proteins. *Antioxid. Redox Signal.*, **7**, 999–1010.

Jang, H.H., Lee, K.O., Chi, Y.H., Jung, B.G., Park, S.K., Park, J.H., Lee, J.R., Lee, S.S., Moon, J.C., Yun, J.W., Choi, Y.O., Kim, W.Y., Kang, J.S., Cheong, G.W., Yun, D. J., Rhee, S.G., Cho, M.J., and Lee, S.Y. 2004. Two enzymes in one; two yeast peroxiredoxins display oxidative stress-dependent switching from a peroxidase to a molecular chaperone function. *Cell*, **117**, 625–635.

Jeong, J.S., Kwon, S.J., Kang, S.W., Rhee, S.G., and Kim, K. 1999. Purification and characterization of a second type thioredoxin peroxidase (type II TPx) from *Saccharomyces cerevisiae*. *Biochemistry*, **38**, 776–783.

Jeong, W., Cha, M.K., and Kim, I.H. 2000. Thioredoxin-dependent hydroperoxide peroxidase activity of bacterioferritin comigratory protein (BCP) as a new member of the thiol-specific antioxidant protein (TSA)/alkyl hydroperoxide peroxidase C (AhpC) family. *J. Biol. Chem.*, **275**, 2924–2930.

Jin, D.Y., Chae, H.Z., Rhee, S.G., and Jeang, K.T. 1997. Regulatory role for a novel human thioredoxin peroxidase in NF-kappaB activation. *J. Biol. Chem.*, **272**, 30952–30961.

Jo, S.H., Son, M.K., Koh, H.J., Lee, S.M., Song, I.H., Kim, Y.O., Lee, Y.S., Jeong, K.S., Kim, W.B., Park, J.W., Song, B.J., Huh, T.L., and Huhe, T.L. 2001. Control of mitochondrial redox balance and cellular defense against oxidative damage by mitochondrial $NADP^+$-dependent isocitrate dehydrogenase. *J. Biol. Chem.*, **276**, 16168–16176.

Jocelyn, P.C. 1967. The standard redox potential of cysteine-cystine from the thiol-disulphide exchange reaction with glutathione and lipoic acid. *Eur. J. Biochem.*, **2**, 327–331.

Jocelyn, P.C. 1975. Some properties of mitochondrial glutathione. *Biochim. Biophys. Acta*, **396**, 427–436.

Johansson, C., Kavanagh, K.L., Gileadi, O., and Oppermann, U. 2006. Reversible sequestration of active site cysteines in a 2FE2S-bridged dimer provides a mechanism for glutaredoxin 2 regulation in human mitochondria. *J. Biol. Chem.*, **282**, 3077–3082.

Johansson, C., Lillig, C.H., and Holmgren, A. 2004. Human mitochondrial glutaredoxin reduces S-glutathionylated proteins with high affinity accepting electrons from either glutathione or thioredoxin reductase. *J. Biol. Chem.*, **279**, 7537–7543.

Jones, D.P. 2006a. Disruption of mitochondrial redox circuitry in oxidative stress. *Chem. Biol. Interact.*, **163**, 38–53.

Jones, D.P. 2006b. Redefining oxidative stress. *Antioxid. Redox Signal.*, **8**, 1865–1879.

Kang, S.W., Baines, I.C., and Rhee, S.G. 1998. Characterization of a mammalian peroxiredoxin that contains one conserved cysteine. *J. Biol. Chem.*, **273**, 6303–6311.

Kato, M., Chuang, J.L., Tso, S.C., Wynn, R.M., and Chuang, D.T. 2005. Crystal structure of pyruvate dehydrogenase kinase 3 bound to lipoyl domain 2 of human pyruvate dehydrogenase complex. *EMBO J.*, **24**, 1763–1774.

Ke, B. 1957. The polarographic behavior of alpha-lipoic acid. *Biochim. Biophys. Acta*, **25**, 650–651.

Kim, K., Kim, I.H., Lee, K.Y., Rhee, S.G., and Stadtman, E.R. 1988. The isolation and purification of a specific "protector" protein which inhibits enzyme inactivation by a thiol/Fe(III)/O2 mixed-function oxidation system. *J. Biol. Chem.*, **263**, 4704–4711.

Knoops, B., Clippe, A., Bogard, C., Arsalane, K., Wattiez, R., Hermans, C., Duconseille, E., Falmagne, P., and Bernard, A. 1999. Cloning and characterization of AOEB166, a novel mammalian antioxidant enzyme of the peroxiredoxin family. *J. Biol. Chem.*, **274**, 30451–30458.

Laurent, T.C., Moore, E.C., and Reichard, P. 1964. Enzymatic synthesis of deoxyribonucleotides. IV. Isolation and characterization of thioredoxin, the hydrogen donor from *Escherichia coli* B. *J. Biol. Chem.*, **239**, 3436–3444.

Lee, S.P., Hwang, Y.S., Kim, Y.J., Kwon, K.S., Kim, H.J., Kim, K., and Chae, H.Z. 2001. Cyclophilin a binds to peroxiredoxins and activates its peroxidase activity. *J. Biol. Chem.*, **276**, 29826–29832.

Lee, S.R., Kim, J.R., Kwon, K.S., Yoon, H.W., Levine, R.L., Ginsburg, A., and Rhee, S.G. 1999. Molecular cloning and characterization of a mitochondrial selenocysteine-containing thioredoxin reductase from rat liver. *J. Biol. Chem.*, **274**, 4722–4734.

Lillig, C.H., Berndt, C., Vergnolle, O., Lonn, M.E., Hudemann, C., Bill, E., and Holmgren, A. 2005. Characterization of human glutaredoxin 2 as iron-sulfur protein: A possible role as redox sensor. *Proc. Natl. Acad. Sci. U S A*, **102**, 8168–8173.

Liu, S., Baker, J.C., and Roche, T.E. 1995. Binding of the pyruvate dehydrogenase kinase to recombinant constructs containing the inner lipoyl domain of the dihydrolipoyl acetyltransferase component. *J. Biol. Chem.*, **270**, 793–800.

Lmoumene, C.E., Conte, D., Jacquot, J.P., and Houee-Levin, C. 2000. Redox properties of protein disulfide bond in oxidized thioredoxin and lysozyme: A pulse radiolysis study. *Biochemistry*, **39**, 9295–9301.

Lundberg, M., Johansson, C., Chandra, J., Enoksson, M., Jacobsson, G., Ljung, J., Johansson, M., and Holmgren, A. 2001. Cloning and expression of a novel human glutaredoxin(Grx2) with mitochondrial and nuclear isoforms. *J. Biol. Chem.*, **276**, 26269–26275.

Luthman, M. and Holmgren, A. 1982. Glutaredoxin from calf thymus. Purification to homogeneity. *J. Biol. Chem.*, **257**, 6686–6690.

Manevich, Y., Feinstein, S.I., and Fisher, A.B. 2004. Activation of the antioxidant enzyme 1-CYS peroxiredoxin requires glutathionylation mediated by heterodimerization with pi GST. *Proc. Natl. Acad. Sci. U S A*, **101**, 3780–3785.

Mannick, J.B. and Schonhoff, C.M. 2002. Nitrosylation: The next phosphorylation? *Arch. Biochem. Biophys.*, **408**, 1–6.

Martin, J.L. 1995. Thioredoxin—a fold for all reasons. *Structure*, **3**, 245–250.

Mato, T.K., Davis, P.A., Odin, J.A., Coppel, R.L., and Gershwin, M.E. 2004. Sidechain biology and the immunogenicity of PDC-E2, the major autoantigen of primary biliary cirrhosis. *Hepatology*, **40**, 1241–1248.

Meister, A. and Anderson, M.E. 1983. Glutathione. *Annu. Rev. Biochem.*, **52**, 711–760.

Meredith, M.J. and Reed, D.J. 1982. Status of the mitochondrial pool of glutathione in the isolated hepatocyte. *J. Biol. Chem.*, **257**, 3747–3753.

Minard, K.I., Jennings, G.T., Loftus, T.M., Xuan, D., and McAlister-Henn, L. 1998. Sources of NADPH and expression of mammalian $NADP^+$-specific isocitrate dehydrogenases in *Saccharomyces cerevisiae*. *J. Biol. Chem.*, **273**, 31486–31493.

Miranda-Vizuete, A., Damdimopoulos, A.E., Pedrajas, J.R., Gustafsson, J.A., and Spyrou, G. 1999a. Human mitochondrial thioredoxin reductase cDNA cloning, expression and genomic organization. *Eur. J. Biochem.*, **261**, 405–412.

Miranda-Vizuete, A., Damdimopoulos, A.E., and Spyrou, G. 1999b. cDNA cloning, expression and chromosomal localization of the mouse mitochondrial thioredoxin reductase gene(1). *Biochim. Biophys. Acta*, **1447**, 113–118.

Murata, H., Ihara, Y., Nakamura, H., Yodoi, J., Sumikawa, K., and Kondo, T. 2003. Glutaredoxin exerts an antiapoptotic effect by regulating the redox state of Akt. *J. Biol. Chem.*, **278**, 50226–50233.

Mustacich, D. and Powis, G. 2000. Thioredoxin reductase. *Biochem. J.*, **346 (Pt 1)**, 1–8.

Nakamura, H., Masutani, H., Tagaya, Y., Yamauchi, A., Inamoto, T., Nanbu, Y., Fujii, S., Ozawa, K., and Yodoi, J. 1992. Expression and growth-promoting effect of adult T-cell leukemia-derived factor. A human thioredoxin homologue in hepatocellular carcinoma. *Cancer*, **69**, 2091–2097.

Neveling, U., Bringer-Meyer, S., and Sahm, H. 1998. Gene and subunit organization of bacterial pyruvate dehydrogenase complexes. *Biochim. Biophys. Acta*, **1385**, 367–372.

Noguera-Mazon, V., Lemoine, J., Walker, O., Rouhier, N., Salvador, A., Jacquot, J.P., Lancelin, J.M., and Krimm, I. 2006. Glutathionylation induces the dissociation of 1-Cys D-peroxiredoxin non-covalent homodimer. *J. Biol. Chem.*, **281**, 31736–31742.

Nonn, L., Williams, R.R., Erickson, R.P., and Powis, G. 2003. The absence of mitochondrial thioredoxin 2 causes massive apoptosis, exencephaly, and early embryonic lethality in homozygous mice. *Mol. Cell Biol.*, **23**, 916–922.

Nordberg, J. and Arner, E.S. 2001. Reactive oxygen species, antioxidants, and the mammalian thioredoxin system. *Free Radic. Biol. Med.*, **31**, 1287–1312.

Nulton-Persson, A.C. and Szweda, L.I. 2001. Modulation of mitochondrial function by hydrogen peroxide. *J. Biol. Chem.*, **276**, 23357–23361.

Nulton-Persson, A.C., Starke, D.W., Mieyal, J.J., and Szweda, L.I. 2003. Reversible inactivation of alpha-ketoglutarate dehydrogenase in response to alterations in the mitochondrial glutathione status. *Biochemistry*, **42**, 4235–4242.

Olafsdottir, K. and Reed, D.J. 1988. Retention of oxidized glutathione by isolated rat liver mitochondria during hydroperoxide treatment. *Biochim. Biophys. Acta*, **964**, 377–382.

Outten, C.E. and Culotta, V.C. 2003. A novel NADH kinase is the mitochondrial source of NADPH in *Saccharomyces cerevisiae*. *EMBO J.*, **22**, 2015–2024.

Outten, C.E. and Culotta, V.C. 2004. Alternative start sites in the *Saccharomyces cerevisiae* GLR1 gene are responsible for mitochondrial and cytosolic isoforms of glutathione reductase. *J. Biol. Chem.*, **279**, 7785–7791.

Packer, L., Kraemer, K., and Rimbach, G. 2001. Molecular aspects of lipoic acid in the prevention of diabetes complications. *Nutrition*, **17**, 888–895.

Park, S.G., Cha, M.K., Jeong, W., and Kim, I.H. 2000. Distinct physiological functions of thiol peroxidase isoenzymes in *Saccharomyces cerevisiae*. *J. Biol. Chem.*, **275**, 5723–5732.

Pe'er, I., Felder, C.E., Man, O., Silman, I., Sussman, J.L., and Beckmann, J.S. 2004. Proteomic signatures: Amino acid and oligopeptide compositions differentiate among phyla. *Proteins*, **54**, 20–40.

Pedrajas, J.R., Kosmidou, E., Miranda-Vizuete, A., Gustafsson, J.A., Wright, A.P., and Spyrou, G. 1999. Identification and functional characterization of a novel mitochondrial thioredoxin system in *Saccharomyces cerevisiae*. *J. Biol. Chem.*, **274**, 6366–6373.

Pedrajas, J.R., Miranda-Vizuete, A., Javanmardy, N., Gustafsson, J.A., and Spyrou, G. 2000. Mitochondria of *Saccharomyces cerevisiae* contain one-conserved cysteine type peroxiredoxin with thioredoxin peroxidase activity. *J. Biol. Chem.*, **275**, 16296–16301.

Pedrajas, J.R., Porras, P., Martinez-Galisteo, E., Padilla, C.A., Miranda-Vizuete, A., and Barcena, J.A. 2002. Two isoforms of *Saccharomyces cerevisiae* glutaredoxin 2 are expressed in vivo and localize to different subcellular compartments. *Biochem. J.*, **364**, 617–623.

Pekkari, K. and Holmgren, A. 2004. Truncated thioredoxin: Physiological functions and mechanism. *Antioxid. Redox Signal.*, **6**, 53–61.

Perham, R.N. 2000. Swinging arms and swinging domains in multifunctional enzymes: Catalytic machines for multistep reactions. *Annu. Rev. Biochem.*, **69**, 961–1004.

Peshenko, I.V. and Shichi, H. 2001. Oxidation of active center cysteine of bovine 1-Cys peroxiredoxin to the cysteine sulfenic acid form by peroxide and peroxynitrite. *Free Radic. Biol. Med.*, **31**, 292–303.

Pick, U., Haramaki, N., Constantinescu, A., Handelman, G.J., Tritschler, H.J., and Packer, L. 1995. Glutathione reductase and lipoamide dehydrogenase have opposite stereospecificities for alpha-lipoic acid enantiomers. *Biochem. Biophys. Res. Commun.*, **206**, 724–730.

Porras, P., Pedrajas, J.R., Martínez-Galisteo, E., Johansson, C., Holmgren, A., Padilla, C.A., and Bárcena, J.A. 2002. Glutaredoxins catalyze the reduction of glutathione by dihydrolipoamide with high efficiency. *Biochem. Biophys. Res. Commun.*, **295**, 1046–1051.

Porras, P., Padilla, C.A., Krayl, M., Voos, W., and Barcena, J.A. 2006. One single in-frame AUG codon is responsible for a diversity of subcellular localizations of glutaredoxin 2 in *Saccharomyces cerevisiae. J. Biol. Chem.*, **281**, 16551–16562.

Rand, J.D. and Grant, C.M. 2006. The thioredoxin system protects ribosomes against stress-induced aggregation. *Mol. Biol. Cell*, **17**, 387–401.

Reed, L.J. and Hackert, M.L. 1990. Structure–function relationships in dihydrolipoamide acyltransferases. *J. Biol. Chem.*, **265**, 8971–8974.

Rhee, S.G. 1999. Redox signaling: Hydrogen peroxide as intracellular messenger. *Exp. Mol. Med.*, **31**, 53–59.

Roche, T.E., Hiromasa, Y., Turkan, A., Gong, X., Peng, T., Yan, X., Kasten, S.A., Bao, H., and Dong, J. 2003. Essential roles of lipoyl domains in the activated function and control of pyruvate dehydrogenase kinases and phosphatase isoform 1. *Eur. J. Biochem.*, **270**, 1050–1056.

Rodriguez-Manzaneque, M.T., Ros, J., Cabiscol, E., Sorribas, A., and Herrero, E. 1999. Grx5 glutaredoxin plays a central role in protection against protein oxidative damage in *Saccharomyces cerevisiae. Mol. Cell Biol.*, **19**, 8180–8190.

Rodriguez-Manzaneque, M.T., Tamarit, J., Belli, G., Ros, J., and Herrero, E. 2002. Grx5 is a mitochondrial glutaredoxin required for the activity of iron/sulfur enzymes. *Mol. Biol. Cell*, **13**, 1109–1121.

Rouhier, N., Gelhaye, E., and Jacquot, J.P. 2002. Glutaredoxin-dependent peroxiredoxin from poplar: Protein–protein interaction and catalytic mechanism. *J. Biol. Chem.*, **277**, 13609–13614.

Sanadi, D.R., Langley, M., and Searls, R.L. 1959. alpha-Ketoglutaric dehydrogenase. VI. Reversible oxidation of dihydrothioctamide by diphosphopyridine nucleotide. *J. Biol. Chem.*, **234**, 178–182.

Schafer, F.Q. and Buettner, G.R. 2001. Redox environment of the cell as viewed through the redox state of the glutathione disulfide/glutathione couple. *Free Radic. Biol. Med.*, **30**, 1191–1212.

Sies, H. 1982. Nicotinamide nucleotide compartmentation. In Sies, H. (Ed.), *Metabolic compartmentation*. Academic Press, London, pp. 205–231.

Skrede, S. 1968. The mechanism of disulphide reduction by mitochondria. *Biochem. J.*, **108**, 693–699.

Soderdahl, T., Enoksson, M., Lundberg, M., Holmgren, A., Ottersen, O.P., Orrenius, S., Bolcsfoldi, G., and Cotgreave, I.A. 2003. Visualization of the compartmentalization of glutathione and protein-glutathione mixed disulfides in cultured cells. *FASEB J.*, **17**, 124–126.

Spyrou, G., Enmark, E., Miranda-Vizuete, A., and Gustafsson, J. 1997. Cloning and expression of a novel mammalian thioredoxin. *J. Biol. Chem.*, **272**, 2936–2941.

Sun, Q.A., Kirnarsky, L., Sherman, S., and Gladyshev, V.N. 2001. Selenoprotein oxidoreductase with specificity for thioredoxin and glutathione systems. *Proc. Natl. Acad. Sci. U S A*, **98**, 3673–3678.

Tanaka, T., Hosoi, F., Yamaguchi-Iwai, Y., Nakamura, H., Masutani, H., Ueda, S., Nishiyama, A., Takeda, S., Wada, H., Spyrou, G., and Yodoi, J. 2002. Thioredoxin-2 (TRX-2) is an essential gene regulating mitochondria-dependent apoptosis. *EMBO J.*, **21**, 1695–1703.

Toledano, M.B., Delaunay, A., Monceau, L., and Tacnet, F. 2004. Microbial H_2O_2 sensors as archetypical redox signaling modules. *Trends Biochem. Sci.*, **29**, 351–357.

Tribble, D.L. and Jones, D.P. 1990. Oxygen dependence of oxidative stress. Rate of NADPH supply for maintaining the GSH pool during hypoxia. *Biochem. Pharmacol.*, **39**, 729–736.

Verdoucq, L., Vignols, F., Jacquot, J.P., Chartier, Y., and Meyer, Y. 1999. In vivo characterization of a thioredoxin h target protein defines a new peroxiredoxin family. *J. Biol. Chem.*, **274**, 19714–19722.

Watabe, S., Kohno, H., Kouyama, H., Hiroi, T., Yago, N., and Nakazawa, T. 1994. Purification and characterization of a substrate protein for mitochondrial ATP-dependent protease in bovine adrenal cortex. *J. Biochem. (Tokyo)*, **115**, 648–654.

Wingert, R.A., Galloway, J.L., Barut, B., Foott, H., Fraenkel, P., Axe, J.L., Weber, G.J., Dooley, K., Davidson, A.J., Schmidt, B., Paw, B.H., Shaw, G.C., Kingsley, P., Palis, J., Schubert, H., Chen, O., Kaplan, J., and Zon, L.I. 2005. Deficiency of glutaredoxin 5 reveals Fe-S clusters are required for vertebrate haem synthesis. *Nature*, **436**, 1035–1039.

Wood, Z.A., Poole, L.B., and Karplus, P.A. 2003a. Peroxiredoxin evolution and the regulation of hydrogen peroxide signaling. *Science*, **300**, 650–653.

Wood, Z.A., Schroder, E., Robin Harris, J., and Poole, L.B. 2003b. Structure, mechanism and regulation of peroxiredoxins. *Trends Biochem. Sci.*, **28**, 32–40.

Yang, K.S., Kang, S.W., Woo, H.A., Hwang, S.C., Chae, H.Z., Kim, K., and Rhee, S.G. 2002. Inactivation of human peroxiredoxin I during catalysis as the result of the oxidation of the catalytic site cysteine to cysteine-sulfinic acid. *J. Biol. Chem.*, **277**, 38029–38036.

Zhao, F., Yan, J., Deng, S., Lan, L., He, F., Kuang, B., and Zeng, H. 2006. A thioredoxin reductase inhibitor induces growth inhibition and apoptosis in five cultured human carcinoma cell lines. *Cancer Lett.*, **236**, 46–53.

Zhong, L., Arner, E.S., and Holmgren, A. 2000. Structure and mechanism of mammalian thioredoxin reductase: The active site is a redox-active selenolthiol/selenenylsulfide formed from the conserved cysteine-selenocysteine sequence. *Proc. Natl. Acad. Sci. U S A*, **97**, 5854–5859.

Zhong, L. and Holmgren, A. 2000. Essential role of selenium in the catalytic activities of mammalian thioredoxin reductase revealed by characterization of recombinant enzymes with selenocysteine mutations. *J. Biol. Chem.*, **275**, 18121–18128.

14 Lipoic Acid as an Inducer of Phase II Detoxification Enzymes through Activation of Nrf2-Dependent Gene Expression

Kate Petersen Shay, Swapna Shenvi, and Tory M. Hagen

CONTENTS

INTRODUCTION

Cells constantly encounter toxic electrophiles, mutagens, and oxidants that arise as by-products of normal metabolism and from the environment. Overlapping arrays of antioxidant and detoxification enzymes, collectively termed phase II detoxification enzymes, defend against both endogenous and exogenous insults. As might be expected from such an important collective defense system, phase II enzymes are highly inducible. The transcription factor, NF E2-related factor 2 (Nrf2), is essential for induction of phase II defenses, as it is the principal regulatory protein for genes containing the antioxidant response element (ARE; also termed the electrophile response element). Because of their potent and diverse means to defend against xenobiotic challenge, there has been significant interest in finding effective but non-toxic inducers of phase II defenses as a prophylaxis against chronic diseases (e.g., diabetes, cancer, and neurodegenerative disorders) and pathologies where reactive oxygen and nitrogen species (ROS, RNS, respectively) are associated. The dithiol compound, α-lipoic acid (LA), may be one such phase II inducing compound. This review will focus on the known means of regulating the phase II detoxification response through induction of Nrf2 as well as both the theoretical and known role of LA to stimulate Nrf2-mediated phase II gene induction. In particular, this role will be discussed in the context of LA to upregulate glutathione (GSH) synthesis genes, quintessential phase II enzymes regulated by Nrf2.

PHASE II GENE INDUCTION: ROLE OF NRF2 NUCLEAR TRANSLOCATION AND ITS BINDING TO THE ANTIOXIDANT RESPONSE ELEMENT

Because of the constant threat of endogenous and exogenous challenges from toxins, mutagens, and oxidants, cellular defenses have evolved that are sufficiently diverse and adaptable to respond to this seemingly limitless set of stress stimuli. Toxicologists have classified these defenses into three broad categories where phase I enzymes typically increase polarity of xenobiotic compounds, while the so-called phase II enzymes are detoxification catalysts that conjugate xenobiotics with other biomolecules (e.g., glutathione, glucuronic acid, etc.) to limit their chemical reactivity. Phase III proteins typically remove the modified toxin from cells for excretion. For Phase II enzymes, upon which this review is focused, many antioxidant enzymes are also included in this category. The reader is referred to Table 14.1, which shows a partial list of phase II enzymes and their detoxification function.

As might be expected from their diverse set of substrates, phase II enzymes are also quite varied in their catalytic structure, substrate specificity, and overall function. However, they share certain common traits, especially with regard to their transcriptional regulation following stress stimuli. Most if not all phase II enzymes contain at least one sequence comprised of TGACnnnnGC in their 5′ flanking region. This core sequence, commonly known as the antioxidant response element (ARE), is often found with an embedded TPA response element as well as different flanking regions around the core sequence. Each of these traits may affect the degree of gene expression following a stimulus. Regardless of sequence diversity, Nrf2 appears to be a critical transcription factor necessary for induced ARE gene transcription following a toxicological stress.

TABLE 14.1
Phase II Enzymes and Their Functions

Gene	Abbreviation	Function	References
NAD(P)H: Quinone oxidoreductase 1	NQO1	2 electron reduction of quinones	[54,59,63]
Glutathione-*S*-transferase	GSTA1, GSTA2	Formation of hydrophilic glutathionyl conjugates	[52,54,64,68]
Glutamate cysteine ligase catalytic and modulatory	GCLC, GCLC	Glutathione synthesis	[60–62,75,76]
Glutathione synthetase	GS	Glutathione synthesis	[24,58]
UDP glucoronosyl transferase	UGT	Conjugation of glucuronic acid with xenobiotics	[70,71,73]
Aldehyde dehydrogenase	ALDH3a1	Oxidation of aldehydes	[57]
Thioredoxin	Trx	Protein disulfide oxidoreductase	[56,72]
Thioredoxin reductase 1	TrxR	Reduction of Trx	[53,69]
Hemoxygenase 1	HO-1	Heme catabolism	[55,65,66]
Ferritin heavy chain	Ferritin (H)	Iron storage	[66,74]
Aldo-keto reductase	AKR1B3	Aldehyde metabolism	[64]
Microsomal epoxide hydrolase	EPHX	Inactivation of epoxides	[71]
γ-Glutamyl transpeptidase	GGT	Metabolism of GSH S-conjugates	[77,78]
Leukotriene B_4 12-hydroxydehydrogenase	LTB_4-12HD	Leukotriene B_4 detoxication	[67]
Glutathione peroxidase	GPx	Reduction of hydroperoxides by means of glutathione	[51]

Nrf2 is part of the basic leucine zipper (bZIP) Cap-n-Collar family of transcription factors that binds to the ARE and regulates both the basal and induced expression of many phase II genes. Following binding, Nrf2 forms heterodimers with other bZIP transcription factors, which either activate (c-Jun, small maf proteins, ATF's) or in some cases suppress (Fra-1, Bach1) gene expression. Thus, the degree of gene induction following an insult can be dynamically regulated not only by the ARE sequence motif of a given gene, but also by differences in heterodimer partners that bind with Nrf2. Regardless, Nrf2 is the principal factor necessary for phase II stress response. This is best illustrated in knockout mice where the Nrf2 gene has been ablated. These animals display lower glutathione (GSH) levels and glutathione-synthesizing enzymes (γ-glutamate-cysteine ligase [GCL] and glutathione synthetase [GS]) are more susceptible to toxicological insult [7,16], and the induction of many phase II enzyme genes is almost completely inhibited [43,58]. This suggests that Nrf2 regulation of the ARE truly represents a stress-response system, inactive when not needed, but ready to be activated by the appropriate stimuli.

KEAP-1 AND ITS ROLE IN NRF2-MEDIATED GENE EXPRESSION

Because of its central importance in governing response to environmental insults, a significant effort has been made toward understanding how stress stimuli induce Nrf2-mediated gene expression. Nrf2 is normally found in the cytosol anchored to the actin cytoskeleton by an inhibitory protein called Keap-1 (also termed iNrf2). Pioneering work done in many labs [14,26,27,29,34,36,37,44,73,74,87] showed that Keap-1 sequesters Nrf2 in the cytoplasm and the Keap-1/Nrf2 complex seems to be critical for regulating Nrf2 action. Currently, it is believed that Keap-1 has two interrelated roles regarding Nrf2 (Figure 14.1). First, it acts as a protein bridge for Nrf2 to interact with Cul3 and Rbx1, the union of which initiates ubiquitylation and subsequent degradation of Nrf2 [12,32,80]. In fact, Nrf2/Keap-1 is rapidly turned over, which only allows a small amount of

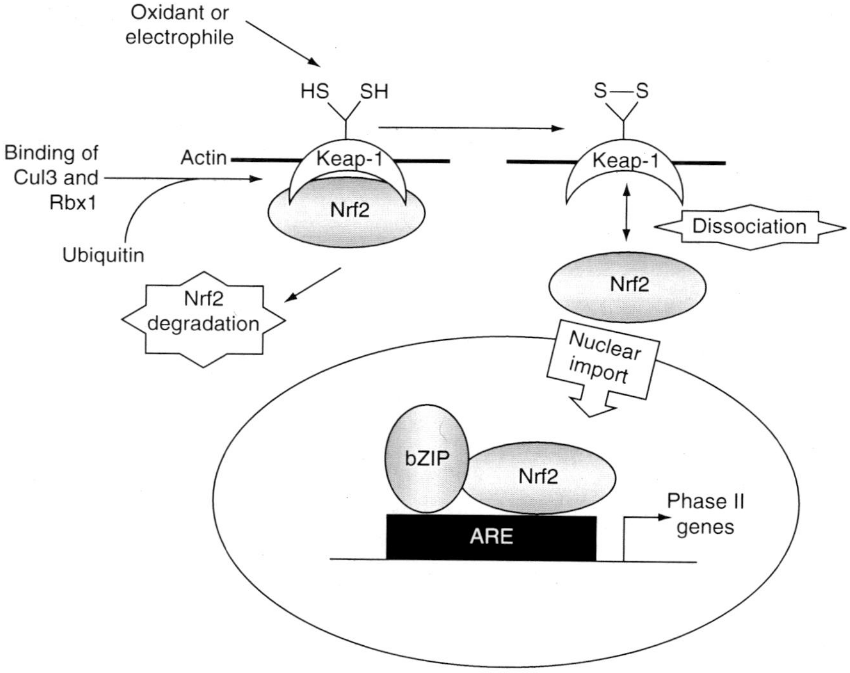

FIGURE 14.1 Keap-1-mediated regulation of Nrf2 and phase II enzyme induction. Nrf2 binds to Keap-1, which recruits Cul3 and Rbx1 to form an E3 ubiquitin ligation complex. In this situation, Nrf2 is rapidly degraded by the proteasome. Critical cysteine residues on Keap-1 can be modified by electrophilic compounds, with the result that Nrf2 degradation is halted. Then, Nrf2 is either immediately released or it remains bound to Keap-1, awaiting a further modification that allows its nuclear localization. In the nucleus, Nrf2 dimerizes with bZIP partner proteins and binds to the ARE of genes for phase II enzymes. Lipoic acid is a compound that may cause the oxidation of Keap-1's critical cysteines.

Nrf2 to be available for basal gene expression. Currently, it is not clear whether all newly synthesized Nrf2 becomes bound to Keap-1, or whether there is a pool of Nrf2 that is free to translocate to the nucleus to perform the task of basal transcription. Kobayashi et al. [33] suggested that newly synthesized Nrf2 under stress conditions may never bind Keap-1, but instead, is transported directly to the nucleus. Thus, not all of the entire cellular pool of Nrf2 may be quickly degraded, implicating Keap-1 as mainly a modulator of Nrf2 levels in the cell.

Second, Keap-1 may act as a sensor of oxidative and toxicological insults. Monofunctional phase II inducers (compounds that do not induce phase I enzymes) include hydroperoxides, isothiocyanates (e.g., sulforaphane), carotenoids, divalent heavy metal cations, Michael addition acceptors (e.g., curcuminoids, coumarins, cinnamates, chalcones), oxidizable diphenols and quinones, and 1,2-dithiole-3-thiones (e.g., oltipraz). Even though their structure and chemical reactivity differ greatly, a common property of these diverse compounds is that many if not all react with critical cysteine thiols. Keap-1 contains redox active cysteines that are highly sensitive to both oxidation as well as electrophilic attack from these known inducers of phase II gene expression. When oxidative or xenobiotic stressors modify specific Keap-1 sulfhydryls, Nrf2 degradation is halted and Keap-1 is thought to release sequestered Nrf2, which is then free to translocate to the nucleus. This idea is supported by observations showing Keap-1 is the limiting factor in the Keap-1-Nrf2 interaction. However, newer evidence suggests that Keap-1 thiol modification results in a far more complex set of actions that ultimately affect Nrf2 nuclear localization. Kobayashi et al. [33] showed that mutation of critical cysteines implicated in Nrf2 release was insufficient to disrupt the Keap-1/Nrf2 complex and revealed that cysteine 273 and 288 modification mainly prevents Nrf2 ubiquitylation, and, therefore, degradation. Lately it has been proposed that thiol modification of Keap-1 switches ubiquitylation from Nrf2 to Keap-1, which would severely limit its availability to tether Nrf2, but not necessarily release the transcription factor from the Nrf2/Keap-1 complex. In this regard, He et al. [22] showed that the half-life of cytosolic Nrf2 increased 10-fold following treatment of cells with arsenic. This was because ubiquitylation of Nrf2 was prevented even though the Nrf2-Keap-1-Cul3 complex remained intact. Thus, nuclear Nrf2 accumulation may not be simply because of its release from Keap-1, but also due to its rapid de novo synthesis and direct nuclear translocation. This may occur particularly when there is not enough Keap-1 available to bind all the Nrf2. This is a developing area of study and undoubtedly, the ultimate role that Keap-1 plays in regulating Nrf2 nuclear translocation may depend upon the stressor and/or cell type under study.

The overall role of regulating Nrf2 activation has taken on added complexity as it has been found that many inducers of Nrf2-mediated gene expression also initiate stress-activated signaling pathways that could directly affect Nrf2-dependent gene transcription. These pathways will now be briefly addressed.

STRESS SIGNALING PATHWAYS THAT INDUCE NRF2 NUCLEAR TRANSLOCATION

Thiol-modifying compounds do not specifically affect only Keap-1; many other stress-response pathways, including protein kinases, may become activated simultaneously with Keap-1. In this regard, it is notable that Nrf2 has multiple putative and proven phosphorylation sites, indicating that the Nrf2 molecule itself may be a target of protein modification and activation independent its interaction with Keap-1. Phosphorylation would be expected to modify the conformation of Nrf2, its interactions with other proteins, and also possibly affect the degree of ARE gene expression. Alternatively, many stress stimuli are known to affect phosphatases that could likewise remove phosphate groups from these domains, thus regulating how Nrf2 phosphorylation ultimately results in its nuclear translocation.

While Nrf2 phosphorylation has been implicated in its nuclear localization, the particular stress-sensing kinases involved and the mechanism(s) they elicit to affect Nrf2 activation, positively or negatively, are not well understood. Sequence analysis suggests that Nrf2 has multiple potential phosphorylation sites, and experiments using mainly cultured transformed cells show that a plethora of stress-activated kinases phosphorylate Nrf2, though not all of the target sites on Nrf2 are known. These kinases include phosphoinositol-3-phosphate kinase/Akt [42], protein kinase C isoforms [3,23,24,62] on Nrf2 Ser40, endoplasmic reticulum-activated stress kinases (PERK) [13], ERK2 and JNK1 [78], p38 MAPK, GSK3-β [64], and the protein tyrosine kinase Fyn on Nrf2 Tyr568 [28]. Thus, stress stimuli that induce any one of these enzymes may ultimately initiate a cellular response through Nrf2 phosphorylation and its activation of ARE-dependent phase II genes. Supporting this idea is research showing that direct Nrf2 phosphorylation on Ser40 by activated PKC may be responsible for the release of non-ubiquitylated Nrf2 from Keap-1, freeing it to translocate to the nucleus [23]. Additionally, inhibitors of the mitogen-activated protein kinases (MAPK) reduce Nrf2 stability [49], buttressing the concept that one or more of the these phosphorylation cascades regulate Nrf2 phosphorylation. In support of this, we recently tested the effects of the MAPK family on Nrf2 nuclear localization using menadione as a standard phase II gene inducer. As shown in Figure 14.2, nuclear levels of Nrf2 increased the following treatment. Menadione also induced phosphorylation of several MAPK family members, including ERK1/2, p38 MAPK, and SAPK/JNK. A pan-specific MAPK inhibitor (U0126, Calbiochem) not only prevented menadione-induced phosphorylation of MAPK family members, but also reduced the nuclear accumulation of Nrf2 to the levels seen in the vehicle-treated sample. These results suggest that phosphorylation of Nrf2 by the MAPK family or their downstream targets may regulate localization of Nrf2. However, whether MAPK-mediated phosphorylation causes either dissociation of Nrf2 from Keap-1 or prevents its conjugation to ubiquitin remains to be determined.

Despite these interesting results, the role of Nrf2 phosphorylation in regulating ARE-dependent stress response is far from elucidated and may be different depending on the cell type and stressor studied. For instance, treatment of HepG2

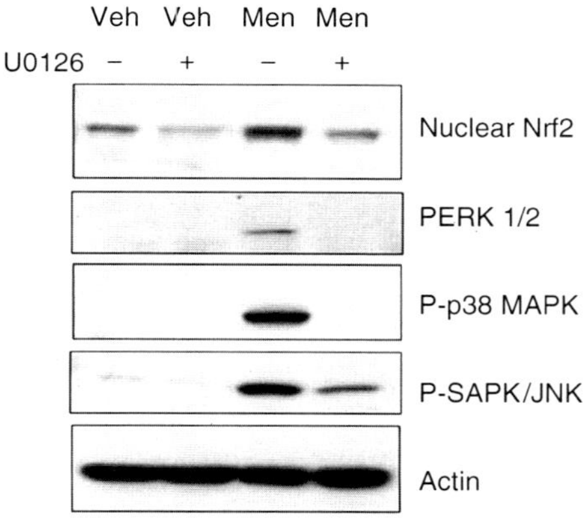

FIGURE 14.2 Menadione induces nuclear accumulation of Nrf2, an effect that is blocked by a MAPK family inhibitor. HepG2 cells were treated with 50 μM menadione for 1 h, or with DMF as the vehicle (veh.), and Nrf2 was observed in the nucleus. Pretreatment (2 h) with U0126 inhibitor prior to menadione resulted in lower levels of nuclear Nrf2 as well as a reduction of menadione-induced phosphorylation of MAPK family kinases ERK 1/2, p38 MAPK, and SAPK/JNK. Actin was used as a loading control.

cells with phorbol ester, *tert*-butylhydroquinone, or beta-napthoflavone results in protein kinase C (PKC)-mediated Nrf2 phosphorylation, and Keap-1 dissociation. However, other studies showed that PKC-dependent phosphorylation did not influence stability of Nrf2 [3]. These differences in action may be due to different PKC orthologs in diverse cell types phosphorylating Nrf2 at multiple sites with markedly different results. Thus, depending on the particular cell or tissue, Nrf2 phosphorylation could result in dissociation from Keap-1, nuclear import or export, and perhaps even a conformational change that renders it vulnerable to degradation.

Finally, phosphorylation of Nrf2 on sites affecting the nuclear export signal (NES) may also provide an additional overlay of regulation to Nrf2-mediated gene induction. Jaiswal et al. [28] recently confirmed a tyrosine phosphorylation site near the predicted NES domain. Phosphorylation at this residue, Tyr568, ultimately affects Nrf2 nuclear steady-state levels and hence may also affect the degree of Nrf2-induced ARE gene expression. This group also showed that Fyn kinase, a member of the Src kinase family, is responsible for tyrosine phosphorylation of Nrf2 at this site. Whether Nrf2 is phosphorylated at additional tyrosine sites remains largely undetermined. Phosphorylation mapping using mass spectrometry as well as design and production of antibodies specifically recognizing these sites are needed to advance the biological importance of Nrf2 phosphorylation.

In summary, release of Nrf2 from Keap-1 has been the subject of intensive research and there is increasing awareness of the complexity of regulation surrounding Nrf2-mediated phase II gene expression. Keap-1 undoubtedly plays a significant role in this process, but phosphorylation of Nrf2 at sites at or near its localization and binding motifs, and under different stress stimuli, may also markedly affect the degree of Nrf2 binding to the ARE and/or its ability to initiate gene expression.

While there is much to learn about the impressive regulatory machinery involved in Nrf2 and its regulation of phase II gene expression, there is also a strong theoretical rationale for LA to be an inducer of this detoxification system. The following section will outline both the theoretical and experimental evidences for LA to activate ARE-mediated gene expression.

THEORETICAL EVIDENCE OF α-LIPOIC ACID AS AN INDUCER OF PHASE II DETOXIFICATION

Endogenous synthesis of LA solely supplies it for mitochondrial α-ketoacid dehydrogenase and this pool does not readily enter the cytosol. Thus, for LA to induce Nrf2-mediated gene expression, it would have to be available from foods or from dietary supplements. Results from both animal and human studies show that LA is indeed readily bioavailable from the diet. It has been reported that 20%–40% of an oral LA dose given to humans is seen in the plasma [4,17,70]. Clinically, oral LA doses ranging from 600 to 1800 mg/day (up to ~26 mg/kg daily dose, based on a 70 kg body weight) have been given with high tolerance [59,60,85,86]. Following its appearance in plasma, LA is rapidly cleared with peak plasma concentrations seen after 1–2 h in humans. LA transiently increases in the liver, heart, and skeletal muscle but is also found in most tissues following ingestion [21]. LA crosses the blood–brain barrier and doses as little as 10 mg/kg given to rats result in its rapid accumulation in cerebral cortex, spinal cord, and peripheral nerves. Once taken into cells and tissues, the dithiolane ring of LA may be rapidly reduced, forming dihydrolipoic acid (DHLA), which creates a powerful redox couple with LA. However, it must be emphasized that nonprotein-bound LA neither accumulates nor remains at high levels in tissues. In fact, no detectable free plasma lipoate is evident in the post-fed state. These traits may actually make LA a better Nrf2-activating agent versus stronger compounds that elicit a sustained phase II induction. Such chronic phase II activators may be detrimental, leading to less cell turnover and increased risk for mutagenesis.

Because of the high electron density of the dithiolane ring [6], nonprotein-bound LA could theoretically react with the redox sensitive cysteine groups of Keap-1, catalyzing Nrf2 release or slowing its degradation. Either scenario would result in a higher Nrf2 nuclear accumulation. This concept is supported by similar mechanisms shown for other known Keap-1 thiol modifiers (e.g., sulforaphane and the dithiol thiones). While LA or its reduced form is generally known as a potent antioxidant, there is evidence that these compounds, like all antioxidants, can also become pro-oxidants under the right conditions, which would similarly strengthen an argument for LA to act as a moderate modifier of Keap-1 thiol redox state. With this similar rationale, many cell signaling pathways, which are known to be involved in Nrf2 nuclear translocation (see above), are induced under mild oxidative insult.

LA activates many kinases also implicated in inducing Nrf2 nuclear accumulation. For instance, a recent paper by O'Connell et al. [53] shows that chemically inhibiting p38 MAPK in human monocytic cells prevents the LA-induced binding of Nrf2 to the ARE and ultimately limits expression of HO-1. Other studies have now also shown that LA/DHLA induces phosphorylation and increased activity of Akt, a major cell survival kinase also putatively involved in Nrf2 phosphorylation [48,83]. Klip and coworkers [79] revealed that LA might work by inducing autophosphorylation of IRS-1 of the insulin signaling pathway, and activating the phosphoinositide signaling cascade, which ultimately phosphorylates Akt. IRS-1 activation resulted from an LA-induced thiol oxidation, which initiated a protein conformational change and subsequent autophosphorylation. Thus, in addition to acting on keap-1, LA is likely to act on the sulfhydryl groups of many different kinases, which would induce Nrf2 nuclear translocation and phase II gene expression.

In contrast to LA-mediated action on kinases, there is evidence that LA may also inhibit phosphatases, which could modulate signaling cascades affecting Nrf2 activation. Packer et al. [10] showed that LA repressed the activity of protein tyrosine phosphatase (PTP) 1B, which halts insulin-generated signals by dephosphorylating proteins of the insulin pathway. In this case, LA may act more as a pro-oxidant by oxidizing the critical cysteine residues of PTP1B.

EXPERIMENTAL EVIDENCE FOR α-LIPOIC ACID AS AN INDUCER OF NRF2

As outlined, both its chemical nature and the signaling pathways affected by LA treatment suggest that it would induce phase II detoxification enzymes through Nrf2 nuclear localization. The actual experimental evidence for this role will now be discussed with particular emphasis on how LA increases GSH levels by its ability to induce Nrf2-mediated expression of genes affecting GSH synthesis.

It has long been established that LA increases intracellular GSH levels in a variety of cell types and tissues [2,5]. Packer and coworkers [20] showed that LA treatment enhanced GSH levels in human cell lines and primary cells, including T cells, erythrocytes, lymphocytes, glial cells, and neuroblastoma cells. These authors concluded that DHLA reduced cystine to cysteine, which is the limiting substrate for GSH synthesis. Additionally, LA may also facilitate increased cellular cysteine levels by this effect by enhancing uptake of cystine from the plasma followed by its reduction to cysteine. Suh et al. also showed that LA reverses the age-related decline in myocardial GSH by increasing cysteine availability [68]. Interestingly, the X_c^- transport protein responsible for cystine uptake is regulated by an ARE in its 5′ flanking region, further strengthening the connection of LA as a regulator of Nrf2 activation.

Aside from enhancing cysteine availability, LA affects intracellular GSH levels by increasing GSH-synthesizing enzymes. Nrf2 is essential for the basal and inducible transcription of genes for both the catalytic (GCLC) and regulatory (GCLM) subunits of γ-glutamylcysteine ligase (GCL), which is the rate-controlling enzyme for glutathione synthesis. To determine whether LA affects GSH synthesizing capacity via Nrf2 induction of GCL expression, we examined whether LA reverses the age-related loss of GCL levels seen in aging rat liver in a Nrf2-dependent manner [67]. Treatment of old F344 rats (24 month) with the *R* enantiomer of LA (*R*-LA; 40 mg/kg body weight, IP) induced Nrf2 nuclear translocation and ARE binding, and increased the transcription of both GCLC and GCLM. Ultimately, higher hepatic GSH concentrations resulted from LA-induced increases in GCL activity (Figure 14.3). These results illustrate the power of LA to markedly augment endogenous cellular antioxidant capacity via inducement of ARE-mediated gene expression. This could arguably be a more important means to maintain cellular antioxidant defenses in times of oxidative insult than if LA merely acted as a transiently accumulating oxidant scavenger. As GSH is the most abundant cellular antioxidant, the effect of LA was to increase the cells' antioxidant capacity to a degree that could never be achieved by an exogenous antioxidant alone.

A recent study by Mervaala et al. [40] found that 6 weeks of dietary supplementation with LA (in chow, 0.5% w/w) had a positive effect on parameters of hypertension and nephrotoxicity in cyclosporine A (CsA)-treated spontaneously hypertensive rats. In addition, liver cysteine, GSH levels, and GCL mRNA were significantly increased by CsA plus LA treatment over both the control and CsA-treated rats; however, Nrf2 expression, as measured by immunohistochemistry, was significantly increased only by CsA, and brought back to control levels in rats treated with CsA plus LA. Nuclear localization and ARE binding of Nrf2 were not measured, but the observed increase in GCL message suggests that LA did induce transcription of GCL, a classic ARE-containing gene. GCL was not increased with CsA treatment alone, in spite of the higher overall levels of Nrf2 in this tissue.

The above-mentioned study by O'Connell et al. [53] used THP-1 human monocytic cells to show that LA induces HO-1, a gene that contains the ARE, through Nrf2. LA treatment caused rapid nuclear localization of Nrf2 that lasted up to 4 h, and although HO-1 mRNA was first measured at 4 h post-induction, Nrf2 binding to the HO-1 ARE was detected by 30 min treatment using electrophoretic mobility supershift assay. Nrf2 was responsible for at least two-thirds of the LA-induced HO-1 mRNA expression, as shown by transfection with a dominant negative Nrf2 construct, a mutant that lacks the transcription activation domain. Like other phase II inducers, LA can affect multiple proteins, confounding results obtained by measuring transcript abundance, but this set of experiments by O'Connell et al. confirms a specific and important role for Nrf2 in the induction of phase II enzymes following LA treatment.

LA is synthesized and biologically active as a cofactor for α-ketoacid dehydrogenase solely as the *R*-isoform, but is industrially synthesized both as the *S*- and *R*,*S*-enantiomers. We determined whether different enantiomers were equally effective at inducing Nrf2 nuclear localization in isolated rat hepatocytes by treating them with *R*-, *S*-, and *R*,*S*-LA. Results showed that *R*-LA treatment

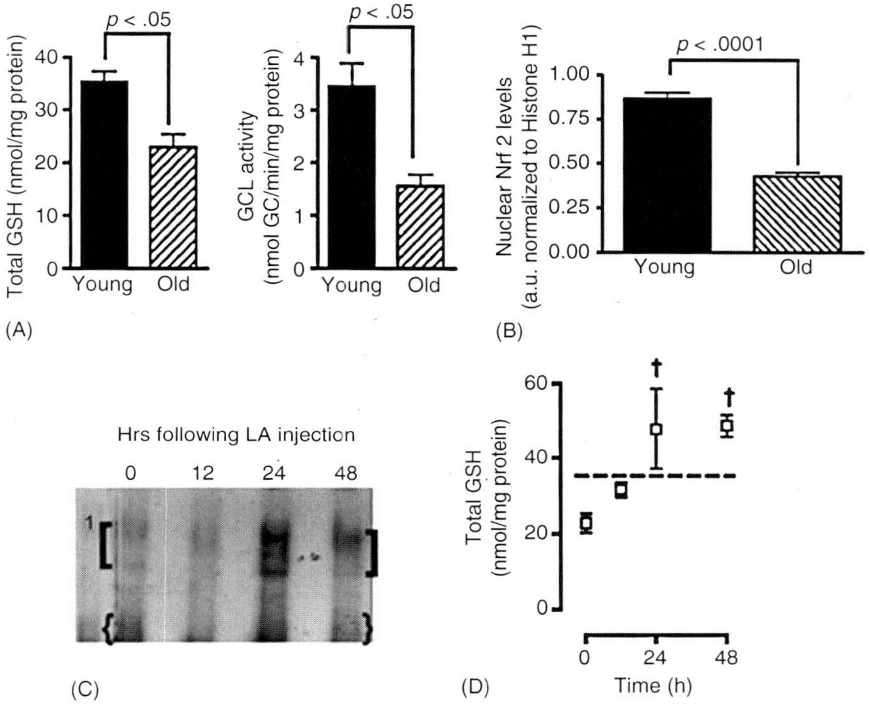

FIGURE 14.3 GSH levels, GCL activity, and nuclear Nrf2 are lower in liver from old rats compared to young, but LA improves Nrf2 binding to the gclc ARE and increases total GSH. (A) Age-related decline in total hepatic GSH is due to loss in GCL activity and expression. Hepatic GSH levels in young (3 month; $n = 4$) and old (24 month; $n = 4$) F344 rats are shown. Results show a 35% decline in total GSH (GSH plus 2 glutathione disulfide [GSSG]) in old relative to young rats. Measurement of GCL activity reveals a significant 54.8% decline with age. Results are expressed as the mean ± SEM. (B) Aged rats display a significant loss in nuclear Nrf2 content and ARE-binding activity. (C) *R*-LA induces nuclear Nrf2 levels and increases its ARE-binding activity. The time-dependent changes in nuclear Nrf2 levels after LA injection (40 mg/kg of body weight) in old rats were determined by supershift assays, which indicate that LA increases Nrf2 binding to the ARE in a time-dependent manner and show maximal binding at 24 h after LA injection. Lane 1 shows a negative control using an antibody against the P65 subunit of NF-κB. Results are representative of three independent experiments. (D) *R*-LA improves hepatic GSH levels. An overall increase in hepatic GSH 24 h post LA injection was observed. Results are expressed as the mean ± SEM. (Adapted from Suh, J.H., Shenvi, S.V., Dixon, B.M., Liu, H., Jaiswal, A.K., Liu, R.M., and Hagen, T.M., *Proc. Natl. Acad. Sci. U S A*, 101, 3381, 2004. Copyright with permission from the National Academy of Sciences, 2007.)

(100 μM) resulted in a high level of nuclear of Nrf2 after 12 h, while *S*-LA did not. The racemic mixture resulted in a small increase of nuclear Nrf2 versus the vehicle (Figure 14.4A). Qualitative real-time PCR analysis showed a pattern of

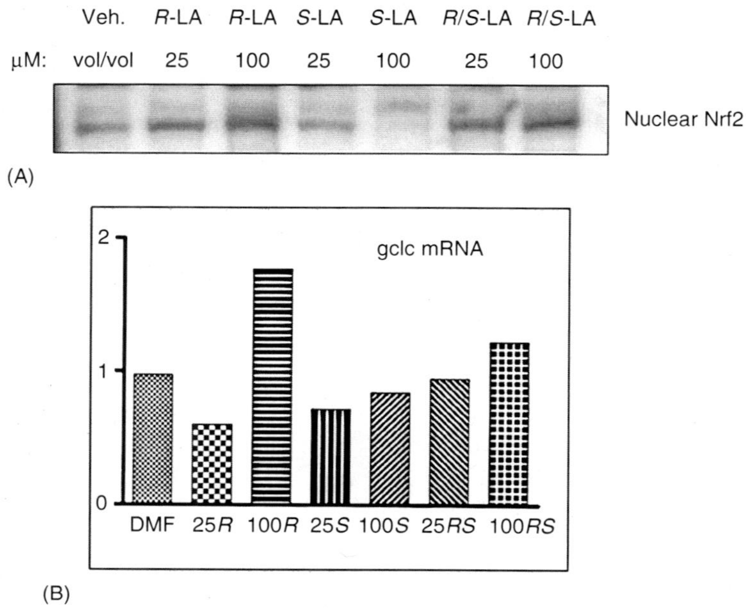

FIGURE 14.4 *R*-LA is more effective than *S*-LA at inducing Nrf2 nuclear localization and transcription of the ARE-containing gene, gclc. Hepatocytes were isolated from young rats and cultured for 24 h before treatment. Cells were then dosed with 0 (veh.), 25, or 100 μM *R*- or *S*-LA in DMF vehicle for 24 h. (A) Nuclear protein was isolated from treated cells and Nrf2 was analyzed by SDS-PAGE and western blotting. (B) mRNA was isolated from treated cells and gclc mRNA was analyzed by quantitative PCR. Data shown is representative of three separate experiments.

GCLC expression consistent with the nuclear Nrf2 levels that the different LA isoforms induced (Figure 14.4B). These results indicate that the chiral nature of LA is an important property of its ability to affect Nrf2-mediated gene transcription. On the basis of the previous reports showing that *S*-LA is less effective at preventing cataracts [41] or improving glucose handling [79] because it is more slowly reduced to DHLA than the *R*-isoform, our results also suggest that the protective effect of *R*-LA may rely upon its reduction to DHLA and its subsequent modification of proteins in the Keap-1/Nrf2 pathway.

To further illustrate the potential benefits of *R*- over *S*-LA to increase endogenous cellular defenses against toxicological insults, we showed that hepatocytes from old rats were markedly more susceptible to *N*-*tert*-butylhydroperoxide (*t*-BuOOH) than cells from young animals [19]. This increased vulnerability was due to the aforementioned age-related decline in endogenous hepatic GSH levels and a resultant compromised detoxification response. Pretreating cells with the *R*- but not the *S*-enantiomer reversed the susceptibility to *t*-BuOOH

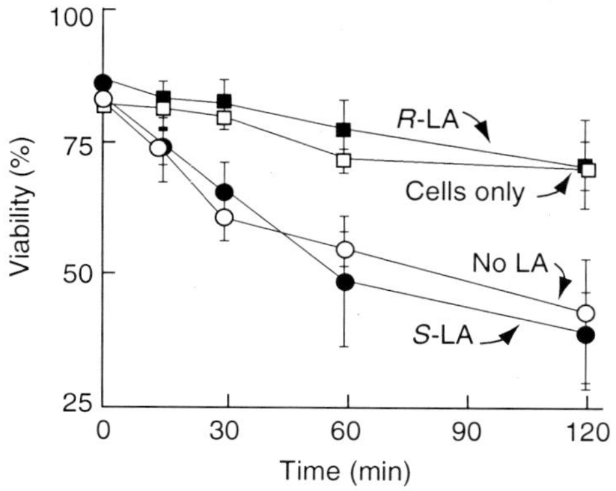

FIGURE 14.5 *R*- but not *S*-LA protects against *t*-BuOOH toxicity in cells from old rats. Isolated hepatocytes from old rats were incubated with 100 μM *R*-LA (■) or *S*-LA (●) in DMF, or DMF alone (○) for 30 min at 37°C and then given a 300 μM dose of *t*-BuOOH. Untreated cells are denoted by □. Cell viability was measured by trypan blue exclusion. Results show that only the cells pretreated with *R*-LA were significantly protected against *t*-BuOOH-mediated cytotoxicity. (Adapted from Hagen, T.M., Vinarsky, V., Wehr, C.M., and Ames, B.N., *Antiox. Redox Signal*, 2(3), 473, 2000. Copyright with permission from Mary Ann Liebert, Inc., 2007.)

(Figure 14.5). Moreover, a two-week *R*-LA (0.5% wt/wt) feeding regimen increased GSH levels such that there was no longer any significant difference between young and old GSH status. A *t*-BuOOH challenge to the isolated hepatocytes showed that *R*-LA obtained through feeding was just as effective for cell protection as direct treatment in vitro. These results indicate that *R*-LA induces GSH-synthesizing enzymes sufficiently to protect cells against toxicological insult while *S*-LA does not afford this protective effect.

As described above, the PI3K/Akt pathway has been mentioned as a potential means how LA may ultimately induce Nrf2 nuclear translocation. However, to our knowledge no direct evidence for such a role has been experimentally defined. We thus explored the role of LA to induce Nrf2 nuclear localization via the PI3K/Akt pathway. Initial results using HepG2 cells show that *R*-LA increases the phosphorylation of Akt (Figure 14.6) and simultaneously induces the nuclear accumulation of Nrf2. Increased nuclear Nrf2 localization is sustained for at least 24 h in this cell type. However, suppression of Akt phosphorylation through chemical inhibition (Calbiochem inhibitor 124018) failed to prevent the accumulation of nuclear Nrf2 induced by LA. The same result was obtained using LY294002, a PI3K inhibitor (data not shown). These results demonstrate that Akt may not be directly involved in LA-induced nuclear localization of Nrf2, and the effects of LA on Akt and Nrf2 may be two separate phenomena. Thus, other potential phosphorylation cascades that "cross-talk" with PI3K/Akt may be ultimately responsible for LA-mediated Nrf2 activation.

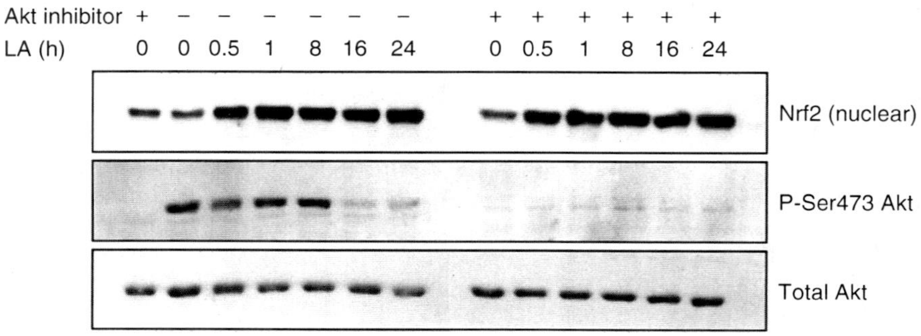

FIGURE 14.6 LA-induced nuclear accumulation of Nrf2 is not blocked by inhibition of Akt in HepG2 cells. Cells were pretreated for 2 h with Akt inhibitor VIII (Calbiochem) or DMF vehicle, then dosed with 50 μM *R*-LA for the indicated times. Proteins were isolated by subcellular fractionation (CelLytic NuCLEAR extraction kit, Sigma-Aldrich) and analyzed by SDS-PAGE and western blotting. Data shown is representative of three separate experiments.

FUTURE DIRECTIONS FOR RESEARCH ON LIPOIC ACID AND THE NRF2-KEAP-1 SYSTEM

Data suggesting that LA stimulates signal transduction pathways opens up new therapeutic possibilities in addition to its long-standing use in diabetes. Because of its chemical nature, LA is not likely to be specific to particular signaling pathways, but probably affects a wide array of proteins, including kinases, phosphatases, and transcription factors containing critical cysteine residues. Therefore, the protective effects of LA described for many cell and tissue types may be attributed not just to the scavenging of ROS and RNS, but to the concomitant increase in GSH, or to the activation of pro-survival kinases.

We have shown that LA increases the binding of Nrf2 to the ARE using nuclei isolated from young and old rat hepatocytes (Figure 14.3C) [67]. Although this study specifically examined the increase in transcription of GCLC and GCLM, Nrf2 will also bind to the AREs of various other genes. Considering that the transcriptional activity that results from Nrf2 binding is dependent upon the heterodimers it forms with partner proteins, it cannot be assumed that every ARE-containing gene is activated in all cases. LA and other phase II inducers may in fact have an effect upon this process as well. In this regard, LA has been shown to induce AP-1 nuclear localization. As there are often AP-1 binding sites embedded in ARE sequences, LA may elicit a transcriptional response through both Nrf2 and its transactivating bZIP partners. However, little is known about the dimerization of other transcription factors with Nrf2 under these circumstances. The effect of LA on Nrf2 binding partners will be another interesting area for future study.

Currently, it is not known how Nrf2 localizes to the nucleus, or whether release from Keap-1 is a prerequisite. Overexpression of wild-type Nrf2 without concurrent overexpression of Keap-1 results in nuclear accumulation of Nrf2, suggesting that Nrf2 passively localizes to the nucleus or is imported without the need for signals from phase II inducers. This evidence also suggests that Keap-1 is the limiting factor, and can be saturated by excess Nrf2. Mesecar [15] has suggested that modification of Keap-1 to halt Nrf2 degradation does not necessarily result in the dissociation of these two proteins, and because Keap-1 has also been found in the nucleus [50], there exists the possibility for co-translocation of this complex. Keap-1 has also been reported to contain an active NES [30,72]. Notably, the pool of cytoplasmic Nrf2, presumably bound to Keap-1, appears to be a much smaller quantity than the pool of Nrf2 that accumulates in the nucleus after induction by LA treatment. Can this entire amount be the result of halting Keap-1-mediated degradation of Nrf2? Is there an increase in translation of Nrf2 protein from existing transcripts, or does post-translational modification of Nrf2 contribute to its stability?

Figure 14.7 shows a schematic of the proposed mechanisms of action for LA on Nrf2 nuclear translocation. LA may oxidize the critical cysteine residues of Keap-1 to halt Nrf2 degradation or catalyze its release, or LA may modify thiol residues of a kinase that becomes activated and, in turn, phosphorylates Nrf2. Similarly, the thiol residues of a phosphatase may be modified by LA, preventing the removal of phosphate groups from kinases that may phosphorylate Nrf2. These roles are not mutually exclusive, as the redox activity of LA/DHLA may modify sulfhydryl groups on both Keap-1 and protein kinases, initiating Nrf2 activation by halting degradation as well as by promoting release. The phosphorylation of Nrf2 on an appropriate site, such as Ser40, can disrupt its binding to Keap-1. Non-Keap-1-bound Nrf2 has been shown to translocate to the nucleus, but transporters or chaperones may play a role here as well, though no such requirement has been described thus far.

The observed ability of LA to cause the phosphorylation of the insulin receptor and its downstream kinases [79] and the repression of phosphatases like PTP1B [10] indicates that there may be another layer of complexity in the regulation of Nrf2 by LA treatment. Given all the potential but heretofore unexamined phosphorylation sites on Nrf2, it will be important to test whether the import, export, and degradation of Nrf2 are regulated by phosphorylation. This will determine the role of kinases and phosphatases in controlling the availability of nuclear Nrf2 and subsequent expression of ARE-containing genes. Our results suggest that phosphorylation of Nrf2 by the MAPK family or their downstream targets may regulate localization of Nrf2. However, whether MAP kinase-mediated phosphorylation causes either its dissociation from Keap-1 or prevents its conjugation to ubiquitin remains to be determined. Because several serine/threonine kinases have been implicated in the regulation of Nrf2 localization, a question that now seems relevant is whether there are specific phosphorylation sites on Nrf2 that either cause the exposure of the NLS or the burying of the NES, or even the availability of PEST sequences that signal ubiquitinylation. Nrf2 has been reported to have a short half-life (but to be constantly synthesized), so halting degradation would be expected to have a significant effect on cellular levels of Nrf2. He et al. [22] showed that the half-life of cytosolic Nrf2 increased

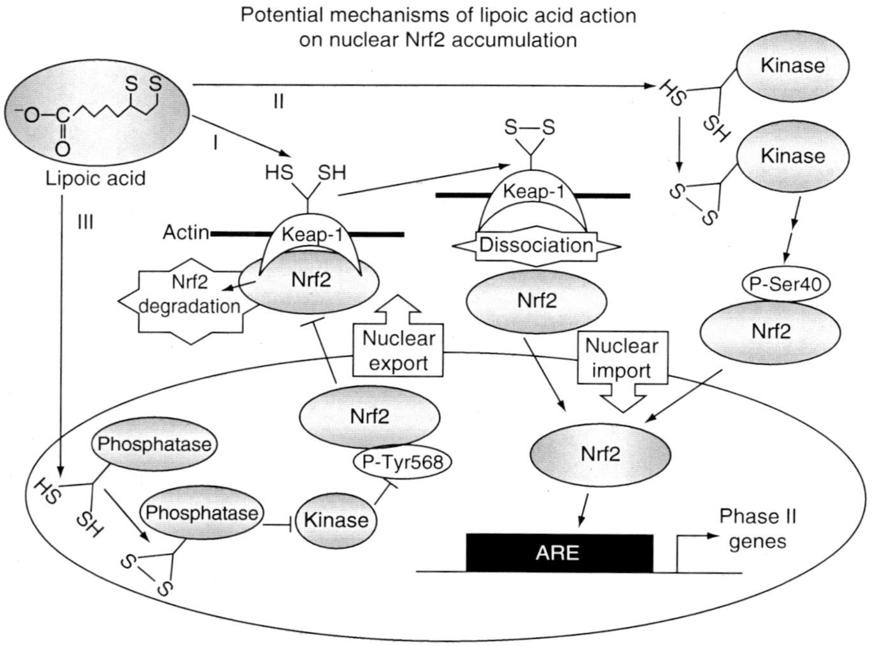

FIGURE 14.7 Proposed mechanisms of LA action on nuclear Nrf2 accumulation. (I) LA may stop Nrf2 degradation by oxidizing the critical sulfhydryl groups of Keap-1, and/or may catalyze Keap-1/Nrf2 dissociation. (II) LA may oxidize sulfhydryl groups of a kinase that becomes activated and, in turn, either phosphorylates Nrf2 directly or initiates a signaling cascade that results in Nrf2 phosphorylation. Ser40 is a strong candidate phosphorylation site. Free Nrf2 as well as Nrf2 dissociated from Keap-1 translocates to the nucleus, but the importance of phosphorylation in this step is not well understood. Once inside the nucleus, Nrf2 and its bZIP partner proteins bind to the ARE for phase II genes. (III) LA may also act to retain Nrf2 in the nucleus: The Tyr568 site on Nrf2 that mediates its export to the cytoplasm for degradation may not become phosphorylated if LA activates a phosphatase that deactivates Fyn or a related kinase.

10-fold following treatment of cells with arsenic, because ubiquitylation of Nrf2 was prevented even though the Nrf2-Keap-1-Cul3 complex remained intact. Thus, nuclear Nrf2 accumulation may not occur simply because of its release from Keap-1, but also due to its rapid de novo synthesis and direct nuclear translocation. This is a developing area of study and undoubtedly, the ultimate role that Keap-1 plays in regulating Nrf2 nuclear translocation may depend upon the stressor and cell type under study.

Experiments are currently underway that will determine the optimal dose of LA for inducing phase II genes and increasing GSH levels in vivo over the long-term. There are still many unknowns in the area of dietary supplementation with LA and the role of Nrf2. Keap-1 knockout is embryonic lethal in mice because constitutive Nrf2 activation caused hyperkeratosis of the esophagus and forestomach [74]. However, adult mice homozygous for a hepatocyte-specific deletion of Keap-1 were normal except for having high levels of nuclear Nrf2, and the animals displayed unusually high resistance to toxic doses of acetaminophen [54]. Although more work needs to be done to examine the effects of elevated Nrf2 on other tissues, these results are helpful in determining the beneficial effects of Nrf2. The mild and transient increase in Nrf2 activity incited by dietary supplementation with LA may act to prime the cellular response to ROS, RNS, and xenobiotics. Due to its use in treating diabetic polyneuropathy, the safety of LA in rats [84] and humans [11,59,60,85,86] has been investigated, and no ill effects are found at normal doses. Next, it will be important to determine the activity of Nrf2 and phase II enzymes with long-term treatment. Also worthy of consideration are the effects of LA on the signaling pathways mentioned above, because of their involvement in cell proliferation. The stronger and more sustained accumulation of nuclear Nrf2 by other phase II inducers will also need to be tested by long-term feeding for safety and organ-specific effects.

REFERENCES

1. Banning, A., S. Deubel, D. Kluth, Z. Zhou, and R. Brigelius-Flohe. 2005. The GI-GPx gene is a target for Nrf2. *Mol Cell Biol* **25:**4914–4923.
2. Bast, A. and G.R. Haenen. 1988. Interplay between lipoic acid and glutathione in the protection against microsomal lipid peroxidation. *Biochim Biophys Acta* **963:**558–561.
3. Bloom, D.A. and A.K. Jaiswal. 2003. Phosphorylation of Nrf2 at Ser40 by protein kinase C in response to antioxidants leads to the release of Nrf2 from INrf2, but is not required for Nrf2 stabilization/accumulation in the nucleus and transcriptional activation of antioxidant response element-mediated NAD(P)H:quinone oxidoreductase-1 gene expression. *J Biol Chem* **278:**44675–44682.
4. Breithaupt-Grogler, K., G. Niebch, E. Schneider, K. Erb, R. Hermann, H.H. Blume, B.S. Schug, and G.G. Belz. 1999. Dose-proportionality of oral thioctic acid—coincidence of assessments via pooled plasma and individual data. *Eur J Pharm Sci* **8:**57–65.
5. Busse, E., G. Zimmer, B. Schopohl, and B. Kornhuber. 1992. Influence of alpha-lipoic acid on intracellular glutathione in vitro and in vivo. *Arzneimittelforschung* **42:**829–831.
6. Bustamante, J., J.K. Lodge, L. Marcocci, H.J. Tritschler, L. Packer, and B.H. Rihn. 1998. Alpha-lipoic acid in liver metabolism and disease. *Free Radic Biol Med* **24:**1023–1039.
7. Chan, K., X.D. Han, and Y.W. Kan. 2001. An important function of Nrf2 in combating oxidative stress: Detoxification of acetaminophen. *Proc Natl Acad Sci U S A* **98:**4611–4616.
8. Chanas, S.A., Q. Jiang, M. McMahon, G.K. McWalter, L.I. McLellan, C.R. Elcombe, C.J. Henderson, C.R. Wolf, G.J. Moffat, K. Itoh, M. Yamamoto, and J.D. Hayes. 2002. Loss of the Nrf2 transcription factor causes a marked reduction in constitutive and inducible expression of the glutathione *S*-transferase Gsta1, Gsta2, Gstm1, Gstm2, Gstm3 and Gstm4 genes in the livers of male and female mice. *Biochem J* **365:**405–416.

9. Chen, Z.H., Y. Saito, Y. Yoshida, A. Sekine, N. Noguchi, and E. Niki. 2005. 4-Hydroxynonenal induces adaptive response and enhances PC12 cell tolerance primarily through induction of thioredoxin reductase 1 via activation of Nrf2. *J Biol Chem* **280:**41921–41927.
10. Cho, K.J., H. Moini, H.K. Shon, A.S. Chung, and L. Packer. 2003. Alpha-lipoic acid decreases thiol reactivity of the insulin receptor and protein tyrosine phosphatase 1B in 3T3-L1 adipocytes. *Biochem Pharmacol* **66:**849–858.
11. Cremer, D.R., R. Rabeler, A. Roberts, and B. Lynch. 2006. Long-term safety of alpha-lipoic acid (ALA) consumption: A 2-year study. *Regul Toxicol Pharmacol* **46:**193–201.
12. Cullinan, S.B., J.D. Gordan, J. Jin, J.W. Harper, and J.A. Diehl. 2004. The Keap1-BTB protein is an adaptor that bridges Nrf2 to a Cul3-based E3 ligase: Oxidative stress sensing by a Cul3-Keap1 ligase. *Mol Cell Biol* **24:**8477–8486.
13. Cullinan, S.B., D. Zhang, M. Hannink, E. Arvisais, R.J. Kaufman, and J.A. Diehl. 2003. Nrf2 is a direct PERK substrate and effector of PERK-dependent cell survival. *Mol Cell Biol* **23:**7198–7209.
14. Dinkova-Kostova, A.T., W.D. Holtzclaw, R.N. Cole, K. Itoh, N. Wakabayashi, Y. Katoh, M. Yamamoto, and P. Talalay. 2002. Direct evidence that sulfhydryl groups of Keap1 are the sensors regulating induction of phase 2 enzymes that protect against carcinogens and oxidants. *Proc Natl Acad Sci U S A* **99:**11908–11913.
15. Eggler, A.L., G. Liu, J.M. Pezzuto, R.B. van Breemen, and A.D. Mesecar. 2005. Modifying specific cysteines of the electrophile-sensing human Keap1 protein is insufficient to disrupt binding to the Nrf2 domain Neh2. *Proc Natl Acad Sci U S A* **102:**10070–10075.
16. Enomoto, A., K. Itoh, E. Nagayoshi, J. Haruta, T. Kimura, T. O'Connor, T. Harada, and M. Yamamoto. 2001. High sensitivity of Nrf2 knockout mice to acetaminophen hepatotoxicity associated with decreased expression of ARE-regulated drug metabolizing enzymes and antioxidant genes. *Toxicol Sci* **59:**169–177.
17. Evans, J.L., C.J. Heymann, I.D. Goldfine, and L.A. Gavin. 2002. Pharmacokinetics, tolerability, and fructosamine-lowering effect of a novel, controlled-release formulation of alpha-lipoic acid. *Endocr Pract* **8:**29–35.
18. Favreau, L.V. and C.B. Pickett. 1991. Transcriptional regulation of the rat NAD(P)H: quinone reductase gene. Identification of regulatory elements controlling basal level expression and inducible expression by planar aromatic compounds and phenolic antioxidants. *J Biol Chem* **266:**4556–4561.
19. Hagen, T.M., V. Vinarsky, C.M. Wehr, and B.N. Ames. 2000. (*R*)-alpha-lipoic acid reverses the age-associated increase in susceptibility of hepatocytes to *tert*-butylhydroperoxide both in vitro and in vivo. *Antioxid Redox Signal* **2:**473–483.
20. Han, D., G. Handelman, L. Marcocci, C.K. Sen, S. Roy, H. Kobuchi, H.J. Tritschler, L. Flohe, and L. Packer. 1997. Lipoic acid increases de novo synthesis of cellular glutathione by improving cystine utilization. *Biofactors* **6:**321–338.
21. Harrison, E.H. and D.B. McCormick. 1974. The metabolism of dl-(1,6–14C)lipoic acid in the rat. *Arch Biochem Biophys* **160:**514–522.
22. He, X., M.G. Chen, G.X. Lin, and Q. Ma. 2006. Arsenic induces NAD(P)H-quinone oxidoreductase I by disrupting the Nrf2 × Keap1 × Cul3 complex and recruiting Nrf2 × Maf to the antioxidant response element enhancer. *J Biol Chem* **281:**23620–23631.
23. Huang, H.C., T. Nguyen, and C.B. Pickett. 2002. Phosphorylation of Nrf2 at Ser-40 by protein kinase C regulates antioxidant response element-mediated transcription. *J Biol Chem* **277:**42769–42774.

24. Huang, H.C., T. Nguyen, and C.B. Pickett. 2000. Regulation of the antioxidant response element by protein kinase C-mediated phosphorylation of NF-E2-related factor 2. *Proc Natl Acad Sci U S A* **97:**12475–12480.
25. Itoh, K., T. Chiba, S. Takahashi, T. Ishii, K. Igarashi, Y. Katoh, T. Oyake, N. Hayashi, K. Satoh, I. Hatayama, M. Yamamoto, and Y. Nabeshima. 1997. An Nrf2/small Maf heterodimer mediates the induction of phase II detoxifying enzyme genes through antioxidant response elements. *Biochem Biophys Res Commun* **236:**313–322.
26. Itoh, K., N. Wakabayashi, Y. Katoh, T. Ishii, K. Igarashi, J.D. Engel, and M. Yamamoto. 1999. Keap1 represses nuclear activation of antioxidant responsive elements by Nrf2 through binding to the amino-terminal Neh2 domain. *Genes Dev* **13:**76–86.
27. Itoh, K., N. Wakabayashi, Y. Katoh, T. Ishii, T. O'Connor, and M. Yamamoto. 2003. Keap1 regulates both cytoplasmic-nuclear shuttling and degradation of Nrf2 in response to electrophiles. *Genes Cells* **8:**379–391.
28. Jain, A.K. and A.K. Jaiswal. 2006. Phosphorylation of tyrosine 568 controls nuclear export of Nrf2. *J Biol Chem* **281:**12132–12142.
29. Kang, M.I., A. Kobayashi, N. Wakabayashi, S.G. Kim, and M. Yamamoto. 2004. Scaffolding of Keap1 to the actin cytoskeleton controls the function of Nrf2 as key regulator of cytoprotective phase 2 genes. *Proc Natl Acad Sci U S A* **101:**2046–2051.
30. Karapetian, R.N., A.G. Evstafieva, I.S. Abaeva, N.V. Chichkova, G.S. Filonov, Y.P. Rubtsov, E.A. Sukhacheva, S.V. Melnikov, U. Schneider, E.E. Wanker, and A.B. Vartapetian. 2005. Nuclear oncoprotein prothymosin alpha is a partner of Keap1: Implications for expression of oxidative stress-protecting genes. *Mol Cell Biol* **25:**1089–1099.
31. Kim, Y.C., Y. Yamaguchi, N. Kondo, H. Masutani, and J. Yodoi. 2003. Thioredoxin-dependent redox regulation of the antioxidant responsive element (ARE) in electrophile response. *Oncogene* **22:**1860–1865.
32. Kobayashi, A., M.I. Kang, H. Okawa, M. Ohtsuji, Y. Zenke, T. Chiba, K. Igarashi, and M. Yamamoto. 2004. Oxidative stress sensor Keap1 functions as an adaptor for Cul3-based E3 ligase to regulate proteasomal degradation of Nrf2. *Mol Cell Biol* **24:**7130–7139.
33. Kobayashi, A., M.I. Kang, Y. Watai, K.I. Tong, T. Shibata, K. Uchida, and M. Yamamoto. 2006. Oxidative and electrophilic stresses activate Nrf2 through inhibition of ubiquitination activity of Keap1. *Mol Cell Biol* **26:**221–229.
34. Kobayashi, M., K. Itoh, T. Suzuki, H. Osanai, K. Nishikawa, Y. Katoh, Y. Takagi, and M. Yamamoto. 2002. Identification of the interactive interface and phylogenic conservation of the Nrf2-Keap1 system. *Genes Cells* **7:**807–820.
35. Kwak, M.K., T.W. Kensler, and R.A. Casero, Jr. 2003. Induction of phase 2 enzymes by serum oxidized polyamines through activation of Nrf2: Effect of the polyamine metabolite acrolein. *Biochem Biophys Res Commun* **305:**662–670.
36. Kwak, M.K., N. Wakabayashi, J.L. Greenlaw, M. Yamamoto, and T.W. Kensler. 2003. Antioxidants enhance mammalian proteasome expression through the Keap1-Nrf2 signaling pathway. *Mol Cell Biol* **23:**8786–8794.
37. Kwak, M.K., N. Wakabayashi, K. Itoh, H. Motohashi, M. Yamamoto, and T.W. Kensler. 2003. Modulation of gene expression by cancer chemopreventive dithiolethiones through the Keap1-Nrf2 pathway. Identification of novel gene clusters for cell survival. *J Biol Chem* **278:**8135–8145.
38. Lee, T.D., H. Yang, J. Whang, and S.C. Lu. 2005. Cloning and characterization of the human glutathione synthetase 5′-flanking region. *Biochem J* **390:**521–528.

39. Li, Y. and A.K. Jaiswal. 1992. Regulation of human NAD(P)H:quinone oxidoreductase gene. Role of AP1 binding site contained within human antioxidant response element. *J Biol Chem* **267:**15097–15104.
40. Louhelainen, M., S. Merasto, P. Finckenberg, R. Lapatto, Z.J. Cheng, and E.M. Mervaala. 2006. Lipoic acid supplementation prevents cyclosporine-induced hypertension and nephrotoxicity in spontaneously hypertensive rats. *J Hypertens* **24:**947–956.
41. Maitra, I., E. Serbinova, H.J. Tritschler, and L. Packer. 1996. Stereospecific effects of *R*-lipoic acid on buthionine sulfoximine-induced cataract formation in newborn rats. *Biochem Biophys Res Commun* **221:**422–429.
42. Martin, D., A.I. Rojo, M. Salinas, R. Diaz, G. Gallardo, J. Alam, C.M. De Galarreta, and A. Cuadrado. 2004. Regulation of heme oxygenase-1 expression through the phosphatidylinositol 3-kinase/Akt pathway and the Nrf2 transcription factor in response to the antioxidant phytochemical carnosol. *J Biol Chem* **279:**8919–8929.
43. McMahon, M., K. Itoh, M. Yamamoto, S.A. Chanas, C.J. Henderson, L.I. McLellan, C.R. Wolf, C. Cavin, and J.D. Hayes. 2001. The Cap'n'Collar basic leucine zipper transcription factor Nrf2 (NF-E2 p45-related factor 2) controls both constitutive and inducible expression of intestinal detoxification and glutathione biosynthetic enzymes. *Cancer Res* **61:**3299–3307.
44. McMahon, M., K. Itoh, M. Yamamoto, and J.D. Hayes. 2003. Keap1-dependent proteasomal degradation of transcription factor Nrf2 contributes to the negative regulation of antioxidant response element-driven gene expression. *J Biol Chem* **278:**21592–21600.
45. Moinova, H.R. and R.T. Mulcahy. 1998. An electrophile responsive element (EpRE) regulates beta-naphthoflavone induction of the human gamma-glutamylcysteine synthetase regulatory subunit gene. Constitutive expression is mediated by an adjacent AP-1 site. *J Biol Chem* **273:**14683–14689.
46. Moinova, H.R. and R.T. Mulcahy. 1999. Up-regulation of the human gamma-glutamylcysteine synthetase regulatory subunit gene involves binding of Nrf-2 to an electrophile responsive element. *Biochem Biophys Res Commun* **261:**661–668.
47. Mulcahy, R.T. and J.J. Gipp. 1995. Identification of a putative antioxidant response element in the 5′-flanking region of the human gamma-glutamylcysteine synthetase heavy subunit gene. *Biochem Biophys Res Commun* **209:**227–233.
48. Muller, C., F. Dunschede, E. Koch, A.M. Vollmar, and A.K. Kiemer. 2003. Alpha-lipoic acid preconditioning reduces ischemia-reperfusion injury of the rat liver via the PI3-kinase/Akt pathway. *Am J Physiol Gastrointest Liver Physiol* **285:**G769–G778.
49. Nguyen, T., P.J. Sherratt, H.C. Huang, C.S. Yang, and C.B. Pickett. 2003. Increased protein stability as a mechanism that enhances Nrf2-mediated transcriptional activation of the antioxidant response element. Degradation of Nrf2 by the 26 S proteasome. *J Biol Chem* **278:**4536–4541.
50. Nguyen, T., P.J. Sherratt, P. Nioi, C.S. Yang, and C.B. Pickett. 2005. Nrf2 controls constitutive and inducible expression of ARE-driven genes through a dynamic pathway involving nucleocytoplasmic shuttling by Keap1. *J Biol Chem* **280:** 32485–32492.
51. Nioi, P., M. McMahon, K. Itoh, M. Yamamoto, and J.D. Hayes. 2003. Identification of a novel Nrf2-regulated antioxidant response element (ARE) in the mouse NAD(P)H:quinone oxidoreductase 1 gene: Reassessment of the ARE consensus sequence. *Biochem J* **374:**337–348.

52. Nishinaka, T. and C. Yabe-Nishimura. 2005. Transcription factor Nrf2 regulates promoter activity of mouse aldose reductase (AKR1B3) gene. *J Pharmacol Sci* **97:**43–51.
53. Ogborne, R.M., S.A. Rushworth, and M.A. O'Connell. 2005. Alpha-lipoic acid-induced heme oxygenase-1 expression is mediated by nuclear factor erythroid 2-related factor 2 and p38 mitogen-activated protein kinase in human monocytic cells. *Arterioscler Thromb Vasc Biol* **25:**2100–2105.
54. Okawa, H., H. Motohashi, A. Kobayashi, H. Aburatani, T.W. Kensler, and M. Yamamoto. 2006. Hepatocyte-specific deletion of the keap1 gene activates Nrf2 and confers potent resistance against acute drug toxicity. *Biochem Biophys Res Commun* **339:**79–88.
55. Prestera, T., P. Talalay, J. Alam, Y.I. Ahn, P.J. Lee, and A.M. Choi. 1995. Parallel induction of heme oxygenase-1 and chemoprotective phase 2 enzymes by electrophiles and antioxidants: Regulation by upstream antioxidant-responsive elements (ARE). *Mol Med* **1:**827–837.
56. Primiano, T., T.W. Kensler, P. Kuppusamy, J.L. Zweier, and T.R. Sutter. 1996. Induction of hepatic heme oxygenase-1 and ferritin in rats by cancer chemopreventive dithiolethiones. *Carcinogenesis* **17:**2291–2296.
57. Primiano, T., Y. Li, T.W. Kensler, M.A. Trush, and T.R. Sutter. 1998. Identification of dithiolethione-inducible gene-1 as a leukotriene B4 12-hydroxydehydrogenase: Implications for chemoprevention. *Carcinogenesis* **19:**999–1005.
58. Ramos-Gomez, M., M.K. Kwak, P.M. Dolan, K. Itoh, M. Yamamoto, P. Talalay, and T.W. Kensler. 2001. Sensitivity to carcinogenesis is increased and chemoprotective efficacy of enzyme inducers is lost in Nrf2 transcription factor-deficient mice. *Proc Natl Acad Sci U S A* **98:**3410–3415.
59. Reljanovic, M., G. Reichel, K. Rett, M. Lobisch, K. Schuette, W. Moller, H.J. Tritschler, and H. Mehnert. 1999. Treatment of diabetic polyneuropathy with the antioxidant thioctic acid (alpha-lipoic acid): A two year multicenter randomized double-blind placebo-controlled trial (ALADIN II). alpha lipoic acid in diabetic neuropathy. *Free Radic Res* **31:**171–179.
60. Ruhnau, K.J., H.P. Meissner, J.R. Finn, M. Reljanovic, M. Lobisch, K. Schutte, D. Nehrdich, H.J. Tritschler, H. Mehnert, and D. Ziegler. 1999. Effects of 3-week oral treatment with the antioxidant thioctic acid (alpha-lipoic acid) in symptomatic diabetic polyneuropathy. *Diabet Med* **16:**1040–1043.
61. Rushmore, T.H., R.G. King, K.E. Paulson, and C.B. Pickett. 1990. Regulation of glutathione *S*-transferase Ya subunit gene expression: Identification of a unique xenobiotic-responsive element controlling inducible expression by planar aromatic compounds. *Proc Natl Acad Sci U S A* **87:**3826–3830.
62. Rushworth, S.A., R.M. Ogborne, C.A. Charalambos, and M.A. O'Connell. 2006. Role of protein kinase C delta in curcumin-induced antioxidant response element-mediated gene expression in human monocytes. *Biochem Biophys Res Commun* **341:**1007–1016.
63. Sakurai, A., M. Nishimoto, S. Himeno, N. Imura, M. Tsujimoto, M. Kunimoto, and S. Hara. 2005. Transcriptional regulation of thioredoxin reductase 1 expression by cadmium in vascular endothelial cells: Role of NF-E2-related factor-2. *J Cell Physiol* **203:**529–537.
64. Salazar, M., A.I. Rojo, D. Velasco, R.M. de Sagarra, and A. Cuadrado. 2006. Glycogen synthase kinase-3beta inhibits the xenobiotic and antioxidant cell response

by direct phosphorylation and nuclear exclusion of the transcription factor Nrf2. *J Biol Chem* **281:**14841–14851.
65. Shelby, M.K. and C.D. Klaassen. 2006. Induction of rat UDP-glucuronosyltransferases in liver and duodenum by microsomal enzyme inducers that activate various transcriptional pathways. *Drug Metab Dispos* **34:**1772–1778.
66. Slitt, A.L., N.J. Cherrington, M.Z. Dieter, L.M. Aleksunes, G.L. Scheffer, W. Huang, D.D. Moore, and C.D. Klaassen. 2006. trans-Stilbene oxide induces expression of genes involved in metabolism and transport in mouse liver via CAR and Nrf2 transcription factors. *Mol Pharmacol* **69:**1554–1563.
67. Suh, J.H., S.V. Shenvi, B.M. Dixon, H. Liu, A.K. Jaiswal, R.M. Liu, and T.M. Hagen. 2004. Decline in transcriptional activity of Nrf2 causes age-related loss of glutathione synthesis, which is reversible with lipoic acid. *Proc Natl Acad Sci U S A* **101:**3381–3386.
68. Suh, J.H., H. Wang, R.M. Liu, J. Liu, and T.M. Hagen. 2004. (*R*)-alpha-lipoic acid reverses the age-related loss in GSH redox status in post-mitotic tissues: Evidence for increased cysteine requirement for GSH synthesis. *Arch Biochem Biophys* **423:**126–135.
69. Tanito, M., H. Masutani, Y.C. Kim, M. Nishikawa, A. Ohira, and J. Yodoi. 2005. Sulforaphane induces thioredoxin through the antioxidant-responsive element and attenuates retinal light damage in mice. *Invest Ophthalmol Vis Sci* **46:**979–987.
70. Teichert, J., J. Kern, H.J. Tritschler, H. Ulrich, and R. Preiss. 1998. Investigations on the pharmacokinetics of alpha-lipoic acid in healthy volunteers. *Int J Clin Pharmacol Ther* **36:**625–628.
71. Thimmulappa, R.K., K.H. Mai, S. Srisuma, T.W. Kensler, M. Yamamoto, and S. Biswal. 2002. Identification of Nrf2-regulated genes induced by the chemopreventive agent sulforaphane by oligonucleotide microarray. *Cancer Res* **62:**5196–5203.
72. Velichkova, M. and T. Hasson. 2005. Keap1 regulates the oxidation-sensitive shuttling of Nrf2 into and out of the nucleus via a Crm1-dependent nuclear export mechanism. *Mol Cell Biol* **25:**4501–4513.
73. Wakabayashi, N., A.T. Dinkova-Kostova, W.D. Holtzclaw, M.I. Kang, A. Kobayashi, M. Yamamoto, T.W. Kensler, and P. Talalay. 2004. Protection against electrophile and oxidant stress by induction of the phase 2 response: Fate of cysteines of the Keap1 sensor modified by inducers. *Proc Natl Acad Sci U S A* **101:**2040–2045.
74. Wakabayashi, N., K. Itoh, J. Wakabayashi, H. Motohashi, S. Noda, S. Takahashi, S. Imakado, T. Kotsuji, F. Otsuka, D.R. Roop, T. Harada, J.D. Engel, and M. Yamamoto. 2003. Keap1-null mutation leads to postnatal lethality due to constitutive Nrf2 activation. *Nat Genet* **35:**238–245.
75. Wasserman, W.W. and W.E. Fahl. 1997. Functional antioxidant responsive elements. *Proc Natl Acad Sci U S A* **94:**5361–5366.
76. Wild, A.C., J.J. Gipp, and T. Mulcahy. 1998. Overlapping antioxidant response element and PMA response element sequences mediate basal and beta-naphthoflavone-induced expression of the human gamma-glutamylcysteine synthetase catalytic subunit gene. *Biochem J* **332 (Pt 2):**373–381.
77. Wild, A.C., H.R. Moinova, and R.T. Mulcahy. 1999. Regulation of gamma-glutamylcysteine synthetase subunit gene expression by the transcription factor Nrf2. *J Biol Chem* **274:**33627–33636.
78. Xu, C., X. Yuan, Z. Pan, G. Shen, J.H. Kim, S. Yu, T.O. Khor, W. Li, J. Ma, and A.N. Kong. 2006. Mechanism of action of isothiocyanates: The induction of ARE-regulated genes is associated with activation of ERK and JNK and the phosphorylation and nuclear translocation of Nrf2. *Mol Cancer Ther* **5:**1918–1926.

79. Yaworsky, K., R. Somwar, T. Ramlal, H.J. Tritschler, and A. Klip. 2000. Engagement of the insulin-sensitive pathway in the stimulation of glucose transport by alpha-lipoic acid in 3T3-L1 adipocytes. *Diabetologia* **43:**294–303.
80. Zhang, D.D., S.C. Lo, J.V. Cross, D.J. Templeton, and M. Hannink. 2004. Keap1 is a redox-regulated substrate adaptor protein for a Cul3-dependent ubiquitin ligase complex. *Mol Cell Biol* **24:**10941–10953.
81. Zhang, H., H. Liu, D.A. Dickinson, R.M. Liu, E.M. Postlethwait, Y. Laperche, and H.J. Forman. 2006. Gamma-glutamyl transpeptidase is induced by 4-hydroxynonenal via EpRE/Nrf2 signaling in rat epithelial type II cells. *Free Radic Biol Med* **40:**1281–1292.
82. Zhang, H., H. Liu, K.E. Iles, R.M. Liu, E.M. Postlethwait, Y. Laperche, and H.J. Forman. 2006. 4-Hydroxynonenal induces rat gamma-glutamyl transpeptidase through mitogen-activated protein kinase-mediated electrophile response element/nuclear factor erythroid 2-related factor 2 signaling. *Am J Respir Cell Mol Biol* **34:**174–181.
83. Zhang, L., G.Q. Xing, J.L. Barker, Y. Chang, D. Maric, W. Ma, B.S. Li, and D.R. Rubinow. 2001. Alpha-lipoic acid protects rat cortical neurons against cell death induced by amyloid and hydrogen peroxide through the Akt signaling pathway. *Neurosci Lett* **312:**125–128.
84. Ziegler, D. 2004. Thioctic acid for patients with symptomatic diabetic polyneuropathy: A critical review. *Treat Endocrinol* **3:**173–189.
85. Ziegler, D., M. Hanefeld, K.J. Ruhnau, H. Hasche, M. Lobisch, K. Schutte, G. Kerum, and R. Malessa. 1999. Treatment of symptomatic diabetic polyneuropathy with the antioxidant alpha-lipoic acid: A 7-month multicenter randomized controlled trial (ALADIN III Study). ALADIN III Study Group. Alpha-lipoic acid in diabetic neuropathy. *Diabetes Care* **22:**1296–1301.
86. Ziegler, D., M. Hanefeld, K.J. Ruhnau, H.P. Meissner, M. Lobisch, K. Schutte, and F.A. Gries. 1995. Treatment of symptomatic diabetic peripheral neuropathy with the anti-oxidant alpha-lipoic acid. A 3-week multicentre randomized controlled trial (ALADIN Study). *Diabetologia* **38:**1425–1433.
87. Zipper, L.M. and R.T. Mulcahy. 2002. The Keap1 BTB/POZ dimerization function is required to sequester Nrf2 in cytoplasm. *J Biol Chem* **277:**36544–36552.

Section III

Clinical Aspects

15 Deficiency Disorders of Components of PDH Complex: E2, E3, and E3BP Deficiencies

Jessie M. Cameron, Mary C. Maj, and Brian H. Robinson

CONTENTS

INTRODUCTION

The pyruvate dehydrogenase complex (PDHc) is a multimeric enzyme comprising several catalytic subunits and cofactors. The enzyme links the glycolytic and tricarboxylic acid pathways by decarboxylating pyruvate to acetyl coenzyme A. The first enzymatic component of the complex is E1 (pyruvate decarboxylase), a heterotetramer of two α and two β subunits. The activity of E1 is regulated by the phosphorylation state of three conserved serine residues of the E1α subunit. This reversible modification of E1α is catalyzed by two enzymes, pyruvate dehydrogenase kinase (PDK) and pyruvate dehydrogenase phosphatase (PDP), which inactivate and activate the complex, respectively (Reed 1981). Thiamine pyrophosphate is non-covalently bound to this first component and is critical for both the decarboxylation of pyruvate and the reductive transfer of the acetyl group to the multi-domained E2 component (dihydrolipoamide transacetylase). The inner catalytic domain of E2 forms the major structural core of the PDH complex; two lipoyl domains catalyze the transfer of the acetyl group to CoA and the E1-binding domain is responsible for the binding of the E1 component to the complex. The E3-binding protein (E3BP) is a minor constituent of the structural core, which is required to anchor the E3 homodimer component (dihydrolipoamide dehydrogenase) to the complex. The E3 dimer contains two flavin adenine dinucleotide (FAD) molecules required to reoxidize the dihydrolipoyl moiety of E2. $FADH_2$ is then oxidized by NAD^+ to complete the catalytic events of the complex (Figure 15.1).

The 3D structure of the mammalian PDHc has been visualized by cryo-electron microscopy at a resolution of ~35 Å. The complex consists of a pentagonal dodecahedral inner core comprising 60 molecules of E2. Trimers of E2 form the 20 vertices of the icosahedron (Zhou, McCarthy et al. 2001). Up to 30 E1 tetramers are bound to the E2 core-scaffold forming an outer shell. Twelve E3-binding proteins, found at each of the icosahedral faces of the E2 core, associate with 6–12 E3 dimers (Zhou, McCarthy et al. 2001; Milne et al. 2002). Recently, a substitution model for human PDHc has been proposed whereby 12 E3BP molecules replace 12 E2 molecules within the core, resulting in 48 E2 molecules and 6 E3BP dimers (Hiromasa et al. 2004).

Clinical Deficiency of PDHc

PDHc deficiency (OMIM 300502) is clinically heterogeneous with phenotypes ranging from fatal infantile lactic acidosis in newborns to chronic neurological dysfunction, or a milder version with intermittent ataxia. Highly aerobic,

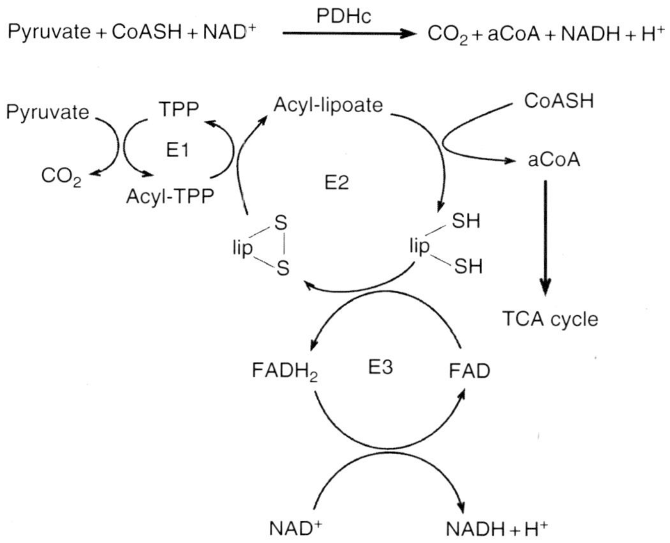

FIGURE 15.1 Catalytic reactions of the pyruvate dehydrogenase complex (PDK).

glucose-dependent tissues such as the brain are most severely affected in patients with PDHc deficiency, since these tissues are most dependent on the normal conversion of pyruvate to acetyl-CoA; affected individuals do not often survive to maturity. The majority of PDHc deficiencies arise from mutations in the X-linked *PDHA1* gene (which encodes the E1α subunit), although mutations have been identified in the genes encoding the E1β, E2, E3, and E3BP components (Kerr et al. 1996; Robinson 2001). PDHc defects arising from aberrant regulation of the complex have recently been discovered: mutations in one of the isoforms of the PDH phosphatase (*PDP1*) have been identified in both humans and dogs, giving rise to developmental delay and exercise intolerance (Maj et al. 2005; Cameron et al. 2007). Figure 15.2 shows examples of Western blots of fibroblast mitochondria derived from patients with PDHc deficiency due to mutation of different subunits. Much attention has been given to PDHc deficiencies that are ascribed to mutations of the *PDHA1* gene (Lissens et al. 2000; Cameron et al. 2004). In light of recent advances in identifying novel causes of PDHc deficiency, we discuss the less common causes of deficiency arising from E2, E3, and E3BP dysfunction. The contribution of these subunits to the structure of PDHc and how their mutation manifests as clinical phenotypes is discussed in the following sections.

E2 (DIHYDROLIPOAMIDE TRANSACETYLASE)

Human mitochondria contain three known α-ketoacid dehydrogenase complexes, all of which are comprised of E1, E2, and E3 components. The E1 and E2 component are specific to each complex whereas the same E3 component is

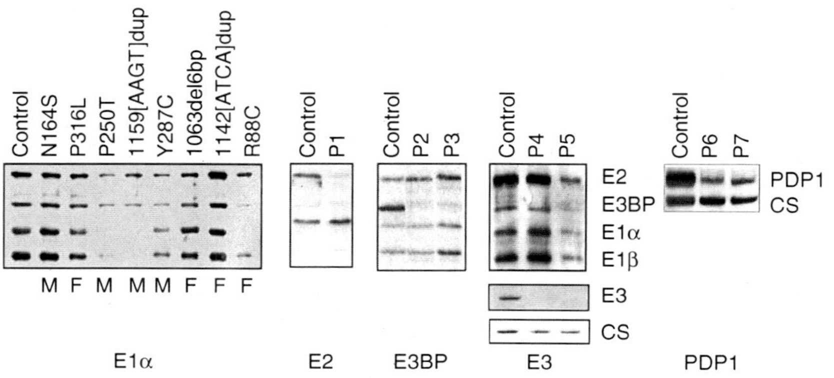

FIGURE 15.2 Western blots of human fibroblast mitochondria derived from patients with PDHc E1α, E2, E3BP, E3, and PDP1 deficiencies. E1α-deficient patients show variation in the levels of E1α and E1β protein that are immunoreactive; usually a loss of E1α will precipitate a concomitant loss of E1β as these subunits bind together in a heterotetramer. E1α patients are identified by the coding mutation that caused the deficiency, and males and females (M and F) are noted.

common to all three. Unlike PDHc, both the E2 component of mammalian 2-oxoglutarate dehydrogenase (α-ketoglutarate dehydrogenase) known as dihydrolipoyl transsuccinylase and E2 for mammalian branched-chain 2-oxoacid dehydrogenase (branched-chain α-keto acid dehydrogenase) form an octahedral core (24 E2 subunits, arranged as eight trimers) (Linn et al. 1972). The dodecahedron formation of the E2 component of mammalian PDHc is thought to be unique among oligomeric enzyme complexes (Zhou, McCarthy et al. 2001). The E2 trimers are connected in a flexible way such that the complex can "breathe" and this size variation is thought to be integral to the function of the complex (Zhou, Liao et al. 2001; Zhou, McCarthy et al. 2001).

Structure of E2 Protein

Each E2 protein is a single polypeptide with four distinct domains. The two N-terminal lipoyl domains that each has a lipoic acid moiety are covalently linked to a lysine residue (L1 and L2 domains). The peripheral subunit binding domain (PSBD) interacts with the E1 heterotetramer. There is a final catalytic inner domain at the C-terminal end that catalyzes the acetyltransferase reaction. These domains are all connected by flexible linker regions (hinges H1-3). It is through the action of these "swinging arms" that the attached lipoyl groups are able to reach the necessary active sites of multiple E1 and E3 components for catalytic activity. The size variability of the E2 core not only may enhance the transfer of reactive intermediates between catalytic sites (Zhou, Liao et al. 2001) but also may allow for selective interaction with the regulatory kinases and phosphatases.

Specific isoforms of PDP and PDK are known to interact primarily with the inner E2 lipoyl domain, but some isoforms interact with the outer lipoyl domain.

These interactions have been shown to enhance the catalytic activity of these regulatory enzymes. There are approximately one to three PDK dimers that are tightly associated with PDHc. It has been suggested that PDK must migrate around the complex using a "hand over hand" mechanism, dissociating and binding to successive lipoyl domains to phosphorylate the regulatory serine residues at the active site channel of the E1 component (Pratt and Roche 1979; Liu et al. 1995; Chen et al. 1996; Ravindran et al. 1996; Baker et al. 2000; Roche et al. 2001; Tuganova et al. 2002; Hiromasa and Roche 2003). The presence of calcium is thought to enhance the binding of PDP1 to the inner E2 lipoyl domain thereby increasing dephosphorylation activity (Chen et al. 1996). As the phospho-serine residues of the E1 component are found on the inner face of the E1 shell, size variability of the E2 core may be an integral regulator of phosphorylation events.

Genetic Mutations Causing E2 Deficiency

Since E2 forms the structural core of the PDHc, it is likely to be the least mutable of the subunits. Loss of the molecule would have a detrimental effect on the assembly of the complex as a whole. This is corroborated by the fact that only three patients have been identified with E2 deficiency, indicative of the importance of its role in the complex. Most mutations within the subunit are probably incompatible with normal function and thus homozygous mutants would be incompatible with life.

The first patient with E2 deficiency (OMIM 608770) was identified in 1990; however, insufficient genomic data precluded the identification of the defect at the molecular level (Robinson et al. 1990). The defect was discovered after a thorough examination of patients with chronic lactic acidemia and total pyruvate dehydrogenase complex deficiency. Separate activity measurements of E1, E2, and E3 in skin fibroblasts demonstrated a low E2 activity compared to normal. Western blotting with an anti-PDHc antibody showed a complete absence of E2 protein, and a reduction in E3BP (P10 in Figure 15.3). This would be expected, as with no E2 to anchor it to the complex, there will be less immunoreactive E3BP present. The patient showed very severe clinical progress with severe lactic acidosis, psychomotor retardation, and microcephaly. She was alive at 3.5 years. Unfortunately, there is no patient material available with which to complete this investigation, now that the genetic sequence of E2 is known.

In light of this early identification of the first E2-deficient patient, why have there been so few further patients identified? Is the disease that rare, or are we missing subtle clinical signs identifying those patients with non-E1α defects? The genetic sequence of E2 (*DLAT*) was determined in 1988 and consists of 14 exons, the gene is present on chromosome 11q23.1 (Thekkumkara et al. 1988).

Only two other patients have subsequently been described with E2 deficiency; they both have homozygous mutations in *DLAT*, presumably from autosomal recessive inheritance (Figure 15.4). The mutations were inherited from consanguineous parents who were confirmed to be heterozygote carriers for the mutations. The patient's clinical presentations are milder than those seen with

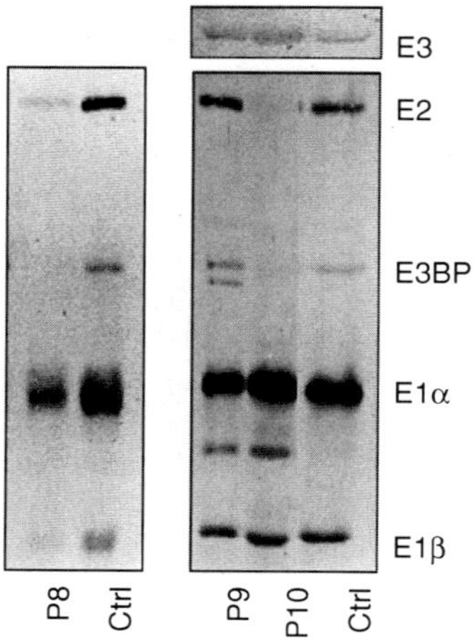

FIGURE 15.3 Western blot of the first identified patients with E3BP (P8 and P9 and E2 deficiency (P10).

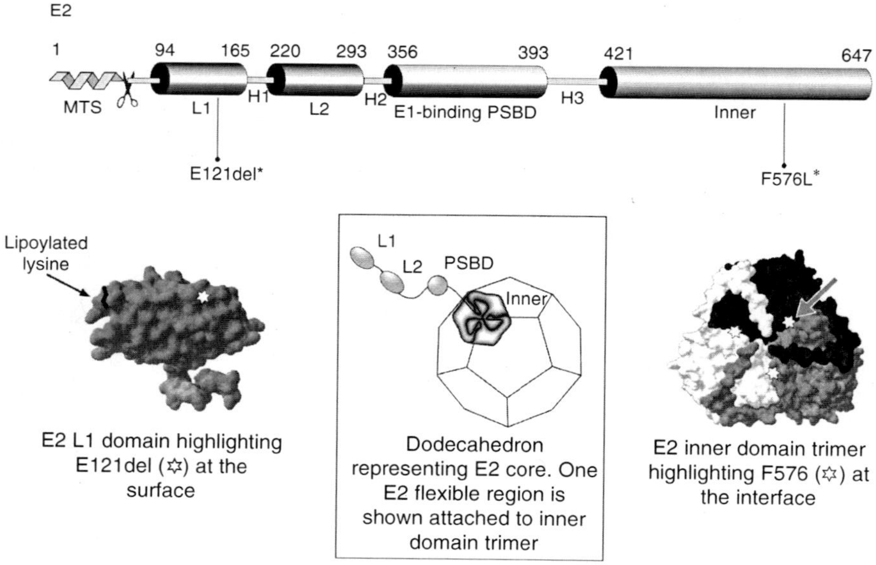

FIGURE 15.4 Protein alterations arising from homozygous mutations in *DLAT*, which encodes the E2 subunit of PDHc. Domains within the E2 protein sequence are shown. Numbers represent amino acids and are based on Entrez protein NP_001922. The approximate location of the mutations within the structural E2 domain models are shown. Structures are based on 2DNE and 1B5S in the RCSB Protein Data Bank. * denotes homozygous mutation.

PDHA1 deficiency, with survival into childhood (Head et al. 2005). One patient (E121del) demonstrated nystagmus, ptosis, and ataxia; lactic acid accumulated in the brain and there were some lesions in the globus pallidus. Importantly, the patient responded extremely well to lipoic acid and thiamine supplementation with implementation of a ketogenic diet, and is now 11 years old. The second patient (F576L) has two normal siblings and is now 8 years old. He has episodes of dystonia and developmental delay, with focal abnormalities in the basal ganglia and globus pallidus. Lactate levels were consistently normal. Both patients showed PDHc activities in fibroblasts of 30%–40% of normal rates and parents had activities of approximately 60%. Western blots for patient 1 (E121del) showed 50% reduction in E2 protein levels relative to E1α, and surprisingly a 20%–25% reduction in the parent's cells. Patient 2 (F576L) had normal E2 protein levels.

Effect of E2 Mutations on PDHc Structure and Function

In one patient, the homozygous mutation results in the deletion of an amino acid in the L1 outer lipoyl binding domain (E121del), and in the second, there is a homozygous missense mutation in the inner catalytic domain (F576L). The mutation in the lipoyl domain resulted in a 50% reduction in E2 protein levels relative to E1α as visualized by Western blot, and a 25% protein reduction in the heterozygote parents. The researchers suggest that the reduction in E2 versus E1α may simply reflect reduced avidity of the monoclonal E2 antibody employed rather than an alteration of the E1 binding. Others have shown that mammalian PDHc will bind from 20–30 E1 tetramers (Wagenknecht et al. 1991; Sanderson et al. 1996), therefore the reduction of E2 relative to E1 is possible. Still it is unclear how the deletion of a residue in the outer lipoyl domain would cause a reduction in the assembly of the E2 core. On the basis of the solution structure of the L1 domain of E2, E121 is located on the outer face in a loop region upstream of β-strand 4. The lipoylated lysine residue of this domain is found within the loop region at the C-terminal end of the same beta-strand. The loss of this residue may affect either the positioning of the lipoyl moiety reducing the efficiency of acetyltransferase activity or may unstabilize L1 domain interaction at any of the E1, E2, or E3 active sites. The L1 domain of E2 has recently been shown to interact with specific isoforms of PDK, though the majority of interaction between E2 and various kinases and phosphatases appears to be with the inner lipoyl domain (Tuganova et al. 2002; Roche et al. 2003). PDHc activities from these patient fibroblasts showed no difference compared to normal cells when exposed to PDH kinase and phosphatase, though the effect remains unknown for other tissues where kinase and phosphatase isoforms show differential expression.

The inner domain of E2, which forms the structural core of PDHc, also catalyzes the acetyltransferase reaction to form acetyl-CoA. As such, this domain should be less tolerant to mutation. The researchers suggest that the amino acid substitution of conserved phenylalanine to leucine in the second patient restricts access of the acetyl group to the E2 active site. The finding that this mutation did not cause any marked degradation of the E2 protein further suggests that the defect is primarily catalytic. Crystallographic and biochemical investigation of E2 inner domain active site residues of *Azotobacter vinelandii* (Hendle et al. 1995) suggests that residues S558 and H610 are involved in the stabilization and proton transfer during catalytic turnover. Homologous residues have also found to be required for the activity of chloramphenicol acetyltransferase (Kleanthous et al. 1985; Lewendon et al. 1990). Modeling of these residues with human F576 in the structure of the catalytic E2 dodecahedral structure of *Bacillus stearothermophilus* indicates that phenylalanine is located near the interface of two subunits and is found within 6.5 Å of both the catalytic serine and histidine residues. Thus, the relatively mild reduction of PDHc activity of the second patient is likely due to reduced efficiency at the catalytic centre. This degree of impairment of PDHc (~40% normal activity) in E1α and E3 deficiency tends to produce later onset neurodegenerative disease, developmental delay, occasionally with ataxia or dystonia. Both E2 deficient patients with reported 30%–40% rates compared to normal have survived and have not had a downward clinical course.

E3 (DIHYDROLIPOAMIDE DEHYDROGENASE)

Dihydrolipoamide dehydrogenase (dihydrolipoamide:NAD^+ oxidoreductase, E3, EC 1.8.1.4) is a component of a number of mitochondrial multienzyme complexes: pyruvate dehydrogenase complex, α-ketoglutarate dehydrogenase, (αKGDHc 2-oxoglutarate dehydrogenase), and branched-chain α-keto acid dehydrogenase (BCKDHc branched-chain 2-oxoacid dehydrogenase). It is known as E3 as it catalyzes the third and last step in a series of reactions performed by each of the complexes. In a series of intricate reactions, the regeneration of E2 lipoamide begins via the transfer of two electron and two proton molecules from dihydrolipoamide to the E3 active site disulfide bridge. This dithiol of E3 is subsequently reoxidized as the electrons and protons are transferred to FAD. $FADH_2$ is then reoxidized by NAD^+ to form NADH and H^+ and the resting E3 component is restored (see Figure 15.1). E3 is also a component of the glycine cleavage system (GCS), in which it is known as L protein. αKGDHc is one of the enzymes in the tricarboxylic acid cycle; BCKDHc catalyzes the catabolism of the essential amino acids leucine, isoleucine, and valine, and GCS degrades glycine. Because of the essential role of E3 in four enzyme complexes essential to normal intermediary metabolism, its deficiency has a devastating clinical effect. Homozygous disruption of the gene to produce knockout mouse models, results in embryonic lethality (Johnson et al. 1997).

Structure of E3 Protein

E3 protein is a homodimer with a molecular mass of 51 kDa. One molecule of FAD is non-covalently attached and one molecule of NAD^+ is transiently bound to each subunit. Each E3 monomer consists of four subdomains, an FAD-binding domain, an NAD^+-binding domain, central and homodimer interface domains. The catalytic disulfide site is found at the bottom of a deep cavity which is in close proximity to the isoalloxazine of FAD. NAD^+ binds near the other side of this FAD ring system. The FAD cofactor is so integral to the function of E3 that its binding is required for correct assembly and folding (Lindsay et al. 2000). Human E3 has been crystallized at 2.5 Å resolution in complex with NAD^+ and FAD (Toyoda et al. 1998; Brautigam et al. 2005). Inspection of the structure reveals that most of the mutations in the E3 component that give rise to disease form close contacts with FAD but are not necessarily found within the FAD-binding domain.

Genetic Mutations Causing E3 Deficiency

E3 (*DLD*) is present on chromosome 7q31–32 and consists of 14 exons (Scherer et al. 1991; Feigenbaum and Robinson 1993). As with the other enzymes in the PDHc, there is a characteristic pattern to the mutational spectrum within this gene: the majority of the defects in E3 are compound heterozygote mutations. Owing to its involvement in several enzyme complexes, the deficiency of this enzyme is also reported as maple syrup urine disease type III, when diagnosed primarily on the basis of BCKDHc deficiency; type III signifies deficiency due to mutation of the E3 subunit.

As is common with autosomal recessive conditions, E3 deficiency shows prevalence in a discrete genetic population, the Ashkenazi Jews. This population is endogamous, and so there is a higher incidence than normal of homozygosity for recessive diseases. E3 deficiency can manifest by compound heterozygosity as well as homozygosity and so the chances of inheriting the disease in this population are high. There is one common mutant allele, G229C, which presents as a milder, adult-onset E3 deficiency in the homozygous form but has also been identified in compound heterozygotes. The carrier rate for the G229C allele has been estimated at 1:94 in Ashkenazi Jews in Israel (Shaag et al. 1999). This would suggest 1:35,000 live births in the Ashkenazi Jewish population will be G229C homozygotes.

The first patient with E3 deficiency was identified in 1977 (Robinson et al. 1977; Taylor et al. 1978). The child died at 7 months of progressive neurological deterioration and persistent metabolic acidosis. The patient had elevated blood pyruvate, lactate, α-ketoglutarate, branched-chain amino acids, and occasional hypoglycemia. PDHc activity was deficient in all tissues tested, as was αKGDHc activity. BCKDHc activity was later found also to be deficient. E1 activities were normal, and E3 activities were deficient suggesting specific deficiency of the E3 subunit. Several other cases of E3 deficiency have been reported that have not been justified at the molecular level (Taylor et al. 1978; Robinson et al. 1981; Munnich et al. 1982; Kuhara et al. 1983; Matalon et al. 1984; Matuda et al. 1984; Otulakowski et al. 1985; Sakaguchi et al. 1986; Elpeleg et al. 1995).

Clinical Outcome of E3 Deficiency

Clinical outcome in E3 deficiency can vary substantially between patients, and this is attributable to the negative contribution from two mutant alleles. The common metabolic findings in E3 deficient patients are increased plasma branched-chain amino acids, increased urinary α-ketoacids, and α-hydroxy organic acids. Lactic acidosis is a common finding, but the extent of neurological dysfunction appears to depend on the nature of the two *DLD* mutations. Affected children usually manifest with vomiting, hypotonia, hepatopathy, and psychomotor retardation very early in life.

The common G229C mutation has been identified in tandem with different heterozygote mutations and as a homozygous mutation. It is predominate in the Ashkenazi Jewish population, but has been identified in non-Jewish patients. In the homozygous state, the E3 deficiency is a much milder presentation with a later age of onset. The reported patients' age of presentation vary from 3 to 36 years (Shaag et al. 1999). These patients had normal psychomotor development between metabolic decompensation episodes with exertional fatigue. In contrast to other mutations, there is no neurological dysfunction but there is a predominance of liver involvement.

Genetic Heterogeneity in E3-Deficient Patients

Mutations have been identified in all regions of the *DLD* coding sequence, including one splice site mutation in an intron (Figure 15.5). The enzymatic data, Western blot data, and clinical phenotypes associated with reported mutations are summarized in Table 15.1. Amino acid numbering in E3 is from the start of the precursor polypeptide, and thus includes the mitochondrial targeting sequence (MTS), which is later cleaved from the mature protein.

The potential effects to the E3 protein for individual mutations are discussed below. As most of the mutations occur in tandem with a second additional mutation, the effects can be detrimental to the protein.

Mutations in FAD Domain (Amino Acids 35–184)

I47T: Alters a large hydrophobic amino acid to a small polar one. Located in β sheet F-3 in FAD domain.

K72E: Alteration results in poor FAD binding, which leads to improper folding and assembly of E3. Activity of the enzyme is normal for the small amounts of enzyme that do properly fold. Thus, K72E mutant enzyme can restore full activity to E3-deficient yeast strain (Lanterman et al. 1996).

G136del: Deletion of this amino acid would affect binding of lipoamide at the disulfide-exchange site (Brautigam et al. 2005).

Mutations in NAD Domain (Amino Acids 185–317)

G229C: Alteration of this amino acid would lead to alteration in the α-helix of the NAD-binding domain that contacts both FAD and NADH. Binding of both FAD and NAD^+/NADH could therefore be affected. Reduction in patient protein levels suggests that the mutation causes instability of the protein. This mutation has shown to result in an increase in the K_m for NAD^+ (Cameron et al. 2006).

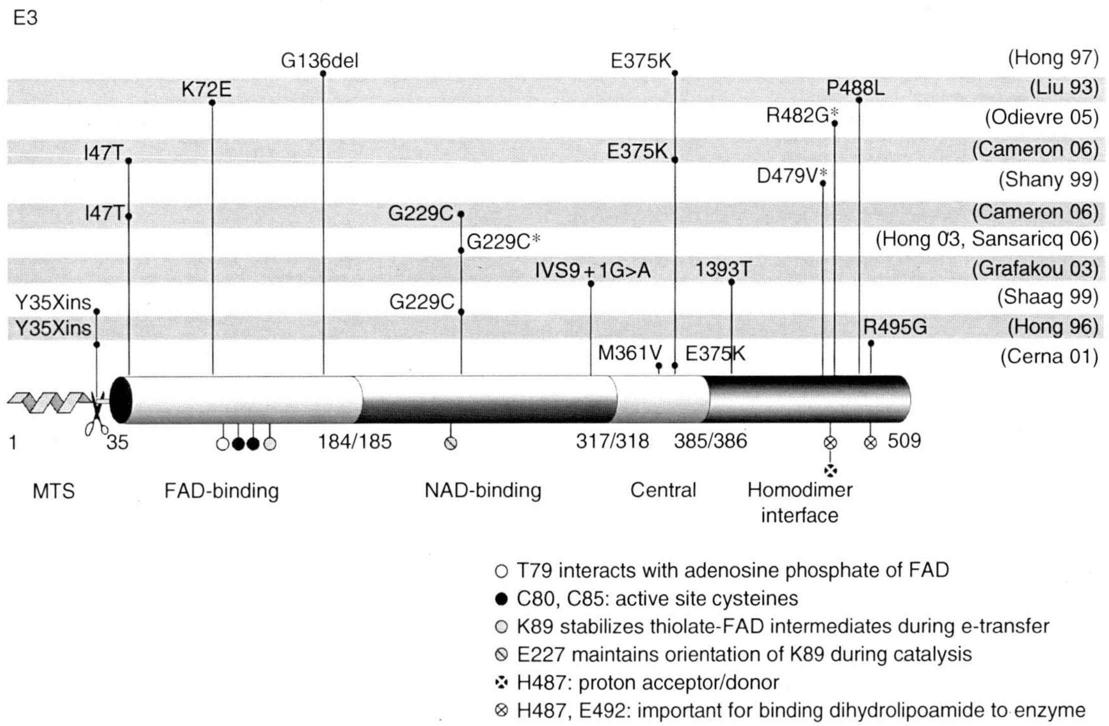

FIGURE 15.5 Protein alterations resulting from genetic mutations identified in *DLD*, which encodes the E3 subunit of PDHc, αKGDH, BCKDHc, and GCS. Domains within the protein sequence are shown based on Brautigam et al. (2005). Numbers represent amino acids and are based on Entrez protein NP_000099. Compound heterozygote mutations are highlighted in rows to illustrate the two contributing mutations, and homozygous mutations are noted (*).

TABLE 15.1
Published Mutations in *DLD* Gene and Clinical Enzyme Profiles of Patients and Parents

Mutations	Enzyme Activities: Blood lymphocytes E3	Blood lymphocytes PDHc	Fibroblasts E3	Fibroblasts PDHc	Fibroblasts αKGDHc	Muscle E3	Muscle PDHc	Muscle αKGDH
K72E/ P488L		10%–30%	6% (60% protein)			0%[a]		
	Clinical phenotype: Hypotonia, minimum dystonic movements							(Liu et al. 1993)
Y35X ins/ R495G	1.5%	26%	14% (50% protein)	11%	20%			
Parents Father/ mother	Y35X: 45% R495G: 35%	Y35X: 54% R495G: 64%	Y35X: 42% (30% protein) R495G: 44% (90% protein)	Y35X: 55% R495G: 52%	Y35X: 75% R495G: 140%			
	Clinical phenotype: DD, hypotonia, LA							(Hong et al. 1996)
ΔG136/ E375K	9%	13%	3% (100% protein)	12%	6%	11% (0% protein)[a]	14%[a]	1%[a]
Parents Father/ mother	E375K: 57% ΔG136: 61%	E375K: 130% ΔG136: 127%	E375K: 22% (100% protein) ΔG136: 38% (100% protein)	E375K: 25% ΔG136: 31%				
	Clinical phenotype: Seizures, hypoventilation, circulatory failure, hepatomegaly, hypoglycemia, LA Parents: Good health, high intellect							(Hong et al. 1997)
Y35X ins/ G229C						(20% protein)	8%–20% (2 patients)	
	Clinical phenotypes: Mild ataxia, attention deficit disorder, muscle hypotonia, impaired vision/hepatomegaly							(Shaag et al. 1999)

G229C/ G229C	12%–31% (8 patients)					(25%–60% protein)	8%–21% (8 patients)	
	Clinical phenotypes: Exertional fatigue/died of sepsis complicated by multiorgan failure/died during metabolic decompensation							(Shaag et al. 1999)
D479V/ D479V						15% (13% protein)	0%	<2%
Parents Father/ mother	D479V: 45%	D479V: 48%						
	Clinical phenotype: LA, microcephaly, low muscle tone, mild HCM							(Shany et al. 1999)
E375K/ M361V	<5%	<5%	<5%	<5%		<5% (40% protein)	<5%	
Parents Father/ mother	E375K: (100% protein)	M361V: (100% protein)						
	Clinical phenotype: Progressive hypotonia, psychomotor retardation, LA							(Cerna et al. 2001)
I393T/ IVS9+ 1G>A				100%		29%	14%	
Parents Father/ mother	IVS9 + 1G > A: (gDNA)	I393T: (gDNA)						
	Clinical phenotype: Hypoglycemia, ataxia, microcephaly, MR, LA, ketoacidemia, Leigh syndrome							(Grafakou et al. 2003)
G229C/ G229C			8%–33% (34%–68% protein) (4 patients)					
	Clinical phenotypes: Vomiting with encephalopathy/ataxia, hepatomegaly, LA/hypoglycemia/LA, hepatic failure							(Hong et al. 2003)
R482G/ R482G			20% (2 patients)	63%	0% (BCKDHc 56%)			
	Clinical phenotypes: Severe motor dysfunction, mild cortical atrophy/HCM/death from neurologic deterioration							(Odievre et al. 2005)

(continued)

TABLE 15.1 (continued)
Published Mutations in *DLD* Gene and Clinical Enzyme Profiles of Patients and Parents

	Enzyme Activities							
Mutations	**Blood lymphocytes**		**Fibroblasts**			**Muscle**		
	E3	**PDHc**	**E3**	**PDHc**	**αKGDHc**	**E3**	**PDHc**	**αKGDH**
I47T/ G229C			10% (0% protein)	69%	44% (BCKDHc 58%)			
Parents Father/ mother	I47T: (gDNA)	G229C: (gDNA)						
	Clinical phenotype: LA, seizures, myocardial dysfunction with recurrent encephalopathy							(Cameron et al. 2006)
I47T/ E375K			9% (0% protein)	59%	25% (BCKDHc 62%)			
Parents Father/ mother	E375K: (gDNA)	I47T: (gDNA)						
	Clinical phenotype: DD, hypotonia, weakness, microcephaly							(Cameron et al. 2006)
G229C/ G229C								
	Clinical phenotype: Hypotonia, tonic-clonic seizures, hepatomegaly							(Sansaricq et al. 2006)

Note: Amino acid numbering is based on the full-length E3 protein sequence including the mitochondrial targeting signal. Protein levels refer to determinations from Western blots, relative to reference protein. Branched-chain α-keto acid dehydrogenase enzyme rates (BCKDHc) are noted if measured.
DD, developmental delay; HCM, hypertrophic cardiomyopathy; LA, lactic acidosis; MR, mental retardation.

[a] Postmortem results.

IVS9 + 1G > A: This splice site mutation produces an unstable mRNA and so the mutant allele cannot be amplified from cDNA. No E3 protein will therefore be translated from this allele.

Mutations in Central Domain (Amino Acids 318–385)

M361V: The backbone carboxy-oxygen of M361 binds NAD^+ and NADH. This residue also interacts with FAD. M361 is found in a loop region between helix 10 and 11, both of which form part of the FAD-binding pocket. Mutants expressed in *E. coli* contained less than 3% bound FAD than normal (Brautigam et al. 2005). The residue is, therefore, also involved in FAD binding, which is necessary for proper E3 folding.

E375K: E375 forms a salt bridge to R482′ in the neighboring monomer. Mutation of this glutamic acid residue (negatively charged) to a lysine (positively charged) will disrupt the ability of E3 to homodimerize due to electrostatic repulsion. This will affect the formation of the disulfide-exchange site and thus lower E3 activity (Brautigam et al. 2005). E375 is also at the surface of the E3 component where the peripheral binding domain of E3BP docks. Further, this mutation has shown to result in an increase in the K_m for NADH (Cameron et al. 2006).

Mutations in Homodimer Interface Domain (Amino Acids 386–509)

The interface domain in PDHc interacts with E3BP, and in αKGDHc and BCKDc it interacts with E2. Mutations in this region will restrict the ability of the molecule to dimerize, consequently affecting enzyme activity as both monomers contribute to the disulfide-exchange sites (Brautigam et al. 2005).

I393T: Mutation of this residue affects NAD^+ and NADH binding (Brautigam et al. 2005). The backbone amide stabilizes E227, which interacts with the nicotinamide region of NAD^+.

D479V: D479 forms a hydrogen bond with Y473′ in the neighboring monomer. The hydroxy-oxygen of the Y473 side chain forms a close contact with the side chain of an arginine residue (R155 as annotated in the structure 2F5Z, deposited at the RCSB Protein Data Bank [Brautigam et al. 2006]) of E3BP peripheral binding domain. Its alteration to valine in patients will disrupt a hydrogen bonding network for both dimerization and interaction with E3BP. The carboxy group of the peptide bond also interacts with the amide bond of R482 (see next).

R482G: This arginine residue forms a salt bridge to E375 of the neighboring subunit. The NE nitrogen of the arginine side chain H-bonds with E478, orienting D479 to form a salt bridge with an arginine residue of the E3BP peripheral binding domain, (R155, see above).

P488L: P488 inserts a bend in a loop region near the C-terminus of the protein. This bend is absolutely required for the correct positioning of the catalytic residues H487 and E492 (Thieme et al. 1981; Liu et al. 1995). These catalytic residues participate in the redox reaction at the disulfides of the neighboring subunit. The mutation of P488 may also affect local interactions with FAD (Brautigam et al. 2005).

R495G: R495 forms a salt bridge with D368 in the neighboring monomer. Its alteration to glycine would affect homodimer formation. This salt bridge is required for the position of the R495 amide backbone to interact with the backbone carboxy oxygen of the catalytic residue E492.

Mutations Creating Null Alleles

Western blot analysis of the PDHc using monoclonal antibodies to the total complex can be misleading in demonstrating the presence of E3, as this protein is often masked by the presence of E3BP migrating to the same position when performing SDS-PAGE electrophoresis. Many other factors can also affect poor visualization of E3 on a Western blot. A mutation in the protein could affect either the epitope required for antibody binding or induce degradation of the protein due to unstable dimerization. In some cases, the mutation in the gene can cause degradation either by the creation of a premature stop codon or splice site mutation. Examples of all these situations have been found in patients with E3 deficiency.

Y35Xins: This mutation results in the insertion of an extra A nucleotide in the final codon of the MTS. This creates a stop codon, which triggers premature translation termination of the precursor E3 polypeptide. No protein is therefore transcribed from alleles with this mutation. The presence of residual E3 protein in a patient with this mutation therefore depends on the second mutation and its effect on the protein sequence. The mutation has been seen twice in patients, with R495G and with G229C. The R495G mutation showed 50% E3 protein in fibroblasts, and the G229C mutations had 20% protein in muscle. In both cases, E3-specific antibody was used for immunoblotting and so immunoreactive epitopes are still available for the antibody to react with.

G136del/E375K: In this patient, normal levels of E3 were detected in the fibroblasts, but no E3 was detected in postmortem liver, heart, or skeletal muscle using an E3-specific antibody. This could be due to instability of the mutant protein under non-cultured conditions (Hong et al. 1997).

IVS9 + 1G > A/I393T: The splice site mutation clearly leads to a degradation of the mutant mRNA, suggesting that no protein is translated from this allele. No Western blots were performed on this patient, however. It is interesting that normal PDHc activities were determined in fibroblasts.

I47T/G229C and I47T/E375K: No E3 protein was identified in fibroblasts from either patient using an E3-specific antibody. These means that the second mutant allele in both patients is unable to produce enough E3 protein to be experimentally identified, presumably in all patients affected with it. Alternatively, since G229C in a homozygous state can produce immunoreactive E3, the I47T mutation must be affecting the stability of the enzyme to a great extent.

Several unpublished mutations in E3 have been recently identified: D479V/D479V (7% PDHc activity in fibroblasts, 13% E3 activity); I257R/E375K

(19% PDHc activity, 15% E3 activity); R109G/R109G (20% PDHc activity, 10% E3 activity); and T45R/A470V (2 patients: 22%, 40% PDHc activity; 7%, 5% E3 activity) (Garry Brown, personal communication).

Clinical Consequences for Heterozygote Carrier Parents

Parents of E3-affected children are usually heterozygote carriers for the mutations present in their child, and often demonstrate reduced enzyme activity, which is not low enough to cause obvious clinical signs. It is assumed that the presence of one normal *DLD* allele will produce enough normal E3 protein to escape harmful clinical consequences. However, when both parents are heterozygous carriers for different E3 mutations, these partial disease causing mutations can result in offspring with severe E3 deficiency.

Parents of one E3 patient had 50%–64% PDHc activity in blood lymphocytes and fibroblasts (patient enzyme activity was 26% and 11%, see Table 15.1) (Hong et al. 1996). E3 activity was 35%–45% of control levels for the parents and 1.5% and 14% for the patient in lymphocytes and fibroblasts, respectively. αKGDH activity in fibroblasts was within the normal range for the parents and 20% for the patient. When E3 protein levels were determined from fibroblasts, the father had 30% E3 compared with E1α and the mother had 90%. The patient inherited both mutations, Y35Xins from father and R495G from mother. The mutation, a single A insertion, results in Y35X that generates a stop codon in one of his alleles. The truncated translation product would therefore be degraded, explaining the protein levels in both the patient (50%) and the father.

A second patient was discovered with both G136del (maternal) and E375K (paternal) mutations. The patient and both parents showed control amounts of E3 protein in fibroblasts. However, E3 activity for the parents was reduced to 57%–61% in lymphocytes and 22%–38% in fibroblasts with overall PDHc activity of 25%–31%. Overall PDHc activity for the patient was 12%–13% in lymphocytes and fibroblasts whereas E3 activity was 9% and 3% of normal rates in the respective tissues (Hong et al. 1997).

In both the cases outlined above, the compound heterozygous mutation resulted in a severe clinical form of E3 deficiency, which resulted in the death of the children in infancy. The parents of the first proband presented in this section have subsequently performed prenatal screening of their next child. E3 enzyme activity was measured from frozen chorionic villus homogenate from the mother (Hong et al. 1997). E3 activity was 45% of controls, and PDHc was 93%. Fibroblasts from the chorionic villus showed 64% and 73% activities, respectively. cDNA prepared from the chorionic villus showed the presence of only the mother's mutation (G136del) confirming that the prenatal diagnosis of the sister was that of a heterozygote carrier.

Therapy for E3 Deficiency

Improvements have been shown in E3-deficient patients treated with D,L-lipoic acid (Matalon et al. 1984), but there are also patients who have not benefited from this treatment (Hong et al. 1997; Cerna et al. 2001). The disease mutations were not identified in the first patient, the latter patients had G136del/E375K and M361V/E375K mutations. Others patients have shown transient effects with lipoic acid, thiamine, carnitine, and dichloroacetate (DCA) (Craigen 1996; Hong et al. 1996).

A ketogenic diet can have varying effects, decreasing lactic acidosis in one case (Robinson et al. 1981) with no effect in another (Matalon et al. 1984). Branched-chain amino acid-restricted formula showed improvements in some cases (Sakaguchi et al. 1986; Liu et al. 1993). Daily supplementation of riboflavin, coenzyme Q10, biotin, and carnitine was favorable in a patient with G229C/G229C compared to fatal outcomes of three similarly affected patients who were not treated (Hong et al. 2003).

High-fat, low-protein diet supplemented with MCT oils (medium chain triglycerides) and DCA resulted in normalization of lactate, amino acids, and organic acids in body fluids but had no effect on psychomotor development in a patient with E375K/M361V mutations (Cerna et al. 2001). Supplementation of carnitine, vitamins, minerals, a ketogenic diet, and protein-restriction reduced episodic deterioration in one patient with IVS9 + 1G > A/I393T mutation (Grafakou et al. 2003).

E3-BINDING PROTEIN

The E3-binding protein (E3BP) plays an important structural role in PDHc. Where 20 E2 trimers form a pentagonal dodecahedron, one E3BP will associate within each of the 12 pentagonal faces. E3BP is unique to eukaryotic PDHc, anchoring E3 to the complex. Two E3BP molecules associate with a single E3 homodimer forming 2:1 stoichiometric "cross-bridges" in human PDHc (Smolle et al. 2006). E3BP has only been identified in eukaryotes, therefore Gram-positive bacteria utilize E2 to associate with both E1 and E3 (Lessard and Perham 1995). E3BP is extremely homologous to E2 but only has a single N-terminal lipoyl domain. Both have segmented multiple domains connected by linker peptide and have retained ~40% sequence identity suggesting that E2 and E3BP are isozymes derived from a common genetic ancestor (Harris et al. 1997).

Structure of E3BP Protein

The E3BP is comprised of three specific domains. Similar to E2, the N-terminal of E3BP is a lipoyl domain (L3) that contains a covalently bound lipoic acid. E3 has only one lipoyl domain compared to two domains in E2. Next is the E3 binding domain (PSBD), followed by the C-terminal inner binding domain. As with the E2 subunit, E3BP domains are linked by ~30 amino acid hinge regions, which allow for remarkable flexibility. In mammals, E2 is thought to retain ability, albeit at very low affinity, to bind E3 if E3BP has been depleted (Ling et al. 1998). However, the E3BP lipoyl domain lacks a conserved transacetylase catalytic histidine residue and is thus unable to demonstrate acetyltransferase activity

(Harris et al. 1997). The protein can be acetylated by ^{14}C pyruvate and it can be hypothesized that this occurs via the catalytic center of E1 (De Marcucci and Lindsay 1985). E3BP therefore appears to serve a purely structural role.

Genetic Mutations Causing E3BP Deficiency

In recent years, there have been very few mutations identified in *PDHX*, the gene encoding E3BP, but more mutations are coming to light due to greater vigilance for the milder clinical course and more careful sequencing of the gene. Over half of the mutations appear to be splice site mutations that would preclude their easy identification by sequencing. A splice site mutation is the one that affects the complex splicing signal that indicates removal of introns when processing pre-mRNA. Introns possess a conserved splice donor at the start of their sequence (5′ GU) and end with a splice acceptor (AG) at the 3′ end. In the intron, there is also a conserved A nucleotide called the branch site, located 20–50 base pairs upstream of the splice acceptor site. These conserved nucleotides together identify the location of introns to spliceosomes, which can remove them enzymatically (Figure 15.6). Thus, mutations within the non-coding intron can create disastrous repercussions in the coding sequence of the mRNA by misdirecting the spliceosomes (Figure 15.7).

These intronic mutations in *PDHX* cannot be identified unless genomic DNA exons are sequenced with included partial flanking intronic sequence. Large deletions caused by splice site mutations may be missed when cDNA is sequenced, due to degradation of the major deletion products. Low levels of normal transcripts escaping the aberrant splicing could provide a false indication of intact *PDHX* genomic sequence. In recent years, investigators have become

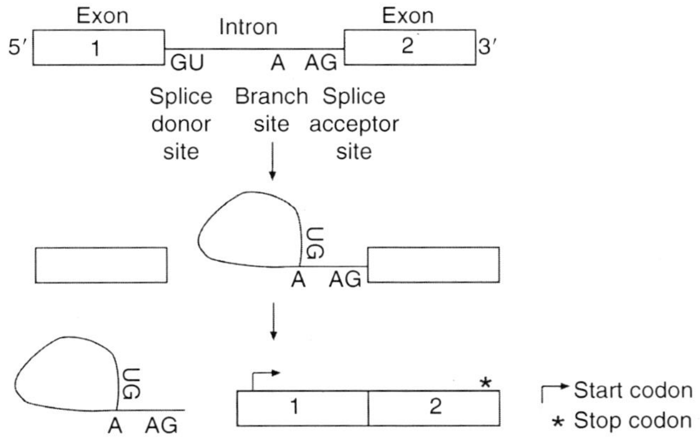

FIGURE 15.6 Splicing of precursor mRNA to mature mRNA. Spliceosomes are small ribonuclear bodies that choreograph precursor mRNA splicing reactions. The adenosine at the branch site combines with the guanine of the donor splice site, releasing the first exon and forming a lariat. Exon 1 then combines with exon 2, releasing the lariat.

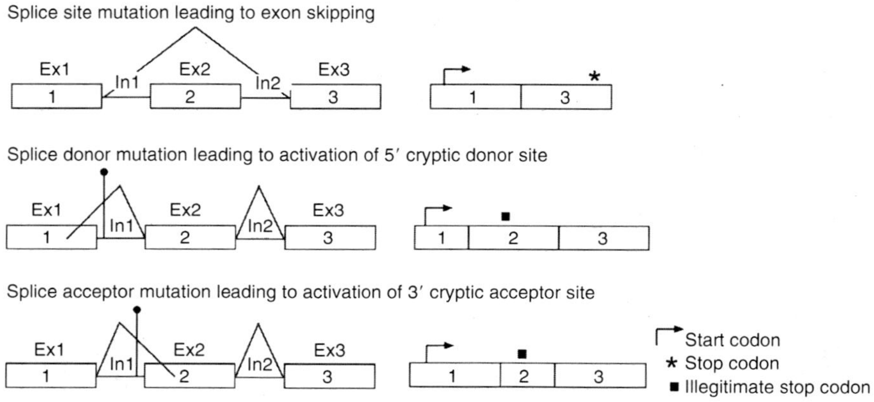

FIGURE 15.7 Examples of splice site mutations resulting in incorrect transcription of pre-mRNA.

more attuned to the nature of these mutations, and *PDHX* mutations are now more easily defined by molecular analysis.

The first two patients with E3BP deficiency (OMIM 245349, then called protein X) were identified in 1990 (Robinson et al. 1990). The defect was discovered after a thorough examination of patients presenting with chronic lactic acidemia who had enzymatically defined total PDHc deficiency. E1, E2, and E3 enzyme measurements were normal, and the specific enzyme defect was highlighted by Western blotting with an anti-PDHc antibody (Figure 15.3). One patient (P9) showed a normal band for E3BP as well as a second band of lower molecular mass. The second patient (P8) had no E3BP, with reduced E2, E1α, and E1β protein levels. The clinical presentation of P9 is extremely indicative of E3BP deficiency: the symptoms are milder than other PDHc deficiencies, and the child survived a number of years. The residual total PDHc activity in fibroblasts was also relatively well maintained, at 55% normal levels, and is presumable attributable to the presence of some normal-sized E3BP protein as seen in the Western blot. Unfortunately, there is no material available for these particular patients with which to complete this investigation, now that the genetic sequence of *PDHX* is known.

Nature of Genetic Mutations Causing E3BP Deficiency

The majority of patients with E3BP deficiency are homozygous for splice site mutations (~50% of cases), only a few cases are compound heterozygotes. The remaining patients have nonsense mutations creating inappropriate stop codons, or insertions/deletions resulting in frameshifting and subsequent premature termination.

Abolition of a splice donor or acceptor site will result in skipping of the previous or next exon. However, the consequences of splice site mutations are extremely variable: there is usually one dominant transcript arising from the aberrant splicing of the pre-mRNA, though sometimes several incorrectly spliced products can be made, as well as some normal transcript. The severity of the clinical phenotype depends on the abundance of aberrantly processed transcripts; the patient usually has a better prognosis when no E3BP is made at all, as is the case when frameshifts in the transcript lead to message or protein degradation (alternative termination).

Mutations identified to date in *PDHX* are shown in Figure 15.8. These include several different genetic mutational mechanisms.

Loss of splice 3′-acceptor site: These mutations usually result in skipping of downstream exons; for example, 965-1G > A in intron 5 results in downstream exon skipping (Dey et al. 2002). This mutation was identified in the homozygous state in two siblings, and as a compound heterozygote with another splice site mutation in another patient. 1024-1G > A in intron 8 results in two products, one of which results in skipping exon 9. The other is described in the following section (Brown et al. 2002).

Loss of splice 3′-acceptor site and activation of cryptic splice acceptor site: The second transcript produced from the mutation 1024-1G > A in intron 8 results in the use of a cryptic splice site downstream of the normal splice acceptor site, causing deletion of 10 base pairs of the beginning of exon 9 (Brown et al. 2002).

Loss of splice 5′-donor site: These mutations usually result in skipping of upstream exons; for example, 640 + 1G > A (Brown et al. 2002). This mutation is unusual in that it allows for the production of multiple mRNA species: in some transcripts exon 5 is deleted, in some exons 5 and 6 are deleted, and in some normal full-length transcript is also produced.

Loss of splice 5′-donor site and activation of cryptic splice donor site: Mutation 160 + 1G > A abolishes normal splicing of intron 1. Instead, the mutation results in the activation of an exon splicing enhancer (ESE) that creates a cryptic splice site that is used preferentially within exon 1, upstream of the 5′-splice site mutation (Dey et al. 2002). Exon enhancers determine splicing efficiency, and are usually purine rich, containing the core sequence $(GAR)_n$ where R = G or A (Berget 1995; Cartegni et al. 2002). An ESE is present in exon 1 with the sequence GAAG, which directs the spliceosomes to the cryptic splice site GT one nucleotide downstream of it.

Frameshift mutations: There are several reports of nonsense mutations that create an inappropriate stop codon (Ramadan et al. 2004; Brown et al. 2006), as well as small insertions (Brown et al. 2006) or deletions (Ling et al. 1998; Dey et al. 2003) which result in a frameshift creating a downstream premature stop codon. Transcripts containing frameshifts caused by nonsense mutations are usually degraded by the process of nonsense-mediated decay (NMD). This is a result of mRNA surveillance that eliminates transcripts with premature termination codons. Transcripts in which the mutant termination codon is located more than 50–55 nucleotides upstream of the 3′-most exon–exon junction are degraded

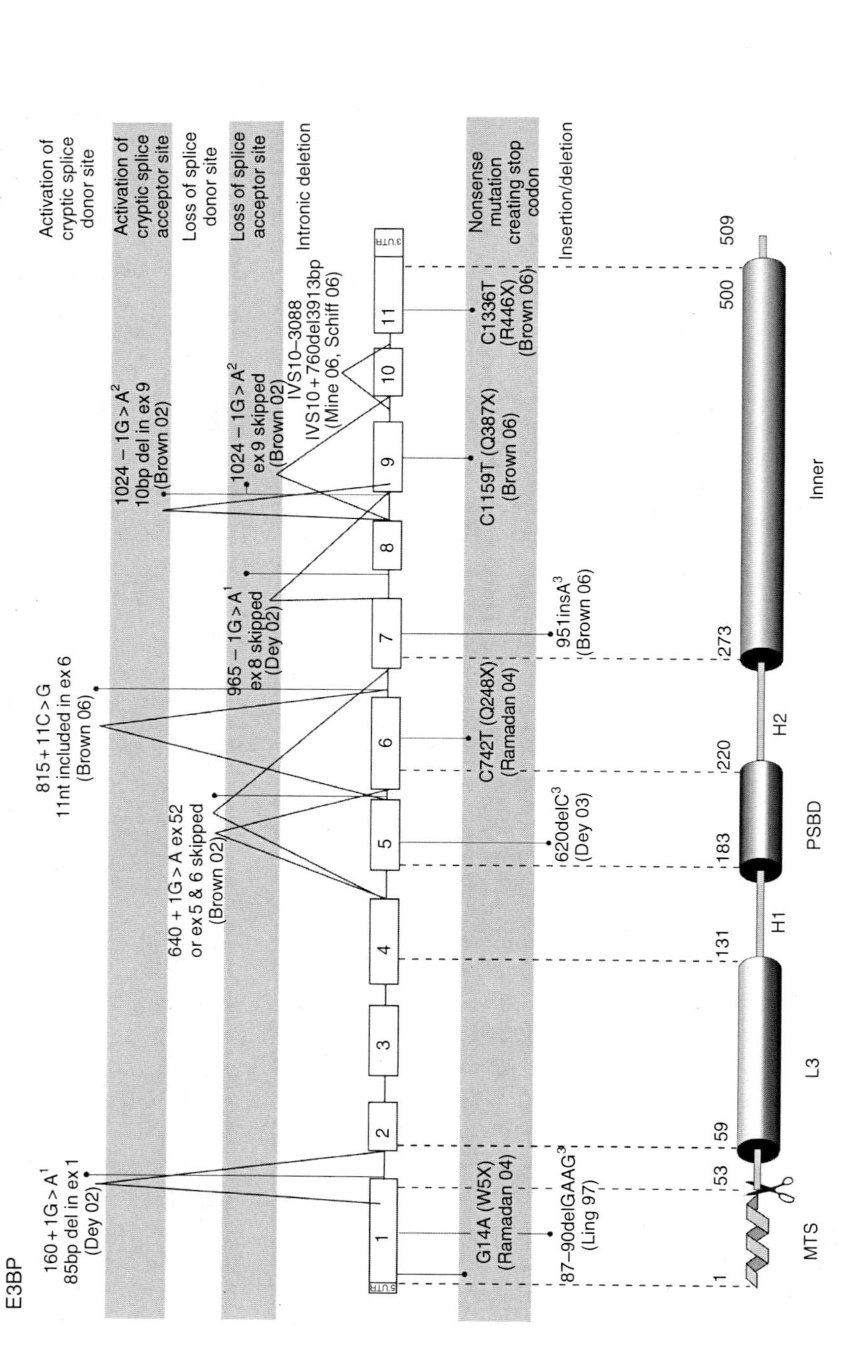

FIGURE 15.8 Genetic mutations in *PDHX* (E3BP) genomic sequence. E3BP protein sequence is shown below depicting structural domains. Numbers represent amino acids and are based on Entrez protein NP_003468.

[1] Heterozygous mutation (all others are homozygous).

[2] Mutation creates several different aberrant products as well as normal product.

[3] Mutation produces a frameshift creating premature stop codon.

(Nagy and Maquat 1998; Maquat 2004). However, naturally intron-less genes do not contain an exon–exon junction and so remaining consistent with this rule, do not elicit NMD.

Mutations causing E3BP are for the most part homozygous, suggesting autosomal recessive inheritance. This is confirmed by the fact that most of the parents of affected children are consanguineous.

E3BP Deficiency Leads to Reduced mRNA Transcripts and Loss of Protein Product

PDHX mRNA transcripts from E3BP-deficient patients usually show reduced normal levels or deletions within the coding sequence. Almost all cases of E3BP deficiency report that there is no immunoreactive E3BP on Western blots (Robinson et al. 1990; Marsac et al. 1993; Geoffroy et al. 1996; Aral et al. 1997; De Meirleir et al. 1998; Ling et al. 1998; Brown et al. 2002; Dey et al. 2002; Dey et al. 2003; Ramadan et al. 2004).

There are unusual cases where there is either lowered E3BP protein or different-sized protein products. Aral et al. (1997) showed minimal residual E3BP product by Western blot. The mRNA size and amount was normal as detected by Northern blot, but instead of no E3BP protein product as is the usual case, there was a very faint band (Aral et al. 1997). The patient had compound heterozygote mutations, and one of these mutations ($965 - 1G > A$) at the $3'$ end of the gene results in very unstable mRNA, which is difficult to amplify by RT-PCR. The $5'$ mutation ($160 + 1G > A$) in contrast is stable and so may produce some normal protein product that is detectable by Western blot (Aral et al. 1997; Rouillac et al. 1999).

Another case reported by the same authors shows a clear loss of E3BP product on Western blot, but also a higher mobility product that could be attributed to a deletion product (Aral et al. 1997). This second band was not present in a second published Western blot, however, and so could be a spurious result (De Meirleir et al. 1998). The *PDHX* mutation in this patient was not identified as the patient appeared to have normal cDNA sequence. This is surprising for if some normal mRNA is transcribed, it could be assumed that some protein product would be visible. Northern blot analysis corroborated a distinct lack of full-length mRNA. A comparable situation was seen by Brown et al. (2002), whereby a patient did have some normally spliced mRNA, but no E3BP protein on Western blot. This patient survived to at least age 24 years, highlighting the value of even miniscule levels of normal *PDHX* transcript (Hargreaves et al. 2003).

Most of the mutations in *PDHX* that have been identified are either splice site mutations that would either severely truncate the protein product when in frame, or result in a frameshift that would terminate translation. Both protein products are most likely to be degraded. The other mutations are nonsense mutations creating premature stop codons, or small insertion/deletions, which again result in frameshifting and premature termination. Thus, in the majority of

E3BP-deficient patients, there is no detectable protein by immunoblotting. It is possible that proteins of abnormal lengths are present, and the epitopes required by the antibody are simply not present, though it is more likely that degradation has occurred. Mutant proteins can cause irreparable damage if they are incorporated into enzyme complexes, the dead-end binding of a nonfunctional protein may lead to dominant-negative effects, gain-of-function, or competitive binding inhibiting the function of any normal protein.

Parental Carrier Status Shows Variable Clinical Phenotypes

Parents of affected children have rarely been tested for carrier status, but do show surprising results when they are. DeMierlier et al. (1998) showed that the father and mother of a patient had 71% and 48% residual PDHc activity in fibroblasts, respectively, as might be expected for a heterozygote carrier and lowered E3BP product compared to E2, by Western blot. In contrast, Geoffroy et al. (1996) reports two parents with normal PDHc activity and also normal E3BP product on Western blot.

Clinical Symptoms of E3BP Deficiency

The clinical course for E3BP-deficient patients shows some unique characteristics, which can help segregate its diagnosis, alleviating potential confusion with E1α-deficient patients, the most common cause of PDHc deficiency. Although both sets of patients have reduced PDHc activities, there are distinct genetic anomalies associated with E1α mutations due to the presence of the *PDHA1* gene on the X chromosome. As a result of X-inactivation, affected girls often show a more severe clinical phenotype than boys, which can also be highly variable due to the presence of normal and mutant X chromosomes in varying degrees in different tissues. Affected girls have been identified with missense and nonsense mutations, but are more likely to have insertion and deletion mutations, usually close to the end of the gene (Lissens et al. 2000; Cameron et al. 2004). The apparent mutant load can be so slight that some girls can have extremely mild phenotypes, or appear asymptomatic, preventing easy diagnosis. Affected boys tend to show a milder phenotype, due to the fact that more severe mutations are incompatible with life. They, therefore, present with missense or nonsense mutations that do not alter E1α protein production. No mutations in boys abolish PDHc activity completely (Robinson 2001).

In contrast to *PDHA1*, the gene encoding E3BP, *PDHX*, is on chromosome 11p13. The disease shows autosomal recessive inheritance, and so parents of affected children often have indications of consanguinity. The disease affects males and females equally, with similar clinical outcomes. Significantly, E3BP deficiency is unique among the PDHc enzyme deficiencies in that affected children most often show a prolonged clinical course, with considerable residual PDHc enzyme activities. The patients reported in the literature have PDHc activities ranging from 10%–55% of normal rates. Of note, the disease is most often associated with complete lack of immunoreactive E3BP protein, making this an important diagnostic tool.

Clinical symptoms can be variable, but usually include lactic acidosis with moderate to high blood lactate levels and hypotonia. Neurological symptoms include absence or thinning of the corpus callosum, and necrotic lesions of the basal ganglia (Leigh syndrome); developmental delay usually leads to mental retardation. However, patients do survive till childhood and in some cases to adulthood, a rare finding with E1α deficiency.

It is not easy to correlate the severity of the E3BP defect with chances of survival and clinical course. One child with 12% PDHc activity developed Leigh syndrome, a particularly severe outcome, as did a child with 55% residual activity (Robinson et al. 1990; Marsac et al. 1993). Patients with intermediate PDHc activities have had much better clinical outcomes, such as the patient with 22% activity that lived to 16 years (Marsac et al. 1993). This is in direct contrast to the patient with neonatal lactic acidosis who had a similar PDHc activity, and yet died in 35 days (Dey et al. 2003). In all of these cases, no immunoreactive E3BP protein was produced. It is possible that differential splicing is occurring in tissues other than fibroblasts, which could affect the prognosis of the patient.

It is apparent from patients with E3BP deficiency, that the majority of cases have low PDHc enzyme activity (10%–20% in fibroblasts) caused by splice site and other mutations that cause total degradation of the protein; this has been confirmed by in vitro studies (Seyda and Robinson 2000). The E2 protein binding domain must, therefore, maintain some ability to associate with E3 without the intermediary help of E3BP, thus resulting in residual PDHc activity (McCartney et al. 1997). Dissociation constants of human E2 with E1 and E3 (0.33 mM vs. 0.58 mM) suggest that E2 will bind E3 when not saturated with E1 (E1 will displace the bound E3) (Lessard et al. 1996). However, it has been shown that E3 does not bind to recombinant E2 when there is no E3BP present (Hiromasa et al. 2004).

CONCLUSION

The defects in the genes encoding E2, E3, and E3BP might on initial consideration be expected to produce patient cohorts with the same symptoms. However, we have seen that as is the case with *PDHA1* and *PDHAB* gene defects, the clinical sequelae are variable and depend very much on the functional consequences of individual mutations and the way in which mutations are combined. As research moves forward, we often find that it is the milder cases with a smaller decrease in total PDHc activity that are more difficult to detect and define. At the same time, the study of the subtlety of the interactions of these mutations with clinical phenotype will continue to provide insights into metabolic pathway flux and normal health and development.

ABBREVIATIONS

BCKDHc = branched-chain α-keto acid dehydrogenase complex
DCA = dichloroacetate

ESE	=	exon splicing enhancer
KGDHc	=	α-ketoglutarate dehydrogenase complex
MTS	=	mitochondrial targeting sequence
NMD	=	nonsense-mediated decay
PDHc	=	pyruvate dehydrogenase complex
PDK	=	pyruvate dehydrogenase kinase
PDP	=	pyruvate dehydrogenase phosphatase
PSBD	=	peripheral subunit binding domain

REFERENCES

Aral, B., C. Benelli, G. Ait-Ghezala, M. Amessou, F. Fouque, C. Maunoury, N. Creau, P. Kamoun, and C. Marsac. 1997. Mutations in PDX1, the human lipoyl-containing component X of the pyruvate dehydrogenase-complex gene on chromosome 11p1, in congenital lactic acidosis. *Am. J. Hum. Genet.* 61 (6):1318–1326.

Baker, J.C., X. Yan, T. Peng, S. Kasten, and T.E. Roche. 2000. Marked differences between two isoforms of human pyruvate dehydrogenase kinase. *J. Biol. Chem.* 275 (21):15773–15781.

Berget, S.M. 1995. Exon recognition in vertebrate splicing. *J. Biol. Chem.* 270 (6):2411–2414.

Brautigam, C.A., J.L. Chuang, D.R. Tomchick, M. Machius, and D.T. Chuang. 2005. Crystal structure of human dihydrolipoamide dehydrogenase: NAD + /NADH binding and the structural basis of disease-causing mutations. *J. Mol. Biol.* 350 (3):543–552.

Brautigam, C.A., R.M. Wynn, J.L. Chuang, M. Machius, D.R. Tomchick, and D.T. Chuang. 2006. Structural insight into interactions between dihydrolipoamide dehydrogenase (E3) and E3 binding protein of human pyruvate dehydrogenase complex. *Structure* 14 (3):611–621.

Brown, R.M., R.A. Head, and G.K. Brown. 2002. Pyruvate dehydrogenase E3 binding protein deficiency. *Hum. Genet.* 110 (2):187–191.

Brown, R.M., R.A. Head, A.A. Morris, J.A. Raiman, J.H. Walter, W.P. Whitehouse, and G.K. Brown. 2006. Pyruvate dehydrogenase E3 binding protein (protein X) deficiency. *Dev. Med. Child Neurol.* 48 (9):756–760.

Cameron, J.M., V. Levandovskiy, N. Mackay, I. Tein, and B.H. Robinson. 2004. Deficiency of pyruvate dehydrogenase caused by novel and known mutations in the E1alpha subunit. *Am. J. Med. Genet. A* 131 (1):59–66.

Cameron, J.M., V. Levandovskiy, N. Mackay, J. Raiman, D.L. Renaud, J.T. Clarke, A. Feigenbaum, O. Elpeleg, and B.H. Robinson. 2006. Novel mutations in dihydrolipoamide dehydrogenase deficiency in two cousins with borderline-normal PDH complex activity. *Am. J. Med. Genet. A* 140 (14):1542–1552.

Cameron, J.M., M.C. Maj, V. Levandovskiy, N. Mackay, G.D. Shelton, and B.H. Robinson. 2007. Identification of a canine model of pyruvate dehydrogenase phosphatase 1 deficiency. *Mol. Genet. Metab.* 90 (1):15–23.

Cartegni, L., S.L. Chew, and A.R. Krainer. 2002. Listening to silence and understanding nonsense: Exonic mutations that affect splicing. *Nat. Rev. Genet.* 3 (4):285–298.

Cerna, L., L. Wenchich, H. Hansikova, S. Kmoch, K. Peskova, P. Chrastina, J. Brynda, and J. Zeman. 2001. Novel mutations in a boy with dihydrolipoamide dehydrogenase deficiency. *Med. Sci. Monit.* 7 (6):1319–1325.

Chen, G., L. Wang, S. Liu, C. Chuang, and T.E. Roche. 1996. Activated function of the pyruvate dehydrogenase phosphatase through Ca^{2+}-facilitated binding to the inner lipoyl domain of the dihydrolipoyl acetyltransferase. *J. Biol. Chem.* 271 (45):28064–28070.

Craigen, W.J. 1996. Leigh disease with deficiency of lipoamide dehydrogenase: Treatment failure with dichloroacetate. *Pediatr. Neurol.* 14 (1):69–71.

De Marcucci, O. and J.G. Lindsay. 1985. Component X. An immunologically distinct polypeptide associated with mammalian pyruvate dehydrogenase multi-enzyme complex. *Eur. J. Biochem.* 149 (3):641–648.

De Meirleir, L., W. Lissens, C. Benelli, C. Marsac, J. De Klerk, J. Scholte, O. van Diggelen, W. Kleijer, S. Seneca, and I. Liebaers. 1998. Pyruvate dehydrogenase complex deficiency and absence of subunit X. *J. Inherit. Metab. Dis.* 21 (1):9–16.

Dey, R., B. Aral, M. Abitbol, and C. Marsac. 2002. Pyruvate dehydrogenase deficiency as a result of splice-site mutations in the PDX1 gene. *Mol. Genet. Metab.* 76 (4):344–347.

Dey, R., M. Mine, I. Desguerre, A. Slama, L. Van Den Berghe, M. Brivet, B. Aral, and C. Marsac. 2003. A new case of pyruvate dehydrogenase deficiency due to a novel mutation in the PDX1 gene. *Ann. Neurol.* 53 (2):273–277.

Elpeleg, O.N., W. Ruitenbeek, C. Jakobs, V. Barash, D.C. De Vivo, and N. Amir. 1995. Congenital lacticacidemia caused by lipoamide dehydrogenase deficiency with favorable outcome. *J. Pediatr.* 126 (1):72–74.

Feigenbaum, A.S. and B.H. Robinson. 1993. The structure of the human dihydrolipoamide dehydrogenase gene (*DLD*) and its upstream elements. *Genomics* 17 (2):376–381.

Geoffroy, V., F. Fouque, C. Benelli, F. Poggi, J.M. Saudubray, W. Lissens, L.D. Meirleir, C. Marsac, J.G. Lindsay, and S.J. Sanderson. 1996. Defect in the X-lipoyl-containing component of the pyruvate dehydrogenase complex in a patient with neonatal lactic acidemia. *Pediatrics* 97 (2):267–272.

Grafakou, O., K. Oexle, L. van den Heuvel, R. Smeets, F. Trijbels, H.H. Goebel, N. Bosshard, A. Superti-Furga, B. Steinmann, and J. Smeitink. 2003. Leigh syndrome due to compound heterozygosity of dihydrolipoamide dehydrogenase gene mutations. Description of the first E3 splice site mutation. *Eur. J. Pediatr.* 162 (10):714–718.

Hargreaves, I.P., S.J. Heales, A. Briddon, P.J. Lee, M.G. Hanna, and J.M. Land. 2003. Primary pyruvate dehydrogenase E3 binding protein deficiency with mild hyperlactataemia and hyperalaninaemia. *J. Inherit. Metab. Dis.* 26 (5):505–506.

Harris, R.A., M.M. Bowker-Kinley, P. Wu, J. Jeng, and K.M. Popov. 1997. Dihydrolipoamide dehydrogenase-binding protein of the human pyruvate dehydrogenase complex. DNA-derived amino acid sequence, expression, and reconstitution of the pyruvate dehydrogenase complex. *J. Biol. Chem.* 272 (32):19746–19751.

Head, R.A., R.M. Brown, Z. Zolkipli, R. Shahdadpuri, M.D. King, P.T. Clayton, and G.K. Brown. 2005. Clinical and genetic spectrum of pyruvate dehydrogenase deficiency: Dihydrolipoamide acetyltransferase (E2) deficiency. *Ann. Neurol.* 58 (2):234–241.

Hendle, J., A. Mattevi, A.H. Westphal, J. Spee, A. de Kok, A. Teplyakov, and W.G. Hol. 1995. Crystallographic and enzymatic investigations on the role of Ser558, His610, and Asn614 in the catalytic mechanism of *Azotobacter vinelandii* dihydrolipoamide acetyltransferase (E2p). *Biochemistry* 34 (13):4287–4298.

Hiromasa, Y. and T.E. Roche. 2003. Facilitated interaction between the pyruvate dehydrogenase kinase isoform 2 and the dihydrolipoyl acetyltransferase. *J. Biol. Chem.* 278 (36):33681–33693.

Hiromasa, Y., T. Fujisawa, Y. Aso, and T.E. Roche. 2004. Organization of the cores of the mammalian pyruvate dehydrogenase complex formed by E2 and E2 plus the E3-binding protein and their capacities to bind the E1 and E3 components. *J. Biol. Chem.* 279 (8):6921–6933.

Hong, Y.S., D.S. Kerr, W.J. Craigen, J. Tan, Y. Pan, M. Lusk, and M.S. Patel. 1996. Identification of two mutations in a compound heterozygous child with dihydrolipoamide dehydrogenase deficiency. *Hum. Mol. Genet.* 5 (12):1925–1930.

Hong, Y.S., D.S. Kerr, T.C. Liu, M. Lusk, B.R. Powell, and M.S. Patel. 1997. Deficiency of dihydrolipoamide dehydrogenase due to two mutant alleles (E340K and G101del). Analysis of a family and prenatal testing. *Biochim. Biophys. Acta* 1362 (2–3):160–168.

Hong, Y.S., S.H. Korman, J. Lee, P. Ghoshal, Q. Wu, V. Barash, S. Kang, S. Oh, M. Kwon, A. Gutman, A. Rachmel, and M.S. Patel. 2003. Identification of a common mutation (Gly194Cys) in both Arab Moslem and Ashkenazi Jewish patients with dihydrolipoamide dehydrogenase (E3) deficiency: Possible beneficial effect of vitamin therapy. *J. Inherit. Metab. Dis.* 26 (8):816–818.

Johnson, M.T., H.S. Yang, T. Magnuson, and M.S. Patel. 1997. Targeted disruption of the murine dihydrolipoamide dehydrogenase gene (*Dld*) results in perigastrulation lethality. *Proc. Natl. Acad. Sci. U S A* 94 (26):14512–14517.

Kerr, D.S., I.D. Wexler, A. Tripatara, and M.S. Patel. 1996. Human defects of the pyruvate dehydrogenase complex. In *Alpha-Keto Acid Dehydrogenase Complexes*, edited by Roche T.E., Patel M.S., and Harris R.A. Basel: Birkhäuser.

Kleanthous, C., P.M. Cullis, and W.V. Shaw. 1985. 3-(Bromoacetyl)chloramphenicol, an active site directed inhibitor for chloramphenicol acetyltransferase. *Biochemistry* 24 (20):5307–5313.

Kuhara, T., T. Shinka, Y. Inoue, M. Matsumoto, M. Yoshino, Y. Sakaguchi, and I. Matsumoto. 1983. Studies of urinary organic acid profiles of a patient with dihydrolipoyl dehydrogenase deficiency. *Clin. Chim. Acta* 133 (2):133–140.

Lanterman, M.M., J.R. Dickinson, and D.J. Danner. 1996. Functional analysis in *Saccharomyces cerevisiae* of naturally occurring amino acid substitutions in human dihydrolipoamide dehydrogenase. *Hum. Mol. Genet.* 5 (10):1643–1648.

Lessard, I.A. and R.N. Perham. 1995. Interaction of component enzymes with the peripheral subunit-binding domain of the pyruvate dehydrogenase multienzyme complex of *Bacillus stearothermophilus*: Stoichiometry and specificity in self-assembly. *Biochem. J.* 306 (Pt 3):727–733.

Lessard, I.A., C. Fuller, and R.N. Perham. 1996. Competitive interaction of component enzymes with the peripheral subunit-binding domain of the pyruvate dehydrogenase multienzyme complex of *Bacillus stearothermophilus*: Kinetic analysis using surface plasmon resonance detection. *Biochemistry* 35 (51):16863–16870.

Lewendon, A., I.A. Murray, W.V. Shaw, M.R. Gibbs, and A.G. Leslie. 1990. Evidence for transition-state stabilization by serine-148 in the catalytic mechanism of chloramphenicol acetyltransferase. *Biochemistry* 29 (8):2075–2080.

Lindsay, H., E. Beaumont, S.D. Richards, S.M. Kelly, S.J. Sanderson, N.C. Price, and J.G. Lindsay. 2000. FAD insertion is essential for attaining the assembly competence of the dihydrolipoamide dehydrogenase (E3) monomer from *Escherichia coli*. *J. Biol. Chem.* 275 (47):36665–36670.

Ling, M., G. McEachern, A. Seyda, N. MacKay, S.W. Scherer, S. Bratinova, B. Beatty, M.L. Giovannucci-Uzielli, and B.H. Robinson. 1998. Detection of a homozygous four base pair deletion in the protein X gene in a case of pyruvate dehydrogenase complex deficiency. *Hum. Mol. Genet.* 7 (3):501–505.

Linn, T.C., J.W. Pelley, F.H. Pettit, F. Hucho, D.D. Randall, and L.J. Reed. 1972. α-Keto acid dehydrogenase complexes. XV. Purification and properties of the component enzymes of the pyruvate dehydrogenase complexes from bovine kidney and heart. *Arch. Biochem. Biophys.* 148 (2):327–342.

Lissens, W., L. De Meirleir, S. Seneca, I. Liebaers, G.K. Brown, R.M. Brown, M. Ito, E. Naito, Y. Kuroda, D.S. Kerr, I.D. Wexler, M.S. Patel, B.H. Robinson, and A. Seyda. 2000. Mutations in the X-linked pyruvate dehydrogenase (E1) alpha subunit gene (PDHA1) in patients with a pyruvate dehydrogenase complex deficiency. *Hum. Mutat.* 15 (3):209–219.

Liu, S., J.C. Baker, and T.E. Roche. 1995. Binding of the pyruvate dehydrogenase kinase to recombinant constructs containing the inner lipoyl domain of the dihydrolipoyl acetyltransferase component. *J. Biol. Chem.* 270 (2):793–800.

Liu, T.C., H. Kim, C. Arizmendi, A. Kitano, and M.S. Patel. 1993. Identification of two missense mutations in a dihydrolipoamide dehydrogenase-deficient patient. *Proc. Natl. Acad. Sci. U S A* 90 (11):5186–5190.

Maj, M.C., N. MacKay, V. Levandovskiy, J. Addis, E.R. Baumgartner, M.R. Baumgartner, B.H. Robinson, and J.M. Cameron. 2005. Pyruvate dehydrogenase phosphatase deficiency: Identification of the first mutation in two brothers and restoration of activity by protein complementation. *J. Clin. Endocrinol. Metab.* 90 (7):4101–4107.

Maquat, L.E. 2004. Nonsense-mediated mRNA decay: Splicing, translation and mRNP dynamics. *Nat. Rev. Mol. Cell Biol.* 5 (2):89–99.

Marsac, C., D. Stansbie, G. Bonne, J. Cousin, P. Jehenson, C. Benelli, J.P. Leroux, and G. Lindsay. 1993. Defect in the lipoyl-bearing protein X subunit of the pyruvate dehydrogenase complex in two patients with encephalomyelopathy. *J. Pediatr.* 123 (6):915–920.

Matalon, R., D.A. Stumpf, K. Michals, R.D. Hart, J.K. Parks, and S.I. Goodman. 1984. Lipoamide dehydrogenase deficiency with primary lactic acidosis: Favorable response to treatment with oral lipoic acid. *J. Pediatr.* 104 (1):65–69.

Matuda, S., A. Kitano, Y. Sakaguchi, M. Yoshino, and T. Saheki. 1984. Pyruvate dehydrogenase subcomplex with lipoamide dehydrogenase deficiency in a patient with lactic acidosis and branched chain ketoaciduria. *Clin. Chim. Acta* 140 (1): 59–64.

McCartney, R.G., S.J. Sanderson, and J.G. Lindsay. 1997. Refolding and reconstitution studies on the transacetylase-protein X (E2/X) subcomplex of the mammalian pyruvate dehydrogenase complex: evidence for specific binding of the dihydrolipoamide dehydrogenase component to sites on reassembled E2. *Biochemistry* 36 (22):6819–6826.

Milne, J.L., D. Shi, P.B. Rosenthal, J.S. Sunshine, G.J. Domingo, X. Wu, B.R. Brooks, R.N. Perham, R. Henderson, and S. Subramaniam. 2002. Molecular architecture and mechanism of an icosahedral pyruvate dehydrogenase complex: A multifunctional catalytic machine. *Embo. J.* 21 (21):5587–5598.

Munnich, A., J.M. Saudubray, J. Taylor, C. Charpentier, C. Marsac, F. Rocchiccioli, O. Amedee-Manesme, F.X. Coude, J. Frezal, and B.H. Robinson. 1982. Congenital lactic acidosis, alpha-ketoglutaric aciduria and variant form of maple syrup urine disease due to a single enzyme defect: Dihydrolipoyl dehydrogenase deficiency. *Acta Paediatr. Scand.* 71 (1):167–171.

Nagy, E. and L.E. Maquat. 1998. A rule for termination-codon position within intron-containing genes: when nonsense affects RNA abundance. *Trends Biochem. Sci.* 23 (6):198–199.

Odievre, M.H., D. Chretien, A. Munnich, B.H. Robinson, R. Dumoulin, S. Masmoudi, N. Kadhom, A. Rotig, P. Rustin, and J.P. Bonnefont. 2005. A novel mutation in the dihydrolipoamide dehydrogenase E3 subunit gene (*DLD*) resulting in an atypical form of alpha-ketoglutarate dehydrogenase deficiency. *Hum. Mutat.* 25 (3):323–324.

Otulakowski, G., W. Nyhan, L. Sweetman, and B.H. Robinson. 1985. Immunoextraction of lipoamide dehydrogenase from cultured skin fibroblasts in patients with combined alpha-ketoacid dehydrogenase deficiency. *Clin. Chim. Acta* 152 (1–2):27–36.

Pratt, M.L. and T.E. Roche. 1979. Mechanism of pyruvate inhibition of kidney pyruvate dehydrogenasea kinase and synergistic inhibition by pyruvate and ADP. *J. Biol. Chem.* 254 (15):7191–7196.

Ramadan, D.G., R.A. Head, A. Al-Tawari, Y. Habeeb, M. Zaki, F. Al-Ruqum, G.T. Besley, J.E. Wraith, R.M. Brown, and G.K. Brown. 2004. Lactic acidosis and developmental delay due to deficiency of E3 binding protein (protein X) of the pyruvate dehydrogenase complex. *J. Inherit. Metab. Dis.* 27 (4):477–485.

Ravindran, S., G.A. Radke, J.R. Guest, and T.E. Roche. 1996. Lipoyl domain-based mechanism for the integrated feedback control of the pyruvate dehydrogenase complex by enhancement of pyruvate dehydrogenase kinase activity. *J. Biol. Chem.* 271 (2):653–662.

Reed, L.J. 1981. Regulation of mammalian pyruvate dehydrogenase complex by a phosphorylation-dephosphorylation cycle. *Curr. Top. Cell Regul.* 18:95–106.

Robinson, B.H. 2001. Lactic academia: Disorders of pyruvate carboxylase, pyruvate dehydrogenase. In *The Metabolic and Molecular Bases of Inherited Disease*, edited by Beaudet A.L., Scriver C.R., Sly W.S., and Valle D. New York: McGraw-Hill.

Robinson, B.H., J. Taylor, and W.G. Sherwood. 1977. Deficiency of dihydrolipoyl dehydrogenase (a component of the pyruvate and alpha-ketoglutarate dehydrogenase complexes): A cause of congenital chronic lactic acidosis in infancy. *Pediatr. Res.* 11 (12):1198–1202.

Robinson, B.H., J. Taylor, S.G. Kahler, and H.N. Kirkman. 1981. Lactic acidemia, neurologic deterioration and carbohydrate dependence in a girl with dihydrolipoyl dehydrogenase deficiency. *Eur. J. Pediatr.* 136 (1):35–39.

Robinson, B.H., N. MacKay, R. Petrova-Benedict, I. Ozalp, T. Coskun, and P.W. Stacpoole. 1990. Defects in the E2 lipoyl transacetylase and the X-lipoyl containing component of the pyruvate dehydrogenase complex in patients with lactic acidemia. *J. Clin. Invest.* 85 (6):1821–1824.

Roche, T.E., J.C. Baker, X. Yan, Y. Hiromasa, X. Gong, T. Peng, J. Dong, A. Turkan, and S.A. Kasten. 2001. Distinct regulatory properties of pyruvate dehydrogenase kinase and phosphatase isoforms. *Prog. Nucleic Acid Res. Mol. Biol.* 70:33–75.

Roche, T.E., Y. Hiromasa, A. Turkan, X. Gong, T. Peng, X. Yan, S.A. Kasten, H. Bao, and J. Dong. 2003. Essential roles of lipoyl domains in the activated function and control of pyruvate dehydrogenase kinases and phosphatase isoform 1. *Eur. J. Biochem.* 270 (6):1050–1056.

Rouillac, C., B. Aral, F. Fouque, D. Marchant, J.M. Saudubray, Y. Dumez, G. Lindsay, M. Abitbol, J.L. Dufier, C. Marsac, and C. Benelli. 1999. First prenatal diagnosis of defects in the HsPDX1 gene encoding protein X, an additional lipoyl-containing subunit of the human pyruvate dehydrogenase complex. *Prenat. Diagn.* 19 (12):1160–1164.

Sakaguchi, Y., M. Yoshino, S. Aramaki, I. Yoshida, F. Yamashita, T. Kuhara, I. Matsumoto, and T. Hayashi. 1986. Dihydrolipoyl dehydrogenase deficiency: A therapeutic trial with branched-chain amino acid restriction. *Eur. J. Pediatr.* 145 (4):271–274.

Sanderson, S.J., S.S. Khan, R.G. McCartney, C. Miller, and J.G. Lindsay. 1996. Reconstitution of mammalian pyruvate dehydrogenase and 2-oxoglutarate dehydrogenase complexes: Analysis of protein X involvement and interaction of homologous and heterologous dihydrolipoamide dehydrogenases. *Biochem. J.* 319 (Pt 1):109–116.

Sansaricq, C., S. Pardo, M. Balwani, M. Grace, and K. Raymond. 2006. Biochemical and molecular diagnosis of lipoamide dehydrogenase deficiency in a North American Ashkenazi jewish family. *J. Inherit. Metab. Dis.* 29 (1):203–204.

Scherer, S.W., G. Otulakowski, B.H. Robinson, and L.C. Tsui. 1991. Localization of the human dihydrolipoamide dehydrogenase gene (*DLD*) to 7q31–q32. *Cytogenet. Cell Genet.* 56 (3–4):176–177.

Seyda, A. and B.H. Robinson. 2000. Expression and functional characterization of human protein X variants in SV40-immortalized protein X-deficient and E2-deficient human skin fibroblasts. *Arch. Biochem. Biophys.* 382 (2):219–223.

Shaag, A., A. Saada, I. Berger, H. Mandel, A. Joseph, A. Feigenbaum, and O.N. Elpeleg. 1999. Molecular basis of lipoamide dehydrogenase deficiency in Ashkenazi Jews. *Am. J. Med. Genet.* 82 (2):177–182.

Shany, E., A. Saada, D. Landau, A. Shaaq, E. Hershkovitz, and O.N. Elpeleg. 1999. Lipoamide dehydrogenase deficiency due to a novel mutation in the interface domain. *Biochem. Biophys. Res. Commun.* 262 (1):163–166.

Smolle, M., A.E. Prior, A.E. Brown, A. Cooper, O. Byron, and J.G. Lindsay. 2006. A new level of architectural complexity in the human pyruvate dehydrogenase complex. *J. Biol. Chem.* 281 (28):19772–19780.

Taylor, J., B.H. Robinson, and W.G. Sherwood. 1978. A defect in branched-chain amino acid metabolism in a patient with congenital lactic acidosis due to dihydrolipoyl dehydrogenase deficiency. *Pediatr. Res.* 12 (1):60–62.

Thekkumkara, T.J., L. Ho, I.D. Wexler, G. Pons, T.C. Liu, and M.S. Patel. 1988. Nucleotide sequence of a cDNA for the dihydrolipoamide acetyltransferase component of human pyruvate dehydrogenase complex. *FEBS Lett.* 240 (1–2):45–48.

Thieme, R., E.F. Pai, R.H. Schirmer, and G.E. Schulz. 1981. Three-dimensional structure of glutathione reductase at 2 A resolution. *J. Mol. Biol.* 152 (4):763–782.

Toyoda, T., K. Suzuki, T. Sekiguchi, L.J. Reed, and A. Takenaka. 1998. Crystal structure of eucaryotic E3, lipoamide dehydrogenase from yeast. *J. Biochem. (Tokyo)* 123 (4):668–674.

Tuganova, A., I. Boulatnikov, and K.M. Popov. 2002. Interaction between the individual isoenzymes of pyruvate dehydrogenase kinase and the inner lipoyl-bearing domain of transacetylase component of pyruvate dehydrogenase complex. *Biochem. J.* 366 (Pt 1):129–136.

Wagenknecht, T., R. Grassucci, G.A. Radke, and T.E. Roche. 1991. Cryoelectron microscopy of mammalian pyruvate dehydrogenase complex. *J. Biol. Chem.* 266 (36):24650–24656.

Zhou, Z.H., W. Liao, R.H. Cheng, J.E. Lawson, D.B. McCarthy, L.J. Reed, and J.K. Stoops. 2001. Direct evidence for the size and conformational variability of the pyruvate dehydrogenase complex revealed by three-dimensional electron microscopy. The "breathing" core and its functional relationship to protein dynamics. *J. Biol. Chem.* 276 (24):21704–21713.

Zhou, Z.H., D.B. McCarthy, C.M. O'Connor, L.J. Reed, and J.K. Stoops. 2001. The remarkable structural and functional organization of the eukaryotic pyruvate dehydrogenase complexes. *Proc. Natl. Acad. Sci. U S A* 98 (26):14802–14807.

16 Relationship between Primary Biliary Cirrhosis and Lipoic Acid

Carlo Selmi, Xiao-Song He, Christopher L. Bowlus, and M. Eric Gershwin

CONTENTS

INTRODUCTION

Primary biliary cirrhosis (PBC) [1] is a progressive autoimmune liver disease with a 9:1 female predominance. Its peak incidence occurs over age 50, and it is uncommon under 25 years of age. The disease is characterized by immune-mediated destruction of intrahepatic small bile ducts, leading to decreased bile secretion and the retention of toxic substances within the liver, further contributing to the development of fibrosis, cirrhosis, and eventual liver failure. Understanding the etiology of PBC remains a challenge. However, recent studies have provided new insights into the events leading to the breakdown of self-tolerance and development of autoimmunity in PBC. In this article we will review emerging data and discuss the potential link between immune system dysregulation and the immunopathology of PBC.

EPIDEMIOLOGY OF PRIMARY BILIARY CIRRHOSIS

Primary biliary cirrhosis has a major genetic basis, with a higher relative risk than nearly any other autoimmune disease [2]. The relative risk of PBC in first-degree relatives of patients is 50–100-fold higher than the general population [3]. However, there is little, if any, association between PBC and MHC class I or class II haplotypes [4,5]. Recently our group evaluated the concordance of PBC in a genetically defined population of twin sets as well as the clinical characteristics in this cohort [6]. We identified 16 pairs of twins within a 1400-family cohort followed up by several centers worldwide, including eight sets each of monozygotic and dizygotic twins. In five of eight sets (0.63 pairwise concordance) of the monozygotic twins, both individuals had PBC. Among the eight pairs of the dizygotic twins, none was found to be concordant for PBC. Interestingly, the age at onset of disease was similar in four of five concordant sets of monozygotic pairs; however, there were differences in natural history and disease severity. Hence, while the concordance rate of PBC in identical twins is among the highest reported for any autoimmune disease, discordant pairs were also identified. These results also emphasize that either epigenetic factors or environment plays a critical role [7]. The importance of a genetic background for the development of PBC is further highlighted from a recent study that demonstrated abnormally elevated levels of the inflammatory chemokines IP-10 and MIG in the liver and peripheral blood from patients with PBC compared to controls [8]. Interestingly, the healthy daughters and sisters of PBC cases also demonstrated increased plasma levels of IP-10 and MIG compared to unrelated healthy controls, suggesting a shared immunogenetic response. The results of epidemiological studies suggest that environmental factors also play a role in triggering or exacerbating this autoimmune liver disease [9–11]. For example, several lines of evidence support an infectious etiology for PBC. These include the clustering of disease cases in unrelated individuals at certain geographical locations and increased incidence of PBC in immigrants moving from areas of low prevalence to high prevalence [12]. Disease recurs in 20% of patients with PBC who receive liver transplantation; this rate of recurrence appears to correlate with increased immunosuppression [13,14]. These findings are reminiscent of the findings with recurrence of hepatitis C virus (HCV) infection following liver transplantation for end stage hepatitis C [15].

SEROLOGICAL FEATURES OF PRIMARY BILIARY CIRRHOSIS

Primary biliary cirrhosis is characterized in over 90% of cases by high-titer serum antimitochondrial antibodies (AMA). The targets of the AMA response are all enzymatic members of the family of the 2-oxo-acid dehydrogenase complexes which include the E2 subunits of the pyruvate dehydrogenase complex, the branched chain 2-oxo-acid dehydrogenase complex, the α-ketoglutaric acid dehydrogenase complex, and the dihydrolipoamide dehydrogenase binding protein. Figure 16.1 illustrates the biochemical pathways of these enzyme complexes. Approximately 90%–95% of PBC sera samples react against the pyruvate

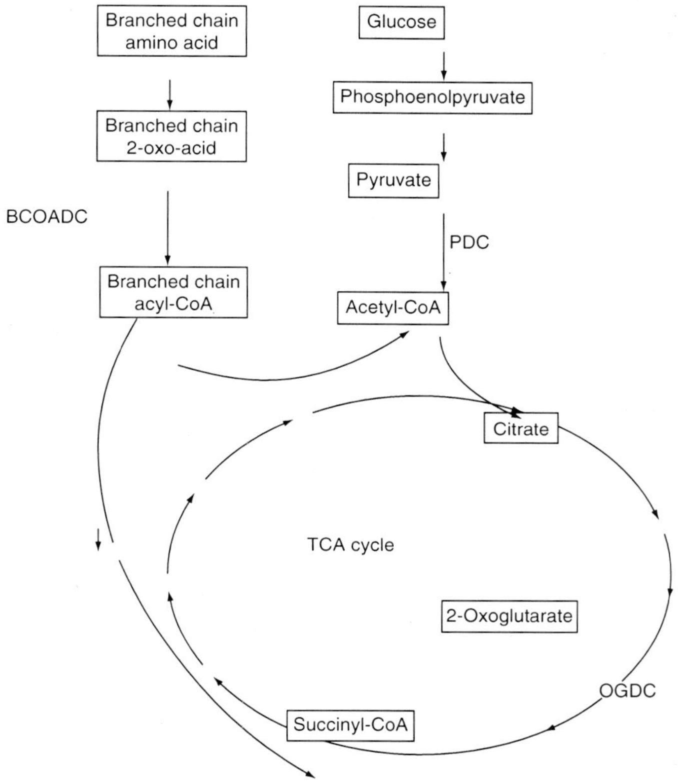

Antigenic subunit	Location of epitope		
	B cell	CD4 T cell	CD8 T cell
PDC-E2	AA167–186	AA163–176	AA159–167 AA165–174
OGDC-E2	AA67–147		
BCOADC-E2	AA1–227		

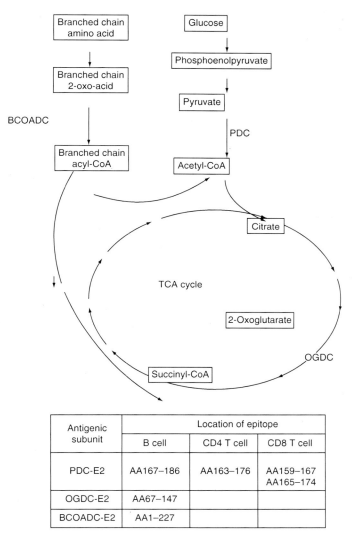

FIGURE 16.1 The major autoantigens in primary biliary cirrhosis (PBC) in their biochemical pathways. The major autoantigens are the enzymatic members of the family of the 2-oxo-acid dehydrogenase complexes: PDC, pyruvate dehydrogenase complex; BCOADC, branched chain 2-oxo-acid dehydrogenase complex; OGDC, oxoglutarate dehydrogenase complex. All the identified B cell and T cell epitopes locate at the inner lipoyl domains of the E2 subunits of these mitochondrial enzyme complexes (see box). Substantial amino acid sequence homology exists among these lipoylated epitopes, which are required in the metabolism of pyruvic acid.

TABLE 16.1
Prevalence of Anti-Mitochondrial Antibodies in Different Biological Samples from Patients with Primary Biliary Cirrhosis

Enzyme Complex	Antigenic Subunit	% Positive Serum	Bile	Saliva
	PDC-E2	90–95	79	87
PDC	PDC-E3BP	90–95		
	PDC-E1α	41–66		
OGDC	OGDC-E2	39–88	5.3	38
BCOADC	BCOADC-E2	53–55	32	28

dehydrogenase E2 complex (PDC-E2) [16]. AMA from some patients react with PDC-E2 alone, whereas AMA from most patients also show reactivity against either some or all of the other E2 subunit members (Table 16.1). In addition to the serum, AMA can also be detected in bile and saliva of patients with PBC (Table 16.1), with the prevalence and antigen specificity correlated with AMA in sera [17–19].

These target antigens are all localized within the inner mitochondrial matrix and catalyze the oxidative decarboxylation of keto acid substrates. The E2 enzymes have a common structure consisting of an N-terminal domain containing a single or multiple lipoyl groups. Previous studies using peptides or recombinant proteins have demonstrated that the dominant epitopes recognized by AMA are all located within the lipoylated domains of these target antigens [16,20–22]. Interestingly, there are only five proteins in mammals that are known to contain lipoic acid (LA), and four of the five are autoantigens in PBC. As mentioned above, particularly when the recombinant forms of the mitochondrial proteins are used diagnostically, a positive test for AMA is virtually diagnostic of PBC, or at least suggests that the person will likely develop PBC over the next 5–10 years [23].

Although AMA are considered the humoral hallmark of PBC, antibodies against various mitochondrial enzymes can be frequently detected in patients with infectious liver diseases [24–31]. However, these AMA responses are not directed at the same epitope as found in patients with PBC and have led to significant misunderstandings regarding the specificity of the AMA response. These issues do, however, raise the possibility that the production of autoantibodies does not cause PBC, but is rather the consequence of liver cell damage caused by distinct etiologies, for example, hepatitis viruses or intestinal bacteria that translocate to the liver. It has also been hypothesized that immune responses against virus-infected cells, mediated by NK cells and virus-specific MHC restricted $CD8^+$ T lymphocytes, lead to the lysis of target cells with the release of cryptic proteins that otherwise have never been seen by the immune system.

These proteins are thus recognized as foreign, which initiates the autoimmune disease process. Alternatively, autoreactive antibody response could be induced by viral antigens with structural homology to host proteins. Thus, autoantibodies could have been initially induced by a B cell response directed at the viral homologs of self-proteins, again implicating viral infection in the induction of autoimmunity [32,33]. This phenomenon has been termed "molecular mimicry."

While AMA are detected in the majority of PBC patients, sera from a subgroup of patients are also positive for antibodies to nuclear components (ANA). A number of nuclear structures have been recognized as specific targets of ANA in PBC, including Sp100, promyelocytic leukemia proteins, and two components of the nuclear pore complex. Autoantibodies against these structures generate different patterns in indirect immunofluorescence staining [34]. The clinical significance of ANA in PBC has been widely investigated recently and data indicate that, unlike AMA, which are not associated with disease severity and may be present many years before other clinical, biochemical, or histological manifestations, the PBC-associated ANA correlate with disease severity and may therefore serve as a marker of poor prognosis [35]. Studies are certainly warranted to elucidate the role of ANA in the immunopathology of PBC.

IMMUNOBIOLOGY OF THE INTRAHEPATIC BILIARY EPITHELIUM

In contrast to systemic autoimmune diseases such as lupus and Sjögren's syndrome, a striking feature of PBC is the specific destruction of small bile ducts, despite the fact that the primary targets of autoimmunity in PBC, including PDC-E2, oxoglutarate dehydrogenase E2 complex (OGDC-E2), and branched chain 2-oxo-acid dehydrogenase E2 complex (BCOADC-E2), are ubiquitous mitochondrial proteins expressed in all nucleated cells. This apparent paradox suggests unique immunopathological characteristics of biliary epithelial cell (BEC), the target tissue of autoimmune destruction in PBC. Staining of small bile ducts with a panel of monoclonal antibodies (mAbs) against the mitochondrial autoantigens showed intense staining at the apical surface of the cells lining the lumen of PBC livers, but not controls [36,37]. This pattern is distinct from the normal cytoplasmic pattern seen with mAbs against other non-PBC-related mitochondrial proteins. The fact that such apical staining is seen only with select PDC-E2 specific mAbs but not all mAbs that react with PDC-E2, and that distinct epitopes were identified with these apically staining mAbs, has led to the hypothesis that the target molecule cross-reactive with PDC-E2 is located at the apical surface of PBC BEC [17,36,37]. However, subsequent research demonstrated that the apical staining was due to a new complex between IgA AMA and PDC-E2 and further raised the potential role of IgA as a participant in the immune-mediated destruction of PBC [38,39]. These findings also highlight the theory that the BEC are not just an "innocent victim," but an active participant in the autoimmune pathology of PBC.

Biliary epithelial cells are actively involved in biliary and mucosal secretion including dimeric IgA [18,19,40]. Using MDCK cells expressing human polymeric immunoglobulin receptor (pIgR) as a model for BEC, a dimeric monoclonal IgA specific for PDC-E2 was imported into the cytoplasmic space within the cells and co-localized with PDC-E2 [39]. These data support the hypothesis that PDC-E2-specific IgA may enter BEC of PBC patients via pIgR and complex with PDC-E2, thereby potentially contributing to the pathology of BECs. When MDCK-pIgR$^+$ cells were incubated with purified IgA from PBC patient sera, the IgA specifically increased caspase activation. The titer of anti-PDC-E2 IgA antibody in the sera from PBC patients directly correlated with the level of caspase activation [41]. These data suggest that during transcytosis through pIgR$^+$ cells, exposure to PDC-E2-specific dimeric IgA results in the initiation of caspase activation. Considering the high concentration of dimeric IgA in biliary and mucosal secretions, constant transcytosis may render the exposed cells more susceptible to apoptosis, resulting in subsequent bile duct damage.

LIPOIC ACID AND PRIMARY BILIARY CIRRHOSIS

Recently, we have shown that foreign compounds that mimic or alter lipoic acid are capable of binding AMA when the lipoate molecule is conjugated to a carrier molecule and that immunization of rabbits with a bovine serum albumin (BSA) conjugate of a lipoic acid mimic (6-bromohexanoic acid) induces AMA. These findings suggest that this autoimmune disease may be secondary to chemical exposure and that lipoic acid or a lipoic acid mimic is important in breaking tolerance to mitochondrial antigens [32,33,42]. A number of lipoic acid mimotopes have been identified with the use of mimotope conjugated carrier molecules and affinity purified anti-PDC-E2 antibodies. Specifically, 79/97 (81%) of sera from AMA-positive PBC cases reacted to lipoylated human serum albumin (HSA-LA) (Table 16.2). The highest mean reactivity was noted in those PBC sera that were reactive with PDC-E2. High reactivity to HSA-LA correlated with the level of reactivity to PDC-E2. When PDC-E2 affinity-purified sera were tested, such sera also reacted with HSA-LA, suggesting that some of the antibodies to HSA-LA are a subset of anti-PDC-E2 specificities. Reactivity against lipoylated PDC-E2 and PDC-E2 is predominantly IgG and IgM whereas IgM is predominant against HSA-LA (Table 16.3). High IgM reactivity was also detected against BSA-LA, lipoylated rabbit serum albumin (RSA-LA), and lipoylated hemocyanin (KLH-LA) (data not shown). No Ig reactivity was observed against HSA, BSA, RSA, and KLH. To confirm whether the antibody reactivity to HSA-LA and PDC-E2 reflects the presence of two or more distinct populations of antibodies, i.e., anti-lipoic acid (hapten) antibodies, anti-PDC-E2 (carrier) specific antibodies, and a lipoic acid-PDC-E2 conjugate (conjugated form of the hapten-carrier complex) specific antibodies; sera from five PBC patients were adsorbed at a predetermined dilution against HSA, HSA-LA, non-lipoylated PDC-E2, and lipoylated PDC-E2. Aliquots were taken at various time points of adsorption and individually tested for the residual reactivity against

TABLE 16.2
Anti-Mitochondrial Antibodies Profile and Reactivity of Primary Biliary Cirrhosis Patient Sera to Lipoylated Human Serum Albumin

	AMA Profile			Reactivity to HSA-LA		
	Antigens			O.D. Mean ± S.D.	O.D. Range	Frequency of Positive Sera[a]
	PDC-E2	BCKD	OGDC			
I	+	+	+	0.429 ± 0.097	0.115–1.574	16/16
II	+	+	−	0.397 ± 0.065	0.062–1.280	22/25
III	+	−	+	0.302 ± 0.063	0.107–0.675	8/8
IV	+	−	−	0.289 ± 0.048	0.000–1.223	27/39
V	−	+	−	0.133 ± 0.020	0.062–0.215	6/9
VI	−	−	−	0.072 ± 0.024	0.000–0.219	3/8

Source: From Long, S.A., Quan, C., Van de Walter, J., Nantz, M.H., Barsky, D., Colvin, M.E., et al., *J. Immunol.*, 167, 2956, 2001.

[a] The positive sera were determined as two standard deviations above the mean O.D. unit for healthy volunteer sera.

HSA-LA, nonlipoylated PDC-E2, and lipoylated PDC-E2. Absorption with HSA alone did not appreciably reduce the reactivity of the sera against HSA-LA, nonlipoylated and lipoylated PDC-E2. When PBC sera were adsorbed with HSA-LA, Ig reactivity to HSA-LA was markedly decreased whereas Ig reactivity to both lipoylated and nonlipoylated PDC-E2 remained virtually unchanged. In contrast, when PBC sera were adsorbed with nonlipoylated PDC-E2, reactivity to both lipoylated and nonlipoylated PDC-E2 decreased with increasing time of adsorption, with a greater reduction in reactivity against nonlipoylated PDC-E2 than lipoylated PDC-E2. This could be interpreted as due to the specific removal of antibodies to nonlipoylated PDC-E2, with the remaining antibodies

TABLE 16.3
Immunoglobulin Class to Lipoic Acid-Protein Conjugates. Data Are Presented as Mean O.D. Unit ± S.E.M

	HSA	HSA-LA	PDC	PDC-LA
IgG	0.012 ± 0.001	0.133 ± 0.021	0.401 ± 0.131	0.394 ± 0.119
IgM	0.143 ± 0.007	0.807 ± 0.061	0.326 ± 0.055	0.351 ± 0.059

Source: From Long, S.A., Quan, C., Van de Walter, J., Nantz, M.H., Barsky, D., Colvin, M.E., et al., *J. Immunol.*, 167, 2956, 2001.

recognizing lipoylated PDC-E2 and the lipoic moiety on lipoylated PDC-E2. Reactivity to HSA-LA remained at 90% of unadsorbed sera even after 6 h of adsorption with nonlipoylated PDC-E2. Finally, adsorption with lipoylated PDC-E2 reduced reactivity against all three antigens with increasing time of adsorption. Specifically, in lipoylated PDC-E2 adsorbed PBC sera, the decrease in reactivity against HSA-LA was less than the decrease in reactivity against nonlipoylated PDC-E2 and lipoylated PDC-E2. This is likely due to the higher number of lipoic acid conjugates on lysine residues of HSA (up to 61) than PDC-E2. These results demonstrate the presence of (a) antibodies with specificity against PDC-E2 that are capable of recognizing both PDC-E2 and lipoic acid and (b) antibodies to lipoic acid that recognize a conjugated form of lipoic acid but not the PDC-E2 backbone in sera of patients with PBC.

ANTIBODIES TO LIPOIC ACID IN THE ABSENCE OF PROTEIN CARRIER

To determine the immunoreactivity of patients' sera with lipoic acid per se, our group also investigated the reactivity against lipoic acid attached to different peptide backbones (including a KLH protein motif) or serum albumin from bovine, human, and rabbit [44]. Absorption of PBC sera with lipoylated polyethylene glycol (PEG) reduced the reactivity of PBC sera against KLH-LA, HSA-LA ($84.9\% \pm 2.0\%$), and BSA-LA ($90.0\% \pm 1.5\%$) compared to unabsorbed sera. A dose dependent response was noted. We also investigated the possible reactivity of PBC sera with compounds (xenobiotics) found in the environment that could mimic lipoic acid [43]. More importantly, similar absorption of PBC sera with control haptens, compounds complexed with PEG, or PEG alone did not result in any appreciable decrease of reactivity. Absorption with lipoylated PEG, PEG-compound 3, or PEG alone did not have influence on sera reactivity against PDC-E2 or lipoylated PDC-E2. Furthermore, to determine whether antibodies to free, i.e., non-conjugated lipoic acid can be detected, 5 PBC sera were absorbed in 3% milk with lipoic acid (0.5 mM) and a number of control haptens including compound 3 (0.5 mM), 4-(bromo)hexanoic acid (0.5 mM), and octanoic acid (0.5 mM). Interestingly, absorption with free lipoic acid also reduced the reactivity of PBC sera against BSA-LA ($49\% \pm 5\%$), HSA-LA ($62\% \pm 5\%$), RSA-LA ($50\% \pm 6\%$), and KLH-LA ($36\% \pm 6\%$) while adsorption with compound 3,4-(bromo)hexanoic acid or octanoic acid did not influence sera reactivity against lipoylated proteins.

SIGNIFICANCE OF REACTIVITY TO LIPOIC ACID

In PBC, the following populations of antibodies to lipoic acid are revealed: (a) antibodies to free lipoic acid (as identified by absorption with free lipoic acid), (b) antibodies to lipoylated PDC-E2, and (c) antibodies to PDC-E2 unrelated carrier bound lipoate. Of particular interest, antibodies to the lipoyl–peptide conjugate have a unique specificity for the lipoyl moiety and appear to bind to lipoic acid and a conjugated form of lipoic acid irrespective of the protein carrier but with low to undetectable reactivity against nonlipoylated carrier proteins, providing support for the view that the reactivity is specific for a conjugated form of lipoic acid moiety. Our data clearly demonstrate that anti-lipoic acid antibodies are present uniquely in sera of patients with PBC; such antibodies were not seen in control sera. Although previous studies have suggested the presence of cross-reactive antibodies between the inner lipoyl domain of PDC-E2 and lipoic acid, the results were not definitive [45,46]. Indeed, it is important to note that the lipoic acid is attached to the ε-amino group of the lysine residue which is part of the signature DKA motif within the inner lipoyl domain of PDC-E2 [47]. In our study [44], the lipoic acid was attached to a number of the ε-amino groups of non-PDC-E2 carrier proteins, via the lysine residue of human albumin and rabbit albumin. The number of lysine residues per mole of human albumin and rabbit albumin is 61 and 58, respectively, and the only DKA motif identified was present in human albumin. Therefore, the PDC-E2 inner lipoyl domain peptide backbone was not likely to serve as a mimic with the use of the rabbit protein conjugates in the studies described herein.

Subpopulations of AMA appear to recognize the conjugated lipoyl moiety in at least two distinct contexts. The first, and perhaps the major population, recognizes the PDC-E2 peptide both in its lipoylated and nonlipoylated form. The second population specifically recognizes a conjugated form of lipoic acid, but not the inner lipoyl domain of PDC-E2. These two populations of antibodies have different affinities for their antigens, with the antibodies directed against conjugated lipoic acid having a much lower concentration for lipoic acid than the anti-PDC-E2/lipoic acid antibodies for PDC-E2. Furthermore, sera reactivity against nonlipoylated and lipoylated PDC-E2 was virtually unchanged when PBC sera were adsorbed by HSA-LA suggesting that antibodies to lipoyl–peptide conjugates may represent only a minor population of AMA. Interestingly, antibodies to PDC-E2 are primarily IgG antibodies whereas there are more IgM than IgG antibodies to the lipoyl-peptide conjugate. Hence, when PBC sera are adsorbed with lipoylated PDC-E2, primarily the IgG autoantibodies are removed and the remaining antibody reactivity to lipoylated peptide conjugates is mainly IgM. The converse is also evident when PBC sera are adsorbed with lipoylated HSA. Thus, antibodies to PDC-E2 LA and antibodies to lipoylated peptide conjugates are not cross-reactive antibodies; they differ not only in their epitope specificity but also in the Ig subclass and affinity to antigens.

While previous studies failed to detect anti-lipoic acid antibodies when free lipoic acid was used as an antigen [46], we demonstrated the presence of antibodies to free lipoic acid in sera of patients with PBC for the reasons provided below [44]. Previous studies using direct binding ELISA assays against lipoic acid were unsuccessful because free lipoic acid does not bind to ELISA plates efficiently. In contrast, using alternative backbones such as HSA, RSA, and KLH in an ELISA, we showed that PBC sera but not control sera recognize the lipoylated forms of HSA, RSA, and KLH (Figure 16.2). In addition, absorption

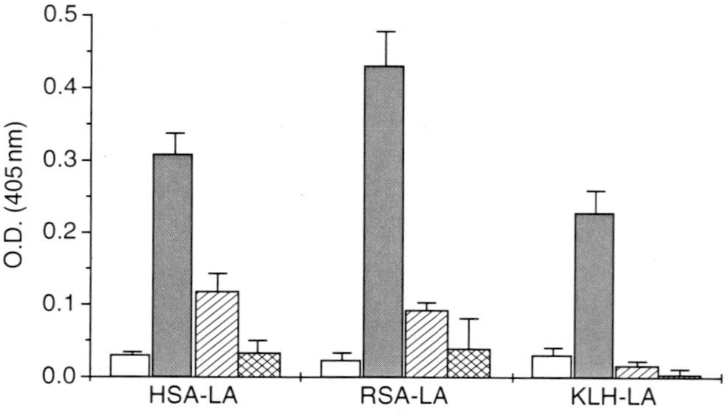

FIGURE 16.2 Reactivity of total sera from healthy volunteers ($n = 42$), anti-mitochondrial antibodies (AMA) positive primary biliary cirrhosis (PBC) patients ($n = 97$), AMA negative PBC patients ($n = 8$), patients with primary sclerosing cholangitis (PSC) ($n = 69$), and patients with rheumatoid arthritis (RA) ($n = 28$), against lipoylated human serum albumin (HSA-LA), lipoylated rabbit serum albumin (RSA-LA), or lipoylated hemocyanin (KLH-LA). Values presented as mean ± SEM of the average O.D. unit of three samples per serum measured by direct ELISA after correction for the background activity on the protein alone. Triplicate dilutions of sera were prepared in 1:1000 in PBST buffer containing 3% milk. Reactivity of total sera on the protein alone, all groups combined, was 0.042 ± 0.008, 0.036 ± 0.003, and 0.095 ± 0.004 for RSA, HSA, and KLH, respectively.

of PBC sera with lipoylated PEG and free lipoic acid reduces the Ig reactivity to the lipoylated conjugates. Since the possibility of noncovalent conjugation of lipoic acid to milk proteins cannot be totally excluded, the reduction in reactivity (to lipoylated conjugates of BSA, HSA, RSA, KLH) may be partly due to binding of anti-LA antibodies to lipoic acid moieties that are noncovalently conjugated to milk proteins during the absorption. Furthermore, the fact that absorption with free lipoic acid and lipoylated PEG only partially removes the reactivity of PBC sera against the various lipoic acid peptide conjugates suggests the antibodies have a higher affinity for lipoylated protein than lipoic acid in the absence of protein carrier. Also, this could suggest a third subpopulation of anti-lipoic antibodies reactive to free lipoic acid and, to a lesser extent, to lipoylated protein. This latter issue while important is difficult to address due to the overwhelming titers against PDC-E2 which precludes identification of such antibodies. Thus, following adsorption of PBC sera with a LA conjugate, one has to probe the adsorbed sera against PDC-E2 and demonstrate a decrease in the titer against PDC-E2, which is difficult to demonstrate.

The identification of autoantibodies to lipoic acid in PBC is particularly interesting since similarities exist in the mechanism of the immune response to iodine in autoimmune thyroiditis. This phenomenon has been demonstrated in a number of other diseases such as autoimmune thyroiditis animal models in chicken, rat, and NOD mice [48–52]. In addition, direct iodine stimulation of immune cells such as macrophages, T cells, dendritic cells, and B cells has been reported to play a significant role in the development of thyroid autoimmune reactivity [53,54]. Iodine also has been shown to play an important role in the precise specificity of the disease associated epitope of thyroglobulin. Thus, T cells from patients with thyroiditis react with iodinated but not non-iodinated human thyroglobulin and the use of selected monoclonal antibodies as a surrogate for the T cell receptor suggest that a specific iodine containing epitope is sometimes involved in the recognition [55].

As explained elsewhere, α-lipoic acid is a naturally occurring disulfide derivative of octanoic acid. Due to its low negative redox potential, lipoic acid fulfills all of the criteria for an ideal antioxidant, namely free-radical quenching, metal chelation, interaction with other antioxidants, absorption, and bioavailability properties [56]. Lipoic acid has been used as a therapeutic agent in the treatment of diabetic neuropathy, in conditions related to a number of liver diseases, and as a nutritional supplement in European countries and the United States [57]. Moreover, the liver has a high capacity for uptake and accumulation of lipoic acid and its metabolites. The detection of anti-lipoic acid antibodies warrants further studies on the safety of lipoic acid supplements, with particular regard to PBC. Indeed, our data provide compelling evidence for lipoic acid as the target of the reactive antibodies, yet no data are available as to the effect of dietary lipoic acid supplementation in PBC. Accordingly, there are no current, clear contraindication to its use by patients with autoimmune diseases, including PBC.

ACKNOWLEDGMENTS

We thank the many patients with primary biliary cirrhosis and their respective families who have been enormously cooperative and generous in providing samples and most importantly their time in participating in the studies reported herein.

REFERENCES

1. Kaplan MM and Gershwin ME. Primary biliary cirrhosis. *N Engl J Med* 2005;353:1261–1273.
2. Tsuji K, Watanabe Y, Van de Water J, Nakanishi T, Kajiyama G, Parikh-Patel A, Coppel R, et al. Familial primary biliary cirrhosis in Hiroshima. *J Autoimmun* 1999;13:171–178.
3. Parikh-Patel A, Gold E, Mackay IR, and Gershwin ME. The geoepidemiology of primary biliary cirrhosis: Contrasts and comparisons with the spectrum of autoimmune diseases. *Clin Immunol* 1999;91:206–218.
4. Invernizzi P, Battezzati PM, Crosignani A, Perego F, Poli F, Morabito A, De Arias AE, et al. Peculiar HLA polymorphisms in Italian patients with primary biliary cirrhosis. *J Hepatol* 2003;38:401–406.
5. Donaldson PT, Baragiotta A, Heneghan MA, Floreani A, Venturi C, Underhill JA, Jones DE, et al. HLA class II alleles, genotypes, haplotypes, and amino acids in primary biliary cirrhosis: A large-scale study. *Hepatology* 2006;44:667–674.

6. Selmi C, Mayo MJ, Bach N, Ishibashi H, Invernizzi P, Gish RG, Gordon SC, et al. Primary biliary cirrhosis in monozygotic and dizygotic twins: Genetics, epigenetics, and environment. *Gastroenterology* 2004;127:485–492.
7. Selmi C, Invernizzi P, Zuin M, Podda M, and Gershwin ME. Genetics and geoepidemiology of primary biliary cirrhosis: Following the footprints to disease etiology. *Semin Liver Dis* 2005;25:265–280.
8. Chuang YH, Lian ZX, Cheng CM, Lan RY, Yang GX, Moritoki Y, Chiang BL, et al. Increased levels of chemokine receptor CXCR3 and chemokines IP-10 and MIG in patients with primary biliary cirrhosis and their first degree relatives. *J Autoimmun* 2005;25:126–132.
9. Triger DR. Primary biliary cirrhosis: An epidemiological study. *Br Med J* 1980;281:772–775.
10. Uibo R and Salupere V. The epidemiology of primary biliary cirrhosis: Immunological problems. *Hepatogastroenterology* 1999;46:3048–3052.
11. Watson RG, Angus PW, Dewar M, Goss B, Sewell RB, and Smallwood RA. Low prevalence of primary biliary cirrhosis in Victoria, Australia. Melbourne liver group. *Gut* 1995;36:927–930.
12. Haydon GH and Neuberger J. PBC: An infectious disease? *Gut* 2000;47:586–588.
13. Polson RJ, Portmann B, Neuberger J, Calne RY, and Williams R. Evidence for disease recurrence after liver transplantation for primary biliary cirrhosis. Clinical and histologic follow-up studies. *Gastroenterology* 1989;97:715–725.
14. Dmitrewski J, Hubscher SG, Mayer AD, and Neuberger JM. Recurrence of primary biliary cirrhosis in the liver allograft: The effect of immunosuppression. *J Hepatol* 1996;24:253–257.
15. Sheiner PA, Schwartz ME, Mor E, Schluger LK, Theise N, Kishikawa K, Kolesnikov V, et al. Severe or multiple rejection episodes are associated with early recurrence of hepatitis C after orthotopic liver transplantation. *Hepatology* 1995;21:30–34.
16. Van de Water J, Gershwin ME, Leung P, Ansari A, and Coppel RL. The autoepitope of the 74-kD mitochondrial autoantigen of primary biliary cirrhosis corresponds to the functional site of dihydrolipoamide acetyltransferase. *J Exp Med* 1988;167:1791–1799.
17. Van de Water J, Turchany J, Leung PS, Lake J, Munoz S, Surh CD, Coppel R, et al. Molecular mimicry in primary biliary cirrhosis. Evidence for biliary epithelial expression of a molecule cross-reactive with pyruvate dehydrogenase complex-E2. *J Clin Invest* 1993;91:2653–2664.
18. Nishio A, Van de Water J, Leung PS, Joplin R, Neuberger JM, Lake J, Björkland A, et al. Comparative studies of antimitochondrial autoantibodies in sera and bile in primary biliary cirrhosis. *Hepatology* 1997;25:1085–1089.
19. Reynoso-Paz S, Leung PS, Van de Water J, Tanaka A, Munoz S, Bass N, Lindor K, et al. Evidence for a locally driven mucosal response and the presence of mitochondrial antigens in saliva in primary biliary cirrhosis. *Hepatology* 2000;31:24–29.
20. Surh CD, Coppel R, and Gershwin ME. Structural requirement for autoreactivity on human pyruvate dehydrogenase-E2, the major autoantigen of primary biliary cirrhosis. Implication for a conformational autoepitope. *J Immunol* 1990;144:3367–3374.
21. Leung PS, Chuang DT, Wynn RM, Cha S, Danner DJ, Ansari A, Coppel RL, et al. Autoantibodies to BCOADC-E2 in patients with primary biliary cirrhosis recognize a conformational epitope. *Hepatology* 1995;22:505–513.
22. Moteki S, Leung PS, Dickson ER, Van Thiel DH, Galperin C, Buch T, Alarcon-Segovia D, et al. Epitope mapping and reactivity of autoantibodies to the E2

component of 2-oxoglutarate dehydrogenase complex in primary biliary cirrhosis using recombinant 2-oxoglutarate dehydrogenase complex. *Hepatology* 1996;23: 436–444.
23. Gershwin ME, Ansari AA, Mackay IR, Nakanuma Y, Nishio A, Rowley MJ, and Coppel RL. Primary biliary cirrhosis: An orchestrated immune response against epithelial cells. *Immunol Rev* 2000;174:210–225.
24. Lenzi M, Ballardini G, Fusconi M, Cassani F, Selleri L, Volta U, Zauli D, et al. Type 2 autoimmune hepatitis and hepatitis C virus infection. *Lancet* 1990;335:258–259.
25. Garson JA, Lenzi M, Ring C, Cassani F, Ballardini G, Briggs M, Tedder RS, et al. Hepatitis C viraemia in adults with type 2 autoimmune hepatitis. *J Med Virol* 1991;34:223–226.
26. Todros L, Touscoz G, D'Urso N, Durazzo M, Albano E, Poli G, Baldi M, et al. Hepatitis C virus-related chronic liver disease with autoantibodies to liver-kidney microsomes (LKM). Clinical characterization from idiopathic LKM-positive disorders. *J Hepatol* 1991;13:128–131.
27. Philipp T, Durazzo M, Trautwein C, Alex B, Straub P, Lamb JG, Johnson EF, et al. Recognition of uridine diphosphate glucuronosyl transferases by LKM-3 antibodies in chronic hepatitis D. *Lancet* 1994;344:578–581.
28. Nishioka M, Morshed SA, Kono K, Himoto T, Parveen S, Arima K, Watanabe S, et al. Frequency and significance of antibodies to P450IID6 protein in Japanese patients with chronic hepatitis C. *J Hepatol* 1997;26:992–1000.
29. Lindgren S, Braun HB, Michel G, Nemeth A, Nilsson S, Thome-Kromer B, and Eriksson S. Absence of LKM-1 antibody reactivity in autoimmune and hepatitis-C-related chronic liver disease in Sweden. Swedish internal medicine liver club. *Scand J Gastroenterol* 1997;32:175–178.
30. Lohse AW, Gerken G, Mohr H, Lohr HF, Treichel U, Dienes HP, and Meyer zum Buschenfelde KH. Relation between autoimmune liver diseases and viral hepatitis: Clinical and serological characteristics in 859 patients. *Z Gastroenterol* 1995;33:527–533.
31. Manns MP and Strassburg CP. Autoimmune hepatitis: Clinical challenges. *Gastroenterology* 2001;120:1502–1517.
32. Amano K, Leung PS, Xu Q, Marik J, Quan C, Kurth MJ, Nantz MH, et al. Xenobiotic-induced loss of tolerance in rabbits to the mitochondrial autoantigen of primary biliary cirrhosis is reversible. *J Immunol* 2004;172:6444–6452.
33. Amano K, Leung PS, Rieger R, Quan C, Wang X, Marik J, Suen YF, et al. Chemical xenobiotics and mitochondrial autoantigens in primary biliary cirrhosis: Identification of antibodies against a common environmental, cosmetic, and food additive, 2-octynoic acid. *J Immunol* 2005;174:5874–5883.
34. Talwalkar JA and Lindor KD. Primary biliary cirrhosis. *Lancet* 2003;362:53–61.
35. Invernizzi P, Selmi C, Ranftler C, Podda M, and Wesierska-Gadek J. Antinuclear antibodies in primary biliary cirrhosis. *Semin Liver Dis* 2005;25:298–310.
36. Migliaccio C, Nishio A, Van de Water J, Ansari AA, Leung PS, Nakanuma Y, Coppel RL, et al. Monoclonal antibodies to mitochondrial E2 components define autoepitopes in primary biliary cirrhosis. *J Immunol* 1998;161:5157–5163.
37. Migliaccio C, Van de Water J, Ansari AA, Kaplan MM, Coppel RL, Lam KS, Thompson RK, et al. Heterogeneous response of antimitochondrial autoantibodies and bile duct apical staining monoclonal antibodies to pyruvate dehydrogenase complex E2: The molecule versus the mimic. *Hepatology* 2001;33: 792–801.

38. Malmborg AC, Shultz DB, Luton F, Mostov KE, Richly E, Leung PS, Benson GD, et al. Penetration and co-localization in MDCK cell mitochondria of IgA derived from patients with primary biliary cirrhosis. *J Autoimmun* 1998;11:573–580.
39. Fukushima N, Nalbandian G, Van de Water J, White K, Ansari AA, Leung P, Kenny T, et al. Characterization of recombinant monoclonal IgA anti-PDC-E2 autoantibodies derived from patients with PBC. *Hepatology* 2002;36:1383–1392.
40. Tanaka A, Nalbandian G, Leung PS, Benson GD, Munoz S, Findor JA, Branch AD, et al. Mucosal immunity and primary biliary cirrhosis: Presence of antimitochondrial antibodies in urine. *Hepatology* 2000;32:910–915.
41. Matsumura S, Van de Water J, Leung P, Odin JA, Yamamoto K, Gores GJ, Mostov K, et al. Caspase induction by IgA antimitochondrial antibody: IgA-mediated biliary injury in primary biliary cirrhosis. *Hepatology* 2004;39:1415–1422.
42. Leung PS, Quan C, Park O, Van de Water J, Kurth MJ, Nantz MH, Ansari AA, et al. Immunization with a xenobiotic 6-bromohexanoate bovine serum albumin conjugate induces antimitochondrial antibodies. *J Immunol* 2003;170:5326–5332.
43. Long SA, Quan C, Van de Water J, Nantz MH, Kurth MJ, Barsky D, Colvin ME, et al. Immunoreactivity of organic mimeotopes of the E2 component of pyruvate dehydrogenase: Connecting xenobiotics with primary biliary cirrhosis. *J Immunol* 2001;167:2956–2963.
44. Bruggraber SF, Leung PS, Amano K, Quan C, Kurth MJ, Nantz MH, Benson GD, et al. Autoreactivity to lipoate and a conjugated form of lipoate in primary biliary cirrhosis. *Gastroenterology* 2003;125:1705–1713.
45. Flannery GR, Burroughs AK, Butler P, Chelliah J, Hamilton-Miller J, Brumfitt W, and Baum H. Antimitochondrial antibodies in primary biliary cirrhosis recognize both specific peptides and shared epitopes of the M2 family of antigens. *Hepatology* 1989;10:370–374.
46. Quinn J, Diamond AG, Palmer JM, Bassendine MF, James OF, and Yeaman SJ. Lipoylated and unlipoylated domains of human PDC-E2 as autoantigens in primary biliary cirrhosis: Significance of lipoate attachment. *Hepatology* 1993;18:1384–1391.
47. Howard MJ, Fuller C, Broadhurst RW, Perham RN, Tang JG, Quinn J, Diamond AG, et al. Three-dimensional structure of the major autoantigen in primary biliary cirrhosis. *Gastroenterology* 1998;115:139–146.
48. Li M, Eastman CJ, and Boyages SC. Iodine induced lymphocytic thyroiditis in the BB/W rat: Early and late immune phenomena. *Autoimmunity* 1993;14:181–187.
49. Maczek C, Neu N, Wick G, and Hala K. Target organ susceptibility and autoantibody production in an animal model of spontaneous autoimmune thyroiditis. *Autoimmunity* 1992;12:277–284.
50. Many MC, Maniratunga S, and Denef JF. The non-obese diabetic (NOD) mouse: An animal model for autoimmune thyroiditis. *Exp Clin Endocrinol Diabetes* 1996;104 Suppl 3:17–20.
51. Sternthal E, Like AA, Sarantis K, and Braverman LE. Lymphocytic thyroiditis and diabetes in the BB/W rat. A new model of autoimmune endocrinopathy. *Diabetes* 1981;30:1058–1061.
52. Wick G, Brezinschek HP, Hala K, Dietrich H, Wolf H, and Kroemer G. The obese strain of chickens: An animal model with spontaneous autoimmune thyroiditis. *Adv Immunol* 1989;47:433–500.
53. Paschke R, Vogg M, Winter J, Wawschinek O, Eber O, and Usadel KH. The influence of iodine on the intensity of the intrathyroidal autoimmune process in Graves' disease. *Autoimmunity* 1994;17:319–325.

54. Toussaint-Demylle D, Many MC, Theisen H, Kraal G, and Denef JF. Effects of iodide on class II-MHC antigen expression in iodine deficient hyperplastic thyroid glands. *Autoimmunity* 1990;7:51–62.
55. Rose NR and Burek CL. Autoantibodies to thyroglobulin in health and disease. *Appl Biochem Biotechnol* 2000;83:245–251; discussion 251–244, 297–313.
56. Packer L, Witt EH, and Tritschler HJ. Alpha-lipoic acid as a biological antioxidant. *Free Radic Biol Med* 1995;19:227–250.
57. Moini H, Packer L, and Saris NE. Antioxidant and prooxidant activities of alpha-lipoic acid and dihydrolipoic acid. *Toxicol Appl Pharmacol* 2002;182:84–90.

17 Effects of Lipoic Acid on Insulin Action in Animal Models of Insulin Resistance

Erik J. Henriksen and Stephan Jacob

CONTENTS

REGULATION AND DYSREGULATION OF WHOLE-BODY GLUCOSE HOMEOSTASIS

Normal Regulation of Glucose Homeostasis

The regulation of glucose homeostasis in the intact organism is mediated by the coordinated functions of several organ systems. The liver is the primary site of glucose production from the glycogenolytic and gluconeogenic processes (Wasserman and Cherrington 1991), whereas the skeletal muscle, which comprises ~40% of the body weight of humans and most other mammalian species, is the predominant site of glucose disposal following a meal or during an exercise bout (DeFronzo et al. 1983; Baron et al. 1988). Both hepatic glucose production and skeletal muscle glucose uptake are under the influence of several neural and hormonal factors, as described below.

Hepatic glucose production is activated acutely via stimulation of glycogen breakdown (glycogenolysis) and is maintained chronically by upregulation of the de novo synthesis of glucose (gluconeogenesis) from several precursor molecules (glycerol, lactate, pyruvate, alanine, and glutamine) derived primarily from adipose tissue and skeletal muscle (Wasserman and Cherrington 1991; Camacho et al. 2004). The regulation of glycogenolysis is mediated acutely by alterations in the secretion and action of the pancreatic hormones insulin (via decreased insulin secretion) and glucagon (via increased glucagons secretion), and this hormonally activated breakdown of glycogen is the primary contributor to the initial acceleration of hepatic glucose production. The required decrease in insulin secretion and increase in glucagon secretion are elicited by acute autonomic neural inputs to the pancreatic islets, possibly mediated by signals emanating from the contracting muscle, from neural activity originating in the splanchnic bed, from subtle changes in glucose availability, and via neural input from the motor centers (Wasserman et al. 2002).

Additionally, epinephrine and cortisol can provide longer-term regulation of hepatic glucose production by accelerating the provision of gluconeogenic precursors and by an enhanced intrahepatic conversion of these substrates to glucose (Wasserman and Cherrington 1991; Camacho et al. 2004). The longer-term regulation of hepatic gluconeogenesis is mediated by altered gene expression of phosphoenolpyruvate carboxykinase (PEPCK) and glucose-6-phosphatase (G6Pase), enzymes active at the rate-limiting steps of the gluconeogenic pathway. Glucagon and cortisol can also enhance the expression of the PEPCK and G6Pase genes, whereas insulin can downregulate the expression of these genes (O'Brien and Granner 1990; Sutherland et al. 1996; Argaud et al. 1996).

Glucose transport activity in skeletal muscle is a critical site of regulation of whole-body glucose disposal (Henriksen 2002). The glucose transport process is regulated acutely by insulin through the activation of a series of intracellular proteins (White 1997; Shepherd and Kahn 1999; Zierath et al. 2000; Henriksen 2002). Insulin binding to its receptor activates tyrosine kinase activity via an autophosphorylation event, and the activated insulin receptor subsequently phosphorylates insulin receptor substrate proteins (IRS-1 and IRS-2 in skeletal muscle) on tyrosine residues. The tyrosine-phosphorylated IRS protein then interacts with the p85 regulatory subunit of phosphotidylinositol-3-kinase (PI-3-kinase), thereby activating the p110 catalytic subunit of this enzyme. p110 produces phosphoinositide moieties that in turn allosterically activate 3-phosphoinositide-dependent kinases (PDK). 3-Phosphoinositide-dependent kinase, a serine/threonine kinase, can phosphorylate and activate Akt and the atypical protein kinase C (PKC) isoform PCK-ζ. The activation of these steps up to and including PI-3-kinase, Akt, and PKC-ζ ultimately results in the translocation of a specific glucose transporter protein isoform (GLUT-4) to the membranes of the sarcolemma and the t-tubules, where glucose transport takes place via a facilitative diffusion process. The amount of GLUT-4 protein incorporated into the sarcolemmal membrane correlates closely with the degree of insulin-stimulated glucose transport, and strongly suggests that this GLUT-4 translocation represents the major mechanism for insulin-stimulated glucose transport in skeletal muscle of rodents (Lund et al. 1993; Gao et al. 1994; Etgen et al. 1996) and humans (Guma et al. 1995).

Glucose transport in skeletal muscle is also stimulated by contractile activity (Jessen and Goodyear 2005). The intracellular signaling mechanisms involved in contraction-dependent glucose transport activity include the allosteric and covalent activation of 5′-adenosine-monophosphate-activated protein kinase (AMP kinase) (Kurth-Kraczek et al. 1999; Mu et al. 2001; Winder 2001; Wright et al. 2004). In addition, this pathway for stimulation of glucose transport in muscle includes the activation of calcium- and calmodulin-dependent protein kinase (Holloszy and Hansen 1996; Wright et al. 2004). Contraction-dependent activation of glucose transport in muscle also depends on the translocation of GLUT-4 to the plasma membrane translocation (Goodyear et al. 1991; Gao et al. 1994).

Dysregulation of Glucose Homeostasis

Overproduction of glucose by the liver can exacerbate the dysregulation of glucose homeostasis in type-2 diabetic humans (Reaven 1995). This accelerated hepatic glucose production can be attributed to several metabolic defects. In type-2 diabetic subjects characterized by central obesity, there is a reduction of insulin-stimulated hepatic glycogen synthase (Cline and Shulman 1991), insulin's ability to suppress hepatic glucose production is suppressed and is associated with a decreased insulin-dependent inhibition of hepatic glycogenolysis (Maggs et al. 1998), and there is an elevation of the rate of hepatic gluconeogenesis (Magnusson et al. 1992). Moreover, the relative decrease in insulin secretion by the pancreatic β-cells, which is an obligatory event in the conversion from the prediabetic condition to a state of overt type-2 diabetes, will also contribute to this enhancement of hepatic glucose production (Reaven 1995).

The reduced action of insulin to stimulate glucose transport in skeletal muscle, termed insulin resistance, is also a major contributor to disruptions of whole-body glucose homeostasis (White 1997; Shepherd and Kahn 1999; Zierath et al. 2000; Henriksen 2002). Studies using a variety of animal models of insulin resistance, as well as clinical investigations on insulin-resistant human subjects, indicate that disruptions in the normal functioning of the insulin-signaling pathway, leading to defective GLUT-4 translocation to the plasma membrane, underlie most conditions of skeletal muscle insulin resistance (Shepherd and Kahn 1999; Zierath et al. 2000; Henriksen 2002). An increasing body of evidence in the literature indicates that altered protein expression and functionality of IR and IRS-1 is associated with insulin resistance (Shulman 2000). Specifically, the protein expression and functionality of these insulin-signaling proteins are downregulated by serine and threonine phosphorylation (Strack et al. 1997; Bossenmaier et al. 2000; Shulman 2000; Aguirre et al. 2000, 2002; Pederson et al. 2001; Yu et al. 2002; Bloch-Damti et al. 2003; Potashnik et al. 2003).

There are several serine/threonine kinases involved in cellular signaling which could mediate the serine phosphorylation of IR and IRS-1, including Akt (Li et al. 1999), p70 S6 kinase (Ming et al. 1994), atypical protein kinase C isoforms (Standaert et al. 1999), glycogen synthase kinase-3 (GSK-3) (Eldar-Finkelman and Krebs 1997; Liberman and Eldar-Finkelman 2005; Henriksen and Teachey 2007), and the mitogen-activated protein kinases (MAPK) (Mothe and Van Obberghen 1996; De Fea and Roth 1997). Serine phosphorylation of IR is associated with decreased tyrosine phosphorylation of IRS-1 and a subsequently diminished activation of PI3-kinase (Strack et al. 1997; Bossenmaier et al. 2000), whereas serine phosphorylation of IRS-1 causes enhanced degradation of this protein in cell lines (Pederson et al. 2001; Potashnik et al. 2003).

One family of MAPKs, p38 MAPK, is an attractive candidate for the investigation of potential molecular mechanisms of insulin resistance in insulin-sensitive tissues. The basal phosphorylation state of p38 MAPK is elevated in adipocytes (Carlson et al. 2003) and skeletal muscle (Koistinen et al. 2003) of insulin-resistant type-2 diabetic humans. Moreover, the protein expression and activation of p38 MAPK is inversely related to insulin-stimulated glucose transport activity in rat skeletal muscle (O'Keefe et al. 2004a,b). These results complement the findings of Ho et al. (2004) demonstrating that overexpression of the p38 MAPK γ-isoform in mouse skeletal muscle causes an increase in basal glucose transport, a decreased GLUT-4 protein expression, and a diminished stimulation of glucose transport by contractions. Finally, the selective inhibition of p38 MAPK in human adipocytes prevents the hyperinsulinemia-induced loss of GLUT-4 protein (Carlson et al. 2003), a molecule critical for normal stimulation of glucose transport activity.

Although insulin resistance of skeletal muscle glucose transport is multifactorial in its etiology, one condition that can negatively impact insulin signaling in muscle is oxidative stress (Henriksen 2000, 2006). Definitive evidence linking oxidative stress and insulin resistance comes from cell culture and isolated muscle incubation studies. Rudich et al. (1997, 1998) have demonstrated in 3T3-L1 adipocytes and L6 myocytes that prolonged exposure to a low-grade oxidant stress (H_2O_2) markedly decreases insulin-stimulated glucose metabolism. This decreased insulin responsiveness is associated with increased GLUT-1 protein and mRNA, and decreased GLUT-4 protein and mRNA (Rudich et al. 1997; Kozlovsky et al. 1997), and with impaired insulin signaling at the level of PI3-kinase (Rudich et al. 1998; Maddux et al. 2001). Moreover, Blair et al. (1999) showed that exposure of L6 myocytes to a low level of H_2O_2 activates p38 MAPK. Thus, these findings are consistent with the hypothesis that oxidative stress can directly and negatively impact insulin action on glucose transport, perhaps through the action of the p38 MAPK. An observation consistent with this deleterious role of oxidative stress was that exposure of hepatoma cells to an oxidant stress (H_2O_2) induces increased phosphorylation at Ser307 and Ser632 of IRS-1 (Bloch-Damti et al. 2003), and that this oxidant exposure is associated degradation of the IRS-1 protein (Potashnik et al. 2003). Preliminary data from our own research group indicates that oxidative stress induced by low levels of

H_2O_2 causes a diminution of IR- and IRS-1-dependent insulin signaling and insulin-dependent glucose transport activity in rat skeletal muscle (Dokken et al. 2007, manuscript submitted).

BRIEF OVERVIEW OF α-LIPOIC ACID AND ITS ROLE AS AN ANTIOXIDANT

Interventions that improve whole-body insulin resistance and insulin resistance of skeletal muscle glucose transport include exercise training or caloric restriction leading to a loss of visceral fat mass, and treatment with a variety of pharmaceutical or nutriceutical compounds (Henriksen 2002, 2006). In the context of nutriceutical interventions, numerous cell-based and animal model studies have addressed the antioxidant properties of α-lipoic acid (Packer et al. 1995; Packer et al. 2001; Henriksen 2006), which is enriched in certain food sources, including vegetables such as spinach, broccoli, and tomato, and in animal tissues such as kidney and liver (Lodge and Packer 1999). Because they possess reactive sulfhydryl groups, α-lipoic acid and its reduced form, dihydrolipoic acid, can function as potent antioxidants, scavenging several reactive oxygen and nitrogen species and also acting to chelate transition metals (Packer et al. 2001). Moreover, α-lipoic acid can interact with the cellular antioxidant network of vitamins C and E and glutathione to facilitate the regeneration of the reduced forms of glutathione and vitamin E (Packer et al. 2001).

Because it possesses a chiral carbon, α-lipoic acid can exist either as an *R*-enantiomer or an *S*-enantiomer. The *R*-enantiomer is the naturally occurring cofactor in several mitochondrial dehydrogenase complexes involved in substrate oxidation (Packer et al. 1995). There is evidence in humans that the oral bioavailability of *R*-α-lipoic acid is superior to that of the *S*-form of the compound (Hermann and Niebsch 1997). In addition, limited evidence indicates that the *R*-form of α-lipoic acid may have greater beneficial effects than the racemic mixture or the *S*-enantiomer on certain pathological conditions, such as insulin resistance and its associated complications, as described in the next section.

BENEFICIAL METABOLIC EFFECTS OF α-LIPOIC ACID IN ANIMAL MODELS OF INSULIN RESISTANCE

The metabolic actions of α-lipoic acid have been studied in a variety of animal models of insulin resistance and of type-1 and type-2 diabetes. The important outcomes from these studies will be reviewed below, first addressing genetic models of insulin resistance, then turning to diet-induced models of insulin resistance, and then finally covering a model of type-1 diabetes.

Genetic Models of Insulin Resistance

The metabolic actions of α-lipoic acid have been studied most extensively in the obese Zucker (fa/fa) rat, a model of prediabetes and the metabolic syndrome.

Due to a point mutation in the gene encoding for the leptin receptor (Iida et al. 1996; Phillips et al. 1996; Takaya et al. 1996), the obese Zucker rat develops central obesity, marked glucose intolerance and skeletal muscle insulin resistance, compensatory hyperinsulinemia, and dyslipidemia (Mathe 1995; Henriksen 2002). Moreover, this animal model displays evidence of oxidative stress in skeletal muscle, heart, and liver (Saengsirisuwan et al. 2001, 2004; Henriksen et al. 2003; Teachey et al. 2003). In isolated skeletal muscle from the obese Zucker rat, racemic α-lipoic acid can stimulate glucose transport activity (Henriksen et al. 1997), likely associated with an activation of p38 MAPK (Saengsirisuwan, Henriksen, unpublished data). Moreover, insulin-stimulated glucose transport activity can be potentiated in isolated skeletal muscle exposure to racemic α-lipoic acid (Henriksen et al. 1997).

In vivo administration of α-lipoic acid to the obese Zucker rat is also associated with significant metabolic improvements. A single acute treatment of obese Zucker rats with α-lipoic acid increases the ability of insulin to activate skeletal muscle glucose transport (Jacob et al. 1996; Streeper et al. 1997), with the *R*-enantiomer being more effective than the *S*-enantiomer (Streeper et al. 1997). We have also shown that chronic treatment of obese Zucker rats with α-lipoic acid causes an enhancement of whole-body glucose tolerance and insulin sensitivity and decreases in hyperinsulinemia and dyslipidemia (Jacob et al. 1996; Streeper et al. 1997; Peth et al. 2000; Saengsirisuwan et al. 2001, 2004). Moreover, the chronic α-lipoic acid treatments (with either the racemic mixture or the pure *R*-enantiomer) enhanced increased stimulation of skeletal muscle glucose transport (Jacob et al. 1996; Streeper et al. 1997; Peth et al. 2000; Saengsirisuwan et al. 2001, 2004) that were associated with upregulation of the IRS-dependent insulin-signaling pathway (Saengsirisuwan et al. 2004) and were correlated with reductions in muscle oxidative stress (Saengsirisuwan et al. 2001) and lipid storage (Saengsirisuwan et al. 2004).

Important investigations of the metabolic actions of chronic α-lipoic acid treatments have been conducted in at least two other genetic rat models of insulin resistance. The Otsuka Long-Evans Tokushima Fatty (OLETF) rat develops obesity at an early age, is insulin-resistant, and at later ages becomes overtly diabetic (Kawano et al. 1992). Chronic dietary treatment of OLETF rats with racemic α-lipoic acid was associated with a decrease in body weight gain and the prevention of hyperglycemia, glucosuria, hyperinsulinemia, and dyslipidemia, as well as with a diminution of lipid markers of oxidative stress and muscle triglycerides (Song et al. 2004). Shorter-term (3 day) dietary administration of α-lipoic acid to these OLETF rats, independent of any body weight loss, caused a significant enhancement of whole-body glucose disposal and skeletal muscle glycogen synthesis that was associated with a reduction in muscle triglyceride concentration and fatty acid oxidation and an activation of the α2-isoform of AMP kinase (Lee et al. 2005). Interestingly, these beneficial effects of α-lipoic acid were prevented when a dominant-negative AMP kinase construct was introduced into the skeletal muscle of these insulin-resistant rats (Lee et al. 2005). These results support the concept that the beneficial effects of α-lipoic acid on muscle insulin sensitivity may be mediated by recruitment of local AMP kinase-dependent signaling pathways and actions.

One final genetic model of insulin resistance that has been used in investigations of the metabolic actions of α-lipoic acid is the Goto–Kakizaki rat. The Goto–Kakizaki rat is a non-obese model of spontaneous type-2 diabetes that was originally characterized by mild fasting hyperglycemia, hyperinsulinemia, impaired glucose tolerance (Goto et al. 1975a, 1975b, 1988). Subsequently it has been shown to display whole-body and skeletal muscle insulin resistance, dyslipidemia, and elevated markers of muscle oxidative stress (Bitar et al. 2004, 2005). Chronic treatment of the Goto–Kakizaki rats with racemic α-lipoic acid for 30 days led to reductions in fasting plasma glucose, insulin, and free fatty acids (Bitar et al. 2004, 2005) and improvements of insulin action on whole-body glucose disposal during a euglycemic, hyperinsulinemic clamp (Bitar et al. 2004), enhanced insulin-stimulated glucose transport activity in skeletal muscle (Bitar et al. 2004, 2005), upregulation of IRS-1-dependent insulin signaling in skeletal muscle (Bitar et al. 2004, 2005), and greater translocation of GLUT-4 to the plasma membrane (Bitar et al. 2005). Moreover, chronic treatment of these diabetic rats with α-lipoic acid was associated with reductions in markers of oxidative stress in muscle (Bitar et al. 2004, 2005) and urine (Bitar et al. 2005).

Nutritional Intervention Models of Insulin Resistance

Diets high in calories from carbohydrates can induce insulin resistance in normal rats. Several investigations have been published on the chronic effects of α-lipoic acid treatment on insulin resistance induced by high carbohydrate diets. The chronic administration of α-lipoic acid to rats made hypertensive and insulin-resistant due to diets enriched in glucose which brought about reductions in systolic blood pressure (El Midaoui and de Champlain 2002, 2005; El Midaoui et al. 2006), an increase in whole-body insulin sensitivity (El Midaoui and de Champlain 2002), and a diminution of markers of oxidative stress (El Midaoui and de Champlain 2002, 2005; El Midaoui et al. 2006). Similar increases in whole-body insulin sensitivity, as well as reductions in plasma lipids, were elicited by α-lipoic acid treatment of rats made insulin-resistant by high fructose feeding (Thirunavukkarasu et al. 2004a, 2004b).

Models of Type-1 Diabetes

A widely used model of type-1diabetes is the streptozotocin-treated rat, in which the ß-cells of the pancreas are selectively destroyed by this agent, rendering the animal insulinopenic and markedly hyperglycemic. Chronic administration of α-lipoic acid to these type-1 diabetic animals led to significant reductions in both fed and fasting plasma glucose levels, and caused increases in insulin-stimulated glucose transport activity that were associated with enhanced GLUT-4 expression (Khamaisi et al. 1997). Moreover, shorter-term treatment of streptozotocin-diabetic rats with α-lipoic acid also elicited a reduction in blood glucose that was accompanied by a significant suppression of hepatic glucoeneogenesis and hepatic fatty acid oxidation (Khamaisi et al. 1999). These results indicate that α-lipoic acid can act to regulate glucose homeostasis by affecting both muscle and liver.

Streptozotocin-diabetic rats provided dietary glucose (a model of combined type-1 diabetes and glucose-induced insulin resistance) also responded favorably to treatment with α-lipoic acid, displaying decreased plasma glucose levels and increased whole-body insulin sensitivity (El Midaoui and de Champlain 2005).

Effects of α-Lipoic Acid Conjugates and Derivatives on Metabolic Regulation

A limited number of investigations have addressed the potential metabolic interactions between α-lipoic acid and other compounds affecting conditions of oxidative stress and insulin resistance. In these studies, unique conjugates of α-lipoic acid and the secondary agent have been derived and used to treat insulin-resistant animals. In one study from our research group (Teachey et al. 2003), insulin-resistant obese Zucker rats were treated chronically with either the *R*-enantiomer of α-lipoic acid alone, a mixture of isomers of conjugated linoleic acid (CLA) alone, or a combination of the *R*-enantiomer of α-lipoic acid and CLA, at lower or higher doses. Individually, the higher doses of the *R*-enantiomer of α-lipoic acid and CLA caused significant increases in insulin-stimulated glucose transport activity and reductions in triglycerides and markers of oxidative stress in skeletal muscle. However, no further improvements were elicited by treatment with the combination of these higher doses of the *R*-enantiomer of α-lipoic acid and CLA. In contrast, the combination of lower, minimally effective doses of the *R*-enantiomer of α-lipoic acid and CLA caused marked improvements in glucose tolerance and whole-body insulin sensitivity and elicited significant increases in insulin-stimulated glucose transport activity in skeletal muscle that were closely associated with reductions in oxidative stress and triglyceride levels in that tissue.

Several studies have investigated the interactions between racemic α-lipoic acid and the n-6 essential fatty acid and prostaglandin precursor γ-linolenic acid (GLA). The combination of α-lipoic acid and GLA was shown to have a synergistic effect in improving neurovascular function in a rat model of type-1 diabetes (Cameron et al. 1998; Hounsom et al. 1998). In this same animal model of type-1 diabetes, Khamaisi et al. (1999) showed that an α-lipoic acid–GLA conjugate reduced fasting blood glucose, enhanced whole-body glucose tolerance and insulin sensitivity, and elicited an improvement in skeletal muscle GLUT-4 protein expression. Finally, our research group demonstrated that this conjugate of the antioxidant α-lipoic acid and the n-6 essential fatty acid GLA elicited significant dose-dependent improvements in whole-body and skeletal muscle insulin action on glucose disposal in insulin-resistant obese Zucker rats, due to the additive effects of the individual components of the conjugate (Peth et al. 2000).

Very recently, derivatives of the antidiabetic agent thiazolidinedione (TZD) were synthesized by coupling the TZD pharmacophore to an α-lipoic acid fragment retaining the reactive sulfhydryl groups (Chittiboyina et al. 2006). Several such TZD–α-lipoic acid derivatives displayed a markedly enhanced ability to activate the peroxisome proliferator-activated receptor-γ, the site of action of TZDs for mediating their antidiabetic effects on fat and muscle cells. Moreover, chronic (4 week) treatment of insulin-resistant, prediabetic obese Zucker rats with one such TZD–α-lipoic acid derivative induced substantial decreases in serum insulin and triglycerides, without an effect on serum glucose (Chittiboyina et al. 2006), indicating an increase in whole-body insulin sensitivity. A more extensive evaluation of the potential metabolic actions of these TZD–α-lipoic acid derivatives in the context of insulin-resistant states is clearly warranted.

SUMMARY AND PERSPECTIVES

The information provided in this chapter indicates that the acute and chronic treatment with the antioxidant α-lipoic acid can induce beneficial metabolic effects in a variety of animal models characterized by insulin resistance of skeletal muscle glucose transport and metabolism. In general, these studies have shown that treatment with α-lipoic acid enhances whole-body glucose tolerance and insulin sensitivity due to an improvement in the action of insulin to stimulate the disposal of glucose in skeletal muscle. In addition, in some models, it has been shown that α-lipoic acid can suppress hepatic glucose production, which would also contribute to this improved glucoregulation. A limited number of studies support a greater effect on glucoregulation from treatment with the *R*-enantiomer of α-lipoic acid compared with the *S*-enantiomer. The enhancement of insulin action on skeletal muscle glucose transport following chronic treatment with α-lipoic acid is associated with a decrease in muscle oxidative stress and with an upregulation of the IRS-1-dependent insulin-signaling pathway. In addition, some evidence supports a role of the engagement of AMP kinase-dependent signaling pathways in skeletal muscle in the enhancement of insulin action by α-lipoic acid in conditions of insulin resistance.

Several studies have also demonstrated that specific conjugates of α-lipoic acid can be effective in enhancing glucose tolerance and insulin action in conditions of insulin resistance associated with both type-1 diabetes and prediabetes. α-Lipoic acid conjugated with either CLA, GLA, or TZDs has greater effects on glucose dysregulation than the individual effects of the respective components of the conjugates, due to the additive influences of these components. In light of the fact that several other pharmaceutical and nutriceutical interventions exist for the treatment of prediabetes and overt type-2 diabetes, the number of possible conjugates that can be created with α-lipoic acid and investigated for their metabolic interaction is quite large.

Although there is an abundance of information on the metabolic benefits of α-lipoic acid treatment in these various animal models of insulin resistance and type-1 and type-2 diabetes, the same cannot be said for insulin resistance in humans. Very few clinical investigations have been published, though the implications from these human studies are important. Jacob and colleagues have demonstrated in insulin-resistant type-2 diabetic subjects that whole-body insulin sensitivity of glucose disposal, as assessed by the isoglycemic, hyperinsulinemic glucose clamp method, can be enhanced by ~30% following either a single intravenous infusion of racemic α-lipoic acid (Jacob et al. 1995) or the chronic intravenous infusion (Jacob et al. 1996) or oral administration (Jacob et al. 1999) of the racemic mixture of this antioxidant. It is clear that metabolic actions of α-lipoic acid in insulin-resistant prediabetic, type-1 diabetic, and type-2 diabetic humans should be more thoroughly investigated.

In closing, the animal model studies cited in this chapter support the role of α-lipoic acid as a beneficial intervention in conditions of insulin resistance. More information is needed on the relationship between oxidative stress and the etiology of insulin resistance, and how antioxidants, such as α-lipoic acid, can impact this relationship. Moreover, more information on the molecular mechanisms of action of α-lipoic acid in the context of insulin resistance is required. Finally, as stated above, large clinical trials involving human subjects with prediabetes or type-2 diabetes must be designed and implemented to more thoroughly evaluate the efficacy of α-lipoic acid to treat human insulin resistance.

REFERENCES

Aguirre V, Uchida T, Yenush L, Davis RJ, and White MF. The c-Jun NH_2-terminal kinase promotes insulin resistance during association with insulin receptor substrate-1 and phosphorylation of Ser^{307}. *J Biol Chem* 2000; 275: 9047–9054.

Aguirre V, Werner ED, Giraud J, Lee YH, Shoelson SE, and White MF. Phosphorylation of Ser^{307} in insulin receptor substrate-1 blocks interactions with the insulin receptor and inhibits insulin action. *J Biol Chem* 2002; 277: 1531–1537.

Argaud D, Zhang Q, Pan W, Maitra S, Pilkis SJ, and Lange AJ. Regulation of rat liver glucose-6-phosphatase gene expression in different nutritional and hormonal states. *Diabetes* 1996; 45: 1563–1571.

Baron AD, Brechtel G, Wallace P, and Edelman SV. Rates and tissue sites of non-insulin- and insulin-mediated glucose uptake in humans. *Am J Physiol Endocrinol Metab* 1988; 255: E769–E774.

Bitar MS, Al-Saleh E, and Al-Mulla F. Oxidative stress-mediated alterations in glucose dynamics in a genetic animal model of type II diabetes. *Life Sci* 2005; 77: 2552–2573.

Bitar MS, Wahid S, Pilcher CW, Al-Saleh E, and Al-Mulla F. Alpha-lipoic acid mitigates insulin resistance in Goto-Kakizaki rats. *Horm Metab Res* 2004; 36: 542–549.

Blair AS, Hajduch E, Litherland GJ, and Hundal HS. Regulation of glucose transport and glycogen synthesis in L6 muscle cells during oxidative stress. Evidence for cross-talk between the insulin and SAPK2/p38 mitogen-activated protein kinase signaling pathways. *J Biol Chem* 1999; 274: 36293–36299.

Bloch-Damti A, Potashnik R, Tanti JF, Le Marchand-Brustel Y, Rudich A, and Bashan N. IRS-1 protein degradation induced by oxidative stress is associated with distinct time-course of phosphorylation on serine 307 and serine 632. *Diabetologia* 2003; 46 (Suppl 2): A209.

Bossenmaier B, Strack V, Stoyanov B, Krutzfeldt J, Beck A, Lehmann R, Kellerer M, Klein H, Ullrich A, Lammers R, and Haring HU. Serine residues 1177/78/82 of the insulin receptor are required for substrate phosphorylation but not autophosphorylation. *Diabetes* 2000; 49: 889–895.

Camacho RC, Galassetti P, Davis SN, and Wasserman DH. Glucoregulation during and after exercise in health and insulin-dependent diabetes. *Exerc Sports Sci Rev* 2004; 33: 17–23.

Cameron NE, Cotter MA, Horrobin DH, and Tritschler HJ. Effects of α-lipoic acid on neurovascular function in diabetic rats: Interaction with essential fatty acids. *Diabetologia* 1998; 41: 390–399.

Carlson CJ, Koterski S, Sciotti RJ, Poccard GB, and Rondinone CM. Enhanced basal activation of mitogen-activated protein kinases in adipocytes from type 2 diabetes: Potential role of p38 in the downregulation of GLUT4 expression. *Diabetes* 2003; 52: 634–641.

Chittiboyina AG, Venkatraman MS, Mizuno CS, Desai PV, Patny A, Benson SC, Ho CI, Kurtz TW, Pershadsingh HA, and Avery MA. Design and synthesis of the first generation of dithiolane thiazolidinedione- and phenylacetic acid-based PPAR-gamma agonists. *J Med Chem* 2006; 49: 4072–4084.

Cline GW and Shulman GI. Quantitative analysis of the pathways of glycogen repletion in periportal and perivenous hepatocytes in vivo. *J Biol Chem* 1991; 266: 4094–4098.

De Fea K and Roth RA. Modulation of insulin receptor substrate-1 tyrosine phosphorylation and function by mitogen-activated protein kinase. *J Biol Chem* 1997; 272: 31400–31406.

DeFronzo RA, Ferrannini E, Hendler R, Felig P, and Wahren J. Regulation of splanchnic and peripheral glucose uptake by insulin and hyperglycemia in man. *Diabetes* 1983; 32: 32–45.

Eldar-Finkelman H and Krebs EG. Phosphorylation of insulin receptor substrate 1 by glycogen synthase kinase 3 impairs insulin action. *Proc Natl Acad Sci* 1997; 94: 9660–9664.

El Midaoui A and de Champlain J. Prevention of hypertension, insulin resistance, and oxidative stress by α-lipoic acid. *Hypertension* 2002; 39: 303–307.

El Midaoui A and de Champlain J. Effects of glucose and insulin on the development of oxidative stress and hypertension in animal models of type 1 and type 2 diabetes. *J Hypertens* 2005; 23: 581–588.

El Midaoui A, Wu L, Wang R, and de Champlain J. Modulation of cardiac and aortic peroxisome proliferator-activated receptor-gamma expression by oxidative stress in chronically glucose-fed rats. *Am J Hypertens* 2006; 19: 407–412.

Etgen GJ, Wilson CM, Jensen J, Cushman SW, and Ivy JL. Glucose transport and cell surface GLUT-4 protein in skeletal muscle of the obese Zucker rat. *Am J Physiol Endocrinol Metab* 1996; 271: E294–E301.

Gao J, Ren J, Gulve EA, and Holloszy JO. Additive effect of contractions and insulin on GLUT-4 translocation into the sarcolemma. *J Appl Physiol* 1994; 77: 1587–1601.

Goodyear LJ, Hirshman MF, and Horton ES. Exercise-induced translocation of skeletal muscle glucose transporters. *Am J Physiol Endocrinol Metab* 1991; 261: E795–E799.

Goto Y, Kakizaki M, Masaki N. Production of spontaneous diabetic rats by repetition of selective breeding. *Tohobu J Exp Med* 1975a; 119: 85–90.

Goto Y, Kakizaki M, Masaki N. Spontaneous diabetes produced by selective breeding of normal Wistar rats. *Proc J Jpn Acad* 1975b; 5: 80–85.

Goto Y, Suzuki KI, Seraki M, Ono T, Abe S. GK rat as a model of non obese, non insulin dependent diabetes: Selective breeding over 35 generations. In: *Frontiers in Diabetes Research. Lessons from Animal Diabetes*, Shafir E, Renold AE (Eds), John Libbey, London, 1988; 301–303.

Guma A, Zierath JR, Wallberg-Henriksson H, and Klip A. Insulin induces translocation of GLUT-4 glucose transporters in human skeletal muscle. *Am J Physiol Endocrinol Metab* 1995; 268: E613–E622.

Henriksen EJ. Oxidative stress and antioxidant treatment: Effects on muscle glucose transport in animal models of type 1 and type 2 diabetes. In: *Antioxidants in Diabetes Management*, Packer L, Rösen P, Tritschler HJ, King GL, Azzi A (Eds.), Marcel Dekker Inc. New York, 2000; 303–318.

Henriksen EJ. Invited Review: Effects of acute exercise and exercise training on insulin resistance. *J Appl Physiol* 2002; 93: 788–796.

Henriksen EJ. Exercise training and the antioxidant alpha-lipoic acid in the treatment of insulin resistance and type 2 diabetes. *Free Rad Biol Med* 2006; 40: 3–12.

Henriksen EJ, Jacob S, Streeper RS, Fogt DL, Hokama JY, and Tritschler HJ. Stimulation by α-lipoic acid of glucose transport activity in skeletal muscle of lean and obese Zucker rats. *Life Sci* 1997; 61: 805–812.

Henriksen EJ and Teachey MK. Short-term in vitro inhibition of glycogen synthase kinase-3 potentiates insulin signaling in skeletal muscle of Zucker diabetic fatty rats. *Metabolism* 2007; 56: 931–938.

Henriksen EJ, Teachey MK, Taylor ZC, Jacob S, Krämer K, and Hasselwander O. Isomer-specific actions of conjugated linoleic acid on muscle glucose transport in the obese Zucker rat. *Am J Physiol Endocrinol Metab* 2003; 285: E98–E105.

Hermann R and Niebsch G. Human pharmacokinetics of lipoic acid. In: *Biothiols in Health and Disease*, Packer L, Cadenas E (Eds.), Marcel Dekker, New York, 1997; 337.

Ho RC, Alcazar O, Fujii N, Hirshman MF, and Goodyear LJ. p38γ MAPK regulation of glucose transporter expression and glucose uptake in L6 myotubes and mouse skeletal muscle. *Am J Physiol Regul Integr Comp Physiol* 2004; 286: R342–R349.

Holloszy JO and Hansen PA. Regulation of glucose transport into skeletal muscle. *Rev Physiol Biochem Pharmacol* 1996; 128: 99–193.

Hounsom L, Horrobin DF, Tritschler H, Corder R, and Tomlinson DR. A lipoic acid-gamma linolenic acid conjugate is effective against multiple indices of experimental diabetic neuropathy. *Diabetologia* 1998; 41: 839–843.

Iida M, Murakami T, Ishida K, Mizuno A, Kuwajima M, and Shima K. Substitution at codon 269 (glutamine –> proline) of the leptin receptor (OB-R) cDNA is the only mutation found in the Zucker fatty (fa/fa) rat. *Biochem Biophys Res Comm* 1996; 224: 597–604.

Jacob S, Henriksen EJ, Schiemann AL, Simon I, Clancy DE, Tritschler HJ, Jung WI, Augustin HJ, and Dietze GJ. Enhancement of glucose disposal in patients with Type 2 diabetes by alpha-lipoic acid. *Drug Res* 1995; 45: 872–874.

Jacob S, Henriksen EJ, Tritschler HJ, Augustin HJ, and Dietze GJ. Improvement of insulin-stimulated glucose disposal in type 2 diabetes after repeated parenteral administration of thioctic acid. *Exp Clin Endocrinol Diab* 1996; 104: 284–288.

Jacob S, Ruus P, Hermann R, Tritschler HJ, Maerker E, Renn W, Augustin HJ, Dietze GJ, and Rett K. Oral administration of rac-α-lipoic acid modulates insulin sensitivity in patients with type 2 diabetes mellitus—a placebo controlled pilot trial. *Free Radic Biol Med* 1999; 27: 309–314.

Jacob S, Streeper RS, Fogt DL, Hokama JY, Tritschler HJ, Dietze GJ, and Henriksen EJ. The antioxidant α-lipoic acid enhances insulin stimulated glucose metabolism in insulin-resistant rat skeletal muscle. *Diabetes* 1996; 45: 1024–1029.

Jessen N and Goodyear LJ. Contraction signaling to glucose transport in skeletal muscle. *J Appl Physiol* 2005; 99: 330–337.

Kawano K, Hirashima T, Mori S, Saitoh Y, Kurosumi M, and Natori T. Spontaneous long-term hyperglycemic rat with diabetic complications. Otsuka Long-Evans Tokushima Fatty (OLETF) strain. *Diabetes* 1992; 41: 1422–1428.

Khamaisi M, Potashnik R, Tirosh A, Demshchak E, Rudich A, Tritschler H, Wessel K, and Bashan N. Lipoic acid reduces glycemia and increases muscle GLUT4 content in streptozotocin-diabetic rats. *Metabolism* 1997; 46: 763–768.

Khamaisi M, Rudich A, Beeri I, Pessler D, Friger M, Gavrilov V, Tritschler H, and Bashan N. Metabolic effects of gamma-linolenic acid-alpha-lipoic acid conjugate in streptozotocin diabetic rats. *Antioxidants Redox Signaling* 1999; 1: 523–535.

Khamaisi M, Rudich A, Potashnik R, Tritschler HJ, Gutman A, and Bashan N. Lipoic acid acutely induces hypoglycemia in fasting nondiabetic and diabetic rats. *Metabolism* 1999; 48: 504–510.

Koistinen HA, Chibalin AV, and Zierath JR. Aberrant p38 mitogen-activated protein kinase signalling in skeletal muscle from type 2 diabetic patients. *Diabetologia* 2003; 46: 1324–1328.

Kozlovsky N, Rudich A, Potashnik R, Ebina Y, Murakami T, and Bashan N. Transcriptional activation of the Glut1 gene in response to oxidative stress in L6 myotubes. *J Biol Chem* 1997; 272: 33367–33372.

Kurth-Kraczek EJ, Hirshman MF, Goodyear LJ, and Winder WW. 5′ AMP-activated protein kinase activation causes GLUT4 translocation in skeletal muscle. *Diabetes* 1999; 48: 1667–1671.

Lee WJ, Song K-H, Koh EH, Won JC, Kim HS, Park H-S, Kim M-S, Kim S-W, Lee K-U, and Park J-Y. α-Lipoic acid increases insulin sensitivity by activating AMPK in skeletal muscle. *Biochem Biophys Res Commun* 2005; 332: 885–891.

Li J, De Fea K, and Roth RA. Modulation of insulin receptor substrate-1 tyrosine phosphorylation by an Akt/phosphatidylinositol 3-kinase pathway. *J Biol Chem* 1999; 274: 9351–9356.

Liberman Z and Eldar-Finkelman H. Serine 332 phosphorylation of insulin receptor substrate-1 by glycogen synthase kinase-3 attenuates insulin signaling. *J Biol Chem* 2005; 280: 4422–4428.

Lodge J and Packer L. Natural source of lipoic acid in plant and animal tissues. In: *Antioxidant Food Supplements in Human Health*, Packer L, Hiramatsu M, Yoshikawa T (Eds.), Academic Press, San Diego, 1999; 121.

Lund S, Holman GD, Schmitz O, and Pedersen O. Glut 4 content in the plasma membrane of rat skeletal muscle: Comparative studies of the subcellular fractionation method and the exofacial photolabelling technique using ATB-BMPA. *FEBS Lett* 1993; 330: 312–318.

Maddux BA, See W, Lawrence JC Jr., Goldfine AL, Goldfine ID, and Evans JL. Protection against oxidative stress-induced insulin resistance in rat L6 muscle cells by micromolar concentrations of α-lipoic acid. *Diabetes* 2001; 50: 404–410.

Maggs DG, Buchanan TA, Burant CF, Cline G, Gumbiner B, Hsueh WA, Inzucchi S, Kelley D, Nolan J, Olefsky JM, Polonsky KS, Silver D, Valiquett TR, and Shulman GI. Metabolic effects of troglitazone monotherapy in type 2 diabetes mellitus. A randomized, double-blind, placebo-controlled trial. *Ann Intern Med* 1998; 128: 176–185.

Magnusson I, Rothman DL, Katz LD, Shulman RG, and Shulman GI. Increased rate of gluconeogenesis in type II diabetes mellitus. *J Clin Invest* 1992; 90: 1323–1327.

Mathe D. Dyslipidemia and diabetes: Animal models. *Diabetes Metab* 1995; 21: 106–111.

Ming XF, Burgering BM, Wennstrom S, Claesson-Welsh L, Heldin CH, Bos JL, Kozma SC, and Thomas J. Activation of p70/p86 S6 kinase by a pathway independent of p21ras. *Nature* 1994; 371: 426–429.

Mothe I and Van Obberghen E. Phosphorylation of insulin receptor substrate-1 on multiple serine residues, 612, 632, 662, and 731, modulates insulin action. *J Biol Chem* 1996; 271: 11222–11227.

Mu J, Brozinick JT Jr., Valladares O, Bucan M, and Birnbaum MJ. A role for AMP-activated protein kinase in contraction- and hypoxia-regulated glucose transport in skeletal muscle. *Molecular Cell* 2001; 7: 1085–1094.

O'Brien RM and Granner DK. PEPCK gene as a model of inhibitory effects of insulin on gene transcription. *Diabetes Care* 1990; 13: 327–339.

O'Keefe MP, Perez FR, Kinnick TR, Tischler ME, and Henriksen EJ. Development of whole-body and skeletal muscle insulin resistance after one day of hindlimb suspension. *Metabolism* 2004a; 53: 1215–1222.

O'Keefe MP, Perez FR, Sloniger JA, Tischler ME, and Henriksen EJ. Enhanced insulin action on glucose transport and insulin signaling in 7-day unweighted rat soleus muscle. *J Appl Physiol* 2004b; 97: 63–71.

Packer L, Kraemer K, and Rimbach G. Molecular aspects of lipoic acid in the prevention of diabetes complications. *Nutrition* 2001; 17: 888–895.

Packer L, Witt EH, and Tritschler HJ. Alpha-lipoic acid as a biological antioxidant. *Free Radic Biol Med* 1995; 19: 227–250.

Pederson TM, Kramer DL, and Rondinone CM. Serine/threonine phosphorylation of IRS-1 triggers its degradation: Possible regulation by tyrosine phosphorylation. *Diabetes* 2001; 50: 24–31.

Peth JA, Kinnick TR, Youngblood EB, Tritschler HJ, and Henriksen EJ. Effects of a unique conjugate of alpha-lipoic acid and gamma-linolenic acid on insulin action in the obese Zucker rat. *Am J Physiol Reg Integrative Comp Physiol* 2000; 278: R453–R459.

Phillips MS, Liu Q, Hammond HA, Dugan V, Hey PJ, Caskey CJ, and Hess JF. Leptin receptor missense mutation in the fatty Zucker rat. *Nature Genet* 1996; 13: 18–19.

Potashnik R, Bloch-Damti A, and Bashan N. IRS1 degradation and increased serine phosphorylation cannot predict the degree of metabolic insulin resistance induced by oxidative stress. *Diabetologia* 2003; 46: 639–648.

Reaven GM. Pathophysiology of insulin resistance in human disease. *Physiol Rev* 1995; 75: 473–486.

Rudich A, Kozlovsky N, Potashnik R, and Bashan N. Oxidant stress reduces insulin responsiveness in 3T3-L1 adipocytes. *Am J Physiol Endocrinol Metab* 1997; 272: E935–E940.

Rudich A, Tirosh A, Potashnik R, Hemi R, Kanety H, and Bashan N. Prolonged oxidative stress impairs insulin-induced GLUT4 translocation in 3T3-L1 adipocytes. *Diabetes* 1998; 47: 1562–1569.

Saengsirisuwan V, Kinnick TR, Schmit MB, and Henriksen EJ. Interactions of exercise training and alpha-lipoic acid on glucose transport in obese Zucker rat. *J Appl Physiol* 2001; 91: 145–153.

Saengsirisuwan V, Perez FR, Sloniger JA, Maier T, and Henriksen EJ. Interactions of exercise training and R-(+)-alpha-lipoic acid on insulin signaling in skeletal muscle of obese Zucker rats. *Am J Physiol Endocrinol Metab* 2004; 287: E529–E536.

Shepherd PR and Kahn BB. Glucose transporters and insulin action. *N Eng J Med* 1999; 341: 248–257.

Shulman GI. Cellular mechanisms of insulin resistance. *J Clin Invest* 2000; 106: 171–176.

Song K-H, Lee WJ, Koh J-M, Kim HS, Youn J-Y, Park H-S, Koh EH, Kim M-S, Youn JH, Lee K-U, and Park J-Y. α-Lipoic acid prevents diabetes mellitus in diabetes-prone obese rats. *Biochem Biophys Res Commun* 2004; 326: 197–202.

Standaert ML, Bandyopadhyay G, Perez L, Price D, Galloway L, Poklepovic A, Sajan MP, Cenni V, Sirri A, Moscat J, Toker A, and Farese RV. Insulin activates protein kinases C-zeta and C-lambda by an autophosphorylation-dependent mechanism and stimulates their translocation to GLUT4 vesicles and other membrane fractions in rat adipocytes. *J Biol Chem* 1999; 274: 25308–25316.

Strack V, Stoyanov B, Bossenmaier B, Mosthaf L, Kellerer M, and Haring H-U. Impact of mutations at different serine residues on the tyrosine kinase activity of the insulin receptor. *Biochem Biophys Res Comm* 1997; 239: 235–239.

Streeper RS, Henriksen EJ, Jacob S, Hokama JY, Fogt DL, Tritschler HJ, and Dietze GJ. Differential effects of stereoisomers of alpha-lipoic acid on glucose metabolism in insulin-resistant rat skeletal muscle. *Am J Physiol Endocrinol Metab* 1997; 273: E185–E191.

Sutherland C, O'Brien RM, and Granner DK. New connections in the regulation of PEPCK gene expression by insulin. *Phil Trans R Soc Lond* 1996; 215: 314–332.

Takaya K, Ogawa Y, Isse N, Okazaki T, Satoh N, Masuzaki H, Mori K, Tamura N, Hosoda K, and Nakao K. Molecular cloning of rat leptin receptor isoform complementary DNAs—identification of a missense mutation in Zucker fatty (fa/fa) rats. *Biochem Biophys Res Comm* 1996; 225: 75–83.

Teachey MK, Taylor ZC, Maier T, Saengsirisuwan V, Sloniger JA, Jacob S, Klatt MJ, Ptock A, Krämer K, Hasselwander O, and Henriksen EJ. Interactions of conjugated linoleic acid and alpha-lipoic acid on insulin action in the obese Zucker rat. *Metabolism* 2003; 52: 1167–1174.

Thirunavukkarasu V, Anitha Nandhini AT, and Anuradha CV. Effect of alpha-lipoic acid on lipid profile in rats fed a high-fructose diet. *Experiment Diabesity Res* 2004a; 5: 195–200.

Thirunavukkarasu V, Anitha Nandhini AT, and Anuradha CV. Lipoic acid attenuates hypertension and improves insulin sensitivity, kallikrein activity and nitrite levels in high fructose-fed rats. *J Comp Physiol B Biochem System Environ Physiol* 2004b; 174: 587–592.

Wasserman DH and Cherrington AD. Hepatic fuel metabolism during muscular work: Role and regulation. *Am J Physiol Endocrinol Metab* 1991; 260: E811–E824.

Wasserman DH, Shi Z, and Vranic M. Metabolic implications of exercise and physical fitness in physiology and diabetes. In: *Diabetes Mellitus: Theory and Practice*, Porte D, Sherwin R, Baron A (Eds.), Appleton and Lange, Stamford, CT, 2002; 453–480.

White MF. The insulin signalling system and the IRS-1 proteins. *Diabetologia* 1997; 40 (Suppl 2): S2–S17.

Winder WW. Energy-sensing and signaling by AMP-activated protein kinase in skeletal muscle. *J Appl Physiol* 2001; 91: 1017–1028.

Wright DC, Hucker KA, Holloszy JO, and Han DH. Ca^{2+} and AMPK both mediate stimulation of glucose transport by muscle contractions. *Diabetes* 2004; 53, 330–335, 2004.

Yu C, Chen Y, Cline GW, Zhang D, Zong H, Wang Y, Bergeron R, Kim JK, Cushman SW, Cooney GJ, Atcheson B, White MF, Kraegen EW, and Shulman GI. Mechanism by which fatty acids inhibit insulin activation of insulin receptor substrate-1 (IRS-1)-associated phosphatidylinositol 3-kinase activity in muscle. *J Biol Chem* 2002; 277: 50230–50236.

Zierath JR, Krook A, and Wallberg-Henriksson H. Insulin action and insulin resistance in human skeletal muscle. *Diabetologia* 2000; 43: 821–835.

18 Activation of Cytoprotective Signaling Pathways by Alpha-Lipoic Acid

Alexandra K. Kiemer and Britta Diesel

CONTENTS

α-LIPOIC ACID AS AN ENDOGENOUS ANTIOXIDANT

α-Lipoic acid (LA; 1,2-dithiocyclopentan-3-valeric acid, thioctic acid) is a biogenic antioxidant which physiologically acts as a coenzyme in the oxidative decarboxylation of α-keto acids, such as from pyruvate into acetyl-CoA (Widlansky et al., 2004). The cyclic disulfide LA is a potent free-radical scavenger that is absorbed from the diet, transported into cells, and reduced to its open chain form dihydrolipoic acid (DHLA), which has even greater antioxidant activity (Biewenga et al., 1997). Both forms can quickly be converted into each other by redox reactions.

In the last couple of years LA has experienced significant scientific and general interest as an anti-aging medicine (Hagen et al., 2002; Liu et al., 2002a,b) and as a fat-reducing compound (Kim et al., 2004; Shen et al., 2005). Decades before this recent interest, LA has been approved as a drug against diabetic polyneuropathy in Germany (Ziegler et al., 2004). In this context LA has been described to act mainly as an antioxidant. The antioxidative potential of LA has also been proposed as the responsible mechanism of action in LA's protective action in mushroom poisoning and protection from CCl_4-induced liver damage (Bustamante et al., 1998).

α-Lipoic acid is optically active, with the *R*-form being the naturally occurring enantiomer. As described for other optically active compounds, LA-mediated effects differ depending on stereochemistry. Indeed, the classical function of LA depends on stereochemistry: the pyruvate dehydrogenase complex has a catalytic preference for the *R*-enantiomer, and even opposing effects between the two enantiomers have been suggested (Hong, 1999; Walgren, 2004). Additionally, there has been described a differential bioavailability between the *R*- and *S*-enantiomer, where both the maximal plasma concentration and the "area-under the curve" levels were significantly higher for the *R*- versus the *S*-form (Breithaupt-Grogler et al., 1999).

The antioxidant reactivity of a compound is typically widely independent of its stereochemistry. As a pharmacological agent LA is mainly employed due to its antioxidant actions. It is therefore interesting to note that several of LA's biological actions have in fact been shown to be stereospecific.

In conditions of insulin resistance, the beneficial metabolic effects of LA have been attributed to the *R*-isomer, with the (*R*1)-enantiomer being more effective than the (*S*2)-enantiomer (Streeper et al., 1997). In bovine aortic endothelial cells exposed to oxidative stress, *R*-LA and racemic LA protected while *S*-LA was ineffective as determined by mitochondrial metabolism (Smith et al., 2005). The *R*-enantiomer also proved to be more efficient to protect against buthionine sulfoximine-induced cataract in newborn rats because of its protective effects on the endogenous lens antioxidants glutathione, ascorbate, and vitamin E. The stereospecificity exhibited was suggested to be due to selective uptake and reduction of *R*-LA by lens cells (Maitra et al., 1996).

These examples of stereospecific action of LA suggest that the biological effects of LA are not solely mediated by its antioxidative properties. We would therefore like to highlight examples of LA's pharmacological action where interaction with specific cell signaling pathways were reported.

LA AS AN ANTI-INFLAMMATORY AGENT

The pathophysiology of numerous diseases is known to be governed by an inflammatory reaction. Macrophages are key players in chronic inflammatory diseases. They are activated by exogenous and endogenous stimuli, such as bacterial lipopolysaccharides (LPS), histamine, or inflammatory cytokines. Receptor binding induces the activation of signaling cascades resulting in the activation of pro-inflammatory transcription factors, such as nuclear factor-κB (NF-κB) and activator protein-1 (AP-1). They induce the production of inflammatory mediators, such as tumor necrosis factor-α (TNF-α), interleukins, interferons, prostaglandins, and nitric oxide (NO). These mediators are part of the hosts' innate immune defense. However, their excessive production can be detrimental to the host tissue.

In addition to exerting direct cytotoxic effects to the inflamed tissue, they activate endothelial cells to express adhesion molecules, recruiting more leukocytes resulting in a vicious cycle of inflammation: a sustained production of cytokines, chemokines, and reactive oxygen species (ROS) is kept up. These pathogenic processes play a role in different diseases, such as atherosclerosis, rheumatoid arthritis, neurodegenerative diseases, or sepsis.

NF-κB is a pivotal transcription factor critical for production of the majority of inflammatory mediators ranging from NO produced by the inducible nitric oxide synthase (iNOS), adhesion molecules and chemokines, to cytokines, among which TNF-α plays the most important role (Calzado et al., 2007).

NF-κB has been proposed to be a redox-sensitive transcription factor and it has even been suggested that ROS play a role as universal second messenger in the signaling events leading to NF-κB activation (Schreck et al., 1992). Antioxidants have therefore been described to inhibit NF-κB activation (Van den Berg et al., 2001). Thiol antioxidants may influence the release of NF-κB from its inhibitor IκB, a process believed to be under redox control. Alternatively, antioxidants may influence the direct binding of NF-κB to DNA by influencing the redox-state of relevant cysteine residues (Matthews et al., 1993).

In this context of antioxidants' action on NF-κB, it might not seem surprising that LA reduces NF-κB-mediated expression of intercellular adhesion molecules (ICAM-1) in human monocytic cells: LA inhibited the binding of NF-κB to the respective DNA region. Since neither vitamin E nor vitamin C had an effect on ICAM-1 expression, however, action of LA on an antioxidant level has to be questioned (Lee and Hughes, 2002).

A similar observation was made in human monocytic THP-1 cells, where LA treatment leads also to an attenuated LPS-induced NF-κB activation and an attenuated consecutive cytokine and chemokine production (Zhang et al., 2007). Most interestingly, the activation of a specific signal transduction pathway which is known to be involved in cellular proliferation and survival processes, the phosphoinositol-3-kinase (PI-3-K)/Akt pathway, is required for this NF-κB inhibition. More and more indications suggest that this pathway also plays an important role as negative feedback regulator of excessive pro-inflammatory responses (Zhang et al., 2007). In fact, the LA-induced activation of PI-3-K/Akt was responsible for significantly improved survival of mice after LPS challenge in a model of sepsis (Zhang et al., 2007).

The activation of the PI-3-K/Akt pathway conferring an inhibitory effect on NF-κB may also explain the observation that LA reduces the LPS-induced, NF-κB-mediated expression of iNOS and TNF-α in rat Kupffer cells, the resident macrophages of the liver, and in murine RAW 264.7 macrophages (Figure 18.1) (Kiemer et al., 2002a).

Other anti-inflammatory effects of LA that may not depend on its direct antioxidative potential have also been described in human aortic endothelial cells. LA inhibits TNF-α or LPS-induced NF-κB-mediated expression of cell adhesion molecules and the chemokine MCP-1 and subsequent monocyte-endothelial interaction (Zhang and Frei, 2001). In this study, LA dose-dependently inhibited TNF-α-induced IκB kinase activation, subsequent degradation of IκB, the cytoplasmic NF-κB inhibitor, and nuclear translocation of NF-κB.

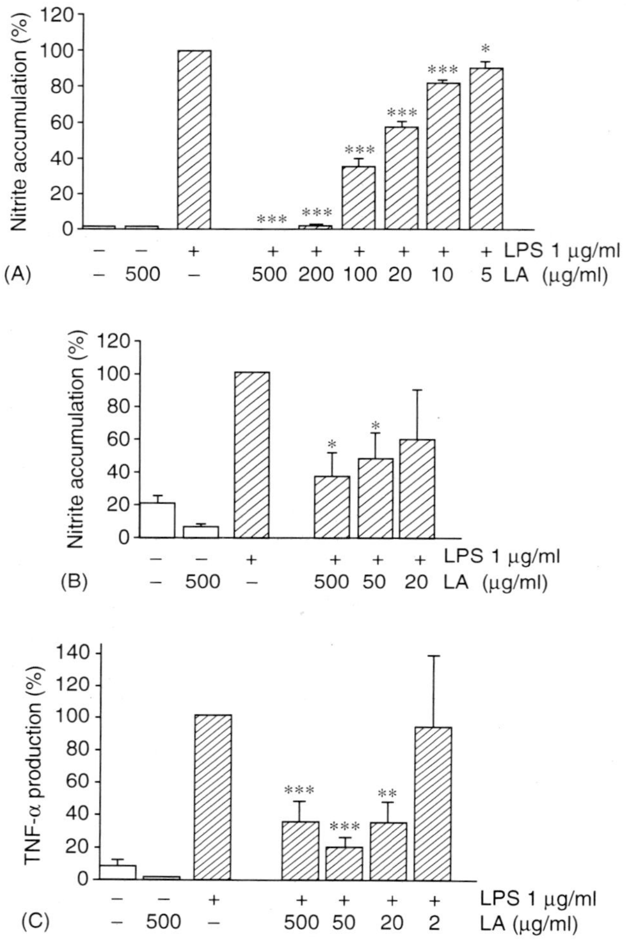

FIGURE 18.1 α-Lipoic acid (LA) inhibits lipopolysaccharide (LPS)-induced production of nitric oxide and tumor necrosis factor-α (TNF-α) release in macrophages. (A), (B) Lipopolysaccharide (LPS)-induced nitric oxide production in RAW 264.7 macrophages and in Kupffer cells. Murine RAW264.7 (A) macrophages or rat Kupffer cells, the resident macrophages of the liver (B), were cultivated for 20 h in either medium alone, in the presence of LA, or in medium containing LPS in the presence or absence of LA (conc. as indicated). The accumulation of nitrite, a stable metabolite of NO, in the supernatant was determined by the Griess assay. Data are expressed as percentage of nitrite accumulation in cells activated with LPS alone (100%). $^{***}P < .001$ and $^{*}P < .05$ represent significant differences compared to the values seen in LPS-activated cells. (C) Release of TNF-α from Kupffer cells, the resident macrophages of the liver. Cells were cultivated in either medium alone or in medium containing LPS or a combination of LPS and various concentrations of LA. Culture supernatants were assayed for TNF-α production using an L929 bioassay. Data are expressed as percentage of TNF-α concentration accumulated in the supernatant of LPS-activated KC (100%). $^{***}P < .001$ and $^{**}P < .01$ represent significant differences compared to the values seen in LPS-activated cells. (From Kiemer, A.K., Muller, C., and Vollmar, A.M., *Immunol. Cell Biol.*, 80, 550, 2002a. With permission.)

This supports that LA inhibits TNF-α-induced endothelial activation by affecting the NF-κB/IκB signaling pathway rather than by preventing DNA binding of NF-κB as discussed by Lee and Hughes (2002). In addition to the PI-3-K/Akt pathway, LA may induce anti-inflammatory responses *via* activation of p38 MAPK and nuclear factor erythroid 2-related factor (Nrf2) in human monocytes (Ogborne et al., 2005). These pathways are responsible for LA's induction of heme oxygenase-1 (HO-1) in human monocytic cells. The induction of HO-1 is known to exert anti-inflammatory, cytoprotective, and antioxidative action (Ogborne et al., 2005).

Further evidence for *in vivo* efficiency of LA in inflammatory diseases is given by the fact that LA inhibits airway inflammation in a mouse model of asthma. LA-treated animals show reduced airway hyperresponsiveness, a lower proportion of eosinophils among cells in bronchioalveolar lavage fluid, and significantly improved pathologic lesion scores of the lungs. LA also significantly reduced inflammatory markers, an effect most likely exerted by attenuated NF-κB DNA-binding activity (Cho et al., 2004).

There have also been reports on beneficial effects of LA on skeletal inflammatory diseases: LA and its reduced form DHLA can inhibit osteoclast formation and bone loss in inflammatory conditions by suppressing prostaglandin E2 synthesis *in vitro* and *in vivo* by inhibiting COX-2 peroxidase activity (Ha et al., 2006). In this context recent evidence suggests LA as an adjunctive treatment for rheumatoid arthritis since it suppresses the development of collagen-induced arthritis and protects against bone destruction in mice (Lee et al., 2007).

Taken together, LA has been described as an anti-inflammatory agent having important potential for the therapy of different inflammatory diseases. Both the antioxidative effects of LA as well as specific activation of signaling pathways like PI-3-K/Akt resulting in NF-κB inhibition, may contribute to these many beneficial effects of LA.

ANTI-ATHEROSCLEROTIC ACTION OF LA

The development and progression of atherosclerosis is increasingly recognized to be an inflammatory disease (Hansson, 2005). Atherosclerosis involves the formation of lesions in the arteries called atherosclerotic plaques resulting in flow-limiting stenoses and severe clinical events following the rupture of a plaque causing sudden thrombotic occlusion of the artery. This may lead to myocardial infarction and heart failure, ischemic stroke, and transient ischemic attacks.

Many of LA's actions might be relevant as anti-atherosclerotic mechanisms and shall be summarized briefly. As in other inflammatory diseases, activated endothelial cells produce chemokines and express leukocyte adhesion molecules, such as VCAM-1 (vascular cell adhesion molecule-1) in atherosclerotic vessels. Monocytes migrate into the arterial tissue by binding to VCAM-1-expressing endothelial cells and respond to the locally produced chemokines. As described above, LA inhibits these initial events of atherosclerosis since it attenuates TNF-α-induced inflammatory activation of endothelial cells *via* inhibition of IκB kinase independent of its antioxidative effects (Zhang and Frei, 2001).

Chemokines, such as M-CSF (macrophage colony-stimulating factor), promote the differentiation of monocytes to macrophages. Other factors, such as TNF-α, activate macrophages to produce pro-inflammatory mediators, such as TNF-α itself, chemokines, and ROS.

Activated endothelial cells allow low-density lipoprotein (LDL) to diffuse from the blood into the innermost layer of the artery and macrophage-derived ROS oxidize LDL-cholesterol. By taking up oxidized LDL and other lipids, macrophages transform to foam cells, representing a highly activated macrophage subtype.

As described above, LA inhibits the activation of macrophages (Kiemer et al., 2002a), a cell type playing a central role in atherosclerosis progression (Ogborne et al., 2005).

Further processes in atherogenesis include an increased proliferation of vascular smooth muscle cells, representing a critical event in restenosis. In a rat carotid artery balloon injury model, LA prevented neointimal hyperplasia. The suppressed smooth muscle cell expression of the chemokine fractalkine exerted by an inhibition of NF-κB activity could be proven responsible for this action (Lee et al., 2006).

Sola et al. evaluated in a clinical study the ability of LA to affect endothelial function and inflammation in patients with the metabolic syndrome (Sola et al., 2005). After a 4 week LA treatment with 300 mg/day, a significantly improved endothelial function was shown as investigated by flow-mediated vasodilatation. Moreover, LA-treated patients showed significantly lower plasma levels of the inflammatory markers IL-6 and plasminogen activator inhibitor-1 (PAI-1) compared with the placebo group. α-Lipoic acid was quite as effective in reducing the parameters mentioned above as irbesartan, an angiotensin receptor blocker. This implicates that LA, similar to an angiotensin receptor blocker, antagonizes the angiotensin II-induced effects in the pathogenesis of vascular inflammation (Frank et al., 2005; Sierra and De La Sierra, 2005). Interestingly, there was no significant antioxidant effect of LA observed in this clinical trial, as determined by plasma levels of 8-isoprostane as an indicator for oxidative stress. This strongly indicates a specific interaction of LA with cellular signaling pathways in vascular disease.

In this context, the activation of cellular signaling pathways by LA is also shown in cell culture models relevant for atherosclerosis processes: LA activates p38 MAPK in human monocytes (Ogborne et al., 2005) and inhibits extracellular regulated protein kinase ERK2 but not ERK1 in microvascular rat endothelial cells (Marsh et al., 2005). Both protein kinases represent members of the mitogen-activated protein kinase family known to be central mediators of inflammatory processes.

In a diabetic mouse model LA prevented the increase in atherosclerosis (Yi and Maeda, 2006). The protective effects of LA were accompanied by a reduction of plasma glucose and an accelerated recovery of insulin-producing cells in the pancreas. The authors suggest that part of LA's effects is attributable to protecting pancreatic beta-cells from damage. The effect might also be explained by observations from Konrad et al. suggesting that LA increases glucose uptake by specifically activating molecular signaling cascades of the cell (Konrad et al., 2001). Since LA has recently been shown to activate the insulin receptor (IR), this pathway will definitely play a major event in these metabolic actions (Diesel et al., 2007).

In addition to inflammatory conditions in the vasculature, an atherosclerotic vessel is characterized by an attenuated production of NO by the endothelial NO synthase (eNOS, NOSIII). This condition plays an important role in the pathogenesis of atherosclerosis (Li et al., 2002; Münzel et al., 2005). Both a decreased expression as well as a reduced activity of eNOS is responsible for this low NO production (Li et al., 2002). eNOS enzyme activity is largely regulated through phosphorylation at Ser1177 mediated through PI-3-Kinase/Akt (Dimmeler et al., 1999) and an activation of cAMP-dependent protein kinase (AMPK) (Chen et al., 1999; Drew et al., 2004; Thors et al., 2004). The latter protein kinase also affects smooth muscle cell proliferation known to contribute to atherogenesis (Nagata et al., 2004).

Nitric oxide produced by endothelial cells from eNOS is seen as an important mediator conferring mainly atheroprotective effects. In contrast, NO production that gets out of control, especially due to induction of the iNOS in macrophages, may result in pro-atherogenic processes.

Interestingly, LA inhibits iNOS expression in macrophages (Kiemer et al., 2002a) and activates eNOS in endothelial cells (Lee et al., 2005a). The activation of AMPK by LA was proven to induce eNOS activity in both human aortic endothelial cells as well as in obese rats (Lee et al., 2005a,b). It remains to be shown how LA affects AMPK, a protein kinase playing an increasingly recognized role in atherogenesis. It is known that a change in AMP/ATP concentration regulates the AMPK activity, usually with higher AMP levels leading to activation. It might therefore be surprising that LA has been proven to increase ATP contents in different cells and organs, among which are rat (Müller et al., 2003) and human livers (Dünschede et al., 2006a). This increased ATP level may, however, also reflect a downstream event of AMPK activation.

In summary, tissue culture, animal experiments, and clinical trials have proven LA as a potent anti-atherosclerotic compound. Anti-inflammatory and metabolic actions mediated by IR activation and activation of AMPK might represent central mechanistic explanations.

PROTECTION FROM APOPTOTIC AND NECROTIC CELL DEATH BY LA

Apoptosis, also known as programmed cell death, is characterized as a defined series of biochemical events, such as the activation of a cascade of cysteine aspartate proteases, named caspases, and endonucleases. Morphological characteristics involve shrinkage of the cell and nuclear condensation and as a result the cell is transferred into closed, sequestered packages of cell material known as apoptotic bodies. These bodies are removed by phagocytes or by neighboring cells in the last step. This apoptotic process contrasts with necrotic cell death, also referred to as accidental cell death. In this process cells rupture and induce inflammation. In pathophysiological conditions the two modes of cell death often occur consecutively or in parallel. This is often the case if apoptotic bodies are not removed properly resulting in lysis of the packages, a process named secondary necrosis.

There is a big variety of apoptotic stimuli, among which are deregulated cell cycle signals or inflammatory mediators like TNF-α and NO. The apoptotic cascade is a tightly controlled process with the group of Bcl proteins comprising several anti-apoptotic (e.g., Bcl-2, Bcl-xL) or pro-apoptotic (e.g., Bax, Bad) members having important regulatory power. Also protein kinases play a central role in the regulation of life and death decisions. They modulate the expression and phosphorylation status of key elements in the execution of apoptosis. Whereas apoptosis-signal-regulating kinase-1 (ASK-1) is a prototypical member of pro-apoptotic protein kinases, the protein kinase downstream of PI-3-K activation, Akt, has been shown *in vitro* and *in vivo* to improve cell survival. This is achieved by phosphorylating and deactivating pro-apoptotic factors, such as Bad, caspase-9, or several transcription factors (Zimmermann and Green, 2001; Choi et al., 2002). Akt has also been reported to promote phosphorylation of IκB and therefore initiate activation of NF-κB, resulting in the transcription of NF-κB dependent pro-survival genes (Bcl-xL and caspase inhibitors).

Numerous diseases have been described to be governed by too high apoptosis rates, many of which involve inflammatory events. Examples are neurodegenerative diseases, viral infections, sepsis, ischemia–reperfusion injury (IRI), or diabetes (Lee and Pervaiz, 2007). Compounds attenuating apoptotic cell death might therefore represent pharmacological agents to prevent or treat respective disorders.

Anti-apoptotic action of LA has often been described in the literature; frequently the protective action of LA is explained by its antioxidative action. α-Lipoic acid inhibits retinal capillary cell death in diabetic rats *in vivo* (Kowluru and Odenbach, 2004) and protects primary neurons of rat cerebral cortex against cell death induced by β-amyloid peptide or hydrogen peroxide (Zhang et al., 2001). Furthermore, LA inhibited apoptosis in a mouse model of Parkinson's disease by abolishing the activation of ASK-1 after treatment with the neurotoxin MPTP (Karunakaran et al., 2007).

α-Lipoic acid also seems to be useful for the prevention of IRI. In a rat model of warm hepatic ischemia, LA led to a significant reduction in necrosis- and apoptosis-related cell death in IRI of the liver and improved animal survival (Figure 18.2A) (Dünschede et al., 2006b; Dünschede et al., 2007b) and liver regeneration (Dünschede et al., 2007a). Also in humans, pretreatment with LA protects from IRI (Figure 18.2B,C) (Dünschede et al., 2006a). Similarly, pretreatment with LA improves cell-survival after freezing of human and rat hepatocytes (Terry et al., 2006). Dietary supplementation of LA combined with vitamin E protects the aged rat heart from ischemia–reperfusion-induced lipid peroxidation.

This protection is associated with improved cardiac performance during reperfusion (Coombes et al., 2000).

In contrast to these reports on anti-apoptotic action of LA, LA did not protect endothelial cells from oxidant-induced apoptosis (Marsh et al., 2005). Marsh et al. report that LA even induced apoptosis when added to the cells alone. This effect was, as in primary hepatocytes (Diesel et al., 2007), only observed in concentrations of several 100 μM up to 1 mM and might therefore not be relevant *in vivo*. Interestingly, lower concentrations of LA have been shown to induce apoptosis rather specifically only in tumor cells: whereas LA concentrations as

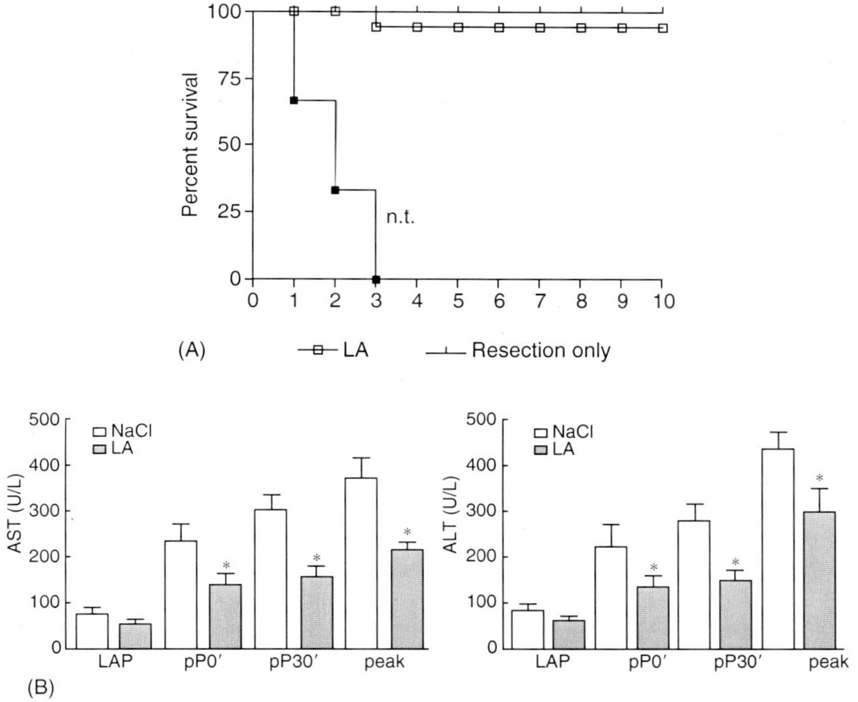

FIGURE 18.2 Beneficial effects of α-lipoic acid (LA) on ischemia–reperfusion injury (IRI) in vivo. (A) Rat survival curves using the method of Kaplan and Meier after 90 min of ischemia and resection of the non-ischemic liver tissue after ischemia with and without LA treatment compared to resection alone. Log rank $P < .01$ comparing the survival curves of nontreated (n.t.) and LA. LA reduces necrotic (B) and apoptotic (C) cell death in human hepatic IRI. In a clinical trial, effects of LA pretreatment on hepatic IRI where assessed. (B) Aspartate aminotransferase (AST) (left) and alanine aminotransferase (ALT) (right) levels in units of enzyme activity/liter (U/L) as markers for liver injury of human livers after laparotomy (LAP) followed by 30 min of ischemia (pP0′), 30 min of reperfusion (pP30′), and after 1–3 days (peak level) with and without LA treatment prior to ischemia. $P < .05$ vs. NaCl.

(*continued*)

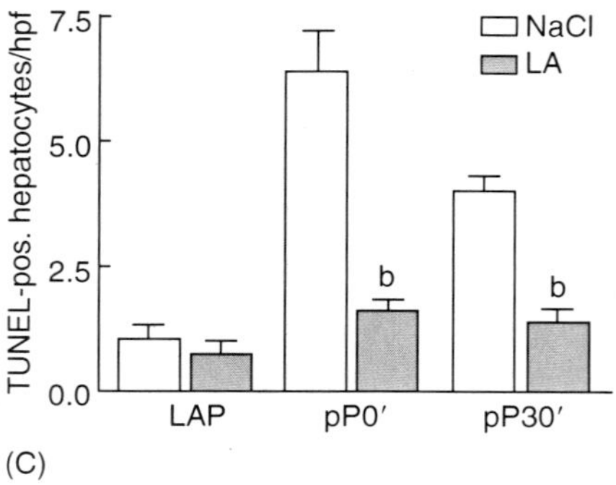

FIGURE 18.2 (continued) (C) Pretreatment with LA reduces post-ischemic apoptotic cell death as measured by TUNEL (terminal deoxynucleotidyl transferase mediated dUTP nick end labeling) assay indicating apoptosis related cell injury. TUNEL-positive hepatocytes in human liver sections after laparotomy (LAP), after 30 min of ischemia (pP0′) and after 30 min of reperfusion (pP30′), with and without LA treatment prior to ischemia. $P < .01$ vs. vehicle (NaCl). (Figure (A) from Dünschede, F., Erbes, K., Kirchner, A., et al., *Shock*, 27, 644, 2007a; Figure (B) and (C) from Dünschede, F., Erbes, K., Kirchner, A., et al., *World J. Gasteroenterol.*, 12, 6812, 2006a. With permission.)

low as 100 μM potentiated Fas-mediated apoptosis of leukemic Jurkat cells, peripheral blood lymphocytes from healthy humans were not affected (Sen et al., 1999). α-Lipoic acid also induces mitochondria-dependent apoptosis in lung epithelial cancer cells, most likely *via* a rapid generation of ROS mainly in the mitochondria (Moungjaroen et al., 2006). LA and DHLA have in fact been shown to inhibit proliferation and induce apoptosis of several other cancer and transformed cell lines while being less active towards normal nontransformed cells (Sen et al., 1999; Pack et al., 2002; Van de Mark et al., 2003; Wenzel et al., 2005). Our own unpublished data reveal that LA only protects primary hepatocytes but not transformed hepatoma cells from TNF-α/actinomycin D-induced apoptosis. α-Lipoic acid was even reported to induce apoptosis in hepatoma cells, an effect that is accompanied by the production of ROS (Simbula et al., 2007).

It has to be noted that the mechanisms of these varying effects of LA on the fate of either normal or tumor cells have never been studied in detail and should attract future research. Concerning the effect on normal cells one might summarize that although LA has a certain potential to induce apoptosis in high concentrations, these concentrations might never be relevant *in vivo*. The safety of LA has in fact been demonstrated by its successful use as a drug without any severe side effects for decades (Ziegler et al., 2004).

The mechanisms how LA mediates its mainly cytoprotective effects need further attention. An increasing number of data show that LA is not only cytoprotective due to its antioxidative properties but also activates specific signaling pathways.

Müller et al. showed for the first time that pretreatment with LA protects rats from IRI *via* the activation of a specific cytoprotective signaling pathway: the activation of PI-3-K/Akt (Figure 18.3) (Müller et al., 2003). Interestingly, in this study, the protection was exerted especially on necrotic processes not affecting apoptosis-related caspase 3-like activity. One has to note, however, that these mechanisms were studied in an *ex vivo* isolated rat liver model which not always reflects the *in vivo* situation. Newer data reveal in fact that LA does attenuate both apoptotic as well as necrotic cell death in hepatic IRI in rats (Figure 18.2) (Dünschede et al., 2006a,b; Dünschede et al., 2007a,b). Besides some anti-oxidative action, LA specifically balances the expression of Bcl proteins towards an anti-apoptotic pattern, i.e., increased Bcl-2 expression with attenuated Bax expression (Dünschede et al., 2007b).

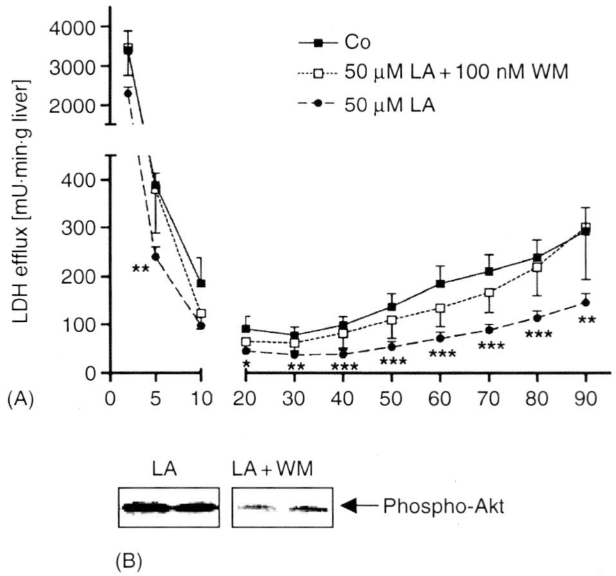

FIGURE 18.3 α-lipoic acid (LA) protects from rat ischemia–reperfusion injury via the activation of PI-3-K/Akt. Isolated livers were perfused for 30 min in the absence (Co) or presence of LA alone or in combination with wortmannin (WM) an inhibitor of PI-3-K upstream of Akt. After ischemia, livers were reperfused for 90 min. Lactate dehydrogenase (LDH) efflux served as an indicator for liver injury of the upstream PI3-K. (A) Analysis of LDH was carried out immediately after collecting perfusate. Data are expressed as enzyme activity in mU·min^{-1}·g liver weight. ***$P < .001$, **$P < .01$, and *$P < .05$ represent significant differences between LA-treated and untreated livers. (B) The efficiency of WM treatment was proven by phospho-Akt Western blots of simultaneously pretreated livers with WM and LA compared with livers preconditioned with LA alone. (From Müller, C., Dünschede, F., Koch, E., et al., *Am. J. Physiol. Gastrointest. Liver Physiol.*, 285, G769, 2003. With permission.)

The activation of the PI-3-K/Akt pathway reported to be important in protection from hepatic IRI is also responsible for anti-apoptotic actions of LA in TNF-α/actinomycin D-treated hepatocytes (Diesel et al., 2007). These apoptosis-reducing properties of LA in hepatocytes had previously been demonstrated by Pierce et al. but were interpreted to be mediated by LA's antioxidant actions (Pierce et al., 2000). Newest data reveal that other antioxidants exert this action either not at all or to a much lesser degree (Diesel et al., 2007). Interestingly, the crucial upstream event for PI-3-K activation by LA is autophosphorylation of the IR.

The IR is a transmembrane tyrosine kinase composed of two extracellular α-subunits and two transmembrane β-subunits. Binding of the ligand to the extracellular domain induces autophosphorylation of the intracellular receptor domains with subsequent phosphorylation of the insulin receptor substrate-1 (IRS-1) inducing PI-3-K/Akt signaling. The phosphorylation of IR followed by the activation of respective downstream targets induced by LA has previously been shown for smooth muscle cells and for adipocytes (Moini et al., 2002; Cho et al., 2003). It has been unknown, however, how LA might lead to the activation of the IR. Again, its redox-modulating potential had been hypothesized to contribute to this action (Konrad, 2005). Such observations may be explained by the fact that oxidants, such as the oxidized form of LA, can inactivate protein tyrosine phosphatases resulting in increased tyrosine phosphorylation of IR or may directly influence the oxidation of critical thiol groups of the IR β-subunit (Konrad, 2005). A respective mechanism of action for LA, however, has in the past been ruled out by experiments measuring LA-induced IR tyrosine phosphorylation in the presence of tyrosine phosphatase inhibitors in the assay buffer (Moini et al., 2002). Other antioxidants, among them Trolox, were either reported to even abolish ROS-induced IR activation (Fang et al., 2004) or to be without effect (Diesel et al., 2007).

Computer modelling studies suggest that the activation of the IR after LA is mediated by a novel mechanism: a direct binding of LA to the tyrosine kinase domain of the IR (Figure 18.4). The model suggests a stabilizing function of LA on the active kinase. α-Lipoic acid binds to the loop A of the active IR tyrosine kinase that is involved in ATP binding. These data were confirmed by the fact that LA induces exogenous substrate phosphorylation downstream of IR in a cell free system after immunoprecipitation of the IR β-chain (Figure 18.5) (Diesel et al., 2007). These findings make LA a unique compound by binding to the IR tyrosine kinase domain.

Taken together, many of LA's biological functions might contribute to anti-atherogenic action. They range, beside positive metabolic effects, from attenuated inflammatory action of macrophages and endothelial cells, attenuated SMC proliferation to the activation of eNOS.

α-LIPOIC ACID MIMICS AND AMPLIFIES INSULIN SIGNALING

Insulin receptor activation by LA has been demonstrated for the first time in adipocytes (Yaworsky et al., 2000). This work investigated LA's actions on

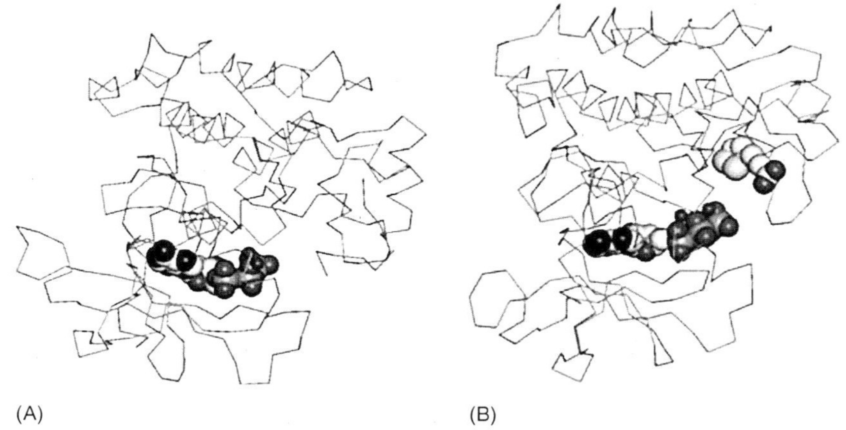

FIGURE 18.4 Computer modelling of binding of α-lipoic acid (LA) to the insulin receptor tyrosine kinase domain. (A) Crystal structure of the insulin receptor tyrosine kinase in its active form (with ATP), (B) in order to identify binding sites of LA, the GRID program was used. The program calculates interaction energies between specific functional groups and the protein and produces a map of possible binding pockets of the ligand. GRID fields were calculated for a sulfur, a methyl, and a carboxylate probe. The distance between the calculated carboxylate field and the lipophilic sulfur and methyl field matches the distance between the acid group of LA and the dithiolan ring (data not shown). LA was placed into this postulated binding pocket of the active kinase and a molecular dynamics simulation was performed in order to investigate whether the three-dimensional structure of the protein remains intact in the presence of LA and whether LA in fact forms stable interactions in the binding pocket. The figure shows the result of the simulation: the three-dimensional structure of the protein remains intact and LA has found a stabile position in the binding pocket. (From Diesel, B., Kulhanek-Heinze, S., Höltje, M., et al., *Biochemistry*, 46, 2146, 2007. With permission.)

glucose metabolism in adipocytes. Although the authors for the first time showed that LA is a non-insulin antidiabetic compound to be able to activate the insulin receptor, the group did not causally show that the action of LA on adipocyte glucose metabolism is in fact IR-dependent. By employing a specific inhibitor of IR tyrosine kinase phosphorylation, HNMPA(AM_3) (Saperstein et al., 1989), LA-induced IR autophosphorylation was only recently causally linked to induce downstream biological effects, i.e., protection from hepatocyte apoptosis (Diesel et al., 2007). These findings make LA an insulin mimetic drug. In fact, similar to LA, insulin has been described to protect hepatocytes against apoptosis *via* activation of the PI-3-K/Akt signaling cascade (Kang et al., 2003; Valverde et al., 2004). A work performed in neonatal rat hepatocytes (Valverde et al., 2004) further investigated potential downstream targets induced by insulin treatment. The group observed an increased expression of anti-apoptotic genes (Bcl-xL) and downregulated pro-apoptotic genes (Bim and nuclear Foxo1)

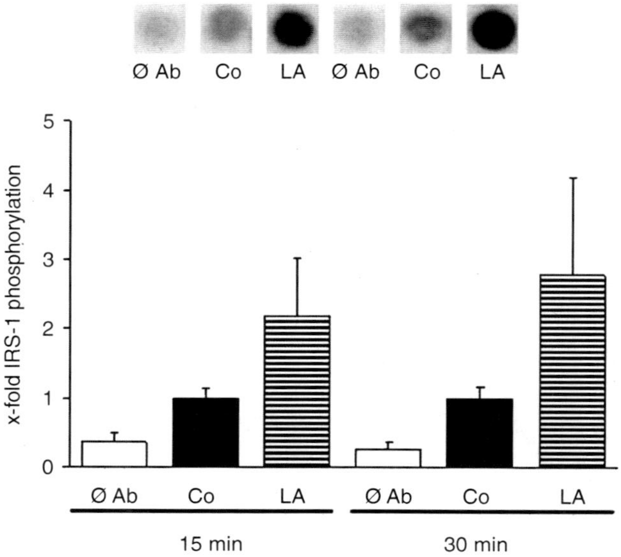

FIGURE 18.5 Kinase assay demonstrating α-lipoic acid (LA) action in a cell-free system. The insulin receptor β-chain was precipitated from untreated HepG2 cells grown for 8 h under serum-free conditions. The precipitates were left either unstimulated (Co) or were treated for 30 min with LA. Control samples were treated like LA-stimulated samples but IR-specific antibody was omitted (ø Ab). Phosphorylation of the insulin receptor substrate IRS-1 (Y608) was determined using the phosphocellulose adsorption method after 15 and 30 min. The upper panel shows phosphor imaging analysis of IRS-1 peptide phosphorylation. The lower panel shows quantification of phosphor imaging analysis. Data are expressed as x-fold IRS-1 phosphorylation relative to nontreated control samples. $*P < .05$ significantly different from nontreated cells. (From Diesel, B., Kulhanek-Heinze, S., Höltje, M., et al., *Biochemistry*, 46, 2146, 2007. With permission.)

showing similarity to the anti-apoptotic Bcl protein expression profile induced by LA in rat livers (Dünschede et al., 2007b). In addition to modulating Bcl protein expression there is evidence that IR-induced Akt activation leads to Ser136 phosphorylation of pro-apoptotic Bad which involves its inactivation and thereby leads to protection from apoptosis (Diesel et al., 2007).

The remarkable decrease of mortality and prevention from sepsis in insulin-treated critically ill patients (Van den Berghe et al., 2001) might be connected with an IR-mediated protection from hepatocyte apoptosis: insulin has been shown to have anti-apoptotic potential in hepatocytes of septic animals (Jeschke et al., 2005). Since hepatocyte apoptosis in sepsis is mediated *via* TNF-α (Shimizu et al., 2005) and LA has been shown to attenuate the production of pro-inflammatory factors contributing to the pathogenesis of sepsis in macrophages, Kupffer cells, and endothelial cells (Kiemer et al., 2002a; Sung et al., 2005; Zhang et al., 2007), these different actions might add up to an interesting pharmacological profile for the treatment of inflammatory diseases. α-Lipoic acid directly activating the IR (Diesel et al., 2007) in combination with its strong antioxidative potential (Konrad, 2005; Maric et al., 2005) may in this context be of special interest. The observation that activation of the PI-3-K/Akt pathway act typically downstream of IR by LA protects the mice from LPS-induced sepsis is an important proof of this concept (Zhang et al., 2007).

Reactive oxygen species generation is commonly assumed to be an event promoting inflammatory processes. With LA showing strong anti-inflammatory action it might seem surprising that there is increasing evidence that both LA and DHLA may exert pro-oxidant properties (Konrad, 2005). As discussed above, the influence of LA on IR autophosphorylation has been suggested to be mediated by a direct pro-oxidant action of LA (Konrad, 2005). Also Dicter et al. describe that LA dose-dependently increases intramuscular ROS production and stimulates glucose uptake into adipocytes by increasing intracellular oxidant levels (Dicter et al., 2002). Also some of LA's pro-apoptotic effects on tumor cells were suggested to be induced by ROS (see above). Scott et al. reported that DHLA may have pro-oxidant effects *via* its ability to reduce iron and generate reactive sulfur-containing radicals that can damage proteins such as α-1-antiproteinase and creatine kinase (Scott et al., 1994).

In the context of activation of signaling pathways by LA, there could be also a specific impact leading to a pro-oxidant cellular state/pathway. The activation of IR by its natural ligand insulin results in activation of nicotinamide adenine dinucleotide phosphate (NADPH) oxidase (Goldstein et al., 2005). It is therefore possible that LA may also activate NADPH oxidase. In the kidney, an elevation of the NADPH oxidase subunits p22 and $p47^{phox}$ by LA has in fact been described (Maric et al., 2005). In the vasculature an activated NADPH oxidase leading to higher ROS production may result in an uncoupling of the eNOS. This leads to the production of superoxide anions by the normally NO-producing enzyme (Li et al., 2002; Münzel et al., 2005). This process is discussed to aggravate endothelial dysfunction. On the other hand, the generation of endothelial ROS does not necessarily result in pathologic effects. The enhancement of basal NADPH oxidase activity by the cardiovascular hormone ANP has been shown to induce cytoprotective action *via* induction of MKP-1, a protein which exerts cytoprotective effects on endothelial cells (Kiemer et al., 2002b; Fürst et al., 2005). Whether LA activates endothelial NADPH oxidase *via* IR and what consequences this effect has on vascular dysfunction should be in the focus of future research.

Besides direct activation of the IR, LA also seems to act like an insulin sensitizing drug. This class of compounds increases sensitivity of cells towards insulin. They may act *via* indirect AMPK activation (Musi et al., 2001). α-Lipoic acid as an activator of AMPK (Lee et al., 2005a,b) has in fact been shown to improve insulin sensitivity in animal models of insulin resistance and obesity, as well as in patients with type-2 diabetes, suggesting a similar mechanism (Pershadsingh, 2007).

Also the nuclear peroxisome proliferator-activated receptor (PPAR) family of transcription factors which modulate genes fundamental for control of glucose and lipid metabolism may be a further specific signaling pathway influenced by

LA. α-Lipoic acid has recently been shown to activate both PPAR-α and PPAR-γ (Pershadsingh, 2005) which both represent typical targets for insulin sensitizers. Also an increase in PPAR-regulated genes by LA was shown (Pershadsingh, 2007).

Taken together, both the activation of AMPK and PPAR make LA a compound employing the same pathways as the highly successful clinically employed insulin sensitizer drugs. This combined action of an activation of AMPK and PPAR-γ might also make LA a promising therapeutic for the treatment of NASH (nonalcoholic steatohepatitis) (Trappoliere et al., 2005). This inflammatory liver disease is associated with adiposity, the metabolic syndrome, and insulin resistance, and is thought to affect 2%–5% of Americans with a dramatically rising incidence. As observed in other inflammatory liver diseases, there is an elevated rate of hepatocyte apoptosis contributing to the pathogenesis of the disease. With LA being anti-inflammatory, anti-apoptotic, and improving metabolism, it might represent an ideal drug candidate for the treatment of respective diseases.

CONCLUSION

Taken together, due to its numerous effects on different signaling molecules, employment of LA does not seem suitable to prove the involvement of ROS in a

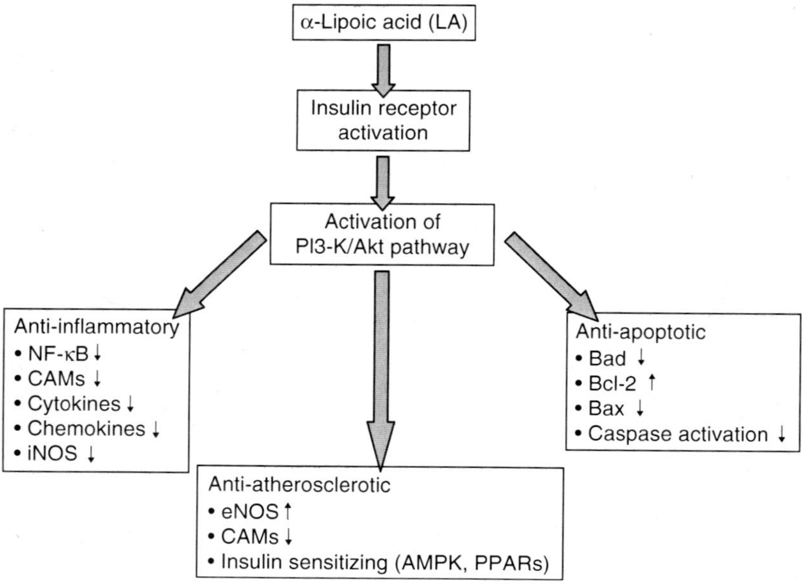

FIGURE 18.6 Anti-inflammatory, anti-atherosclerotic, and anti-apoptotic effects of α-lipoic acid which may be mediated by specifically activating the insulin receptor. PI-3-K, phosphatidylinositol-3-kinase; CAMs, cell adhesion molecules; iNOS, inducible nitric oxide synthase; eNOS, endothelial nitric oxide synthase; AMPK, AMP-activated kinase; PPAR, peroxisome proliferator-activated receptor.

specific system. In contrast, an increasing number of data shows that LA specifically interacts with cytoprotective cell signaling pathways. Most of its actions can be explained by the activation of the IR leading to different pharmacological profiles. In this article we highlighted respective anti-inflammatory, anti-atherosclerotic, and anti-apoptotic actions of LA (Figure 18.6). These involve (i) an attenuated activation of the inflammatory transcription factor NF-κB with subsequent reduced expression of cytokines, chemokines, or adhesion molecules during inflammation, (ii) an activation of endothelial NO synthase in the atherosclerotic vessel, and (iii) the modulated expression and phosphorylation of apoptotic regulators towards anti-apoptotic action. These effects might only be some of the events induced by IR-mediated PI-3-K/Akt activation triggered by direct interaction of LA with the IR. α-Lipoic acid is therefore both suitable to serve as a safe and orally available drug itself but also as a model compound for the development of structurally similar insulin mimetics.

REFERENCES

Biewenga, G.P., G.R. Haenen, and A. Bast. 1997. The pharmacology of the antioxidant lipoic acid. *General Pharmacology* 29:315–331.

Breithaupt-Grogler, K., G. Niebch, E. Schneider et al. 1999. Dose-proportionality of oral thioctic acid—coincidence of assessments via pooled plasma and individual data. *European Journal of Pharmaceutical Sciences* 8:57–65.

Bustamante, J., J.K. Lodge, L. Marcocci et al. 1998. Alpha-lipoic acid in liver metabolism and disease. *Free Radical Biology and Medicine* 24:1023–1039.

Calzado, M.A., S. Bacher, and M.L. Schmitz. 2007. NF-kappaB inhibitors for the treatment of inflammatory diseases and cancer. *Current Medicinal Chemistry*. 14:367–376.

Chen, Z.P., K.I. Mitchelhill, B.J. Michell et al. 1999. AMP-activated protein kinase phosphorylation of endothelial NO synthase. *FEBS Letters* 443:285–289.

Cho, K.J., H.E. Moon, A.S. Chung et al. 2003. Alpha-lipoic acid inhibits adipocyte differentiation by regulating pro-adipogenic transcription factors via mitogen-activated protein kinase pathways. *Journal of Biological Chemistry* 278:34823–34833.

Cho, Y.S., J. Lee, T.H. Lee et al. 2004. Alpha-lipoic acid inhibits airway inflammation and hyperresponsiveness in a mouse model of asthma. *The Journal of Allergy and Clinical Immunology* 114:429–435.

Choi, B.M., H.O. Pae, S.I. Jang et al. 2002. Nitric oxide as a pro-apoptotic as well as anti-apoptotic modulator. *Journal of Biochemistry and Molecular Biology* 35:116–126.

Coombes, J.S., S.K. Powers, K.L. Hamilton et al. 2000. Improved cardiac performance after ischemia in aged rats supplemented with vitamin E and alpha-lipoic acid. *American Journal of Physiology. Regulatory, Integrative and Comparative Physiology* 279:R2149–R2155.

Dicter, N., Z. Madar, and O. Tirosh. 2002. Alpha-lipoic acid inhibits glycogen synthesis in rat soleus muscle via its oxidative activity and the uncoupling of mitochondria. *The Journal of Nutrition* 132:3001–3006.

Diesel, B., S. Kulhanek-Heinze, M. Höltje et al. 2007. Alpha-lipoic acid as a directly binding activator of the insulin receptor-protection from hepatocyte apoptosis. *Biochemistry* 46:2146–2155.

Dimmeler, S., C. Hermann, A.M. Zeiher et al. 1999. Activation of nitric oxide synthase in endothelial cells by Akt-dependent phosphorylation. *Nature* 399:601–605.

Drew, B.G., N.H. Fidge, G. Gallon-Beaumier et al. 2004. High-density lipoprotein and apolipoprotein AI increase endothelial NO synthase activity by protein association and multisite phosphorylation. *Proceedings of the National Academy of Sciences of the United States of America* 101:6999–7004.

Dünschede, F., K. Erbes, A. Kircher et al. 2007a. Protection from hepatic ischemia/reperfusion injury and improvement of liver regeneration by alpha-lipoic acid. *Shock* 27:644–651.

Dünschede, F., S. Westermann, N. Riegler et al. 2007b. Protective effects of ischemic preconditioning and application of lipoic acid prior to 90 minutes of rat liver ischemia. *World Journal of Gastroenterology* 13:3692–3698.

Dünschede, F., K. Erbes, A. Kircher et al. 2006a. Reduction of ischemia–reperfusion injury after liver resection and hepatic inflow occlusion by alpha-lipoic acid in humans. *World Journal of Gastroenterology* 12:6812–6817.

Dünschede, F., S. Westermann, N. Riegler et al. 2006b. Different protection mechanisms after pretreatment with glycine or alpha-lipoic acid in a rat model of warm hepatic ischemia. *European Surgical Research* 38:503–512.

Fang, Y., S.I. Han, C. Mitchell et al. 2004. Bile acids induce mitochondrial ROS, which promote activation of receptor tyrosine kinases and signaling pathways in rat hepatocytes. *Hepatology* 40:961–971.

Frank, G.D., S. Eguchi, and E.D. Motley. 2005. The role of reactive oxygen species in insulin signaling in the vasculature. *Antioxidants and Redox Signaling* 7:1053–1061.

Fürst, R., C. Brueckl, W.M. Kuebler et al. 2005. Atrial natriuretic peptide induces mitogen-activated protein kinase phosphatase-1 in human endothelial cells via Rac1 and NAD(P)H oxidase/Nox2-activation. *Circulation Research* 96:43–53.

Goldstein, B.J., K. Mahadev, X. Wu et al. 2005. Role of insulin-induced reactive oxygen species in the insulin signaling pathway. *Antioxidants and Redox Signaling* 7:1021–1031.

Ha, H., J.H. Lee, H.N. Kim et al. 2006. Alpha-lipoic acid inhibits inflammatory bone resorption by suppressing prostaglandin E2 synthesis. *Journal of Immunology* 176:111–117.

Hagen, T.M., J. Liu, J. Lykkesfeldt et al. 2002. Feeding acetyl-L-carnitine and lipoic acid to old rats significantly improves metabolic function while decreasing oxidative stress. *Proceedings of the National Academy of Sciences of the United States of America* 99:1870–1875.

Hansson, G.K. 2005. Mechanisms of disease: Inflammation, atherosclerosis, and coronary artery disease. *New England Journal of Medicine* 352:1685–1695.

Hong, Y.S., S.J. Jacobia, L. Packer et al. 1999. The inhibitory effects of lipoic compounds on mammalian pyruvate dehydrogenase complex and its catalytic components. *Free Radical Biology & Medicine* 26:685–694.

Jeschke, M.G., H. Rensing, D. Klein et al. 2005. Insulin prevents liver damage and preserves liver function in lipopolysaccharide-induced endotoxemic rats. *Journal of Hepatology* 42:870–879.

Kang, S., J. Song, H. Kang et al. 2003. Insulin can block apoptosis by decreasing oxidative stress via phosphatidylinositol-3-kinase and extracellular signal-regulated protein kinase-dependent signaling pathways in HepG2 cells. *European Journal of Endocrinology* 148:147–155.

Karunakaran, S., L. Diwakar, U. Saeed et al. 2007. Activation of apoptosis signal regulating kinase 1 (ASK1) and translocation of death-associated protein, Daxx, in substantia nigra pars compacta in a mouse model of Parkinson's disease: Protection by alpha-lipoic acid. *FASEB Journal* 21:2226–2236.

Kiemer, A.K., C. Müller, and A.M. Vollmar. 2002a. Inhibition of LPS-induced nitric oxide and TNF-alpha production by alpha-lipoic acid in rat Kupffer cells and in RAW 264.7 murine macrophages. *Immunology and Cell Biology* 80:550–557.

Kiemer, A.K., N.C. Weber, R. Fürst et al. 2002b. Inhibition of p38 MAPK activation via induction of MKP-1: Atrial natriuretic peptide reduces TNF-alpha-induced actin polymerization and endothelial permeability. *Circulation Research* 90:874–881.

Kim, M.S., J.Y. Park, I.S. Namgoong et al. 2004. Anti-obesity effects of alpha-lipoic acid mediated by suppression of hypothalamic AMP-activated protein kinase. *Nature Medicine* 10:727–733.

Konrad, D. 2005. Utilization of the insulin-signaling network in the metabolic actions of alpha-lipoic acid-reduction or oxidation? *Antioxidants and Redox Signaling* 7:1032–1039.

Konrad, D., R. Somwar, G. Sweeney et al. 2001. The antihyperglycemic drug alpha-lipoic acid stimulates glucose uptake via both GLUT4 translocation and GLUT4 activation: Potential role of p38 mitogen-activated protein kinase in GLUT4 activation. *Diabetes* 50:1464–1471.

Kowluru, R.A. and S. Odenbach. 2004. Effect of long-term administration of alpha-lipoic acid on retinal capillary cell death and the development of retinopathy in diabetic rats. *Diabetes* 53:3233–3238.

Lee, E.Y., C.K. Lee, K.U. Lee et al. 2007. Alpha-lipoic acid suppresses the development of collagen-induced arthritis and protects against bone destruction in mice. *Rheumatology International* 27:225–233.

Lee, H.A. and D.A. Hughes. 2002. Alpha-lipoic acid modulates NF-kappaB activity in human monocytic cells by direct interaction with DNA. *Experimental Gerontology* 37:401–410.

Lee, K.M., K.G. Park, Y.D. Kim et al. 2006. Alpha-lipoic acid inhibits fractalkine expression and prevents neointimal hyperplasia after balloon injury in rat carotid artery. *Atherosclerosis* 189:106–114.

Lee, S.C. and S. Pervaiz. 2007. Apoptosis in the pathophysiology of diabetes mellitus. *The International Journal of Biochemistry and Cell Biology* 39:497–504.

Lee, W.J., I.K. Lee, H.S. Kim et al. 2005a. Alpha-lipoic acid prevents endothelial dysfunction in obese rats via activation of amp-activated protein kinase. *Arteriosclerosis, Thrombosis, and Vascular Biology* 25:2488–2494.

Lee, W.J., K.H. Song, E.H. Koh et al. 2005b. Alpha-lipoic acid increases insulin sensitivity by activating AMPK in skeletal muscle. *Biochemical and Biophysical Research Communications* 332:885–891.

Li, H., T. Wallerath, T. Münzel et al. 2002. Regulation of endothelial-type NO synthase expression in pathophysiology and in response to drugs. *Nitric Oxide-Biology and Chemistry* 7:149–164.

Liu, J., E. Head, A.M. Gharib et al. 2002a. Memory loss in old rats is associated with brain mitochondrial decay and RNA/DNA oxidation: Partial reversal by feeding acetyl-L-carnitine and/or *R*-alpha-lipoic acid. *Proceedings of the National Academy of Sciences of the United States of America* 99:2356–2361.

Liu, J., D.W. Killilea, and B.N. Ames. 2002b. Age-associated mitochondrial oxidative decay: Improvement of carnitine acetyltransferase substrate-binding affinity and activity in brain by feeding old rats acetyl-L-carnitine and/or *R*-alpha-lipoic acid. *Proceedings of the National Academy of Sciences of the United States of America* 99:1876–1881.

Maitra, I., E. Serbinova, H.J. Tritschler et al. 1996. Stereospecific effects of *R*-lipoic acid on buthionine sulfoximine-induced cataract formation in newborn rats. *Biochemical and Biophysical Research Communications* 221:422–429.

Maric, C., F. Bhatti, R.W. Mankhey et al. 2005. Mechanisms of antioxidant and pro-oxidant effects of alpha-lipoic acid in the diabetic and nondiabetic kidney. *Kidney International* 67:1371–1380.

Marsh, S.A., J.S. Coombes, B.K. Pat et al. 2005. Evidence for a non-antioxidant, dose-dependent role of alpha-lipoic acid in caspase-3 and ERK2 activation in endothelial cells. *Apoptosis* 10:657–665.

Matthews, J.R., W. Kaszubska, G. Turcatti et al. 1993. Role of cysteine62 in DNA recognition by the P50 subunit of NF-kappa B. *Nucleic Acids Research* 21:1727–1734.

Moini, H., O. Tirosh, Y.C. Park et al. 2002. *R*-Alpha-lipoic acid action on cell redox status, the insulin receptor, and glucose uptake in 3T3-L1 adipocytes. *Archives of Biochemistry and Biophysics* 397:384–391.

Moungjaroen, J., U. Nimmannit, P.S. Callery et al. 2006. Reactive oxygen species mediate caspase activation and apoptosis induced by lipoic acid in human lung epithelial cancer cells through Bcl-2 down-regulation. *The Journal of Pharmacology and Experimental Therapeutics* 319:1062–1069.

Müller, C., F. Dünschede, E. Koch, A.M. Vollmar, and A.K. Kiemer. 2003. Alpha-lipoic acid preconditioning reduces ischemia–reperfusion injury of the rat liver via the PI3-kinase/Akt pathway. *American Journal of Physiology. Gastrointestinal and Liver Physiology* 285:G769–G778.

Münzel, T., A. Daiber, A. Mülsch et al. 2005. Vascular consequences of endothelial nitric oxide synthase uncoupling for the activity and expression of the soluble guanylyl cyclase and the cGMP-dependent protein kinase. *Arteriosclerosis, Thrombosis, and Vascular Biology* 25:1551–1557.

Musi, N., T. Hayashi, N. Fujii et al. 2001. AMP-activated protein kinase activity and glucose uptake in rat skeletal muscle. *American Journal of Physiology. Endocrinology and Metabolism* 280:E677–E684.

Nagata, D., R. Takeda, M. Sata et al. 2004. AMP-activated protein kinase inhibits angiotensin II-stimulated vascular smooth muscle cell proliferation. *Circulation* 110:444–451.

Ogborne, R.M., S.A. Rushworth, and M.A. O'Connell. 2005. Alpha-lipoic acid-induced heme oxygenase-1 expression is mediated by 2 and p38 mitogen-activated protein kinase in human monocytic cells. *Arteriosclerosis, Thrombosis, and Vascular Biology* 25:2100–2105.

Pack, R.A., K. Hardy, M.C. Madigan et al. 2002. Differential effects of the antioxidant alpha-lipoic acid on the proliferation of mitogen-stimulated peripheral blood lymphocytes and leukaemic T cells. *Molecular Immunology* 38:733–745.

Pershadsingh, H.A. 2005. Alpha-lipoic acid is a weak dual PPARa/g agonist an ester derivative with increased PPARa/g efficacy and antioxidant activity. *Journal of Applied Research* 5:510–523.

Pershadsingh, H.A. 2007. Alpha-lipoic acid: Physiologic mechanisms and indications for the treatment of metabolic syndrome. *Expert Opinion on Investigational Drugs* 16:291–302.

Pierce, R.H., J.S. Campbell, A.B. Stephenson et al. 2000. Disruption of redox homeostasis in tumor necrosis factor-induced apoptosis in a murine hepatocyte cell line. *American Journal of Pathology* 157:221–236.

Saperstein, R., P.P. Vicario, H.V. Strout et al. 1989. Design of a selective insulin receptor tyrosine kinase inhibitor and its effect on glucose uptake and metabolism in intact cells. *Biochemistry* 28:5694–5701.

Schreck, R., K. Albermann, and P.A. Baeuerle. 1992. Nuclear factor kappa B: An oxidative stress-responsive transcription factor of eukaryotic cells (a review). *Free Radical Research Communications*. 17:221–237.

Scott, B.C., O.I. Aruoma, P.J. Evans et al. 1994. Lipoic and dihydrolipoic acids as antioxidants. A critical evaluation. *Free Radical Research* 20:119–133.

Sen, C.K., S. Roy, and L. Packer. 1999. Fas mediated apoptosis of human Jurkat T-cells: Intracellular events and potentiation by redox-active alpha-lipoic acid. *Cell Death and Differentiation* 6:481–491.

Shen, Q.W., C.S. Jones, N. Kalchayanand et al. 2005. Effect of dietary alpha-lipoic acid on growth, body composition, muscle pH, and AMP-activated protein kinase phosphorylation in mice. *Journal of Animal Science* 83:2611–2617.

Shimizu, S., Y. Yamada, M. Okuno et al. 2005. Liver injury induced by lipopolysaccharide is mediated by TNFR-1 but not by TNFR-2 or Fas in mice. *Hepatology Research* 31:136–142.

Sierra, C. and A. De La Sierra. 2005. Antihypertensive, cardiovascular, and pleiotropic effects of angiotensin-receptor blockers. *Current Opinion in Nephrology and Hypertension* 14:435–441.

Simbula, G., A. Columbano, G.M. Ledda-Columbano et al. 2007. Increased ROS generation and p53 activation in alpha-lipoic acid-induced apoptosis of hepatoma cells. *Apoptosis* 12:113–123.

Smith, J.R., H.V. Thiagaraj, B. Seaver et al. 2005. Differential activity of lipoic acid enantiomers in cell culture. *Journal of Herbal Pharmacotherapy* 5:43–54.

Sola, S., M.Q. Mir, F.A. Cheema et al. 2005. Irbesartan and lipoic acid improve endothelial function and reduce markers of inflammation in the metabolic syndrome: Results of the Irbesartan and lipoic acid in endothelial dysfunction (ISLAND) study. *Circulation* 111:343–348.

Streeper, R.S., E.J. Henriksen, S. Jacob et al. 1997. Differential effects of lipoic acid stereoisomers on glucose metabolism in insulin-resistant skeletal muscle. *American Journal of Physiology* 273:E185–E191.

Sung, M.J., W. Kim, S.Y. Ahn et al. 2005. Protective effect of alpha-lipoic acid in lipopolysaccharide-induced endothelial fractalkine expression. *Circulation Research* 97:880–890.

Terry, C., A. Dhawan, R.R. Mitry et al. 2006. Preincubation of rat and human hepatocytes with cytoprotectants prior to cryopreservation can improve viability and function upon thawing. *Liver Transplantation* 12:165–177.

Thors, B., H. Halldrsson, and G. Thorgeirsson. 2004. Thrombin and histamine stimulate endothelial nitric-oxide synthase phosphorylation at Ser1177 via an AMPK mediated pathway independent of PI3K-Akt. *FEBS Letters* 573:175–180.

Trappoliere, M., C. Tuccillo, A. Federico et al. 2005. The treatment of NAFLD. *European Review for Medical and Pharmacological Sciences* 9:299–304.

Valverde, A.M., I. Fabregat, D.J. Burks, M.F. White, and M. Benito. 2004. IRS-2 mediates the antiapoptotic effect of insulin in neonatal hepatocytes. *Hepatology* 40:1285–1294.

Van de Mark, K., J.S. Chen, K. Steliou et al. 2003. Alpha-lipoic acid induces p27Kip-dependent cell cycle arrest in non-transformed cell lines and apoptosis in tumor cell lines. *Journal of Cellular Physiology* 194:325–340.

Van den Berghe, G., P. Wouters, F. Weekers et al. 2001. Intensive insulin therapy in critically ill patients. *New England Journal of Medicine* 345:1359–1367.

Van den Berg, R., G.R. Haenen, H. van den Berg et al. 2001. Transcription factor NF-kappaB as a potential biomarker for oxidative stress. *The British Journal of Nutrition* 86 Suppl 1:S121–S127.

Walgren, J.L., Z. Amani, J.M. McMillan et al. 2004. Effect of R(+)alpha-lipoic acid on pyruvate metabolism and fatty acid oxidation in rat hepatocytes. *Metabolism* 53:165–173.

Wenzel, U., A. Nickel, and H. Daniel. 2005. Alpha-lipoic acid induces apoptosis in human colon cancer cells by increasing mitochondrial respiration with a concomitant O2-generation. *Apoptosis* 10:359–368.

Widlansky, M., A.R. Smith, S.V. Shenvi et al. 2004. Lipoic acid as a potential therapy for chronic diseases associated with oxidative stress. *Current Medicinal Chemistry* 11:1135–1146.

Yaworsky, K., R. Somwar, T. Ramlal et al. 2000. Engagement of the insulin-sensitive pathway in the stimulation of glucose transport by alpha-lipoic acid in 3T3-L1 adipocytes. *Diabetologia* 43:294–303.

Yi, X. and N. Maeda. 2006. Alpha-lipoic acid prevents the increase in atherosclerosis induced by diabetes in apolipoprotein E-deficient mice fed high-fat/low-cholesterol diet. *Diabetes* 55:2238–2244.

Zhang, L., G.Q. Xing, J.L. Barker et al. 2001. Alpha-lipoic acid protects rat cortical neurons against cell death induced by amyloid and hydrogen peroxide through the Akt signaling pathway. *Neuroscience Letters* 312:125–128.

Zhang, W.J. and B. Frei. 2001. Alpha-lipoic acid inhibits TNF-alpha-induced NF-kappaB activation and adhesion molecule expression in human aortic endothelial cells. *FASEB Journal* 15:2423–2432.

Zhang, W.J., H. Wei, T. Hagen et al. 2007. Alpha-lipoic acid attenuates LPS-induced inflammatory responses by activating the phosphoinositide 3-kinase/Akt signaling pathway. *Proceedings of the National Academy of Sciences of the United States of America* 104:4077–4082.

Ziegler, D., H. Nowak, P. Kempler et al. 2004. Treatment of symptomatic diabetic polyneuropathy with the antioxidant alpha-lipoic acid: A meta-analysis. *Diabetic Medicine* 21:114–121.

Zimmermann, K.C. and D.R. Green. 2001. How cells die: Apoptosis pathways. *The Journal of Allergy and Clinical Immunology* 108:S99–S103.

19 Selenotrisulfide Derivatives of Alpha-Lipoic Acid: Potential Use as a Novel Topical Antioxidant

William T. Self

CONTENTS

INTRODUCTION

Damage caused by reactive oxygen species (ROS) has been linked to cancer due to oxidative modifications to DNA, and it is also well established that ROS can contribute to degenerative diseases and aging. The use of small molecule antioxidants to prevent the development of cancer as well as to postpone the natural effects of aging has expanded in recent years, both in model cell culture systems as well as in animal studies. α-Lipoic acid (LA) has shown great promise as a lipid soluble antioxidant, especially in studies focused on oxidative stress and aging in the brain [1,2]. Recent analysis of the safety of LA indicates that oral administration yields no significant side effects in long-term exposure studies in animals [3,4].

Selenium is required in mammals as a micronutrient and present in the form of selenocysteine in key enzymes involved in defense against ROS [5]. Several isoenzymes of selenoproteins, thioredoxin reductase (TrxR) and glutathione peroxidase (Gpx), represent a large portion of the complement of "antioxidant" enzymes in mammals [6]. Clinical trials supplementing selenium in the diet, primarily in the form of selenomethionine, have uncovered an inverse correlation between serum selenium levels and increased cancer rates of the prostate and potentially other organs, as summarized in a recent review [5]. The cumulative data reported from several clinical trials suggest that one's selenium status can be a key determinant for predisposition for cancer and possibly degenerative diseases. Development of novel selenium-containing small molecules that can effectively supply selenium for selenoprotein synthesis and subsequently increase the TrxR and Gpx activities in tissues remains a goal towards reducing the severity of ROS-associated diseases. Towards this goal, a novel selenium-containing derivative of α-lipoic acid or lipoamide (LASe and LNSe, Figure 19.1), has recently been developed and tested as a topical antioxidant [7,8].

In this chapter, I will discuss the evidence that lipoic acid is a potent and clinically relevant antioxidant, the role of that selenotrisulfides play in selenium metabolism, as well as the development, testing, and potential use of selenotrisulfide derivative of α-lipoic acid (LASe) as a novel topical antioxidant to aid in the prevention of skin cancer and aging of skin.

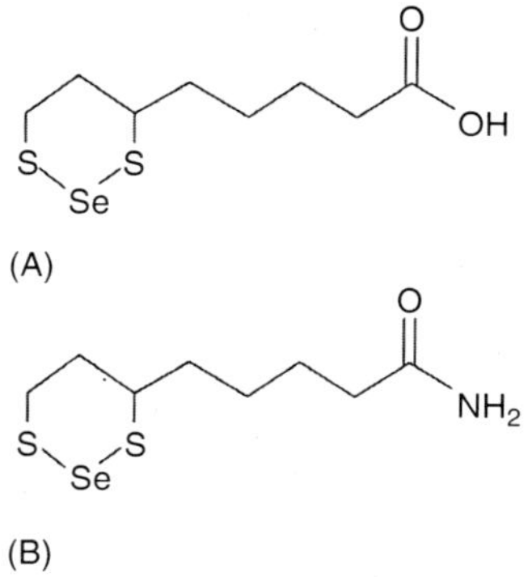

FIGURE 19.1 Selenotrisulfide derivatives of (A) lipoic acid (LASe) and (B) lipoamide (LNSe). (From Self, W.T., Tsai, L., and Stadtman, T.C., *Proc. Natl. Acad. Sci. U.S.A.*, **97,** 12481, 2000.)

α-LIPOIC ACID: USE AS AN ANTIOXIDANT

Gunsalus first reported that an unidentified factor present in yeast extract was required as a cofactor in pyruvate oxidation [9]. This cofactor, α-lipoic acid, was later isolated in a crystalline form by Lester Reed in a Herculean effort in 1951 [10]. Since these seminal studies, which were centered around the role of LA as a cofactor in α-keto acid dehydrogenases, LA has also found utility as a lipid soluble antioxidant given the redox active dithiol of the compound [1,2,11–17]. Early work by Packer revealed that oxidized LA could be reduced by several oxidoreductases using either NADH or NADPH as a reductant [13]. In cytosolic extracts of hepatocytes, NADPH was more effective as a reductant, whereas in cell extracts derived from heart, brain, and kidney NADH-dependent enzymes acted more efficiently on oxidized LA. Regardless of the source of reducing potential, these studies indicated that LA undergoes oxidation/reduction in the cytosol and therefore could serve as a thiol-based antioxidant in the cell. One key study indicated that administration of LA to cells in culture resulted in higher levels of reduced glutathione (GSH) [12]. Han et al. [13] showed that the likely molecular mechanism for this observed effect was the efficient reduction of cystine to cysteine in the culture medium, leading to more efficient synthesis of GSH in the cell. Taken together, these studies indicate that administration of LA may act either directly as an antioxidant or perhaps indirectly by stimulating GSH biosynthesis. In either case, administration of LA could lead to an enhanced pool of reduced thiols (RSH) to aid in the prevention of oxidative damage.

A modified LA derivative, *N,N*-dimethyl,*N′*-2amidoethyl-lipoate (LA-plus), was developed by Packer et al. [15]. A net positive charge at physiological pH facilitated uptake of this thioctic acid analogue [15]. *N,N*-dimethyl,*N′*-2amidoethyl-lipoate was more efficiently taken up by human T cells, was more efficiently reduced to DHLA-plus, and was able to prevent peroxide-dependent activation of NF-κB. Interestingly, both LA and LA-plus acted synergistically with selenium (given in the form of selenite) to prevent peroxide-dependent NF-κB activation. A similar synergism between selenium and LA-plus was observed in a neuronal cell line (HT4), this time in preventing glutamate-induced excitotoxicity [17]. In this case clear changes in the steady-state level of ROS were confirmed by dichlorofluorescein fluorescence.

More recently, oral administration of LA has been used successfully to prevent, or perhaps even reverse oxidative damage to mitochondrial enzymes in aged rats [2]. It is well established that damage from protein oxidation can accumulate during the normal aging process (as assessed by protein carbonyl levels) and this damage can alter the kinetic properties of metabolic enzymes, such as decreased affinity for substrates (higher apparent K_m) or reduce efficiency of catalysis (lower K_{cat}). Rats fed LA showed improvements in kinetic parameters for carnitine acetyltransferase versus untreated controls [2]. Similar studies with LA-fed aged rats confirmed that LA administration reduced the age-associated decreases in mitochondrial membrane potential. Tissues from LA-treated animals also exhibited a reduced level of ROS compared to untreated control [11].

Aged rats fed LA also showed improved memory during behavior tests and lower levels of DNA damage (oxo8dG) in the brain [1]. Lipoic acid administration to older animals reversed the age-related decline in Nrf2 synthesis, a key transcriptional regulator of genes encoding enzymes responsible for GSH biosynthesis [16]. This may indeed be the underlying mechanism behind the significant improvements in biochemical and neurological parameters in aged rats upon LA administration.

These and several other studies clearly demonstrated the potential of LA as a therapeutic, and suggest that further studies are needed to evaluate LA in the treatment of degenerative diseases. One underlying concern would be safety of high dose and/or long term treatment with LA. Two recent studies also show that oxidized LA can be safely administered orally. In a short-term regimen, there was no toxicity to LA in rats and the lowest observed adverse effect level (LOAEL) was determined to be 121 mg/kg bw/day [4]. Lipoic acid was not mutagenic in the Ames test, and displayed no genotoxicity using the mouse micronucleus test [4]. Long-term oral administration of LA (2 years) to rats resulted in no apparent changes in clinical chemistry or histopathology when animals were treated with 20, 60, or 180 mg/kg bw/day [3]. There was no change in spontaneous tumor formation rate above vehicle treatment. However, there was a slight decrease in food intake in animals treated with the highest dose of LA (180 mg/kg bw/day). On the basis of these studies, the NOAEL was determined to be 60 mg/kg bw/day for oral administration of LA. Given the promise LA administration has shown in animal studies and the recent data supporting its safety, additional animal studies are warranted to evaluate the effect of oral administration of LA to prevent or treat conditions for which oxidative stress is a known contributor.

SELENOTRISULFIDES AND THEIR ROLE IN SELENOPROTEIN SYNTHESIS

Chemistry of Selenotrisulfides

The reduction of selenite (SeO_3^{2-}) to selenide (Se^{2-}) in vivo is believed to occur through a chemical reaction with thiols known as the Painter reaction. Painter first described the sequential reduction of selenite to selenide by studying the reaction with cysteine [18]. Following up on Painter's studies, Ganther examined the reaction of thiols and selenite using CoA, 2-mercaptoethanol, or GSH [19]. For each thiol Ganther showed that an intermediate in the complete reduction to selenide was the formation of a selenotrisulfide (RSSeSR), and that these compounds exhibited a unique UV absorption peak with a maximum at 265 nm [19]. All selenotrisulfides analyzed in the above-mentioned studies were stabilized by acidic conditions and were sensitive to excess thiols. An optimal molar ratio of thiols to selenite was found to be 4:1 for production of selenotrisulfide [19]. Further studies revealed a protein-bound selenotrisulfide using RNaseA as a model substrate, with the cysteine thiols being utilized within an active enzyme [20]. Again, the optimal ratio for selenotrisulfide formation was 4:1 with respect to protein thiols (Cys). Biologists studying selenium metabolism have for many years pointed to these studies as defining the molecular mechanism for the reduction of selenite to selenide in vivo. However, until recently little biochemical evidence existed that selenotrisulfides were present in living biological systems.

Selenoprotein Synthesis

The first bacterial selenoprotein, selenoprotein A, was identified in *Clostridium sticklandii* [21]. Since this seminal work, the specific incorporation of selenium into selenoproteins has been best defined in *Escherichia coli* [22]. Early bacteriological studies suggested a role for selenium in formate dehydrogenase of *E. coli* [23]. The key discovery occurred when a series of formate dehydrogenase mutants were isolated [24]. Complementation of these mutations with wild type DNA identified the genes encoding the protein and RNA components needed for selenoprotein biosynthesis. Figure 19.2 is a schematic overview of the pathway for insertion of selenium into selenoproteins in *E. coli*. The *selC* gene, which encodes the selenocysteine-specific tRNA, is first charged with serine by the SerS protein [25]. The *selD* gene encodes selenophosphate synthetase (SPS), activates selenium from a reduced form (selenide, or Se^{2-}) to selenophosphate in an ATP-dependent manner [26–29]. The enzyme encoded by the *selA* gene ligates selenophosphate to the seryl-tRNA, converting it to the selenocysteinyl-tRNA needed for insertion into the polypeptide chain of selenoproteins. The *selB* gene encodes the specific elongation factor SelB [30], which binds to the stem-loop structure within the mRNA (SECIS element) and mediates the efficient translation of the UGA codon for insertion of selenocysteine. It should be noted that even in the well-studied *E. coli* system, very little is known about the uptake of selenium and no specific transport protein has been identified in any organism. Nonetheless, the uptake of the inorganic form selenite is known to be efficient in *E. coli* and results in optimum selenoprotein synthesis. Many of corresponding proteins in selenoprotein synthesis have been identified and studied in mammalian systems [31], and the overall metabolism of selenium appears to be similar, and this is beyond the scope of this current chapter. The key point to be made regarding

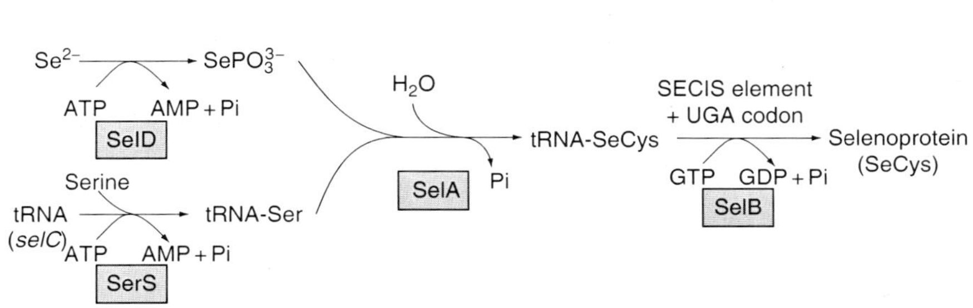

FIGURE 19.2 An overview of selenoprotein synthesis in the model system of *Escherichia coli*.

selenoprotein synthesis in the context of this chapter is the fact that selenium transported into the cell must be first converted to the reduced form selenide (Se^{2-}) before entry into the pathway (Figure 19.2) in both prokaryotes and eukaryotes.

Selenotrisulfides as Intermediates in Selenium Metabolism

Since the metabolism of selenium upstream of SPS (SelD, Figure 19.2) is poorly understood, Lacourciere attempted to decipher the proteins that interact with selenium using ^{75}Se radioisotope labeled cells which lacked both selenophosphate synthetase and O-acetylserine sulphydrylase [32]. Because of these mutations, this strain could not incorporate selenium specifically into selenoproteins (e.g., formate dehydrogenase) or nonspecifically into the cysteine or methionine pool. The rationale for these experiments was that if selenium could not be metabolized for either specific or nonspecific protein synthesis, it would build up a pool of the intermediate form(s) of selenium that are used for selenoprotein synthesis. Several selenium-binding proteins were identified, including glyceraldehyde-3-phosphate dehydrogenase (GAPDH). Direct coupling of the selenium loaded GAPDH to SPS [33], demonstrated that reactive Cys residues are likely the key players in binding of selenium after reduction to selenide. These two studies, when taken together, strongly suggest that GAPDH may play a key role in selenium metabolism. The form of selenium presumed to be present was $RSSe^{-}$—with selenium bound as a *perselenide* to a reactive cysteine thiol.

Another selenium-binding protein was also uncovered in *Methanococcus vannielii* [34,35]. This protein was identified by Stadtman in a selenium-bound form using biochemical techniques and ^{75}Se radioisotope labeling. This protein-bound Se is likely in the form of a perselenide ($RSSe^{-}$). This and other selenium-binding proteins identified so far may sequester selenium for specific selenoprotein synthesis. Yet there is insufficient biochemical data to date that shows these particular proteins play a direct role in the metabolism of selenium. Indeed, the absence of a selenium-binding protein in the pool of mutants obtained in the model organism *E. coli* suggests that more than one protein(s) can function in delivery of selenium to SPS.

Effect of Selenotrisulfides on Cellular Targets

Although selenotrisulfides synthesized in vitro are generally unstable [19,20,33,36], several studies have shown the ability of pyridine nucleotide-dependent oxidoreductases to reduce and release selenium from selenotrisulfides. Hsieh and Ganther [37] showed that glutathione reductase could more efficiently reduce GSSeSG to selenide than GSH alone. It is interesting to note that arsenite was shown to inhibit both the chemical and enzymatic reduction. GSSeSG, in the presence of TrxR, also decreased 15 lipoxygenase activity [38,39]. This inhibition was tied to production of the reactive reduced form of selenium, selenide (Se^{2-}).

GSSeSG also inhibited binding of the transcription factor AP-1 to DNA, likely by interacting with reactive Cys residues in both Fos (Cys154) and Jun (Cys272) proteins [40]. GSSeSG also inactivated protein kinase C, apparently by reducing binding of ATP [41]. In each of the above studies, the investigators implicated the mechanism of inhibition as part of the potential anticancer effect of selenium supplementation in the diet. However, this is speculative since at the time of these publications little evidence existed that selenotrisulfides are formed and are stable in vivo.

Three separate reports documented the inhibition of protein synthesis by GSSeSG using cell culture models [42–44]. GSSeSG appeared to target elongation factor 2 in inhibition of the translation of mRNA using polyribosomes from rat liver [42]. GSSeSG also inhibited protein synthesis when added to the culture medium of 3T3 cells, yet had no significant effect on DNA synthesis [43]. Given the sensitivity of selenotrisulfides to excess thiols, it is possible that these inhibitory effects may have been downstream of the further chemical or enzymatic reduction of selenotrisulfides, but unfortunately this line of research was not continued beyond these initial studies.

Uptake of Selenotrisulfides and Identification In Vivo

With the rationale that selenite would only exist in an oxidizing or acidic environment and that reduced forms of selenium are more likely to be present in the gut for selenium uptake, Vendeland et al. [45] studied the uptake of selenium in brush-border membranes when present as either Cys or GSH selenotrisulfide derivatives. Notably, a tenfold increase in selenium uptake was seen with either selenotrisulfide [45]. This suggests that a specific mechanism for uptake of reduced selenium compounds may exist and further suggests that such compounds may indeed be stable in tissues.

Further support for this possibility has been gained from studies using mass spectrometry techniques. Lindemann and Hintelmann [46] identified selenotrisulfides as a significant fraction of the selenium metabolites in yeast extracts. Both GSH and Cys derivatives (and mixed) were identified. A more recent study using ICP-MS techniques also confirmed the presence of selenotrisulfides in epithelial cell homogenates [47]. These studies reveal that mammalian cells have a specific system to transport selenium in this form, but that these compounds are stable in the cytosol of cells. The identification of selenotrisulfides has only been made possible with the advent of more advanced mass spectrometry [45–47].

SELENOTRISULFIDE DERIVATIVE OF α-LIPOIC ACID

Synthesis and Initial Characterization

Since selenotrisulfides and similar forms of selenium bound to protein thiols ($RSSe^-$) are likely key intermediates in selenium metabolism, an attempt was made to synthesize stable derivatives of lipoic acid using the Painter reaction [8] to better understand the nature of these compounds. Dihydrolipoic acid (DHLA) was reacted with selenite under acidic conditions in 50% ethanol using varying ratios of reduced thiols. A stable selenotrisulfide derivative of LA was formed (Figure 19.1) and confirmed by mass spectrometry [8]. A similar derivative of lipoamide was also synthesized, to further confirm that the chemistry needed was solely due to the dithiol and the unique absorption peak at 288 nm was due to this thiol-Se derivative [8]. Both compounds were easily separated from oxidized lipoic acid or lipoamide by reverse phase HPLC. Notably, this study was undertaken primarily to obtain a perselenide derivative ($RSSe^-$) which is the putative active site form of selenium in a unique class of molybdenum hydroxylases [48–52]. However, these studies did not yield stable perselenide derivatives. Nonetheless, the properties of the relatively stable ring compound formed by addition of selenium across the disulfide link in oxidized lipoic acid were quite intriguing.

Selenotrisulfide derivative of α-lipoic acid was an excellent substrate for reduction by both glutathione reductase and thioredoxin reductase, releasing selenium in a form that was not detected by reverse phase HPLC [8]. It is likely that selenide was formed in this reduction, based on the evidence from previous studies [38,39], and was subsequently oxidized to elemental selenium that is notorious for becoming tightly bound to hydrophobic media in chromatographic columns used in small molecule or protein separations. Radioisotope labeling of LASe revealed a molar extinction coefficient at 288 nm of 1500, and this characteristic has facilitated quantization of subsequent preparations of LASe. One of the more interesting features of this selenotrisulfide was the fact that LASe was more stable than its GSH counterpart (GSSeSG) at physiological pH [8]. Considerable interest in the compound from a therapeutic perspective arose after publication, and this interest has led to the subsequent study of LASe as a novel selenium containing antioxidant.

Rationale for Topical Application of Selenium

A series of in vitro cell culture studies have shown that selenium supplementation of keratinocytes and melanocytes is effective to protect cells against ROS induced by UV radiation [53–57]. Selenite addition to the culture medium proved more protective, as compared to L-selenomethionine, for protection against UVB-induced cell death in keratinocytes [56]. Both selenite and selenomethionine were effective in prevention of 8-hydroxy-2-deoxyguanosine modifications to DNA [55]. However, selenite was still more effective at lower concentrations. It was suggested that selenodiglutathionine, formed after Painter reaction with selenite, could be responsible for the more efficient protective effect observed in these cell culture experiments [53].

Topical studies of selenium as a protective antioxidant have been limited to the use of L-selenomethionine (SeMet, [58]). Burke et al. demonstrated that topical SeMet administration, either with or without vitamin E, protected skin in hairless mice from blistering induced by UV irradiation. Treatment with SeMet also reduced the number and size of tumors, especially when administered in conjunction with vitamin E. Unfortunately, SeMet did not synergize with vitamin E nor did it significantly improve the performance above vitamin E administration alone. Taken together with the in vitro cell culture work of McKenzie [53–55,57,58], this would suggest that conversion of selenium to a more efficient nutritional form (i.e., conversion to selenide) could have impaired the efficacy of selenium in the hairless mice study [58].

Efficiency of Selenium Utilization and Delivery from LASe

Penetration of human skin represents a significant barrier for development of topical antioxidants [59,60]. Given that selenium treatment to skin and cells in culture has shown benefit against UV induced damage, a study was undertaken to determine the comparative efficiency of selenium adsorption into skin [7]. As compared to SeMet, LASe (Figure 19.2) was far more efficiently absorbed into pig skin, a good model for human skin [7]. The conjugation of selenium to the hydrophobic LA indeed resulted in levels of selenium within skin at several orders of magnitude above basal selenium levels, as assessed by atomic absorption spectroscopy. Other forms of selenium (organic and inorganic) were far less efficiently absorbed in this study. These results were encouraging and suggested that LASe could be a good candidate to topically administer selenium into skin.

To determine whether this form of selenium could also be used efficiently for selenoprotein synthesis, LASe was tested in a comparative analysis against selenite, L-selenocysteine (SeCys), SeMet, and selenate using a keratinocyte cell culture model [7]. L-Selenomethionine was a poor selenium source in this model, as was selenate. However, LASe, SeCys, and selenite were equally effective as a source of selenium for selenoprotein biosynthesis. In addition, pretreatment of HaCat with LASe also protected keratinocytes from UVA/UVB-induced cell death [7]. This was the first study that has focused on the use of a selenotrisulfide as a potential therapeutic, with the goal being optimal selenium nutrition to enable the highest level of selenoenzyme biosynthesis.

CONCLUSIONS AND FUTURE DIRECTIONS

Topical application of selenium compounds may prove to be a safe and effective way to reduce oxidative damage in a tissue that is inundated with UV and γ-radiation. However, not enough animal studies have been carried out to determine whether this approach is feasible and sound in terms of actual changes in selenoenzyme activities within the affective dermal layers. Further animal studies are needed to determine whether topical application of lipophilic selenium compounds, such as LASe, can be effective at prevention of radiation-induced oxidative damage to skin. It has also yet to be determined whether the protective effects of selenium compounds are due to increases in selenoenzymes such as TrxR and Gpx. Given the existing literature on the protective effects of selenium and lipoic acid, LASe indeed may prove to be a potent topical antioxidant.

ABBREVIATIONS

CoA, coenzyme A
Cys, L-cysteine
DHLA, dihydrolipoic acid
Gpx, glutathione peroxidase
GSH, glutathione
LA, α-lipoic acid
LA-plus, *N,N*-dimethyl,*N*′-2amidoethyl-lipoate
LASe, selenotrisulfide derivative of α-lipoic acid
Nrf2, nuclear factor erythroid-2 related factor
ROS, reactive oxygen species
TrxR, thioredoxin reductase

REFERENCES

1. Liu, J., E. Head, A.M. Gharib, W. Yuan, R.T. Ingersoll, T.M. Hagen, C.W. Cotman, and B.N. Ames. 2002. Memory loss in old rats is associated with brain mitochondrial decay and RNA/DNA oxidation: Partial reversal by feeding acetyl-L-carnitine and/or R-alpha-lipoic acid. *Proc Natl Acad Sci U S A* **99:**2356–2361.
2. Liu, J., D.W. Killilea, and B.N. Ames. 2002. Age-associated mitochondrial oxidative decay: Improvement of carnitine acetyltransferase substrate-binding affinity and activity in brain by feeding old rats acetyl-L-carnitine and/or R-alpha-lipoic acid. *Proc Natl Acad Sci U S A* **99:**1876–1881.
3. Cremer, D.R., R. Rabeler, A. Roberts, and B. Lynch. 2006. Long-term safety of alpha-lipoic acid (ALA) consumption: A 2-year study. *Regul Toxicol Pharmacol* **46:**193–201.
4. Cremer, D.R., R. Rabeler, A. Roberts, and B. Lynch. 2006. Safety evaluation of alpha-lipoic acid (ALA). *Regul Toxicol Pharmacol* **46:**29–41.
5. Rayman, M.P. 2005. Selenium in cancer prevention: A review of the evidence and mechanism of action. *Proc Nutr Soc* **64:**527–542.
6. Kryukov, G.V., S. Castellano, S.V. Novoselov, A.V. Lobanov, O. Zehtab, R. Guigo, and V.N. Gladyshev. 2003. Characterization of mammalian selenoproteomes. *Science* **300:**1439–1443.
7. Alonis, M. and W.T. Self. 2006. Bioavailability of selenium from the selenotrisulphide derivative of lipoic acid. *Photodermatol Photoimmunol Photomed* **22:**315–323.
8. Self, W.T., L. Tsai, and T.C. Stadtman. 2000. Synthesis and characterization of selenotrisulfide-derivatives of lipoic acid and lipoamide. *Proc Natl Acad Sci U S A* **97:**12481–12486.
9. O'Kane, D.J. and I.C. Gunsalus. 1948. Pyruvic acid metabolism: A factor required for oxidation by *Streptococcus faecalis*. *J Bacteriol* **56:**499–506.
10. Reed, L.J., B.B. De, I.C. Gunsalus, and C.S. Hornberger, Jr. 1951. Crystalline alpha-lipoic acid; a catalytic agent associated with pyruvate dehydrogenase. *Science* **114:**93–94.
11. Hagen, T.M., R.T. Ingersoll, J. Lykkesfeldt, J. Liu, C.M. Wehr, V. Vinarsky, J.C. Bartholomew, and A.B. Ames. 1999. (R)-alpha-lipoic acid-supplemented old rats

have improved mitochondrial function, decreased oxidative damage, and increased metabolic rate. *FASEB J* **13:**411–418.

12. Han, D., G. Handelman, L. Marcocci, C.K. Sen, S. Roy, H. Kobuchi, H.J. Tritschler, L. Flohe, and L. Packer. 1997. Lipoic acid increases de novo synthesis of cellular glutathione by improving cystine utilization. *Biofactors* **6:**321–338.
13. Haramaki, N., D. Han, G.J. Handelman, H.J. Tritschler, and L. Packer. 1997. Cytosolic and mitochondrial systems for NADH- and NADPH-dependent reduction of alpha-lipoic acid. *Free Radic Biol Med* **22:**535–542.
14. Perricone, N., K. Nagy, F. Horvath, G. Dajko, I. Uray, and I. Zs-Nagy. 1999. Alpha lipoic acid (ALA) protects proteins against the hydroxyl free radical-induced alterations: Rationale for its geriatric topical application. *Arch Gerontol Geriatr* **29:**45–56.
15. Sen, C.K., O. Tirosh, S. Roy, M.S. Kobayashi, and L. Packer. 1998. A positively charged alpha-lipoic acid analogue with increased cellular uptake and more potent immunomodulatory activity. *Biochem Biophys Res Commun* **247:**223–228.
16. Suh, J.H., S.V. Shenvi, B.M. Dixon, H. Liu, A.K. Jaiswal, R.M. Liu, and T.M. Hagen. 2004. Decline in transcriptional activity of Nrf2 causes age-related loss of glutathione synthesis, which is reversible with lipoic acid. *Proc Natl Acad Sci U S A* **101:**3381–3386.
17. Tirosh, O., C.K. Sen, S. Roy, M.S. Kobayashi, and L. Packer. 1999. Neuroprotective effects of alpha-lipoic acid and its positively charged amide analogue. *Free Radic Biol Med* **26:**1418–1426.
18. Painter, H.E. 1941. The chemistry and toxicity of selenium compounds, with sepcial reference to the selenium problem. *Chem Rev* **28:**179–213.
19. Ganther, H.E. 1968. Selenotrisulfides. Formation by the reaction of thiols with selenious acid. *Biochemistry* **7:**2898–2905.
20. Ganther, H.E. and C. Corcoran. 1969. Selenotrisulfides. II. Cross-linking of reduced pancreatic ribonuclease with selenium. *Biochemistry* **8:**2557–2563.
21. Turner, D.C. and T.C. Stadtman. 1973. Purification of protein components of the clostridial glycine reductase system and characterization of protein A as a selenoprotein. *Arch Biochem Biophys* **154:**366–381.
22. Bock, A., K. Forchhammer, J. Heider, W. Leinfelder, G. Sawers, B. Veprek, and F. Zinoni. 1991. Selenocysteine: The 21st amino acid. *Mol Microbiol* **5:**515–520.
23. Pinsent, J. 1954. The need for selenite and molybdate in the formation of formic dehydrogenase by members of the coli-aerogenes group of bacteria. *Biochem J* **57:**10–16.
24. Leinfelder, W., K. Forchhammer, F. Zinoni, G. Sawers, M.A. Mandrand-Berthelot, and A. Bock. 1988. *Escherichia coli* genes whose products are involved in selenium metabolism. *J Bacteriol* **170:**540–546.
25. Leinfelder, W., T.C. Stadtman, and A. Bock. 1989. Occurrence in vivo of selenocysteyl-tRNA(SERUCA) in *Escherichia coli*. Effect of sel mutations. *J Biol Chem* **264:**9720–9723.
26. Ehrenreich, A., K. Forchhammer, P. Tormay, B. Veprek, and A. Bock. 1992. Selenoprotein synthesis in *E. coli*. Purification and characterisation of the enzyme catalysing selenium activation. *Eur J Biochem* **206:**767–773.
27. Glass, R.S., W.P. Singh, W. Jung, Z. Veres, T.D. Scholz, and T.C. Stadtman. 1993. Monoselenophosphate: Synthesis, characterization, and identity with the prokaryotic biological selenium donor, compound SePX. *Biochemistry* **32:**12555–12559.

28. Leinfelder, W., K. Forchhammer, B. Veprek, E. Zehelein, and A. Bock. 1990. In vitro synthesis of selenocysteinyl-tRNA(UCA) from seryl-tRNA(UCA): Involvement and characterization of the selD gene product. *Proc Natl Acad Sci U S A* **87:**543–547.
29. Veres, Z., I.Y. Kim, T.D. Scholz, and T.C. Stadtman. 1994. Selenophosphate synthetase. Enzyme properties and catalytic reaction. *J Biol Chem* **269:**10597–10603.
30. Forchhammer, K., W. Leinfelder, and A. Bock. 1989. Identification of a novel translation factor necessary for the incorporation of selenocysteine into protein. *Nature* **342:**453–456.
31. Allmang, C. and A. Krol. 2006. Selenoprotein synthesis: UGA does not end the story. *Biochimie* **88:**1561–1571.
32. Lacourciere, G.M., R.L. Levine, and T.C. Stadtman. 2002. Direct detection of potential selenium delivery proteins by using an *Escherichia coli* strain unable to incorporate selenium from selenite into proteins. *Proc Natl Acad Sci U S A* **99:**9150–9153.
33. Ogasawara, Y., G.M. Lacourciere, K. Ishii, and T.C. Stadtman. 2005. Characterization of potential selenium-binding proteins in the selenophosphate synthetase system. *Proc Natl Acad Sci U S A* **102:**1012–1016.
34. Patteson, K.G., N. Trivedi, and T.C. Stadtman. 2005. *Methanococcus vannielii* selenium-binding protein (SeBP): Chemical reactivity of recombinant SeBP produced in *Escherichia coli. Proc Natl Acad Sci U S A* **102:**12029–12034.
35. Self, W.T., R. Pierce, and T.C. Stadtman. 2004. Cloning and heterologous expression of a *Methanococcus vannielii* gene encoding a selenium-binding protein. *IUBMB Life* **56:**501–507.
36. Ganther, H.E. 1971. Reduction of the selenotrisulfide derivative of glutathione to a persulfide analog by glutathione reductase. *Biochemistry* **10:**4089–4098.
37. Hsieh, H.S. and H.E. Ganther. 1975. Acid-volatile selenium formation catalyzed by glutathione reductase. *Biochemistry* **14:**1632–1636.
38. Bjornstedt, M., S. Kumar, and A. Holmgren. 1995. Selenite and selenodiglutathione: Reactions with thioredoxin systems. *Methods Enzymol* **252:**209–219.
39. Bjornstedt, M., B. Odlander, S. Kuprin, H.E. Claesson, and A. Holmgren. 1996. Selenite incubated with NADPH and mammalian thioredoxin reductase yields selenide, which inhibits lipoxygenase and changes the electron spin resonance spectrum of the active site iron. *Biochemistry* **35:**8511–8516.
40. Spyrou, G., M. Bjornstedt, S. Kumar, and A. Holmgren. 1995. AP-1 DNA-binding activity is inhibited by selenite and selenodiglutathione. *FEBS Lett* **368:**59–63.
41. Gopalakrishna, R., U. Gundimeda, and Z.H. Chen. 1997. Cancer-preventive selenocompounds induce a specific redox modification of cysteine-rich regions in Ca(2+)-dependent isoenzymes of protein kinase C. *Arch Biochem Biophys* **348:**25–36.
42. Vernie, L.N., W.S. Bont, H.B. Ginjaar, and P. Emmelot. 1975. Elongation factor 2 as the target of the reaction product between sodium selenite and glutathione (GSSeSG) in the inhibiting of amino acid incorporation in vitro. *Biochim Biophys Acta* **414:**283–292.
43. Vernie, L.N., J.G. Collard, A.P. Eker, A. de Wildt, and I.T. Wilders. 1979. Studies on the inhibition of protein synthesis by selenodiglutathione. *Biochem J* **180:**213–218.
44. Vernie, L.N., H.B. Ginjaar, I.T. Wilders, and W.S. Bont. 1978. Amino acid incorporation in a cell-free system derived from rat liver studied with the aid of selenodiglutathione. *Biochim Biophys Acta* **518:**507–517.
45. Vendeland, S.C., J.T. Deagen, and P.D. Whanger. 1992. Uptake of selenotrisulfides of glutathione and cysteine by brush border membranes from rat intestines. *J Inorg Biochem* **47:**131–140.

46. Lindemann, T. and H. Hintelmann. 2002. Identification of selenium-containing glutathione S-conjugates in a yeast extract by two-dimensional liquid chromatography with inductively coupled plasma MS and nanoelectrospray MS/MS detection. *Anal Chem* **74:**4602–4610.
47. Gabel-Jensen, C., B. Gammelgaard, L. Bendahl, S. Sturup, and O. Jons. 2006. Separation and identification of selenotrisulfides in epithelial cell homogenates by LC-ICP-MS and LC-ESI-MS after incubation with selenite. *Anal Bioanal Chem* **384:**697–702.
48. Dilworth, G.L. 1982. Properties of the selenium-containing moiety of nicotinic acid hydroxylase from *Clostridium barkeri. Arch Biochem Biophys* **219:**30–38.
49. Gladyshev, V.N., S.V. Khangulov, and T.C. Stadtman. 1996. Properties of the selenium- and molybdenum-containing nicotinic acid hydroxylase from *Clostridium barkeri. Biochemistry* **35:**212–223.
50. Schrader, T., A. Rienhofer, and J.R. Andreesen. 1999. Selenium-containing xanthine dehydrogenase from *Eubacterium barkeri. Eur J Biochem* **264:**862–871.
51. Self, W.T. 2002. Regulation of purine hydroxylase and xanthine dehydrogenase from *Clostridium purinolyticum* in response to purines, selenium, and molybdenum. *J Bacteriol* **184:**2039–2044.
52. Self, W.T., M.D. Wolfe, and T.C. Stadtman. 2003. Cofactor determination and spectroscopic characterization of the selenium-dependent purine hydroxylase from *Clostridium purinolyticum. Biochemistry* **42:**11382–11390.
53. McKenzie, R.C. 2000. Selenium, ultraviolet radiation and the skin. *Clin Exp Dermatol* **25:**631–636.
54. Rafferty, T.S., G.J. Beckett, C. Walker, Y.C. Bisset, and R.C. McKenzie. 2003. Selenium protects primary human keratinocytes from apoptosis induced by exposure to ultraviolet radiation. *Clin Exp Dermatol* **28:**294–300.
55. Rafferty, T.S., M.H. Green, J.E. Lowe, C. Arlett, J.A. Hunter, G.J. Beckett, and R.C. McKenzie. 2003. Effects of selenium compounds on induction of DNA damage by broadband ultraviolet radiation in human keratinocytes. *Br J Dermatol* **148:**1001–1009.
56. Rafferty, T.S., R.C. McKenzie, J.A. Hunter, A.F. Howie, J.R. Arthur, F. Nicol, and G.J. Beckett. 1998. Differential expression of selenoproteins by human skin cells and protection by selenium from UVB-radiation-induced cell death. *Biochem J* **332** (Pt 1):231–236.
57. Rafferty, T.S., C. Walker, J.A. Hunter, G.J. Beckett, and R.C. McKenzie. 2002. Inhibition of ultraviolet B radiation-induced interleukin 10 expression in murine keratinocytes by selenium compounds. *Br J Dermatol* **146:**485–489.
58. Burke, K.E., J. Clive, G.F. Combs, Jr., and R.M. Nakamura. 2003. Effects of topical L-selenomethionine with topical and oral vitamin E on pigmentation and skin cancer induced by ultraviolet irradiation in Skh:2 hairless mice. *J Am Acad Dermatol* **49:**458–472.
59. Lin, J. Y., M.A. Selim, C.R. Shea, J.M. Grichnik, M.M. Omar, N.A. Monteiro-Riviere, and S.R. Pinnell. 2003. UV photoprotection by combination topical antioxidants vitamin C and vitamin E. *J Am Acad Dermatol* **48:**866–874.
60. Pinnell, S.R. 2003. Cutaneous photodamage, oxidative stress, and topical antioxidant protection. *J Am Acad Dermatol* **48:**1–19; quiz 20–2.

20 Alpha-Lipoic Acid: A Potent Mitochondrial Nutrient for Improving Memory Deficit, Oxidative Stress, and Mitochondrial Dysfunction

Jiankang Liu

CONTENTS

INTRODUCTION

Cognitive function declines with age. Increasing evidence shows that mitochondrial dysfunction due to the oxidation of lipids, proteins, and nucleic acids plays an important role in brain aging and age-related neurodegenerative diseases, such as Alzheimer's disease (AD), Parkinson's disease (PD), amyotrophic lateral sclerosis, and Huntington's disease. Mitochondrial decay may be a principal underlying event in aging, including brain aging (Ames et al., 1993; Shigenaga et al., 1994; Ames et al., 1995; Beckman and Ames, 1998; Harman, 2002; Ames, 2003; Roubertoux et al., 2003) and is also associated withwith the onset and development of neurodegenerative diseases (Beal, 1992; Tritschler et al., 1994; Liu and Mori, 1999; Albers and Beal, 2000; Perry et al., 2000; Aliev et al., 2002; Rao and Balachandran, 2002).

We have identified a group of mitochondrial targeting nutrients and named them as "mitochondrial nutrients," which can (1) prevent the generation of oxidants, (2) scavenge oxidants or inhibit oxidant reactivity, (3) repair oxidative damage to lipids, proteins/enzymes, and RNA/DNA by enhancing antioxidant defense systems, and (4) elevate cofactors of defective enzymes (increased K_m) in mitochondria to stimulate enzyme activity, and also protect enzymes from further oxidation (cofactor function). Mitochondrial dysfunction could possibly be reversed in aged animals by feeding mitochondrial nutrients. Thus, neurodegenerative diseases such as AD and PD may be delayed or ameliorated by treatment with mitochondrial nutrients, which could delay or repair mitochondrial damage, thus improving mitochondrial function.

Clinical trials to determine whether micronutrients will delay aging and improve memory in the elderly or will prevent or treat AD and PD have been promising. A few reviews have summarized the effects of different nutrients and antioxidants on aging and neurological diseases including PD and AD (Butterfield et al., 2002a; Butterfield et al., 2002b; Price, 2002; Baker and Tarnopolsky, 2003; Beal, 2003; Black, 2003; Marriage et al., 2003; McDaniel et al., 2003). In the present review, we survey the recently published literature on α-lipoic acid (LA) and its derivatives on age-associated cognitive and mitochondrial dysfunction and on oxidative damage in the nervous system.

α-LIPOIC ACID: A MITOCHONDRIAL NUTRIENT?

α-Lipoic acid is a coenzyme involved in mitochondrial metabolism. The reduced form of LA, dihydrolipoic acid, is a powerful mitochondrial antioxidant (Packer et al., 1995; Packer et al., 1997a; Packer et al., 1997b; Moini et al., 2002). It recycles other cellular antioxidants, including CoQ, vitamins C and E, glutathione, and chelates iron and copper (Packer et al., 1995; Packer et al., 1997a; Packer et al., 1997b; Moini et al., 2002). LA readily crosses the blood–brain barrier and is accepted by human cells as a substrate where it is reduced to dihydrolipoic acid by NADH-dependent mitochondrial dihydrolipoamide dehydrogenase (Moini et al., 2002).

LA plays a fundamental role in mitochondrial metabolism. Biologically, it exists in proteins where it is linked covalently to a lysyl residue as a lipoamide. The mitochondrial E3 enzyme, dihydrolipoyl dehydrogenase, reduces lipoate to dihydrolipoate at the expense of NADH. Lipoate is also a substrate for the NADPH-dependent enzyme glutathione reductase (Fuchs et al., 1997; Packer et al., 1997a; Bustamante et al., 1998). In recent years, LA has gained considerable attention as an antioxidant (Packer et al., 1995; Fuchs et al., 1997; Sen et al., 1999). The reduced form of LA, dihydrolipoic acid, reacts with oxidants such as superoxide radicals, hydroxyl radicals, hypochlorous acid, peroxyl radicals, and singlet oxygen. It also protects membranes by reducing oxidized vitamin C and glutathione, which may in turn recycle vitamin E. Administration of LA is beneficial to a number of oxidative stress models such as diabetes, cataract, HIV activation, neurodegeneration, and radiation injury in animals. Furthermore, LA functions as a redox regulator of proteins such as myoglobin, prolactin, thioredoxin, and NF-κB transcription factor (Packer et al., 1995; Fuchs et al., 1997; Bustamante et al., 1998). LA has neuroprotective effects in neuronal cells. One possible mechanism for the antioxidant effect of LA is its metal-chelating activity (Ou et al., 1995). LA can increase ambulatory activity and partially restore age-associated mitochondrial decay in liver and heart (Hagen et al., 1999; Hagen et al., 2000; Suh et al., 2001).

On the basis of our definition for mitochondrial nutrients, we consider LA satisfies all criteria for a mitochondrial nutrient. Therefore, LA is a mitochondrial nutrient and it is also one of the mostly studied mitochondrial nutrients with regard to mitochondrial function in cellular and animal models related to brain aging and neurodegeneration.

MITOCHONDRIAL DYSFUNCTION: A CONSEQUENCE OF AGING?

Mitochondria provide energy for basic metabolic processes, produce oxidants as inevitable byproducts, and decay with age-impairing cellular metabolism and leading to cellular decline. Mitochondrial membrane potential, respiratory control ratios, and cellular oxygen consumption decline with age, and oxidant production increases (Harman, 1972; Shigenaga et al., 1994; Hagen et al., 1998). Oxidative damage to DNA, RNA, proteins, and lipid membranes in mitochondria may be involved. Mutations in mitochondrial genes compromises mitochondria by altering components of the electron transport chain, resulting in inefficient electron transport and increased superoxide production (Shigenaga et al., 1994; Sohal and Weindruch, 1996). The resulting oxidative damage to mitochondria compromises their ability to meet cellular energy demands. Mitochondrial enzymes are especially susceptible to inactivation by superoxide and hydroxyl radicals, as these oxidants are generated in mitochondria (Cadenas and Davies, 2000). Oxidized proteins accumulate with age (Stadtman and Levine, 2000) that cause mitochondrial inefficiencies, leading to more oxidant formation. Mitochondrial membrane fluidity also declines with age (Chen and Yu, 1994; Choi and Yu, 1995), which may lead to deformation of membrane proteins and cause mitochondrial dysfunction. The significant age-related loss of cardiolipin, a phospholipid that occurs primarily in the mitochondrial inner membrane, may be in part because of greater oxidative damage or reduced biosynthesis. Loss of cardiolipin, coupled with oxidation of critical thiol groups in key proteins, adversely affects transport of substrates and cytochrome *c* oxidase activity (Paradies et al., 1994) necessary for mitochondrial function. These changes could directly impact the ability of mitochondria to maintain their membrane potential.

AGE-ASSOCIATED COGNITIVE DYSFUNCTION AND NEURODEGENERATIVE DISEASES: PREVENTION AND AMELIORATION BY LA?

LA shows improvement on cognitive function in normal old mice. LA improved long-term memory of aged female NMRI mice in habituation in the open field test and also alleviated age-related NMDA receptor deficits (Bmax) without changing muscarinic, benzodiazepine, and α_2-adrenergic receptor deficits (Stoll et al., 1993).

LA shows improvement on cognitive function in senescence-accelerated mice. The senescence-accelerated mouse strain 8 (SAMP8), at the age of 12 months old, exhibits age-related deterioration in memory and learning, has increased levels of β-amyloid and oxidative damage to proteins and lipids. Chronic administration of LA improved cognition of 12 month old SAMP8 mice in both the T-maze foot-shock avoidance paradigm and the lever press appetitive task without inducing nonspecific effects on motor activity, motivation to avoid shock, or body weight, and reduced oxidative damage to proteins and lipids (Farr et al., 2003). The LA treatment also significantly increased the expressions of three brain proteins (neurofilament triplet L protein, α-enolase, and ubiquitous mitochondrial creatine kinase) and significantly decreased specific carbonyl levels of three brain proteins (lactate dehydrogenase B, dihydropyrimidinase-like protein 2, and α-enolase) in the aged SAMP8 mice (Poon et al., 2005a; Poon et al., 2005b).

LA shows improvement on cognitive function in chemical-induced aging accelerated mice. Chronic systemic exposure of D-galactose to mice induced a spatial memory deficit, an increase in cell karyopyknosis, apoptosis, and caspase-3 protein levels in hippocampal neurons, a decrease in the number of new neurons in the subgranular zone in the dentate gyrus, a reduction of migration of neural progenitor cells, and an increase in death of newly formed neurons in granular cell layer (Cui et al., 2006). The D-galactose exposure also induced an increase in peripheral oxidative stress, including an increase in malondialdehyde, a decrease in total antioxidative capabilities, total superoxide dismutase, and glutathione peroxidase activities (Cui et al., 2006). A concomitant treatment with LA ameliorated cognitive dysfunction and neurodegeneration in the hippocampus, and also reduced peripheral oxidative damage by decreasing malondialdehyde and increasing total antioxidative capabilities and total superoxide dismutase, without an effect on glutathione peroxidase (Cui et al., 2006).

LA shows improvement on hippocampal-dependent memory deficits of Tg2576 mice, a transgenic model of cerebral amyloidosis associated with AD. LA-treated Tg2576 mice exhibited significantly improved learning and memory retention in the Morris water maze task and significantly more context freezing compared to untreated Tg2576 mice (Quinn et al., 2007).

LA shows improvement on cognitive function in x-irradiation-induced memory impairment in mice. Whole body x-irradiation of mice substantially impaired the reference memory and motor activities of mice and treatment with LA before irradiation significantly attenuated such cognitive dysfunction (Manda et al., 2007). LA pretreatment also exerted a significant protection against radiation-induced increase in oxidative damage to proteins and lipids in mice cerebellum (Manda et al., 2007). LA pretreatment also inhibited radiation-induced deficit of total nonprotein- and protein-bound sulfhydryl contents of cerebellum and plasma ferric reducing power. In addition, LA-treated mice showed an intact cytoarchitecture of cerebellum, and higher counts of intact Purkinje cells and granular cells in comparison to untreated irradiated mice (Manda et al., 2007).

LA shows effect on improving diabetic peripheral and cardiac autonomic neuropathy (Ziegler and Gries, 1997; Evans and Goldfine, 2000). LA prevents cognitive impairment and oxidative stress induced by intracerebroventricular streptozotocin administration in rats. Intracerebroventricular streptozotocin induced cognitive impairment in rats, which is characterized by a progressive deterioration of memory, cerebral glucose and energy metabolism, and oxidative stress while LA-treated rats showed significantly less cognitive impairment as compared to the vehicle-treated rats (Sharma and Gupta, 2003).

LA prevents the development of multiple sclerosis. LA dose dependently prevented the development of clinical signs in a rat model for multiple sclerosis and acute experimental allergic encephalomyelitis. LA has a protective effect on encephalomyelitis development not only by affecting the migratory capacity of monocytes, but also by stabilizing the blood–brain barrier (Schreibelt et al., 2006).

LA has also been studied as a treatment option for Alzheimer-type dementia. A dose of 600 mg LA, given daily to nine patients with AD and related dementias for about 1 year, led to a stabilization of cognitive functions in the study group, as shown by constant scores in two neuropsychological tests (Hager et al., 2001). Though the study was small and not randomized, it suggests that treatment with LA is a possible neuroprotective therapy option for AD and related dementias.

IMPROVING MITOCHONDRIAL FUNCTION AND REDUCING OXIDATIVE DAMAGE: POSSIBLE MECHANISMS OF LA ON COGNITION?

Various mechanisms of the LA effects on improving cognitive function have been suggested, including the improvement of memory-related signaling pathways, reducing oxidative stress, and improving mitochondrial function.

LA may restore the activity of acetylcholinesterase and Na^+/K^+ ATPase. The activity of acetylcholinesterase was found to be significantly decreased in the cerebral cortex, cerebellum, striatum, hippocampus, and hypothalamus in aged rats while administration of LA reversed the decrease in activity in discrete brain regions (Arivazhagan et al., 2006). In aged rats, the level of lipofuscin was increased, and the activity of Na^+/K^+ ATPase was decreased. Administration of LA to aged rats led to a duration-dependent reduction in lipofuscin and elevation of enzyme activity, respectively, in the cortex, cerebellum, striatum, hippocampus, and hypothalamus of the brain (Arivazhagan and Panneerselvam, 2004).

Treatment with LA protected cortical neurons against cytotoxicity induced by β-amyloid or hydrogen peroxide and induced an increase in the level of Akt, an effector immediately downstream of phosphatidylinositol kinase, suggesting that the neuroprotective effects of the LA are partly mediated through activation of the PKB/Akt-signaling pathway (Zhang et al., 2001).

The glutamate receptors mediate excitatory neurotransmission in the brain and are important in memory acquisition, learning, and are implicated in some neurodegenerative disorders (Choi, 1988; Beal, 1992; Choi, 1992). This receptor family is classified in three groups: *N*-methyl-D-aspartate (NMDA), α-amino-3-hydroxy-5-methyl-4-isoxazolepropionate (AMPA)-kainate, and metabotropic receptors. Excessive activation of the NMDA receptor leads to a large influx of calcium into neurons and subsequent generation of oxidants and oxidative stress by the stimulation of phospholipase A_2 (Dumuis et al., 1988; Lafon et al., 1993; Liu and Mori, 1999). Increased intracellular calcium may cause mitochondrial dysfunction, which can result in localized oxidant formation within mitochondria and an inability to handle free calcium (Henneberry et al., 1989; Beal, 1992). Intact mitochondrial function appears to be essential for neuronal resistance to excitotoxic insults. It is believed that the reduced levels of ATP that accompany abnormal mitochondrial function are insufficient to drive the ion pumps that maintain neuronal membrane polarization. With depolarization of the neuronal membrane, the magnesium that normally blocks the NMDA receptor ion channel is extruded, and ambient extracellular levels of glutamate may become lethal via NMDA receptor mechanism. On the basis of this mechanism, it seems likely that LA may play its memory-improving effect by enhancing mitochondrial function, scavenging free radicals to decrease oxidative damage, or increasing the levels of the antioxidants glutathione (GSH) and ascorbate to enhance the antioxidant defense. We have examined the effects of LA on neurotoxin- or oxidant-induced toxicity in HT4 cells and HT22 cells. The HT4 cell line was constructed by McKay et al. in 1989 and was derived from mouse neuronal tissue. Morimoto and Koshland have shown that HT4 cells possess NMDA receptors (Morimoto and Koshland, 1990). The HT22 cell line is a subclone of HT4. HT22 (the immortalized mouse cell line) lacks ionotropic glutamate receptors and responds to oxidative glutamate toxicity with a form of programmed cell death distinct from classical apoptosis.

We have found that dose-dependent cell injury in HT4 and HT22 cells is caused by glutamate (an excitotoxin), thapsigargin (an apoptosis-inducing agent), hydrogen peroxide (a typical oxidant), homocysteic acid (a cysteine uptake inhibitor), diethyl maleate (a pro-oxidant which depletes intracellular glutathione), apomorphine (a memory-impairing agent), SIN-1 (a generator of peroxynitrate), and 6-hydroxydopamine (an oxidant generator in brain) (Liu et al., 2002a). Mitochondrial dysfunction was a key event in the excitotoxicity-independent component of neuronal cell death. Reactive oxygen species accumulation and glutathione depletion were prominent in glutamate-treated cells (Tirosh et al., 2000). LA showed effects on reducing most of glutamate- and oxidant-induced cell death, decreasing oxidative damage, increasing antioxidant defense, and improving mitochondrial function (Liu et al., 2002a). These results, together with previous studies (Wolz and Krieglstein, 1996; Aksenova et al., 1998; Marangon et al., 1999; Tirosh et al., 2000) suggest that LA is an effective neuroprotective agent for ameliorating excitotoxins and oxidant-induced and age-associated neurodegeneration.

Iron may play a role in cognitive dysfunction, and LA may act as an iron chelator to prevent oxidant generation. It was found that the cerebral iron levels in 24–28 month old rats were increased by 80% relative to 3 month old rats and the iron accumulation correlated with a decline in GSH and the GSH/GSSG ratio (Suh et al., 2005). LA-treated old rats showed a decrease in cerebral iron and an improvement of the antioxidant status and thiol redox state, compared with untreated old rats (Suh et al., 2005), confirming that LA is a potent chelator of divalent metal ions (Ou et al., 1995) in the brain.

Another mechanism of the protective effect of LA, like the thiol-reactive compound sulforaphane (Gao et al., 2001), is mediated through induction of the phase 2 enzyme response via the transcription factor Nrf2. LA was shown to induce Nrf2, Nrf2 binding to antioxidant response element, and consequently, higher glutamylcysteine ligase activity (Smith et al., 2004). Phase 2 enzymes (e.g., glutathione transferases, NAD(P)H:quinone reductase) and glutathione synthesis are part of the elaborate system for protection against the toxicity of reactive oxygen and nitrogen species and electrophiles, which are constant dangers to the integrity of mammalian DNA (Gao et al., 2001). Induction of phase 2 enzymes, which neutralize reactive electrophiles and act as indirect antioxidants, appears to be effective in achieving protection against a variety of carcinogens and other oxidative damage in animals and humans (Ramos-Gomez et al., 2001). This mechanism appears as an indirect mitochondrial protection because the induced phase 2 enzymes reduce cytosolic oxidative stress and enhance the cellular antioxidant defense, thus indirectly relieving oxidative stress to mitochondria.

COMBINATION OF LA AND OTHER MITOCHONDRIAL NUTRIENTS OR COMPOUNDS: MORE EFFECTIVE THAN LA ALONE IN IMPROVING COGNITIVE FUNCTION?

It was suggested that a combination of nonsteroidal anti-inflammatory drugs (NSAIDs) and appropriate levels and types of micronutrients might be more effective than the individual agents in the prevention and in the treatment of AD (Prasad et al., 2002). On the basis of epidemiologic, laboratory, and clinical studies, we propose that using optimal combinations of LA and other mitochondrial nutrients to target mitochondrial dysfunction may provide an effective strategy in delaying aging, preventing, and treating cognitive dysfunction, including AD and PD.

Combinations of a number of nutritional cofactors have been tested in different mitochondrial disorders for additive or synergistic effects, including riboflavin plus carnitine to improve muscle weakness and exercise capacity in complex I deficient myopathy; riboflavin plus nicotinamide to improve encephalopathic symptoms and nerve conduction; vitamin K_3 plus ascorbate to clinically improve exercise capacity in patients with complex III defect; CoQ plus vitamin K_3, ascorbate, thiamin, riboflavin, and niacin to reduce mortality in mitochondrial myopathy and encephalomyopathies (Marriage et al., 2003); and carnitine plus choline and caffeine to reduce body fat and serum leptin concentrations (Hongu and Sachan, 2000).

LA and vitamin E have shown synergistic effects against lipid peroxidation by oxidant radicals in several pathological conditions such as a thromboembolic stroke model in rats for neurological functions, glial reactivity, and neuronal remodeling (Gonzalez-Perez et al., 2002). We observed that LA-plus acetyl-L-carnitine (ALCAR) was more effective than LA and ALCAR used alone to ameliorate mitochondrial decay in old rats, possibly by playing different roles in restoring mitochondrial function, including the complementary effect of LA on ALCAR by inhibiting oxidative stress (Hagen et al., 2002; Liu et al., 2002b; Liu et al., 2002c).

We first examined the effects of ALCAR and LA separately as well as their combined use on memory with the the Morris water maze test (Morris, 1984), and with the Skinner box test (fixed-interval performance in the peak procedure) (Gharib et al., 2001). Old rats showed an age-associated decline in spatial memory in the Morris water maze test and treatments with ALCAR, LA, or a combination improved the age-associated spatial memory decline. ALCAR, LA, or their combination improved the spatial memory by (1) reducing the time to find the hidden escape platform, (2) increasing the time at the platform position and the percentage time in the quadrant where the escape platform was formerly contained during the 60 s transfer (no platform) test, and (3) reducing the time to find the visible escape platform. ALCAR showed a greater effect than LA. The combination of ALCAR and LA showed a synergistic effect on reversing the decay of spatial memory in old animals (Liu et al., 2002b). Figure 20.1 shows an example performance curve of young, old, old + ALCAR, old + LA, and old + ALCAR + LA rats in the first four days in Morris water maze test. The Skinner box test reflects the internal clock and memory. Old rats had a much lower response rate, and LA increased the response rate in old rats. Although ALCAR does not show any effect (comparing the ALCAR group to the old controls), LA seems to slightly increase the peak rate in old rats. The combination of LA and ALCAR showed a larger and significant increase in response rate

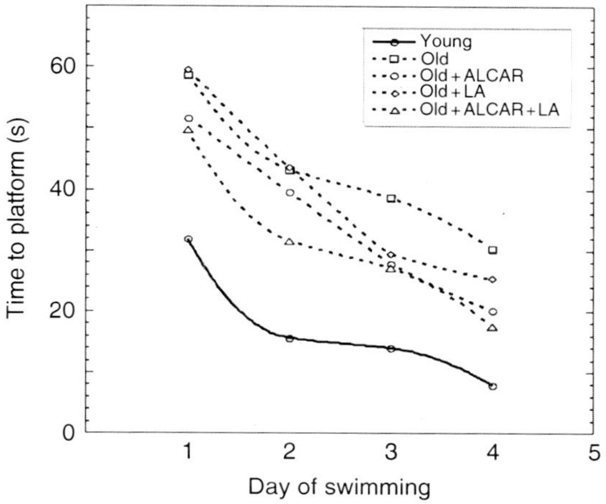

FIGURE 20.1 An example performance curve of young, old, old + acetyl-L-carnitine (ALCAR), old + LA, and old + ALCAR + LA rats in the first four days in Morris water maze test.

and also peak rate in old animals compared to LA alone, suggesting a synergistic action of LA and ALCAR (Liu et al., 2002b). Figure 20.2 shows the response rate to sound signal in the Skinner box test.

The memory-improving effect of LA-plus ALCAR was found to be accompanied by a decrease in lipid peroxidation, protein oxidation, oxidative RNA/DNA damage, and mitochondrial dysfunction in the brain of rats (Hagen

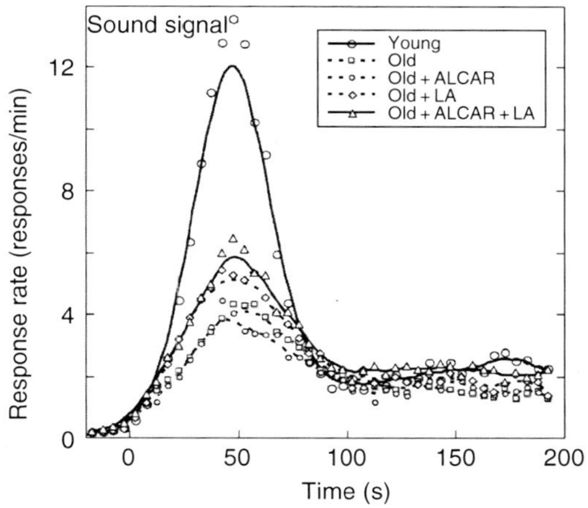

FIGURE 20.2 The response rate (to sound signal) of young, old, old + acetyl-L-carnitine (ALCAR), old + LA, and old + ALCAR + LA rats in the Skinner box test.

et al., 2002; Liu et al., 2002b,c). These results strongly suggest that LA-plus ALC improve cognitive dysfunction by improving mitochondrial dysfunction and decreasing oxidative damage in the brain. Similar results on reducing oxidative damage have been obtained by administrating LA and carnitine to rats by Muthuswamy et al. (2006). Aged rats had a significant decline in the antioxidant status and increase in lipid peroxidation, protein carbonyl and DNA protein cross-links as compared to young rats in the cerebral cortex, striatum, and hippocampus; co-supplementation of LA and carnitine was effective in reducing brain regional lipid peroxidation, protein carbonyl, and DNA protein cross-links and in increasing the activities of enzymatic antioxidants in aged rats to near normalcy (Muthuswamy et al., 2006). In addition, immunohistochemical staining assay showed that brains of old rats had a significant increase in nitrotyrosine administrations of ALCAR, LA, especially ALCAR + LA reduced nitrotyrosine in old rat brain (data not shown).

Using electron microscopy we have also observed the mitochondrial structural changes in the hippocampus (Liu et al., 2002b). It was found that structural abnormalities develop with age. Compared with young rats, old rats showed some disruption and loss of cristae in about half of the mitochondria in the dentate gyrus area, indicating structural decay. Animals treated with 0.5% ALCAR or 0.2% LA showed less structural disruption and loss of cristae. In addition, old rats had more lipofuscin in the cytoplasm of granule cells of the dentate gyrus, and the combined treatment rats also seemed to have less lipofuscin. Figure 20.3 shows a few representative electron microscopic images obtained from the brain of young, old, old + ALCAR, old + LA, and old + ALCAR + LA rats. We should emphasize that the mitochondrial structures are heterogous in the brains of young and old rats.

In a cellular oxidative stress model, LA-plus ALC is found to mediate prosurvival-signaling mechanism in aldehyde-induced oxidative toxicity. 4-Hydroxy-2-nonenal (HNE) is a highly reactive product of lipid peroxidation of unsaturated lipids, and induces oxidative toxicity as a model of oxidative stress-induced neurodegeneration. Pretreatment of primary cortical neuronal cultures with ALCAR and LA significantly attenuated HNE-induced cytotoxicity, protein oxidation, lipid peroxidation, and apoptosis, also led to elevated cellular GSH and heat shock protein (HSP) levels, and activation of phosphoinositol-3-kinase (PI3K), PKG, and ERK1/2 pathways (Abdul and Butterfield, 2007).

LA may be more active and effective when combined with CoQ because dihydrolipoic acid (the reduced form of LA), which is soluble in the aqueous compartment, has been found to reduce CoQ to ubiquinol by the transfer of a pair of electrons, thereby increasing the antioxidant capacity in biomembranes (Kozlov et al., 1999). Packer et al. (1997b) have suggested that LA can recycle most of the antioxidants, including CoQ, ascorbic acid, glutathione, and thioredoxin. It may be useful to add LA to other antioxidants to recycle the oxidized antioxidants in vivo.

A mixture with LA and other antioxidants and mitochondrial enzyme cofactors could slow age-dependent cognitive decline in dogs. A mixture of LA, carnitine, and other antioxidants vitamin E, ascorbic acid, and also spinach

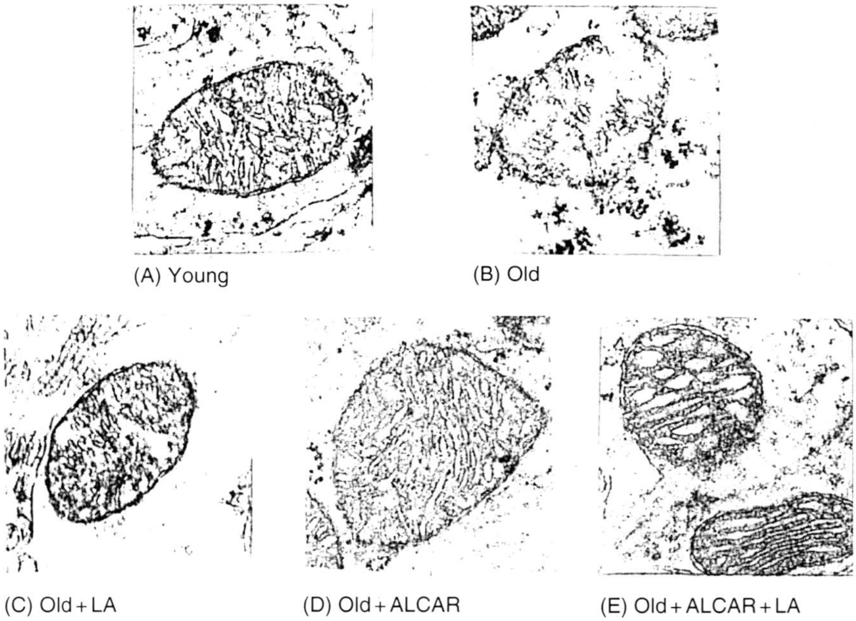

FIGURE 20.3 A few representative electron microscopic images obtained from the brains of (A) young, (B) old, (C) old + LA, (D) old + acetyl-L-carnitine (ALCAR), and (E) old + ALCAR + LA rats.

flakes, tomato pomace, grape pomace, carrot granules, and citrus pulp was fed to aged beagle dogs for 2 years. At 1 and 2 years, the mixture-fed group with a behavioral enrichment showed more accurate learning than the other aged groups. Discrimination learning was significantly improved by behavioral enrichment; reversal learning was improved by both behavioral enrichment and mixture feeding (Milgram et al., 2005). Oxidative stress biomarkers, i.e., protein carbonyls, 3-nitrotyrosine (3-NT), and the lipid peroxidation product 4-hydroxynonenal (HNE), were decreased by the mixture feeding, the behavioral enrichment, and the combination of both when compared to control, with the most significant effects found in the combination of the mixture feeding and behavioral enrichment (Opii et al., 2006). The combined treatment significantly reduced the specific protein carbonyl levels of glutamate dehydrogenase [NAD (P)], glyceraldehyde-3-phosphate dehydrogenase (GAPDH), α-enolase, neurofilament triplet L protein, glutathione-*S*-transferase (GST), and fascin actin bundling protein, significantly increased the expression of Cu/Zn superoxide dismutase, fructose-bisphosphate aldolase C, creatine kinase, glutamate dehydrogenase, and glyceraldehyde-3-phosphate dehydrogenase, and also significantly increased the enzymatic activities of glutathione-*S*-transferase (GST) and total superoxide dismutase (SOD) (Opii et al., 2006).

In one study, LA administration has been found to increase oxidative stress in rat brain. Kayali et al. (2006) found that LA administration caused an increase in protein carbonyl and nitrotyrosine levels and a decrease in total thiol, non-protein thiol, and lipid hydroperoxide levels in the brain tissue of aged rats. The authors assumed pro-oxidative effects of LA in the brain tissue of aged rats may be due to the pro-oxidant effects of LA (Kayali et al., 2006). However, the results should be further confirmed with more convincing parameters and methods.

LA DERIVATIVES: MORE POTENT THAN LA FOR TREATING NEURODEGENERATIVE DISEASES?

One active direction in LA studies is to develop LA derivatives with more potent effect or with more functions or specificities. Packer and coworkers (Tirosh et al., 1999) have developed a positively charged water soluble LA amide analogue, 2-(*N*,*N*-dimethylamine) ethylamidolipoate HCl and named it LA-plus. Compared to LA, LA-plus was found to be more effective in (1) protecting cells against glutamate-induced neurotoxicity, (2) preventing glutamate-induced loss of intracellular GSH, and (3) disallowing increase of intracellular peroxide level following the glutamate challenge (Tirosh et al., 1999). LA-plus was also shown to provide protection against HNE-induced inactivation of pyruvate dehydrogenase in the mitochondria (Korotchkina et al., 2001) and to inhibit NO production in RAW 264.7 macrophages (Guo et al., 2001). Therefore, the authors considered that LA-plus is a potent protector of neuronal cells against glutamate-induced cytotoxicity and associated oxidative stress.

LA has been connected with melatonin. The conjugate was named melatoninolipoamide (Venkatachalam et al., 2006). It was found that the melatonin moiety of the conjugate reacts preferably with oxidizing radicals and the LA moiety exhibits preferential reaction with reducing radical. Using γ-radiation-induced lipid peroxidation in liposomes and hemolysis of erythrocytes as a model, the radioprotection ability was compared with those of melatonin and LA and the results suggest that the conjugate can be explored as a probable radioprotector (Venkatachalam et al., 2006).

In order to target the Au nanoparticles (AuNP) to folate receptor positive tumor cells via receptor-mediated endocytosis, a polyethylene glycol (PEG) construct with LA and folic acid coupled on opposite ends of the polymer chain was synthesized (Dixit et al., 2006). The folate-PEG grafted AuNPs was shown to have a selective uptake by KB cells, a folate receptor positive cell line that overexpress the folate receptor. This is helpful for developing methods that use targeted metal nanoparticles for tumor imaging and ablation (Dixit et al., 2006).

For PD treatment, one big problem is the pro-oxidant effect associated with L-dopa therapy. In order to overcome this problem, a series of multifunctional codrugs, obtained by joining L-dopa and dopamine with (*R*)-LA, were synthesized and evaluated as potential drugs with antioxidant and iron-chelating properties (Di Stefano et al., 2006). It was found that there are potential advantages of using some of these codrugs rather than L-dopa in treating PD due to their "in vivo" dopaminergic activity and a sustained release of the parent drug in human plasma (Di Stefano et al., 2006).

Conjugated LA and γ-linolenic acid improves functional deficits in the peripheral and central nervous system in streptozotocin-diabetic rats. As we know, diabetes mellitus can lead to functional and structural deficits in both the peripheral and central nervous system. A 12 week of treatment with the conjugate in streptozotocin-diabetic rats showed that the conjugate treatment improved long-term potentiation in the hippocampus (Biessels et al., 2001).

CONCLUSION

LA is used to improve age-associated decline of cognitive function, and also to treat or prevent peripheral neuropathy and cardiac autonomic neuropathy, insulin resistance in type-2 diabetes, retinopathy and cataract, glaucoma, HIV/AIDS, cancer, liver disease, Wilson's disease, cardiovascular disease, and lactic acidosis caused by inborn errors of metabolism, and also Alzheimer-type dementia. Though further investigations are needed for studying molecular and cellular mechanisms and specially, more clinical trials on LA supplementations, evidence has demonstrated that LA has beneficial effects and show promising applications for delaying, preventing, and repairing mitochondrial decay; improving cognitive function; and preventing/ameliorating age-related degenerative diseases. It is also an important direction of developing LA derivatives with more functions and specificities for treating various degenerative diseases.

ACKNOWLEDGMENTS

This book chapter is adapted from a review for *Neurochemical Research*. This study was supported by NIH grant R21 AT001918, and grant AG023265-01.

REFERENCES

Abdul, H.M. and Butterfield, D.A. 2007. Involvement of PI3K/PKG/ERK1/2 signaling pathways in cortical neurons to trigger protection by cotreatment of acetyl-L-carnitine and alpha-lipoic acid against HNE-mediated oxidative stress and neurotoxicity: Implications for Alzheimer's disease. *Free Radic Biol Med 42*, 371–384.

Aksenova, M.V., Aksenov, M.Y., Carney, J.M., and Butterfield, D.A. 1998. Protein oxidation and enzyme activity decline in old brown Norway rats are reduced by dietary restriction. *Mech Ageing Dev 100*, 157–168.

Albers, D.S. and Beal, M.F. 2000. Mitochondrial dysfunction and oxidative stress in aging and neurodegenerative disease. *J Neural Transm Suppl 59*, 133–154.

Aliev, G., Smith, M.A., Seyidov, D., Neal, M.L., Lamb, B.T., Nunomura, A., Gasimov, E.K., Vinters, H.V., Perry, G., LaManna, J.C., and Friedland, R.P. 2002. The role of oxidative stress in the pathophysiology of cerebrovascular lesions in Alzheimer's disease. *Brain Pathol 12*, 21–35.

Ames, B.N. 2003. Delaying the mitochondrial decay of aging-a metabolic tune-up. *Alzheimer Dis Assoc Disord 17 Suppl 2*, S54–S57.

Ames, B.N., Shigenaga, M.K., and Hagen, T.M. 1993. Oxidants, antioxidants, and the degenerative diseases of aging. *Proc Natl Acad Sci U S A 90*, 7915–7922.

Ames, B.N., Shigenaga, M.K., and Hagen, T.M. 1995. Mitochondrial decay in aging. *Biochim Biophys Acta 1271*, 165–170.

Arivazhagan, P., Ayusawa, D., and Panneerselvam, C. 2006. Protective efficacy of alpha-lipoic acid on acetylcholinesterase activity in aged rat brain regions. *Rejuvenation Res 9*, 198–201.

Arivazhagan, P., and Panneerselvam, C. 2004. Alpha-lipoic acid increases Na^+ K^+ ATPase activity and reduces lipofuscin accumulation in discrete brain regions of aged rats. *Ann N Y Acad Sci 1019*, 350–354.

Baker, S.K. and Tarnopolsky, M.A. 2003. Targeting cellular energy production in neurological disorders. *Expert Opin Investig Drugs 12*, 1655–1679.

Beal, M.F. 1992. Does impairment of energy metabolism result in excitotoxic neuronal death in neurodegenerative illnesses? *Ann Neurol 31*, 119–130.

Beal, M.F. 2003. Bioenergetic approaches for neuroprotection in Parkinson's disease. *Ann Neurol 53 Suppl 3*, S39–S47; discussion S8.

Beckman, K.B. and Ames, B.N. 1998. The free radical theory of aging matures. *Physiol Rev 78*, 547–581.

Biessels, G.J., Smale, S., Duis, S.E., Kamal, A., and Gispen, W.H. 2001. The effect of gamma-linolenic acid-alpha-lipoic acid on functional deficits in the peripheral and central nervous system of streptozotocin-diabetic rats. *J Neurol Sci 182*, 99–106.

Black, M.M. 2003. Micronutrient deficiencies and cognitive functioning. *J Nutr 133*, 3927S–3931S.

Bustamante, J., Lodge, J.K., Marcocci, L., Tritschler, H.J., Packer, L., and Rihn, B.H. 1998. Alpha-lipoic acid in liver metabolism and disease. *Free Radic Biol Med 24*, 1023–1039.

Butterfield, D., Castegna, A., Pocernich, C., Drake, J., Scapagnini, G., and Calabrese, V. 2002a. Nutritional approaches to combat oxidative stress in Alzheimer's disease. *J Nutr Biochem 13*, 444.

Butterfield, D.A., Castegna, A., Drake, J., Scapagnini, G., and Calabrese, V. 2002b. Vitamin E and neurodegenerative disorders associated with oxidative stress. *Nutr Neurosci 5*, 229–239.

Cadenas, E. and Davies, K.J. 2000. Mitochondrial free radical generation, oxidative stress, and aging. *Free Radic Biol Med 29*, 222–230.

Chen, J.J. and Yu, B.P. 1994. Alterations in mitochondrial membrane fluidity by lipid peroxidation products. *Free Radic Biol Med 17*, 411–418.

Choi, D.W. 1988. Glutamate neurotoxicity and diseases of the nervous system. *Neuron 1*, 623–634.

Choi, D.W. 1992. Excitotoxic cell death. *J Neurobiol 23*, 1261–1276.

Choi, J.H. and Yu, B.P. 1995. Brain synaptosomal aging: free radicals and membrane fluidity. *Free Radic Biol Med 18*, 133–139.

Cui, X., Zuo, P., Zhang, Q., Li, X., Hu, Y., Long, J., Packer, L., and Liu, J. 2006. Chronic systemic D-galactose exposure induces memory loss, neurodegeneration, and oxidative damage in mice: Protective effects of R-alpha-lipoic acid. *J Neurosci Res 83*, 1584–1590.

Di Stefano, A., Sozio, P., Cocco, A., Iannitelli, A., Santucci, E., Costa, M., Pecci, L., Nasuti, C., Cantalamessa, F., and Pinnen, F. 2006. L-dopa- and dopamine-(R)-alpha-lipoic acid conjugates as multifunctional codrugs with antioxidant properties. *J Med Chem 49*, 1486–1493.

Dixit, V., Van den Bossche, J., Sherman, D.M., Thompson, D.H., and Andres, R.P. 2006. Synthesis and grafting of thioctic acid-PEG-folate conjugates onto Au nanoparticles for selective targeting of folate receptor-positive tumor cells. *Bioconjug Chem 17*, 603–609.

Dumuis, A., Sebben, M., Haynes, L., Pin, J.P., and Bockaert, J. 1988. NMDA receptors activate the arachidonic acid cascade system in striatal neurons. *Nature 336*, 68–70.

Evans, J.L. and Goldfine, I.D. 2000. Alpha-lipoic acid: A multifunctional antioxidant that improves insulin sensitivity in patients with type 2 diabetes. *Diabetes Technol Ther 2*, 401–413.

Farr, S.A., Poon, H.F., Dogrukol-Ak, D., Drake, J., Banks, W.A., Eyerman, E., Butterfield, D.A., and Morley, J.E. 2003. The antioxidants alpha-lipoic acid and N-acetylcysteine reverse memory impairment and brain oxidative stress in aged SAMP8 mice. *J Neurochem 84*, 1173–1183.

Fuchs, J., Packer, L., and Zimmer, G. 1997. *Lipoic Acid in Health and Disease*, Vol 5 New York: Marcel Dekker, Inc.

Gao, X., Dinkova-Kostova, A.T., and Talalay, P. 2001. Powerful and prolonged protection of human retinal pigment epithelial cells, keratinocytes, and mouse leukemia cells against oxidative damage: The indirect antioxidant effects of sulforaphane. *Proc Natl Acad Sci U S A 98*, 15221–15226.

Gharib, A., Derby, S., and Roberts, S. 2001. Timing and the control of variation. *J Exp Psychol Anim Behav Process 27*, 165–178.

Gonzalez-Perez, O., Gonzalez-Castaneda, R.E., Huerta, M., Luquin, S., Gomez-Pinedo, U., Sanchez-Almaraz, E., Navarro-Ruiz, A., and Garcia-Estrada, J. 2002. Beneficial effects of alpha-lipoic acid plus vitamin E on neurological deficit, reactive gliosis and neuronal remodeling in the penumbra of the ischemic rat brain. *Neurosci Lett 321*, 100–104.

Guo, Q., Tirosh, O., and Packer, L. 2001. Inhibitory effect of alpha-lipoic acid and its positively charged amide analogue on nitric oxide production in RAW 264.7 macrophages. *Biochem Pharmacol 61*, 547–554.

Hagen, T.M., Ingersoll, R.T., Lykkesfeldt, J., Liu, J., Wehr, C.M., Vinarsky, V., Bartholomew, J.C. and Ames, A.B. 1999. (R)-alpha-lipoic acid-supplemented old rats have improved mitochondrial function, decreased oxidative damage, and increased metabolic rate. *FASEB J 13*, 411–418.

Hagen, T.M., Ingersoll, R.T., Wehr, C.M., Lykkesfeldt, J., Vinarsky, V., Bartholomew, J.C., Song, M.H., and Ames, B.N. 1998. Acetyl-L-carnitine fed to old rats partially restores mitochondrial function and ambulatory activity. *Proc Natl Acad Sci U S A 95*, 9562–9566.

Hagen, T.M., Liu, J., Lykkesfeldt, J., Wehr, C.M., Ingersoll, R.T., Vinarsky, V., Bartholomew, J.C., and Ames, B.N. 2002. Feeding acetyl-L-carnitine and lipoic acid to old rats significantly improves metabolic function while decreasing oxidative stress. *Proc Natl Acad Sci U S A 99*, 1870–1875.

Hagen, T.M., Vinarsky, V., Wehr, C.M., and Ames, B.N. 2000. (R)-alpha-lipoic acid reverses the age-associated increase in susceptibility of hepatocytes to tert-butylhydroperoxide both in vitro and in vivo. *Antioxid Redox Signal 2*, 473–483.

Hager, K., Marahrens, A., Kenklies, M., Riederer, P., and Munch, G. 2001. Alpha-lipoic acid as a new treatment option for Azheimer type dementia. *Arch Gerontol Geriatr 32*, 275–282.

Harman, D. 1972. The biologic clock: The mitochondria? *J Am Geriatr Soc 20*, 145–147.

Harman, D. 2002. *Increasing Healthy Life Span*, Vol 959 New York: The New York Academy of Sciences.

Henneberry, R.C., Novelli, A., Cox, J.A., and Lysko, P.G. 1989. Neurotoxicity at the N-methyl-D-aspartate receptor in energy-compromised neurons. An hypothesis for cell death in aging and disease. *Ann N Y Acad Sci 568*, 225–233.

Hongu, N. and Sachan, D.S. 2000. Caffeine, carnitine and choline supplementation of rats decreases body fat and serum leptin concentration as does exercise. *J Nutr 130*, 152–157.

Kayali, R., Cakatay, U., Akcay, T., and Altug, T. 2006. Effect of alpha-lipoic acid supplementation on markers of protein oxidation in post-mitotic tissues of ageing rat. *Cell Biochem Funct 24*, 79–85.

Korotchkina, L.G., Yang, H., Tirosh, O., Packer, L., and Patel, M.S. 2001. Protection by thiols of the mitochondrial complexes from 4-hydroxy-2-nonenal. *Free Radic Biol Med 30*, 992–999.

Kozlov, A.V., Gille, L., Staniek, K., and Nohl, H. 1999. Dihydrolipoic acid maintains ubiquinone in the antioxidant active form by two-electron reduction of ubiquinone and one-electron reduction of ubisemiquinone. *Arch Biochem Biophys 363*, 148–154.

Lafon, C.M., Pietri, S., Culcasi, M., and Bockaert, J. 1993. NMDA-dependent superoxide production and neurotoxicity. *Nature 364*, 535–537.

Liu, J., Atamna, H., Kuratsune, H., and Ames, B.N. 2002a. Delaying brain mitochondrial decay and aging with mitochondrial antioxidants and metabolites. *Ann N Y Acad Sci 959*, 133–166.

Liu, J., Head, E., Gharib, A.M., Yuan, W., Ingersoll, R.T., Hagen, T.M., Cotman, C.W., and Ames, B.N. 2002b. Memory loss in old rats is associated with brain mitochondrial decay and RNA/DNA oxidation: Partial reversal by feeding acetyl-L-carnitine and/or R-alpha-lipoic acid. *Proc Natl Acad Sci U S A 99*, 2356–2361.

Liu, J., Killilea, D.W., and Ames, B.N. 2002c. Age-associated mitochondrial oxidative decay: Improvement of carnitine acetyltransferase substrate-binding affinity and activity in brain by feeding old rats acetyl-L-carnitine and/or R-alpha-lipoic acid. *Proc Natl Acad Sci U S A 99*, 1876–1881.

Liu, J. and Mori, A. 1999. Stress, aging, and brain oxidative damage. *Neurochem Res 24*, 1479–1497.

Manda, K., Ueno, M., Moritake, T., and Anzai, K. 2007. Radiation-induced cognitive dysfunction and cerebellar oxidative stress in mice: Protective effect of alpha-lipoic acid. *Behav Brain Res 177*, 7–14.

Marangon, K., Devaraj, S., Tirosh, O., Packer, L., and Jialal, I. 1999. Comparison of the effect of alpha-lipoic acid and alpha-tocopherol supplementation on measures of oxidative stress. *Free Radic Biol Med 27*, 1114–1121.

Marriage, B., Clandinin, M.T., and Glerum, D.M. 2003. Nutritional cofactor treatment in mitochondrial disorders. *J Am Diet Assoc 103*, 1029–1038.

McDaniel, M.A., Maier, S.F., and Einstein, G.O. 2003. "Brain-specific" nutrients: A memory cure? *Nutrition 19*, 957–975.

Milgram, N.W., Head, E., Zicker, S.C., Ikeda-Douglas, C.J., Murphey, H., Muggenburg, B., Siwak, C., Tapp, D., and Cotman, C.W. 2005. Learning ability in aged beagle dogs is preserved by behavioral enrichment and dietary fortification: A two-year longitudinal study. *Neurobiol Aging 26*, 77–90.

Moini, H., Packer, L., and Saris, N.E. 2002. Antioxidant and prooxidant activities of alpha-lipoic acid and dihydrolipoic acid. *Toxicol Appl Pharmacol 182*, 84–90.

Morimoto, B.H., and Koshland, D.E., Jr. 1990. Excitatory amino acid uptake and N-methyl-D-aspartate-mediated secretion in a neural cell line. *Proc Natl Acad Sci U S A 87*, 3518–3521.

Morris, R. 1984. Developments of a water-maze procedure for studying spatial learning in the rat. *J Neurosci Methods 11*, 47–60.

Muthuswamy, A.D., Vedagiri, K., Ganesan, M., and Chinnakannu, P. 2006. Oxidative stress-mediated macromolecular damage and dwindle in antioxidant status in aged rat brain regions: Role of L-carnitine and DL-alpha-lipoic acid. *Clin Chim Acta 368*, 84–92.

Opii, W.O., Joshi, G., Head, E., Milgram, N.W., Muggenburg, B.A., Klein, J.B., Pierce, W.M., Cotman, C.W., and Butterfield, D.A. 2006. Proteomic identification of brain proteins in the canine model of human aging following a long-term treatment with antioxidants and a program of behavioral enrichment: Relevance to Alzheimer's disease. *Neurobiol Aging*, Epub 2006 Oct 20.

Ou, P., Tritschler, H.J., and Wolff, S.P. 1995. Thioctic (lipoic) acid: A therapeutic metal-chelating antioxidant? *Biochem Pharmacol 50*, 123–126.

Packer, L., Roy, S., and Sen, C.K. 1997a. Alpha-lipoic acid: A metabolic antioxidant and potential redox modulator of transcription. *Adv Pharmacol 38*, 79–101.

Packer, L., Tritschler, H.J., and Wessel, K. 1997b. Neuroprotection by the metabolic antioxidant alpha-lipoic acid. *Free Radic Biol Med 22*, 359–378.

Packer, L., Witt, E.H., and Tritschler, H.J. 1995. Alpha-Lipoic acid as a biological antioxidant. *Free Radic Biol Med 19*, 227–250.

Paradies, G., Ruggiero, F.M., Petrosillo, G., Gadaleta, M.N., and Quagliariello, E. 1994. Effect of aging and acetyl-L-carnitine on the activity of cytochrome oxidase and adenine nucleotide translocase in rat heart mitochondria. *FEBS Lett 350*, 213–215.

Perry, G., Nunomura, A., Friedlich, A.L., Boswell, M.V. 2000. Factors controlling oxidative damage in Alzheimer disease: Metals and mitochondria, In *Free Radicals in Chemistry, Biology and Medicine*, T. Yoshikawa, S. Toyokuni, Y. Yamamato, Y. Naito (Eds.), (OICA International), pp. 417–423.

Poon, H.F., Farr, S.A., Thongboonkerd, V., Lynn, B.C., Banks, W.A., Morley, J.E., Klein, J.B., and Butterfield, D.A. 2005a. Proteomic analysis of specific brain proteins in aged SAMP8 mice treated with alpha-lipoic acid: Implications for aging and age-related neurodegenerative disorders. *Neurochem Int 46*, 159–168.

Poon, H.F., Frasier, M., Shreve, N., Calabrese, V., Wolozin, B., and Butterfield, D.A. 2005b. Mitochondrial associated metabolic proteins are selectively oxidized in A30P alpha-synuclein transgenic mice—a model of familial Parkinson's disease. *Neurobiol Dis 18*, 492–498.

Prasad, K.N., Cole, W.C., and Prasad, K.C. 2002. Risk factors for Alzheimer's disease: Role of multiple antioxidants, non-steroidal anti-inflammatory and cholinergic agents alone or in combination in prevention and treatment. *J Am Coll Nutr 21*, 506–522.

Price, M.C. 2002. Longevity Report 91: The role of enzymic cofactors in aging. www.quantiumcwenet/lr91htm#_Toc22709815.

Quinn, J.F., Bussiere, J.R., Hammond, R.S., Montine, T.J., Henson, E., Jones, R.E., and Stackman, R.W., Jr. 2007. Chronic dietary alpha-lipoic acid reduces deficits in hippocampal memory of aged Tg2576 mice. *Neurobiol Aging 28*, 213–225.

Ramos-Gomez, M., Kwak, M.K., Dolan, P.M., Itoh, K., Yamamoto, M., Talalay, P., and Kensler, T.W. 2001. Sensitivity to carcinogenesis is increased and chemoprotective efficacy of enzyme inducers is lost in nrf2 transcription factor-deficient mice. *Proc Natl Acad Sci U S A 98*, 3410–3415.

Rao, A.V. and Balachandran, B. 2002. Role of oxidative stress and antioxidants in neurodegenerative diseases. *Nutr Neurosci 5*, 291–309.

Roubertoux, P.L., Sluyter, F., Carlier, M., Marcet, B., Maarouf-Veray, F., Cherif, C., Marican, C., Arrechi, P., Godin, F., Jamon, M., et al. 2003. Mitochondrial DNA modifies cognition in interaction with the nuclear genome and age in mice. *Nat Genet 35*, 65–69.

Schreibelt, G., Musters, R.J., Reijerkerk, A., de Groot, L.R., van der Pol, S.M., Hendrikx, E.M., Dopp, E.D., Dijkstra, C.D., Drukarch, B., and de Vries, H.E. 2006. Lipoic acid affects cellular migration into the central nervous system and stabilizes blood–brain barrier integrity. *J Immunol 177*, 2630–2637.

Sen, C.K., Roy, S., Khanna, S., and Packer, L. 1999. Determination of oxidized and reduced lipoic acid using high-performance liquid chromatography and coulometric detection. *Methods Enzymol 299*, 239–246.

Sharma, M. and Gupta, Y.K. 2003. Effect of alpha lipoic acid on intracerebroventricular streptozotocin model of cognitive impairment in rats. *Eur Neuropsychopharmacol 13*, 241–247.

Shigenaga, M.K., Hagen, T.M., and Ames, B.N. 1994. Oxidative damage and mitochondrial decay in aging. *Proc Natl Acad Sci U S A 91*, 10771–10778.

Smith, A.R., Shenvi, S.V., Widlansky, M., Suh, J. H., and Hagen, T.M. 2004. Lipoic acid as a potential therapy for chronic diseases associated with oxidative stress. *Curr Med Chem 11*, 1135–1146.

Sohal, R.S. and Weindruch, R. 1996. Oxidative stress, caloric restriction, and aging. *Science 273*, 59–63.

Stadtman, E.R. and Levine, R.L. 2000. Protein oxidation. *Ann N Y Acad Sci 899*, 191–208.

Stoll, S., Hartmann, H., Cohen, S.A., and Muller, W.E. 1993. The potent free radical scavenger alpha-lipoic acid improves memory in aged mice: Putative relationship to NMDA receptor deficits. *Pharmacol Biochem Behav 46*, 799–805.

Suh, J.H., Moreau, R., Heath, S.H., and Hagen, T.M. 2005. Dietary supplementation with (R)-alpha-lipoic acid reverses the age-related accumulation of iron and depletion of antioxidants in the rat cerebral cortex. *Redox Rep 10*, 52–60.

Suh, J.H., Shigeno, E.T., Morrow, J.D., Cox, B., Rocha, A.E., Frei, B., and Hagen, T.M. 2001. Oxidative stress in the aging rat heart is reversed by dietary supplementation with (R)-(alpha)-lipoic acid. *FASEB J 15*, 700–706.

Tirosh, O., Sen, C.K., Roy, S., Kobayashi, M.S., and Packer, L. 1999. Neuroprotective effects of alpha-lipoic acid and its positively charged amide analogue. *Free Radic Biol Med 26*, 1418–1426.

Tirosh, O., Sen, C.K., Roy, S., and Packer, L. 2000. Cellular and mitochondrial changes in glutamate-induced HT4 neuronal cell death. *Neuroscience 97*, 531–541.

Tritschler, H.J., Packer, L., and Medori, R. 1994. Oxidative stress and mitochondrial dysfunction in neurodegeneration. *Biochem Mol Biol Int 34*, 169–181.

Venkatachalam, S.R., Salaskar, A., Chattopadhyay, A., Barik, A., Mishra, B., Gangabhagirathi, R., and Priyadarsini, K.I. 2006. Synthesis, pulse radiolysis, and in vitro radioprotection studies of melatoninolipoamide, a novel conjugate of melatonin and alpha-lipoic acid. *Bioorg Med Chem 14*, 6414–6419.

Wolz, P. and Krieglstein, J. 1996. Neuroprotective effects of alpha-lipoic acid and its enantiomers demonstrated in rodent models of focal cerebral ischemia. *Neuropharmacology 35*, 369–375.

Zhang, L., Xing, G.Q., Barker, J.L., Chang, Y., Maric, D., Ma, W., Li, B.S., and Rubinow, D.R. 2001. Alpha-lipoic acid protects rat cortical neurons against cell death induced by amyloid and hydrogen peroxide through the Akt signalling pathway. *Neurosci Lett 312*, 125–128.

Ziegler, D. and Gries, F.A. 1997. Alpha-lipoic acid in the treatment of diabetic peripheral and cardiac autonomic neuropathy. *Diabetes 46 Suppl 2*, S62–S66.

21 Effects of Alpha-Lipoic Acid on AMP-Activated Protein Kinase in Different Tissues: Therapeutic Implications for the Metabolic Syndrome

Eun Hee Koh, Eun Hee Cho, Min-Seon Kim, Joong-Yeol Park, and Ki-Up Lee

CONTENTS

INTRODUCTION

Metabolic syndrome is a cluster of cardiovascular risk factors, including abdominal obesity, glucose intolerance, dyslipidemia, and hypertension [1]. The incidence of metabolic syndrome is increasing, and it is estimated to affect 15%–30% of individuals in industrialized countries [2]. The metabolic syndrome is associated with a two- to four-fold increase in cardiovascular morbidity and mortality, and a five- to nine-fold increase in the risk of developing type 2 diabetes mellitus [3,4]. Although life style modifications (e.g., exercise and calorie restriction to reduce weight) are considered first-line therapy for metabolic syndrome, they are often ineffective by themselves.

Recent studies from our group and others have found that dysregulation of intracellular fatty acid metabolism is an important contributor of metabolic syndrome [5–8]. We also found that α-lipoic acid (LA) exerts beneficial effects on metabolic syndrome by affecting 5′-adenosine monophosphate (5′-AMP)-activated protein kinase (AMPK), the metabolic master switch in response to cellular energy change, in various tissues [9–11]. In this chapter, we will review the role of skeletal muscle fatty acid metabolism in the genesis of insulin resistance and the role of AMPK in fuel metabolism. We will then focus on the effects of LA on AMPK activity in various tissues, in particular the peripheral tissues and the hypothalamus.

ROLE OF SKELETAL MUSCLE FATTY ACID METABOLISM IN THE GENESIS OF INSULIN RESISTANCE

Insulin resistance is considered a common pathogenic factor for metabolic syndrome [12]. Skeletal muscle is responsible for 70%–80% of whole-body glucose uptake and is the major regulator of whole-body energy balance. In a state of insulin resistance, there is a striking decrease in glucose utilization in the skeletal muscle [13], making decreased glucose utilization in the skeletal muscle a major determinant of whole-body insulin sensitivity [14]. Defects in glycogen synthase

[15] or insulin-signaling pathways, including insulin-stimulated phosphorylation of insulin receptor substrates (IRS) and phosphatidylinositol-3-kinase (PI-3-kinase) [16–18], have been demonstrated in the skeletal muscle of obese and type 2 diabetes patients. However, it is not yet clear whether these defects are primary or secondary to other abnormalities inherent in these disorders.

Plasma concentrations of free fatty acid (FFA) are increased in obese individuals, and the reciprocal relationship between glucose and fatty acid metabolism [19] has gained special attention. Triglyceride accumulation in skeletal muscle is also increased in insulin-resistant subjects [5,20]. Recent studies have suggested, however, that defective intracellular fatty acid metabolism in skeletal muscle, rather than simple oversupply of fatty acid fuel, is causally related to the development of insulin resistance [21,22].

Glucose–Fatty Acid Cycle

Over 40 years ago, it was hypothesized that the increased availability of FFA leads to decreased glucose utilization in muscle [19]. Increased FFA availability and oxidation lead to an increase in the ratio of intra-mitochondrial acetyl-CoA to CoA, thus inactivating pyruvate dehydrogenase (PDH) and decreasing glucose oxidation. In addition, an increase in intracellular citrate concentration inhibits phosphofructokinase (PFK) and glycolytic flux, and this in turn reduces glucose uptake by increasing glucose-6-phosphate (G-6-P) levels (Figure 21.1) [23].

Since the initial proposal of the "glucose–fatty acid cycle" [19], many investigators have demonstrated the reciprocal relationship between fatty acid and glucose metabolism in vivo and in vitro. Acute elevation of plasma FFA levels by lipid infusion [24,25] or chronic elevation of plasma FFA levels by a high-fat diet [26,27] has been consistently shown to decrease glucose utilization in skeletal muscle by affecting both glucose oxidation and glycogen synthesis [28–30]. In addition, FFA has been found to alter insulin-signaling pathways. For example, elevations in plasma FFA levels by lipid infusion abolish insulin-stimulated IRS-1-associated PI-3-kinase activity in skeletal muscle [31]. Lipid infusion also activates protein kinase C-βII and protein kinase-δ [32,33], decreasing tyrosine phosphorylation of IRS-1 and translocation of glucose transporter-4 (GLUT-4) [33]. Taken together, these results suggest that increased availability of FFA leads to insulin resistance in skeletal muscle by affecting insulin-signaling pathways.

Increased Intramyocellular Triglyceride Accumulation

Although increased FFA availability inhibits glucose utilization in skeletal muscle, increased plasma concentrations of FFA cannot wholly explain insulin resistance in obesity or metabolic syndrome. For example, the correlation coefficient between plasma FFA concentration and insulin sensitivity indices has been reported to be less than 0.6 [34], indicating that the contribution of plasma FFA to insulin resistance is not high. In addition, fatty acid oxidation in

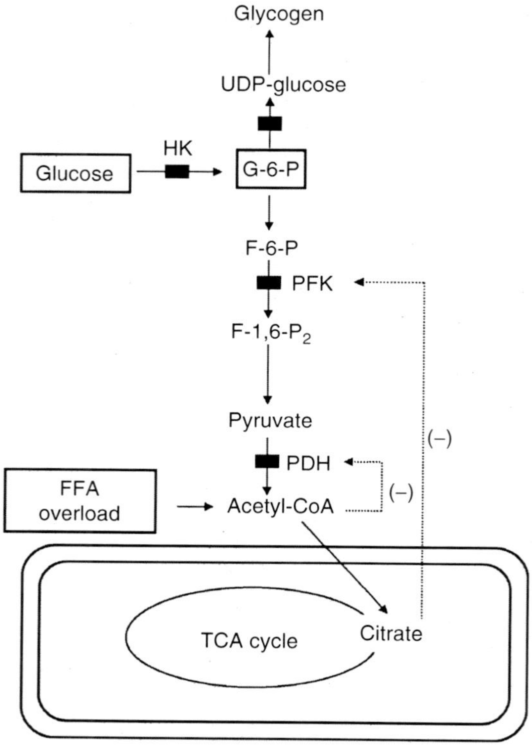

FIGURE 21.1 The glucose–fatty acid cycle describes reciprocal relationship between carbohydrate and fatty acid metabolism. Increased free fatty acid (FFA) availability and FFA oxidation in the skeletal muscle lead to an increase in mitochondrial ratio of acetyl-CoA/CoA, which inhibits the pyruvate dehydrogenase (PDH). Accumulation of citrate inhibits PFK, the rate limiting enzyme of glycolysis. This in turn accumulates G-6-P, and inhibits hexokinase (HK) and reduces muscle glucose uptake. F-1,6-P_2, fructose-1,6-biphosphate; F-6-P, fructose-6-phosphate; TCA; tricarboxylic acid.

the skeletal muscle of patients with type 2 diabetes was decreased, rather than increased [35]. Using the limb-balance technique, the respiratory quotient across the leg was found to be elevated in type 2 diabetes, denoting increased glucose oxidation and decreased fatty acid oxidation. Further, whole-body calorimetry disclosed that reliance on fatty acid oxidation was decreased in obesity and that this was related to decreased lipoprotein lipase activity in the skeletal muscle [36].

Recent studies have repeatedly demonstrated that the triglyceride content in skeletal muscle is increased in subjects with obesity or type 2 diabetes and that this is associated with insulin resistance [5,37]. The rate of fatty acid oxidation in skeletal muscle is determined not only by plasma FFA released from adipose tissue but also by fatty acids released from stored triglycerides in the skeletal muscle itself. Thus local lipolysis of intramyocellular pools of fatty acid could provide a missing link in the glucose–fatty acid cycle hypothesis [19] and may explain the apparently low correlation between plasma FFA concentrations and insulin resistance [34].

Decreased Skeletal Muscle Lipolysis in Insulin Resistance

To determine whether local lipolysis of intramyocellular triglycerides contributes to insulin resistance by activating the glucose–fatty acid cycle, we directly measured the rate of lipolysis by using a microdialysis technique [38]. This technique, which utilizes a semiquantitative "no net flux" protocol [39], can measure local lipolysis in various tissues. Contrary to expectations, lipolysis in skeletal muscle was decreased in rats fed a high-fat diet (Figure 21.2) [39], suggesting that intracellular lipid accumulation in an insulin-resistant state does not cause insulin resistance by activating the glucose–fatty acid cycle. These results are similar to those showing that skeletal muscle FFA utilization is decreased, rather than increased, in insulin-resistant subjects [34,40]. Taken together, these observations suggest that intracellular triglyceride accumulation in an insulin-resistant state is the consequence of a diminished fatty acid oxidation capacity rather than the cause of insulin resistance. Indeed, diminished fatty acid oxidation in the mitochondria would increase cytosolic long-chain acyl-CoA (LCAC). Accumulation of LCAC has been shown to impair glycogen synthase or hexokinase activity in skeletal muscle [41,42] and to affect insulin signaling [43]. In accordance with this concept, short-term interventions that ameliorate insulin resistance in rats fed a high-fat diet, such as a low-fat meal, overnight fasting, or acute exercise, were all associated with a reduction of in LCAC levels [44].

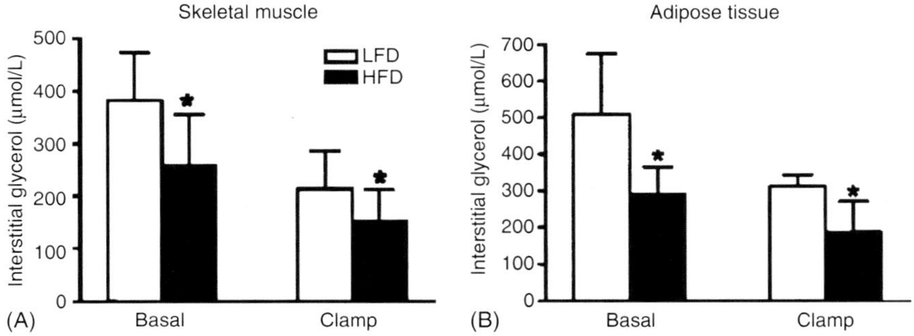

FIGURE 21.2 Interstitial glycerol concentrations in (A) skeletal muscle and (B) adipose tissue under basal and hyperinsulinemic euglycemic clamp conditions. $*P < .05$ versus LFD group. These results do not support the idea that increased fatty acid oxidation resulting from increased lipolysis of intramyocellular triglycerides is responsible for the insulin resistance in rats on a diet high in fat. (Adapted from Kim, C.H., Kim, M.S., Lee, K.U., et al., *Metabolism*, 52, 1586, 2003.)

Impaired Mitochondrial Function in Insulin Resistance

Mitochondria play a central role in the oxidation of fatty acids derived from lipid breakdown, a very efficient source of metabolic energy [45]. Skeletal muscle represents about 40% of the total body mass and has high oxidative metabolism linked to a high mitochondrial number [46]. Recent studies have demonstrated that insulin resistance in the elderly or diabetic offspring is related to a reduction in mitochondrial oxidative phosphorylation capacity [21,47]. A population-based epidemiologic study showed that mitochondrial DNA content was decreased in the peripheral blood of diabetic subjects, as well as in subjects who developed diabetes mellitus within 2 years [48]. Exercise training, which increases insulin sensitivity, also increases peak oxygen uptake, the activity of muscle mitochondrial enzymes and mRNA levels of mitochondrial genes and genes involved in mitochondrial biogenesis [49]. Skeletal muscle from type 2 diabetes patients showed decreased mitochondrial respiratory chain activity [20] and a deficiency of subsarcolemmal mitochondria [50].

Collectively, skeletal muscle mitochondrial dysfunction is one of the major factors in the pathophysiology of insulin resistance, which leads to an accumulation of lipid metabolites (e.g., LCAC) and impairment of insulin signaling (Figure 21.3).

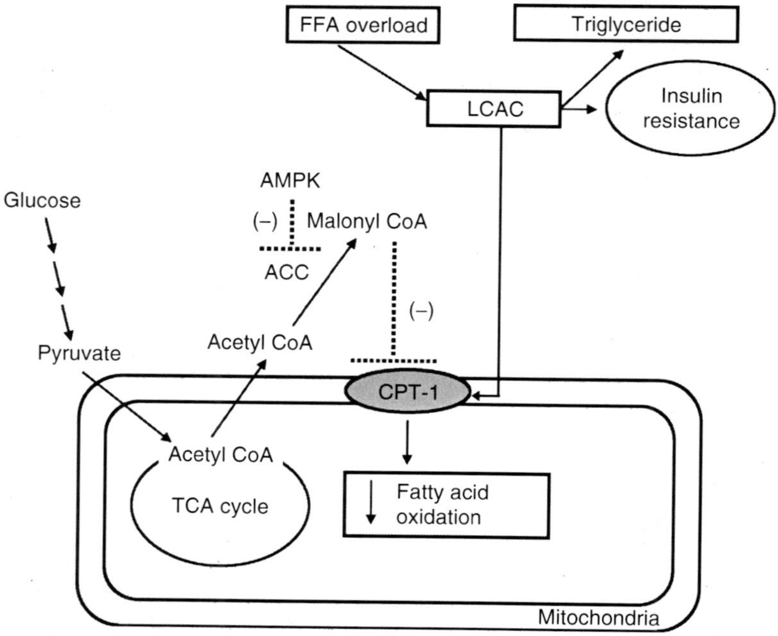

FIGURE 21.3 Diagrams showing effects of impaired mitochondrial fatty acid oxidation on insulin resistance in skeletal muscle. Decreased mitochondrial capacity to oxidize fatty acids may be the primary event leading to excess accumulation of intramuscular lipid metabolites (LCAC) and insulin resistance.

ADENOSINE MONOPHOSPHATE-ACTIVATED PROTEIN KINASE

AMPK: Structure, Regulation, and Functions

AMPK was first described as an enzyme capable of phosphorylating and inactivating hydroxymethylglutaryl-CoA (HMG-CoA) reductase and acetyl-CoA carboxylase (ACC), key enzymes in the synthesis of cholesterol and fatty acids [51,52]. Since its activity is highly dependent on the presence of 5′-AMP, this enzyme was later named AMPK [53].

AMPK is a heterotrimeric enzyme consisting of a catalytic α-subunit and two regulatory subunits, β and γ [54]. The α-subunit contains a typical serine/threonine protein kinase catalytic domain in its N-terminus. The β-subunit functions as a scaffold for the binding of α- and γ-subunits, as well as possessing a glycogen binding domain. The γ-subunit has four CBS domains, which bind to the adenosine of AMP. Multiple isoforms of mammalian AMPK exist (α1, α2, β1, β2, γ1–3), each encoded by different genes. The α1-subunit appears to be more ubiquitous and is located mainly in the cytosol, whereas the α2-subunit is mostly restricted to the heart, muscle, and liver and is present in both cytosol and nuclei [55–57].

AMPK is activated by an increase in the intracellular ratio of AMP to ATP [58]. AMP activates AMPK allosterically and induces phosphorylation of a threonine residue (Thr172) within the α2-subunit by an upstream kinase, the tumor suppressor LKB1. AMP also inhibits the dephosphorylation of Thr172 by protein phosphatase, whereas a high concentration of ATP inhibits the activation of AMPK [58,59]. Although elevation in the AMP/ATP ratio is the most well-known activator of AMPK, hyperosmotic stress does not alter this ratio, suggesting that other mechanisms are involved in the regulation of the AMPK pathway. Recently, calmodulin-dependent protein kinase kinase (CaMKK) was shown to be an additional upstream kinase of AMPK [60]. Activation of AMPK by CaMKK is triggered by a rise in intracellular calcium ions, without detectable changes in the AMP/ATP ratio.

Once activated, AMPK exerts effects on specific enzymes and transcriptional regulators, stimulating multiple events that enhance ATP generation and inhibiting events that consume ATP but are not acutely necessary for survival (Figure 21.4) [61]. In addition, AMPK increases mitochondrial biogenesis in response to chronic energy depletion [62,63]. Chronic feeding of β-guanadinopropionic acid (β-GPA), which reduces the intramuscular AMP/ATP ratio in rats, resulted in chronic AMPK activation and increased cytochrome *c* protein expression and mitochondrial density in the skeletal muscle [62]. Interestingly, under these conditions, AMPK activation promotes mitochondrial biogenesis through nuclear respiratory factor-1 and peroxisome proliferator-activated receptor-γ coactivator-1α (PGC-1α), master regulators of mitochondrial biogenesis [64,65]. Conversely, feeding β-GPA to transgenic mice expressing a dominant-negative mutant form of AMPK in skeletal muscle had no effect on mitochondrial content [63].

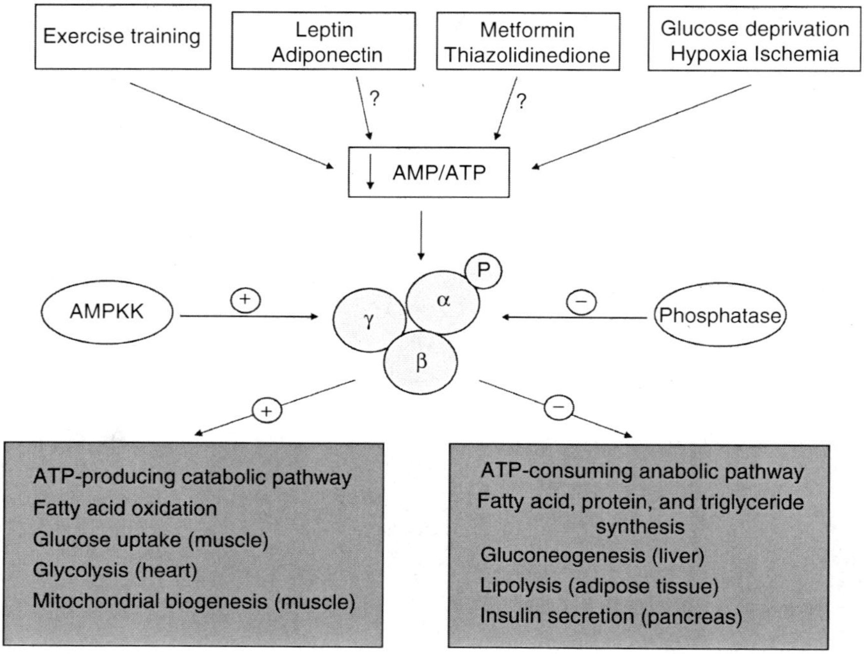

FIGURE 21.4 Role of AMPK in the regulation of cellular energy homeostasis. AMPK is the metabolic master switch in response to cellular energy change. AMPK activation in many tissues induces the ATP-producing catabolic pathways, while it suppresses the ATP-consuming anabolic pathways.

AMPK and Metabolic Syndrome

Recent findings have suggested that dysregulation of AMPK may play a major role in the development of metabolic syndrome. For example, metformin and thiazolidinedione, widely used to treat type-2 diabetes, were shown to increase AMPK activity. The primary hypoglycemic effect of metformin is through its inhibition of hepatic glucose production, which is mediated by AMPK activation through LKB1, irrespective of any increase in the cellular AMP:ATP ratio [66]. By contrast, improvement of insulin sensitivity by thiazolidinediones is partly attributed to AMPK activation in the muscle, which is through the increase in the cellular AMP:ATP ratio [67]. Moreover, adiponectin, a fat cell hormone that increases insulin sensitivity and reduces atherosclerotic processes, was shown to activate AMPK in C2Cl2 myocytes, hepatocytes, and endothelial cells [68,69]. In agreement with this concept, we and others observed that AMPK activity was decreased in tissues of obese animals [7,70], although others have reported no changes.

Roles of AMPK in Peripheral Tissues

Skeletal Muscle

Skeletal muscle is the major site of insulin-stimulated glucose disposal [71], and insulin resistance in this target tissue has long been considered a major factor in the pathogenesis of type 2 diabetes. Exercise training increases insulin sensitivity, but exercise-induced increases in glucose transport are insulin-independent. Muscle contraction activates AMPK [72], and administration of the AMPK activator 5-aminoimidazole-4-carboxamide (AICAR) enhances glucose uptake by increasing GLUT-4 translocation [73]. Therefore, AMPK activation was thought to be responsible for exercise-induced increases in glucose transport. AMPK activation, however, may not be necessary for contraction to stimulate glucose transport. Thus the molecular mechanism by which contraction increases glucose transport remains to be determined.

AMPK also regulates fatty acid oxidation in skeletal muscle. AMPK activation increases fatty acid oxidation by inhibiting ACC [74]. Malonyl-CoA is an intermediate in the de novo synthesis of long-chain fatty acids, and is an inhibitor of carnitine palmitoyltransferase-1 (CPT-1), an enzyme necessary for mitochondrial fatty acid uptake. By inhibiting ACC, the rate limiting enzyme of malonyl-CoA synthesis, activation of AMPK leads to increased mitochondrial fatty acid oxidation [75]. Moreover, the adipocyte-derived hormones, leptin and adiponectin, were shown to increase fatty acid oxidation through AMPK activation in skeletal muscle [68,76,77].

Inactivation of ACC, however, may not be the sole mechanism by which AMPK activation increases fatty acid oxidation in skeletal muscle. During prolonged exercise, ACC phosphorylation (inactivation) peaks after 1 h and returns to resting levels at exhaustion, while AMPK phosphorylation and AMPK $\alpha 2$ activity progressively increase [78], indicating that a mechanism other than ACC inactivation is involved in increasing AMPK-induced fatty acid oxidation. This discrepancy may be due to (1) activation of malonyl-CoA decarboxylase (MCD), an enzyme that degrades malonyl-CoA [79], or (2) transcriptional activation of mitochondrial fatty acid oxidative enzymes by peroxisome proliferator-activated receptors α (PPARα) and PGC-1 [80,81]. Furthermore, activation of muscle AMPK is required for increased expression of PGC-1 and mitochondrial biogenesis in response to chronic energy deficiency in vivo [63]. Thus AMPK activation increases fatty acid oxidation in cells through at least three distinct pathways: inactivation of ACC and activation of MCD, increase of mitochondrial biogenesis, and transcriptional activation of mitochondrial fatty acid oxidation genes (Figure 21.5).

Liver

In addition to skeletal muscle, AMPK exerts a wide range of metabolic effects in various tissues. In the liver, AMPK inactivates ACC and stimulates fatty

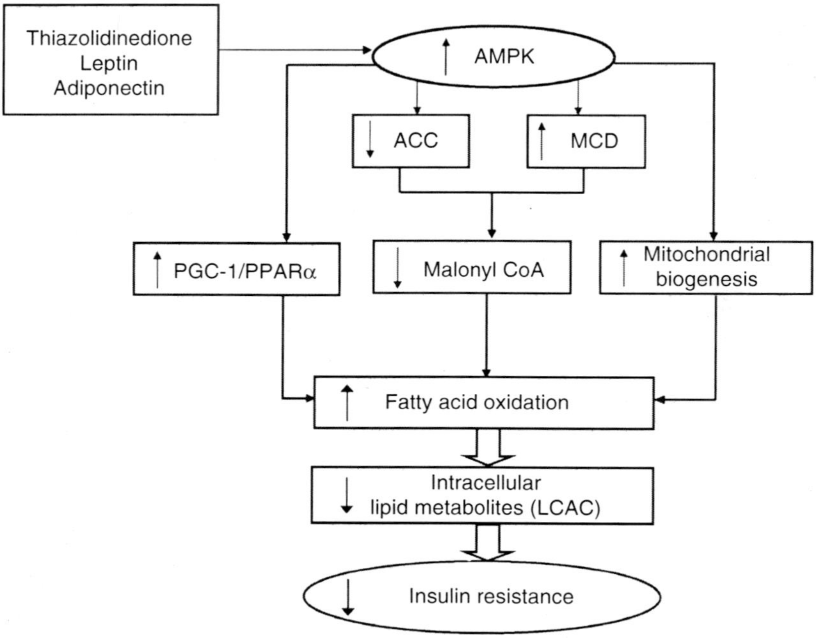

FIGURE 21.5 Possible mechanisms of AMP-activated protein kinase (AMPK) in the cellular fatty acid metabolism and insulin resistance.

acid oxidation [82]. Furthermore, AMPK activation suppresses the key enzymes involved in fatty acid and cholesterol synthesis, fatty acid synthase (FAS) and HMG-CoA reductase, respectively, as well as the mRNA expression of sterol element-binding protein-1 and hepatocyte nuclear factor 4α, the family of transcription factors that stimulate sterol and fatty-acid biosynthesis [82–84].

In insulin-resistant states, excessive hepatic glucose production is the major contributor to both fasting and postprandial hyperglycemia. Hepatic glucose production is the sum of glycogenolysis (breakdown of glycogen) and gluconeogenesis from noncarbohydrate precursors such as lactate, amino acids, and glycerol [85]. Although increased fatty acid oxidation in the liver is usually associated with increased gluconeogenesis [12], AMPK activation by AICAR, metformin, and adiponectin was shown to decrease hepatic gluconeogenesis [82,86,87]. The rate of gluconeogenesis is controlled principally by the activities of enzymes such as phosphoenolpyruvate carboxykinase and glucose-6-phosphatase [85]. The genes encoding these gluconeogenic enzymes are controlled at the transcriptional level by the activation of AMPK [88,89].

Vascular Endothelial Cell

To date, little has been known about the functional role of AMPK in endothelial cells. AMPK activation induces the phosphorylation of endothelial nitric oxide synthase (eNOS) and increases the activity of eNOS, thus regulating the vascular tone by endothelial NO generation [90]. Incubation of human umbilical endothelial cells (HUVECs) in a glucose-deprived medium or hydrogen peroxide (H_2O_2) was shown to increase AMPK activity [91,92]. In contrast, incubation of HUVECs with high glucose for 72 h caused significant apoptosis [93]. AICAR completely prevented high-glucose-mediated cell apoptosis, suggesting that AMPK plays an important role in the regulation of endothelial cell function.

Pancreatic Islet β-Cells

In pancreatic β-cells, high concentrations of glucose decrease AMPK activity [55]. Conversely, AMPK activation suppresses glucose-induced increases in glycolysis, mitochondrial oxidative metabolism, Ca^{2+} influx, and insulin secretion [94]. AMPK activation also inhibits the effect of glucose on the promoter activities of L-type pyruvate kinase (L-PK) and preproinsulin (PPI) [55]. Furthermore, antibodies to α2-AMPK mimic the effects of elevated glucose on L-PK and PPI promoters. These findings suggest that AMPK antagonizes the effects of glucose on insulin secretion and gene expression in β-cells. AMPK activation may also have adverse effects in pancreatic β-cells by inducing apoptosis, as observed in MIN6 β-cells [95].

AMPK activation, however, may act to favor the preservation of β-cell function under lipid overloading conditions. Accumulation of triglyceride in β-cells reduces insulin secretion and β-cell apoptosis, an important mechanism of β-cell loss in type-2 diabetes mellitus [8]. Similarly, overexpression of a constitutively active form of sterol response element-binding protein 1c (SREBP1c) in MIN6 β-cells leads to the activation of FAS gene expression, accumulation of triacylglycerol, a decrease in total islet ATP contents, and a profound inhibition of glucose-stimulated insulin release [96]. These effects of SREBP1c overexpression in pancreatic β-cells are reversed by AICAR [96].

Role of AMPK in the Hypothalamus

Nutrient Sensing in the Hypothalamus and Feeding Regulation

Neuronal cells have been shown to sense nutrient availability and to signal relevant neuronal pathways that lead to feeding behavior. Sustained intracerebroventricular (ICV) infusion of brain fuels, such as glucose, glycerol, and β-hydroxybutyrate, decreases body weight and food intake [97]. ICV administration of long-chain fatty acids also inhibits food intake and hepatic glucose production [98]. Conversely, central administration of 2-deoxyglucose (2-DG, a non-metabolizable glucose analogue) or mercaptoacetate (an inhibitor of fatty acid oxidation) elicits feeding [99,100].

Several signaling pathways are thought to be involved in mediating nutrient-induced feeding signaling. Central administration of the FAS inhibitors cerulenin and C75 reduces food intake, and this can be prevented by the coadministration of the ACC inhibitor TOFA [101]. These results suggest that malonyl-CoA, an intermediate metabolite between ACC and FAS, may be an anorexic signal. On the other hand, inhibition of hypothalamic CPT-1, which increases cytosolic LCAC concentrations in hypothalamic neurons, reduces food intake [102]. FAS inhibitor-induced increases in intracellular malonyl-CoA concentrations can also suppress CPT-1 activity [75], suggesting that LCAC may serve as a signaling molecule in the regulation of food intake by the hypothalamus.

Role of Hypothalamic AMPK in Regulating Food Intake and Energy Expenditure

We and others [11,103,104] have shown that AMPK activity in hypothalamic neurons is altered by various factors and mediates their feeding effects, and that this activity is affected by nutritional availability (Figure 21.6). Administration of the glucose antimetabolite 2-DG increased hypothalamic AMPK activity, while coadministration of an AMPK inhibitor, compound C, inhibited the 2-DG-induced glucoprivic feeding [11]. Conversely, ICV administration of glucose or restoration of food intake decreases hypothalamic AMPK activity [103].

Interestingly, hypothalamic AMPK activity is also altered by physiologic feeding regulators. ICV administration of the anorexigenic hormones (such as insulin and leptin) reduces AMPK activity, while the orexigenic hormone ghrelin increases this activity [103,104]. AMPK activity in the hypothalamic paraventricular nucleus is decreased by the melanocortin receptor agonist MT-II but increased by the melanocortin receptor antagonist Agouti-related protein (AGRP). The FAS inhibitor C75 has potent anorexic action and suppresses AMPK activity in neurons [105,106]. Taken together, these findings indicate that AMPK is a common signaling pathway by which various factors regulate feeding behavior.

If hypothalamic AMPK activity is an important determinant of food intake, alteration in hypothalamic AMPK activity may cause the dysregulation of feeding observed in metabolic disorders. We recently found that hypothalamic AMPK activity is enhanced in rats with uncontrolled diabetic mellitus [107], and that this was suppressed by hypothalamic AMPK inhibitors, suggesting that hypothalamic AMPK activation is responsible for the development of diabetic hyperphagia. Plasma leptin and insulin concentrations are much lower in diabetic than in normal animals, and ICV leptin and insulin reversed diabetes-induced increases in food intake and hypothalamic AMPK activity. Thus, leptin and insulin deficiencies in diabetic animals may cause increased hypothalamic AMPK activity.

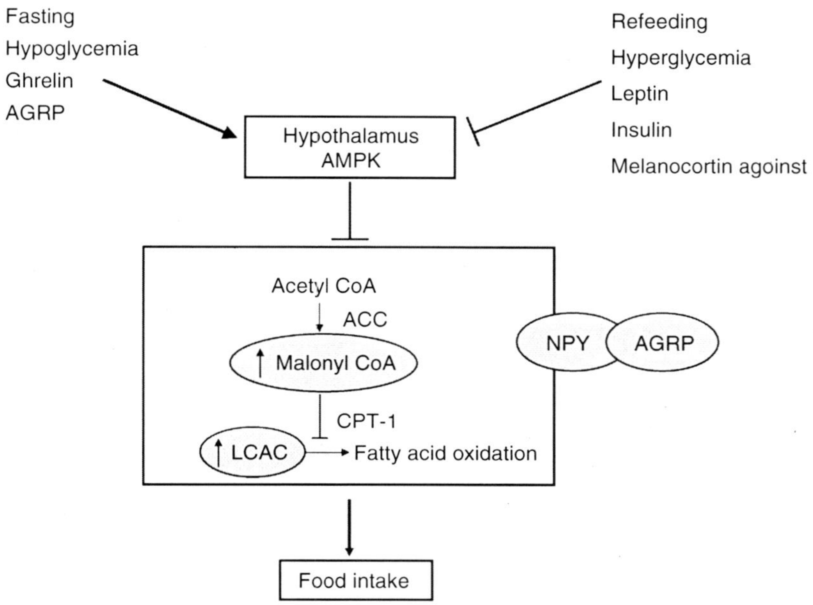

FIGURE 21.6 The role of hypothalamic AMP-activated protein kinase (AMPK) in the regulation food intake. Hypothalamic AMPK activity is regulated by several hormonal signals. Anorexigenic hormones, insulin and leptin decrease AMPK activity, while the orexigenic hormone, ghrelin increases it. Hypothalamic AMPK is also affected by nutritional status. Energy depletion such as fasting and hypoglycemia causes activation of hypothalamic AMPK, whereas refeeding and hyperglycemia decrease hypothalamic AMPK activity and food intake. Decrease in hypothalamus AMPK activity increases intracellular malonyl-CoA and long-chain acyl-CoA (LCAC), which were shown to decrease appetite.

The mechanism by which AMPK activity in hypothalamic neurons affects feeding behavior is still not fully understood. However, hypothalamic AMPK controls feeding behavior, at least in part, through the regulation of orexigenic neuropeptide Y (NPY) and AGRP expression. Overexpression of dominant-negative-AMPK in the medial hypothalamus decreased NPY and AGRP mRNA expression in ad libitum fed rats, whereas overexpression of constitutively active-AMPK augmented the fast-induced increases in NPY and AGRP expression [103]. Similarly, in neuronal cell lines and ex vivo hypothalamic cultures, low glucose, 2-DG, and AICAR increased both hypothalamic AMPK activity and AGRP expression [108]. Alternatively, changes in AMPK activity may affect feeding via changes in intracellular malonyl-CoA or LCAC concentrations [102,103]. The effect of AMPK may also be through alterations in ion channel activity, including K_{ATP} channels, or cytosolic Ca^{2+} concentrations [109,110].

LA AND METABOLIC SYNDROME

Beneficial Effect of LA in Metabolic Syndrome

LA is a strong antioxidant, and this activity has been shown to mediate beneficial effects in metabolic syndrome. Oxidative stress plays a major role in insulin resistance and β-cell damage [111,112]. For example, direct exposure of L6 muscle cells, 3T3L1 adipocytes and other cell lines to a low-grade oxidant (H_2O_2) markedly impaired insulin-stimulated glucose metabolism and insulin-signaling pathway, perhaps through activation of serine/threonine kinase cascades such as c-Jun N-terminal kinases and nuclear factor-κB (NF-κB) [113–115]. In 3T3L1 adipocytes and L6 myocytes, micromolar concentrations of LA were shown to protect the insulin-signaling system from oxidative stress, and to enhance glucose uptake and GLUT4 translocation [116,117]. Acute administration of LA in vivo enhanced glucose transport into the skeletal muscles of lean and insulin-resistant obese animals [118]. Both in insulin-resistant rats and in patients with type-2 diabetes, chronic administration of LA increased skeletal muscle glucose transport and improved insulin sensitivity [119,120]. However, the precise mechanisms responsible for the effect of LA on insulin sensitivity remain unclear.

Another target of oxidative stress is the β-cell [112]. Chronic exposure to high FFA and glucose results in the generation of reactive oxygen species (ROS) and consequently increases β-cell dysfunction [121,122]. In addition, triglycerides and lipid metabolites accumulate in pancreatic islets of obese animals. Accumulation of LCAC may induce apoptosis of pancreatic β-cells by sequentially activating ceramide and the inducible nitric oxide synthase pathway [8]. Dihydrolipoic acid (DHLA), a reduced form of LA, has been shown to protect pancreatic islets from damage by macrophage cytotoxicity [123]. Furthermore, LA blocked IL-1β mediated inhibition of glucose-stimulated insulin secretion from islet cells [124]. We found that chronic LA treatment reversed the decreased β-cell mass, disrupted islets, and accumulation of triglyceride content in obese, diabetes-prone OLETF rats. In addition, LA decreased plasma levels of oxidative stress markers, such as malondialdehyde and 8-hydroxy-deoxyguanosine [7].

Oxidative stress also plays an important role in atherogenesis. Reactive oxygen species induce vascular dysfunction by (1) reducing the bioavailability of nitric oxide, (2) impairing endothelium-dependent vasodilatation and endothelial cell growth, (3) causing apoptosis, (4) stimulating endothelial cell migration, and (5) activating adhesion molecules and inflammatory reactions [125–128]. Many studies have shown that LA may prevent atherosclerosis. For example, LA inhibited the expression of adhesion molecules, as well as monocyte adhesion, by inhibiting the IκB/NF-κB signaling pathway in human endothelial cells [129]. LA treatment prevented neointimal hyperplasia by reducing fractalkine expression, and in cultured vascular smooth cells, LA inhibited TNF-α stimulated expression of vascular cell adhesion molecule-1 and monocyte chemotactic protein-1 [130]. In healthy subjects, LA treatment reversed hypertriglyceridemia-induced endothelial dysfunction and decreased superoxide formation [131].

Effect of LA on AMPK in Peripheral Tissues

Skeletal Muscle

Chronic administration of LA to OLETF rats completely prevented the development of diabetes mellitus [10]. LA enhanced both non-oxidative and oxidative glucose metabolism in skeletal muscle, as evidenced by improvements in whole body and skeletal muscle glycogen synthesis and by reductions in plasma lactate concentrations. Interestingly, α2-AMPK phosphorylation level in the skeletal muscle was lower in obese rats than in control rats. Similar to leptin and adiponectin [68,76], LA increased α2-AMPK phosphorylation and fatty acid oxidation, and decreased triglyceride contents, in skeletal muscles of obese rats. Overexpression of dominant-negative α2-AMPK gene reversed the effects of LA on fatty acid oxidation, triglyceride content, and insulin-stimulated glucose uptake. These results suggest that LA improves insulin sensitivity in skeletal muscle by reducing lipid accumulation, which is mediated by AMPK activation.

Pancreatic Islet β-Cells

Acute or chronic treatment with LA led to dose-dependent increases in the phosphorylation of the AMPK α-subunit and ACC in islets and MIN6 cells [132]. In agreement with this, LA increased fatty acid oxidation and AMPK activation, leading to reduced triglyceride accumulation in pancreatic islets [7].

Vascular Endothelial Cell

AMPK is expressed in vascular endothelial cells and AMPK dysregulation contributes to endothelial dysfunction [10]. Endothelium-dependent vascular relaxation was impaired, and the number of apoptotic endothelial cells was higher, in the aorta of OLETF rats compared with control rats. In addition, triglyceride and lipid peroxide levels were higher, and NO synthesis and AMPK activity were lower, in the endothelium of OLETF rats. All of these alterations in endothelial cells and vascular dysfunction were substantially improved by LA treatment. In human aortic endothelial cells treated with linoleic acid, LA prevented the linoleic acid-induced decrease in AMPK phosphorylation, an effect associated with normalization of endothelial apoptosis and ROS generation in the presence of linoleic acid. Dominant-negative AMPK nearly completely reversed the effects of LA. These results suggest that reduced AMPK activity in endothelial cells contributes to endothelial dysfunction in obesity, and that LA improves endothelial dysfunction in obese rats by activating AMPK in endothelial cells.

ALA, AMPK, and Antioxidant Action

In summary, LA activates AMPK in peripheral tissues, thereby increasing fatty acid oxidation and reducing lipid accumulation. LA, which is present in mitochondria, is an essential cofactor of mitochondrial respiratory enzymes.

Administration of LA to aged rats restored aging-associated decreases in mitochondrial function [133]. LA is also a strong antioxidant, directly scavenging free radicals, chelating transition metal ions (e.g., iron and copper), increasing cytosolic glutathione and vitamin C levels, and preventing toxicities associated with their loss [134]. In addition, we demonstrated that the antioxidant action of LA is mediated, at least in part, by its effect on AMPK. Major sites of intracellular ROS generation are the mitochondria and cell membrane NAD(P)H oxidase [135]. The mitochondrial respiratory chain generates ROS when the electrochemical gradient between the mitochondrial inner membrane is high and the rate of electron transport is limited [136]. Consistent with this concept, recent studies from our group [137,138] and others [139] have shown that high glucose or linoleic acid levels lead to significant increases in mitochondrial membrane potential (hyperpolarization) and ROS generation. LA treatment nearly completely prevented hyperpolarization induced by linoleic acid, and pretreatment with Ad-dominant negative-AMPK reversed the effects of LA. LCAC has been found to impair the flow of electrons through the electron transfer chain [6], and the reduction of LCAC levels by AMPK activation may be responsible for improvements in electron transfer and reductions in ROS generation [10].

Effect of LA on AMPK in the Hypothalamus

We recently found that LA decreases hypothalamic AMPK activity and causes profound weight loss in rodents by reducing food intake and enhancing energy expenditure [11]. Activation of hypothalamic AMPK by AICAR or constitutively active AMPK reversed the effects of LA on food intake and energy expenditure. Conversely, 2-DG-induced hyperphagia was reversed by inhibiting hypothalamic AMPK. These data are quite surprising because LA increases glucose uptake and fatty acid oxidation by activating AMPK in skeletal muscle or endothelial cells [9,10]. Thus, LA exerts opposite effects on AMPK activity in different tissues, namely the hypothalamus versus the peripheral tissues. Similarly, leptin increases α2-AMPK activity in skeletal muscle [76] but decreases it in the hypothalamus [76,104].

Although the precise mechanism by which AMPK activity is differentially regulated in different tissues is not yet known, hypothalamic AMPK may sense whole-body energy status, and enhancement of its activity during energy depletion may help restore energy deficiency by increasing food intake. LA and leptin may signal the brain of excess whole-body energy status, thereby suppressing hypothalamic AMPK activity. In peripheral tissues, however, these molecules may increase AMPK activity and energy expenditure to decrease lipid depots in the body.

Finally, it should be noted that LA reduced food intake and body weight in both $Lep^{-/-}$ and $Lepr^{-/-}$ mice. We also showed that LA effectively reduced adiposity and visceral fat mass in genetically obese OLETF rats, which are leptin resistant [140]. Because most obese people are resistant to leptin [141], this agent is ineffective in treating human obesity. Thus, LA may be a promising anti-obesity drug for treatment of leptin-resistant human obesity and related diseases.

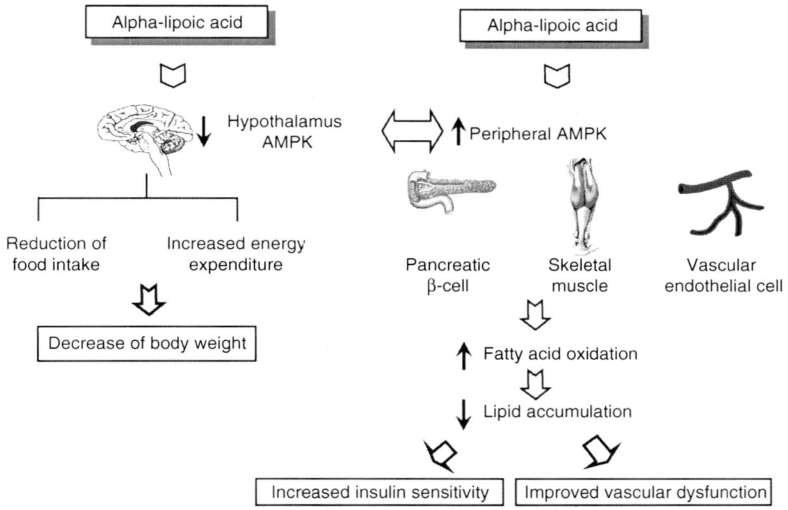

FIGURE 21.7 Differential effects of α-lipoic acid (LA) on AMP-activated protein kinase (AMPK) activity in peripheral tissues and hypothalamus.

CONCLUSION

There is considerable evidence that defective mitochondrial fatty acid oxidation in various tissues underlies the pathogenesis of the metabolic syndrome. AMPK is a fuel-sensing enzyme that increases intracellular fatty acid oxidation by multiple actions, including increased mitochondrial biogenesis. Administration of LA to obese rats prevented the development of diabetes and vascular dysfunction, and these effects were attributed to AMPK activation in skeletal muscle, pancreatic islets, and vascular endothelium. In contrast, LA decreased hypothalamic AMPK activity and caused profound weight loss in rodents by reducing food intake and enhancing energy expenditure (Figure 21.7). Although the precise mechanism by which AMPK activity is differentially regulated in different tissues is not yet known, LA may be a promising new drug for treatment of human obesity and metabolic syndrome.

REFERENCES

1. National Institutes of Health, Third Report of the National Cholesterol Education Program Expert Panel on Detection, Evaluation, and Treatment of High Blood Cholesterol in Adults (Adult Treatment Panel III). Bethesda, MD: National Institutes of Health. 2001; 01–3670.
2. Kim ES, Han SM, Kim YI, et al. Prevalence and clinical characteristics of metabolic syndrome in a rural population of South Korea. *Diabet Med* 2004; 21:1141–1143.
3. Lakka HM, Laaksonen DE, Lakka TA, et al. The metabolic syndrome and total and cardiovascular disease mortality in middle-aged men. *JAMA* 2002; 288:2709–2716.

4. Boyko EJ, de Courten M, Zimmet PZ, et al. Features of the metabolic syndrome predict higher risk of diabetes and impaired glucose tolerance: A prospective study in Mauritius. *Diabetes Care* 2000; 23:1242–1248.
5. Pan DA, Lillioja S, Storlien LH, et al. Skeletal muscle triglyceride levels are inversely related to insulin action. *Diabetes* 1997; 46:983–988.
6. Bakker SJ, IJzerman RG, Teerlink T, et al. Cytosolic triglycerides and oxidative stress in central obesity: The missing link between excessive atherosclerosis, endothelial dysfunction, and beta-cell failure? *Atherosclerosis* 2000; 148:17–21.
7. Song KH, Lee WJ, Koh JM, et al. Alpha-lipoic acid prevents diabetes mellitus in diabetes-prone obese rats. *Biochem Biophys Res Commun* 2005; 326:197–202.
8. Unger RH, Zhou YT. Lipotoxicity of beta-cells in obesity and in other causes of fatty acid spillover. *Diabetes* 2001; 50:S118–S121.
9. Lee WJ, Song KH, Koh EH, et al. Alpha-lipoic acid increases insulin sensitivity by activating AMPK in skeletal muscle. *Biochem Biophys Res Commun* 2005; 332:885–891.
10. Lee WJ, Lee IK, Kim HS, et al. Alpha-lipoic acid prevents endothelial dysfunction in obese rats via activation of AMP-activated protein kinase. *Arterioscler Thromb Vasc Biol* 2005; 25:2488–2494.
11. Kim MS, Park JY, Namkoong C, et al. Anti-obesity effects of alpha-lipoic acid mediated by suppression of hypothalamic AMP-activated protein kinase. *Nat Med* 2004; 10:727–733.
12. Reaven GM. Banting lecture 1988. Role of insulin resistance in human disease. *Diabetes* 1988; 37:1595–1607.
13. Simoneau JA, Colberg SR, Thaete FL, et al. Skeletal muscle glycolytic and oxidative enzyme capacities are determinants of insulin sensitivity and muscle composition in obese women. *FASEB J* 1995; 9:273–278
14. Henry RR, Abrams L, Nikoulina S, et al. Insulin action and glucose metabolism in nondiabetic control and NIDDM subjects. Comparison using human skeletal muscle cell cultures. *Diabetes* 1995; 44:936–946.
15. Wright KS, Beck-Nielsen H, Kolterman OG, et al. Decreased activation of skeletal muscle glycogen synthase by mixed-meal ingestion in NIDDM. *Diabetes* 1998; 37:436–440.
16. Kim YB, Nikoulina SE, Ciaraldi TP, et al. Normal insulin-dependent activation of Akt/protein kinase B, with diminished activation of phosphoinositide 3-kinase, in muscle in type 2 diabetes. *J Clin Invest* 1999; 104:733–741.
17. Goodyear LJ, Giorgino F, Sherman LA, et al. Insulin receptor phosphorylation, insulin receptor substrate-1 phosphorylation, and phosphatidylinositol 3-kinase activity are decreased in intact skeletal muscle strips from obese subjects. *J Clin Invest* 1995; 95:2195–2204.
18. Bjornholm M, Kawano Y, Lehtihet M, et al. Insulin receptor substrate-1 phosphorylation and phosphatidylinositol 3-kinase activity in skeletal muscle from NIDDM subjects after in vivo insulin stimulation. *Diabetes* 1997; 46:524–527.
19. Randle PJ, Garland PB, Hales CN, et al. The glucose–fatty acid cycle: Its role in insulin sensitivity and the metabolic disturbances of diabetes mellitus. *Lancet* 1963; 1:785–789.
20. Kelly DE, Goodpaster BH. Skeletal muscle triglyceride: An aspect of regional adiposity and insulin resistance. *Diabetes Care* 2001; 4:933–941.

21. Petersen KF, Dufour S, Befroy D, et al. Impaired mitochondrial activity in the insulin-resistant offspring of patients with type-2 diabetes. *N Engl J Med* 2004; 350:664–671.
22. Kelley DE, He J, Menshikova EV, et al. Dysfunction of mitochondria in human skeletal muscle in type-2 diabetes. *Diabetes* 2002; 51:2944–2950.
23. Newsholme EA, Sugden PH, Williams T. Effect of citrate on the activities of 6-phosphofructokinase from nervous and muscle tissues from different animals and its relationships to the regulation of glycolysis. *Biochem J* 1977; 166:123–129.
24. Ferrannini E, Barrett EJ, Bevilacqua S, et al. Effect of fatty acids on glucose production and utilization in man. *J Clin Invest* 1983; 72:1737–1747.
25. Lee KU, Lee HK, Koh CS, et al. Artificial induction of intravascular lipolysis by lipid-heparin infusion leads to insulin resistance in man. *Diabetologia* 1988; 31:285–290.
26. Kraegen EW, James DE, Storlien LH, et al. In vivo insulin resistance in individual peripheral tissues of the high fat fed rat: Assessment by euglycemic clamp plus deoxyglucose administration. *Diabetologia* 1986; 29:192–198.
27. Kim JK, Wi JK, Youn JH. Metabolic impairment precedes insulin resistance in skeletal muscle during high-fat feeding in rats. *Diabetes* 1996; 45:651–658.
28. Roden M, Price TB, Perseghin G, et al. Mechanism of free fatty acid-induced insulin resistance in humans. *J Clin Invest* 1996; 97:2859–2865.
29. Boden G. Role of fatty acids in the pathogenesis of insulin resistance and NIDDM. *Diabetes* 1997; 46:3–10.
30. Kim CH, Youn JH, Lee KU, et al. Effects of high-fat diet and exercise training on intracellular glucose metabolism in rats. *Am J Physiol Endocrinol Metab* 2000; 278: E977–E984.
31. Dresner A, Laurent D, Shulman GI, et al. Effects of free fatty acids on glucose transport and IRS-1-associated phosphatidylinositol 3-kinase activity. *J Clin Invest* 1999; 103:253–259.
32. Itani SI, Ruderman NB, Schmieder F, et al. Lipid-induced insulin resistance in human muscle is associated with changes in diacylglycerol, protein kinase C, and IkappaB-alpha. *Diabetes* 2002; 51:2005–2011.
33. Griffin ME, Marcucci MJ, Shulman GI, et al. Free fatty acid-induced insulin resistance is associated with activation of protein kinase C theta and alterations in the insulin signaling cascade. *Diabetes* 1999; 48:1270–1274.
34. Golay A, Chen N, Chen YD, et al. Effect of central obesity on regulation of carbohydrate metabolism in obese patients with varying degrees of glucose tolerance. *J Clin Endocrinol Metab* 1990; 71:1299–1304.
35. Kelley DE, Simoneau JA. Impaired free fatty acid utilization by skeletal muscle in non-insulin-dependent diabetes mellitus. *J Clin Invest* 1994; 94:2349–2356.
36. Ferro R, Eckel R, Larson E, et al. Relationship between skeletal muscle lipoprotein lipase activity and 24-h macronutrient oxidation. *J Clin Invest* 1993; 92:441–445.
37. Oakes ND, Cooney GJ, Camilleri S, et al. Mechanisms of liver and muscle insulin resistance induced by chronic high-fat feeding. *Diabetes* 1997; 46: 1768–1774.
38. Rosdahl H, Ungerstedt U, Jorfeldt L, et al. Interstitial glucose and lactate balance in human skeletal muscle and adipose tissue studied by microdialysis. *J Physiol* 1993; 471:637–657.

39. Kim CH, Kim MS, Lee KU, et al. Lipolysis in skeletal muscle is decreased in high-fat-fed rats. *Metabolism* 2003; 52:1586–1592.
40. Simoneau JA, Veerkamp JH, Turcotte LP, et al. Markers of capacity to utilize fatty acids in human skeletal muscle: Relation to insulin resistance and obesity and effects of weight loss. *FASEB J* 1999; 13:2051–2060.
41. Wititsuwannakul D, Kim KH. Mechanism of palmityl coenzyme A inhibition of liver glycogen synthase. *J Biol Chem* 1977; 252:7812–7817.
42. Thompson AL, Cooney GJ. Acyl-CoA inhibition of hexokinase in rat and human skeletal muscle is a potential mechanism of lipid-induced insulin resistance. *Diabetes* 2000; 49:1761–1765.
43. Schmitz-Peiffer C, Browne CL, Biden TJ, et al. Alterations in the expression and cellular localization of protein kinase C isozymes epsilon and theta are associated with insulin resistance in skeletal muscle of the high-fat-fed rat. *Diabetes* 1997; 46:169–178.
44. Oakes ND, Bell KS, Kraegen EW, et al. Diet-induced muscle insulin resistance in rats is ameliorated by acute dietary lipid withdrawal or a single bout of exercise: Parallel relationship between insulin stimulation of glucose uptake and suppression of long-chain fatty acyl-CoA. *Diabetes* 1997; 46:2022–2028.
45. Ruderman NB, Saha AK, Vavvas D, et al. Malonyl-CoA, fuel sensing, and insulin resistance. *Am J Physiol* 1999; 276:E1–E18.
46. Rimbert V, Boirie Y, Bedu M, et al. Muscle fat oxidative capacity is not impaired by age but by physical inactivity: Association with insulin sensitivity. *FASEB J* 2004; 18:737–739.
47. Iossa S, Mollica MP, Lionetti L, et al. A possible link between skeletal muscle mitochondrial efficiency and age-induced insulin resistance. *Diabetes* 2004 53:2861–2866.
48. Song J, Oh JY, Sung YA, et al. Peripheral blood mitochondrial DNA content is related to insulin sensitivity in offspring of type-2 diabetic patients. *Diabetes Care* 2001; 24:865–869.
49. Short KR, Vittone JL, Nair KS, et al. Impact of aerobic exercise training on age-related changes in insulin sensitivity and muscle oxidative capacity. *Diabetes* 2003; 52:1888–1896.
50. Ritov VB, Menshikova EV, He J, et al. Deficiency of subsarcolemmal mitochondria in obesity and type-2 diabetes. *Diabetes* 2005; 54:8–14.
51. Beg ZH, Allmann DW, Gibson DM. Modulation of 3-hydroxy-3-methylglutaryl coenzyme A reductase activity with cAMP and with protein fractions of rat liver cytosol. *Biochem Biophys Res Commun* 1973; 54:1362–1369.
52. Carlson CA, Kim KH. Regulation of hepatic acetyl coenzyme A carboxylase by phosphorylation and dephosphorylation. *J Biol Chem* 1973; 248:378–390.
53. Yeh LA, Lee KH, Kim KH. Regulation of rat liver acetyl-CoA carboxylase. Regulation of phosphorylation and inactivation of acetyl-CoA carboxylase by the adenylate energy charge. *J Biol Chem* 1980; 255:2308–2314.
54. Stapleton D, Woollatt E, Mitchelhill KI, et al. AMP-activated protein kinase isoenzyme family: Subunit structure and chromosomal location. *FEBS Lett* 1997; 409:452–456.
55. da Silva Xavier G, Leclerc I, Salt IP, et al. Role of AMP-activated protein kinase in the regulation by glucose of islet beta cell gene expression. *Proc Natl Acad Sci U S A* 2000; 97:4023–4028.

56. Woods A, Azzout-Marniche D, Foretz M, et al. Characterization of the role of AMP-activated protein kinase in the regulation of glucose-activated gene expression using constitutively active and dominant negative forms of the kinase. *Mol Cell Biol* 2000; 20:6704–6711.
57. Chen Z, Heierhorst J, Mann RJ, et al. Expression of the AMP-activated protein kinase beta1 and beta2 subunits in skeletal muscle. *FEBS Lett* 1999; 460:343–348.
58. Hardie DG, Scott JW, Pan DA, et al. Management of cellular energy by the AMP-activated protein kinase system. *FEBS Lett* 2003; 546:113–120.
59. Long YC, Zierath JR. AMP-activated protein kinase signaling in metabolic regulation. *J Clin Invest* 2006; 116:1776–1783.
60. Hawley SA, Pan DA, Mustard KJ, et al. Calmodulin-dependent protein kinase kinase-beta is an alternative upstream kinase for AMP-activated protein kinase. *Cell Metab* 2005; 2:9–19.
61. Luo Z, Saha AK, Xiang X, et al. AMPK, the metabolic syndrome and cancer. *Trends Pharmacol Sci* 2005; 26:69–76.
62. Bergeron R, Ren JM, Cadman KS, et al. Chronic activation of AMP kinase results in NRF-1 activation and mitochondrial biogenesis. *Am J Physiol Endocrinol Metab* 2001; 281:E1340–E1346.
63. Zong H, Ren JM, Young LH, et al. AMP kinase is required for mitochondrial biogenesis in skeletal muscle in response to chronic energy deprivation. *Proc Natl Acad Sci U S A* 2002; 99:15983–15987.
64. Kelly DP, Scarpullar RC. Transcriptional regulatory circuits controlling mitochondrial biogenesis and function. *Genes Dev* 2004; 18:357–368.
65. Wu Z, Puigserver P, Andersson U, et al. Mechanisms controlling mitochondrial biogenesis and respiration through the thermogenic coactivator PGC-1. *Cell* 1999; 98:115–124.
66. Shaw RJ, Lamia KA, Vasquez D, et al. The kinase LKB1 mediates glucose homeostasis in liver and therapeutic effects of metformin. *Science* 2005; 310:1642–1646.
67. Fryer LG, Parbu-Patel A, Carling D. The anti-diabetic drugs rosiglitazone and metformin stimulate AMP-activated protein kinase through distinct signaling pathways. *J Biol Chem* 2002; 277:25226–25232.
68. Yamauchi T, Kamon J, Minokoshi Y, et al. Adiponectin stimulates glucose utilization and fatty-acid oxidation by activating AMP-activated protein kinase. *Nat Med* 2002; 8:1288–1295.
69. Chen H, Montagnani M, Funahashi T, et al. Adiponectin stimulates production of nitric oxide in vascular endothelial cells. *J Biol Chem* 2003; 278:45021–45026.
70. Lessard SJ, Chen ZP, Watt MJ, et al. Chronic rosiglitazone treatment restores AMPK alpha 2 activity in insulin-resistant rat skeletal muscle. *Am J Physiol Endocrinol Metab* 2006; 290:E251–E257.
71. Newsholme EA, Sugden PH, Williams T. Effect of citrate on the activities of 6-phosphofructokinase from nervous and muscle tissues from different animals and its relationships to the regulation of glycolysis. *Biochem J* 1977; 166:123–129.
72. Marsin AS, Bouzin C, Bertrand L, et al. The stimulation of glycolysis by hypoxia in activated monocytes is mediated by AMP-activated protein kinase and inducible 6-phosphofructo-2-kinase. *J Biol Chem* 2002; 277:30778–83.
73. Hayashi T, Hirshman MF, Kurth EJ, et al. Evidence for 5′ AMP-activated protein kinase mediation of the effect of muscle contraction on glucose transport. *Diabetes* 1998; 47:1369–1373.

74. Saha AK, Schwarsin AJ, Ruderman NB, et al. Activation of malonyl-CoA decarboxylase in rat skeletal muscle by contraction and the AMP-activated protein kinase activator 5-aminoimidazole-4-carboxamide-1-beta-D-ribofuranoside. *J Biol Chem* 2000; 275:24279–24283.
75. Ruderman NB, Saha AK, Kraegen EW. Minireview: Malonyl-CoA, AMP-activated protein kinase, and adiposity. *Endocrinology* 2003; 144:5166–5171.
76. Minokoshi Y, Kim YB, Peroni OD, et al. Leptin stimulates fatty-acid oxidation by activating AMP-activated protein kinase. *Nature* 2002; 415:339–343.
77. Tomas E, Tsao TS, Saha AK, et al. Enhanced muscle fat oxidation and glucose transport by ACRP30 globular domain: Acetyl-CoA carboxylase inhibition and AMP-activated protein kinase activation. *Proc Natl Acad Sci U S A* 2002; 99:16309–16313.
78. Wojtaszewski JF, Mourtzakis M, Hillig T, et al. Dissociation of AMPK activity and ACC beta phosphorylation in human muscle during prolonged exercise. *Biochem Biophys Res Commun* 2002; 98:309–316.
79. Park H, Kaushik VK, Saha AK, et al. Coordinate regulation of malonyl-CoA decarboxylase, sn-glycerol-3-phosphate acyltransferase, and acetyl-CoA carboxylase by AMP-activated protein kinase in rat tissues in response to exercise. *J Biol Chem* 2002; 277:32571–32577.
80. Suwa M, Nakano H, Kumagai S. Effects of chronic AICAR treatment on fiber composition, enzyme activity, UCP3, and PGC-1 in rat muscles. *J Appl Physiol* 2003; 95:960–968.
81. Tunstall RJ, Mehan KA, Cameron-Smith D, et al. Exercise training increases lipid metabolism gene expression in human skeletal muscle. *Am J Physiol Endocrinol Metab* 2002; 283:E66–E72.
82. Zhou G, Myers R, Li Y, Chen Y, et al. Role of AMP-activated protein kinase in mechanism of metformin action. *J Clin Invest* 2001; 108:1167–1174.
83. Leclerc I, Lenzner C, Gourdon L, et al. Hepatocyte nuclear factor-4alpha involved in type 1 maturity-onset diabetes of the young is a novel target of AMP-activated protein kinase. *Diabetes* 2001; 50:1515–1521.
84. Carlson CA, Kim KH. Regulation of hepatic acetyl coenzyme A carboxylase by phosphorylation and dephosphorylation. *J Biol Chem* 1973; 248:378–80.
85. Postic C, Dentin R, Girard J. Role of the liver in the control of carbohydrate and lipid homeostasis. *Diabetes Metab* 2004; 30:398–408.
86. Vincent MF, Erion MD, Gruber HE, et al. Hypoglycemic effect of AICA riboside in mice. *Diabetologia* 1996; 39:1148–1155.
87. Rajala MW, Scherer PE. Minireview: The adipocyte—at the crossroads of energy homeostasis, inflammation, and atherosclerosis. *Endocrinology* 2003; 144: 3765–3773.
88. Koo SH, Flechner L, Qi L, et al. The CREB coactivator TORC2 is a key regulator of fasting glucose metabolism. *Nature* 2005; 437:1109–1111.
89. Shaw RJ, Lamia KA, Vasquez D, et al. The kinase LKB1 mediates glucose homeostasis in liver and therapeutic effects of metformin. *Science* 2005; 310:1642–1646.
90. Chen ZP, Mitchelhill KI, Michell BJ, et al. AMP-activated protein kinase phosphorylation of endothelial NO synthase. *FEBS Lett* 1999; 443:285–289.
91. Dagher Z, Ruderman N, Tornheim K, et al. Acute regulation of fatty acid oxidation and AMP-activated protein kinase in human umbilical vein endothelial cells. *Circ Res* 2001; 88:1276–1282.

92. Ruderman NB, Cacicedo JM, Itani S, et al. Malonyl-CoA and AMP-activated protein kinase (AMPK): Possible links between insulin resistance in muscle and early endothelial cell damage in diabetes. *Biochem Soc Trans* 2003; 31:202–206.
93. Ido Y, Carling D, Ruderman N. Hyperglycemia-induced apoptosis in human umbilical vein endothelial cells: Inhibition by the AMP-activated protein kinase activation. *Diabetes* 2002; 51:159–167.
94. da Silva Xavier G, Leclerc I, Varadi A, et al. Role for AMP-activated protein kinase in glucose-stimulated insulin secretion and preproinsulin gene expression. *Biochem J* 2003; 371:761–774.
95. Kefas BA, Heimberg H, Vaulont S, et al. AICA-riboside induces apoptosis of pancreatic beta cells through stimulation of AMP-activated protein kinase. *Diabetologia* 2003; 46:250–254.
96. Diraison F, Parton L, Ferre P, et al. Over-expression of sterol-regulatory-element-binding protein-1c (SREBP1c) in rat pancreatic islets induces lipogenesis and decreases glucose-stimulated insulin release: Modulation by 5-aminoimidazole-4-carboxamide ribonucleoside (AICAR). *Biochem J* 2004; 378:769–778.
97. Davis JD, Wirtshafter D, Asin KE, et al. Sustained intracerebroventricular infusion of brain fuels reduces body weight and food intake in rats. *Science* 1981; 212:81–3.
98. Obici S, Feng Z, Morgan K, et al. Central administration of oleic acid inhibits glucose production and food intake. *Diabetes* 2002; 51:271–5.
99. Thompson DA, Campbell RG. Hunger in humans induced by 2-deoxy-D-glucose: Glucoprivic control of taste preference and food intake. *Science* 1977; 198: 1065–1068.
100. Sergeyev V, Broberger C, Gorbatyuk O, et al. Effect of 2-mercaptoacetate and 2-deoxy-D-glucose administration on the expression of NPY, AGRP, POMC, MCH and hypocretin/orexin in the rat hypothalamus. *Neuroreport* 2000; 11:117–121.
101. Loftus TM, Jaworsky DE, Frehywot GL, et al. Reduced food intake and body weight in mice treated with fatty acid synthase inhibitors. *Science* 2000; 288: 2379–2381.
102. Obici S, Feng Z, Arduini A, Conti R, et al. Inhibition of hypothalamic carnitine palmitoyltransferase-1 decreases food intake and glucose production. *Nat Med* 2003; 9:756–761.
103. Minokoshi Y, Alquier T, Furukawa N, et al. AMP-kinase regulates food intake by responding to hormonal and nutrient signals in the hypothalamus. *Nature* 2004; 428:569–574.
104. Andersson U, Filipsson K, Abbott CR, et al. AMP-activated protein kinase plays a role in the control of food intake. *J Biol Chem* 2004; 279:12005–12008.
105. Kim EK, Miller I, Aja S, et al. C75, a fatty acid synthase inhibitor, reduces food intake via hypothalamic AMP-activated protein kinase. *J Biol Chem* 2004; 279:19970–19976.
106. Landree LE, Hanlon AL, Strong DW, et al. C75, a fatty acid synthase inhibitor, modulates AMP-activated protein kinase to alter neuronal energy metabolism. *J Biol Chem* 2004; 279:3817–3827.
107. Namkoong C, Kim MS, Jang PG, et al. Enhanced hypothalamic AMP-activated protein kinase activity contributes to hyperphagia in diabetic rats. *Diabetes* 2005; 54:63–68.

108. Lee K, Li B, Xi X, Suh Y, et al. Role of neuronal energy status in the regulation of adenosine 5′-monophosphate-activated protein kinase, orexigenic neuropeptides expression, and feeding behavior. *Endocrinology* 2005; 146:3–10.
109. Light PE, Wallace CH, Dyck JR. Constitutively active adenosine monophosphate-activated protein kinase regulates voltage-gated sodium channels in ventricular myocytes. *Circulation* 2003; 107:1962–1965.
110. Hallows KR, Raghuram V, Kemp BE, et al. Inhibition of cystic fibrosis transmembrane conductance regulator by novel interaction with the metabolic sensor AMP-activated protein kinase. *J Clin Invest* 2000; 105:1711–1721.
111. Evans JL, Maddux BA, Goldfine ID. The molecular basis for oxidative stress-induced insulin resistance. *Antioxid Redox Signal* 2005; 7:1040–1052.
112. Evans JL, Goldfine ID, Maddux BA, et al. Are oxidative stress-activated signaling pathways mediators of insulin resistance and beta-cell dysfunction? *Diabetes* 2003; 52:1–8.
113. Maddux BA, See W, Lawrence JC Jr, et al. Protection against oxidative stress-induced insulin resistance in rat L6 muscle cells by micromolar concentrations of alpha-lipoic acid. *Diabetes* 2001; 50:404–410.
114. Rudich A, Tirosh A, Potashnik R, et al. Prolonged oxidative stress impairs insulin-induced GLUT4 translocation in 3T3-L1 adipocytes. *Diabetes* 1998; 47:1562–1569.
115. Gardner CD, Eguchi S, Reynolds CM, et al. Hydrogen peroxide inhibits insulin signaling in vascular smooth muscle cells. *Exp Biol Med (Maywood)* 2003; 228: 836–842.
116. Rudich A, Tirosh A, Potashnik R, et al. Lipoic acid protects against oxidative stress induced impairment in insulin stimulation of protein kinase B and glucose transport in 3T3-L1 adipocytes. *Diabetologia* 1999; 42:949–957.
117. Jacob S, Streeper RS, Fogt DL, et al. The antioxidant alpha-lipoic acid enhances insulin-stimulated glucose metabolism in insulin-resistant rat skeletal muscle. *Diabetes* 1996; 45:1024–1029.
118. Saengsirisuwan V, Perez FR, Sloniger JA, et al. Interactions of exercise training and alpha-lipoic acid on insulin signaling in skeletal muscle of obese Zucker rats. *Am J Physiol Endocrinol Metab* 2004; 287:E529–E536.
119. Saengsirisuwan V, Kinnick TR, Schmit MB, et al. Interactions of exercise training and lipoic acid on skeletal muscle glucose transport in obese Zucker rats. *J Appl Physiol* 2001; 91:145–153.
120. Jacob S, Henriksen EJ, Tritschler HJ, et al. Improvement of insulin-stimulated glucose-disposal in type-2 diabetes after repeated parenteral administration of thioctic acid. *Exp Clin Endocrinol Diabetes* 1996; 104:284–288.
121. Robertson RP, Zhang HJ, Pyzdrowski KL, et al. Preservation of insulin mRNA levels and insulin secretion in HIT cells by avoidance of chronic exposure to high glucose concentrations. *J Clin Invest* 1992; 90:320–325.
122. Zhou YP, Grill VE. Long-term exposure of rat pancreatic islets to fatty acids inhibits glucose-induced insulin secretion and biosynthesis through a glucose fatty acid cycle. *J Clin Invest* 1994; 93:870–876.
123. Burkart V, Koike T, Brenner HH, et al. Dihydrolipoic acid protects pancreatic islet cells from inflammatory attack. *Agents Actions* 1993; 38:60–65.
124. Schroeder MM, Belloto RJ Jr, Hudson RA, et al. Effects of antioxidants coenzyme Q10 and lipoic acid on interleukin-1 beta-mediated inhibition of glucose-stimulated insulin release from cultured mouse pancreatic islets. *Immunopharmacol Immunotoxicol* 2005; 27:109–122.

125. Kunsch C, Medford RM. Oxidative stress as a regulator of gene expression in the vasculature. *Circ Res* 1999; 85:753–766.
126. Choy JC, Granville DJ, Hunt DW, et al. Endothelial cell apoptosis: Biochemical characteristics and potential implications for atherosclerosis. *J Mol Cell Cardiol* 2001; 33:1673–1690.
127. Ross R. Atherosclerosis—an inflammatory disease. *N Engl J Med* 1999; 340: 115–126.
128. Gryglewski RJ, Palmer RM, Moncada S. Superoxide anion is involved in the breakdown of endothelium-derived vascular relaxing factor. *Nature* 1986; 320:454–456.
129. Zhang WJ, Frei B. Alpha-lipoic acid inhibits TNF-alpha-induced NF-kappaB activation and adhesion molecule expression in human aortic endothelial cells. *FASEB J* 2001; 15:2423–2432.
130. Lee KM, Park KG, Kim YD, et al. Alpha-lipoic acid inhibits fractalkine expression and prevents neointimal hyperplasia after balloon injury in rat carotid artery. *Atherosclerosis* 2006; 189:106–114.
131. Park KG, Kim MJ, Kim HS, et al. Prevention and treatment of macroangiopathy: Focusing on oxidative stress. *Diabetes Res Clin Pract* 2004; 66:S57–S62.
132. Targonsky ED, Dai F, Koshkin V, et al. Alpha-lipoic acid regulates AMP-activated protein kinase and inhibits insulin secretion from beta cells. *Diabetologia* 2006; 49:1587–1598.
133. Hagen TM, Ingersoll RT, Lykkesfeldt J, et al. *R*-Alpha-lipoic acid-supplemented old rats have improved mitochondrial function, decreased oxidative damage, and increased metabolic rate. *FASEB J* 1999; 13:411–418.
134. Smith AR, Shenvi SV, Widlansky M, et al. Lipoic acid as a potential therapy for chronic diseases associated with oxidative stress. *Curr Med Chem* 2004; 11:1135–1146.
135. Schafer M, Schafer C, Ewald N, et al. Role of redox signaling in the autonomous proliferative response of endothelial cells to hypoxia. *Circ Res* 2003; 92:1010–1015.
136. Arsenijevic D, Onuma H, Pecqueur C, et al. Disruption of the uncoupling protein-2 gene in mice reveals a role in immunity and reactive oxygen species production. *Nat Genet* 2000; 26:435–439.
137. Park JY, Park KG, Kim HJ, et al. The effects of the overexpression of recombinant uncoupling protein 2 on proliferation, migration and plasminogen activator inhibitor 1 expression in human vascular smooth muscle cells. *Diabetologia* 2005; 48: 1022–1028.
138. Lee KU, Lee IK, Han J, et al. Effects of recombinant adenovirus-mediated uncoupling protein 2 overexpression on endothelial function and apoptosis. *Circ Res* 2005; 96:1200–1207.
139. Russell JW, Golovoy D, Vincent AM, et al. High glucose-induced oxidative stress and mitochondrial dysfunction in neurons. *FASEB J* 2002; 16:1738–1748.
140. Niimi M, Sato M, Yokote R, et al. Effects of central and peripheral injection of leptin on food intake and on brain Fos expression in the Otsuka Long-Evans Tokushima Fatty rat with hyperleptinemia. *J Neuroendocrinol* 1999; 11:605–611.
141. Maffei M, Halaas J, Ravussin E, et al. Leptin levels in human and rodent: Measurement of plasma leptin and ob RNA in obese and weight-reduced subjects. *Nat Med* 1995; 1:1155–1161.

Index

A

B

C

D

E

F

G

H

I

J

K

L

M

N

O

P

R

S

T

U

V